BIRKHÄUSER

Jacques Helmstetter
Artibano Micali

Quadratic Mappings and Clifford Algebras

Birkhäuser
Basel · Boston · Berlin

Authors:

Jacques Helmstetter
Institut Fourier (Mathématiques)
Université Grenoble I
B.P. 74
38402 Saint-Martin d'Hères Cedex
France
e-mail: jacques.helmstetter@ujf-grenoble.fr

Artibano Micali
Département des Sciences mathématiques
Université Montpellier II
Place Eugène Bataillon, CC 051
34095 Montpellier Cedex 5
France
e-mail: micali@math.univ-montp2.fr

2000 Mathematical Subject Classification: 15A66, 15A69, 15A63, 15A75, 16H05

Library of Congress Control Number: 2007942636

Bibliographic information published by Die Deutsche Bibliothek. Die Deutsche Bibliothek lists this publication in the Deutsche Nationalbibliografie; detailed bibliographic data is available in the Internet at http://dnb.ddb.de

ISBN 978-3-7643-8605-4 Birkhäuser Verlag AG, Basel - Boston - Berlin

Basel · Boston · Berlin
P.O. Box 133, CH-4010 Basel, Switzerland
Part of Springer Science+Business Media
Printed on acid-free paper produced from chlorine-free pulp. TCF∞
Printed in Germany

ISBN 978-3-7643-8605-4

e-ISBN 978-3-7643-8606-1

9 8 7 6 5 4 3 2 1

www.birkhauser.ch

Contents

Introduction

The first purpose of an introduction is to explain what distinguishes the newly written book from other books that might as well have the same title. This book deals with quadratic mappings between modules over an arbitrary ring K (commutative, associative, with unit element); therefore it requires an effective mastery of some little part of commutative algebra. It is especially interested in quadratic forms and in their Clifford algebras. The most common object under consideration is a *quadratic module* (M, q), that is any module M provided with a quadratic form $q : M \to K$, and the deepest results are obtained when (M, q) is a *quadratic space*, in other words, when M is a finitely generated projective module and q induces a bijection from M onto the dual module M^*. In particular the study of Clifford algebras of quadratic spaces shall (very progressively) lead to sophisticated theories involving noncommutative algebras over the ring K (Azumaya algebras, Morita theory, separability).

This book is almost never interested in results that would follow from some special properties of the basic ring K; therefore much more emphasis has been put on a serious study of Clifford algebras than on sophisticated properties of quadratic forms which always depend on subtle hypotheses on the ring K. Here, when K is not an arbitrary ring, it is a local ring, or even a field; the consideration of such particular rings is justified by the importance of localization and globalization in many chapters, and the important role of residue fields at some critical moments. Besides, many useful applications of Clifford algebras outside mathematics involve quadratic spaces over fields.

Another essential feature of this book is the narrowness of the set of prerequisites, and its constancy from the beginning to the end. These prerequisites are made precise below, and although they are not elementary, they are much less difficult and fewer than would be required for a pioneering or scholarly work. All essential properties of Clifford algebras have been reached by elementary means in the first five chapters before more difficult theories are presented in Chapter 6. The concern of the authors about teaching has led them to limit the amount of prerequisites, and to prove all results in the core of the book (almost the whole book) on the basis of these prerequisites; for all these results the complete path leading to their proof (sometimes by new simpler means) is explained.

Of course it has not been possible to impose the above-mentioned features on the whole book. We thought it sensible to present interesting examples involving theories outside the scope of the book, and to give information about related topics which do not appear in the core of the book. Thus for the proof of several statements it has been necessary to refer to other publications. For instance, quadratic forms over the ring of integers often afford illuminating applications of general theories; but since this book does not deal with arithmetic, it just mentions which arithmetical knowledge is indispensable.

Readers are assumed already to know elementary algebra (rings, fields, groups, quotients,...), and also linear and multilinear algebra over fields, especially tensor products and exterior algebras. They are assumed to know the usual properties of quadratic forms over the usual fields $\mathbb{R}$ and $\mathbb{C}$, which should enable a rapid understanding of the properties of more general quadratic mappings. Even some knowledge of linear algebra over rings (over commutative, associative rings with unit) is required: exact sequences, projective and flat modules,.... Most of these prerequisites are briefly recalled, especially in Chapter **1**. A self-contained yet concise exposition of commutative algebra is provided; it only covers the small part that is needed, essentially localization and globalization, and finitely generated modules. Homological algebra is never involved, except in isolated allusions.

Many pages are devoted to "exercises"; their purposes are varied. Some of them are training exercises, in other words, direct applications. Others present still more results, which have seemed less important to the authors, but which nevertheless deserve to be stated with indications about how to prove them. Others present examples enlightening the reader on some particular features or some unexpected difficulties. There are also developments showing applications in other domains, and some few extracts from the existing literature. The levels of difficulty are varied; when an exercise has seemed to be very difficult, or to require some knowledge that is not treated in the book, an asterisk has been put on its number, and often a hint has been supplied.

In the opinion of the authors, many applications of Clifford algebras outside algebra, and even outside mathematics, raise problems that are universally interesting, even for algebraists. It is the duty of algebraists to find clear concepts and effective treatments, especially in places that are usually obscured by a lot of cumbersome calculations. In many applications of Clifford algebras there are interior multiplications; here (in Chapter **4**) it is explained that they can be derived from the comultiplication that makes every Clifford algebra become a comodule over the exterior algebra (treated as a coalgebra). In many applications of Clifford algebras the calculations need two multiplications, a Clifford multiplication and an exterior one; here (in Chapter **4**) this practice is related to the concept of "deformation of Clifford algebra", which allows an elaborate presentation of a well-known result stated for instance in [Chevalley 1954], §2.1, and with more generality in [Bourbaki 1959, *Algèbre*, Chap. 9] (see Proposition 3 in §9, n^o3). But the true meaning of this essential result only appears when it is stated that it gives isomorphisms of comodules over exterior algebras, and not merely isomorphisms of K-modules.

Spinor spaces in quantum mechanics raise problems for which insightful algebraic interpretation and smart proof eschewing tedious calculations are still objects of discussion. Spinor spaces are often said to be Clifford modules although they are actually *graded* Clifford modules (see Example (6.2.2) in this book); the word "graded" refers to a parity grading which distinguishes even and odd elements. Whereas the theory of Clifford modules is a long sequence of particular cases, graded Clifford modules come under a unified and effective theory. The last exercises of Chapter **6** propose a smart and effective path to the essential algebraic properties that are needed in quantum mechanics.

This comment about spinor spaces is just one example of the constant emphasis put on parity gradings (from Chapter **3** to the end), in full agreement with C.T.C. Wall and H. Bass. In many cases the reversion of two odd factors must be compensated by a multiplication by -1, and here this rule is systematically enforced in all contexts in which it is relevant; indeed only a systematic treatment of parity gradings can avoid repeated hesitations about such multiplications by -1. For instance if f and g are linear forms on M, their exterior product can be defined as the linear form on $\bigwedge^2(M)$ that takes this value on the exterior product of two elements x and y of M :

$$(f \wedge g)(x \wedge y) \;=\; -f(x)\, g(y) \;+ f(y)\, g(x) \;;$$

the sign $-$ before $f(x)g(y)$ comes from the reversion of the odd factors g and x; but in $f(y)g(x)$ the odd factor y has jumped over two odd factors x and g, whence the sign $+$.

Lipschitz, the forgotten pioneer

Rudolf O.S. Lipschitz (1832–1903) discovered Clifford algebras in 1880, two years after William K. Clifford (1845–1879) and independently of him, and he was the first to use them in the study of orthogonal transformations. Up to 1950 people mentioned "Clifford-Lipschitz numbers" when they referred to this discovery of Lipschitz. Yet Lipschitz's name suddenly disappeared from the publications involving Clifford algebras; for instance Claude Chevalley (1909–1984) gave the name "Clifford group" to an object that is never mentioned in Clifford's works, but stems from Lipschitz's. The oblivion of Lipschitz's role is corroborated by [Weil], a letter that A. Weil first published anonymously, probably to protest against authors who discovered again some of Lipschitz's results in complete ignorance of his priority. Pertti Lounesto (1945–2002) contributed greatly to recalling the importance of Lipschitz's role: see his historical comment in [Riesz, 1993].

This extraordinary oblivion has generated two different controversies, a historical one and a mathematical one. On one side, some people claimed that the name "Clifford group" was historically incorrect and should be replaced with "Lipschitz group"; their action at least convinced other mathematicians to make correct references to Lipschitz when they had to invent *new* terms for objects that still had no name, even when they were reluctant to forsake the name "Clifford group". On the other side, some people were not satisfied with Chevalley's presentation of the so-called Clifford group, and completed it with additional developments that meant a return to Lipschitz's ideas; this is especially flagrant in [Sato, Miwa, Jimbo 1978], where the authors discovered again some of Lipschitz's results and gave them much more generality and effectiveness; the same might be said about [Helmstetter 1977, 1982]; but since the Japanese team showed applications of his cliffordian ideas to difficult problems involving differential operators (the "holonomic quantum fields"), the necessity of going beyond Chevalley's ideas became obvious for external reasons too. The fact that all these authors at that time completely ignored Lipschitz's contribution proves that the mathematical controversy is independent of the historical one.

The part of this book devoted to orthogonal transformations can be understood as a modernization of Lipschitz's theory. Whereas Lipschitz only considered real positive definite quadratic forms for which Clifford–Lipschitz groups may look quite satisfying, with more general quadratic forms it becomes necessary to attach importance to "Lipschitz monoids" from which "Lipschitz groups" are derived. Here the historically incorrect "Clifford groups" are still accepted (with the usual improved definition that pays due attention to the parity grading), but they only play an incidental role. They coincide with the Lipschitz groups in the classical case of quadratic spaces; but when beyond this classical case Clifford groups and Lipschitz groups no longer coincide, the latter prove to be much more interesting. Thus the mathematical controversy happens to prevail over the historical one.

Contraction and expansion are opposite and equally indispensable stages in all scientific research. At Chevalley's time it was opportune to contract the arguments and to exclude developments that no longer looked useful; but in Chevalley's works there is at least one part (in [Chevalley 1954], Chapter 3) that should have led him to reinstate Lipschitz if he had continued developing it. Our Chapter **7** is an expansion of this part of Chevalley's work, which for a long time has remained as he left it. This expansion involves the contributions of both Lipschitz and Chevalley, and should give evidence that it is much better to accept the *whole* heritage from *all* pioneers without prolonging inopportune exclusions. Besides, Lipschitz's ideas also proved to be very helpful in the cliffordian treatment of Weyl algebras.

Weyl algebras

Weyl algebras represent for alternate bilinear forms the same structure as Clifford algebras for quadratic forms, and in some publications they are even called "symplectic Clifford algebras". In [Dixmier 1968] you can find a concise exposition of what was known about them before cliffordian mathematicians became interested in them. Revoy was probably the first to propose a cliffordian treatment of Weyl algebras; see [Nouazé, Revoy 1972] and [Revoy 1978]. Later and independently, Crumeyrolle in France and the Japanese team Sato–Miwa–Jimbo produced some publications developing the cliffordian treatment of Weyl algebras, although they ignored (at least in their first publications) that these algebras had been already studied, and had received H. Weyl's name. Revoy's isolated work was hardly noticed, the cliffordian ideas of the Japanese team (which the renewal of Lipschitz's ideas mentioned above) were inserted in a very long and difficult work devoted to differential operators, which discouraged many people, and Crumeyrolle's statements bumped up against severe and serious objections. That is why the cliffordian treatment of Weyl algebras has not yet won complete acknowledgement.

There is no systematic presentation of Weyl algebras in this book, which already deals with a large number of other subjects. But at the end of Chapters **4**, **5** and **7**, many exercises about them have been proposed; Weyl algebras are defined in (4.ex.18). These exercises explain the cliffordian treatment of Weyl algebras as long as it is an imitation, or at least an adaptation, of the analogous treatment of Clifford algebras. For the most difficult results that require Fourier analysis and related theories, a short summary has been supplied; it should help readers to understand the purposes and the achievements of this new theory.

Acknowledgements

A. Micali is pleased to recall that he has worked for a long time with Orlando E. Villamayor (1923–1998), who efficiently contributed with some other authors to much progress that is now common knowledge and presented as such in this book. J. Helmstetter declares that he is indebted to Chevalley for his interest in Clifford algebras. Both authors have also learned much from H. Bass's publications. For some difficult topics we have also consulted [Knus 1991].

During the writing of our text, we took advantage of the services of the Mathematical Department (Fourier Institute) of the University of Grenoble; our text was prepared in this Institute and its Library was very often visited. Therefore we are grateful to the Director and to the Librarian for their help. Several colleagues in this Institute suggested some good ideas, or tried to answer embarrassing questions; they also deserve our gratitude.

The authors also thank the Director and the Librarian of the Mathematical Department of the University of Montpellier, who always gave us a helpful welcome. Our colleague Philippe Revoy followed our work and gave much good advice, for which we are indebted to him.

The authors are also grateful to Max-Albert Knus (Zürich, ETH-Zentrum) for his critical reading of a large part of the book, and for his judicious suggestions.

Finally we thank all the persons that have worked on our text to improve its literary quality and its presentation; despite our efforts, we yet needed their help. Without the benevolence of the editorial board of Birkhäuser Verlag, our project would not have been completed.

Grenoble, September 2007.

Jacques Helmstetter
Université Grenoble I
Institut Fourier (Mathématiques)
B.P. 74
F-38402 Saint-Martin d'Hères
France

Artibano Micali
Université Montpellier II
Département des Sciences Mathématiques
Place Eugène Bataillon
F-34095 Montpellier Cedex 5
France

Chapter 1

Algebraic Preliminaries

This preliminary chapter is devoted to the following three subjects, which together allows us to review a great part of the prerequisites, and to add some more specialized knowledge:

(a) a very simple presentation of the notion of universal property, with many examples; Sections **1.2** to **1.4** are devoted to this subject.

(b) additional information about categories of modules; Sections **1.5** to **1.9** contain reminders about exact sequences, usual functors, projective modules and changes of basic rings; but in **1.8** finitely presented modules are treated with more detail.

(c) a self-contained presentation of rings and modules of fractions, localization and globalization; although this material is already well treated in the existing literature, a concise exposition of the exact part that is here actually useful should prevent beginners from wandering in too-specialized topics.

Complete knowledge of *all* this chapter is not indispensable, because precise references will always be given when the most difficult or specialized results are needed in the following chapters.

1.1 Some general notation and definitions

The following notation and definitions will be used *in all chapters.*

As usual, $\mathbb{N}$, $\mathbb{Z}$, $\mathbb{Q}$, $\mathbb{R}$, $\mathbb{C}$, $\mathbb{H}$ are respectively the set of all integers ≥ 0, the ring of all integers, the fields of rational, real and complex numbers, and the division ring of real quaternions.

A ring A is always an associative ring with a unit element 1_A unless otherwise stated; the group of all invertible elements of A is denoted by $A^\times$. If a mapping $f : A \to B$ between two rings is called a ring morphism, it must be understood that the equality $f(1_A) = 1_B$ is also required. By definition a subring of A must contain 1_A.

Every module over the ring A is a left module, unless it is stated that it is a right module; the category of all left A-modules is denoted by $\mathcal{M}od(A)$. Right A-modules are often treated as left modules over the opposite algebra A^o (defined in **3.1**) or over the twisted opposite algebra A^{to} (defined in **3.2**) and therefore their category is denoted by $\mathcal{M}od(A^o)$ or $\mathcal{M}od(A^{to})$. If M and N are two modules over A, the set of all A-linear mappings from M into N is denoted by $\mathrm{Hom}_A(M,N)$; the ring of all A-linear endomorphisms of M is denoted by $\mathrm{End}_A(M)$. When $A = \mathbb{Z}$, $\mathcal{M}od(\mathbb{Z})$ is also the category of all additive groups.

The letter K always refers to a commutative ring, the unit element of which is denoted by 1; it is silently assumed that K is not reduced to 0 (in other words, $1 \neq 0$). When gradings (or filtrations) get involved, K is always trivially graded (or filtered); in other words, all its nonzero elements have degree 0. At some places it will be assumed that K is a local ring (in which there is only one maximal ideal $\mathfrak{m}$) or even a field. When there is no mention of another ring, all algebraic notions such as linearity, tensor products,..., refer to this ring K; consequently notation like $\mathrm{Hom}(M,N)$, $M \otimes N$, $\bigwedge(M)$,..., must be understood as $\mathrm{Hom}_K(M,N)$, $M \otimes_K N$, $\bigwedge_K(M)$,....

Unless otherwise stated, an algebra A over K is a ring provided with a ring morphism $K \to A$ that maps K into the center of A. If this ring morphism is injective, K can be identified with a subring of A, and thus 1_A is identified with 1. Every ring is an algebra over the ring $\mathbb{Z}$ of integers. The category of all K-algebras is denoted by $\mathcal{A}lg(K)$; thus the notation $\mathrm{Hom}_{\mathcal{A}lg(K)}(A,B)$ means the set of all K-linear ring morphisms from A into B. The subcategory of all commutative algebras in $\mathcal{A}lg(K)$ is denoted by $\mathcal{C}om(K)$.

Internal references must be understood in this way: the notation **4.2** means the second section of the fourth chapter; inside each section, all emphasized statements or formulas are numbered in a single file; for instance (4.2.3) means the third statement (theorem, or definition, or remark, or example, or formula, or anything else) in **4.2**. References to other works (listed in the bibliography at the end of the book) are indicated by brackets: for instance [Lounesto 1981].

1.2 Universal objects in a category

There are different ways to introduce universal properties, but here the simplest way, based on the notion of universal object in a category, is already sufficient. A category $\mathcal{C}$ consists of objects and morphisms (also called homomorphisms or arrows); each morphism relates two objects, called the source and the target; the set of morphisms relating the objects M and N in the category $\mathcal{C}$ is denoted by $\mathrm{Hom}_{\mathcal{C}}(M,N)$. A morphism f from M to N and a morphism g from N to P can be linked together to give a morphism from M to P denoted by $g \circ f$ or gf, and it is required that the equality $h(gf) = (hg)f$ is true whenever it is meaningful (in other words, whenever the targets of f and g are respectively the sources of g and h). In the definition of a category is also mentioned the existence of an identity morphism id_N for each object N, with the requirement that the equalities $\mathrm{id}_N\ f = f$ and $g\ \mathrm{id}_N = g$ hold whenever they are meaningful. A morphism $f : M \to N$ is called an isomorphism if there exists $g : N \to M$ such that $gf = \mathrm{id}_M$ and $fg = \mathrm{id}_N$; this morphism g is unique and is called the reciprocal isomorphism.

An object U in a category $\mathcal{C}$ is called an *initial universal object* (resp. a *final universal object*) if for every object M in $\mathcal{C}$ the set $\mathrm{Hom}_{\mathcal{C}}(U, M)$ (resp. $\mathrm{Hom}_{\mathcal{C}}(M, U)$) contains exactly one element. This definition implies that id_U is the only element in $\mathrm{Hom}_{\mathcal{C}}(U, U)$. Here almost all universal objects under consideration will be initial ones. We must keep in mind the following evident theorem.

(1.2.1) **Theorem.** *If a category contains two initial universal objects (resp. two final universal objects) U and V, the only morphism from U to V is an isomorphism.*

Proof. There is one morphism f from U to V and also one morphism g from V to U; since gf (resp. fg) is a morphism from U (resp. V) to itself, it must be equal to id_U (resp. id_V); therefore f and g are reciprocal isomorphisms. □

The most evident category is the category of all sets in which any mapping is a morphism; the empty set is the only initial universal object, and any set containing exactly one element is a final universal object.

In the category $\mathcal{M}od(K)$ of all modules over K, the morphisms are the K-linear mappings; the notation $\mathrm{Hom}(M, N)$ (or $\mathrm{Hom}_K(M, N)$ when more precision is necessary) is the usual abbreviation for $\mathrm{Hom}_{\mathcal{M}od(K)}(M, N)$. In this category any module reduced to zero is both an initial universal object and a final one.

Now let us consider the category $\mathcal{A}lg(K)$ of all algebras over the ring K, in which the morphisms are the K-linear ring morphisms. The algebras reduced to zero are the only ones in which the unit element is equal to 0, and they are final universal objects. Since the morphisms must respect the unit elements (see **1.1**), the ring K itself is an initial universal object. The category $\mathcal{A}lg(\mathbb{Z})$ is the category of all rings; for any ring A, the only morphism $\mathbb{Z} \to A$ determines the characteristic of A : it is the integer $n \geq 0$ such that $n\mathbb{Z}$ is the kernel of this morphism.

Before more interesting examples are given in the next two sections, it must be recalled that sometimes the objects of a category are morphisms in another category. In particular, with each category $\mathcal{C}$ is associated a category $\mathcal{C}_{\mathrm{hom}}$ in which the objects are all the morphisms of $\mathcal{C}$; if $u : M_1 \to M_2$ and $v : N_1 \to N_2$ are two objects of $\mathcal{C}_{\mathrm{hom}}$, a morphism from u to v is a couple (f_1, f_2) of morphisms $f_1 : M_1 \to N_1$ and $f_2 : M_2 \to N_2$ such that $vf_1 = f_2u$. The composition of two morphisms (f_1, f_2) and (g_1, g_2) in $\mathcal{C}_{\mathrm{hom}}$ is given by the formula $(g_1, g_2) \circ (f_1, f_2) = (g_1f_1,\ g_2f_2)$ whenever it is meaningful. If U is an initial universal object in $\mathcal{C}$, then id_U is an initial universal object in $\mathcal{C}_{\mathrm{hom}}$.

Sometimes it is convenient to associate with a given category $\mathcal{C}$ a *dual category* $\mathcal{C}^*$ containing the same objects and the same morphisms; but the target (resp. the source) of a morphism in $\mathcal{C}^*$ is by definition its source (resp. its target) in $\mathcal{C}$. If $gf : M \to N \to P$ is a product of morphisms in $\mathcal{C}$, it must be written as $fg : P \to N \to M$ in $\mathcal{C}^*$. Every final universal object in $\mathcal{C}$ is an initial one in $\mathcal{C}^*$, and this explains why both kinds of universal objects are considered as dual to each other.

1.3 Examples of universal objects

Later in **3.1** Clifford algebras are introduced by means of a *universal property*, meaning that they afford an initial universal object in some category; because of (1.2.1), this property characterizes a Clifford algebra up to isomorphy. To prove that a universal property can often distinguish an interesting object, we show several already known objects, the interest of which actually depends on some universal property.

Quotient modules

Let M be a module over K, and N a submodule of M; when the quotient module M/N and the quotient morphism $\varphi : M \to M/N$ are involved in an argument, the following proposition is often referred to.

(1.3.1) **Proposition.** *For every linear mapping $f : M \to P$ such that $f(N) = 0$, there exists a unique linear mapping $f' : M/N \to P$ such that $f = f'\varphi$. Moreover if M is a K-algebra and N an ideal of M, then M/N is also a K-algebra and φ is an algebra morphism.*

This property of M/N and φ is said to be a universal property because it means that φ is an initial universal object in the category of all linear mappings f from M into any module P such that $f(N) = 0$. If $f : M \to P$ and $g : M \to Q$ are objects in this category, a morphism from f to g is a linear mapping $u : P \to Q$ such that $g = uf$. When M is a K-algebra and N an ideal, we may require f and u to be algebra morphisms. The proof of (1.3.1) is based on the fact that the image and kernel of φ are exactly M/N and N; therefore these properties characterize M/N and φ up to isomorphy.

The null morphism $M \to 0$ with a trivial target is obviously a final universal object, but quite uninteresting. In the following examples such trivial final universal objects will not even be mentioned.

Freely generated modules

Let S be a set. The K-module freely generated by S is the set $K^{(S)}$ of all mappings $\varepsilon : S \to K$ such that $\varepsilon(s)$ vanishes for all $s \in S$ except a finite number; it is a K-module in an evident way; it is reduced to 0 if S is empty. Let φ be the mapping from S to $K^{(S)}$ which maps each $s \in S$ to the mapping $e_s : S \to K$ such that

$e_s(s) = 1$ and $e_s(t) = 0$ whenever $s \neq t$. It is clear that $K^{(S)}$ is a free K-module in which the family of all e_s is a basis; this property explains the name given to $K^{(S)}$. It also implies that φ is an initial universal object in the category of all mappings f from S into any K-module P; if $f : S \to P$ and $g : S \to Q$ are objects in this category, a morphism from f to g is a linear mapping $u : P \to Q$ such that $g = uf$. Indeed every mapping f from S into a K-module P determines a unique linear mapping f' from $K^{(S)}$ into P such that $f = f'\varphi$, or equivalently, $f'(e_s) = f(s)$ for all $s \in S$.

Let P be any K-module, and S any subset of P; the universal property of $K^{(S)}$ implies the existence of a canonical linear mapping $K^{(S)} \to P$ which maps every e_s to s. It is surjective if and only if P is generated by S; therefore every generating subset S makes P isomorphic to a quotient of the free module $K^{(S)}$. If P is finitely generated, it is isomorphic to the quotient of a free module with a finite basis.

Tensor products of modules

Let M and N be K-modules. The canonical bilinear mapping φ from $M \times N$ into the tensor product $M \otimes N$ (that is $(x, y) \longmapsto x \otimes y$) is also an initial universal object because of the following universal property.

(1.3.2) **Proposition.** *For any bilinear mapping $f : M \times N \to P$ there exists a unique linear mapping $f' : M \otimes N \to P$ such that $f = f'\varphi$, or equivalently, $f'(x \otimes y) = f(x, y)$ for all $(x, y) \in M \times N$.*

The objects of the category under consideration are the bilinear mappings f defined on $M \times N$; if $f : M \times N \to P$ and $g : M \times N \to Q$ are two objects, a morphism from f to g is a linear mapping $u : P \to Q$ such that $g = uf$.

The precise definition of $M \otimes N$ is very seldom needed, since it is characterized by (1.3.2) up to isomorphy. For instance (1.3.2) is sufficient to explain that two linear mappings $v : M \to M'$ and $w : N \to N'$ determine a linear mapping $v \otimes w$ from $M \otimes M'$ into $N \otimes N'$; indeed $v \otimes w$ is the only linear mapping such that $(v \otimes w)(x \otimes y) = v(x) \otimes w(y)$ for all $(x, y) \in M \times N$.

Of course it is possible to state an analogous universal property for a tensor product of several modules. The so-called commutativity property of tensor products refers to the evident isomorphisms $M \otimes N \longleftrightarrow N \otimes M$. There is also an associativity property:

$$(M \otimes_K N) \otimes_L P \;\cong\; M \otimes_K (N \otimes_L P);$$

it is valid when M is a K-module, P an L-module, and N a module over K and L such that $\kappa(\lambda y) = \lambda(\kappa y)$ for all $\kappa \in K$, $\lambda \in L$ and $y \in N$.

Tensor products of algebras

When A and B are K-algebras, the tensor product $A\otimes B$ is also an algebra; indeed the quadrilinear mapping $(x,y,x',y') \longmapsto xx'\otimes yy'$ from $A\times B\times A\times B$ into $A\otimes B$ determines a linear mapping from $A\otimes B\otimes A\otimes B$ into $A\otimes B$; thus $A\otimes B$ is provided with a multiplication such that

$$(x\otimes y)\ (x'\otimes y') \ = \ xx'\otimes yy';$$

it is easy to prove that it is associative and that $1_A\otimes 1_B$ is a unit element. Observe that the mappings $x \longmapsto x\otimes 1_B$ and $y \longmapsto 1_A\otimes y$ are algebra morphisms φ_1 and φ_2 from respectively A and B into $A\otimes B$, and that $\varphi_1(x)$ and $\varphi_2(y)$ always commute in $A\otimes B$:

$$(x\otimes 1_B)(1_A\otimes y) \ = \ x\otimes y \ = \ (1_A\otimes y)(x\otimes 1_B).$$

As an algebra, $A\otimes B$ has the following universal property.

(1.3.3) **Proposition.** *If $f_1 : A\to P$ and $f_2 : B\to P$ are algebra morphisms, and if $f_1(x)$ and $f_2(y)$ commute in the algebra P for all $(x,y)\in A\times B$, there exists a unique algebra morphism $f' : A\otimes B\to P$ such that $f_1 = f'\varphi_1$ and $f_2 = f'\varphi_2$, or in other words, $f_1(x) = f'(x\otimes 1_B)$ for all $x\in A$ and $f_2(y) = f'(1_A\otimes y)$ for all $y\in B$.*

From (1.3.3) it should be easy to deduce the category in which (φ_1,φ_2) is an initial universal object. Nevertheless it has become usual to say that (1.3.3) is a universal property of the target $A\otimes B$.

Direct products and direct sums

With every family $(M_j)_{j\in J}$ of modules over K are associated a direct product $\prod_j M_j$ and a direct sum $\bigoplus_j M_j$ which coincide whenever the set J of indices is finite; the direct product is the ordinary cartesian product consisting of all families (x_j) such that $x_j\in M_j$ for all $j\in J$, whereas the direct sum is the submodule of all families (x_j) in which all x_j vanish except a finite number. Let us realize that the direct product is the source of a final universal object in some category $\mathcal{C}'$, whereas the direct sum is the target of an initial universal object in another category $\mathcal{C}''$.

The objects of $\mathcal{C}'$ are the families (f_j) of linear mappings $P\to M_j$ with the same source P, and a morphism from (f_j) to $(g_j : Q\to M_j)$ is a morphism $u : Q\to P$ such that $g_j = f_j u$ for all $j\in J$. The following proposition means that the family of all projections $\psi_j : \prod_{i\in J} M_i \to M_j$ is a final universal object.

(1.3.4) **Proposition.** *Every family of linear mappings $f_j : P\to M_j$ determines a unique linear mapping $f' : P\to\prod_j M_j$ such that $f_j = \psi_j f'$ for all $j\in J$; in other words there is a canonical (and linear) bijection*

$$\mathrm{Hom}(P,\prod_j M_j) \longrightarrow \prod_j \mathrm{Hom}(P,M_j), \qquad f' \longmapsto (\psi_j f')_{j\in J}.$$

The objects of $\mathcal{C}''$ are the families (f_j) of linear mappings $M_j \to P$ with the same target P, and a morphism from (f_j) to $(g_j : M_j \to Q)$ is a morphism $u : P \to Q$ such that $g_j = uf_j$ for all $j \in J$. The next proposition means that the family of all natural injections $\varphi_j : M_j \to \bigoplus_i M_i$ is an initial universal object.

(1.3.5) **Proposition.** *Every family of linear mappings $f_j : M_j \to P$ determines a unique linear mapping $f' : \bigoplus_j M_j \to P$ such that $f_j = f'\varphi_j$ for all $j \in J$; in other words there is a canonical (and linear) bijection*

$$\mathrm{Hom}\left(\bigoplus_j M_j,\ P\right) \longrightarrow \prod_j \mathrm{Hom}(M_j, P), \qquad f' \longmapsto (f'\varphi_j)_{j\in J}.$$

For a family of two modules M and N the direct product and the direct sum coincide; but the notation $M \oplus N$ is generally preferred when it is the source of a linear mapping, whereas $M \times N$ is preferred when it is the source of a bilinear mapping or the target of a mapping of any kind.

Obviously the categories $\mathcal{C}'$ and $\mathcal{C}''$ have been derived from the category $\mathcal{C} = \mathcal{M}od(K)$ and the family (M_j) in such a way that no special properties of $\mathcal{M}od(K)$ have been needed; consequently it is possible to repeat the same construction of $\mathcal{C}'$ and $\mathcal{C}''$ by starting from any category $\mathcal{C}$ and any family (M_j) of objects of $\mathcal{C}$. If $\mathcal{C}'$ contains a final universal object, its source is called the direct product of the family (M_j) in $\mathcal{C}$; and if $\mathcal{C}''$ contains an initial universal object, its target is called its direct sum in $\mathcal{C}$.

When the objects of $\mathcal{C}$ are sets provided with some common structure, and when its morphisms are the mappings respecting this structure, it often happens that every cartesian product of objects is still an object, that the canonical projections from this cartesian product onto its components are morphisms in $\mathcal{C}$, and that finally this cartesian product is the direct product in $\mathcal{C}$. This is true for the category of all sets, the category $\mathcal{A}lg(K)$ and its subcategory $\mathcal{C}om(K)$.

The concepts of direct product and direct sum may be understood as dual to each other if we remember the dual category $\mathcal{C}^*$ defined in **1.2**. Indeed if we replace $\mathcal{C}$ with $\mathcal{C}^*$ in the construction of $\mathcal{C}'$ (without changing the family (M_j)), we get the dual category of $\mathcal{C}''$; and conversely. Therefore a direct sum in $\mathcal{C}$ is a direct product in $\mathcal{C}^*$, and conversely.

Despite this duality, the existence of direct sums is often a more difficult problem than the existence of direct products when the objects of $\mathcal{C}$ are sets provided with some structure as above. Besides the previous example with $\mathcal{C} = \mathcal{M}od(K)$, we will also look for direct sums of two objects in two other categories. First, in the category of all sets, the direct sum of two sets M and N is the so-called *disjoint union.* When M and N are actually disjoint, it is merely $M \cup N$; if not, it may be the union of two disjoints sets M' and N' respectively isomorphic to M and N.

Secondly the direct sum of two commutative algebras A and B in $\mathcal{C}om(K)$ is their tensor product $A \otimes B$; indeed (1.3.3) implies that there is a canonical bijection

$$\mathrm{Hom}_{\mathcal{C}om(K)}(A \otimes B,\ P) \longrightarrow \mathrm{Hom}_{\mathcal{C}om(K)}(A, P) \times \mathrm{Hom}_{\mathcal{C}om(K)}(B, P).$$

1.4 Tensor algebras and symmetric algebras

Let M be a K-module. We consider the category of all linear mappings from M into a K-algebra P; a morphism from $f : M \to P$ to $g : M \to Q$ is an algebra morphism $u : P \to Q$ such that $g = uf$. If there is an initial universal object $\varphi : M \to U$, its target U (which is thus defined up to isomorphy because of (1.2.1)) should be called *the algebra freely generated by* M. Two properties are understood in this name. First U coincides with the subalgebra U' generated by $\varphi(M)$ in U; indeed there must be an algebra morphism $u : U \to U'$ such that $\varphi = u\varphi$, and then $U \to U' \to U$ must be equal to id_U, whence $U = U'$ (since $U' \to U$ is the natural injection). Secondly, if the algebra P is generated by the image of $f : M \to P$, the unique algebra morphism $f' : U \to P$ satisfying $f = f'\varphi$ is surjective and makes P become a quotient of U; in other words, any algebra P generated by a linear image $f(M)$ of M is determined (up to isomorphy) by the knowledge of $\mathrm{Ker}(f')$ (or that of any subset generating it as an ideal). This is what you must understand when one says that P can be deduced from U by adding more relations between the generators; and these relations correspond to the elements of $\mathrm{Ker}(f')$. This point of view leads to an effective construction of U, starting with a K-algebra V of noncommutative polynomials in noncommuting indeterminates e_x (indexed by $x \in M$), and then taking the quotient of V by the ideal W generated by all $e_x + e_y - e_{x+y}$ and $\lambda e_x - e_{\lambda x}$ with $x, y \in M$ and $\lambda \in K$, so that a linear mapping $\varphi : M \to V/W$ can be defined.

Nonetheless the above name of U is very seldom used, because there is another construction of U that has become more popular, and that has led to it being called *the tensor algebra of* M. Let us set

$$\mathrm{T}^0(M) = K\,, \quad \mathrm{T}^1(M) = M\,, \quad \mathrm{T}^2(M) = M \otimes M\,, \quad \mathrm{T}^3(M) = M \otimes M \otimes M, \dots;$$

because of the so-called associativity of tensor products, for all $(i, j) \in \mathbb{N}^2$ there is a canonical isomorphism from $\mathrm{T}^i(M) \otimes \mathrm{T}^j(M)$ onto $\mathrm{T}^{i+j}(M)$, whence a multiplication mapping $\mathrm{T}^i(M) \times \mathrm{T}^j(M) \longrightarrow \mathrm{T}^{i+j}(M)$; this multiplication can be extended by linearity to the direct sum $\mathrm{T}(M)$ of all the tensor powers $\mathrm{T}^k(M)$, and thus $\mathrm{T}(M)$ is an associative algebra (even a graded algebra). The next proposition means that the natural injection $\varphi : M \to \mathrm{T}(M)$ is the desired universal object.

(1.4.1) **Proposition.** *Any linear mapping f from M into an algebra P extends in a unique way to an algebra morphism $f' : \mathrm{T}(M) \to P$. For every sequence $(x_1, x_2, \dots, x_k)$ of elements of M (of any length $k \geq 1$),*

$$f'(x_1 \otimes x_2 \otimes \cdots \otimes x_k) = f(x_1)f(x_2)\cdots f(x_k).$$

Since there is an algebra freely generated by the module M in the category $\mathcal{A}lg(K)$, it is sensible to look for a commutative algebra freely generated by M in the subcategory $\mathcal{C}om(K)$. Since the former can be constructed as a quotient of an algebra of noncommutative polynomials, the latter can be constructed in the same way by means of commutative polynomials. As explained above, we can also obtain the latter by imposing more relations (relations of commutation) between the generators of the former; in other words, it suffices to take the quotient of the tensor algebra $\mathrm{T}(M)$ by the ideal R generated by all $x \otimes y - y \otimes x$ with $x, y \in M$. The resulting algebra is called the *symmetric algebra of* M, and denoted by $\mathrm{S}(M)$; the symbol $\vee$ is often used to mean the product of two elements of $\mathrm{S}(M)$.

This algebra $\mathrm{S}(M)$ inherits the grading of $\mathrm{T}(M)$, because the ideal R is the direct sum of the intersections $R^k = R \cap \mathrm{T}^k(M)$; thus $\mathrm{S}(M)$ can be identified with the direct sum of the quotients $\mathrm{S}^k(M) = \mathrm{T}^k(M)/R^k$. Obviously $R^0 = R^1 = 0$, whence $\mathrm{S}^0(M) = K$ and $\mathrm{S}^1(M) = M$. The natural injection $M \to \mathrm{S}(M)$ is an initial universal object in the category of all linear mappings $M \to P$ with target a commutative algebra, as stated in the next proposition.

(1.4.2) **Proposition.** *Any linear mapping from M into a commutative algebra P extends in a unique way to an algebra morphism $\mathrm{S}(M) \to P$.*

In the algebra $\mathrm{T}(M)$ each component $\mathrm{T}^k(M)$ has a universal property, because it is a tensor power of M: for every k-linear mapping f from $M^k = M \times M \times \cdots \times M$ into any module P, there is a unique linear mapping $f' : \mathrm{T}^k(M) \to P$ such that

$$f'(x_1 \otimes x_2 \otimes \cdots \otimes x_k) = f(x_1, x_2, \ldots, x_k)$$

for all $x_1, x_2, \ldots, x_k \in M$. The components of $\mathrm{S}(M)$ also have their particular universal property.

(1.4.3) **Proposition.** *For every symmetric k-linear mapping $f : M^k \to P$ there is a unique linear mapping $f'' : \mathrm{S}^k(M) \to P$ such that*

$$f''(x_1 \vee x_2 \vee \cdots \vee x_k) = f(x_1, x_2, \ldots, x_k)$$

for all $x_1, x_2, \ldots, x_k \in M$.

Proof. We suppose $k \geq 2$, otherwise the symmetry hypothesis is empty. First f induces a linear mapping $f' : \mathrm{T}^k(M) \to P$. As above, we treat $\mathrm{S}^k(M)$ as the quotient $\mathrm{T}^k(M)/R^k$, where R^k is spanned by the products

$$x_1 \otimes x_2 \otimes \cdots \otimes x_{j-1} \otimes (x_j \otimes x_{j+1} - x_{j+1} \otimes x_j) \otimes x_{j+2} \otimes \cdots \otimes x_k$$

with $j \in \{1, 2, \ldots, k-1\}$; if the k-linear mapping f is symmetric, the mapping f' vanishes on all these products, whence a linear mapping f'' from $\mathrm{T}^k(M)/R^k = \mathrm{S}^k(M)$ into P with the desired property. Since $\mathrm{S}^k(M)$ is spanned by the products $x_1 \vee x_2 \vee \cdots \vee x_k$, the unicity of f'' is obvious. □

1.5 Functors

Let $\mathcal{C}$ and $\mathcal{D}$ be two categories. A *covariant functor* $\mathcal{F}$ from $\mathcal{C}$ to $\mathcal{D}$ associates with each object M of $\mathcal{C}$ an object $\mathcal{F}(M)$ of $\mathcal{D}$, and with each morphism f of $\mathcal{C}$ a morphism $\mathcal{F}(f)$ of $\mathcal{D}$ in such a way that the following three conditions are always fulfilled:

if $f \in \mathrm{Hom}_{\mathcal{C}}(M,N)$, then $\mathcal{F}(f) \in \mathrm{Hom}_{\mathcal{D}}(\mathcal{F}(M),\mathcal{F}(N))$;

$\mathcal{F}(gf) = \mathcal{F}(g)\mathcal{F}(f)$ whenever gf exists;

if f is an identity morphism, then so is $\mathcal{F}(f)$.

A *contravariant functor* $\mathcal{F}$ from $\mathcal{C}$ to $\mathcal{D}$ is a covariant functor from the dual category of $\mathcal{C}$ (see **1.2**) to the category $\mathcal{D}$; this means that:

if $f \in \mathrm{Hom}_{\mathcal{C}}(N,M)$, then $\mathcal{F}(f) \in \mathrm{Hom}_{\mathcal{D}}(\mathcal{F}(M),\mathcal{F}(N))$;

$\mathcal{F}(fg) = \mathcal{F}(g)\mathcal{F}(f)$ whenever fg exists;

if f is an identity morphism, then so is $\mathcal{F}(f)$.

Many universal properties give rise to a functor. For instance there is a functor T associated with tensor algebras; it is a covariant functor from $\mathcal{M}od(K)$ to $\mathcal{A}lg(K)$; the tensor algebra $\mathrm{T}(M)$ of a module M has been defined in **1.4**; if f is a linear mapping from M into N, then $\mathrm{T}(f)$ is the unique algebra morphism $\mathrm{T}(M) \to \mathrm{T}(N)$ extending the linear mapping $M \to \mathrm{T}(N)$ defined by $x \longmapsto f(x)$. It is easy to verify that a functor has been defined in this way. In the same way we can define a functor S from $\mathcal{M}od(K)$ to $\mathcal{C}om(K)$ by means of symmetric algebras.

The covariant functors between two categories are themselves the objects of a category. When $\mathcal{F}$ and $\mathcal{G}$ are covariant functors from $\mathcal{C}$ to $\mathcal{D}$, a morphism Φ from $\mathcal{F}$ to $\mathcal{G}$ associates with each object M of $\mathcal{C}$ a morphism $\Phi(M)$ from $\mathcal{F}(M)$ to $\mathcal{G}(M)$ in $\mathcal{D}$, in such a way that $\mathcal{G}(f)\circ\Phi(M) = \Phi(N)\circ\mathcal{F}(f)$ for every $f \in \mathrm{Hom}_{\mathcal{C}}(M,N)$. This morphism $\Phi : \mathcal{F} \longrightarrow \mathcal{G}$ is an isomorphism if and only if all morphisms $\Phi(M)$ are isomorphisms.

Let $\mathcal{F}$ be a covariant functor from $\mathcal{C}$ to $\mathcal{D}$, and M and N two objects of $\mathcal{C}$; we assume that $\mathcal{C}$ contains morphisms $\varphi_1 : M \to P$ and $\varphi_2 : N \to P$ for which P is the direct sum of M and N in this category, and that $\mathcal{D}$ contains morphisms $\psi_1 : \mathcal{F}(M) \to Q$ and $\psi_2 : \mathcal{F}(N) \to Q$ for which Q is the direct sum of $\mathcal{F}(M)$ and $\mathcal{F}(N)$ in this category. Now $\mathcal{F}(\varphi_1)$ and $\mathcal{F}(\varphi_2)$ have the same sources as ψ_1 and ψ_2, but their common target is $\mathcal{F}(P)$; since (ψ_1,ψ_2) is universal, all this results in a morphism $Q \to \mathcal{F}(P)$. In other words, the existence of direct sums in $\mathcal{C}$ and in $\mathcal{D}$ implies the existence of canonical morphisms

$$\mathcal{F}(M) \oplus_{\mathcal{D}} \mathcal{F}(N) \longrightarrow \mathcal{F}(M \oplus_{\mathcal{C}} N).$$

Some functors $\mathcal{F}$ have the nice property that these canonical morphisms are isomorphisms. For instance the functor S defined by means of symmetric algebras has this property, as stated in the next theorem; remember that the direct sum of two objects in $\mathcal{C}om(K)$ is their tensor product (see **1.3**).

(1.5.1) **Theorem.** *The symmetric algebra of the module $M \oplus N$ is canonically isomorphic to the tensor product* $\mathrm{S}(M) \otimes \mathrm{S}(N)$.

Proof. For every $x \in M$ and $y \in N$ the canonical algebra morphism $\mathrm{S}(M) \otimes \mathrm{S}(N) \to \mathrm{S}(M \oplus N)$ resulting from the previous argument maps $x \otimes 1$ and $1 \otimes y$ respectively to $(x, 0)$ and $(0, y)$ in $M \oplus N$. Conversely, because of the universal property of $\mathrm{S}(M \oplus N)$, the linear mapping $(x, y) \longmapsto x \otimes 1 + 1 \otimes y$ extends to an algebra morphism $\mathrm{S}(M \oplus N) \to \mathrm{S}(M) \otimes \mathrm{S}(N)$. Then it is easy to prove that in this way two reciprocal isomorphisms have been constructed. □

The definition of a functor of several variables can be easily guessed, and we will present at once the evident example of the functor $\mathrm{Hom}_{\mathcal{C}}$ from $\mathcal{C} \times \mathcal{C}$ to the category of all sets, which is contravariant in the first variable and covariant in the second variable. The set $\mathrm{Hom}_{\mathcal{C}}(M, N)$ has been defined in **1.2**; now let us consider two morphisms in $\mathcal{C}$: $f_1 : M' \to M$ and $f_2 : N \to N'$; from them we derive a mapping denoted by $\mathrm{Hom}_{\mathcal{C}}(f_1, f_2)$:

$$\mathrm{Hom}_{\mathcal{C}}(M, N) \longrightarrow \mathrm{Hom}_{\mathcal{C}}(M', N'), \qquad u \longmapsto f_2 u f_1.$$

Thus we have got a functor because

$\mathrm{Hom}_{\mathcal{C}}(f_1 g_1,\ g_2 f_2) = \mathrm{Hom}_{\mathcal{C}}(g_1, g_2) \circ \mathrm{Hom}_{\mathcal{C}}(f_1, f_2)$ whenever $f_1 g_1$ and $g_2 f_2$ exist,

$\mathrm{Hom}_{\mathcal{C}}(\mathrm{id}_M, \mathrm{id}_N)$ is the identity mapping of $\mathrm{Hom}_{\mathcal{C}}(M, N)$.

By fixing the object N, we get a contravariant functor $\mathrm{Hom}_{\mathcal{C}}(\dots, N)$ of the first variable (whence the notation $\mathrm{Hom}_{\mathcal{C}}(f_1, N)$ meaning $\mathrm{Hom}_{\mathcal{C}}(f_1, \mathrm{id}_N)$), and by fixing M we get a covariant functor $\mathrm{Hom}_{\mathcal{C}}(M, \dots)$ of the second variable (whence the notation $\mathrm{Hom}_{\mathcal{C}}(M, f_2)$ meaning $\mathrm{Hom}_{\mathcal{C}}(\mathrm{id}_M, f_2)$). When $\mathcal{C}$ is the category $\mathcal{M}od(K)$, the functor Hom_K takes its values in the category $\mathcal{M}od(K)$.

Here we are also interested in the functor $\otimes$ which is a twice covariant functor from $\mathcal{M}od(K) \times \mathcal{M}od(K)$ to $\mathcal{M}od(K)$. The notations $M \otimes N$ and $v \otimes w$ have been explained in **1.3**, and it is easy to verify that they correspond to a functor; the notations $v \otimes N$ and $v \otimes \mathrm{id}_N$ are synonymous. This functor has also a nice behaviour relatively to direct sums, even infinite direct sums; as explained above, the universal property of direct sums implies the existence of morphisms like this one:

$$\bigoplus_{j \in J} (M \otimes N_j) \longrightarrow M \otimes \left(\bigoplus\nolimits_{j \in J} N_j\right);$$

to get a reciprocal morphism, it suffices to consider the evident bilinear mapping

$$M \times \left(\bigoplus\nolimits_{j \in J} N_j\right) \longrightarrow \bigoplus_{j \in J} (M \otimes N_j);$$

because of the universal property of tensor products, we derive from it the reciprocal isomorphism.

If N is a free module, it is a direct sum of submodules all isomorphic to K, and since $M \otimes K$ is canonically isomorphic to M, the tensor product $M \otimes N$ is isomorphic to a direct sum of submodules all isomorphic to M. If M and N are free modules, $M \otimes N$ is also a free module, and when x runs through a basis of M and y through a basis of N, the tensor products $x \otimes y$ constitute a basis of $M \otimes N$.

1.6 Exact sequences

Let us consider a sequence of two linear mappings u and v:

$$M' \xrightarrow{u} M \xrightarrow{v} M'';$$

it is called an *exact sequence* if $\mathrm{Im}(u) = \mathrm{Ker}(v)$. The inclusion $\mathrm{Im}(u) \subset \mathrm{Ker}(v)$ is equivalent to the equality $vu = 0$; when we must prove the exactness of a sequence, in most cases the equality $vu = 0$ will be obvious and the converse inclusion $\mathrm{Im}(u) \supset \mathrm{Ker}(v)$ will be the core of the proof. A sequence of several mappings, or an infinite sequence of mappings, is said to be exact if all the subsequences of two consecutive mappings are exact; for instance the sequence

$$0 \longrightarrow M' \xrightarrow{u} M \xrightarrow{v} M'' \longrightarrow 0 \tag{1.6.1}$$

is exact if u is injective, v surjective, and $\mathrm{Im}(u) = \mathrm{Ker}(v)$.

An exact sequence like (1.6.1) is called a *splitting exact sequence* if the median module M is the direct sum of $M_1 = \mathrm{Im}(u) = \mathrm{Ker}(v)$ and some submodule M_2; since the injection u determines an isomorphism $M' \to M_1$, and since the surjection v then determines an isomorphism $M_2 \to M''$, the splitting of the exact sequence (1.6.1) makes M become isomorphic to $M' \oplus M''$. Still under the assumption that the sequence (1.6.1) is exact and splits, there exist two mappings $u' : M \to M'$ and $v' : M'' \to M$ such that

$$M_2 = \mathrm{Ker}(u') = \mathrm{Im}(v'), \quad u'u = \mathrm{id}_{M'} \quad \text{and} \quad vv' = \mathrm{id}_{M''};$$

all this implies $\mathrm{id}_M = uu' + v'v$ and $u'v' = 0$.

Conversely if the sequence (1.6.1) is exact and if there exists $u' : M \to M'$ such that $u'u = \mathrm{id}_{M'}$, it is easy to prove that M is the direct sum of $\mathrm{Im}(u)$ and $\mathrm{Ker}(u')$, and this means that this exact sequence splits. In an analogous way the exact sequence (1.6.1) is also splitting if there exists $v' : M'' \to M$ such that $vv' = \mathrm{id}_{M''}$, because this equality implies $M = \mathrm{Im}(v') \oplus \mathrm{Ker}(v)$.

Besides, when the sequence (1.6.1) is no longer assumed to be exact, and when the only hypothesis is the existence of four mappings u, v, u', v' satisfying the five equalities

$$vu = 0, \quad u'v' = 0, \quad u'u = \mathrm{id}_{M'}, \quad vv' = \mathrm{id}_{M''} \quad \text{and} \quad uu' + v'v = \mathrm{id}_M, \tag{1.6.2}$$

then we get two splitting exact sequences:

$$0 \longrightarrow M' \xrightarrow{u} M \xrightarrow{v} M'' \longrightarrow 0,$$

$$0 \longleftarrow M' \xleftarrow{u'} M \xleftarrow{v'} M'' \longleftarrow 0.$$

Let $\mathcal{F}$ be an *additive functor* from the category $\mathcal{M}od(K)$ into itself; the additiveness of $\mathcal{F}$ means that for every couple of modules (M, N) the mapping $u \longmapsto \mathcal{F}(u)$ is a group morphism from $\mathrm{Hom}(M, N)$ into $\mathrm{Hom}(\mathcal{F}(M), \mathcal{F}(N))$ if $\mathcal{F}$ is covariant, into $\mathrm{Hom}(\mathcal{F}(N), \mathcal{F}(M))$ if $\mathcal{F}$ is contravariant. With this hypothesis, any equality $vu = 0$ implies $\mathcal{F}(v)\mathcal{F}(u) = 0$ or $\mathcal{F}(u)\mathcal{F}(v) = 0$; but this is not sufficient to conclude that every exact sequence is transformed by $\mathcal{F}$ into an exact sequence; $\mathcal{F}$ is called an *exact functor* if it transforms every exact sequence into an exact sequence. It is not difficult to prove that $\mathcal{F}$ is exact if and only if it transforms every exact sequence like (1.6.1) into an exact sequence.

Every splitting exact sequence is transformed into a splitting exact sequence, because four mappings u, v, u', v' satisfying the five equalities (1.6.2) are transformed by $\mathcal{F}$ into four mappings satisfying the analogous five equalities that prove the exactness and the splitting of the transformed sequence. Therefore if a functor is not exact, its lack of exactness can be observed only on exact sequences that do not split.

Unfortunately the most usual additive functors Hom and $\otimes$ are not exact for all rings K; the former is only *left exact* (for both variables), and the latter is *right exact*. This means that for all modules P and all exact sequences

$$\begin{array}{l} 0 \longrightarrow M' \longrightarrow M \longrightarrow M'', \\ \qquad\ N' \longrightarrow N \longrightarrow N'' \longrightarrow 0\,, \end{array}$$

we get these exact sequences:

$$\begin{array}{cccccccc} 0 & \longrightarrow & \mathrm{Hom}(N'', P) & \longrightarrow & \mathrm{Hom}(N, P) & \longrightarrow & \mathrm{Hom}(N', P) & , \\ 0 & \longrightarrow & \mathrm{Hom}(P, M') & \longrightarrow & \mathrm{Hom}(P, M) & \longrightarrow & \mathrm{Hom}(P, M'') & , \\ & & P \otimes N' & \longrightarrow & P \otimes N & \longrightarrow & P \otimes N'' & \longrightarrow \quad 0. \end{array}$$

From the right exactness of the functor $\otimes$ the following statement can be immediately derived: the exactness of the two sequences

$$M' \longrightarrow M \longrightarrow M'' \longrightarrow 0 \quad \text{and} \quad N' \longrightarrow N \longrightarrow N'' \longrightarrow 0$$

implies the exactness of the sequence

$$(M' \otimes N) \oplus (M \otimes N') \longrightarrow M \otimes N \longrightarrow M'' \otimes N'' \longrightarrow 0; \tag{1.6.3}$$

this is proved in Exercise (1.ex.10).

1.7 Projective modules and flat modules

A K-module P is called *injective* if the functor $\mathrm{Hom}(\dots, P)$ is exact; it is called *projective* if the functor $\mathrm{Hom}(P, \dots)$ is exact; and it is called *flat* if the functor $P \otimes \cdots$ is exact. Because of the left exactness of the functor Hom and the right exactness of the functor $\otimes$ (see **1.6**), we get at once the following statements: P is injective if and only if the mapping $\mathrm{Hom}(N, P) \to \mathrm{Hom}(N', P)$ is surjective whenever $N' \to N$ is injective; P is projective if and only if the mapping $\mathrm{Hom}(P, M) \to \mathrm{Hom}(P, M'')$ is surjective whenever $M \to M''$ is surjective; and P is flat if and only if the mapping $P \otimes N' \to P \otimes N$ is injective whenever $N' \to N$ is injective.

Here we never need injective modules; therefore only classical properties of projective or flat modules are recalled in this section.

A (finite or infinite) direct sum of modules is projective if and only if all the components are projective. Since K itself is projective, this proves that every free module is projective.

The following four statements are equivalent:

(a) P is projective;
(b) when the morphism $M \to M''$ is surjective, every morphism $P \to M''$ can be factorized through M by means of some morphism $P \to M$;
(c) every exact sequence $0 \to M' \to M \to P \to 0$ is splitting if it contains P at the fourth place;
(d) there exists a module P' such that $P \oplus P'$ is a free module.

Moreover if P is a finitely generated projective module, there exists a module P' such that $P \oplus P'$ is a free module with finite bases.

A (finite or infinite) direct sum of modules is flat if and only if all components are flat. Consequently, since K is flat, every projective module is flat.

Every tensor product of projective modules (resp. flat modules) is projective (resp. flat). When M is projective, $\mathrm{Hom}(M, N)$ is projective (resp. flat) whenever N is projective (resp. flat).

A module P is called *faithfully flat* if it is flat and if every equality $P \otimes M = 0$ implies $M = 0$.

When P is merely flat, every linear mapping $f : M \to N$ gives an exact sequence

$$0 \longrightarrow P \otimes \mathrm{Ker}(f) \longrightarrow P \otimes M \longrightarrow P \otimes N \longrightarrow P \otimes \mathrm{Coker}(f) \longrightarrow 0$$

which proves that the kernel and cokernel of $P \otimes f$ can be identified with $P \otimes \mathrm{Ker}(f)$ and $P \otimes \mathrm{Coker}(f)$. When P is faithfully flat, then f *is injective (resp. surjective) if and only if* $P \otimes f$ *is injective (resp. surjective)*; indeed the vanishing of $\mathrm{Ker}(f)$ is equivalent to the vanishing of $P \otimes \mathrm{Ker}(f)$, and the same for $\mathrm{Coker}(f)$.

More generally, when P is faithfully flat, *a sequence* $M' \to M \to M''$ *is exact if and only if the sequence* $P \otimes M' \longrightarrow P \otimes M \longrightarrow P \otimes M''$ *is exact.* Indeed the mapping $M' \longrightarrow \mathrm{Ker}(M \to M'')$ is surjective if and only if its gives a surjective mapping by the functor $P \otimes \cdots$.

Let N and N' be submodules of M. When P is merely flat, $P \otimes N$ and $P \otimes N'$ can be identified with submodules of $P \otimes M$, and

$$(P \otimes N) + (P \otimes N') = P \otimes (N + N'),$$
$$(P \otimes N) \cap (P \otimes N') = P \otimes (N \cap N');$$

indeed the former equality is obvious, and the latter one follows from it because of the exact sequence

$$0 \longrightarrow N \cap N' \longrightarrow N \oplus N' \longrightarrow N + N' \longrightarrow 0$$

in which the second arrow is $c \longmapsto (c, -c)$ and the third one is $(a, b) \longmapsto a + b$. Obviously every inclusion $N' \subset N$ in M gives a similar inclusion in $P \otimes M$. *When P is faithfully flat, the inclusion $N' \subset N$ is equivalent to $P \otimes N' \subset P \otimes N$.* Indeed these inclusions are respectively equivalent to the surjectiveness of the mappings $N \to N + N'$ and $P \otimes N \to P \otimes (N + N')$.

The finitely generated projective modules afford the most convenient frame to generalize the classical properties of vector spaces of finite dimension. For instance we may associate with each module M the dual module $M^* = \mathrm{Hom}(M, K)$, and there is a canonical mapping from M into the bidual module $(M^*)^*$; if P is finitely generated and projective, the mapping $P \to (P^*)^*$ is an isomorphism; indeed this is obviously true when P is a free module with a finite basis; when P is not free, there exists a module P' such that $P \oplus P'$ is free with a finite basis; whence an isomorphism $P \oplus P' \to ((P \oplus P')^*)^*$; but $(P \oplus P')^*$ is canonically isomorphic to $P^* \oplus P'^*$, and finally we get an isomorphism

$$P \oplus P' \longrightarrow (P^*)^* \oplus (P'^*)^*;$$

since this isomorphism maps each of the two components on the left side into the corresponding component on the right side, it gives two isomorphisms, among which is the announced isomorphism.

1.8 Finitely presented modules

When M is a K-module, every subset S that spans M, provides a surjective morphism $N \to M$ defined on the module $N = K^{(S)}$ freely generated by S (see **1.3**); if M is finitely generated, we can require S to be a finite subset, and N to have finite bases. Unfortunately in many cases it is not sufficient that N contains finite bases; we must also consider the kernel of the morphism $N \to M$, which is called the module of relations between the generators. A finite presentation of M is a finite subset of generators that gives a finitely generated module of relations between these generators. Nonetheless the existence of finite presentations is ensured by a much weaker definition.

(1.8.1) **Definition.** A module M is called *finitely presented* if it is finitely generated and if there exists a surjective morphism $f : P \to M$ such that P is projective and $\text{Ker}(f)$ finitely generated.

According to this definition every finitely generated projective module is finitely presented. Now the next theorem implies that in a finitely presented module every finite subset of generators actually gives a finitely generated module of relations.

(1.8.2) **Theorem.** *Let M be a finitely presented module, and $f : P \to M$ a surjective morphism from a projective module P onto M; the kernel of f is finitely generated if and only if P is finitely generated.*

This theorem is a consequence of the following lemma.

(1.8.3) **Schanuel's lemma.** *Let $f : P \to M$ and $f' : P' \to M$ be surjective morphisms from two projective modules P and P' onto the same module M; the modules $P \oplus \text{Ker}(f')$ and $P' \oplus \text{Ker}(f)$ are isomorphic. Consequently if P and $\text{Ker}(f')$ are finitely generated, so are P' and $\text{Ker}(f)$ (and conversely).*

Proof of (1.8.3). Since P is projective and f' surjective, there exists $u : P \to P'$ such that $f = f'u$. Let Q be the submodule of $P \oplus P'$ containing all pairs (x, x') such that $f(x) = f'(x')$. We get reciprocal isomorphisms between Q and $P \oplus \text{Ker}(f')$ if we map every $(x, x') \in Q$ to $(x,\ x' - u(x))$, and conversely every $(x, y') \in P \oplus \text{Ker}(f')$ to $(x,\ y' + u(x))$. In an analogous way we get a pair of reciprocal isomorphisms between Q and $P' \oplus \text{Ker}(f)$. □

Proof of (1.8.2). Since M is finitely generated, there exists a surjective mapping $f' : P' \to M$ defined on a finitely generated projective module P'; and since M is finitely presented, there exists a surjective mapping $f'' : P'' \to M$ such that P'' is projective and $\text{Ker}(f'')$ finitely generated. Now if P is finitely generated, since $\text{Ker}(f'')$ is also finitely generated, P'' and $\text{Ker}(f)$ are finitely generated because of (1.8.3). Conversely if $\text{Ker}(f)$ is finitely generated, since P' is finitely generated, P and $\text{Ker}(f')$ are also finitely generated. □

When the basic ring K is noetherian, every submodule of a finitely generated module is also finitely generated; therefore every finitely generated module is also finitely presented.

Suppose that M is finitely presented; every finite subset of generators gives a surjective morphism $f : M_0 \to M$ defined on a free module M_0 with a finite basis; every finite subset of generators of $\text{Ker}(f)$ gives a surjective morphism $M_1 \to \text{Ker}(f)$ defined on a free module M_1 with a finite basis; we can link together the exact sequences

$$M_1 \longrightarrow \text{Ker}(f) \longrightarrow 0 \quad \text{and} \quad 0 \longrightarrow \text{Ker}(f) \longrightarrow M_0 \longrightarrow M \longrightarrow 0,$$

in order to get the exact sequence that symbolizes the finite presentation of M :

$$M_1 \longrightarrow M_0 \longrightarrow M \longrightarrow 0.$$

A tensor product of finitely presented modules is a finitely presented module, because from the two exact sequences

$$M_1 \longrightarrow M_0 \longrightarrow M \longrightarrow 0 \quad \text{and} \quad N_1 \longrightarrow N_0 \longrightarrow N \longrightarrow 0,$$

by means of (1.6.3) we can derive the exact sequence

$$(M_1 \otimes N_0) \oplus (M_0 \otimes N_1) \longrightarrow M_0 \otimes N_0 \longrightarrow M \otimes N \longrightarrow 0.$$

Moreover $\mathrm{Hom}(P, M)$ is a finitely presented module whenever P is a finitely generated projective module, and M a finitely presented module; indeed from the above exact sequence $M_1 \to M_0 \to M \to 0$ we can derive the exact sequence

$$\mathrm{Hom}(P, M_1) \longrightarrow \mathrm{Hom}(P, M_0) \longrightarrow \mathrm{Hom}(P, M) \longrightarrow 0;$$

there exists P' such that $P \oplus P'$ is free with finite bases; thus $\mathrm{Hom}(P \oplus P', M_0)$ is also free with finite bases, and consequently $\mathrm{Hom}(P, M_0)$ is finitely generated and projective; and the same for $\mathrm{Hom}(P, M_1)$.

1.9 Changes of basic rings

When we meet additive groups that are modules over two commutative rings K and L, it is necessary to use precise notations like $\mathrm{Hom}_L(M, N)$, $M \otimes_L N$, $\mathrm{T}_L(M)$, $\mathrm{S}_L(M)$,... indicating which basic ring is referred to. Here we study what happens when there is a ring morphism $f : K \to L$; in this case every L-module M is also a K-module: for all $\kappa \in K$ and all $x \in M$ the product κx is by definition $f(\kappa)x$. Such a ring morphism $f : K \to L$ is called an *extension* of the ring K, even when it is not injective. It may even occur that f is surjective; for instance if M is not a faithful K-module, the ideal containing all $\kappa \in K$ such that $\kappa M = 0$ is not reduced to 0, and M becomes a faithful module over the quotient of K by this ideal.

First let us suppose that M and N are L-modules, and therefore also K-modules; there is obviously a canonical injection $\mathrm{Hom}_L(M, N) \to \mathrm{Hom}_K(M, N)$ and a canonical surjection $M \otimes_K N \to M \otimes_L N$; the kernel of the latter is spanned by the elements $\lambda x \otimes y - x \otimes \lambda y$ with λ running through L and x and y through M and N; both mappings are isomorphisms when f is surjective. If we consider tensor and symmetric algebras, we find canonical morphisms of graded algebras

$$\mathrm{T}_K(M) \longrightarrow \mathrm{T}_L(M) \quad \text{and} \quad \mathrm{S}_K(M) \longrightarrow \mathrm{S}_L(M);$$

in degree 0 we find merely the ring morphism $f : K \to L$; in degree 1 we find the identity mapping of $M = \mathrm{T}^1_K(M) = \mathrm{T}^1_L(M) = \mathrm{S}^1_K(M) = \mathrm{S}^1_L(M)$, and for each degree $k \geq 2$ the mappings $\mathrm{T}_K^k(M) \to \mathrm{T}_L^k(M)$ and $\mathrm{S}_K^k(M) \to \mathrm{S}_L^k(M)$ are surjective, and even bijective when f is surjective.

Let us now suppose that M is merely a K-module; we may derive from it two L-modules, called the *extensions* of M, which are $L \otimes_K M$ and $\mathrm{Hom}_K(L, M)$; an element μ of L multiplies an element $\lambda \otimes x$ of $L \otimes_K M$ or an element ξ of $\mathrm{Hom}_K(L, M)$ in this way:

$$\mu(\lambda \otimes x) = (\mu\lambda) \otimes x \quad \text{and} \quad (\mu\xi)(\lambda) = \xi(\mu\lambda).$$

There are canonical K-linear mappings

$$M \longrightarrow L \otimes_K M \quad \text{and} \quad \mathrm{Hom}_K(L, M) \longrightarrow M$$

defined by $x \longmapsto 1_L \otimes x$ and $\xi \longmapsto \xi(1_L)$; the former is not always injective, and the latter is not always surjective. When $L = K$, both are bijective and usually $K \otimes_K M$ and $\mathrm{Hom}_K(K, M)$ are identified with M. But in general these two extensions are different, and isomorphisms can be found between them only under restrictive hypotheses. The next lemma gives more details in a particular case.

(1.9.1) **Lemma.** *Let us suppose that $f : K \to L$ is surjective; let $\mathfrak{a}$ be its kernel. Then $L \otimes_K M$ is canonically isomorphic to the quotient $M/\mathfrak{a}M$, and $\mathrm{Hom}_K(L, M)$ to the submodule of all elements $x \in M$ such that $\mathfrak{a}x = 0$.*

Proof. The right exactness of the functor $\otimes$ gives the exact sequence

$$\mathfrak{a} \otimes_K M \longrightarrow K \otimes_K M \longrightarrow L \otimes_K M \longrightarrow 0;$$

when $K \otimes_K M$ is identified with M, the image of $\mathfrak{a} \otimes_K M$ in M is $\mathfrak{a}M$, and thus $L \otimes_K M = M/\mathfrak{a}M$. The left exactness of the functor Hom allows us to prove the statement involving $\mathrm{Hom}_K(L, M)$. □

Here we shall only use extensions like $L \otimes_K M$, but the other extension $\mathrm{Hom}_K(L, M)$ may appear in other contexts. Let us begin with these isomorphisms:

$$\mathrm{T}_L(L \otimes_K M) \cong L \otimes_K \mathrm{T}_K(M), \tag{1.9.2}$$

$$\mathrm{S}_L(L \otimes_K M) \cong L \otimes_K \mathrm{S}_K(M). \tag{1.9.3}$$

Most of the here mentioned isomorphisms are easy consequences of the definitions and universal properties of the objects under consideration, and the following explanations about (1.9.2) should be a sufficient model for all the others. Observe that the L-algebras $\mathrm{T}_L(L \otimes_K M)$ and $L \otimes_K \mathrm{T}_K(M)$ are generated by the elements which in each algebra are written $1_L \otimes x$ (with $x \in M$). The mapping which maps every $\lambda \otimes x$ in $L \otimes_K M$ to the corresponding $\lambda \otimes x$ in $L \otimes_K \mathrm{T}_K(M)$, is L-linear, and therefore extends to a morphism of L-algebras $\mathrm{T}_L(L \otimes_K M) \to L \otimes_K \mathrm{T}_K(M)$. Conversely the mapping $x \longmapsto 1_L \otimes x$ extends to a morphism of K-algebras $\mathrm{T}_K(M) \to \mathrm{T}_L(L \otimes_K M)$, and by combining it with the canonical morphism $L \to \mathrm{T}_L(L \otimes_K M)$, the image of which lies in the center of $\mathrm{T}_L(L \otimes_K M)$, we get a morphism of K-algebras $L \otimes_K \mathrm{T}_K(M) \to \mathrm{T}_L(L \otimes_K M)$. This converse morphism too is L-linear, and the behaviour of both morphisms on the elements $1_L \otimes x$ shows that they are reciprocal isomorphisms of L-algebras. □

When M is a K-module as previously, and N an L-module, there are canonical isomorphisms

$$\mathrm{Hom}_K(M, N) \cong \mathrm{Hom}_L(L \otimes_K M,\ N), \tag{1.9.4}$$
$$M \otimes_K N \cong (L \otimes_K M) \otimes_L N. \tag{1.9.5}$$

To be complete, let us also mention the isomorphisms

$$\mathrm{Hom}_K(N, M) \ \cong\ \mathrm{Hom}_L(N,\ \mathrm{Hom}_K(L, M)).$$

The isomorphisms (1.9.4) and (1.9.5) lead to an easy proof of this statement: *the extension $L \otimes_K M$ is L-projective (resp. L-flat) whenever M is K-projective (resp. K-flat).* When M is injective, its injectiveness is inherited by the other extension $\mathrm{Hom}_K(L, M)$.

When both M and N are K-modules, we find the canonical isomorphisms

$$L \otimes_K (M \otimes_K N) \ \cong\ (L \otimes_K M) \otimes_L (L \otimes_K N). \tag{1.9.6}$$

To be complete, let us also mention the isomorphisms

$$\mathrm{Hom}_K(L,\ \mathrm{Hom}_K(M, N)) \ \cong\ \mathrm{Hom}_L(L \otimes_K M,\ \mathrm{Hom}_K(L, N)).$$

Sometimes we also need the following morphism which is not always bijective:

$$L \otimes_K \mathrm{Hom}_K(M, N) \longrightarrow \mathrm{Hom}_L(L \otimes_K M,\ L \otimes_K N),$$
$$\lambda \otimes g \longmapsto (\mu \otimes x \longmapsto \lambda\mu \otimes g(x)).$$

(1.9.7) **Proposition.** *When M is projective, the above canonical morphism from $L \otimes_K \mathrm{Hom}_K(M, N)$ into $\mathrm{Hom}_L(L \otimes_K M,\ L \otimes_K N)$ is an isomorphism.*

Proof. When M is a direct sum, this morphism is bijective if and only if it is bijective when M is successively replaced with each direct summand; consequently it suffices to prove (1.9.7) when $M = K$; in this case it is obviously bijective. □

The extension $K \to L$ is called *flat* (resp. *faithfully flat*) if L is a flat (resp. faithfully flat) K-module. When the extensions $K \to L$ and $L \to L'$ are flat (resp. faithfully flat), it is clear that $K \to L'$ is flat (resp. faithfully flat). Let $K \to K'$ and $K \to L$ be two extensions of K, and $L' = K' \otimes_K L$; when the extension $K \to L$ is flat (resp. faithfully flat), then $K' \to L'$ is also flat (resp. faithfully flat); indeed, for every K'-module M',

$$L' \otimes_{K'} M' \ \cong\ (K' \otimes_K L) \otimes_{K'} M' \ \cong\ L \otimes_K K' \otimes_{K'} M' \ \cong\ L \otimes_K M'.$$

Other properties of flatness or faithful flatness are stated in **1.7**.

(1.9.8) **Proposition.** *When $K \to L$ is a faithfully flat extension and M a K-module, M is finitely generated (resp. finitely presented) if and only if the L-module $L \otimes_K M$ is finitely generated (resp. finitely presented).*

Proof. Without any hypothesis on $K \to L$, every exact sequence $M_1 \to M_0 \to M \to 0$ is transformed by the functor $L \otimes \cdots$ into an exact sequence, and this proves that the mentioned properties of M are inherited by $L \otimes_K M$. Conversely if $L \otimes_K M$ is finitely generated, there exist $x_1, x_2, \dots, x_n \in M$ such that $L \otimes M$ is generated by all $1_L \otimes x_i$ with $i = 1, 2, \dots, n$; let M_0 be a free K-module with basis $(e_1, e_2, \dots, e_n)$, and f the K-linear mapping $M_0 \to M$ such that $f(e_i) = x_i$ for $i = 1, 2, \dots, n$; this mapping f is surjective because $L \otimes f$ is surjective and L faithfully flat; therefore M is finitely generated. The kernel of $L \otimes f$ is canonically isomorphic to $L \otimes \mathrm{Ker}(f)$ because L is flat; if $L \otimes M$ is finitely presented, this kernel is finitely generated; consequently $L \otimes \mathrm{Ker}(f)$ is finitely generated; by a similar argument $\mathrm{Ker}(f)$ is also finitely generated; thus M is finitely presented. □

(1.9.9) **Proposition.** *When $K \to L$ is a flat extension, and M a finitely presented K-module, then for each K-module N the canonical morphism*

$$L \otimes_K \mathrm{Hom}_K(M, N) \longrightarrow \mathrm{Hom}_L(L \otimes_K M,\ L \otimes_K N)$$

is an isomorphism.

Proof. It is obviously an isomorphism when M is a free module with a finite basis. When M is merely finitely presented, there is an exact sequence $M_1 \to M_0 \to M \to 0$ in which M_1 and M_0 are free modules with finite bases; it leads to the following diagram:

$$\begin{array}{ccccc} 0 \to L \otimes \mathrm{Hom}(M, N) & \to & L \otimes \mathrm{Hom}(M_0, N) & \to & L \otimes \mathrm{Hom}(M_1, N) \\ \downarrow & & \downarrow & & \downarrow \\ 0 \to \mathrm{Hom}_L(L \otimes M,\ L \otimes N) & \to & \mathrm{Hom}_L(L \otimes M_0,\ L \otimes N) & \to & \mathrm{Hom}_L(L \otimes M_1,\ L \otimes N) \end{array}$$

The two lines are exact because of the flatness of L and the left exactness of the functors Hom_K and Hom_L ; the second and third vertical arrows are isomorphisms; now it is easy to prove that the first vertical arrow too is an isomorphism. □

(1.9.10) **Proposition.** *When $K \to L$ is a faithfully flat extension and P a K-module, P is a finitely generated projective K-module if and only if $L \otimes_K P$ is a finitely generated projective L-module.*

Proof. When P is finitely generated and projective over K, so is $L \otimes_K P$ over L because of (1.9.4). The hypothesis about $K \to L$ is only needed when we conversely suppose that $L \otimes_K P$ is a finitely generated projective L-module; then according to (1.9.8), P is finitely presented; the projectiveness of P is equivalent to the exactness of the functor $\mathrm{Hom}_K(P, \dots)$, and since L is faithfully flat, it is equivalent to the exactness of the functor $L \otimes_K \mathrm{Hom}_K(P, \dots)$; because of (1.9.9) it is also equivalent to the exactness of the functor $\mathrm{Hom}_L(L \otimes_K P,\ L \otimes_K \cdots)$; now the projectiveness of $L \otimes_K P$ implies the projectiveness of P. □

1.10 Rings and modules of fractions

A *multiplicative subset* of K (also called a multiplicatively closed subset) is a subset S that contains 1 and all products of two elements of S. We consider the category $\mathcal{C}$ of all ring morphisms $f : K \to L$ such that $f(s)$ is invertible in L for all $s \in S$; a morphism from f to $f' : K \to L'$ is a ring morphism $u : L \to L'$ such that $f' = uf$. The morphism $K \to 0$ is a final universal object in $\mathcal{C}$; it is its unique object (up to isomorphy) when S contains 0, and it is preferable also to accept this quite degenerate case. If $\mathcal{C}$ contains an initial universal object $f : K \to U$, it is unique up to isomorphy (see (1.2.1)), and U is called the *ring of fractions of K with denominator in S* and denoted by $S^{-1}K$.

Let us prove the existence of $S^{-1}K$. Two elements (λ, s) and (λ', s') of $K \times S$ are said to be equivalent if there exists $t \in S$ such that $t(s'\lambda - s\lambda') = 0$; it is easy to prove that an equivalence relation has been defined in this way; let $S^{-1}K$ be the set of equivalence classes; the image of (λ, s) in $S^{-1}K$ is written as a fraction λ/s. This set is given a ring structure with the following operations:

$$\frac{\lambda}{s} + \frac{\mu}{t} = \frac{t\lambda + s\mu}{st} \qquad \text{and} \qquad \frac{\lambda}{s}\frac{\mu}{t} = \frac{\lambda\mu}{st};$$

it is easy to prove that this addition and this multiplication are well defined on the set of equivalence classes and satisfy the required properties for $S^{-1}K$ to be a ring; the zero and unit elements are respectively the fractions $0/1$ and $1/1$. Moreover the mapping $f : K \longrightarrow S^{-1}K$ which maps every λ to the fraction $\lambda/1$ is a ring morphism, and it is easy to prove that it is an initial universal object in $\mathcal{C}$, as stated in the following theorem.

(1.10.1) **Theorem.** *For every ring morphism $f : K \to L$ such that $f(s)$ is invertible for all $s \in S$, there exists a unique ring morphism $f' : S^{-1}K \to L$ such that $f(\lambda) = f'(\lambda/1)$ for all $\lambda \in K$.*

It is clear that $f'(\lambda/s) = f(\lambda)f(s)^{-1}$ for all fractions λ/s. Moreover an element $\lambda \in K$ belongs to the kernel of the canonical morphism $f : K \to S^{-1}K$ if and only if there exists $t \in S$ such that $t\lambda = 0$.

If S' is a multiplicative subset containing S, the universal property of $S^{-1}K$ implies that the canonical mapping $K \to S'^{-1}K$ can be factorized through $S^{-1}K$, and thus we get a ring morphism $S^{-1}K \to S'^{-1}K$.

When K is an *integral domain* (a ring without divisors of zero, and not reduced to 0), the set of all nonzero elements of K is a multiplicative subset; the corresponding ring of fractions is a field, which is called the *field of fractions of K*; all rings of fractions of K and K itself can be identified with subrings of this field.

Now let M be any K-module, and $\mathcal{D}$ the category in which the objects are the K-linear mappings $g : M \to P$ from M into any $(S^{-1}K)$-module P; a morphism from g to $g' : M \to P'$ is a $(S^{-1}K)$-linear mapping $u : P \to P'$ such that

$g' = ug$. If $\mathcal{D}$ contains an initial universal object $f_M : M \longrightarrow V$, it is unique up to isomorphy, V is called the *module of fractions of M with denominator in S* and denoted by $S^{-1}M$.

Let us prove the existence of $S^{-1}M$. Two elements (x, s) and (x', s') of $M \times S$ are said to be equivalent if there exists $t \in S$ such that $t(s'x - sx') = 0$; it is easy to prove that an equivalence relation has been defined; the set of equivalence classes is the wanted module $S^{-1}M$ and the image of (x, s) in $S^{-1}M$ is by definition the fraction x/s. It is easy to prove that the following operations are well defined and make $S^{-1}M$ become an $(S^{-1}K)$-module:

$$\frac{x}{s} + \frac{y}{t} = \frac{tx + sy}{st} \quad \text{and} \quad \frac{\lambda}{s}\frac{y}{t} = \frac{\lambda y}{st}.$$

Moreover the mapping $f_M : M \longrightarrow S^{-1}M$ which maps each x to $x/1$, is a K-linear mapping, and it is easy to prove that it is an initial universal object in $\mathcal{D}$, in other words: *for every K-linear mapping $g : M \to P$ into an $(S^{-1}K)$-module P, there exists a unique $(S^{-1}K)$-linear mapping $g' : S^{-1}M \to P$ such that $g(x) = g'(x/1)$ for all $x \in M$.*

An element $x \in M$ belongs to the kernel of the canonical morphism $x \longmapsto x/1$ if and only if there exists $t \in S$ such that $tx = 0$.

The canonical morphism $x \longmapsto x/1$ is an isomorphism from M onto $S^{-1}M$ if and only if the endomorphism $x \longmapsto sx$ is bijective from M onto M for all $s \in S$; this condition is necessary and sufficient for M to have a structure of $(S^{-1}K)$-module compatible with its structure of K-module.

If S' is a multiplicative subset of K containing S, the universal property of $S^{-1}M$ gives a canonical $(S^{-1}K)$-linear mapping $S^{-1}M \to S'^{-1}M$.

Two kinds of multiplicative subsets will be used later. First from any element $s \in K$ we can derive the multiplicative subset $S = \{1, s, s^2, s^3, \dots\}$ of all powers of s; then the ring $S^{-1}K$ is usually denoted by K_s; it is reduced to 0 if and only if s is nilpotent. Similarly M_s is the module of fractions of M with denominator a power of s.

An ideal $\mathfrak{p}$ of K is called a *prime ideal* when the following four equivalent statements are true:

(a) $\mathfrak{p} \neq K$, and whenever $\mathfrak{p}$ contains a product xy of elements of K, it contains x or y;

(b) $\mathfrak{p} \neq K$, and whenever $\mathfrak{p}$ contains a product $\mathfrak{a}\mathfrak{b}$ of ideals of K, it contains $\mathfrak{a}$ or $\mathfrak{b}$;

(c) the quotient $K/\mathfrak{p}$ is an integral domain (without divisors of zero and not reduced to 0);

(d) the complementary subset $S = K \setminus \mathfrak{p}$ is a multiplicative subset.

The corresponding ring $S^{-1}K$ and modules $S^{-1}M$ are denoted by $K_\mathfrak{p}$ and $M_\mathfrak{p}$, and are called the *localizations* of K and M at the prime ideal $\mathfrak{p}$.

Every ring K (not reduced to 0) contains prime ideals; indeed Zorn's Lemma implies the existence of *maximal ideals*, and maximal ideals are prime, because the following two assertions are equivalent:

(a) the ideal $\mathfrak{m}$ is maximal (it is contained in no ideal other than $\mathfrak{m}$ and K, and $\mathfrak{m} \neq K$);

(b) the quotient $K/\mathfrak{m}$ is a field.

Zorn's Lemma even implies that *every ideal other than K is contained in a maximal ideal.*

Let us also recall that for any ring K the following two statements are equivalent:

(a) K contains exactly one maximal ideal;

(b) K is not reduced to 0, and a sum of noninvertible elements is never invertible.

When these statements are true, K is called a *local ring*. In a local ring the unique maximal ideal $\mathfrak{m}$ is the subset of all noninvertible elements. The quotient $K/\mathfrak{m}$ is called the *residue field* of the local ring K.

When $\mathfrak{p}$ is a prime ideal of K, the elements of $K_\mathfrak{p}$ which are not invertible, are the elements of $\mathfrak{p}K_\mathfrak{p}$ (this notation is meaningful since $K_\mathfrak{p}$ is a K-module); this proves that the localized ring $K_\mathfrak{p}$ is a local ring.

Here is a first application of these notions.

(1.10.2) **Theorem.** *The radical of K, which is the subset* $\mathrm{Rad}(K)$ *of all its nilpotent elements, is also the intersection of all its prime ideals.*

Proof. It is clear that every prime ideal contains all nilpotent elements. Conversely we prove that when s is not nilpotent, there exists a prime ideal that does not contain it. Indeed let us consider the ring K_s of fractions with denominator in the set of powers of s, and the canonical morphism $f : K \to K_s$. Since s is not nilpotent, K_s is not reduced to 0. If $\mathfrak{m}$ is a maximal ideal of K_s, $f^{-1}(\mathfrak{m})$ is a prime ideal $\mathfrak{p}$ of K because $K/\mathfrak{p}$ is isomorphic to a subring of the field $K_s/\mathfrak{m}$. Since $f(s)$ is invertible and cannot belong to $\mathfrak{m}$, we are sure that $s \notin \mathfrak{p}$. □

Every multiplicative subset S affords a functor from $\mathcal{M}od(K)$ toward $\mathcal{M}od(S^{-1}K)$; with every K-module M we associate the $(S^{-1}K)$-module $S^{-1}M$, and with every K-linear mapping $f : M \longrightarrow N$ we associate the $(S^{-1}K)$-linear mapping $S^{-1}f : S^{-1}M \to S^{-1}N$ defined in this way: because of the universal property of $S^{-1}M$, the mapping $M \longrightarrow N \to S^{-1}N$ can be factorized in a unique way through $S^{-1}M$.

(1.10.3) **Theorem.** *The functor $M \longmapsto S^{-1}M$ is exact. Moreover it is equivalent to the functor $M \longmapsto S^{-1}K \otimes M$ corresponding to the extension $K \to S^{-1}K$ of the basic ring; in other words, for each K-module M there is a canonical $(S^{-1}K)$-linear isomorphism $S^{-1}M \to S^{-1}K \otimes M$, and for each K-linear mapping $f : M \longrightarrow N$ the canonical isomorphisms $S^{-1}M \to S^{-1}K \otimes M$ and $S^{-1}N \to S^{-1}K \otimes N$ intertwine $S^{-1}f$ and $S^{-1}K \otimes f$.*

Proof. Let us consider an exact sequence $M' \xrightarrow{u} M \xrightarrow{v} M''$, and let us prove the exactness of $S^{-1}M' \to S^{-1}M \to S^{-1}M''$. It is clear that $S^{-1}v \circ S^{-1}u = 0$; therefore we must prove that every fraction $x/s \in S^{-1}M$ that belongs to the kernel of $S^{-1}v$, must belong to the image of $S^{-1}u$. Indeed there exists $t \in S$ such that $tv(x) = 0$, in other words, $tx \in \mathrm{Ker}(v)$; consequently there exists $x' \in M'$ such that $u(x') = tx$, whence $x/s = S^{-1}u(x'/st)$ as desired.

The proof of the second part of (1.10.3) is still easier: there is a mapping $S^{-1}M \to S^{-1}K \otimes M$ resulting from the universal property of $S^{-1}M$, and there is a converse mapping $(\lambda/s) \otimes x \longmapsto (\lambda x)/s$ resulting from the universal property of the tensor product; obviously they are reciprocal isomorphisms. The statement about $S^{-1}f$ and $S^{-1}K \otimes f$ is also evident. □

(1.10.4) **Corollary.** *The ring extension $K \to S^{-1}K$ is flat.*

Indeed the exactness of the functor $M \longmapsto S^{-1}K \otimes M$ is an immediate consequence of the exactness of the functor $M \longmapsto S^{-1}M$. □

(1.10.5) **Corollary.** *For all K-modules M and M' there is a canonical isomorphism*

$$S^{-1}(M \otimes_K M') \;\cong\; S^{-1}M \otimes_{S^{-1}K} S^{-1}M'.$$

Indeed, according to (1.9.6), there is a canonical isomorphism

$$S^{-1}K \otimes_K (M \otimes_K M') \;\cong\; (S^{-1}K \otimes_K M) \otimes_{S^{-1}K} (S^{-1}K \otimes_K M'). \qquad \square$$

(1.10.6) **Corollary.** *Let $\mathfrak{a}$ be the ideal of K generated by the elements $s_1, s_2, \dots, s_n$. The direct product L of the rings K_{s_i} (where $i = 1, 2, \dots, n$) is faithfully flat if and only if $\mathfrak{a} = K$.* When $\mathfrak{a} = K$, this ring L is called a *Zariski extension* of K.

Proof. Since L is a direct sum of flat modules, it is flat. Let us suppose that $L \otimes M = 0$; therefore for each $x \in M$ there exists a positive integer k such that $s_i^k x = 0$ for $i = 1, 2, \dots, n$. If $\mathfrak{a} = K$, there exist λ_1, $\lambda_2, \dots, \lambda_n$ such that $1 = \sum_{i=1}^n \lambda_i s_i$; let us set $m = n(k-1)+1$; from the equality $1 = (\sum_{i=1}^n \lambda_i s_i)^m$ we can deduce the existence of $\mu_1, \mu_2, \dots, \mu_n$ such that $1 = \sum_{i=1}^n \mu_i s_i^k$; this shows that the equalities $s_i^k x = 0$ imply $x = 0$; therefore $M = 0$. Conversely if $\mathfrak{a} \neq K$, the equality $(K/\mathfrak{a}) \otimes L = 0$ (a consequence of (1.9.1)) shows that L is not faithfully flat. □

Among the consequences of Theorem (1.10.3) there is the fact that $S^{-1}N$ can be considered as a submodule of $S^{-1}M$ whenever N is a submodule of M; indeed the sequence $0 \longrightarrow S^{-1}N \longrightarrow S^{-1}M$ is exact. The proof of the following lemma is left to the reader.

(1.10.7) **Lemma.** *When N and N' are submodules of M, then*

$$S^{-1}(N+N') \;=\; S^{-1}N + S^{-1}N' \quad \textit{and} \quad S^{-1}(N \cap N') \;=\; S^{-1}N \,\cap\, S^{-1}N'.$$

Whereas the functor $M \longmapsto S^{-1}M$ has a nice behaviour relatively to tensor products (see (1.10.5)), it requires more caution when the functor Hom is involved.

(1.10.8) **Proposition.** *For all K-modules M and N there is a canonical morphism*

$$S^{-1}\mathrm{Hom}_K(M,N) \longrightarrow \mathrm{Hom}_{S^{-1}K}(S^{-1}M,\ S^{-1}N)$$

which is an isomorphism whenever M is a finitely presented module.

Each element g/s of $S^{-1}\mathrm{Hom}(M,N)$ is mapped to $x/t \longmapsto g(x)/st$. This proposition is an immediate corollary of (1.9.9) because the extension $K \to S^{-1}K$ is flat, as stated in (1.10.4). □

Let us come back to the ring extension $f : K \to S^{-1}K$. Later it shall be necessary to know the prime ideals of $S^{-1}K$. As explained above for the submodules of any K-module M, to each ideal $\mathfrak{a}$ of K corresponds an ideal $S^{-1}\mathfrak{a}$ of $S^{-1}K$; it is the ideal generated by $f(\mathfrak{a})$. Conversely with every ideal $\mathfrak{b}$ of $S^{-1}K$ we associate the ideal $f^{-1}(\mathfrak{b})$ of K; it is clear that $f^{-1}(\mathfrak{b}) \cap S = \emptyset$ whenever $\mathfrak{b} \neq S^{-1}K$. The proof of the following lemma is left to the reader.

(1.10.9) **Lemma.** *The mapping $\mathfrak{q} \longmapsto f^{-1}(\mathfrak{q})$ is a bijection from the set of prime ideals $\mathfrak{q}$ of $S^{-1}K$ onto the set of prime ideals $\mathfrak{p}$ of K such that $\mathfrak{p} \cap S$ is empty; the converse bijection is $\mathfrak{p} \longmapsto S^{-1}\mathfrak{p}$.*

More generally $S^{-1}f^{-1}(\mathfrak{b}) = \mathfrak{b}$ for every ideal $\mathfrak{b}$ of $S^{-1}K$, but if $\mathfrak{a}$ is an ideal of K, the ideal $f^{-1}(S^{-1}\mathfrak{a})$ may be larger than $\mathfrak{a}$, since it is the set of all $\lambda \in K$ such that $s\lambda \in \mathfrak{a}$ for some $s \in S$.

1.11 Localization and globalization

The *spectrum* of the ring K, denoted by $\mathrm{Spec}(K)$, is the set of all its prime ideals. With every ideal $\mathfrak{a}$ of K we associate the subset $\mathcal{V}(\mathfrak{a})$ of all $\mathfrak{p} \in \mathrm{Spec}(K)$ such that $\mathfrak{p} \supset \mathfrak{a}$.

(1.11.1) **Lemma.** *The mapping $\mathfrak{a} \longmapsto \mathcal{V}(\mathfrak{a})$ has the following properties:*

(a) $\mathcal{V}(K) = \emptyset$ *and* $\mathcal{V}(0) = \mathrm{Spec}(K)$*;*

(b) *an inclusion $\mathfrak{a} \subset \mathfrak{b}$ implies $\mathcal{V}(\mathfrak{a}) \supset \mathcal{V}(\mathfrak{b})$;*

(c) *when $(\mathfrak{a}_j)_{j\in J}$ is any family of ideals of K, then*

$$\bigcap_{j\in J} \mathcal{V}(\mathfrak{a}_j) = \mathcal{V}(\sum_{j\in J} \mathfrak{a}_j);$$

(d) *for all ideals $\mathfrak{a}$ and $\mathfrak{b}$ of K, we can write*

$$\mathcal{V}(\mathfrak{a}) \cup \mathcal{V}(\mathfrak{b}) = \mathcal{V}(\mathfrak{a} \cap \mathfrak{b}) = \mathcal{V}(\mathfrak{a}\mathfrak{b}).$$

All these statements are evident except perhaps the last one, a consequence of the inclusions

$$\mathcal{V}(\mathfrak{a}) \cup \mathcal{V}(\mathfrak{b}) \subset \mathcal{V}(\mathfrak{a} \cap \mathfrak{b}) \subset \mathcal{V}(\mathfrak{a}\mathfrak{b}) \subset \mathcal{V}(\mathfrak{a}) \cup \mathcal{V}(\mathfrak{b}).$$

Lemma (1.11.1) proves that there is a topology on Spec(K) for which the closed subsets are the subsets $\mathcal{V}(\mathfrak{a})$; it is called the *Zariski topology* of Spec(K). For every $\mathfrak{p} \in$ Spec(A) the topological closure of $\{\mathfrak{p}\}$ is $\mathcal{V}(\mathfrak{p})$; thus the point $\{\mathfrak{p}\}$ is closed if and only if $\mathfrak{p}$ is a maximal ideal; this topology is almost never a Hausdorff topology.

As explained in **1.10**, with each $\mathfrak{p} \in$ Spec(K) is associated a localized ring $K_\mathfrak{p}$ with maximal ideal $\mathfrak{p}K_\mathfrak{p}$, and a residue field $K_\mathfrak{p}/\mathfrak{p}K_\mathfrak{p}$. The kernel of the ring morphism $K \to K_\mathfrak{p} \to K_\mathfrak{p}/\mathfrak{p}K_\mathfrak{p}$ is exactly $\mathfrak{p}$. Therefore there is an injective morphism from the integral domain $K/\mathfrak{p}$ into the residue field, which extends to a morphism from the field of fractions of $K/\mathfrak{p}$ into the residue field. This morphism is surjective, consequently bijective (since every field morphism is injective). Therefore the residue field $K_\mathfrak{p}/\mathfrak{p}K_\mathfrak{p}$ can be identified with the field of fractions of $K/\mathfrak{p}$. At every maximal ideal $\mathfrak{m}$ the residue field is $K/\mathfrak{m}$.

Every module M gives a localized module $M_\mathfrak{p}$ at the point $\mathfrak{p}$, and a vector space $F_\mathfrak{p} \otimes_K M$ over the residue field $F_\mathfrak{p} = K_\mathfrak{p}/\mathfrak{p}K_\mathfrak{p}$. The associativity of tensor products allows us to write

$$F_\mathfrak{p} \otimes_K M \;=\; F_\mathfrak{p} \otimes_{K_\mathfrak{p}} K_\mathfrak{p} \otimes_K M \;=\; F_\mathfrak{p} \otimes_{K_\mathfrak{p}} M_\mathfrak{p};$$

and then (1.9.1) allows us to identify $F_\mathfrak{p} \otimes M$ with $M_\mathfrak{p}/\mathfrak{p}M_\mathfrak{p}$. For a maximal ideal $\mathfrak{m}$ we can write $F_\mathfrak{m} = K/\mathfrak{m}$ and $F_\mathfrak{m} \otimes M = M/\mathfrak{m}M$.

Let $f : K \to L$ be any ring morphism. When $\mathfrak{q}$ is a prime ideal of L, then $f^{-1}(\mathfrak{q})$ is a prime ideal of K, because the quotient $K/f^{-1}(\mathfrak{q})$ is isomorphic to a subring of $L/\mathfrak{q}$. This defines a mapping Spec(f) from Spec(L) into Spec(K), and it is easy to prove that it is continuous. Thus we have constructed a contravariant functor from the category $\mathcal{C}om(\mathbb{Z})$ of all commutative rings to the category of topological spaces.

(1.11.2) **Example.** Let F be an algebraically closed field, and $K = F[x_1, x_2, \ldots, x_n]$ a ring of polynomials over F; K can be identified with the ring of polynomial functions on F^n and its field of fractions $F(x_1, x_2, \ldots, x_n)$ with the field of rational functions on F^n; every ring of fractions of K is a subring of this field. At every point $a = (a_1, a_2, \ldots, a_n) \in F^n$ there is a valuation morphism $K \to F$ defined by $f \longmapsto f(a)$; if $\mathfrak{m}_a$ is its kernel, the quotient $K/\mathfrak{m}_a$ is isomorphic to F and consequently $\mathfrak{m}_a$ is a maximal ideal; a theorem of Hilbert (Nullstellensatz) proves that the image of the mapping $a \longmapsto \mathfrak{m}_a$ is the set of all maximal ideals. We identify the point $a \in F^n$ with the point $\mathfrak{m}_a \in$ Spec(K). The localized ring K_a at the maximal ideal $\mathfrak{m}_a$ is the subring of all rational functions g that are defined at the point a, and the image of g in the residue field (which is isomorphic to $K/\mathfrak{m}_a = F$) can be identified with the value $g(a)$ of g at the point a. Now let $\mathfrak{p}$ be any prime ideal of K, and $\mathcal{V}'(\mathfrak{p})$ the subset of all $a \in F^n$ such that $f(a) = 0$ for all $f \in \mathfrak{p}$; a subset like $\mathcal{V}'(\mathfrak{p})$ is called an *irreducible algebraic submanifold* of

F^n, and its points are the maximal ideals belonging to $\mathcal{V}(\mathfrak{p})$. The localized ring at $\mathfrak{p}$ is the subring of all rational functions that are defined at least at one point of $\mathcal{V}'(\mathfrak{p})$ (consequently at almost all points of $\mathcal{V}'(\mathfrak{p})$). The ideal $\mathfrak{p}$ is the kernel of the ring morphism that maps every polynomial function $f \in K$ to its restriction to the subset $\mathcal{V}'(\mathfrak{p})$, and thus $K/\mathfrak{p}$ is identified with a ring of functions on $\mathcal{V}'(\mathfrak{p})$, the so-called *regular functions*. The residue field at $\mathfrak{p}$ is the field of fractions of $K/\mathfrak{p}$ and its elements are called the *rational functions* on $\mathcal{V}'(\mathfrak{p})$. If g belongs to the localized ring $K_\mathfrak{p}$, its image in the residue field is its restriction to $\mathcal{V}'(\mathfrak{p})$.

Let s be a nonzero element in the previous ring $K = F[x_1, \dots, x_n]$. It is (in an essentially unique way) a product of prime elements (each one generates a prime ideal in K); with each prime divisor of s is associated an irreducible algebraic hypersurface in F^n, and $\mathcal{V}'(Ks)$ (which is the subset of all $a \in F^n$ such that $s(a) = 0$) is the union of all these hypersurfaces. Let U'_s be the complementary subset of $\mathcal{V}'(Ks)$ in F^n. The elements of K_s are the rational functions on F^n that are defined at all points of U'_s; their restrictions to U'_s are called the *regular functions* on U'_s. According to Lemma (1.10.9) the mapping $\mathrm{Spec}(K_s) \to \mathrm{Spec}(K)$ is a bijection from $\mathrm{Spec}(K_s)$ onto the subset U_s of all prime ideals of K that do not contain s; this subset U_s is open because the complementary subset is $\mathcal{V}(Ks)$. The elements of U_s correspond to the irreducible algebraic submanifolds $\mathcal{V}'(\mathfrak{p})$ contained in U'_s.

Let us examine what happens when $\mathrm{Spec}(K)$ is not a connected topological space.

(1.11.3) **Theorem.** *When K is a direct sum of ideals $K_1, K_2, \dots, K_n$, each K_i is generated by an idempotent e_i and $\sum_{i=1}^n e_i = 1$; moreover the mapping* $\mathrm{Spec}(K_i) \to \mathrm{Spec}(K)$ *associated with the projection $K \to K_i$ induces a bijection from* $\mathrm{Spec}(K_i)$ *onto an open subset U_i of* $\mathrm{Spec}(K)$, *and* $\mathrm{Spec}(K)$ *is the disjoint union of the open subsets U_1, U_2, …,U_n. Conversely when* $\mathrm{Spec}(K)$ *is a disjoint union of open subsets $U_1, U_2, \dots, U_n$, then K is a direct sum of ideals K_1, $K_2, \dots, K_n$ in such a way that each open subset U_i is the image of the mapping* $\mathrm{Spec}(K_i) \to \mathrm{Spec}(K)$.

Proof. Let us suppose that $K = \bigoplus_{i=1}^n K_i$ and let e_i be the component of 1 in K_i for $i = 1, 2, \dots, n$. The equality $\lambda = \sum_{i=1}^n \lambda e_i$ holds for every $\lambda \in K$, and proves that λe_i is the projection of λ in K_i; in particular $e_i^2 = e_i$ and $K_i = Ke_i$. A prime ideal cannot contain all the n idempotents e_i because their sum is 1; if it does not contain e_i, it must contain all e_j such that $j \neq i$, because $e_i e_j = 0$. Therefore each prime ideal contains all the n idempotents e_i except one, and $\mathrm{Spec}(K)$ is the disjoint union of the n subsets U_i defined in this way: U_i is the set of all prime ideals containing e_j whenever $j \neq i$. Since U_i is equal both to $\mathcal{V}(K(1 - e_i))$ and to the subset complementary to $\mathcal{V}(Ke_i)$, it is open and closed. The continuous mapping $\mathrm{Spec}(K_i) \to \mathrm{Spec}(K)$ maps each prime ideal $\mathfrak{p}_i$ of K_i to the prime ideal of K which is the direct sum of $\mathfrak{p}_i$ and all K_j with $j \neq i$; thus we get a bijection $\mathrm{Spec}(K_i) \to U_i$.

Conversely let us suppose that $\mathrm{Spec}(K)$ is a disjoint union of n open subsets U_i. For each i let us choose an ideal $\mathfrak{a}_i$ such that U_i is the complementary subset of $\mathcal{V}(\mathfrak{a}_i)$. Since the intersection of all $\mathcal{V}(\mathfrak{a}_i)$ is empty, the sum of all the ideals $\mathfrak{a}_i$ is K, and therefore we can write

$$1 = \sum_{i=1}^{n} \varepsilon_i \quad \text{with} \quad \varepsilon_i \in \mathfrak{a}_i \quad \text{for} \quad i = 1, 2, \ldots, n.$$

Since $\mathrm{Spec}(K)$ is the union of $\mathcal{V}(\mathfrak{a}_i)$ and $\mathcal{V}(\mathfrak{a}_j)$ whenever $i \neq j$, the intersection $\mathfrak{a}_i \cap \mathfrak{a}_j$ is contained in all prime ideals, and therefore in $\mathrm{Rad}(K)$ (see Theorem (1.10.2)); consequently there exists a positive integer k such that $(\varepsilon_i \varepsilon_j)^k = 0$ whenever $i \neq j$. Let us set $m = n(k-1)+1$; the equality $1 = (\sum_{i=1}^{n} \varepsilon_i)^m$ shows that we can write

$$1 = \sum_{i=1}^{n} e_i \quad \text{with} \quad e_i \in K\varepsilon_i^k \subset \mathfrak{a}_i \quad \text{for} \quad i = 1, 2, \ldots, n.$$

Now observe that $e_i e_j = 0$ whenever $i \neq j$; this proves that every e_i is an idempotent:

$$e_i = e_i \sum_{j=1}^{n} e_j = \sum_{j=1}^{n} e_i e_j = e_i^2.$$

Therefore K is the direct sum of the n ideals Ke_i as above. It remains to prove that $U_i = \mathcal{V}(K(1-e_i))$. It is clear that $\mathcal{V}(\mathfrak{a}_i) \subset \mathcal{V}(Ke_i)$, whence the inclusions $U_i \supset \mathcal{V}(K(1-e_i))$. Since $\mathrm{Spec}(K)$ is the disjoint union of the n subsets U_i and also the disjoint union of the n subsets $\mathcal{V}(K(1-e_i))$, all this enforces the equalities $U_i = \mathcal{V}(K(1-e_i))$. □

After having explained the localization process, we must face the converse problem: which properties of a module (or a morphism) can be derived from an examination of the corresponding localized modules (or morphisms)?

(1.11.4) **Globalization lemma.** *Let $M_{\max}$ be the direct product of all the localized modules $M_{\mathfrak{m}}$ at all maximal ideals $\mathfrak{m}$ of K. By mapping every $x \in M$ to the family of all its images $x/1$ in all these localized modules, we get an injective mapping $M \to M_{\max}$.*

Proof. Let x be an element of M such that $x/1$ vanishes in all the localized modules $M_{\mathfrak{m}}$; for each maximal ideal $\mathfrak{m}$ there exists an element $t_{\mathfrak{m}} \in K$ such that $t_{\mathfrak{m}} x = 0$ and $t_{\mathfrak{m}} \notin \mathfrak{m}$. Let $\mathfrak{a}$ be the ideal generated by all $t_{\mathfrak{m}}$; no maximal ideal $\mathfrak{m}$ can contain $\mathfrak{a}$ because $\mathfrak{a}$ contains an element $t_{\mathfrak{m}}$ outside $\mathfrak{m}$. Therefore $\mathfrak{a} = K$, and from the set of all elements $t_{\mathfrak{m}}$ we can extract a finite sequence $(t_1, t_2, \ldots, t_n)$ such that

$$1 = \sum_{i=1}^{n} \lambda_i t_i \quad \text{for suitable} \quad \lambda_1, \lambda_2, \ldots, \lambda_n \in K.$$

Consequently the equalities $t_i x = 0$ for $i = 1, 2, \ldots, n$ imply that $x = 0$. □

A property involving modules or morphisms (or anything that can be localized) is called a *local property* when its fulfillment for the object under consideration is equivalent to its fulfillment for all the localizations of this object. The following corollaries of (1.11.4) show several examples of local properties.

(1.11.5) **Corollary.** *For a K-module M the following assertions are equivalent:*

(a) $M = 0$;
(b) $M_{\mathfrak{p}} = 0$ *for every prime ideal* $\mathfrak{p}$;
(c) $M_{\mathfrak{m}} = 0$ *for every maximal ideal* $\mathfrak{m}$.

(1.11.6) **Corollary.** *When N and N' are submodules of M, the following assertions are equivalent:*

(a) $N' \subset N$;
(b) $N'_{\mathfrak{p}} \subset N_{\mathfrak{p}}$ *for every prime ideal* $\mathfrak{p}$;
(c) $N'_{\mathfrak{m}} \subset N_{\mathfrak{m}}$ *for every maximal ideal* $\mathfrak{m}$.

We get another triplet of equivalent assertions if we replace the inclusions with equalities.

Proof. It is clear that (a)⇒(b)⇒(c). Let us suppose that $N'_{\mathfrak{m}} \subset N_{\mathfrak{m}}$, whence $N_{\mathfrak{m}} = N_{\mathfrak{m}} + N'_{\mathfrak{m}} = (N+N')_{\mathfrak{m}}$ (see (1.10.7)), and consequently $(N+N')_{\mathfrak{m}}/N_{\mathfrak{m}} = 0$. Because of the exactness of the localization functors (see (1.10.3)) this implies $((N+N')/N)_{\mathfrak{m}} = 0$. When this equality holds for all maximal ideals, then $(N+N')/N = 0$, and $N' \subset N$. □

Consequently the property that an element x of M belongs to a submodule N is a local property (indeed $x \in N$ if and only if $Kx \subset N$). And the property that two submodules are supplementary in M is a local property too.

(1.11.7) **Corollary.** *For a K-linear mapping $f : M \to N$ the following assertions are equivalent:*

(a) *f is injective;*
(b) *the localized mapping $f_{\mathfrak{p}} : M_{\mathfrak{p}} \to N_{\mathfrak{p}}$ is injective for every prime ideal $\mathfrak{p}$;*
(c) *the localized mapping $f_{\mathfrak{m}} : M_{\mathfrak{m}} \to N_{\mathfrak{m}}$ is injective for every maximal ideal $\mathfrak{m}$.*

We get another triplet of equivalent assertions if we replace the word "injective" with "surjective".

Proof. It is clear that (a)⇒(b)⇒(c). Let us suppose that all mappings $f_{\mathfrak{m}}$ are injective; since all localization functors are exact, from the exact sequence $0 \to \mathrm{Ker}(f) \to M \to N$ we derive exact sequences $0 \to (\mathrm{Ker}(f))_{\mathfrak{m}} \to M_{\mathfrak{m}} \to N_{\mathfrak{m}}$ that allow us to identify $\mathrm{Ker}(f_{\mathfrak{m}})$ with $(\mathrm{Ker}(f))_{\mathfrak{m}}$; consequently $(\mathrm{Ker}(f))_{\mathfrak{m}} = 0$ for all maximal ideals $\mathfrak{m}$, whence $\mathrm{Ker}(f) = 0$.

When the mappings $f_{\mathfrak{m}}$ are assumed to be surjective, there is a similar argument with the exact sequence $M \to N \to \mathrm{Coker}(f) \to 0$. □

(1.11.8) **Corollary.** *For a K-module M the following assertions are equivalent:*

(a) *M is flat;*
(b) *$M_{\mathfrak{p}}$ is a flat $K_{\mathfrak{p}}$-module for every prime ideal $\mathfrak{p}$;*
(c) *$M_{\mathfrak{m}}$ is a flat $K_{\mathfrak{m}}$-module for every maximal ideal $\mathfrak{m}$.*

Proof. The module M is flat if and only if the mapping $M \otimes N' \to M \otimes N$ is injective whenever it comes from an injective mapping $N' \to N$. According to Corollary (1.10.5) there is a canonical isomorphism

$$(M \otimes_K N)_{\mathfrak{p}} \;\cong\; M_{\mathfrak{p}} \otimes_{K_{\mathfrak{p}}} N_{\mathfrak{p}},$$

and the same when N is replaced with N', or $\mathfrak{p}$ with $\mathfrak{m}$. Thus the implication (a)⇒(b) follows from the exactness of localization functors and from the fact that $N_{\mathfrak{p}} = N$ when N is already a $K_{\mathfrak{p}}$-module (and the same for N'). Then the implication (b)⇒(c) is trivial, and the implication (c)⇒(a) is a consequence of (1.11.7). □

1.12 Finitely generated modules

The localization and globalization process allows us to reduce problems about modules to problems about modules over local rings, and when the modules under consideration are finitely generated, the following lemma will be repeatedly used.

(1.12.1) **Nakayama's lemma.** *Let K be a local ring with maximal ideal $\mathfrak{m}$, and M a finitely generated K-module; when $M = \mathfrak{m}M$, then $M = 0$.*

Proof. Let M be generated by the finite family $(x_1, x_2, \dots, x_n)$. If $M \neq 0$, there exists $k \in \{1, 2, \dots, n\}$ such that M is generated by $(x_k, x_{k+1}, \dots, x_n)$ but not by $(x_{k+1}, \dots, x_n)$; since $M = \mathfrak{m}M$, there exist $\mu_k, \mu_{k+1}, \dots, \mu_n$ all in $\mathfrak{m}$ such that

$$x_k = \mu_k x_k + \mu_{k+1} x_{k+1} + \cdots + \mu_n x_n;$$

since $1 - \mu_k$ is invertible, this equality shows that x_k belongs to the submodule generated by $(x_{k+1}, \dots, x_n)$; this causes a contradiction. We conclude that $M = 0$. □

(1.12.2) **Corollary.** *Let M be a finitely generated module over a local ring K with maximal ideal $\mathfrak{m}$; when N is a submodule such that $M = N + \mathfrak{m}M$, then $M = N$.*

Indeed $M/N = \mathfrak{m}(M/N)$, whence $M/N = 0$. □

(1.12.3) **Corollary.** *Let M be a finitely generated module over a local ring K with maximal ideal $\mathfrak{m}$, and $(x_1, x_2, \dots, x_n)$ a finite sequence of elements of M; the following two assertions are equivalent:*

(a) *it is a minimal family of generators of M;*
(b) *its image in $M/\mathfrak{m}M$ is a basis of $M/\mathfrak{m}M$ over the residue field $K/\mathfrak{m}$.*

Proof. Since the bases of $M/\mathfrak{m}M$ are its minimal families of generators, it suffices to prove that $(x_1, x_2, \ldots, x_n)$ generates M if and only if its image in $M/\mathfrak{m}M$ generates it. Obviously the image of a family of generators of M is a family of generators of its quotient. Conversely suppose that the images of $x_1, x_2, \ldots, x_n$ generates $M/\mathfrak{m}M$, and let N be the submodule of M generated by $x_1, x_2, \ldots, x_n$ themselves; since $M = N + \mathfrak{m}M$, we conclude that $M = N$. □

The next theorem too follows from Nakayama's lemma.

(1.12.4) **Theorem.** *Let M be a finitely presented module over a local ring K with maximal ideal $\mathfrak{m}$; the following four assertions are equivalent:*

(a) *M is free;*
(b) *M is projective;*
(c) *M is flat;*
(d) *the mapping $\mathfrak{m} \otimes M \to M$ defined by $\mu \otimes x \longmapsto \mu x$ is injective.*

Moreover when M is free, every minimal family of generators of M is a basis of M.

Proof. The implications (a)⇒(b)⇒(c) are evident, and (c)⇒(d) follows from the fact that the tensor product by the flat module M transforms the natural injection $\mathfrak{m} \to K$ into an injection $\mathfrak{m} \otimes M \longrightarrow M$ if $K \otimes M$ is identified with M as usual. We must prove (d)⇒(a). Let $(x_1, x_2, \ldots, x_n)$ be a minimal family of generators of M; its image in $M/\mathfrak{m}M$ is a basis (see (1.12.3)). Let N be a free module with basis $(e_1, e_2, \ldots, e_n)$, and f the surjective morphism $N \to M$ that maps each e_i to x_i; let us consider $R = \mathrm{Ker}(f)$. Since the kernel of $N \to M \to M/\mathfrak{m}M$ is exactly $\mathfrak{m}N$, we know that $R \subset \mathfrak{m}N$. Let us prove that the injectiveness of the morphism $\mathfrak{m} \otimes M \to M$ implies $R = \mathfrak{m}R$, since this immediately leads to the awaited conclusions: $R = 0$, therefore f is an isomorphism, M is free, and the minimal generating family $(x_1, x_2, \ldots, x_n)$ is a basis. Indeed since $R \subset \mathfrak{m}N$, for every $y \in R$ there exist $\mu_1, \mu_2, \ldots, \mu_n$ all in $\mathfrak{m}$ such that $y = \sum_{i=1}^n \mu_i e_i$. By definition of R we can write $\sum_{i=1}^n \mu_i x_i = 0$, and the injectiveness of $\mathfrak{m} \otimes M \to M$ implies that $\sum_{i=1}^n \mu_i \otimes x_i$ vanishes in $\mathfrak{m} \otimes M$. Now let us consider the exact sequence

$$\mathfrak{m} \otimes R \longrightarrow \mathfrak{m} \otimes N \longrightarrow \mathfrak{m} \otimes M \longrightarrow 0;$$

since $\sum_{i=1}^n \mu_i \otimes e_i$ belongs to the kernel of the second morphism of this sequence, it must belong to the image of the first morphism; since $y = \sum_{i=1}^n \mu_i e_i$, this shows that y belongs to $\mathfrak{m}R$, and thus the proof is ended. □

Unlike flatness which is a local property (see (1.11.8)), projectiveness is not a local property. Nevertheless for finitely presented modules there is a statement analogous to (1.11.8).

(1.12.5) **Proposition.** *Let P be a finitely presented module over some ring K; the following three assertions are equivalent:*

(a) *P is projective;*

(b) $P_\mathfrak{p}$ *is a free* $K_\mathfrak{p}$*-module for all prime ideals* $\mathfrak{p}$*;*

(c) $P_\mathfrak{m}$ *is a free* $K_\mathfrak{m}$*-module for all maximal ideals* $\mathfrak{m}$.

Proof. The module P is projective if and only if the mapping $\mathrm{Hom}(P, M) \to \mathrm{Hom}(P, M'')$ is surjective whenever it comes from a surjective mapping $M \to M''$. According to Proposition (1.10.8) there is a canonical isomorphism

$$(\mathrm{Hom}_K(P, M))_\mathfrak{p} \quad \longrightarrow \quad \mathrm{Hom}_{K_\mathfrak{p}}(P_\mathfrak{p}, M_\mathfrak{p}),$$

and the same when M is replaced with M'', or $\mathfrak{p}$ with $\mathfrak{m}$. Since projectiveness and freedom are equivalent in the case of finitely generated $K_\mathfrak{p}$-modules (see (1.12.4)), the implication (a)⇒(b) follows from the exactness of localization functors, and the fact that $M = M_\mathfrak{p}$ when M is already a $K_\mathfrak{p}$-module (and the same for M''). Then the implication (b)⇒(c) is trivial, and the implication (c)⇒(a) is a consequence of (1.11.7). □

At the end of **1.7** it is proved that the canonical mapping $M \to M^{**}$ is an isomorphism when M is a finitely generated projective module; the proof starts with the case of a free module with finite bases. Now there is another proof starting in the same way but using (1.10.8), (1.11.7) and (1.12.4): indeed for all $\mathfrak{p} \in \mathrm{Spec}(K)$ there are isomorphisms $M_\mathfrak{p} \to (M_\mathfrak{p})^{**} \leftarrow (M^{**})_\mathfrak{p}$, and consequently all localizations of the above canonical mapping are bijective.

When M is a finitely generated module over any ring K, and $\mathfrak{p}$ any prime ideal of K, Corollary (1.12.3) shows that all minimal families of generators of $M_\mathfrak{p}$ have the same length, which is the dimension of $F_\mathfrak{p} \otimes M = M_\mathfrak{p}/\mathfrak{p}M_\mathfrak{p}$ over the residue field $F_\mathfrak{p}$; this common length is called the *rank* of M at the prime ideal $\mathfrak{p}$ and denoted by $\mathrm{rk}(\mathfrak{p}, M)$. Thus with M is associated a function on $\mathrm{Spec}(K)$ with values in the set of nonnegative integers; it is clear that this function is constant when M is free.

(1.12.6) **Proposition.** *For all finitely generated modules* M *and* N *and for all prime ideals* $\mathfrak{p}$ *of* K *the following equalities hold:*

$$\begin{aligned} \mathrm{rk}(\mathfrak{p},\ M \oplus N) &= \mathrm{rk}(\mathfrak{p}, M) + \mathrm{rk}(\mathfrak{p}, N), \\ \mathrm{rk}(\mathfrak{p},\ M \otimes N) &= \mathrm{rk}(\mathfrak{p}, M)\ \mathrm{rk}(\mathfrak{p}, N); \end{aligned}$$

when M *is moreover projective, then* $\mathrm{Hom}_K(M, N)$ *is also finitely generated and*

$$\mathrm{rk}(\mathfrak{p},\ \mathrm{Hom}(M, N)) \;=\; \mathrm{rk}(\mathfrak{p}, M)\ \mathrm{rk}(\mathfrak{p}, N).$$

Proof. Since $\mathrm{rk}(\mathfrak{p}, M)$ is the dimension of $F_\mathfrak{p} \otimes M$ over the residue field $F_\mathfrak{p}$, the first conclusions follow from the isomorphisms

$$\begin{aligned} F_\mathfrak{p} \otimes_K (M \oplus N) &\cong (F_\mathfrak{p} \otimes_K M) \oplus (F_\mathfrak{p} \otimes_K N), \\ F_\mathfrak{p} \otimes_K (M \otimes_K N) &\cong (F_\mathfrak{p} \otimes_K M) \otimes_{F_\mathfrak{p}} (F_\mathfrak{p} \otimes_K N). \end{aligned}$$

When M is a finitely generated projective module, there exists M' such that $M \oplus M'$ is free with finite bases, therefore $\mathrm{Hom}(M \oplus M',\ N)$ is finitely generated whenever N is finitely generated, and $\mathrm{Hom}(M, N)$ too. Besides, $(\mathrm{Hom}_K(M, N))_\mathfrak{p}$ is canonically isomorphic to $\mathrm{Hom}_{K_\mathfrak{p}}(M_\mathfrak{p}, N_\mathfrak{p})$ because of (1.10.8); since $M_\mathfrak{p}$ is free, it is easy to calculate the rank of $\mathrm{Hom}_{K_\mathfrak{p}}(M_\mathfrak{p}, N_\mathfrak{p})$. □

(1.12.7) **Proposition.** *Let M be a finitely generated module, and $\mathfrak{p}$ a prime ideal of K; there exists an open subset U of* $\mathrm{Spec}(K)$ *containing $\mathfrak{p}$ such that* $\mathrm{rk}(\mathfrak{q}, M) \leq \mathrm{rk}(\mathfrak{p}, M)$ *for all $\mathfrak{q} \in U$. If M is a finitely generated projective module, the function $\mathfrak{p} \longmapsto \mathrm{rk}(\mathfrak{p}, M)$ is locally constant.*

Proof. Let us set $r = \mathrm{rk}(\mathfrak{p}, M)$ and let $(x_1, x_2, \ldots, x_n)$ be a family of generators of M. Out of this family we can pick out r elements which give a minimal family of generators of $M_\mathfrak{p}$ and we can suppose that they are $x_1, x_2, \ldots, x_r$; consequently there exists a family $(\kappa_{i,j})$ of elements of $K_\mathfrak{p}$ such that

$$\frac{x_j}{1} = \sum_{i=1}^{r} \kappa_{i,j} \frac{x_i}{1} \quad \text{for} \quad j = r+1, r+2, \ldots, n;$$

we can write all the fractions $\kappa_{i,j}$ with a common denominator s (outside $\mathfrak{p}$); consequently there exists a family $(\lambda_{i,j})$ of elements of K such that the following equalities hold for a suitable t also lying outside $\mathfrak{p}$:

$$(st)x_j = \sum_{i=1}^{r} \lambda_{i,j}\, x_i \quad \text{for} \quad j = r+1, r+2, \ldots, n;$$

if the prime ideal $\mathfrak{q}$ does not belong to $\mathcal{V}(Kst)$, then st is invertible in $K_\mathfrak{q}$ and therefore $x_1, x_2, \ldots, x_r$ still give a family of generators of $M_\mathfrak{q}$ (not necessarily a minimal family); this proves that $\mathrm{rk}(\mathfrak{q}, M) \leq r$ for all $\mathfrak{q}$ in the open subset complementary to $\mathcal{V}(Kst)$.

When M is a finitely generated projective module, there exists a finitely generated projective module M' such that $M \oplus M'$ is free, of constant rank $r + r'$ if r and r' are the ranks of M and M' at $\mathfrak{p}$. There are also open subsets U and U' containing $\mathfrak{p}$ such that

$$\mathrm{rk}(\mathfrak{q}, M) \leq r \quad \text{for all} \quad \mathfrak{q} \in U \qquad \text{and} \qquad \mathrm{rk}(\mathfrak{q}, M') \leq r' \quad \text{for all} \quad \mathfrak{q} \in U';$$

since the sum of the ranks of M and M' is everywhere equal to $r + r'$, all this implies that both ranks remain constant on $U \cap U'$. □

(1.12.8) **Corollary.** *Let us suppose that the rank of the finitely generated projective module M is not constant and takes different values $r_1, r_2, \ldots, r_k$. There are idempotents $e_1, e_2, \ldots, e_k$ in K satisfying these properties: first $e_1 + e_2 + \cdots + e_k = 1$ and $e_i e_j = 0$ whenever $i \neq j$; secondly, for each $i \in \{1, 2, \ldots, k\}$, the equality* $\mathrm{rk}(\mathfrak{p}, M) = r_i$ *holds for all $\mathfrak{p} \in \mathcal{V}(K(1 - e_i))$, or equivalently, $e_i M$ is a (Ke_i)-module of constant rank r_i.*

Indeed $\mathrm{Spec}(K)$ is a disjoint union of k open subsets U_i such that $\mathrm{rk}(\mathfrak{p}, M) = r_i$ for all $p \in U_i$, and then it suffices to remember (1.11.3). □

The next theorem improves (1.12.5); its proof may be skipped.

(1.12.9) **Theorem.** *For a K-module P the following five assertions are equivalent:*

(a) *P is a finitely generated projective module;*

(b) *P is a finitely presented module, and $P_{\mathfrak{m}}$ is a free $K_{\mathfrak{m}}$-module for each maximal ideal $\mathfrak{m}$;*

(c) *P is a finitely generated module, $P_{\mathfrak{p}}$ is a free $K_{\mathfrak{p}}$-module for each $\mathfrak{p} \in \mathrm{Spec}(K)$, and the rank function of P is locally constant on $\mathrm{Spec}(K)$;*

(d) *for each maximal ideal $\mathfrak{m}$ there exists $s \in K \setminus \mathfrak{m}$ such that P_s is a free K_s-module of finite rank;*

(e) *there exists a finite sequence $(s_1, s_2, \dots, s_k)$ generating K as an ideal, such that P_{s_i} is a free K_{s_i}-module of finite rank for $i = 1, 2, \dots, k$.*

Proof. From Propositions (1.12.5) and (1.12.7) we derive (b)⇔(a)⇒(c), and it is obvious that (d)⇔(e), because for a subset of elements s of K the following three assertions are equivalent:

- it is contained in no maximal ideal $\mathfrak{m}$;
- it generates K as an ideal;
- it contains a finite subset generating K as an ideal.

Therefore it suffices to prove (c)⇒(d) and (e)⇒(a).

Let us prove (c)⇒(d). Let $\mathfrak{m}$ be a maximal ideal, r the rank of P at $\mathfrak{m}$, and $(x_1, x_2, \dots, x_n)$ a finite family of generators of P such that $(x_1/1, x_2/1, \dots, x_r/1)$ is a minimal family of generators of $P_{\mathfrak{m}}$. From the proof of (1.12.7) we know that there exists $s' \in K \setminus \mathfrak{m}$ such that $(x_1/1, x_2/1, \dots, x_r/1)$ is a family of generators of $P_{\mathfrak{q}}$ at every prime ideal $\mathfrak{q}$ not belonging to $\mathcal{V}(Ks')$. The hypothesis (c) implies the existence of a closed subset $\mathcal{V}(\mathfrak{a})$ not containing $\mathfrak{m}$ such that the rank of P is r at every $\mathfrak{q}$ not belonging to $\mathcal{V}(\mathfrak{a})$; let s'' be any element of $\mathfrak{a} \cap (K \setminus \mathfrak{m})$; we set $s = s's''$. For every prime ideal $\mathfrak{q}$ not belonging to $\mathcal{V}(Ks)$, the image of $(x_1, x_2, \dots, x_r)$ in $P_{\mathfrak{q}}$ is a minimal family of generators, and since $P_{\mathfrak{q}}$ is supposed to be free, it is a basis of $P_{\mathfrak{q}}$ (see (1.12.4)). With the ring morphism $K \longrightarrow K_s$ is associated a continuous mapping $\mathrm{Spec}(K_s) \longrightarrow \mathrm{Spec}(K)$; Lemma (1.10.9) shows that this mapping is injective and that its image is exactly the open subset U_s complementary to $\mathcal{V}(Ks)$. Moreover for any K-module M, the localized modules of the K_s-module M_s can be identified with the localized modules $M_{\mathfrak{q}}$ at all points $\mathfrak{q} \in U_s$. Let N be a free K_s-module with basis $(e_1, e_2, \dots, e_r)$, and f the K_s-linear mapping $N \longrightarrow P_s$ such that $f(e_i) = x_i/1$ for $i = 1, 2, \dots, r$; since the image of $(x_1, x_2, \dots, x_r)$ in $P_{\mathfrak{q}}$ is a basis for all $\mathfrak{q} \in U_s$, all the localizations of f are isomorphisms; consequently f is itself an isomorphism (see (1.11.7)) and P_s is a free K_s-module.

Let us prove (e)⇒(a). As in Corollary (1.10.6), let us consider the direct product L of the rings K_{s_i}; since each P_{s_i} is a free K_{s_i}-module, it is easy to prove that it is a projective L-module; now $L \otimes_K P$ is the direct product (or direct sum) of these P_{s_i}; therefore $L \otimes_K P$ is L-projective. Moreover it is finitely generated, since it is a sum of finitely generated submodules. Because of (1.9.10) P is a finitely generated projective K-module. □

Invertible modules and Picard groups

A module M is said to be *invertible* if there exists a module N such that $M \otimes N$ is isomorphic to K.

(1.12.10) **Theorem.** *A module M is invertible if and only if it is a finitely generated projective module of constant rank* 1. *Moreover every isomorphism $M \otimes N \to K$ induces an isomorphism from N onto $M^* = \mathrm{Hom}(M, K)$.*

Proof. Let f be an isomorphism $M \otimes N \to K$. There are elements $x_1, x_2, \ldots, x_n$ of M, and elements $y_1, y_2, \ldots, y_n$ of N, such that $\sum_i f(x_i \otimes y_i) = 1$. The mapping $x' \longmapsto \sum_i f(x', y_i)x_i$ is an automorphism of M because the mappings $x' \longmapsto \sum_i x_i \otimes y_i \otimes x'$ and $x \otimes y \otimes x' \longmapsto f(x, y)x'$ are reciprocal isomorphisms between M and $M \otimes N \otimes M$, and the mapping $x \otimes y \otimes x' \longmapsto x' \otimes y \otimes x$ is an automorphism of $M \otimes N \otimes M$. Consequently M is generated by the n elements x_i. This set of generators of M gives a surjective mapping $K^n \to M$ and a surjective mapping $K^n \otimes N \to M \otimes N$. Since $M \otimes N$ is projective, it is isomorphic to a direct summand of $K^n \otimes N$, and therefore $M \otimes N \otimes M$ is isomorphic to a direct summand of $K^n \otimes N \otimes M$. Now

$$K^n \otimes N \otimes M \cong K^n \otimes M \otimes N \cong K^n \quad \text{and} \quad M \otimes N \otimes M \cong M;$$

since M is isomorphic to a direct summand of K^n, it is projective. For the same reasons N is finitely generated and projective. Since the rank of $M \otimes N$ is the product of the ranks of M and N, all these ranks are everywhere equal to 1.

Conversely if M is a finitely generated projective module of constant rank 1, for all prime ideal $\mathfrak{p}$ we can write $(M_\mathfrak{p})^* = (M^*)_\mathfrak{p}$ because of (1.9.7), and this allows us to prove by localization the bijectiveness of the mapping $M \otimes M^* \to K$ defined by $x \otimes h \longmapsto h(x)$. Besides, every mapping $f : M \otimes N \to K$ induces a mapping $N \to M^*$ defined by $y \longmapsto (x \longmapsto f(x, y))$; by localization we can derive the bijectiveness of $N \to M^*$ from the bijectiveness of f; indeed when $M_\mathfrak{p}$ and $N_\mathfrak{p}$ are respectively generated by $x_1/1$ and $y_1/1$, the bijectiveness of $f_\mathfrak{p}$ means that $f(x_1, y_1)/1$ is invertible in $K_\mathfrak{p}$. □

The isomorphy classes of invertible modules constitute a commutative group with the operation derived from the tensor product; this group is called the *Picard group* of K and denoted by $\mathrm{Pic}(K)$; its unit element is the isomorphy class of the module K itself.

The following lemma is sometimes useful when invertible modules are involved; since the localization of tensor products raises no difficulty (see (1.10.5)), its proof is omitted.

(1.12.11) **Lemma.** *When at every maximal ideal of K the localization of M is generated by one element, the equality $x \otimes y = y \otimes x$ holds in $M \otimes M$ for all x and y in M.*

Ranks of extensions of modules

(1.12.12) **Proposition.** *Let $f : K \to L$ be an extension of a ring as in* **1.9**, *let $\mathfrak{q}$ be a prime ideal of L, and $\mathfrak{p} = f^{-1}(\mathfrak{q})$ its image in* Spec(K). *For every finitely generated K-module M we can write*

$$\mathrm{rk}(\mathfrak{q},\ L \otimes M) \;=\; \mathrm{rk}(\mathfrak{p}, M).$$

Proof. Let $F = K_\mathfrak{p}/\mathfrak{p}K_\mathfrak{p}$ and $G = L_\mathfrak{q}/\mathfrak{q}L_\mathfrak{q}$ be the residue fields; the ring morphism $K \to L$ induces a field morphism $F \to G$. The rank of M at $\mathfrak{p}$ is the dimension of $F \otimes M$ over F, and the rank of $L \otimes M$ is the dimension of $G \otimes_L (L \otimes_K M)$ over G. Now

$$\begin{aligned} G \otimes_L (L \otimes_K M) &\cong (G \otimes_L L) \otimes_K M \cong G \otimes_K M \\ &\cong (G \otimes_F F) \otimes_K M \cong G \otimes_F (F \otimes_K M); \end{aligned}$$

it suffices to remember that for every finite dimensional vector space V over F, the dimension of $G \otimes_F V$ over G is the dimension of V over F. □

Thus we know the rank of $L \otimes M$ at every prime ideal of L when we know the rank of M at every prime ideal of K. In particular $L \otimes M$ is an invertible L-module whenever M is an invertible K-module, whence a group morphism $\mathrm{Pic}(K) \to \mathrm{Pic}(L)$ which shows that Pic is a functor from the category $\mathcal{C}om(\mathbb{Z})$ to the category of commutative groups. Conversely does the knowledge of the ranks of $L \otimes M$ lead to the knowledge of the ranks of M? The answer is positive when the extension $K \to L$ is faithfully flat, because of the following lemma.

(1.12.13) **Lemma.** *When $f : K \to L$ is a flat extension of K, these three assertions are equivalent:*

(a) *L is a faithfully flat K-algebra;*
(b) *for every prime ideal $\mathfrak{p}$ of K there exists a prime ideal $\mathfrak{q}$ of L such that $\mathfrak{p} = f^{-1}(\mathfrak{q})$;*
(c) *for every maximal ideal $\mathfrak{m}$ of K there exists a maximal ideal $\mathfrak{n}$ of L such that $\mathfrak{m} = f^{-1}(\mathfrak{n})$.*

Proof. First we prove (a)⇒(b). We consider the ring $L_\mathfrak{p} = K_\mathfrak{p} \otimes L$ and the diagram

$$\begin{array}{cccc} f : & K & \longrightarrow & L \\ & \downarrow & & \downarrow \\ f_\mathfrak{p} : & K_\mathfrak{p} & \longrightarrow & L_\mathfrak{p} \end{array}$$

When L is a faithfully flat extension of K, then $L_{\mathfrak{p}}$ is a faithfully flat extension of $K_{\mathfrak{p}}$ (see **1.9**); consequently from the strict inclusion $\mathfrak{p}K_{\mathfrak{p}} \neq K_{\mathfrak{p}}$ we can deduce $\mathfrak{p}L_{\mathfrak{p}} \neq L_{\mathfrak{p}}$ since

$$L_{\mathfrak{p}}/\mathfrak{p}L_{\mathfrak{p}} \cong (K_{\mathfrak{p}}/\mathfrak{p}K_{\mathfrak{p}}) \otimes_{K_{\mathfrak{p}}} L_{\mathfrak{p}} \neq 0.$$

Consequently there is a maximal ideal $\mathfrak{n}'$ in $L_{\mathfrak{p}}$ that contains $\mathfrak{p}L_{\mathfrak{p}}$. Let $\mathfrak{q}$ be the image of $\mathfrak{n}'$ by the mapping $\mathrm{Spec}(L_{\mathfrak{p}}) \to \mathrm{Spec}(L)$; thus $f^{-1}(\mathfrak{q})$ is the image of $f_{\mathfrak{p}}^{-1}(\mathfrak{n}')$ by the mapping $\mathrm{Spec}(K_{\mathfrak{p}}) \to \mathrm{Spec}(K)$. Obviously $f_{\mathfrak{p}}^{-1}(\mathfrak{n}')$ contains $\mathfrak{p}K_{\mathfrak{p}}$, and since this is the maximal ideal of $K_{\mathfrak{p}}$, we get $f_{\mathfrak{p}}^{-1}(\mathfrak{n}') = \mathfrak{p}K_{\mathfrak{p}}$. At last it is well known that the mapping $\mathrm{Spec}(K_{\mathfrak{p}}) \to \mathrm{Spec}(K)$ maps $\mathfrak{p}K_{\mathfrak{p}}$ to $\mathfrak{p}$.

The implication (b)⇒(c) is easy, and it remains to prove (c)⇒(a). When (c) is true, then $\mathfrak{m}L \neq L$ for all maximal ideals $\mathfrak{m}$ of K; consequently $\mathfrak{a}L \neq L$ for every ideal $\mathfrak{a}$ other than K itself. Let M be a K-module containing a nonzero element z; thus the submodule Kz is isomorphic to $K/\mathfrak{a}$ where $\mathfrak{a}$ is the ideal of all $\lambda \in K$ such that $\lambda z = 0$; since L is flat, $L \otimes (Kz)$ is a submodule of $L \otimes M$; therefore to prove $L \otimes M \neq 0$, it suffices to prove $L \otimes (K/\mathfrak{a}) \neq 0$; this is clear, since $(K/\mathfrak{a}) \otimes L = L/\mathfrak{a}L$. □

Now let M be a finitely generated L-module, therefore also a K-module because of the extension $f : K \to L$. We suppose that $\mathfrak{p} = f^{-1}(\mathfrak{q})$ as in (1.12.12), and that L is a finitely generated K-module. The following formula is suggested by a statement that is well known when K and L are fields:

$$\mathrm{rk}(\mathfrak{p}, M) = \mathrm{rk}(\mathfrak{q}, M)\,\mathrm{rk}(\mathfrak{p}, L);$$

unfortunately the example (1.12.14) just beneath shows that it is not always true; in (1.ex.22) it proves to be true when M is a finitely generated projective L-module of constant rank.

(1.12.14) **Example.** Let K be a field, and f the ring morphism $\lambda \longmapsto (\lambda, \lambda)$ from K into $L = K \times K$. Any couple (r, s) of nonnegative integers determines a projective module $M = K^r \times K^s$ over L; you must understand that $(\lambda, \mu)(x, y) = (\lambda x, \mu y)$ for all $x \in K^r$ and $y \in K^s$. Obviously L and M have dimension (or constant rank) 2 and $r + s$ over K. Now $\mathrm{Spec}(L)$ contains two prime ideals, $\mathfrak{q} = 0 \times K$ and $\mathfrak{q}' = K \times 0$; we can identify $L_{\mathfrak{q}}$ with $K \times 0$, $M_{\mathfrak{q}}$ with $K^r \times 0$, $L_{\mathfrak{q}'}$ with $0 \times K$ and $M_{\mathfrak{q}'}$ with $0 \times K^s$; thus the ranks of M at $\mathfrak{q}$ and $\mathfrak{q}'$ are r and s. In the present case the above formula would give the two equalities $r + s = 2r = 2s$; they are only true when $r = s$, in other words, when M has constant rank.

1.13 Some applications

Some properties which are not local properties (see **1.11**) can nevertheless be tested with localizations provided that some suitable hypotheses are fulfilled. Remember that a submodule N of M is called a *direct summand* if it admits a supplementary submodule; the next lemma shows that this property can be tested with localizations under suitable hypotheses.

(1.13.1) **Lemma.** *When N is a submodule of M, and both M and N are finitely presented, these assertions are equivalent:*

(a) *N is a direct summand of M;*
(b) *the natural mapping $\mathrm{Hom}(M, N) \to \mathrm{Hom}(N, N)$ is surjective;*
(c) *$N_\mathfrak{p}$ is a direct summand of $M_\mathfrak{p}$ for every prime ideal $\mathfrak{p}$;*
(d) *$N_\mathfrak{m}$ is a direct summand of $M_\mathfrak{m}$ for every maximal ideal $\mathfrak{m}$.*

Proof. When N is a direct summand, it is clear that the injection $u : N \longrightarrow M$ induces a surjection $\mathrm{Hom}(M, N) \to \mathrm{Hom}(N, N)$. Conversely when this mapping $\mathrm{Hom}(u, N)$ is surjective, there exists $v \in \mathrm{Hom}(M, N)$ such that $v \circ u = \mathrm{id}_N$ and consequently N is a direct summand of M. Thus the equivalence (a)⇔(b) is true without any assumption on M or N. This mapping $\mathrm{Hom}(u, N)$ is surjective if and only if its localizations are surjective. Since M and N are finitely presented, the canonical mappings

$$\mathrm{Hom}_K(M, N)_\mathfrak{p} \longrightarrow \mathrm{Hom}_{K_\mathfrak{p}}(M_\mathfrak{p}, N_\mathfrak{p}) \quad \text{and} \quad \mathrm{Hom}_K(N, N)_\mathfrak{p} \longrightarrow \mathrm{Hom}_{K_\mathfrak{p}}(N_\mathfrak{p}, N_\mathfrak{p})$$

are isomorphisms (see (1.9.9)). Consequently the surjectiveness of the localized mapping $\mathrm{Hom}(u, N)_\mathfrak{p}$ means that $N_\mathfrak{p}$ is a direct summand of $M_\mathfrak{p}$. □

(1.13.2) **Corollary.** *If the K-algebra A is a finitely generated projective K-module, the image $K1_A$ of the canonical morphism $K \to A$ is a direct summand of A.*

Proof. We can suppose $A \neq 0$. When K is a local ring with maximal ideal $\mathfrak{m}$, then 1_A cannot belong to $\mathfrak{m}A$ because of Nakayama's lemma; consequently there is a basis of $A/\mathfrak{m}A$ containing the image of 1_A, and there is a basis of A containing 1_A; this settles this particular case. In the general case, there is an idempotent $e \in K$ such that $A_\mathfrak{p} \neq 0$ if and only if $\mathfrak{p} \in \mathcal{V}(K(1-e))$ (see (1.12.8)); by localization we realize that the canonical morphism $K \to A$ induces an isomorphism $Ke \to K1_A$. Consequently $K1_A$ is projective and the conclusion follows from (1.13.1). □

The module M is said to be *faithful* if every equality $\lambda M = 0$ with $\lambda \in K$ implies $\lambda = 0$. This property can also be tested by localization under suitable hypotheses.

(1.13.3) **Lemma.** *When M is a finitely generated module, these three assertions are equivalent:*

(a) *M is a faithful module;*
(b) *$M_\mathfrak{p}$ a is faithful $K_\mathfrak{p}$-module for every prime ideal $\mathfrak{p}$;*
(c) *$M_\mathfrak{m}$ is a faithful $K_\mathfrak{m}$-module for every maximal ideal $\mathfrak{m}$.*

When M is a finitely generated projective module, they are still equivalent to these two assertions:

(d) $\mathrm{rk}(\mathfrak{p}, M) \neq 0$ *for every prime ideal $\mathfrak{p}$;*
(e) $\mathrm{rk}(\mathfrak{m}, M) \neq 0$ *for every maximal ideal $\mathfrak{m}$.*

Proof. If $\lambda M = 0$ and $\lambda \neq 0$, there is a localization $K_\mathfrak{m}$ in which $\lambda/1 \neq 0$ (see (1.11.4)) and thus $M_\mathfrak{m}$ is not a faithful $K_\mathfrak{m}$-module. Consequently (c)⇒(a), and of course (b)⇒(c). Now let λ/s be an element of $K_\mathfrak{p}$ that annihilates $M_\mathfrak{p}$; this means that $\lambda M_\mathfrak{p} = 0$, and if $(x_1, x_2, \ldots, x_n)$ is a family of generators of M, this implies $t_i \lambda x_i = 0$ for $i = 1, 2, \ldots, n$ and for some t_i outside $\mathfrak{p}$; it follows that $t\lambda M = 0$ if $t = t_1 t_2 \cdots t_n$. If M is faithful, we conclude that $t\lambda = 0$, whence $\lambda/s = 0$. This proves (a)⇒(b).

When M is finitely generated and projective, all its localizations are free modules of finite rank; such a module is faithful if and only if its rank is not equal to 0. □

In (1.13.3) the assertions (d) and (e) mean that under somewhat stronger hypotheses the faithfulness of M can be tested with extensions $K \to K_\mathfrak{p}/\mathfrak{p}K_\mathfrak{p}$ to residue fields; indeed the nonvanishing of $\mathrm{rk}(\mathfrak{p}, M)$ is equivalent to the faithfulness of the vector space $M_\mathfrak{p}/\mathfrak{p}M_\mathfrak{p}$ the dimension of which it is. There are other properties that can be tested with extensions to residue fields. The next two lemmas show that the surjectiveness or the bijectiveness of a linear mapping can be tested in this way; for the sake of briefness only extensions $K \to K/\mathfrak{m}$ have been mentioned. After these positive statements the example (1.13.6) shows that in general injectiveness cannot be tested in this way.

(1.13.4) **Lemma.** *If the target of $f : M \to N$ is finitely generated, these assertions are equivalent:*

(a) *f is surjective;*
(b) *f induces a surjective mapping $M/\mathfrak{m}M \to N/\mathfrak{m}N$ for every maximal ideal $\mathfrak{m}$ of K.*

Proof. The surjectiveness of f is equivalent to the surjectiveness of all localized mappings $f_\mathfrak{m}$ (see (1.11.7)). Now $N/\mathfrak{m}N$ is the same thing as $N_\mathfrak{m}/\mathfrak{m}N_\mathfrak{m}$, and because of (1.12.2) a submodule of $N_\mathfrak{m}$ (for instance $\mathrm{Im}(f_\mathfrak{m})$) is equal to $N_\mathfrak{m}$ if and only if it is mapped onto $N_\mathfrak{m}/\mathfrak{m}N_\mathfrak{m}$ by the quotient mapping. Therefore all $f_\mathfrak{m}$ are surjective if and only if the assertion (b) is true. □

(1.13.5) **Lemma.** *Let $f : M \to N$ be a linear mapping between finitely generated modules; N is even assumed to be projective. The following three assertions are equivalent:*

(a) *f is bijective;*
(b) *f induces a bijective mapping $M/\mathfrak{m}M \to N/\mathfrak{m}N$ for every maximal ideal $\mathfrak{m}$ of K;*
(c) *f is surjective, and M and N have the same rank at every maximal ideal.*

Proof. It is clear that (a)⇒(b)⇒(c). When the assertion (c) is true, the exact sequence $0 \to \mathrm{Ker}(f) \to M \to N \to 0$ splits and M is isomorphic to $N \oplus \mathrm{Ker}(f)$; since M and N everywhere have the same rank, we conclude that $\mathrm{Ker}(f) = 0$. □

(1.13.6) **Example.** Let F be a field and $K = F[[t]]$ the ring of formal series; it is a local ring and its maximal ideal $\mathfrak{m}$ is the ideal generated by the indeterminate t; the residue field $K/\mathfrak{m}$ can be identified with F. The multiplication by t is an injective mapping $K \to K$; but after the extension $K \to F$ it gives the null morphism $F \to F$. On the contrary, for every integer $k \geq 1$, the quotient mapping $K \to K/\mathfrak{m}^k$ is not injective, but after the extension $K \to F$ it gives the injective mapping id_F.

Standard involutions of algebras of constant rank 2

The remainder of this section is devoted to involutions of algebras, which some authors rather call anti-involutions, since they reserve the name "involution" for what is here called an involutive automorphism.

(1.13.7) **Definitions.** An *involution* of an algebra A is an involutive linear mapping $\varphi : A \to A$ such that $\varphi(1_A) = 1_A$ and $\varphi(xy) = \varphi(y)\,\varphi(x)$ for all x, $y \in A$. When the canonical morphism $K \to A$ is injective and its image $K1_A$ is identified with K, an involution φ of A is called a *standard involution* if $x\varphi(x)$ belongs to K for all $x \in A$; then $x\varphi(x)$ is called the *norm* of x, often denoted by $\mathcal{N}(x)$, whereas $x+\varphi(x)$ (that is equal to $\mathcal{N}(x+1) - \mathcal{N}(x) - 1$) is called its *trace*, often denoted by $\mathrm{tr}(x)$.

The injectiveness of $K \to A$ is equivalent to the faithfulness of the module A. When φ is a standard involution, every $x \in A$ commutes with $\varphi(x)$ since it commutes with $x + \varphi(x) \in K$. Easy calculations show that these equalities hold for all x, $y \in A$:

$$\mathcal{N}(xy) = \mathcal{N}(x)\mathcal{N}(y) \quad \text{and} \quad x^2 = \mathrm{tr}(x)x - \mathcal{N}(x).$$

Besides, the equality $\varphi(x) = \mathrm{tr}(x) - x$ shows that φ leaves invariant every subalgebra, and induces a standard involution on it by restriction.

(1.13.8) **Lemma.** *We suppose that the algebra A is a faithful and finitely generated projective module, and that φ is an anti-automorphism of A such that $x\varphi(x)$ belongs to K for all $x \in A$. Then φ is involutive, it is the only standard involution of A, and consequently commutes with all automorphisms and anti-automorphisms of A.*

Proof. Since all these properties can be tested by localization, we can suppose that K is a local ring with maximal ideal $\mathfrak{m}$. As in the proof of (1.13.2), there is a basis $(1, e_1, e_2, \dots, e_n)$ of A containing 1. From the hypotheses we deduce that $x + \varphi(x)$ is an element λ of K such that $x^2 - \lambda x$ belongs to K; indeed

$$x + \varphi(x) = (x+1)\varphi(x+1) - x\varphi(x) - 1 \quad \text{and} \quad x^2 - (x + \varphi(x))x = -x\varphi(x).$$

This determines $\varphi(x)$ in a unique way when $x = e_j$ for $j = 1, 2, \dots, n$. And since $\varphi(1) = 1$, we have proved the unicity of the anti-automorphism φ such that $x\varphi(x)$ always belongs to K.

Now $x\,\varphi^{-1}(x)$ also belongs to K because it is equal to $\varphi^{-1}(x\varphi(x))$; consequently $\varphi^{-1} = \varphi$, and φ is involutive. At last, if ψ is an automorphism (resp. anti-automorphism) of A, then $\psi\varphi\psi^{-1}$ is still an anti-automorphism, and $x\,\psi\varphi\psi^{-1}(x)$ always belongs to K because it is equal to $\psi(y\varphi(y))$ if $y = \psi^{-1}(x)$ (resp. $y = \varphi\psi^{-1}(x)$); therefore $\psi\varphi\psi^{-1} = \varphi$. □

Let us suppose that the algebra A is a free module of rank 2, and contains an element z such that $(1, z)$ is a basis of A. Then there are β and γ in K such that $z^2 = \beta z - \gamma$, and the algebra morphism $K[Z] \to A$ that maps the indeterminate Z to z determines an isomorphism from the quotient $K[Z]/(Z^2 - \beta Z + \gamma)$ onto A, since its image is A, and its kernel the ideal generated by $Z^2 - \beta Z + \gamma$. This proves that A is a commutative algebra; according to a general definition later stated in **3.4**, A is called a *quadratic extension* if the discriminant $\beta^2 - 4\gamma$ of the polynomial $Z^2 - \beta Z + \gamma$ is invertible in K.

Since the polynomial $(\beta - Z)^2 - \beta(\beta - Z) + \gamma$ is equal to $Z^2 - \beta Z + \gamma$, there is an automorphism φ of A that maps z to $\beta - z$. Since $z + \varphi(z) = \beta$ and $z\varphi(z) = \gamma$, it is easy to verify that $x\varphi(x)$ belongs to K for all $x \in A$; consequently φ is a standard involution.

Later in **3.4** it is important to know that a standard involution still exists when A is merely projective of constant rank 2. An elementary proof of this capital fact must be provided here; but hurried readers are advised to skip it; it begins with a preliminary lemma.

(1.13.9) **Lemma.** *Let $(s_1, s_2, \dots, s_n)$ be a sequence of elements of K that generates it as an ideal, and $L = \prod_i K_{s_i}$ the resulting Zariski extension of K. The canonical morphism $K \to L$ is injective, and an element $(\lambda_1, \lambda_2, \dots, \lambda_n)$ of L belongs to the image of K if and only if for every pair of indices (i, j) the image of λ_i by the canonical morphism $K_{s_i} \to K_{s_is_j}$ is equal to the image of λ_j by the analogous morphism $K_{s_j} \to K_{s_is_j}$.*

Proof. It is clear that the morphism $K \to L$ is injective and that every element in its image satisfies the announced property. Conversely let us assume that $(\lambda_1, \dots, \lambda_n)$ satisfies this property. There is an exponent $m \geq 0$ and there are elements $\kappa_1, \dots, \kappa_n \in K$ such that $\lambda_i = \kappa_i/s_i^m$ for $i = 1, 2, \dots, n$. Since κ_i/s_i^m and κ_j/s_j^m have the same image in $K_{s_is_j}$, we can write $(s_is_j)^r(s_i^m\kappa_j - s_j^m\kappa_i) = 0$ for some exponent r, and we can still assume that r is suitable for all pairs (i, j). Since $s_1, \dots, s_n$ generate K as an ideal, we can write $\sum_i s_i\mu_i = 1$ for suitable coefficients $\mu_i \in K$, and if p is any exponent greater than $(m + r - 1)n$, from $(\sum_i s_i\mu_i)^p = 1$ we deduce that $\sum_i s_i^{m+r}\nu_i = 1$ for some coefficients $\nu_i \in K$. Now let us set $\kappa = \sum_i s_i^r\kappa_i\nu_i$, and let us prove that every fraction κ_j/s_j^m is the image of κ in K_{s_j}; it suffices to verify that $s_j^r(s_j^m\kappa - \kappa_j) = 0$. As a matter of fact,

$$\begin{aligned} s_j^r(s_j^m\kappa - \kappa_j) &= \sum\nolimits_i (s_is_j)^r s_j^m \kappa_i\nu_i - s_j^r\kappa_j \;=\; \sum\nolimits_i (s_is_j)^r s_i^m \kappa_j\nu_i - s_j^r\kappa_j \\ &= s_j^r\kappa_j\Big(\sum\nolimits_i s_i^{m+r}\nu_i - 1\Big) \;=\; 0. \end{aligned}$$

□

(1.13.10) **Theorem.** *If the algebra A is a projective module of constant rank 2, then A is a commutative algebra that admits a standard involution.*

Proof. Because of (1.12.9) and (1.13.2) there are elements $s_1, s_1, \dots, s_n$ of K, generating K as an ideal, such that each module of fractions A_{s_i} is a free module of rank 2, with a basis containing the unit element; therefore each algebra A_{s_i} is commutative and admits a standard involution φ_i. Therefore the algebra $B = \prod_i A_{s_i}$ (which is a free module over the Zariski extension L defined as in (1.13.9)) is also commutative and admits a standard involution $\psi = (\varphi_1, \varphi_2, \dots, \varphi_n)$. Since the denominators s_i generate K as an ideal, the natural algebra morphism $A \to B$ is also injective, and consequently A too is a commutative algebra. Besides, if we prove that $\psi(z)$ belongs to the image of A whenever z is an element of B in the image of A, then we can claim that ψ induces a standard involution φ on A. Since $z+\psi(z)$ belongs to L, it suffices to prove that it belongs to the image of K in L, and thus we are led to the previous lemma (1.13.9). By the extension $K_{s_i} \to K_{s_i s_j}$ the standard involution φ_i induces a standard involution $\varphi_{i,j}$ of $A_{s_i s_j}$; and similarly φ_j induces a standard involution $\varphi_{j,i}$ of $A_{s_i s_j}$. According to (1.13.9) we must prove that $y + \varphi_{i,j}(y)$ and $y + \varphi_{j,i}(y)$ are the same element of $K_{s_i s_j}$ whenever y is an element of $A_{s_i s_j}$ in the image of A. This follows from the fact that $A_{s_i s_j}$ admits at most one standard involution (see (1.13.8)), and that $\varphi_{i,j}$ and $\varphi_{j,i}$ must be equal. □

Exercises

Warning: when diagrams are given, they are silently assumed to be "commutative"; if there are several paths from some module to another one, they give the same morphism.

(1.ex.1) Let N be a submodule of M, and $\mathrm{S}(M) \vee N$ the ideal generated by N in the symmetric algebra of M; prove that there is an exact sequence of modules

$$0 \longrightarrow \mathrm{S}(M) \vee N \longrightarrow \mathrm{S}(M) \longrightarrow \mathrm{S}(M/N) \longrightarrow 0.$$

Hint. Define reciprocal algebra morphisms between $\mathrm{S}(M/N)$ and the quotient of $\mathrm{S}(M)$ by $\mathrm{S}(M) \vee N$.

(1.ex.2) Let M and N be vector spaces over the field of complex numbers $\mathbb{C} = \mathbb{R} \oplus \mathbb{R}i$ with $i = \sqrt{-1}$. With N is associated a conjugate space N^c which, as a vector space over $\mathbb{R}$, is isomorphic to N through a canonical bijection $y \longmapsto y^c$, and which becomes a vector space over $\mathbb{C}$ according to the formula $\lambda y^c = (\bar{\lambda} y)^c$ (where $\bar{\lambda}$ means the conjugate complex number). The extension $\mathbb{R} \to \mathbb{C}$ determines two $\mathbb{R}$-linear mappings from $M \otimes_{\mathbb{R}} N$ respectively onto $M \otimes_{\mathbb{C}} N$ and $M \otimes_{\mathbb{C}} N^c$. Prove the bijectiveness of the resulting mapping

$$M \otimes_{\mathbb{R}} N \longrightarrow (M \otimes_{\mathbb{C}} N) \oplus (M \otimes_{\mathbb{C}} N^c), \qquad x \otimes_{\mathbb{R}} y \longmapsto (x \otimes_{\mathbb{C}} y,\ x \otimes_{\mathbb{C}} y^c).$$

(1.ex.3) Let M and N be modules over $K^2 = K \times K$. The ring K^2 is provided with a "swap automorphism" σ such that $\sigma(\lambda, \mu) = (\mu, \lambda)$, and with N is associated a conjugate module N^c which, as an additive group, is isomorphic to N through a canonical bijection $y \longmapsto y^c$, and which becomes a K^2-module according to the formula $(\lambda, \mu)y^c = ((\mu, \lambda)y)^c$. We define a ring morphism $K \to K^2$ and two ring morphisms $K^2 \to K$ in this way:

$$\varphi(\lambda) = (\lambda, \lambda), \quad \varpi(\lambda, \mu) = \lambda \quad \text{and} \quad \varpi'(\lambda, \mu) = \mu;$$

these ring morphisms allow us to define a canonical morphism $M \otimes_K N \to M \otimes_{K^2} N$ and two canonical morphisms $M \otimes_{K^2} N \to M \otimes_K N$; and since $\varphi = \sigma\varphi$, we also get a morphism $M \otimes_K N \to M \otimes_{K^2} N^c$ and two morphisms $M \otimes_{K^2} N^c \to M \otimes_K N$. From these six morphisms derive canonical isomorphisms

$$M \otimes_K N \longleftrightarrow (M \otimes_{K^2} N) \oplus (M \otimes_{K^2} N^c).$$

(1.ex.4) Prove that the following additive groups are not isomorphic:

$$\mathbb{Q} \otimes_{\mathbb{Z}} \prod_{n \in \mathbb{N}} (\mathbb{Z}/n\mathbb{Z}) \quad \text{and} \quad \prod_{n \in \mathbb{N}} (\mathbb{Q} \otimes_{\mathbb{Z}} (\mathbb{Z}/n\mathbb{Z})).$$

Hint. $\prod_{n \in \mathbb{N}} (\mathbb{Z}/n\mathbb{Z})$ contains a subgroup isomorphic to $\mathbb{Z}$, and $\mathbb{Q}$ is flat.

(1.ex.5) Prove that the following additive groups are not isomorphic:

$$\mathbb{Q} \otimes_{\mathbb{Z}} \mathrm{Hom}_{\mathbb{Z}}(\mathbb{Q},\ \mathbb{Q}/\mathbb{Z}) \quad \text{and} \quad \mathrm{Hom}_{\mathbb{Q}}(\mathbb{Q} \otimes_{\mathbb{Z}} \mathbb{Q},\ \mathbb{Q} \otimes_{\mathbb{Z}} (\mathbb{Q}/\mathbb{Z})).$$

(1.ex.6)* When a linear mapping $f : M \longrightarrow N$ is injective, it is not always true that $f \otimes f$ is injective, even when M is a free module; here is a counter-example. Let K be the quotient of the ring of polynomials $\mathbb{Z}[X, Y, Z, X', Y', Z']$ by the ideal generated by all monomials of degree 3, all monomials of degree 2 in (X, Y), all monomials of degree 2 in (X', Y', Z'), and also XX', YX', YY', $XZ' - ZX'$ and $YZ' - ZY'$; let us write x, y, z, x', y', z' for the images of the six indeterminates in K; thus K is a free $\mathbb{Z}$-module with basis

$$(1,\ x,\ y,\ z,\ x',\ y',\ z',\ xz,\ yz,\ z^2,\ xy',\ zx',\ zy',\ zz')$$

and moreover $xz' = zx'$ and $yz' = zy'$. Let N be the quotient of K^3 by the K-submodule generated by (x, y, z), and (e_1, e_2, e_3) the image in N of the canonical basis of K^3 (whence $xe_1 + ye_2 + ze_3 = 0$). Let $f : K^2 \longrightarrow N$ be the linear mapping $(\lambda, \mu) \longmapsto \lambda e_1 + \mu e_2$.

(a) Prove that f is injective.

(b) Prove that $f \otimes f : K^2 \otimes K^2 \longrightarrow N \otimes N$ is not injective, because

$$(xe_1 + ye_2 + ze_3) \otimes (x'e_1 + y'e_2 + z'e_3) - (x'e_1 + y'e_2 + z'e_3) \otimes (xe_1 + ye_2 + ze_3) = 0.$$

For more information about tensor powers or exterior powers of injective mappings, see [Flanders 1967].

(1.ex.7) In the following diagram f is surjective, and h injective:

$$\begin{array}{ccccc} M & \longrightarrow & N & \longrightarrow & P \\ \downarrow{\scriptstyle f} & & \downarrow{\scriptstyle g} & & \downarrow{\scriptstyle h} \\ M' & \longrightarrow & N' & \longrightarrow & P' \end{array}$$

(a) When the first line is exact and g is surjective, then the second line is also exact.

(b) When the second line is exact and g is injective, then the first line is also exact.

(1.ex.8) In the following diagram both lines are exact, f is surjective, and j injective:

$$\begin{array}{ccccccc} M & \longrightarrow & N & \longrightarrow & P & \longrightarrow & Q \\ \downarrow{\scriptstyle f} & & \downarrow{\scriptstyle g} & & \downarrow{\scriptstyle h} & & \downarrow{\scriptstyle j} \\ M' & \longrightarrow & N' & \longrightarrow & P' & \longrightarrow & Q' \end{array}$$

(a) When g is injective, then h too is injective.

(b) When h is surjective, then g too is surjective.

(1.ex.9) Derive an exact sequence $0 \to \mathrm{Ker}(f) \to \mathrm{Ker}(g) \to \mathrm{Ker}(h)$ from the following diagram in which both lines are exact:

$$\begin{array}{ccccccc} 0 & \longrightarrow & M & \longrightarrow & N & \longrightarrow & P \\ & & \downarrow{\scriptstyle f} & & \downarrow{\scriptstyle g} & & \downarrow{\scriptstyle h} \\ 0 & \longrightarrow & M' & \longrightarrow & N' & \longrightarrow & P' \end{array}$$

Derive an exact sequence $\mathrm{Coker}(f) \to \mathrm{Coker}(g) \to \mathrm{Coker}(h) \to 0$ from the following diagram in which both lines are exact:

$$\begin{array}{ccccccc} M & \longrightarrow & N & \longrightarrow & P & \longrightarrow & 0 \\ \downarrow{\scriptstyle f} & & \downarrow{\scriptstyle g} & & \downarrow{\scriptstyle h} & & \\ M' & \longrightarrow & N' & \longrightarrow & P' & \longrightarrow & 0 \end{array}$$

(1.ex.10) Each of the following two diagrams contains an exact line with 3 arrows, and two exact columns with 2 or 3 arrows:

$$\begin{array}{ccccccc} & & & & 0 & & 0 \\ & & & & \downarrow & & \downarrow \\ 0 & \to & P' & \to & P & \to & P'' \\ & & & & \downarrow & & \downarrow \\ & & & & Q & \to & Q'' \\ & & & & \downarrow & & \\ & & & & R & & \end{array} \qquad \begin{array}{ccccccc} & & P & & & & \\ & & \downarrow & & & & \\ Q' & \to & Q & & & & \\ \downarrow & & \downarrow & & & & \\ R' & \to & R & \to & R'' & \to & 0 \\ \downarrow & & \downarrow & & & & \\ 0 & & 0 & & & & \end{array}$$

From the first diagram derive an exact sequence $0 \to P' \to Q \to Q'' \times R$.
From the second diagram derive an exact sequence $P \oplus Q' \to Q \to R'' \to 0$.

Now let $\mathcal{F}$ be a twice covariant functor from the category $\mathcal{M}od(K)\times\mathcal{M}od(K)$ to $\mathcal{M}od(K)$. First we assume that $\mathcal{F}$ is left exact in both variables; prove that two exact sequences $0 \to M' \to M \to M''$ and $0 \to N' \to N \to N''$ give an exact sequence

$$0 \longrightarrow \mathcal{F}(M', N') \longrightarrow \mathcal{F}(M, N) \longrightarrow \mathcal{F}(M'', N) \times \mathcal{F}(M, N'').$$

Then we assume that $\mathcal{F}$ is right exact in both variables. Prove that two exact sequences $M' \to M \to M'' \to 0$ and $N' \to N \to N'' \to 0$ give an exact sequence

$$\mathcal{F}(M', N) \oplus \mathcal{F}(M, N') \longrightarrow \mathcal{F}(M, N) \longrightarrow \mathcal{F}(M'', N'') \longrightarrow 0.$$

Comment. This proves the exactness of the sequence (1.6.3).

(1.ex.11) Let e be an idempotent of K other than 0 or 1, and let M be a K-module. Prove that M is projective if and only if eM is a projective (Ke)-module, and $(1-e)M$ a projective $(K(1-e))$-module. Which conditions must fulfil eM and $(1-e)M$ for M to be a free module?

(1.ex.12) Let P be a finitely generated projective module, and P' a module such that $P \oplus P'$ is free; prove that P' contains a submodule P'' such that $P \oplus P''$ is free with finite bases.

(1.ex.13) Let M be a finitely generated module, N a finitely presented module, and $f : M \to N$ a surjective morphism; prove that $\mathrm{Ker}(f)$ is finitely generated.

(1.ex.14) If $M \otimes N$ is a free module of finite nonzero rank r, then both M and N are faithful and finitely generated projective modules.
Hint. Remember the beginning of the proof of (1.12.10), and prove that M^r is isomorphic to a direct summand of some free module K^{nr}.
Comment. Conversely, when M is a faithful and finitely generated projective module, there exists N such that $M \otimes N$ is a free module of finite nonzero rank, but the proof of this statement is much more difficult.

(1.ex.15) Let $f : K \to L$ be a ring morphism, and $\mathfrak{a}$ an ideal of K. Prove that the ring $L \otimes_K (K/\mathfrak{a})$ is isomorphic to $L/f(\mathfrak{a})L$. Prove that $f(\mathfrak{a})L \neq L$ if and only if there exists a prime ideal $\mathfrak{q}$ of L such that $\mathfrak{a} \subset f^{-1}(\mathfrak{q})$.

Rings of fractions, localizations

(1.ex.16) Let M and N be K-modules, and S a multiplicative subset of K; prove the bijectiveness of the following three canonical morphisms of $(S^{-1}K)$-modules:

$$S^{-1}(M \otimes_K N) \longrightarrow S^{-1}M \otimes_K N \longrightarrow S^{-1}M \otimes_K S^{-1}N \longrightarrow S^{-1}M \otimes_{S^{-1}K} S^{-1}N.$$

(1.ex.17)

(a) Let M be a finitely generated K-module. Assume that $M/\mathfrak{m}M = 0$ (or equivalently $M_\mathfrak{m}/\mathfrak{m}M_\mathfrak{m} = 0$) for all maximal ideals $\mathfrak{m}$ of K; prove that $M = 0$.

(b) Calculate $N_\mathfrak{p}/\mathfrak{p}N_\mathfrak{p}$ for every prime ideal $\mathfrak{p}$ when $K = \mathbb{Z}$ and $N = \mathbb{Q}$ or $N = \mathbb{Q}/\mathbb{Z}$.

(c) Let K be an integral domain that is not a field, and L its field of fractions. Prove that $L/\mathfrak{m}L = 0$ for all maximal ideals $\mathfrak{m}$ of K, and conclude that L is not a finitely generated K-module.

(1.ex.18) Let $\mathfrak{a}$ be an ideal of K.

(a) Prove that $\mathrm{rk}(\mathfrak{p}, K/\mathfrak{a})$ is equal to 1 or 0 according as $\mathfrak{p}$ contains $\mathfrak{a}$ or not.

(b) The radical $\mathfrak{r}$ of $\mathfrak{a}$ is the set of all $\lambda \in K$ such that λ^k belongs to $\mathfrak{a}$ for some k (depending on λ); deduce from (1.10.2) that $\mathfrak{r}$ is the intersection of all prime ideals containing $\mathfrak{a}$, and consequently $K/\mathfrak{a}$ and $K/\mathfrak{r}$ have the same rank at every $\mathfrak{p}$.

(c) Prove that $K/\mathfrak{a}$ is a projective K-module if and only if $\mathfrak{a} = Ke$ for some idempotent $e \in K$.

(1.ex.19) Prove that the following five assertions are equivalent when M is a finitely generated K-module:

all localizations $M_\mathfrak{p}$ at all prime ideals are generated by one element;
all localizations $M_\mathfrak{m}$ at all maximal ideals are generated by one element;
the natural morphism $\mathrm{T}(M) \to \mathrm{S}(M)$ is bijective;
the tensor algebra $\mathrm{T}(M)$ is a commutative algebra;
$x \otimes y = y \otimes x$ in $M \otimes M$ for all $x, y \in M$.

(1.ex.20)

(a) Let $\mathrm{Reg}(K)$ be the subset of all nonzero elements of K that are not divisors of zero; prove that $\mathrm{Reg}(K)$ is a multiplicative subset of K, and that it is the largest multiplicative subset S such that the canonical morphism $K \to S^{-1}K$ is injective; the corresponding ring $\mathrm{Reg}(K)^{-1}K$ is called the *total ring of fractions of* K and denoted by $\mathrm{Fr}(K)$.

(b) Now we suppose that $\mathfrak{p}$ and $\mathfrak{q}$ are prime ideals of K, and that neither contains the other; we denote the canonical morphisms from K onto the quotients $K/\mathfrak{p}$, $K/\mathfrak{q}$ and $K/(\mathfrak{p} \cap \mathfrak{q})$ by f, g and h. Prove that $\mathrm{Reg}(K/(\mathfrak{p} \cap \mathfrak{q}))$ and $h(\mathfrak{p} \cup \mathfrak{q})$ are complementary subsets in $K/(\mathfrak{p} \cap \mathfrak{q})$. This allows you to define a canonical morphism $\mathrm{Fr}(K/(\mathfrak{p}\cap\mathfrak{q})) \longrightarrow \mathrm{Fr}(K/\mathfrak{p}) \times \mathrm{Fr}(K/\mathfrak{q})$; prove that it is an isomorphism.

Hint. Choose a in $\mathfrak{p} \setminus (\mathfrak{p} \cap \mathfrak{q})$ and b in $\mathfrak{q} \setminus (\mathfrak{p} \cap \mathfrak{q})$, and observe that every fraction $f(x)/f(s)$ in $\mathrm{Fr}(K/\mathfrak{p})$ is the image of $h(bx)/h(a + bs)$.

(c)* Generalize the previous result to a finite family of prime ideals $\mathfrak{p}_1, \mathfrak{p}_2, \ldots, \mathfrak{p}_n$ of K, such that no one contains another one.

Hint. An ideal that is contained in none of the prime ideals $\mathfrak{p}_1, \mathfrak{p}_2, \dots, \mathfrak{p}_n$, is not contained in their union; this classical proposition is proved by induction on n; consequently, for $j = 1, 2, \dots, n$, there exists an a_j that belongs to $\mathfrak{p}_j$ but not to $\mathfrak{p}_i$ when $i \neq j$.

(1.ex.21) Let P be a K-module; prove that P is a finitely generated projective module of constant rank if and only if there exists a faithfully flat extension $K \to L$ such that $L \otimes P$ is a free L-module of finite rank.
Hint. If P is a finitely generated projective module, consider the sequence $(s_1, \dots, s_k)$ mentioned in (1.12.9)(e) and the Zariski extension mentioned in (1.10.6). For the converse argument remember (1.9.10), (1.12.12), (1.12.13).

(1.ex.22) Let $f : K \to L$ be a ring extension in which L is a finitely generated K-module, and let P be a finitely generated projective L-module of constant rank n. Prove that for every prime ideal $\mathfrak{p}$ in the image of $\mathrm{Spec}(L) \to \mathrm{Spec}(K)$, the rank of P at $\mathfrak{p}$ is $n\ \mathrm{rk}(\mathfrak{p}, L)$.
Hint. According to (1.ex.21) there is a faithfully flat extension $L \to L'$ such that $L' \otimes_L P$ is free over L'.

Projective modules of constant rank 1

(1.ex.23) Let K be the ring of all continuous (resp. derivable, resp. infinitely derivable...) functions $f : \mathbb{R} \to \mathbb{R}$ such that $f(t+1) = f(t)$ for all $t \in \mathbb{R}$, and let M be the set of all continuous (resp. derivable, resp. infinitely derivable...) functions $u : \mathbb{R} \to \mathbb{R}$ such that $u(t+1) = -u(t)$ for all $t \in \mathbb{R}$. Thus M is a K-module containing the functions r and s defined by $r(t) = \cos(\pi t)$ and $s(t) = \sin(\pi t)$. Prove that the mapping $K^2 \to M^2$ defined by $(f, g) \longmapsto (fr + gs,\ gr - fs)$ is an isomorphism of K-modules. Then prove that M is a projective module of constant rank 1 generated by r and s, but that it is not free.

Prove that the mapping $u \otimes v \longmapsto uv$ determines an isomorphism $M \otimes M \to K$. (*Hint:* (1.13.5)).
Comment. With each $a \in \mathbb{R}$ is associated the maximal ideal $\mathfrak{m}_a$ consisting of all $f \in K$ such that $f(a) = 0$; a classical argument, based on the compactness of the segment $[0, 1]$, proves that every prime ideal of K is an ideal $\mathfrak{m}_a$.

(1.ex.24) Let F be a field of characteristic $\neq 2$, and K the quotient of $F[X, Y]$ by the ideal generated by $X^2 + Y^2 - 1$; this is a prime ideal; the images of the indeterminates X and Y in K are denoted by x and y (whence $x^2 + y^2 = 1$). Let $\mathfrak{a}$ be the ideal of K generated by $1 + x$ and y.

(a) Prove that $\mathfrak{a}^2$ is the principal ideal generated by $1 + x$.
(b) Prove that the mapping $(f, g) \longmapsto ((1 + x)f + yg,\ (1 + x)g - yf))$ is an isomorphism from $K \oplus K$ onto $\mathfrak{a} \oplus \mathfrak{a}$. Therefore $\mathfrak{a}$ is a projective module of constant rank 1.
(c) Prove that $\mathfrak{a} \otimes \mathfrak{a}$ is isomorphic to K.

Hint. Deduce from (1.13.5) the bijectiveness of the natural mapping $\mathfrak{a} \otimes \mathfrak{a} \to \mathfrak{a}^2$.

(d) First suppose that F contains a square root i of -1. Prove that $\mathfrak{a}$ is the principal ideal generated by $1 + x + iy$. Therefore it is a free module.
Prove that every invertible element of K is equal to $\lambda(x + iy)^k$ for some $\lambda \in F^\times$ and some $k \in \mathbb{Z}$.

(e) Then suppose that F contains no square root of -1. Prove that $\mathfrak{a}$ is not a free module.
Hint. If $\mathfrak{a}$ were generated by an element f, in the algebra $K[i] = K \oplus Ki$ the equality $f = \lambda(x + iy)^k(1 + x + iy)$ should be true for some $\lambda \in F[i]^\times$ and some $k \in \mathbb{Z}$.
Comment. It has been proved that the group $\mathrm{Pic}(K)$ has order 1 or 2 according as K contains a square root of -1 or not.

(1.ex.25) Let S be a multiplicative subset of K that contains no divisors of zero; thus K can be identified with a subring of $S^{-1}K$. When M and N are K-submodules of $S^{-1}K$, the notation MN means the K-submodule generated by all products xy with $x \in M$ and $y \in N$. Observe that $S^{-1}M$ can be identified with $(S^{-1}K)M$.

(a) A K-submodule M of $S^{-1}K$ is said to be *nondegenerate* if $S^{-1}M = S^{-1}K$. Prove that M is nondegenerate if and only if $M \cap S$ is not empty.

(b) A K-submodule M of $S^{-1}K$ is said to be *invertible* if there exists a K-submodule N of $S^{-1}K$ such that $MN = K$. Prove that the equality $MN = K$ implies that M is nondegenerate, and that N is equal to $(K : M)$, that is the subset of all $y \in S^{-1}K$ such that $yM \subset K$.

(c)* Prove that every invertible K-submodule M of $S^{-1}K$ is a finitely generated projective K-module of constant rank 1. Consequently the equality $MN = K$ implies that the natural morphism $M \otimes N \longrightarrow MN$ is an isomorphism.
Hint. If $MN = K$, there exist $a_1, \dots, a_n \in M$ and $b_1, \dots, b_n \in N$ such that $\sum_i a_i b_i = 1$; this allows you to construct two morphisms $M \to K^n$ and $K^n \to M$ giving id_M by composition; consequently M is a finitely generated projective module. Now let $\mathfrak{p}$ be a prime ideal of K, r the rank of M at $\mathfrak{p}$, and ST the set of all products st with $s \in S$ and $t \in T = K \setminus \mathfrak{p}$; prove that $(ST)^{-1}K \neq 0$, and that $(ST)^{-1}M$ is a free module of rank r over $(ST)^{-1}K$; then remember that $S^{-1}M = S^{-1}K$.
More information on this topic in [Bourbaki 1961, *Algèbre commutative*, Chap. 2, §5].

(1.ex.26)* Let F be a field in which 2 is invertible, and K the subring of $F[t]$ containing all polynomials f such that $f'(0) = 0$ (in other words, $f(t) = a_0 + a_2t^2 + a_3t^3 + \cdots$).

(a) Prove that K and $F[t]$ have the same field of fractions $F(t)$.

(b) With every $a \in F$ we associate the ideal $M(a)$ generated by $t^2 - a^2$ and $t^3 - a^3$ in K; prove that $M(a)$ is a maximal ideal.

(c) Prove that $M(a)M(-a)$ is the principal ideal $(t^2 - a^2)K$ when $a \neq 0$. Then, with the help of (1.ex.25), prove that $M(a)$ is a projective K-module of constant rank 1 when $a \neq 0$.

Comment. From the surjective algebra morphism $F[X, Y] \to K$ that maps X to t^2 and Y to t^3, it is possible to deduce that K is isomorphic to the quotient of $F[X, Y]$ by the principal and prime ideal generated by $Y^2 - X^3$; when F is algebraically closed, we are in the situation described in (1.11.2): indeed K is then the ring of regular functions on the curve defined in K^2 by the equation $y^2 - x^3 = 0$, the ideals $M(a)$ are all the prime ideals other than 0, and they are in bijection with the points (a^2, a^3) of this curve. This allows us to study $M(a)$ without using (1.ex.25). The ideal $M(0)$ is exceptional because $(0, 0)$ is a singular point on this curve (see (e) beneath).

(d) Prove that $M(a)$ is not a free module when $a \neq 0$.

(e) Verify that $\mathrm{rk}(M(0), M(0)) = 2$, whereas $\mathrm{rk}(M(a), M(b)) = 1$ whenever $a \neq b$. Conclude that $M(0)$ is not a projective K-module.

Direct limits and projective limits

(1.ex.27)* Let J be an ordered set, and D the set of all $(i, j) \in J \times J$ such that $i \leq j$. A family of modules and morphisms over (J, D) is a family $\big((M_j)_{j\in J},\ (f_{j,i})_{(i,j)\in D}\big)$ consisting of modules M_j and linear mappings $f_{j,i} : M_i \to M_j$ satisfying these two conditions: $f_{j,j}$ is the identity of M_j for all $j \in J$, and $f_{k,j}f_{j,i} = f_{k,i}$ whenever (i, j) and (j, k) are in D; this family $(M_j,\ f_{j,i})$ is denoted by (M_j) when there is no ambiguity. A morphism from $(M_j,\ f_{j,i})$ to $(N_j,\ g_{j,i})$ is a family of linear mappings $u_j : M_j \to N_j$ such that $g_{j,i}u_i = u_jf_{j,i}$ whenever $(i, j) \in D$. Obviously a category (depending on J, D and the basic ring K) has been defined. From each module P is derived a constant family (P) in which all modules are equal to P and all mappings to id_P; a morphism between two constant families (P) and (Q) is the same thing as a linear mapping $P \to Q$.

From the previous family (M_j) we derive two modules $\varprojlim(M_j)$ and $\varinjlim(M_j)$. The former is the submodule of all $(x_j) \in \prod_j M_j$ such that $f_{j,i}(x_i) = x_j$ for all $(i, j) \in D$. The latter is the quotient of $\bigoplus_j M_j$ by the submodule generated by all elements $x - y$ defined in this way for all $(i, k) \in D$: x is the natural image in $\bigoplus_j M_j$ of some $x_i \in M_i$ and y is the natural image of $f_{k,i}(x_i)$. If (J, D) satisfies some condition stated beneath in (1.ex.29) (resp. (1.ex.28)), $\varprojlim(M_j)$ (resp. $\varinjlim(M_j)$) is called the *projective limit* (resp. the *direct limit*) of the family (M_j). If the order on J is the equality (in other words, if D is the diagonal of $J \times J$), we get again the direct product and the direct sum of the family (M_j). Of course all these definitions also work when D is replaced with the subset D^* of all $(i, j) \in J \times J$ such that $i \geq j$. Here and in the following exercises we use these abbreviated notations:

$$M_\alpha = \varprojlim(M_j) \quad \text{and} \quad M_\omega = \varinjlim(M_j).$$

(a) For every $j \in J$ construct canonical morphisms $f_{j,\alpha} : M_\alpha \to M_j$ and $f_{\omega,j} : M_j \to M_\omega$ in such a way that these universal properties are true:

the family $(f_{j,\alpha})$ is a morphism from the constant family (M_α) to the given family (M_j), and if (v_j) is another morphism from a constant family (P) to (M_j), there is a unique morphism $v : P \to M_\alpha$ such that $v_j = f_{j,\alpha} v$ for all $j \in J$;

in a dual way the family $(f_{\omega,j})$ is a morphism from (M_j) to the constant family (M_ω), and if (v_j) is another morphism from (M_j) to a constant family (P), there is a unique morphism $v : M_\omega \to P$ such that $v_j = v f_{\omega,j}$ for all $j \in J$.

(b) Prove that every morphism $(u_j) : (M_j) \longrightarrow (N_j)$ induces two linear mappings $u_\alpha : M_\alpha \to N_\alpha$ and $u_\omega : M_\omega \to N_\omega$, so as to define two functors $\varprojlim$ and $\varinjlim$.

(c) A sequence of morphisms between families over (J, D) is said to be exact if for every $j \in J$ it gives an ordinary exact sequence of modules. Suppose that the sequences

$$(0) \longrightarrow (M'_j) \longrightarrow (M_j) \longrightarrow (M''_j) \quad \text{and} \quad (N'_j) \longrightarrow (N_j) \longrightarrow (N''_j) \longrightarrow (0)$$

are exact; prove the exactness of the sequences

$$0 \longrightarrow M'_\alpha \longrightarrow M_\alpha \longrightarrow M''_\alpha \quad \text{and} \quad N'_\omega \longrightarrow N_\omega \longrightarrow N''_\omega \longrightarrow 0.$$

Hint. See (1.ex.9). Indeed M_α is the kernel of some morphism from $\prod_j Mj$ into $\prod_{(i,j)\in D} M_j$ whereas M_ω is the cokernel of some morphism from $\bigoplus_{(j,k)\in D} M_j$ into $\bigoplus_j M_j$.

(d) Deduce from (a) that there are canonical isomorphisms

$$\mathrm{Hom}(P,\ \varprojlim(M_j)) \longleftrightarrow \varprojlim(\mathrm{Hom}(P, M_j)),$$
$$\mathrm{Hom}(\varinjlim(M_j),\ P) \longleftrightarrow \varprojlim(\mathrm{Hom}(M_j, P));$$

because of the contravariance of the functor $\mathrm{Hom}(\dots, P)$, the modules $\mathrm{Hom}(M_j, P)$ constitute a family over (J, D^*). Prove that there are also canonical isomorphisms

$$\varinjlim(P \otimes M_j) \longleftrightarrow P \otimes \varinjlim(M_j).$$

(1.ex.28)* The notations are those of Exercise (1.ex.27). We suppose that for all $(i, j) \in J \times J$ there exists $k \in J$ such that (i, k) and (j, k) are in D; in this case the functor $\varinjlim$ is called *direct limit* (or *inductive limit*); you may understand it as a limit when j becomes greater and greater. Prove that this assumption on (J, D) brings the following five improvements (from (a) to (e)).

(a) Every x in M_ω can be written $f_{\omega,j}(x_j)$ for some x_j in some M_j. Besides, if some $f_{\omega,j}(x_j)$ vanishes, there exists $k \in J$ such that $f_{k,j}(x_j)$ already vanishes.

(b) When all $f_{j,i}$ are injective (resp. surjective), then all $f_{\omega,j}$ are injective (resp. surjective).
When all M_j are submodules of a same module, and when for all $(i,j) \in D$ the inclusion $M_i \subset M_j$ holds and $f_{j,i}$ is the natural injection, then M_ω is the union of all M_j.

(c) When the sequence $(N'_j) \rightarrow (N_j) \rightarrow (N''_j)$ is exact, so is the sequence $N'_\omega \rightarrow N_\omega \rightarrow N''_\omega$.

(d) A direct limit of flat modules is flat.

(e) Let us assume that all M_j are algebras (associative with units), and all $f_{j,i}$ are algebra morphisms; there exists a unique algebra structure on M_ω such that all $f_{\omega,j}$ are algebra morphisms. The morphism u_ω mentioned in (1.ex.27)(b) is an algebra morphism if all objects M_j and N_j and all morphisms $f_{i,j}$, $g_{i,j}$ and u_j belong to the category $\mathcal{A}lg(K)$.

(f) Here is an example with $J = \mathbb{N}$ and $K = \mathbb{Z}$. Let p be a prime integer; for every $j \in \mathbb{N}$ we set $M_j = \mathbb{Z}/p^j\mathbb{Z}$; then $f_{k,j}$ is the group morphism induced by the multiplication by p^{k-j} when $j \le k$. Prove that M_ω is isomorphic to $(S^{-1}\mathbb{Z})/\mathbb{Z}$ if S is the subset of all powers of p.

(1.ex.29)* The notations are those of Exercise (1.ex.27).

(a) Let us assume that all M_j are algebras (associative with units), and all $f_{j,i}$ are algebra morphisms; there exists a unique algebra structure on M_α such that all $f_{j,\alpha}$ are algebra morphisms. The morphism u_α mentioned in (1.ex.27)(b) is an algebra morphism if all objects M_j and N_j and all morphisms between them belong to the category $\mathcal{A}lg(K)$.

(b) We suppose that for all $(j,k) \in J \times J$ there exists $i \in J$ such that (i,j) and (i,k) belong to D; in this case the functor $\varprojlim$ is called *projective limit* (or *inverse limit*); you may understand it as a limit when j becomes smaller and smaller. This assumption on (J,D) does not improve the situation very much; nonetheless prove that all $f_{j,\alpha}$ are injective when all $f_{j,i}$ are injective.
When all M_j are submodules of a same module, and when for all $(i,j) \in D$ the inclusion $M_i \subset M_j$ holds and $f_{j,i}$ is the natural injection, then M_α is the intersection of all M_j.

(c) Here is an example with $J = \mathbb{N}$. For every $j \in \mathbb{N}$ let M_j be the quotient of the ring of polynomials $K[t]$ by the ideal generated by t^j. When $i \ge j$, let $f_{j,i} : M_i \rightarrow M_j$ be the canonical algebra morphism induced by the identity mapping of $K[t]$. Thus we get a family of K-modules over (J, D^*), and consequently the projective limit is a limit when j approaches $+\infty$. Observe that M_j is also the quotient of the ring of formal power series $K[[t]]$ by the ideal generated by t^j, whence algebra morphisms $K[[T]] \rightarrow M_j$ which induce an algebra morphism $K[[t]] \rightarrow M_\alpha$ because of the universal property of M_α; prove that it is an isomorphism.

(d) Here is another classical example with $J = \mathbb{N}$ as above, $K = \mathbb{Z}$, p a prime integer ≥ 2, and $M_j = \mathbb{Z}/p^j\mathbb{Z}$ for all $j \in \mathbb{N}$. The ring morphisms $f_{j,i}$ (with $i \geq j$) are induced by $\mathrm{id}_{\mathbb{Z}}$. For every $j \in \mathbb{N}$ there is a ring morphism $\mathbb{Z}[[t]] \to M_j$ which maps every formal series $\sum_{i=0}^{+\infty} a_i t^i$ to $\sum_{i=0}^{j-1} a_i p^i$ modulo $p^j\mathbb{Z}$, and which induces a ring morphism from $\mathbb{Z}[[t]]/(t-p)$ (the quotient by the ideal generated by $t - p$) onto M_j; thus we get a natural ring morphism $\mathbb{Z}[[t]]/(t-p) \to M_\alpha$; prove that it is an isomorphism.

This ring $\varprojlim(\mathbb{Z}/p^j\mathbb{Z})$ is called the *ring of p-adic integers*. A p-adic integer is usually written as a series $\sum_{i=0}^{+\infty} a_i p^i$; it has a unique canonical expression with all coefficients a_i in the set $\{0, 1, 2, \ldots, p-1\}$.

Chapter 2

Quadratic Mappings

This chapter presents quadratic mappings in general, and quadratic forms in particular. It affords many results that will be useful in the next chapter devoted to Clifford algebras; for instance the concept of "hyperbolic space" presented in **2.5** and the theorems of orthogonal decomposition expounded in **2.6** will be especially helpful in **3.7**, where their effectiveness also depends on an insightful use of localization and globalization. Besides, the concept of "hyperbolic space" is also essential in the last Sections **2.7** and **2.8** devoted to Witt rings.

2.1 Generalities

Let M and N be two K-modules. We say that a mapping $q : M \to N$ is a *K-quadratic mapping* (or a *quadratic mapping over K*) if $q(\lambda x) = \lambda^2 q(x)$ for all λ in K and all x in M, and if the mapping $\mathrm{b}_q : M \times M \to N$ defined by $\mathrm{b}_q(x, y) = q(x + y) - q(x) - q(y)$ is K-bilinear; this mapping b_q is called the *associated bilinear mapping.* When $N = K$, q is called a *quadratic form* on M, and (M, q) is called a *quadratic module.*

The associated bilinear mapping b_q is symmetric: $\mathrm{b}_q(x, y) = \mathrm{b}_q(y, x)$ for all x and y in M. Besides, the equality $q(2x) = 4q(x)$ implies $\mathrm{b}_q(x, x) = 2q(x)$ for all x in M. Consequently, when the mapping $y \longmapsto 2y$ is surjective from M onto M, then q is determined by b_q because the equality $x = 2y$ implies $q(x) = 4q(y) = 2\mathrm{b}_q(y, y)$. Similarly q is determined by b_q when the mapping $z \longmapsto 2z$ is injective from N into N. When this mapping $z \longmapsto 2z$ is injective, we can even prove this stronger statement: if b is a symmetric bilinear mapping $M \times M \to N$ such that $b(x, x)$ is divisible by 2 in N for all $x \in M$, then the mapping $q : M \to N$ defined by $2q(x) = b(x, x)$ is a quadratic mapping such that $\mathrm{b}_q = b$.

The set $\mathrm{Quad}_K(M, N)$ of all K-quadratic mappings $M \to N$ is obviously a K-module, like the set $\mathrm{Bil}_K(M, N)$ of all symmetric K-bilinear mappings $M \times M \to N$, and the mapping $\mathrm{Quad}_K(M, N) \to \mathrm{Bil}_K(M, N)$ defined by $q \longmapsto \mathrm{b}_q$

is obviously K-linear. When it is clear that the basic ring must be K, we omit the repeated mentions of K, and rather write $\mathrm{Quad}(M,N)$ and $\mathrm{Bil}(M,N)$. The above mapping $\mathrm{Quad}(M,N) \to \mathrm{Bil}(M,N)$ is bijective if the mapping $y \longmapsto 2y$ is bijective from M onto M, or if the mapping $z \longmapsto 2z$ is bijective from N onto N; but often it is neither injective nor surjective.

(2.1.1) **Example.** Let K be a field of characteristic 2 and M a finite-dimensional vector space over K; if f is a nonzero linear form on M, its square $x \longmapsto f(x)^2$ is a quadratic form on M that vanishes only on a hyperplane of M; nevertheless the identity $f(x+y)^2 = f(x)^2 + f(y)^2$ shows that the associated bilinear form vanishes everywhere. Besides, if q is a quadratic form on M, the identity $\mathrm{b}_q(x,x) = 2q(x) = 0$ shows that the symmetric bilinear form b_q is alternate; by means of a basis of M it is easy to prove that the image of $\mathrm{Quad}(M,K)$ in $\mathrm{Bil}(M,K)$ is the subspace of all alternate forms.

(2.1.2) **Example.** Let g be any K-bilinear mapping from $M \times M$ into N, and let us set $q(x) = g(x,x)$ for all x in M; it is easy to prove that q is a quadratic mapping such that $\mathrm{b}_q(x,y) = g(x,y) + g(y,x)$. Nevertheless it is not true that every quadratic mapping $q : M \to N$ can always be derived from a bilinear mapping in this way.

Many arguments will need the following technical lemma.

(2.1.3) **Lemma.** *Let M be a free module with basis $(e_j)_{j\in J}$ and N any K-module; let $(y_{i,j})$ be a family of elements of N (with indices (i,j) in $J \times J$) such that $y_{i,j} = y_{j,i}$ for all $(i,j) \in J \times J$. There exists a unique q in* $\mathrm{Quad}(M,N)$ *such that $q(e_j) = y_{j,j}$ for all $j \in J$ and $\mathrm{b}_q(e_i,e_j) = y_{i,j}$ for all $(i,j) \in J \times J$ such that $i \neq j$.*

Proof. If g is any bilinear mapping $M \times M \to N$, the mapping $x \longmapsto g(x,x)$ is quadratic from M into N, and the associated bilinear mapping is $(x,y) \longmapsto g(x,y) + g(y,x)$ (see (2.1.2)). Let us put a total order on the set J (this is always possible) and let $g : M \times M \longrightarrow N$ be the K-bilinear mapping such that $g(e_i,e_j) = y_{i,j}$ for all $(i,j) \in J \times J$ such that $i \leq j$, and $g(e_i,e_j) = 0$ for all (i,j) such that $i > j$. The quadratic mapping $x \longmapsto g(x,x)$ satisfies the conditions of the lemma.

To prove the uniqueness of q, it is sufficient to prove that q must vanish everywhere when all $y_{i,j}$ vanish. Obviously b_q is zero (remember that $\mathrm{b}_q(e_i,e_i) = 2q(e_i)$); consequently the subset of all $x \in M$ such that $q(x) = 0$ is a submodule of M, necessarily equal to M since it contains all e_i. □

When $q_1 : M \to N_1$ is a quadratic mapping and $v : N_1 \to N_2$ a linear mapping, then $q_2 = v \circ q_1$ is a quadratic mapping $M \to N_2$. Therefore it is sensible to consider the category of all quadratic mappings defined on M, like q_1 and q_2 above; a morphism between two objects q_1 and q_2 of this category is a linear mapping v such that $q_2 = v \circ q_1$. The next proposition means that this category contains an initial universal object $\gamma : M \to \Gamma^2(M)$ according to the definition given in **1.2**; γ is called the *universal quadratic mapping* on M.

(2.1.4) **Proposition.** *For every module M there exist a module $\Gamma^2(M)$ and a quadratic mapping $\gamma : M \to \Gamma^2(M)$ which for every module N determine an isomorphism* $\mathrm{Hom}(\Gamma^2(M), N) \to \mathrm{Quad}(M, N)$ *in this way: for every quadratic mapping $q : M \to N$ there exists a unique linear mapping $\bar{q} : \Gamma^2(M) \to N$ such that $q = \bar{q} \circ \gamma$.*

Proof. Let $K^{(M)}$ be the free module with basis $(e_x)_{x\in M}$ (see (**1.3**)), and R the submodule of $P = K^{(M)} \oplus (M \otimes M)$ generated by all elements

$$(e_{x+y} - e_x - e_y \ , \ -x \otimes y) \quad \text{and} \quad (e_{\lambda x} - \lambda^2 e_x \ , \ 0)$$

with x, y in M and λ in K. Then $\Gamma^2(M)$ is the quotient module P/R, and γ is obtained by composing the mapping $x \longmapsto (e_x, 0)$ from M into P and the quotient mapping $P \to \Gamma^2(M)$. This mapping γ is quadratic, and the associated bilinear mapping b_γ maps every $(x, y) \in M \times M$ to the image of $(0,\ x \otimes y)$ in the quotient $\Gamma^2(M)$. For every quadratic mapping $q : M \to N$ there exists a unique linear mapping $\bar{q} : \Gamma^2(M) \to N$ such that $\bar{q} \circ \gamma = q$; indeed $\bar{q}$ must be defined in this way: there is a linear mapping $P \to N$ that maps $(e_x,\ y \otimes z)$ to $q(x) + \mathrm{b}_q(y, z)$ for all x, y, $z \in M$, and since it vanishes on R, it determines the wanted mapping $\bar{q} : \Gamma^2(M) \to N$. □

Let γ_1 and γ_2 be the universal quadratic mappings on the modules M_1 and M_2, and u a linear mapping from M_1 into M_2. Since $\gamma_2 \circ u$ is a quadratic mapping from M_1 into $\Gamma^2(M_2)$, there exists a unique linear mapping $\Gamma^2(u)$ from $\Gamma^2(M_1)$ into $\Gamma^2(M_2)$ such that $\Gamma^2(u) \circ \gamma_1 = \gamma_2 \circ u$. It is clear that a covariant functor Γ^2 has been defined in this way.

(2.1.5) **Proposition.** *As a module, $\Gamma^2(M)$ is generated by the subset $\gamma(M)$. When M is generated by a subset S, then $\Gamma^2(M)$ is generated by $\gamma(S)$ and the elements $\mathrm{b}_\gamma(x, y)$ with x and y running through S. If M is free and $(e_j)_{j\in J}$ is a basis of M indexed by a totally ordered set J, then the elements $\gamma(e_j)$ (with $j \in J$) and $\mathrm{b}_\gamma(e_i, e_j)$ (with $i < j$) constitute a basis of $\Gamma^2(M)$.*

Proof. The first statement is a common property of many initial universal objects (when the arrows of the category under consideration are mappings satisfying some properties); indeed let P be the submodule of $\Gamma^2(M)$ generated by $\gamma(M)$; on one side there is a natural injection $P \longrightarrow \Gamma^2(M)$; on the other side the universal property of $\Gamma^2(M)$ determines a mapping from $\Gamma^2(M)$ into P; then the composition $\Gamma^2(M) \to P \to \Gamma^2(M)$ must be the identity mapping, whence $P = \Gamma^2(M)$. The second statement is an immediate consequence of the first one. The third statement is a consequence of (2.1.3). Indeed a quadratic mapping q from the free module M into any module N is completely determined by the values $q(e_j)$ (with $j \in J$) and $\mathrm{b}_q(e_i, e_j)$ (with $i < j$), and the family of all these values may be arbitrarily chosen in N. If every linear mapping defined on a module P is determined by its values on some subset of P, and if these values can be chosen arbitrarily, this subset is a basis of P. □

(2.1.6) **Proposition.** *For any direct sum* $M_1 \oplus M_2$ *the algebra* $\Gamma^2(M_1 \oplus M_2)$ *is canonically isomorphic to*

$$\Gamma^2(M_1) \oplus \Gamma^2(M_2) \oplus (M_1 \otimes M_2).$$

Proof. Let γ_1, γ_2 and γ be the universal quadratic mappings on M_1, M_2 and $M_1 \oplus M_2$. We get three quadratic mappings on $M_1 \oplus M_2$ if we map each (x_1, x_2) first to $\gamma_1(x_1)$ in $\Gamma^2(M_1)$, then to $\gamma_2(x_2)$ in $\Gamma^2(M_2)$, and finally to $x_1 \otimes x_2$ in $M_1 \otimes M_2$. Together they give a quadratic mapping $(x_1, x_2) \longmapsto (\gamma_1(x_1), \gamma_2(x_2), x_1 \otimes x_2)$ from which the universal property of $\Gamma^2(M_1 \oplus M_2)$ allows us to derive a linear mapping from this module into the direct product (or direct sum) of $\Gamma^2(M_1)$, $\Gamma^2(M_2)$ and $M_1 \otimes M_2$. Conversely each of these three modules has a universal property that allows us to define a natural linear mapping from itself into $\Gamma^2(M_1 \oplus M_2)$; to be precise, we first map every $\gamma_1(x_1)$ to $\gamma(x_1, 0)$, then every $\gamma_2(x_2)$ to $\gamma(0, x_2)$, and finally every $x_1 \otimes x_2$ to $b_\gamma((x_1, 0), (0, x_2))$, that is $\gamma(x_1, x_2) - \gamma(x_1, 0) - \gamma(0, x_2)$; thus we get a linear mapping defined on the direct sum of $\Gamma^2(M_1)$, $\Gamma^2(M_2)$ and $M_1 \otimes M_2$. It is easy to verify that two reciprocal isomorphisms have been defined in this way. □

Because of the universal property of $\mathrm{S}^2(M)$ (see (1.4.3)), there is a unique linear mapping $\mathrm{S}^2(M) \to \Gamma^2(M)$ that maps every $x \vee y$ (with x and y in M) to $\mathrm{b}_\gamma(x, y)$. Conversely because of the universal property of $\Gamma^2(M)$ there is a unique linear mapping $\Gamma^2(M) \to \mathrm{S}^2(M)$ that maps every $\gamma(x)$ to $x \vee x$. It is easy to verify that the mappings

$$\mathrm{S}^2(M) \longrightarrow \Gamma^2(M) \longrightarrow \mathrm{S}^2(M) \quad \text{and} \quad \Gamma^2(M) \longrightarrow \mathrm{S}^2(M) \longrightarrow \Gamma^2(M)$$

are the multiplications by 2 respectively in $\mathrm{S}^2(M)$ and $\Gamma^2(M)$. Let us suppose that the mapping $x \longmapsto 2x$ is bijective from M onto M, and let us write $x \longmapsto x/2$ for the reciprocal mapping (even if 2 is not invertible in K); since the functor S^2 or Γ^2 transforms the multiplication by 2 in M into the multiplication by 4 in $\mathrm{S}^2(M)$ or $\Gamma^2(M)$, the multiplication by 2 in $\mathrm{S}^2(M)$ or $\Gamma^2(M)$ is also bijective, and consequently we have got two isomorphisms $\mathrm{S}^2(M) \to \Gamma^2(M)$ and $\Gamma^2(M) \to \mathrm{S}^2(M)$. The former is more interesting, since later in **4.5** it proves to be the restriction of an algebra isomorphism $\mathrm{S}(M) \to \Gamma(M)$; thus we are led to the next statement.

(2.1.7) **Proposition.** *If the mapping* $x \longmapsto 2x$ *is bijective from* M *onto* M, *the canonical morphism* $\mathrm{S}^2(M) \to \Gamma^2(M)$ *defined by* $x \vee y \longmapsto \mathrm{b}_\gamma(x, y)$ *is an isomorphism, and the converse isomorphism maps every* $\gamma(x)$ *to* $x \vee x/2$.

As a particular case of (2.1.4), the K-module $\mathrm{Quad}(M, K)$ of quadratic forms over the module M is isomorphic to $\Gamma^2(M)^* = \mathrm{Hom}(\Gamma^2(M), K)$. Nevertheless in general the modules $\Gamma^2(M)^*$ and $\Gamma^2(M^*)$ are not isomorphic. According to (2.2.7) below, there are canonical linear mappings $\Gamma^2(M^*) \to \mathrm{S}^2(M)^*$ and $\Gamma^2(M) \to \mathrm{S}^2(M^*)^*$, which are isomorphisms when M is a finitely generated projective module.

2.2 Changes of basic ring and localizations

When $\varphi : M \times M \to N$ is a symmetric K-bilinear mapping, the kernel of φ, denoted by $\mathrm{Ker}(\varphi)$, is the submodule of all $x \in M$ such that $\varphi(x,y) = 0$ for all $y \in M$. When $q : M \to N$ is a K-quadratic mapping, besides $\mathrm{Ker}(\mathrm{b}_q)$ which is the submodule of all $x \in M$ such that $\mathrm{b}_q(x,y) = 0$ for all $y \in M$, we also define the *kernel of* q, denoted by $\mathrm{Ker}(q)$, which is the submodule of all $x \in \mathrm{Ker}(\mathrm{b}_q)$ such that $q(x) = 0$.

Let R be a submodule of M; q is constant modulo R (and consequently determines a K-quadratic mapping $M/R \to N$) if and only if R is a submodule of $\mathrm{Ker}(q)$. Proposition (2.2.1) shows that the inclusion $\mathrm{Ker}(q) \subset \mathrm{Ker}(\mathrm{b}_q)$ is often an equality, but Example (2.2.2) shows that this inclusion is sometimes strict.

(2.2.1) **Proposition.** *The equality* $\mathrm{Ker}(q) = \mathrm{Ker}(\mathrm{b}_q)$ *is valid when the mapping* $y \longmapsto 2y$ *is surjective from* M *onto* M*; it is also valid when the mapping* $z \longmapsto 2z$ *is injective from* N *into* N.

Indeed, when x belongs to $\mathrm{Ker}(\mathrm{b}_q)$, the equalities $2q(x) = \mathrm{b}_q(x,x) = 0$ imply $q(x) = 0$ if the mapping $z \longmapsto 2z$ is injective; and if $x = 2y$ for some $y \in M$, then $q(x) = 4q(y) = 2\mathrm{b}_q(y,y) = \mathrm{b}_q(x,y) = 0$. □

(2.2.2) **Example.** Let K be a field of characteristic 2, M a vector space of finite dimension over K, and f a linear form on M; let us consider $q(x) = f(x)^2$ as in Example (2.1.1); the kernel of q is the kernel of f, whereas the kernel of b_q is M.

Here is the main result of this section.

(2.2.3) **Theorem.** *Let* $f : K \to K'$ *be a ring morphism,* M *and* N *two* K*-modules and* $q : M \longrightarrow N$ *a* K*-quadratic mapping. There exists a unique* K'*-quadratic mapping* $q' : K' \otimes_K M \to K' \otimes_K N$ *such that* $q'(1' \otimes x) = 1' \otimes q(x)$ *for all* $x \in M$ *(if* $1'$ *is the unit element of* K'*).*

This property of q' implies that $\mathrm{b}_{q'}(1' \otimes x,\ 1' \otimes y) = 1' \otimes \mathrm{b}_q(x,y)$ for all x, $y \in M$. It can be written $q' \circ (f \otimes M) = (f \otimes N) \circ q$ if we accept to identify M with $K \otimes_K M$ in the left-hand member, and N with $K \otimes_K N$ in the right-hand member. Sometimes q' is denoted by $K' \otimes q$.

Proof. The uniqueness of q' is obvious and we have to prove its existence. First we suppose that M is a free K-module with basis $(e_j)_{j \in J}$, and we set $y_{j,j} = q(e_j)$ for all j in J and $y_{i,j} = \mathrm{b}_q(e_i, e_j)$ for all (i,j) in $J \times J$ such that $i \neq j$. Since $K' \otimes M$ is a free K'-module with basis $(1' \otimes e_j)_{j \in J}$, according to Lemma (2.1.3) there exists a K'-quadratic mapping $q' : K' \otimes M \to K' \otimes N$ such that $q'(1' \otimes e_j) = 1' \otimes y_{j,j}$ for all $j \in J$ and $\mathrm{b}_{q'}(1' \otimes e_i,\ 1' \otimes e_j) = 1' \otimes y_{i,j}$ for all $(i,j) \in J \times J$ such that $i \neq j$. Obviously this q' satisfies the required conditions.

When M is not free, there is a surjective mapping $g : L \to M$ which makes M become the quotient of a free module L by the submodule $R = \mathrm{Ker}(g)$. If we set $q_0 = q \circ g$, we get a K-quadratic form q_0 on L, and consequently there exists a

K'-quadratic mapping $q_0' : K' \otimes L \to K' \otimes N$ such that $q_0' \circ (f \otimes L) = (f \otimes N) \circ q_0$. Let R' be the kernel of $K' \otimes g : K' \otimes L \to K' \otimes M$; since R' is the image of $K' \otimes R$ in $K' \otimes L$, it is clear that $R' \subset \mathrm{Ker}(q_0')$; in other words, q_0' is constant modulo R'. This proves the existence of a K'-quadratic mapping $q' : K' \otimes M \to K' \otimes N$ such that $q_0' = q' \circ (K' \otimes g)$. The desired equality $q' \circ (f \otimes M) = (f \otimes N) \circ q$ now follows from the surjectiveness of g and these calculations:

$$\begin{aligned} q' \circ (f \otimes M) \circ (K \otimes g) &= q' \circ (K' \otimes g) \circ (f \otimes L) = q_0' \circ (f \otimes L) \\ &= (f \otimes N) \circ q_0 = (f \otimes N) \circ q \circ g \,. \end{aligned} \qquad \square$$

A particular case of change of basic ring will be especially useful later on. Let S be a multiplicative subset of K (see **1.10**), $K' = S^{-1}K$ the ring of fractions of K with denominator in S and $f : K \to K'$ the canonical ring morphism $\lambda \longmapsto \lambda/1$. If $q : M \longrightarrow N$ is a K-quadratic mapping, Theorem (2.2.3) implies that *there exists a unique K'-quadratic mapping $q' : S^{-1}M \longrightarrow S^{-1}N$ such that $(f \otimes N) \circ q = q' \circ (f \otimes M)$* , or equivalently,

$$q'\left(\frac{x}{s}\right) = \frac{q(s)}{s^2} \quad \text{for all} \quad \frac{x}{s} \in S^{-1}M \,.$$

When $\mathfrak{p}$ is a prime ideal of K and $S = K \setminus \mathfrak{p}$, then q' is a quadratic mapping from $M_\mathfrak{p}$ into $N_\mathfrak{p}$, it is called the *localization of q at the prime ideal* $\mathfrak{p}$ and denoted by $q_\mathfrak{p}$; thus $q_\mathfrak{p}(x/s) = q(x)/s^2$ for all x/s in $M_\mathfrak{p}$.

Here are some applications to the functor Γ^2.

(2.2.4) **Proposition.** *Let $f : K \to K'$ be a ring morphism, and M a K-module; the K'-module $\Gamma^2_{K'}(K' \otimes M)$ is canonically isomorphic to $K' \otimes \Gamma^2_K(M)$.*

Proof. Let γ and γ' be the universal quadratic mappings on M and $K' \otimes M$. Since the mapping $x \longmapsto \gamma'(1' \otimes x)$ is K-quadratic from M into $\Gamma^2_{K'}(K' \otimes M)$, it determines a K-linear mapping $\Gamma^2_K(M) \to \Gamma^2_{K'}(K' \otimes M)$ and then a K'-linear mapping $K' \otimes \Gamma^2_K(M) \to \Gamma^2_{K'}(K' \otimes M)$. Conversely, by Theorem (2.2.3) the K-quadratic mapping γ determines a K'-quadratic mapping from $K' \otimes M$ into $K' \otimes \Gamma^2_K(M)$, whence a K'-linear mapping $\Gamma^2_{K'}(K' \otimes M) \to K' \otimes \Gamma^2_K(M)$. It is easy to verify that two reciprocal isomorphisms have been defined in this way. $\square$

(2.2.5) **Corollary.** *Let $f : K \to K'$ be a ring morphism, M a K-module and N' a K'-module; the K'-modules* $\mathrm{Quad}_{K'}(K' \otimes M,\ N')$ *and* $\mathrm{Quad}_K(M, N')$ *are canonically isomorphic.*

Proof. Because of (2.2.4) there are canonical isomorphisms

$$\mathrm{Quad}_{K'}(K' \otimes M,\ N') \cong \mathrm{Hom}_{K'}(\Gamma^2_{K'}(K' \otimes M),\ N') \cong \mathrm{Hom}_{K'}(K' \otimes \Gamma^2(M),\ N') \,;$$

and because of (1.9.4) there are canonical isomorphisms

$$\mathrm{Hom}_{K'}(K' \otimes \Gamma^2(M),\ N') \cong \mathrm{Hom}(\Gamma^2_K(M),\ N') \cong \mathrm{Quad}_K(M, N') \,. \qquad \square$$

As an evident corollary of (2.2.4) we mention the isomorphism

(2.2.6) $$\Gamma^2_{S^{-1}K}(S^{-1}M) \;\cong\; S^{-1}\Gamma^2_K(M)$$

which holds for every ring of fractions of K, in particular for every localization of K.

(2.2.7) **Proposition.** *For every K-module M there are canonical K-linear mappings*

$$\Gamma^2(M^*) \longrightarrow \mathrm{S}^2(M)^* \quad \textit{and} \quad \Gamma^2(M) \longrightarrow \mathrm{S}^2(M^*)^*.$$

When M is a finitely generated projective module, both mappings are isomorphisms, and their sources and targets are finitely generated projective modules.

Proof. Let γ and γ' be the universal quadratic mappings on M and M^*. The following equalities (in which x and y belong to M, and u and v to M^*) define two quadratic mappings $q : M^* \to \mathrm{S}^2(M)^*$ and $q' : M \to \mathrm{S}^2(M^*)^*$:

$$q(u)(x \vee y) = u(x)u(y) \quad \text{and} \quad q'(x)(u \vee v) = u(x)v(x).$$

Then q and q' determine unique linear mappings $\bar{q} : \Gamma^2(M^*) \to \mathrm{S}^2(M)^*$ and $\bar{q}' : \Gamma^2(M) \to \mathrm{S}^2(M^*)^*$ such that $\bar{q} \circ \gamma' = q$ and $\bar{q}' \circ \gamma = q'$.

If M is a free module of finite rank, it is easy to prove that both canonical mappings $\bar{q}$ and $\bar{q}'$ are isomorphisms of free modules. If M is a finitely generated projective module, all its localizations are free modules (see (1.12.4)); from (2.2.6) and (1.9.3) we deduce that, for all $\mathfrak{p} \in \mathrm{Spec}(K)$,

$$\Gamma^2_{K_\mathfrak{p}}(M_\mathfrak{p}) \cong (\Gamma^2_K(M))_\mathfrak{p} \quad \text{and} \quad \mathrm{S}^2_{K_\mathfrak{p}}(M_\mathfrak{p}) \cong (\mathrm{S}^2_K(M))_\mathfrak{p} \;,$$

and the same with M^*; consequently all localizations of both canonical mappings are isomorphisms; by (1.11.7) they are themselves isomorphisms. Their sources and targets are obviously finitely generated, and to prove that they are projective, it suffices to notice that their localizations are free of constant rank whenever the rank of M is constant (see (1.12.9) and (1.12.8)). □

This proof using localization and globalization is more direct than the other one that uses a module M' such that $M \oplus M'$ is free with finite bases, and needs (2.1.6) to decompose $\Gamma^2(M \oplus M')$ and $\Gamma^2(M^* \oplus M'^*)$, and (1.5.1) to decompose $\mathrm{S}^2(M \oplus M')$ and $\mathrm{S}^2(M^* \oplus M'^*)$.

2.3 Nondegenerate quadratic mappings

With every symmetric bilinear mapping $\varphi : M \times M \to N$ we associate the linear mapping d_φ from M into $\mathrm{Hom}(M,N)$ defined by $\mathrm{d}_\varphi(x)(y) = \varphi(x,y)$. And with every quadratic mapping $q : M \to N$ we associate $\mathrm{d}_q : M \to \mathrm{Hom}(M,N)$ defined by $\mathrm{d}_q(x)(y) = \mathrm{b}_q(x,y)$. We say that q (resp. φ) is *nondegenerate* if d_q

(resp. d_φ) is bijective from M onto $\mathrm{Hom}(M,N)$, and we say that q (resp. φ) is *weakly nondegenerate* if d_q (resp. d_φ) is merely injective.

When K is a field, and M a finite-dimensional vector space, every weakly nondegenerate quadratic form $M \to K$ is nondegenerate; but if $\dim(M)$ is infinite, no quadratic form on M is nondegenerate, even if it is weakly nondegenerate. In the most usual cases q or φ may be nondegenerate only if the target N is an invertible module, as it is now explained.

(2.3.1) **Proposition.** *Let M be a faithful and finitely generated projective module. When there is a nondegenerate symmetric bilinear mapping $M \times M \to N$, then N is a finitely generated projective module of constant rank* 1.

Proof. The mapping $\mathrm{Hom}(M,N)\otimes M \to N$ defined by $f\otimes x \longmapsto f(x)$ is surjective; this can be proved by localization, because (1.9.7) allows us to localize $\mathrm{Hom}(M,N)$ without problems, and all localizations of M are free and faithful modules (see (1.12.4) and (1.13.3)). Since $\mathrm{Hom}(M,N)$ is isomorphic to M and therefore finitely generated, N too is finitely generated. Since the rank of $\mathrm{Hom}(M,N)$ is the product of the ranks of M and N (see (1.12.6)), the isomorphism $M \to \mathrm{Hom}(M,N)$ forces the rank of N to be everywhere 1. The faithfulness of M implies the faithfulness of N, and the faithfulness of all its localizations (again (1.13.3)). Consequently all localizations of N are free of rank 1, and the conclusion follows from (1.12.9). □

The next two propositions show that the nondegeneracy property can be tested with faithfully flat extensions and with localizations, provided that the source is a finitely presented module. Under a stronger hypothesis (both source and target are finitely generated and projective), it can even be tested with extensions $K \to K/\mathfrak{m}$ to residue fields, as explained in (2.3.4). For the sake of briefness only nondegenerate quadratic forms are mentioned, but it is clear that (2.3.2) and (2.3.3) remain true when the word "nondegenerate" is replaced with "weakly nondegenerate".

(2.3.2) **Proposition.** *Let $K \to K'$ be an extension of the basic ring, $q : M \to N$ a quadratic mapping, and $q' : K' \otimes M \to K' \otimes N$ the corresponding extension of q.*

(a) *When q is nondegenerate, and M is finitely generated and projective, then q' is also nondegenerate.*

(b) *When the extension $K \to K'$ is faithfully flat, and M is finitely presented, then q' is nondegenerate if and only if q is nondegenerate.*

Proof. Let us consider the linear mappings

$$\mathrm{d}_q : M \longrightarrow \mathrm{Hom}(M,N) \quad \text{and} \quad \mathrm{d}_{q'} : K' \otimes M \longrightarrow \mathrm{Hom}_{K'}(K' \otimes M,\ K' \otimes N)$$

derived from q and q'. When M is finitely generated and projective, or when M is finitely presented and K' flat over K, then the canonical mapping

$$K' \otimes \mathrm{Hom}(M,N) \longrightarrow \mathrm{Hom}_{K'}(K' \otimes M,\ K' \otimes N)$$

is an isomorphism (see (1.9.7) and (1.9.9)); consequently $\mathrm{d}_{q'}$ can be identified with

$$K' \otimes \mathrm{d}_q \ : \ K' \otimes M \longrightarrow K' \otimes \mathrm{Hom}(M,N) \ .$$

Consequently if d_q is an isomorphism, the same is true for $\mathrm{d}_{q'}$. Moreover when K' is faithfully flat over K, the bijectiveness of $\mathrm{d}_{q'}$ conversely implies the bijectiveness of d_q. □

(2.3.3) **Proposition.** *Let $q : M \to N$ be a quadratic mapping defined on a finitely presented module M. The following three assertions are equivalent:*

(a) *the K-quadratic mapping $q : M \to N$ is nondegenerate;*
(b) *for every prime ideal $\mathfrak{p}$ the quadratic mapping $q_\mathfrak{p} : M_\mathfrak{p} \to N_\mathfrak{p}$ is nondegenerate;*
(c) *for every maximal ideal $\mathfrak{m}$ the quadratic mapping $q_\mathfrak{m} : M_\mathfrak{m} \to N_\mathfrak{m}$ is nondegenerate.*

Proof. Since the localized rings $K_\mathfrak{p}$ are flat, the proof of (2.3.2) already shows that the $K_\mathfrak{p}$-linear mapping $M_\mathfrak{p} \to \mathrm{Hom}_{K_\mathfrak{p}}(M_\mathfrak{p}, N_\mathfrak{p})$ induced by $q_\mathfrak{p}$ can be identified with $K_\mathfrak{p} \otimes \mathrm{d}_q = (\mathrm{d}_q)_\mathfrak{p}$. Thus the conclusion follows from (1.11.7). □

Proposition (2.3.2)(a) can be applied to all extensions $K \to K/\mathfrak{a}$ where $\mathfrak{a}$ is an ideal of K. In this case, $(K/\mathfrak{a}) \otimes M$ and $(K/\mathfrak{a}) \otimes N$ can be replaced with $M/\mathfrak{a}M$ and $N/\mathfrak{a}N$ (see (1.9.1)), and thus we get a $(K/\mathfrak{a})$-quadratic mapping $q/\mathfrak{a} : M/\mathfrak{a}M \to N/\mathfrak{a}N$. In this context we get the following converse statement.

(2.3.4) **Proposition.** *Let M and N be finitely generated projective modules, and $q : M \longrightarrow N$ a quadratic mapping; if the $(K/\mathfrak{m})$-quadratic mappings $q/\mathfrak{m} : M/\mathfrak{m}M \longrightarrow N/\mathfrak{m}N$ are nondegenerate for all maximal ideals $\mathfrak{m}$ of K, then q itself is nondegenerate.*

Proof. Let us set $R = \mathrm{Ker}(\mathrm{d}_q)$ and $C = \mathrm{Coker}(\mathrm{d}_q)$ to get the exact sequence

$$0 \longrightarrow R \longrightarrow M \longrightarrow \mathrm{Hom}(M,N) \longrightarrow C \longrightarrow 0.$$

For every maximal ideal $\mathfrak{m}$ we derive from it the exact sequence

$$(K/\mathfrak{m}) \otimes M \longrightarrow (K/\mathfrak{m}) \otimes \mathrm{Hom}(M,N) \longrightarrow (K/\mathfrak{m}) \otimes C \longrightarrow 0$$

in which the first mapping on the left side can be identified with $\mathrm{d}_{q/\mathfrak{m}}$ (as explained above for $\mathrm{d}_{q'}$ in the proof of (2.3.2)). From the assumption that $\mathrm{d}_{q/\mathfrak{m}}$ is bijective, we deduce that $(K/\mathfrak{m}) \otimes C = 0$, whence $C = \mathfrak{m}C$ and $C_\mathfrak{m} = \mathfrak{m}C_\mathfrak{m}$. Since $\mathrm{Hom}(M,N)$ is finitely generated, so is C, and Nakayama's lemma implies that $C_\mathfrak{m} = 0$ for every maximal ideal $\mathfrak{m}$, whence $C = 0$. There remains an exact sequence $0 \to R \to M \to \mathrm{Hom}(M,N) \to 0$. Since M and N are finitely generated projective modules, $\mathrm{Hom}(M,N)$ is projective, this exact sequence is splitting, R is finitely generated, and we get exact sequences

$$0 \longrightarrow (K/\mathfrak{m}) \otimes R \longrightarrow (K/\mathfrak{m}) \otimes M \longrightarrow (K/\mathfrak{m}) \otimes \mathrm{Hom}(M,N) \longrightarrow 0 \ ;$$

now the assumption that $\mathrm{d}_{q/\mathfrak{m}}$ is bijective, implies that $(K/\mathfrak{m}) \otimes R = 0$ for every maximal ideal $\mathfrak{m}$; a similar argument leads to $R = 0$. Thus $R = C = 0$ and d_q is bijective. □

(2.3.5) **Remark.** The propositions (2.3.2), (2.3.3) and (2.3.4) are also valid for symmetric bilinear mappings. By any extension $K \to K'$ it is clear that every bilinear mapping $\varphi : M \times M \to N$ gives a bilinear mapping φ' from $(K' \otimes M) \times (K' \otimes M)$ into $K' \otimes N$. Indeed the knowledge of φ is equivalent to the knowledge of a linear mapping $M \otimes M \to N$, and the same for φ'; therefore the canonical isomorphism $(K' \otimes M) \otimes_{K'} (K' \otimes M) \longleftrightarrow K' \otimes (M \otimes M)$ (see (1.9.6)) shows how to derive φ' from φ.

Orthogonality

If $\varphi : M \times M \to N$ is a symmetric bilinear mapping, or if $q : M \to N$ is a quadratic mapping, two elements x and y of M are said to be *orthogonal* (with respect to φ or q) if $\varphi(x, y)$ or $\mathrm{b}_q(x, y)$ vanishes. With every subset P of M is associated the submodule $P^\perp$ of all elements of M orthogonal to all elements of P; it is called *the submodule orthogonal to* P, although it is exactly the largest submodule orthogonal to P. Obviously $P^\perp$ only depends on the submodule generated by P, but in most cases P is already assumed to be a submodule. It is clear that $P^\perp$ always contains $\mathrm{Ker}(\varphi)$ or $\mathrm{Ker}(\mathrm{b}_q)$. The next lemma gives elementary properties of orthogonal modules like $P^\perp$ and of *orthogonal closures* like $P^{\perp\perp} = (P^\perp)^\perp$; the proof is left to the reader.

(2.3.6) **Lemma.** *The following six assertions are valid for all submodules P or for all pairs (P, Q) of submodules of M :*

$$\begin{aligned} P \subset Q &\Rightarrow P^\perp \supset Q^\perp ; & (P+Q)^\perp &= P^\perp \cap Q^\perp ; \\ P^{\perp\perp} &\supset P ; & (P \cap Q)^\perp &\supset (P^\perp + Q^\perp)^{\perp\perp} ; \\ P^{\perp\perp\perp} &= P^\perp ; & (P^\perp \cap Q^\perp)^\perp &= (P+Q)^{\perp\perp} . \end{aligned}$$

The next two propositions involve a symmetric bilinear mapping $\varphi : M \times M \to N$; of course they might also involve a quadratic mapping $q : M \to N$.

(2.3.7) **Proposition.** *When $\varphi : M \times M \to N$ is nondegenerate, and P is a direct summand of M, then*

(a) *$P^\perp$ too is a direct summand, and every equality $M = P \oplus Q$ implies $M = P^\perp \oplus Q^\perp$;*
(b) *P is orthogonally closed: $P = P^{\perp\perp}$;*
(c) *these four mappings induced by d_φ are bijective:*

$$\begin{aligned} P^\perp &\to \mathrm{Hom}(M/P,\ N) , & M/P^\perp &\to \mathrm{Hom}(P, N) , \\ P &\to \mathrm{Hom}(M/P^\perp,\ N) , & M/P &\to \mathrm{Hom}(P^\perp, N) ; \end{aligned}$$

(d) *the equality $\mathrm{rk}(\mathfrak{p}, M) = \mathrm{rk}(\mathfrak{p}, P) + \mathrm{rk}(\mathfrak{p}, P^\perp)$ holds for every prime ideal $\mathfrak{p}$ provided that M is finitely generated and projective.*

Proof. Let us assume that $M = P \oplus Q$. We can identify $\mathrm{Hom}(M,N)$ with the direct sum of $\mathrm{Hom}(P,N)$ and $\mathrm{Hom}(Q,N)$; this means that $\mathrm{Hom}(P,N)$ is identified with the submodule of all morphisms $M \to N$ vanishing on Q, and the same for $\mathrm{Hom}(Q,N)$. Since $\mathrm{d}_\varphi : M \to \mathrm{Hom}(M,N)$ is bijective, the converse bijection maps $\mathrm{Hom}(Q,N)$ onto $P^\perp$ and $\mathrm{Hom}(P,N)$ onto $Q^\perp$. Consequently d_φ induces bijections $P^\perp \to \mathrm{Hom}(Q,N)$ and $Q^\perp \to \mathrm{Hom}(P,N)$, and moreover $M = P^\perp \oplus Q^\perp$. From this last equality we could deduce in the same way that M is the direct sum of $P^{\perp\perp}$ and $Q^{\perp\perp}$, and since they contain respectively P and Q, they must be equal to them. Thus we have proved (a) and (b). The first two morphisms mentioned in (c) are the above isomorphisms $P^\perp \to \mathrm{Hom}(Q,N)$ and $Q^\perp \to \mathrm{Hom}(P,N)$. Since (a) and (b) show that P and $P^\perp$ can play the same role, all morphisms in (c) are bijective. Now let M be finitely generated and projective; if it is a faithful module, N is an invertible module (see (2.3.1)), and (d) follows from (c). If M is not faithful, there is an idempotent $e \in K$ such that eM is a faithful module over Ke, whereas $(1-e)M = 0$ (see (1.12.8)); therefore (d) is still true. □

(2.3.8) **Proposition.** *Let P be a submodule of M. If the restriction of $\varphi : M \times M \to N$ to $P \times P$ is nondegenerate, then M is the direct sum of P and $P^\perp$.*

Proof. Let ψ be the restriction of φ to $P \times P$, let $j : P \to M$ be the natural injection, and $j' : \mathrm{Hom}(M,N) \to \mathrm{Hom}(P,N)$ the derived restriction morphism. Since d_ψ is bijective, we can consider

$$w = \mathrm{d}_\psi^{-1} \circ j' \circ \mathrm{d}_\varphi \quad : \quad M \to \mathrm{Hom}(M,N) \to \mathrm{Hom}(P,N) \to P.$$

It is easy to realize that $w \circ j = \mathrm{id}_P$; therefore $M = P \oplus \mathrm{Ker}(w)$. Since $\mathrm{Ker}(w) = \mathrm{Ker}(j' \circ \mathrm{d}_\varphi)$, we conclude that $\mathrm{Ker}(w) = P^\perp$. □

2.4 Operations on quadratic mappings and symmetric bilinear mappings

The K-quadratic mappings are the objects of a category $\mathcal{C}_K$; a morphism from $q : M \to N$ to $q' : M' \to N'$ is a couple (u,v) of linear mappings $u : M \to M'$ and $v : N \to N'$ such that $q' \circ u = v \circ q$. In this category an object $q : M \to N$ is usually denoted by (M,q,N), although the knowledge of q implies the knowledge of M and N. Two operations on the objects of $\mathcal{C}_K$ are especially interesting, and are expounded below. As a matter of fact, the first operation is defined inside each subcategory $\mathcal{C}_K(N)$ of quadratic mappings with a same target N; in this subcategory an object $q : M \to N$ is usually denoted by (M,q), and a morphism from (M,q) to (M',q') is a linear mapping $u : M \to M'$ such that (u, id_N) is a morphism in $\mathcal{C}_K$. Analogous definitions might be stated with symmetric bilinear mappings instead of quadratic mappings; for instance a morphism from (M,φ,N) to (M',φ',N') is a couple (u,v) of linear mappings such that $v(\varphi(x,y)) = \varphi'(u(x),u(y))$ for all x, $y \in M$.

Orthogonal sums of quadratic mappings

Let (M, q) and (M', q') be two objects of $\mathcal{C}_K(N)$; their *orthogonal sum* $(M, q) \perp (M', q')$ is the couple $(M \oplus M',\ q \perp q')$, where $q \perp q'$ is the quadratic mapping on $M \oplus M'$ defined in this way: $(q \perp q')(x, x') = q(x) + q'(x')$. It is clear that

$$\mathrm{b}_{q\perp q'}((x,x'),(y,y')) = \mathrm{b}_q(x,y) + \mathrm{b}_{q'}(x',y')$$

for all $x, y \in M$ and all $x', y' \in M'$. The natural injections $j : M \to M \oplus M'$ and $j' : M' \to M \oplus M'$ determine morphisms in the category $\mathcal{C}_K(N)$, and their images $M \oplus 0$ and $0 \oplus M'$ are orthogonal for $q \perp q'$. The next proposition states a universal property of $(M, q) \perp (M', q')$ which means that (j, j') is an initial universal object in some category.

(2.4.1) **Proposition.** *Let $g : (M, q) \to (M'', q'')$ and $g' : (M', q') \to (M'', q'')$ be two morphisms in $\mathcal{C}_K(N)$ that have orthogonal images in M''. There exists a unique morphism g'' from $(M, q) \perp (M', q')$ into (M'', q'') such that $g = g'' \circ j$ and $g' = g'' \circ j'$.*

Proof. It suffices to set $g''(x, x') = g(x) + g'(x')$ for all $x \in M$ and all $x' \in M'$; the orthogonality of $g(x)$ and $g'(x')$ implies $q''(g''(x, x')) = (q \perp q')(x, x')$; consequently g'' is a morphism in $\mathcal{C}_K(N)$. □

(2.4.2) **Proposition.** *Let (M, q), (M', q') and (M'', q'') be objects of $\mathcal{C}_K(N)$. The canonical isomorphisms*

$$\begin{aligned} M \oplus M' &\cong M' \oplus M\ , \\ (M \oplus M') \oplus M'' &\cong M \oplus (M' \oplus M'')\ , \\ \{0\} \oplus M &\cong M\ , \end{aligned}$$

give isomorphisms in $\mathcal{C}_K(N)$ when M, M', M'' and $\{0\}$ are replaced with (M, q), (M', q'), (M'', q'') and the null quadratic mapping $\{0\} \to N$.

Of course Proposition (2.4.2) is an obvious statement, but if we treat the operation $\perp$ as a functor from $\mathcal{C}_K(N) \times \mathcal{C}_K(N)$ into $\mathcal{C}_K(N)$, the three isomorphisms mentioned in (2.4.2) afford three isomorphisms between functors, and the existence of such isomorphisms is the exact meaning of this statement: $\mathcal{C}_K(N)$ is a *monoidal category with neutral object.*

(2.4.3) **Proposition.** *The quadratic mapping $q \perp q'$ is nondegenerate if and only if both q and q' are nondegenerate.*

Proof. Let w be the canonical isomorphism

$$\mathrm{Hom}(M, N) \oplus \mathrm{Hom}(M', N) \longrightarrow \mathrm{Hom}(M \oplus M',\ N)\ ;$$

it suffices to observe that $\mathrm{d}_{q\perp q'} = w \circ (\mathrm{d}_q \oplus \mathrm{d}_{q'})$; if d_q and $\mathrm{d}_{q'}$ are isomorphisms, the same holds for $\mathrm{d}_{q\perp q'}$; and conversely. □

(2.4.4) **Remark.** We can also define the *orthogonal sum* of two symmetric bilinear mappings $\varphi : M \times M \longrightarrow N$ and $\varphi' : M' \times M' \longrightarrow N$; it is the symmetric bilinear mapping $\varphi \perp \varphi'$ from $(M \oplus M') \times (M \oplus M')$ into N that maps every $((x, x'), (y, y'))$ to $\varphi(x, y) + \varphi'(x', y')$. When q and q' are quadratic mappings taking their values in N, it is clear that $\mathrm{b}_{q \perp q'} = \mathrm{b}_q \perp \mathrm{b}_{q'}$. There are statements analogous to the propositions (2.4.1), (2.4.2), (2.4.3), with symmetric bilinear mappings instead of quadratic mappings.

Tensor products of quadratic mappings and symmetric bilinear mappings

If $\varphi : M \times M \longrightarrow N$ and $\varphi' : M' \times M' \longrightarrow N'$ are two symmetric bilinear mappings, the construction of their tensor product $\varphi \otimes \varphi'$ raises no problem; it is the unique element of $\mathrm{Bil}(M \otimes M',\ N \otimes N')$ such that

$$(\varphi \otimes \varphi')(x \otimes x',\ y \otimes y') \ = \ \varphi(x, y) \otimes \varphi'(x', y')$$

for all x, $y \in M$ and all x', $y' \in M'$. The next step is the definition of the quadratic mapping $q \otimes \varphi'$, tensor product of a quadratic mapping q and a symmetric bilinear mapping φ'.

(2.4.5) **Proposition.** *Let $q : M \to N$ be a quadratic mapping and $\varphi' : M' \times M' \longrightarrow N'$ a symmetric bilinear mapping. There exists a unique quadratic mapping $q'' : M \otimes M' \to N \otimes N'$ such that $\mathrm{b}_{q''} = \mathrm{b}_q \otimes \varphi'$ and, for all $x \in M$ and all $x' \in M'$,*

$$q''(x \otimes x') \ = \ q(x) \otimes \varphi'(x', x').$$

Proof. Since $M \otimes M'$ is generated by all elements $x \otimes x'$, the unicity of q'' is a consequence of (2.1.5). Let us prove its existence first when M and M' are free modules with bases $(e_i)_{i \in I}$ and $(e'_j)_{j \in J}$ respectively. We define a family $(y_{(i,j),(k,l)})$ of elements of $N \otimes N'$, with (i, j) and (k, l) running through $I \times J$, in the following way:

$$\begin{aligned} y_{(i,j),(k,l)} &= \mathrm{b}_q(e_i, e_k) \otimes \varphi'(e'_j, e'_l) \qquad \text{if } (i,j) \neq (k,l)\ , \\ y_{(i,j),(i,j)} &= q(e_i) \otimes \varphi'(e'_j, e'_j)\ . \end{aligned}$$

By Lemma (2.1.3) there exists a unique quadratic mapping $q'' : M \otimes M' \to N \otimes N'$ such that

$$\begin{aligned} q''(e_i \otimes e'_j) &= y_{(i,j),(i,j)} \qquad \text{for all } (i,j) \in I \times J\ , \\ \mathrm{b}_{q''}(e_i \otimes e'_j,\ e_k \otimes e'_l) &= y_{(i,j),(k,l)} \qquad \text{whenever } (i,j) \neq (k,l)\ ; \end{aligned}$$

this is the wanted mapping q'' in this case.

When M and M' are arbitrary modules, we treat them as quotients of free modules L and L'; let $g : L \to M$ and $g' : L' \to M'$ be the corresponding surjective mappings. According to the first part of the proof, the quadratic mapping $q \circ g : L \to N$ and the symmetric bilinear mapping $\varphi' \circ (g' \times g') : L' \times L' \to N'$ determine a quadratic mapping $q''' : L \otimes L' \to N \otimes N'$ satisfying analogous conditions. According to (1.6.3) the kernel R of the surjective mapping $L \otimes L' \to M \otimes M'$ is the image of the linear mapping

$$(\mathrm{Ker}(g) \otimes L') \oplus (L \otimes \mathrm{Ker}(g')) \longrightarrow L \otimes L' \ ;$$

obviously $R \subset \mathrm{Ker}(q''')$, whence a quadratic mapping $q'' : M \otimes M' \to N \otimes N'$ such that $q'' \circ (g \otimes g') = q'''$; this is the wanted mapping q''. □

By definition the mapping q'' defined in (2.4.5) is the *tensor product* of the quadratic mapping q and the symmetric bilinear mapping φ', and we write $q'' = q \otimes \varphi'$. In the same way we can construct a tensor product like $\varphi \otimes q'$, and thus we come to the last step, the tensor product of two quadratic mappings q and q'. One may use the detailed notation $(M, q, N) \otimes_K (M', q', N')$ or the abbreviation $q \otimes q'$.

(2.4.6) **Proposition.** *When $q : M \to N$ and $q' : M' \to N'$ are two quadratic mappings, the quadratic mappings $q \otimes \mathrm{b}_{q'}$ and $\mathrm{b}_q \otimes q'$ are equal; either one is the tensor product $q \otimes q'$. Thus $q \otimes q'$ is characterized by these two equalities: first $\mathrm{b}_{q\otimes q'} = \mathrm{b}_q \otimes \mathrm{b}_{q'}$ and secondly, for all $x \in M$ and all $x' \in M'$,*

$$(q \otimes q')(x \otimes x') \ = \ 2\, q(x) \otimes q'(x') \ .$$

The proof is just a straightforward verification.

Later in **2.7** we need especially the following statements.

(2.4.7) **Proposition.** *Let q, q' and q'' be K-quadratic mappings; in the category $\mathcal{C}_K$ there is an isomorphism between $q \otimes q'$ and $q' \otimes q$, and an isomorphism between $(q \otimes q') \otimes q''$ and $q \otimes (q' \otimes q'')$. If 2 is invertible in K, there is a neutral object q_0 (such that $q_0 \otimes q$ is always isomorphic to q), which is the quadratic mapping $K \to K$ defined by $\lambda \longmapsto \lambda^2/2$.*

The first statement in (2.4.7) is an immediate consequence of the well-known properties of commutativity and associativity of tensor products; they make $\mathcal{C}_K$ become a monoidal category for this operation. When 2 is invertible in K, the bilinear mapping associated with $q_0 : \lambda \longmapsto \lambda^2/2$ is the multiplication mapping $(\lambda, \mu) \longmapsto \lambda\mu$, and thus the canonical isomorphisms $K \otimes M \to M$ and $K \otimes N \to N$ yield a canonical isomorphism between $q_0 \otimes q$ and q. □

Of course the category of all symmetric K-bilinear mappings is also a monoidal category for the tensor product, and the multiplication mapping $K \times K \to K$ is always a neutral object.

(2.4.8) **Proposition.** *Let $q : M \to N$ and $q' : M' \to N'$ be two quadratic mappings, and let us assume either that M and M' are finitely generated projective modules, or that M and N are finitely generated projective modules, or that M' and N' are finitely generated projective modules. If q and q' are nondegenerate, their tensor product $q \otimes q'$ is also nondegenerate. The same conclusion holds when q is replaced* *with a nondegenerate symmetric bilinear mapping* $\varphi : M \times M \to N$, *and when* q' *too is replaced with a nondegenerate symmetric bilinear mapping.*

Proof. Let w be the canonical linear mapping

$$\mathrm{Hom}(M,N) \otimes \mathrm{Hom}(M',N') \longrightarrow \mathrm{Hom}(M \otimes M',\ N \otimes N')\ ;$$

any one of the three assumptions proposed at the beginning of (2.4.8) ensures that w is bijective; for the first assumption, this can be proved by localization with the help of (1.10.8); for all three assumptions, this can be proved by embedding the finitely generated projective modules under consideration as direct summands in free modules of finite ranks. It is clear that $\mathrm{d}_{q\otimes q'} = w \circ (\mathrm{d}_q \otimes \mathrm{d}_{q'})$; consequently if d_q and $\mathrm{d}_{q'}$ are bijective, the same holds for $\mathrm{d}_{q\otimes q'}$. □

(2.4.9) **Proposition.** *Let* (M,q) *be an element of* $\mathcal{C}_K(N)$, *and* (M',q') *and* (M'',q'') *two elements of* $\mathcal{C}_K(N')$. *In* $\mathcal{C}_K(N\otimes N')$ *there are canonical isomorphisms*

$$(M,q) \otimes \big((M',q') \perp (M'',q'')\big) \longleftrightarrow (M,q) \otimes (M',q')\ \perp\ (M,q) \otimes (M'',q'').$$

These reciprocal isomorphisms are defined in this way: every $x \otimes (x',x'')$ is mapped to $(x\otimes x',\ x\otimes x'')$, and conversely $(x_1\otimes x',\ x_2\otimes x'')$ to $x_1\otimes(x',0)+x_2\otimes(0,x'')$. □

2.5 Hyperbolic and metabolic spaces

From now on, we are especially interested in quadratic forms, which are objects of the category $\mathcal{C}_K(K)$ according to the notation of **2.4**. This subcategory $\mathcal{C}_K(K)$ is provided with two operations, orthogonal sum and tensor product, because of the canonical isomorphism $K \otimes K \to K$.

(2.5.1) **Definitions**. A *quadratic module* (M,q) is a module M provided with a quadratic form $q : M \to K$; it is called a *quadratic space* if M is finitely generated and projective, and q nondegenerate. A *bilinear module* (M,φ) is a module M provided with a symmetric bilinear form $\varphi : M \times M \to K$; it is called a *bilinear space* if M is finitely generated and projective, and φ nondegenerate.

A morphism from (M,q) to (M',q') is a linear mapping $u : M \to M'$ such that $q'(u(x)) = q(x)$ for all $x \in M$. A morphism u from (M,φ) to (M',φ') must satisfy the condition $\varphi'(u(x),u(y)) = \varphi(x,y)$ for all x and y. Isomorphisms between quadratic or bilinear modules are often called *isometries*.

With every K-module P we associate the dual module $P^* = \mathrm{Hom}_K(P,K)$, and the quadratic form $q_P : P^* \oplus P \to K$ defined by $(f,x) \longmapsto f(x)$; thus we obtain a quadratic module $(P^* \oplus P,\ q_P)$ which is called the *hyperbolic module* associated with P and denoted by $\mathbf{H}[P]$. The bilinear form b_P associated with q_P is this one: $((f,x),\ (g,y)) \longmapsto f(y) + g(x)$. The bilinear module $(P^* \oplus P,\ \mathrm{b}_P)$ is called the *hyperbolic bilinear module* associated with P and denoted by $\mathbf{H}(P)$.

More generally, from a bilinear module (P, ψ) over K we derive the symmetric bilinear form

$$\psi_P : (P^* \oplus P) \times (P^* \oplus P) \longrightarrow K\ , \quad ((f, x), (g, y)) \longmapsto f(y) + g(x) + \psi(x, y)\ ;$$

the bilinear module $(P^* \oplus P,\ \psi_P)$ is called the *metabolic module* associated with (P, ψ) and denoted by $\mathbf{M}(P, \psi)$. When $\psi = 0$, it is clear that we get again the hyperbolic bilinear module $\mathbf{H}(P)$.

There is a canonical mapping $c_P : P \longrightarrow P^{**}$ defined by $x \longmapsto (f \longmapsto f(x))$; the module P is said to be *reflexive* when c_P is bijective. If P is reflexive, $\mathbf{H}[P^*]$ is obviously isomorphic to $\mathbf{H}[P]$. Finitely generated projective modules are reflexive (see **1.7**), and their reflexiveness is the key to the next proposition.

(2.5.2) **Proposition.** *When P is a finitely generated projective module, $\mathbf{H}(P)$ is a quadratic space, and $\mathbf{M}(P, \psi)$ is a bilinear space whatever the symmetric bilinear form ψ may be.*
Therefore we call them respectively the *hyperbolic space* associated with P and the *metabolic space* associated with (P, ψ).

Proof. Obviously $P^* \oplus P$ is a finitely generated projective module. Since b_P is equal to ψ_P when $\psi = 0$, it suffices to prove that ψ_P is always nondegenerate. The linear mapping $P^* \oplus P \to (P^* \oplus P)^*$ derived from ψ_P is this one:

$$P^* \oplus P \longrightarrow P^{**} \oplus P^*\ , \quad (f, x) \longmapsto (c_P(x),\ \mathrm{d}_\psi(x) + f)\ ;$$

it is bijective whenever P is reflexive, in particular when P is finitely generated and projective. □

The next proposition (2.5.5) characterizes hyperbolic spaces and metabolic spaces by means of very special submodules. An element x of a quadratic module (M, q) (resp. a bilinear module (M, φ)) is said to be *isotropic* if $q(x) = 0$ (resp. $\varphi(x, x) = 0$), and a submodule N of M is said to be *totally isotropic* (or *totally singular*) if all its elements are isotropic. The inclusion $N \subset N^\perp$ holds for every totally isotropic submodule N. Conversely in a bilinear module (M, φ) (and also in a quadratic module (M, q) when 2 is invertible in K) the inclusion $N \subset N^\perp$ means that N is totally isotropic.

For quadratic spaces we get a stronger result than for bilinear spaces because of the following lemma (which still plays a capital role later in **4.8**) and its corollary.

(2.5.3) **Lemma.** *Let q be a K-quadratic form on a projective module M; there exists a K-bilinear form $\beta : M \times M \to K$ such that $q(x) = \beta(x, x)$ for all x in M.*

Proof. Let us first suppose that M is a free module with basis $(e_j)_{j \in J}$; we can suppose that the set J of indices is totally ordered; we get a suitable bilinear form β on M by assigning the following values to $\beta(e_i, e_j)$ when i and j run through J :

$$\begin{aligned} \beta(e_i, e_j) &= \mathrm{b}_q(e_i, e_j) && \text{if } i < j \\ &= q(e_i) && \text{if } i = j \\ &= 0 && \text{if } i > j\ . \end{aligned}$$

When M is merely a projective module, there exists a module M' such that $M \oplus M'$ is free; then we consider the orthogonal sum $(M, q) \perp (M', 0)$; there exists a bilinear form β'' on $M \oplus M'$ such that $\beta''((x, x'), (x, x')) = q(x)$ for all $x \in M$ and all $x' \in M'$; by restriction to $M \oplus 0$ we get a suitable bilinear form β on M. □

(2.5.4) **Corollary.** *Let (M, q) be a quadratic space containing a totally isotropic direct summand N such that $N = N^\perp$. There exists a totally isotropic submodule P such that $M = N \oplus P$.*

Proof. Let P be any submodule supplementary to N, not necessarily totally isotropic; any other submodule supplementary to N is determined by a linear mapping $u : P \to N$, since it is the subset of all elements $y + u(y)$ with $y \in P$. We must find a linear mapping $u : P \longrightarrow N$ such that $q(y + u(y)) = 0$ for all $y \in P$. Let us apply the previous lemma to the restriction of q to P : there exists a bilinear form β on P such that $q(y) = \beta(y, y)$ for all $y \in P$; we also consider $\mathrm{d}_\beta : P \to P^*$ defined by $\mathrm{d}_\beta(y)(z) = \beta(y, z)$. The condition required from u is equivalent to this one:

$$\forall y \in P, \qquad \beta(y, y) + \mathrm{b}_q(u(y), y) \;=\; 0\;;$$

let us look for a mapping u satisfying this stronger condition:

$$\forall y \in P, \;\; \forall z \in P, \qquad \beta(y, z) + \mathrm{b}_q(u(y), z) \;=\; 0\;;$$

this stronger condition determines a unique mapping $u : P \to N$; indeed from (2.3.7) we deduce an isomorphism $\mathrm{d}'_q : N \to (M/N)^* \to P^*$ which maps every $x \in N$ to the restriction of $\mathrm{d}_q(x)$ to P; the stronger condition required from u means that $\mathrm{d}_\beta + \mathrm{d}'_q \circ u = 0$, whence $u = -\mathrm{d}'^{-1}_q \circ \mathrm{d}_\beta$. □

(2.5.5) **Proposition.**

(a) *A quadratic space (M, q) is hyperbolic if and only if there exists a totally isotropic direct summand N such that $N = N^\perp$. If such a submodule N exists, (M, q) is isomorphic to $\mathbf{H}[N]$, and also to $\mathbf{H}[M/N]$.*

(b) *A bilinear space (M, φ) is metabolic if and only if there exists a totally isotropic direct summand N such that $N = N^\perp$. If such a submodule N exists, if P is any submodule supplementary to N, and ψ the restriction of φ to $P \times P$, then (M, φ) is isomorphic to $\mathbf{M}(P, \psi)$.*

Proof. In a hyperbolic module $\mathbf{H}[P]$, both direct summands $0 \oplus P$ and $P^* \oplus 0$ are totally isotropic; the former is equal to its orthogonal submodule, and the latter too if the canonical mapping $P \to P^{**}$ is injective. It is actually injective when $\mathbf{H}[P]$ is a quadratic space, since in this case P is finitely generated and projective, therefore reflexive. Similarly in a metabolic space $\mathbf{M}(P, \psi)$ the direct summand $P^* \oplus 0$ is totally isotropic, and equal to its orthogonal submodule; but $0 \oplus P$ is totally isotropic if and only if $\psi = 0$.

Conversely we suppose that (M,q) contains a totally isotropic direct summand N such that $N = N^\perp$. From (2.5.4) we deduce the existence of another totally isotropic submodule P such that $M = N \oplus P$. From (2.3.7) we deduce an isomorphism $\mathrm{d}'_q : N \to (M/N)^* \to P^*$ which maps every $x \in N$ to the restriction of $\mathrm{d}_q(x)$ to P; this proves that $N \cap P^\perp = 0$, whence $P = P^\perp$. Consequently N and P play similar roles, and there is another isomorphism $\mathrm{d}''_q : P \to (M/P)^* \longrightarrow N^*$. The mapping $(M,q) \to \mathbf{H}[N]$ defined by $x + y \longmapsto (\mathrm{d}''_q(y),\, x)$ (for all $x \in N$ and $y \in P$) is an isomorphism of quadratic spaces. Of course (M,q) is also isomorphic to $\mathbf{H}[P]$ by the mapping $x + y \longmapsto (\mathrm{d}'_q(x),\, y)$.

For a bilinear space (M,φ) containing a direct summand N such that $N = N^\perp$, the proof is still simpler, since any supplementary module P will do. □

(2.5.6) **Examples.** First K is the field $\mathbb{Z}/2\mathbb{Z}$ and we consider the nondegenerate symmetric bilinear form φ on $(\mathbb{Z}/2\mathbb{Z})^2$ defined by $\varphi((x_1,x_2),(x'_1,x'_2)) = x_1x'_1 + x_2x'_2$; this bilinear space is called G_2 later in **2.8**. The equality $N = N^\perp$ holds for the line N generated by $(1,1)$; yet there is no other totally isotropic line in G_2; this bilinear space is metabolic but not hyperbolic. Now K is the ring $\mathbb{Z}$ and we provide $\mathbb{Z}^2$ with the $\mathbb{Z}$-bilinear form ψ defined by $\psi((z_1,z_2),(z'_1,z'_2)) = z_1z'_1 - z_2z'_2$; thus we get the $\mathbb{Z}$-bilinear space later denoted by $G_{1,1}$. The diagonal submodule Δ of all elements (z,z) is totally isotropic, it is supplementary to $\mathbb{Z}\times\{0\}$, and obviously $\Delta^\perp = \Delta$. Consequently $G_{1,1}$ is metabolic. But it is not hyperbolic; indeed if it were hyperbolic, the ring extension $\mathbb{Z} \to \mathbb{Z}/2\mathbb{Z}$ would yield a hyperbolic bilinear space over $\mathbb{Z}/2\mathbb{Z}$, whereas it actually yields the above metabolic space G_2 which is not hyperbolic.

Nevertheless when 2 is invertible in K, every K-bilinear module can also be studied as a quadratic module, and from (2.5.5) we deduce that every metabolic space is hyperbolic.

Now we come to the properties that are most useful later in **2.7**. The next proposition is evident, and its proof is omitted.

(2.5.7) **Proposition.** *Let M and N be finitely generated projective K-modules.*

(a) *The natural bijection from $\mathbf{H}[M \oplus N]$ onto $\mathbf{H}[M] \perp \mathbf{H}[N]$ is an isomorphism of quadratic spaces.*

(b) *For all symmetric bilinear forms φ and ψ respectively on M and N, the natural bijection from $\mathbf{M}((M,\varphi) \perp (N,\psi))$ onto $\mathbf{M}(M,\varphi) \perp \mathbf{M}(N,\psi)$ is an isomorphism of bilinear spaces.*

(2.5.8) **Proposition.**

(a) *For every quadratic space (M,q) there is an isomorphism*

$$(M,q) \perp (M,-q) \;\cong\; \mathbf{H}[M]\ .$$

(b) *For every bilinear space (M,φ) there is an isomorphism*

$$(M,\varphi) \perp (M,-\varphi) \;\cong\; \mathbf{M}(M,\varphi)\ .$$

(c) *For every metabolic space* $\mathbf{M}(P,\psi)$*, there is an isomorphism*

$$\mathbf{H}(P) \perp \mathbf{M}(P,-\psi) \;\cong\; \mathbf{M}(P,\psi) \perp \mathbf{M}(P,-\psi)\ .$$

Proof. In the quadratic space $(M,q) \perp (M,-q)$ the submodule Δ of all elements (x,x) with $x \in M$ is isomorphic to M, it is totally isotropic, and it is a direct summand, because $M \oplus 0$ is a supplementary submodule. Consequently $\Delta^\perp$ is generated by Δ and all $(y,0) \in M \oplus 0$ that are orthogonal to Δ; yet this condition means that y must belong to $\mathrm{Ker}(\mathrm{b}_q)$, whence $y=0$ and $\Delta^\perp = \Delta$. Because of (2.5.5) we have a hyperbolic space.

For a bilinear space (M,φ) the proof is similar. When 2 is invertible in K, instead of the supplementary submodule $M \oplus 0$ we might use the other diagonal Δ' containing all elements $(x,-x)$, which is also totally isotropic; this would prove again that $(M,\varphi) \perp (M,-\varphi)$ is hyperbolic. But in the general case we only get a metabolic space (see (2.5.6) above).

The bilinear spaces $\mathbf{H}(P) \perp \mathbf{M}(P,-\psi)$ and $\mathbf{M}(P,\psi) \perp \mathbf{M}(P,-\psi)$ have the same underlying module $P^* \oplus P \oplus P^* \oplus P$; the following mapping is an isomorphism from the former onto the latter:

$$(f,x,g,y) \longmapsto (\,f-g+\mathrm{d}_\psi(y)\,,\ x,\ g,\ x+y\,)\ . \qquad \square$$

(2.5.9) **Proposition.** *Let P be a finitely generated projective module, sometimes provided with a symmetric bilinear form ψ, and M a finitely generated projective module provided either with a nondegenerate symmetric bilinear form φ or with a nondegenerate quadratic form q. There are isomorphisms of the following four types:*

$$\begin{aligned}
\mathbf{M}(P,\psi)\otimes(M,\varphi) &\cong \mathbf{M}(P\otimes M,\ \psi\otimes\varphi)\ ,\\
\mathbf{M}(P,\psi)\otimes(M,q) &\cong \mathbf{H}[P\otimes M]\ ,\\
\mathbf{H}[P]\otimes(M,\varphi) &\cong \mathbf{H}[P\otimes M]\ ,\\
\mathbf{H}[P]\otimes(M,q) &\cong \mathbf{H}[P\otimes M]\ .
\end{aligned}$$

Proof. The bilinear space $\mathbf{M}(P,\psi)\otimes(M,\varphi)$ is the module $(P^*\otimes M)\oplus(P\otimes M)$ provided with the bilinear form χ such that

$$\begin{aligned}
&\chi\,\big((f\otimes x,\ z\otimes y),\ (f'\otimes x',\ z'\otimes y')\big)\\
&\qquad = f(z')\,\varphi(x,y')\ + f'(z)\,\varphi(y,x')\ + \psi(z,z')\,\varphi(y,y')\ ;
\end{aligned}$$

it is clear that $(P^*\otimes M)\oplus 0$ is a totally isotropic submodule, and that $0\oplus(P\otimes M)$ is a supplementary submodule; if the former submodule were not equal to its orthogonal submodule, there should exist a nonzero element in $0\oplus(P\otimes M)$ that would be orthogonal to it; this would mean the existence of a nonzero element of $P\otimes M$ that should be annihilated by all linear forms $f'\otimes \mathrm{d}_\varphi(x')$; but this is impossible, because all the modules under consideration are finitely generated and projective, and the linear mapping

$$P^*\otimes M \longrightarrow (P\otimes M)^*\ ,\quad f'\otimes x' \longmapsto f'\otimes \mathrm{d}_\varphi(x')$$

is an isomorphism; we conclude that $(P^* \otimes M) \oplus 0$ is equal to its orthogonal submodule; thus from (2.5.5) we deduce the existence of an isomorphism of the first type.

For the second and third types a similar argument shows that $(P^* \otimes M) \oplus 0$ is still a totally isotropic submodule which is equal to its orthogonal submodule; here it leads to an isomorphism onto $\mathbf{H}[P^* \otimes M]$. Nevertheless we must remember that $\mathbf{H}[P^* \otimes M]$ is isomorphic to $\mathbf{H}[(P^* \otimes M)^*]$, and that $(P^* \otimes M)^*$ is isomorphic to $P \otimes M^*$, therefore to $P \otimes M$ because of the isomorphism d_q or d_φ from M onto M^*. To treat the fourth type, remember that $\mathbf{H}[P] \otimes (M, q)$ is the same thing as $\mathbf{H}[P] \otimes (M, \mathrm{b}_q)$. □

2.6 Orthogonal decompositions of quadratic spaces

A submodule P of a quadratic or bilinear module is called an *orthogonal summand* if $P^\perp$ contains a supplementary submodule. When M is said to be the orthogonal sum of subspaces P and Q, it must be understood that this sum is direct and that P and Q are orthogonal; in this case the quadratic or bilinear form on M is nondegenerate if and only if its restrictions to P and Q are both nondegenerate (see (2.4.3)); thus the next proposition is an immediate consequence of (2.3.8).

(2.6.1) **Proposition.** *Let (M, q) be a quadratic space (see Definition (2.5.1)), M' a submodule of M, and q' the restriction of q to M'. These two assertions are equivalent:*

(a) *(M', q') is a quadratic space;*

(b) *(M', q') is an orthogonal summand of (M, q).*

Besides, the analogous statement for bilinear spaces is also true.

Quadratic spaces over local rings

(2.6.2) **Theorem.** *Let K be a local ring with maximal ideal $\mathfrak{m}$, and (M, q) a quadratic space over K.*

(a) *If 2 is invertible in K, then (M, q) has an orthogonal basis (a basis in which the elements are pairwise orthogonal).*

(b) *If 2 belongs to $\mathfrak{m}$, the rank of M is even and (M, q) is an orthogonal sum of submodules of rank 2. In every one of these submodules there is an element x such that $q(x)$ is invertible.*

Proof. Let r be the rank of M; we can suppose $r > 0$. Let us first assume that 2 is invertible in K. If $q(x)$ belonged to $\mathfrak{m}$ for all $x \in M$, then $\mathrm{b}_q(x, y)$ should belong to $\mathfrak{m}$ for all x and y, whence the inclusion $\mathrm{Im}(\mathrm{d}_q) \subset \mathfrak{m}M^*$ contrary to the nondegeneracy of q. Consequently there is some e_1 in M such that $q(e_1)$ is not in $\mathfrak{m}$. Let M' be the submodule generated by e_1; the restriction of q to M' is nondegenerate because $\mathrm{b}_q(e_1, e_1) = 2q(e_1)$ is invertible; thus (2.6.1) implies that

M is the orthogonal sum of M' and a submodule M''. We complete the proof by induction on r; when $M'' \neq 0$, by the induction hypothesis there is an orthogonal basis $(e_2, e_3, \dots, e_r)$ in M'', whence the orthogonal basis $(e_1, e_2, \dots, e_r)$ in M.

Now let us assume that 2 belongs to $\mathfrak{m}$. Let $(e_1, e_2, \dots, e_r)$ be any basis of M; it is impossible that all $\mathrm{b}_q(e_i, e_j)$ belong to $\mathfrak{m}$ (for the same reason as above); consequently there exists (i, j) such that $\mathrm{b}_q(e_i, e_j)$ is not in $\mathfrak{m}$; and moreover $i \neq j$ because $\mathrm{b}_q(e_i, e_i) = 2q(e_i)$ belongs to $\mathfrak{m}$; consequently the rank of M is not 1, and we can suppose that $i = 1$ and $j = 2$. Let M' be the submodule generated by e_1 and e_2, and q' the restriction of q to M'; the matrix of $\mathrm{d}_{q'}$ relatively to (e_1, e_2) and the dual basis is made of the four entries $\mathrm{b}_q(e_i, e_j)$ with i and j in $\{1, 2\}$; consequently its determinant $4q(e_1)q(e_2) - \mathrm{b}_q(e_1, e_2)^2$ is invertible; this proves that q' is nondegenerate. Thus (2.6.1) implies that M is the orthogonal sum of M' and a submodule M'' of rank $r - 2$. As above, an induction on r proves that M is an orthogonal sum of submodules of rank 2. Let us prove that, for instance, the submodule generated by e_1 and e_2 contains an element x such that $q(x)$ does not belong to $\mathfrak{m}$; if $q(e_1)$ and $q(e_2)$ both belong to $\mathfrak{m}$, then $q(e_1 + e_2)$ does not belong to it, because it is equal to $q(e_1) + q(e_2) + \mathrm{b}_q(e_1, e_2)$. □

When K is any ring in which 2 is invertible, with every invertible $a \in K^\times$ we associate the quadratic space $\langle a \rangle$ defined in this way: it is the module K provided with the quadratic form $x \longmapsto ax^2/2$; thus the associated bilinear form is $(x, y) \longmapsto axy$. If $a_1, a_2, \dots, a_n$ are invertible elements of K, the notation $\langle a_1, a_2, \dots, a_n \rangle$ means $\langle a_1 \rangle \perp \langle a_2 \rangle \perp \cdots \perp \langle a_n \rangle$; it is the quadratic space K^n with quadratic form $(x_1, x_2, \dots, x_n) \longmapsto \sum_{i=1}^n a_i x_i^2/2$; the associated matrix is the diagonal matrix in which the entries on the diagonal are the coefficients $a_1, a_2, \dots, a_n$. Theorem (2.6.2)(a) says that every quadratic space over a local ring K in which 2 is invertible, is isomorphic to some $\langle a_1, a_2, \dots, a_n \rangle$, with $a_1, a_2, \dots, a_n$ all in $K^\times$.

But when 2 is not invertible in the local ring K, according to Theorem (2.6.2)(b) we need quadratic spaces K^2 with quadratic forms $(x, y) \longmapsto ax^2 + bxy + cy^2$ such that b is invertible (because $4ac - b^2$ must be invertible); we can require that a is invertible; but when the residue field $K/\mathfrak{m}$ has cardinal 2, we cannot always require that a and c are both invertible. Yet we can require $b = 1$; indeed

$$ax^2 + bxy + cy^2 = ax^2 + xy' + c'y'^2 \quad \text{if} \quad y' = by \text{ and } c' = cb^{-2}.$$

For bilinear spaces there is a theorem parallel to (2.6.2).

(2.6.3) **Theorem.** *Let (M, φ) be a bilinear space over a local ring K with maximal ideal $\mathfrak{m}$.*

(a) *If $\varphi(x, x)$ belongs to $\mathfrak{m}$ for all $x \in M$, then* 2 *is* not *invertible in K, the rank of M is even, and (M, φ) is an orthogonal sum of bilinear subspaces of rank* 2.

(b) *If $\varphi(x, x)$ is invertible for some $x \in M$, then (M, φ) admits an orthogonal basis.*

Proof. If $\varphi(x,x)$ belongs to $\mathfrak{m}$ for all $x \in M$, the equality

$$2\,\varphi(x,y) \;=\; \varphi(x+y,\,x+y) - \varphi(x,x) - \varphi(y,y)$$

shows that $\varphi(x,y)$ would belong to $\mathfrak{m}$ for all $x,\,y \in M$ if 2 were invertible in K, in contradiction with the nondegeneracy of φ. Therefore 2 cannot be invertible. In this case the argument presented in the second part of the proof of (2.6.2) shows that (M,φ) is an orthogonal sum of subspaces of rank 2.

If M contains an element e_1 such that $\varphi(e_1,e_1)$ is invertible, we proceed as in the first part of the proof of (2.6.2), because the hyperplane orthogonal to e_1 allows us to make an induction on the rank r of M, provided that it contains an element e_2 such that $\varphi(e_2,e_2)$ is invertible. Unfortunately if 2 is not invertible in K (and $r>1$), it may happen that $\varphi(y,y)$ is never invertible when y runs through this hyperplane. Nevertheless this hyperplane certainly contains two elements e_2 and e_3 such that $\varphi(e_2,e_3)$ is invertible, whereas $\varphi(e_2,e_2)$ and $\varphi(e_3,e_3)$ belong to $\mathfrak{m}$. Let us set $e'_1 = e_1+e_2$ and $e'_2 = \varphi(e_2,e_3)e_1 - \varphi(e_1,e_1)e_3$, so that

$$\varphi(e'_1,e'_2)=0\,, \qquad \varphi(e'_1,e'_1) = \varphi(e_1,e_1)+\varphi(e_2,e_2) \in K^\times\,,$$
$$\varphi(e'_2,e'_2) \;=\; \varphi(e_2,e_3)^2\varphi(e_1,e_1) + \varphi(e_1,e_1)^2\varphi(e_3,e_3) \in K^\times.$$

Since $\varphi(e'_1,e'_1)$ is invertible, and since the hyperplane orthogonal to e'_1 contains an element e'_2 such that $\varphi(e'_2,e'_2)$ is invertible, the induction on r is still possible. □

Zariski extensions

Let $\{s_1,s_2,\dots,s_n\}$ be a set of elements of K that generates K as an ideal; remember that K_{s_i} (for $i=1,2,\dots,n$) is the ring of fractions of K with denominator a power of s_i, and that the direct product $L=\prod_{i=1}^n K_{s_i}$ is a faithfully flat extension of K (see (1.10.6)), which is called a *Zariski extension* of K.

If $K \longrightarrow K'$ is a faithfully flat extension of K, then (M,q) is a K-quadratic space if and only if $K'\otimes_K (M,q)$ is a K'-quadratic space. Indeed M is a finitely generated projective K-module if and only if $K'\otimes M$ is a finitely generated projective K'-module (see (1.9.10)); and (2.3.2) states that q is nondegenerate if and only if its extension q' to $K'\otimes M$ is nondegenerate.

We shall also need the following lemma.

(2.6.4) **Lemma.** *Let (M,q) be a quadratic module, and $e_1,e_2,\dots,e_r$ some elements of M, and let δ be the determinant of the matrix $(\mathrm{b}_q(e_j,e_k))$ (with j and k in $\{1,2,\dots,r\}$). If δ is not a divisor of zero, then $e_1,e_2,\dots,e_r$ are linearly independent over K. If δ is invertible in K, the free module generated by $e_1,\dots,e_r$ is an orthogonal summand of M.*

Proof. Let $\lambda_1,\lambda_2,\dots,\lambda_r$ be elements of K such that $\sum_{j=1}^r \lambda_j e_j = 0$. If in the matrix $(\mathrm{b}_q(e_j,e_k))$ we multiply the first line by λ_1, this line becomes a linear combination of the others; consequently its determinant $\lambda_1\delta$ vanishes. When δ is not a divisor of zero, we realize that $\lambda_1 = 0$. Similarly $\lambda_j = 0$ for $j = 2, \dots, r$. When δ is invertible, the restriction of q to the free submodule generated by $e_1, \dots, e_r$ is nondegenerate, and (2.6.1) implies that it is an orthogonal summand of M. $\square$

The next theorem is a global version of (2.6.2) for a ring K that has no longer to be a local ring.

(2.6.5) **Theorem.** *Let (M, q) be a quadratic space of* constant rank.

(a) *If 2 is invertible in K, there exists a Zariski extension L of K such that the L-quadratic space $L \otimes (M, q)$ has an orthogonal basis.*

(b) *If 2 is not invertible in K, the rank of M is even, and there exists a Zariski extension L of K such that the L-quadratic space $L \otimes (M, q)$ is an orthogonal sum of* free *submodules of rank 2; moreover, if q'' means the extension of q to $L \otimes M$, in each of these submodules there exists a basis (x, y) such that $q''(x)$ is invertible in L.*

Proof. Since M has a constant rank r over K, according to (1.12.12) every extension $K' \otimes M$ has the same constant rank r over K'. If we find in $K' \otimes M$ a family of vectors $(e'_1, e'_2, \dots, e'_r)$ such that the determinant of the matrix $\mathrm{b}_{q'}(e'_j, e'_k)$ is invertible, this family generates a free submodule of rank r that it is an orthogonal summand (see (2.6.4)), therefore equal to $K' \otimes M$; in other words, this family is a basis of $K' \otimes M$.

We first suppose that 2 is invertible in K. For every maximal ideal $\mathfrak{m}$ of K, $M_{\mathfrak{m}}$ has an orthogonal basis made of fractions e_j/t_j (where $j = 1, 2, \dots, r$); this implies that each $q(e_j)$ is outside $\mathfrak{m}$, and that the equality $t_{j,k}\mathrm{b}_q(e_j, e_k) = 0$ holds for some $t_{j,k}$ outside $\mathfrak{m}$ whenever $j < k$. Let us set

$$s_{\mathfrak{m}} = \prod_{j=1}^{r} q(e_j) \prod_{j<k} t_{j,k} \ ;$$

we consider the ring $K_{s_{\mathfrak{m}}}$ and the module of fractions $M_{s_{\mathfrak{m}}}$; according to the above explanations, the r fractions $e_j/1$ make up an orthogonal basis of this module. Now K is generated as an ideal by the family of all factors $s_{\mathfrak{m}}$; consequently there exists a finite subfamily $(s_1, s_2, \dots, s_n)$ that generates K as an ideal; for $i = 1, 2, \dots, n$ there exists a family $(e_{i,1}, e_{i,2}, \dots, e_{i,r})$ of elements of M such that the fractions $e_{i,j}/1$ constitute an orthogonal basis of M_{s_i}. Let us set

$$L = \prod_{i=1}^{n} K_{s_i} \ , \quad \text{whence} \quad L \otimes M = \prod_{i=1}^{n} M_{s_i} \ ;$$

for $j = 1, 2, \dots, r$, let e'_j be the element of $L \otimes M$ equal to $(e_{1,j}/1,\ e_{2,j}/1, \dots, e_{n,j}/1)$; the r elements e'_j constitute an orthogonal basis of $L \otimes M$.

When 2 is not invertible in K, there is at least one maximal ideal that does not contain 2, and consequently the constant rank r of M must be even; let us set $r = 2r'$. If $\mathfrak{m}$ is any maximal ideal of K, from Theorem (2.6.2) we deduce the existence of a family $(e_1, e_2, \dots, e_r)$ of elements of M satisfying all the following properties; first for $j' = 1, 2, \dots, r'$, the elements

$$q(e_{2j'-1}) \quad \text{and} \quad 4\, q(e_{2j'-1})q(e_{2j'}) \;-\; \mathrm{b}_q(e_{2j'-1}, e_{2j'})^2$$

are outside $\mathfrak{m}$; when 2 is not in $\mathfrak{m}$, we can even require that $\mathrm{b}_q(e_{2j'-1}, e_{2j'})$ is annihilated by some element of K outside $\mathfrak{m}$, and consequently $q(e_{2j'})$ is also outside $\mathfrak{m}$; but when 2 belongs to $\mathfrak{m}$, the previous assertion means that $\mathrm{b}_q(e_{2j'-1}, e_{2j'})$ is outside $\mathfrak{m}$; secondly, whenever $j < k$ and (j, k) is not a pair $(2j' - 1,\ 2j')$ with $j' = 1, 2, \dots, r'$, there exists $t_{j,k}$ outside $\mathfrak{m}$ such that $t_{j,k}\mathrm{b}_q(e_j, e_k) = 0$. For convenience we set $t_{j,k} = 1$ when $(j, k) = (2j' - 1,\ 2j')$. Let us also set

$$s_{\mathfrak{m}} \;=\; \prod_{j'=1}^{r'} \left(q(e_{2j'-1})\ (4\, q(e_{2j'-1})q(e_{2j'}) - \mathrm{b}_q(e_{2j'-1}, e_{2j'})^2\right) \prod_{j<k} t_{j,k}\ ;$$

the family of all elements $s_{\mathfrak{m}}$ generates K as an ideal, and contains a finite subfamily $(s_1, s_2, \dots, s_n)$ that already generates K; the proof ends in the same way as in the previous case. □

Free quadratic extensions

A free quadratic extension of K is an algebra isomorphic to the quotient of $K[Z]$ by the ideal generated by some polynomial $Z^2 - \beta Z + \gamma$ the discriminant of which (that is $\beta^2 - 4\gamma$) is invertible in K. It admits a basis $(1, z)$ in which z (the image of Z) satisfies the equality $z^2 = \beta z - \gamma$. It is obviously a faithfully flat extension. Quadratic extensions (whether free or not) shall be presented in **3.4** with more generality.

If the polynomial $Z^2 - \beta Z + \gamma$ admits a root κ in K, then $\kappa' = \beta - \kappa$ is also a root, and the difference $\kappa - \kappa'$ is invertible because it is a square root of $\beta^2 - 4\gamma$. By setting $\varepsilon = (z - \kappa)(\kappa' - \kappa)^{-1}$ and $\varepsilon' = (z - \kappa')(\kappa - \kappa')^{-1}$ we get another basis $(\varepsilon, \varepsilon')$ of $K[z]$ because $\varepsilon + \varepsilon' = 1$ and $\kappa'\varepsilon + \kappa\varepsilon' = z$. Since $\varepsilon\varepsilon' = 0$, these ε and ε' are idempotents, and there is an algebra isomorphism $K[z] \to K \times K$ mapping ε and ε' respectively to $(1, 0)$ and $(0, 1)$. Conversely if there is an isomorphism $K[z] \to K \times K$, this isomorphism maps z to an element $(\kappa, \kappa') \in K \times K$ such that $(\kappa, \kappa')^2 = \beta(\kappa, \kappa') - (\gamma, \gamma)$; thus κ and κ' are roots of $Z^2 - \beta Z + \gamma$.

With free quadratic extensions we can improve the previous results.

(2.6.6) **Proposition.** *Let (M, q) be a quadratic space, with M a finitely generated projective module of* even constant rank. *There exists a faithfully flat extension L of K such that $L \otimes (M, q)$ is isomorphic to the hyperbolic space $\mathbf{H}[P]$ derived from a free L-module P.*

Proof. According to (2.6.5), after a Zariski extension of K we get an orthogonal sum of free modules of rank 2. Since successive faithfully flat extensions $K \to K'$ and $K' \to K''$ give a faithfully flat extension $K \to K''$, and since any extension of a hyperbolic space is still a hyperbolic space, it suffices to prove (2.6.6) when M is a free K-module of rank 2 with a basis (e_1, e_2) such that $q(e_1)$ is invertible in K. Let us set

$$a = q(e_1)\ , \quad b = \mathrm{b}_q(e_1, e_2)\ , \quad c = q(e_2)\ ;$$

$b^2 - 4ac$ is invertible in K because q is nondegenerate. Let $K[z]$ be the free quadratic extension with an element z such that $z^2 = bz - ac$. In the extension $M' = K[z] \otimes M$ we consider the two elements

$$e'_1 \ = \ z \otimes e_1 \ - a \otimes e_2 \quad \text{and} \quad e'_2 \ = \ (b - z) \otimes e_1 \ - a \otimes e_2\ ;$$

let q' be the extension of q to M'; it is easy to verify that

$$q'(e'_1) = q'(e'_2) = 0 \quad \text{and} \quad \mathrm{b}_{q'}(e'_1, e'_2) \ = \ a(4ac - b^2)\ ;$$

since $a(4ac - b^2)$ is invertible, e'_1 and e'_2 generate a free submodule of rank 2 which is an orthogonal summand (see (2.6.4)), and therefore equal to M'; consequently (e'_1, e'_2) is a basis of M', and M' is a hyperbolic plane.

We conclude that the extension L mentioned in (2.6.6) can be constructed by means of a Zariski extension followed by at most r' quadratic extensions if $2r'$ is the rank of M. □

(2.6.7) **Corollary.** *Let (M, q) be a quadratic space, with M a finitely generated projective module of* odd constant rank. *There exists a faithfully flat extension L of K such that $L \otimes (M, q)$ is the orthogonal sum of a free L-hyperbolic space and an L-quadratic space of rank* 1, *generated by a vector on which the quadratic form takes the value* 1.

Proof. As stated in (2.6.5)(a), after a Zariski extension we get a quadratic space with an orthogonal basis; consequently we can already suppose that (M, q) has an orthogonal basis. Thus M is the orthogonal sum of a quadratic space M' of constant even rank and a quadratic space M'' of rank 1. According to (2.6.6) there is a faithfully flat extension $K \to L_1$ such that $L_1 \otimes M'$ is hyperbolic. Then $L_1 \otimes M''$ is generated by a vector e'' on which the quadratic form takes a value γ invertible in L_1; let $L = L_1[z]$ be the quadratic extension of L_1 with an element z such that $z^2 = \gamma$; after the extension $L_1 \to L$, the extended quadratic form takes the value 1 on $z\gamma^{-1} \otimes e''$, and the conclusion follows. □

2.7 Witt rings

The quadratic and bilinear Witt rings of K shall be constructed as quotients of a semiring by an ideal, according to the definitions expounded just beneath. This construction allows us to avoid the Witt–Grothendieck rings. Contrary to the general conventions of **1.1**, the quadratic Witt ring does not always contain a unit element.

Quotient of a semiring by an ideal

A nonempty set M is called an *additive monoid* (or *additive semigroup*) if it is provided with an associative and commutative addition $M \times M \to M$. It is an additive group if it contains a zero element and if every $x \in M$ admits an opposite $-x$. Of course every property of additive monoids is also valid for all commutative monoids, but here we only mention additive monoids because we are especially interested in semirings. A *semiring* is an additive monoid M provided with an associative and distributive multiplication $M \times M \to M$; here we will also require the commutativity of the multiplication, although it is not mentioned in the general definition of a semiring. By definition a zero element 0 of a semiring M must satisfy both conditions $x + 0 = x$ and $0x = 0$ for all $x \in M$; the latter condition is not a consequence of the former.

A *submonoid* of M is a nonempty subset that is stable by addition. A submonoid N is said to be *absorbent* if for every $a \in M$ there exists $b \in M$ such that $a + b$ belongs to N. When M is a semiring, a submonoid N is called an *ideal* if every product belongs to N whenever at least one factor belongs to it.

Let N be a submonoid of M; two elements a and $a' \in M$ are said to be *equivalent modulo* N if there exist b and $b' \in N$ such that $a+b = a'+b'$; it is easy to prove that an equivalence has been defined in this way; the set of equivalence classes is called the *quotient of* M *by* N and denoted by M/N.

(2.7.1) **Proposition.** *Let M be an additive monoid, N a submonoid, and f the canonical mapping $M \to M/N$. There exists a unique addition on M/N such that $f(a) + f(b) = f(a + b)$ for all a, b in M; it makes M/N become an additive monoid with zero element, and $f(b) = 0$ for all $b \in N$. When N is absorbent, M/N is even an additive group. When M is a semiring, and N an ideal, there is a unique multiplication on M/N that makes M/N become a semiring with zero element, and f a morphism of semirings.*

The kernel of the quotient mapping $f : M \to M/N$ (that is the submonoid $f^{-1}(0)$) is the subset of all $a \in M$ such that $a + b$ belongs to N for some $b \in N$; it may be larger than N, and it always contains the zero element of M if such an element exists in M. The quotient M/N admits a universal property, stated in the next proposition.

(2.7.2) **Proposition.** *Let $g : M \to M'$ be a morphism from an additive monoid (resp. semiring) M into an additive monoid (resp. semiring) M' with zero element, and let N be a submonoid (resp. an ideal) of M such that $g(b) = 0$ for all $b \in N$. There exists a unique monoid morphism (resp. semiring morphism) $\bar{g} : M/N \to M'$ such that $g = \bar{g} \circ f$.*

(2.7.3) **Corollary.** *Let M and M' be additive monoids (resp. semirings), N and N' submonoids (resp. ideals) of M and M' respectively, $f : M \to M/N$ and $f' : M' \longrightarrow M'/N'$ the quotient mappings, and let $g : M \to M'$ be a morphism such that $g(N) \subset N'$. There exists a unique morphism $\bar{g} : M/N \to M'/N'$ such that $\bar{g} \circ f = f' \circ g$.*

From every commutative monoid M is derived a *universal group* $\mathrm{G}(M)$ as it is now explained. When M is a semiring, then $\mathrm{G}(M)$ is a ring; but contrary to the conventions accepted in all other contexts, here the name "ring" does not require the existence of unit elements, it only means that a more regular object has been derived from some semiring.

(2.7.4) **Proposition.** *Let M be an additive monoid; there is an additive group $\mathrm{G}(M)$ and a monoid morphism $f : M \to \mathrm{G}(M)$ (unique up to isomorphism) such that, whatever the additive group G' and the monoid morphism $g : M \to G'$ may be, there exists a unique group morphism $\bar{g} : \mathrm{G}(M) \to G'$ satisfying $g = \bar{g} \circ f$. Two elements a and b of M have the same image in $\mathrm{G}(M)$ if and only if the equality $a + c = b + c$ holds for some $c \in M$. When M is a semiring, there is a unique multiplication on $\mathrm{G}(M)$ that makes $\mathrm{G}(M)$ become a ring, and f a semiring morphism. When g too is a semiring morphism, then $\bar{g}$ is a ring morphism.*

Proof. It is possible to construct $\mathrm{G}(M)$ as the quotient of the free additive group $\mathbb{Z}^{(M)}$ with basis $(e_a)_{a\in M}$ (see **1.3**) by the subgroup generated by all elements $e_{a+b} - e_a - e_b$ with $a, b \in M$. But another construction of $\mathrm{G}(M)$ has become more popular because it is more practical. Two elements (a, b) and (a', b') of $M \times M$ are said to be equivalent if $a + b' + x = a' + b + x$ for some $x \in M$; it is easy to prove that an equivalence has been defined in this way. Let $\mathrm{G}(M)$ be the set of equivalence classes, and $(\!(a, b)\!)$ the equivalence class of (a, b). It is easy to prove the existence of an addition on $\mathrm{G}(M)$ such that $(\!(a, b)\!) + (\!(c+d)\!) = (\!(a+c,\ b+d)\!)$ for all a, b, c, $d \in M$; thus $\mathrm{G}(M)$ is a group with zero element $0 = (\!(x, x)\!)$ (with any $x \in M$), and moreover $-(\!(a, b)\!) = (\!(b, a)\!)$ for all a and $b \in M$. Then the morphism f is defined by $f(a) = (\!(a + x,\ x)\!)$ (with any $x \in M$); thus $(\!(a, b)\!) = f(a) - f(b)$. All the remainder of the proof is now a matter of easy verification. Indeed when the morphism g is given, then $\bar{g}$ necessarily maps every $(\!(a, b)\!)$ to $g(a) - g(b)$. And when M is a semiring, the equalities $(\!(a, b)\!) = f(a) - f(b)$ and $(\!(c, d)\!) = f(c) - f(d)$ imply that the only suitable multiplication on $\mathrm{G}(M)$ must be defined in this way:

$$(\!(a, b)\!)\ (\!(c, d)\!) = (\!(ac + bd,\ ad + bc)\!). \qquad \square$$

It is also easy to prove that (2.7.4) leads to a functor from the category of additive monoids (resp. semirings) to the category of additive groups (resp. commutative rings with or without unit element).

When N is an absorbent submonoid (resp. an absorbent ideal) of M, then M/N is a group (resp. a ring), and the universal property of $\mathrm{G}(M)$ allows us to factorize the quotient mapping $M \to M/N$ through $\mathrm{G}(M)$. It is easy to prove that the morphism $\mathrm{G}(M) \to M/N$ induces an isomorphism $\mathrm{G}(M)/N_{\mathrm{G}} \to M/N$ if N_{G} is the subgroup generated by the image of N in $\mathrm{G}(M)$.

The Witt rings $\mathrm{WQ}(K)$ and $\mathrm{WB}(K)$

Let $\mathrm{WIQ}(K)$ (resp. $\mathrm{WIB}(K)$) be the set of isomorphy classes of K-quadratic spaces (resp. K-bilinear spaces), and $\mathrm{H}(K)$ (resp. $\mathrm{M}(K)$) the subset of isomorphy classes of hyperbolic spaces (resp. metabolic spaces). The construction of orthogonal sums and tensor products provides $\mathrm{WIQ}(K)$ (resp. $\mathrm{WIB}(K)$) with an addition and a multiplication, and from various statements in **2.3** it follows that it is a semiring with zero element. The semiring $\mathrm{WIB}(K)$ always contains a unit element (the isomorphy class of the bilinear space K with bilinear form $(\lambda,\mu) \longmapsto \lambda\mu$) whereas $\mathrm{WIQ}(K)$ contains a unit element only if 2 is invertible in K (the isomorphy class of the quadratic space K with quadratic form $\lambda \longmapsto \lambda^2/2$). Moreover from various statements in **2.5** it follows that $\mathrm{H}(K)$ is an ideal of $\mathrm{WIQ}(K)$, and $\mathrm{M}(K)$ an ideal of $\mathrm{WIB}(K)$; these ideals are absorbent because of (2.5.8).

By definition the *quadratic Witt ring* $\mathrm{WQ}(K)$ is the quotient $\mathrm{WIQ}(K)/\mathrm{H}(K)$, and the *bilinear Witt ring* $\mathrm{WB}(K)$ is the quotient $\mathrm{WIB}(K)/\mathrm{M}(K)$. The latter always contains a unit element, whereas the former does not always.

According to this definition the Witt class of a hyperbolic quadratic space (resp. a metabolic bilinear space) is zero. If a morphism from $\mathrm{WIB}(K)$ into an additive group maps all isomorphy classes of hyperbolic bilinear spaces to 0, the statement (c) in (2.5.8) shows that it also maps to 0 all isomorphy classes of metabolic spaces; consequently if we intend to make the quotient of $\mathrm{WIB}(K)$ by a submonoid so as to get a group in which all images of hyperbolic bilinear spaces vanish, we must make the quotient by a submonoid at least as large as $\mathrm{M}(K)$.

Conversely the following question deserves some attention: when the Witt class of a quadratic (resp. bilinear) space vanishes, is it hyperbolic (resp. metabolic)? This question is equivalent to this one: when the quadratic (resp. bilinear) spaces $A \perp B$ and B are hyperbolic (resp. metabolic), is A itself hyperbolic (resp. metabolic)? The answer is not positive for all rings K.

The construction of the Witt rings is functorial with respect to K. Let $f : K \longrightarrow K'$ be a ring morphism; it is easy to prove that the corresponding change of basic rings commutes (up to isomorphy) with the two previous operations on quadratic or bilinear spaces; for instance for quadratic modules we can write

$$(K' \otimes (M,q)) \perp (K' \otimes (M',q')) \cong K' \otimes (M \oplus M',\ q \perp q')\ ,$$
$$(K' \otimes (M,q)) \otimes_{K'} (K' \otimes (M',q')) \cong K' \otimes (M \otimes M',\ q \otimes q')\ ;$$

moreover a hyberbolic (resp. metabolic) K-space gives a hyperbolic (resp. metabolic) K'-space. Because of (2.7.3) we get a ring morphism $\mathrm{WQ}(K) \to \mathrm{WQ}(K')$ (resp. $\mathrm{WB}(K) \to \mathrm{WB}((K'))$; this morphism may be neither injective nor surjective. Thus we have defined two functors WQ and WB from the category $\mathcal{C}om(\mathbb{Z})$ of commutative rings (with unit element as in **1.1**) to the category of commutative rings (with or without unit element).

(2.7.5) **Proposition.** *By associating with every quadratic space* (M,q) *the bilinear space* (M,b_q) *we get a ring morphism* $\mathrm{WQ}(K) \to \mathrm{WB}(K)$ *; it is an isomorphism when* 2 *is invertible in* K.

Proof. The existence of this morphism results from the fact that b_q is hyperbolic whenever q is hyperbolic, and from the following assertions (see **2.4**):

$$\mathrm{b}_{q\perp q'} = \mathrm{b}_q \perp \mathrm{b}_{q'} \quad \text{and} \quad \mathrm{b}_{q\otimes q'} = \mathrm{b}_q \otimes \mathrm{b}_{q'}.$$

When 2 is invertible in K, every symmetric bilinear form φ is associated with a unique quadratic form, namely $x \longmapsto \varphi(x,x)/2$, because for every quadratic form q we can write $q(x) = \mathrm{b}_q(x,x)/2$. □

When this mapping $\mathrm{WQ}(K) \to \mathrm{WB}(K)$ is an isomorphism of rings, we simply write $\mathrm{W}(K)$ for both $\mathrm{WQ}(K)$ and $\mathrm{WB}(K)$, and we say that $\mathrm{W}(K)$ is the *Witt ring* of K.

(2.7.6) **Remark.** Proposition (2.7.5) implies that $\mathrm{WB}(K)$ is a $\mathrm{WQ}(K)$-module. Nevertheless, since $\mathrm{WQ}(K)$ does not always contain a unit element, it is more interesting to observe that $\mathrm{WQ}(K)$ *is a module over the ring* $\mathrm{WB}(K)$ *with unit element.* Indeed by Proposition (2.4.5) with each couple (φ, q') combining a symmetric bilinear form and a quadratic form is associated a tensor product $\varphi \otimes q'$ which is a quadratic form, whence a mapping $\mathrm{WB}(K)\times\mathrm{WQ}(K) \to \mathrm{WQ}(K)$ which makes $\mathrm{WQ}(K)$ become a $\mathrm{WB}(K)$-module. The equality $q\otimes q' = \mathrm{b}_q\otimes q'$ means that the operation of $\mathrm{WB}(K)$ in the ring $\mathrm{WQ}(K)$ is compatible with the ring morphism $\mathrm{WQ}(K) \to \mathrm{WB}(K)$.

No construction of a universal group in the sense of (2.7.4) is necessary to obtain the Witt groups. Nevertheless it is sometimes useful to consider the universal groups $\mathrm{G}(\mathrm{WIQ}(K))$ and $\mathrm{G}(\mathrm{WIQ}(K))$; they are called the Witt–Grothendieck rings of K and denoted by $\mathrm{WGQ}(K)$ and $\mathrm{WGB}(K)$; of course the notation $\mathrm{WG}(K)$ can also be used when 2 is invertible in K. The Witt groups can also be constructed as quotients of the Witt–Grothendieck groups by the subgroups generated by the classes of hyperbolic spaces; the fact that hyperbolic bilinear spaces are sufficient to generate the correct subgroup of $\mathrm{WGB}(K)$ follows from the statement (c) in (2.5.8).

Two quadratic (resp. bilinear) spaces A and B have the same class in $\mathrm{WGQ}(K)$ (resp. $\mathrm{WGB}(K)$) if and only if there is a quadratic (resp. bilinear) space C such that $A \perp C$ and $B \perp C$ are isomorphic (see (2.7.4)). Therefore the mapping $\mathrm{WIQ}(K) \to \mathrm{WGQ}(K)$ is injective if and only if the following *cancellation property* holds for every family (A, B, A', B') of quadratic spaces: if $A \perp B$ and $A' \perp B'$ are isomorphic, and also A and A', then B and B' too are isomorphic.

In (2.7.7) below it is stated that this cancellation property holds when K is a field. But in **2.8** we shall realize that it does not hold when $K = \mathbb{Z}$; only indefinite quadratic spaces (M, q) over $\mathbb{Z}$ (in which q is sometimes positive, sometimes negative) are determined (up to isomorphy) by their Witt–Grothendieck class. When 2 is not invertible in K, it is hard for the mapping $\mathrm{WIB}(K) \to \mathrm{WGB}(K)$ to be injective, since every metabolic space has the same class in $\mathrm{WGB}(K)$ as some hyperbolic space; for instance in **2.8** it shall *not* be injective when K is the field $\mathbb{Z}/2\mathbb{Z}$, and when $K = \mathbb{Z}$.

The quadratic Witt ring of a field

First let us state the cancellation property; it is equivalent to the injectiveness of the canonical mapping $\mathrm{WIQ}(K) \to \mathrm{WGQ}(K)$. It shall be proved more generally in (8.1.1) and (8.2.1) when K is a local ring.

(2.7.7) **Theorem.** *Let A, B, A', B' be four quadratic spaces over a field K; if $A \perp B$ and $A' \perp B'$ are isomorphic, and A and A' too, then B and B' are also isomorphic.*

The cancellation theorem is also a consequence of this powerful Witt theorem, a proof of which can be read in [Chevalley 1954] Chapter I, or in [Bourbaki 1959, *Algèbre*, Chap. 9] §4, n^o3.

(2.7.8) **Theorem.** *Let (M, q) be a quadratic space over a field K, N any linear subspace of M, and $g : N \to M$ an injective linear mapping such that $q(g(x)) = q(x)$ for all $x \in N$. Then g extends to an automorphism of (M, q).*

To derive (2.7.7) from (2.7.8), we can assume that $A \perp B$ and $A' \perp B'$ are equal to the same quadratic space (M, q), and then we extend the isomorphism $A \to A'$ to an automorphism of (M, q), which must map $B = A^\perp$ onto $B' = A'^\perp$.

Like (2.7.7), the next proposition (2.7.9) and its corollary (2.7.10) remain valid when K is a local ring, provided that the word "dimension" is replaced with "rank".

(2.7.9) **Proposition.** *Two hyperbolic spaces are isomorphic if and only if they have the same dimension. A quadratic space has a neutral Witt class if and only if it is hyperbolic. When two quadratic spaces A and B have the same Witt class, and* $\dim(A) \geq \dim(B)$, *there is a hyperbolic space H such that A is isomorphic to $B \perp H$.*

Proof. The isomorphy class of the hyperbolic space $\mathbf{H}[P]$ is determined by the isomorphy class of the linear space P, which depends on $\dim(P)$. This proves the first statement in (2.7.9), and the second one follows from the third one when

$B = 0$. Let us assume the existence of hyperbolic spaces H_1 and H_2 such that $A \perp H_1$ and $B \perp H_2$ are isomorphic, so that A and B have the same Witt class; if $\dim(A) \geq \dim(B)$, then H_2 is the orthogonal sum of two hyperbolic subspaces H and H' such that $\dim(H') - \dim(H_1)$; this implies that H_1 and H' are isomorphic. From (2.7.7) we deduce that A and $B \perp H$ are isomorphic. □

By mapping every quadratic space to its dimension, we get a semiring morphism $\mathrm{WIQ}(K) \to \mathbb{N}$ which induces a ring morphism $\mathrm{WGB}(K) \to \mathbb{Z}$, the image of which is $\mathbb{Z}$ or $2\mathbb{Z}$ according as 2 is invertible in K or not. The notation $\dim(a)$ is still used when a is an element of $\mathrm{WGQ}(K)$ or $\mathrm{WGB}(K)$. Since all hyperbolic spaces have even dimension, we still get a ring morphism $\mathrm{WQ}(K) \to \mathbb{Z}/2\mathbb{Z}$. The next statement is an evident corollary of (2.7.9); it means that a quadratic space is determined (up to isomorphy) by its Witt class and its dimension.

(2.7.10) **Corollary.** *The canonical ring morphisms* $\mathrm{WGQ}(K) \to \mathrm{WQ}(K)$ *and* $\mathrm{WGQ}(K) \to \mathbb{Z}$ *determine an injective ring morphism from* $\mathrm{WGQ}(K)$ *into* $\mathrm{WQ}(K) \times \mathbb{Z}$.

When K is a field, a quadratic form over K is said to be *anisotropic* if the equality $q(x) = 0$ implies $x = 0$. When 2 is invertible in K, such a quadratic form is weakly nondegenerate. An *anisotropic space* is a quadratic space provided with an anisotropic quadratic form. Because of (2.7.9) two anisotropic spaces have the same Witt class if and only if they are isomorphic. Proposition (2.7.11) means that the canonical mapping $\mathrm{WIQ}(K) \to \mathrm{WQ}(K)$ induces a bijection from the set of isomorphy classes of anisotropic spaces onto $\mathrm{WQ}(K)$.

(2.7.11) **Proposition.** *When K is a field, every quadratic space (M, q) is the orthogonal sum of an anisotropic subspace and a hyperbolic subspace.*

Proof. If M contains a nonzero isotropic x, there exists $y \in M$ such that $\mathrm{b}_q(x, y) \neq 0$ because $\mathrm{Ker}(\mathrm{b}_q) = 0$. The restriction of q to the plane $H = Kx \oplus Ky$ is nondegenerate, even hyperbolic (see (2.5.5)), and $M = H \oplus H^\perp$ (see (2.6.1)). Thus the proof ends with an induction on $\dim(M)$. □

The case of a field K of characteristic $\neq 2$ in which every element has a square root (for instance $\mathbb{C}$) can be settled at once: two quadratic spaces are isomorphic if and only if they have the same dimension, and consequently $\mathrm{WI}(K)$, $\mathrm{WG}(K)$ and $\mathrm{W}(K)$ are respectively isomorphic to $\mathbb{N}$, $\mathbb{Z}$ and $\mathbb{Z}/2\mathbb{Z}$.

2.8 Examples of Witt rings

Three examples of calculations of Witt rings and Witt–Grothendieck rings are presented here; they involve the field $\mathbb{R}$ of real numbers, the finite field $\mathbb{Z}/2\mathbb{Z}$ and the ring $\mathbb{Z}$. Some information about $\mathrm{W}(\mathbb{Q})$ is later presented in **8.3**, and the Witt rings of all finite fields are calculated in (8.ex.11).

First example: $\mathrm{W}(\mathbb{R})$

Let q be a quadratic form on a vector space M of finite dimension over $\mathbb{R}$; it is said to be *positive definite* if $q(x) > 0$ whenever $x \neq 0$; this implies that q is anisotropic, consequently nondegenerate, and that M contains an orthogonal basis $(e_1, e_2, \dots, e_r)$ such that $q(e_j) = 1$ for $j = 1, 2, \dots, r$; thus the isomorphy class of this positive definite space (M, q) is determined by its dimension r. And the same for a *negative definite* quadratic form. We say that q is *indefinite* if it is nondegenerate and neither positive definite nor negative definite. The existence of orthogonal bases shows that every quadratic space is the orthogonal sum of a positive definite subspace and a negative definite subspace (perhaps reduced to 0 when the given quadratic form is definite); the following theorem (Sylvester's theorem) shows that the isomorphy class of (M, q) is determined by the dimensions of these two definite subspaces.

(2.8.1) **Theorem.** *Let (M, q) be a real quadratic space, and m (resp. n) the maximal dimension of a positive definite (resp. negative definite) subspace; in each orthogonal basis of M there are always m vectors (resp. n vectors) on which q is positive (resp. negative).*

Proof. This is a consequence of (2.7.8), but it can also be proved in the following way. If M^+ and M^- are respectively a positive definite subspace of maximal dimension m and a negative definite subspace of maximal dimension n, then $M^+ \cap M^- = 0$, and consequently $m+n \leq r$. Now let m' and n' be the numbers of vectors in an orthogonal basis on which q is respectively positive or negative. The former m' vectors span a positive definite subspace, whence $m' \leq m$, and the latter n' vectors span a negative definite subspace, whence $n' \leq n$. Since $m' + n' = r$, we conclude that $m' = m$ and $n' = n$. □

Let us denote by $G_{m,n}$ the vector space $\mathbb{R}^{m+n}$ provided with the quadratic form

$$(x_1, x_2, x_3, \dots, x_{m+n}) \longmapsto x_1^2 + x_2^2 + \dots + x_m^2 - x_{m+1}^2 - x_{m+2}^2 - \dots - x_{m+n}^2 ;$$

Sylvester's theorem means that every real quadratic space is isomorphic to some $G_{m,n}$ for a unique couple (m, n). Consequently the semiring $\mathrm{WIQ}(\mathbb{R})$ is isomorphic to the set $\mathbb{N} \times \mathbb{N}$ provided with the following operations:

$$(m, n) + (m', n') = (m + m',\ n + n'),$$
$$(m, n)\ (m', n') = (mm' + nn',\ mn' + nm'),$$

and $\mathrm{WG}(\mathbb{R})$ is isomorphic to $\mathbb{Z} \times \mathbb{Z}$ provided with the same operations.

Since $G_{n,n}$ is hyperbolic (see (2.5.8)), every hyperbolic space is isomorphic to $G_{n,n}$ for some $n \in \mathbb{N}$ (see (2.7.9)). Therefore $\mathrm{W}(\mathbb{R})$ is isomorphic to the quotient of $\mathbb{N} \times \mathbb{N}$ by the ideal J containing all couples (n, n). By means of (2.7.2) it is easy to prove that the mapping $(m, n) \longmapsto m - n$ determines and isomorphism from $(\mathbb{N} \times \mathbb{N})/J$ onto the ring $\mathbb{Z}$; consequently $\mathrm{W}(\mathbb{R})$ is isomorphic to $\mathbb{Z}$, and the

Witt class of the previous quadratic space (M, q) is determined by the difference $s = m - n$ which is called its *signature.* The integers r and s have the same parity, and conversely $m = (r+s)/2$ and $n = (r-s)/2$.

Second example: $\mathrm{WQ}(\mathbb{Z}/2\mathbb{Z})$ and $\mathrm{WB}(\mathbb{Z}/2\mathbb{Z})$

Now we consider $K = \mathbb{Z}/2\mathbb{Z}$. Since 2 is not invertible in K, every quadratic space over K has even dimension. Every hyperbolic quadratic space of dimension $2m$ over K is isomorphic to K^{2m} provided with the quadratic form

$$(x_1, x_2, \dots, x_{2m}) \longmapsto \sum_{i=1}^{m} x_{2i-1}x_{2i} \ ;$$

let us denote by H_{2m} the space K^{2m} provided with this hyperbolic quadratic form. It is easy to prove that every quadratic plane is either hyperbolic (and consequently isomorphic to H_2) or anisotropic, and consequently isomorphic to K^2 provided with the quadratic form

$$(x_1, x_2) \longmapsto x_1^2 + x_1x_2 + x_2^2 \ ;$$

let us call A_2 the plane K^2 provided with this anisotropic quadratic form. Because of (2.5.8) we know that $A_2 \perp A_2$ is hyperbolic. Let (M, q) be a quadratic space of nonzero dimension $2m$; since it is an orthogonal sum of planes, we realize that it is isomorphic either to H_{2m} or to $A_2 \perp H_{2m-2}$.

Now we must prove that these two quadratic spaces are not isomorphic; this is a consequence of (2.7.8), but it can also be proved by comparing the numbers $N(2m)$ and $N'(2m)$ of isotropic elements respectively in H_{2m} and $A_2 \perp H_{2m-2}$; the number of the other elements, on which the quadratic form takes the value 1, is respectively $2^{2m} - N(2m)$ and $2^{2m} - N'(2m)$. This leads to these induction formulas:

$$N(2m+2) = N(2)\ N(2m) + (2^2 - N(2))\ (2^{2m} - N(2m)),$$
$$N'(2m+2) = N'(2)\ N(2m) + (2^2 - N'(2))\ (2^{2m} - N(2m));$$

since $N(2) = 3$ and $N'(2) = 1$, we get after some calculations

$$N(2m) = \ 2^{2m-1} + 2^{m-1} \neq \ N'(2m) = 2^{2m-1} - 2^{m-1}.$$

(2.8.2) **Proposition.** *A quadratic space (M, q) of nonzero dimension $2m$ over $\mathbb{Z}/2\mathbb{Z}$ is either hyperbolic and isomorphic to H_{2m}, or isomorphic to $A_2 \perp H_{2m-2}$; this depends on whether the number of isotropic elements is larger or smaller than 2^{2m-1}.*

This proves that $\mathrm{WQ}(\mathbb{Z}/2\mathbb{Z})$ has only two elements, the neutral class and the class of A_2. Moreover $A_2 \otimes A_2$ is hyperbolic because its quadratic form obviously vanishes on the 10 elements that can be written as tensor products $u \otimes v$. Consequently $\mathrm{WQ}(\mathbb{Z}/2\mathbb{Z})$ *is isomorphic to the additive group* $\mathbb{Z}/2\mathbb{Z}$ *provided with the null multiplication.*

Then the injective ring morphism mentioned in (2.7.10) gives an isomorphism from $\mathrm{WGQ}(\mathbb{Z}/2\mathbb{Z})$ onto $\mathrm{WQ}(\mathbb{Z}/2\mathbb{Z}) \times (2\mathbb{Z})$. Later we shall need precise information about H_4.

(2.8.3) **Proposition.** *The hyperbolic space* H_4 *contains exactly two anisotropic planes, and* H_4 *is their orthogonal sum.*

Proof. Since H_4 is isomorphic to $A_2 \perp A_2$, it is the orthogonal sum of two anisotropic planes P and P'. Since $N(4) = 10$, the 6 nonzero elements of $P \cup P'$ are the only elements of H_4 that are not isotropic; consequently no other anisotropic plane can exist in H_4. □

Now let us consider bilinear spaces over K. For every r and m in $\mathbb{N}$ we call G_r and U_{2m} the spaces K^r and K^{2m} provided with these bilinear forms:

$$((x_1, x_2, \dots, x_r),\ (x'_1, x'_2, \dots, x'_r)) \longmapsto \sum_{i=1}^{r} x_i x'_i\ ,$$

$$((x_1, x_2, \dots, x_{2m}),\ (x'_1, x'_2, \dots, x'_{2m})) \longmapsto \sum_{i=1}^{m} (x_{2i-1} x'_{2i} + x_{2i} x'_{2i-1})\ .$$

It is clear that every bilinear space admitting an orthogonal basis is isomorphic to some G_r, and that every orthogonal sum of bilinear spaces of rank 2 that are all provided with an alternate form, is isomorphic to some U_{2m}. Thus the following proposition is an immediate consequence of (2.6.3).

(2.8.4) **Proposition.** *Let* (M, φ) *be a bilinear space of dimension* r *over* $\mathbb{Z}/2\mathbb{Z}$. *When* φ *is an alternate bilinear form (in other words, when* $\varphi(x, x) = 0$ *for all* $x \in M$), *then* r *is even and* (M, φ) *is isomorphic to* U_r. *But when* φ *is not alternate,* (M, φ) *is isomorphic to* G_r.

It follows from (2.8.4) that $U_{2m} \perp G_1$ and $G_{2m} \perp G_1$ are isomorphic. Therefore the Witt–Grothendieck class of a bilinear space is determined by its dimension, and there is an isomorphism $\mathrm{WGB}(\mathbb{Z}/2\mathbb{Z}) \to \mathbb{Z}$.

It is clear that U_{2m} is hyperbolic, and from (2.5.6) we deduce that G_{2m} is metabolic. Therefore $\mathrm{WB}(\mathbb{Z}/2\mathbb{Z})$ is isomorphic to $\mathbb{Z}/2\mathbb{Z}$; its unit element is the class of G_1.

For every quadratic form q over $\mathbb{Z}/2\mathbb{Z}$, the associated bilinear form b_q is alternate; therefore the canonical morphism $\mathrm{WQ}(\mathbb{Z}/2\mathbb{Z}) \to \mathrm{WB}(\mathbb{Z}/2\mathbb{Z})$ is null. Besides, the action of the ring $\mathrm{WB}(\mathbb{Z}/2\mathbb{Z})$ (resp. $\mathrm{WGB}(\mathbb{Z}/2\mathbb{Z})$) in $\mathrm{WQ}(\mathbb{Z}/2\mathbb{Z})$ (resp. $\mathrm{WGQ}(\mathbb{Z}/2\mathbb{Z})$) is determined by the fact that the action of the unit element is the identity.

Later we shall also need the following lemma.

(2.8.5) **Lemma.** *Let (M, φ) be a bilinear space over $\mathbb{Z}/2\mathbb{Z}$; there exists a unique $\xi \in M$ such that $\varphi(x, x) = \varphi(\xi, x)$ for all $x \in M$.*

Proof. The bilinear form associated with the quadratic form $x \longmapsto \varphi(x, x)$ is 2φ which here is null. Consequently this quadratic form is also a linear form. Lemma (2.8.5) means that this linear form is equal to $\mathrm{d}_\varphi(\xi)$ for a unique $\xi \in M$; indeed d_φ is a bijection from M onto $\mathrm{Hom}(M, K)$. □

Third example: $\mathrm{WQ}(\mathbb{Z})$ and $\mathrm{WB}(\mathbb{Z})$

The Witt rings of $\mathbb{Z}$ raise more difficulties and require the two theorems (2.8.6) and (2.8.7), the proof of which is wildly too long to be given here.

(2.8.6) **Theorem.** *Every projective module over $\mathbb{Z}$ (or more generally over any principal domain) is free. And if it has a finite rank r, all its submodules are free of rank $\leq r$.*

Let φ be a *weakly* nondegenerate symmetric bilinear form on a free $\mathbb{Z}$-module M of finite rank r. By the extension $\mathbb{Z} \to \mathbb{R}$ we get a nondegenerate bilinear form on $\mathbb{R} \otimes M$; its class in $\mathrm{W}(\mathbb{R})$ is determined by its signature s, and we set $m = (r+s)/2$ and $n = (r-s)/2$; the notation $(r, s; m, n)$ will be used up to the end; φ is called *definite* (resp. *indefinite*) if $mn = 0$ (resp. $mn \neq 0$). The words "definite" and "indefinite" are here only used for weakly nondegenerate forms (either symmetric bilinear or quadratic forms) on free modules of finite rank.

Besides, let $(e_1, \ldots, e_r)$ and $(e'_1, \ldots, e'_r)$ be two bases of M, let Λ be the matrix of the numbers $\lambda_{i,j}$ such that $e'_j = \sum_i \lambda_{i,j} e_i$, and Φ (resp. Φ') the matrix of the numbers $\varphi(e_i, e_j)$ (resp. $\varphi(e'_i, e'_j)$). The equality $\Phi' = {}^t\Lambda\Phi\Lambda$ is well known (see (2.ex.3)); since the only invertible elements of $\mathbb{Z}$ are ±1, here it implies that Φ and Φ' have the same determinant. Consequently $\det(\varphi)$, by definition the determinant of Φ, is a well-defined number, and *φ is nondegenerate if and only if* $\det(\varphi) = \pm1$.

Let q be an indefinite quadratic form on M; does M contain a nonzero isotropic element x (such that $q(x) = 0$)? Let us consider the following quadratic form q on $\mathbb{Z}^3$: $q(x_1, x_2, x_3) = x_1^2 + x_2^2 - 3x_3^2$; by means of the extension $\mathbb{Z} \longrightarrow \mathbb{Z}/9\mathbb{Z}$ it is possible to prove that the answer is negative for this particular quadratic form. The answer to the general question is given for instance in J.P. Serre's course in Arithmetic, referred to as [Serre]. Here we only need the two sufficient conditions stated in the next theorem; the very long proof is expounded in [Serre], where several difficult theorems in the beginning chapters lead to a final argument in Chapter V. §3.1.

(2.8.7) **Theorem.** *Let q be an indefinite quadratic form on the free $\mathbb{Z}$-module M of rank r. If one of these two conditions is fulfilled, there exists a nonzero x in M such that $q(x) = 0$:*

– *if* $r \geq 5$ (Meyer's theorem, 1883) ;
– *if* $r \geq 3$ *and* $\det(\mathrm{b}_q) = \pm 2^k$ *for some* $k \geq 0$.

For a module of rank 2 we use the following easy lemma.

(2.8.8) **Lemma.** *Let q be a quadratic form on a free $\mathbb{Z}$-module M of rank 2. There exists a nonzero x such that $q(x) = 0$ if and only if $-\det(\mathrm{b}_q)$ is a square.*

Proof. We can suppose that $M = \mathbb{Z}^2$ and $q(x,y) = ax^2 + bxy + cy^2$, whence $-\det(\mathrm{b}_q) = b^2 - 4ac$. If $b^2 - 4ac$ is the square of d, then $q(-b \pm d,\ 2a) = 0$; this gives the desired result when $a \neq 0$; but when $a = 0$, we have trivially $q(1,0) = 0$. Conversely we use the equality $4(b^2 - 4ac)\,y^2 = (2ax + by)^2 - 4a\,q(x,y)$; it shows that $b^2 - 4ac$ is a square when $q(x,y) = 0$ and $y \neq 0$; when $q(x,0) = 0$ and $x \neq 0$, then $a = 0$ and the conclusion is trivial. □

When q runs through $\mathrm{Quad}(M,\mathbb{Z})$, the mapping $q \longmapsto \mathrm{b}_q$ is a bijection from $\mathrm{Quad}(M,\mathbb{Z})$ onto the set of all symmetric bilinear forms φ such that $\varphi(x,x)$ is even for all x in M; such a symmetric bilinear form is said to be of *even type*, and the other ones are said to be of *odd type*. Now we shall almost forget the quadratic forms and replace them with symmetric bilinear forms of even type. If $(e_1,\dots,e_r)$ is a basis of M, φ has even type if and only if the r numbers $\varphi(e_i,e_i)$ are even.

After these preliminaries we define some particular bilinear spaces over $\mathbb{Z}$. First $G_{m,n}$ and H_{2m} are the modules $\mathbb{Z}^{m+n}$ and $\mathbb{Z}^{2m}$ provided with these symmetric bilinear forms φ and ψ :

$$\varphi(x,x') = x_1x_1' + x_2x_2' + \cdots + x_mx_m' - x_{m+1}x_{m+1}' - x_{m+2}x_{m+2}' - \cdots - x_{m+n}x_{m+n}' ,$$
$$\psi(x,x') = x_1x_2' + x_2x_1' + x_3x_4' + x_4x_3' + \cdots + x_{2m-1}x_{2m}' + x_{2m}x_{2m-1}' ;$$

in these equalities x and x' mean $(x_1,\dots,x_r)$ and $(x_1',\dots,x_r')$ with r equal to $m+n$ or $2m$. The space $G_{m,n}$ has odd type; it is metabolic if and only if $m = n$ (see (2.5.6)). The space H_{2m} has even type and is hyperbolic.

Now let r be a positive integer divisible by 4, $(e_1,\dots,e_r)$ the canonical basis of $\mathbb{R}^r$, $\tilde{\varphi}$ the $\mathbb{R}$-bilinear form on $\mathbb{R}^r$ defined by $\tilde{\varphi}(x,x') = \sum_{i=1}^r x_ix_i'$, and P_r the subgroup of all elements $(x_1,\dots,x_r) \in \mathbb{R}^r$ satisfying these three conditions: the r numbers $2x_i$ are integers, they all have the same parity, and $\sum_{i=1}^r x_i$ is an even integer.

(2.8.9) **Lemma.** *The following r elements $b_1,\dots,b_r$ constitute a basis of the free additive group P_r :*

$$b_1 = \frac{1}{2}(e_1 - e_2 - e_3 - \cdots - e_{r-1} + e_r) , \qquad b_2 = e_1 + e_2$$
$$\textit{and} \quad b_i = e_{i-1} - e_{i-2} \quad \textit{for} \quad i = 3,4,\dots,r.$$

Moreover $\tilde{\varphi}$ determines by restriction to P_r a $\mathbb{Z}$-bilinear form $\varphi : P_r \times P_r \to \mathbb{Z}$ which is nondegenerate, positive definite, and which has even type if and only if r is divisible by 8.

Proof. Let $(\lambda_{i,j})$ be the matrix defined by $b_j = \sum_i \lambda_{i,j} e_i$; an easy calculation shows that its determinant is -1. Obviously all b_j are in P_r. If x is any element of P_r, then $x - (2x_r)b_1$ is another element y such that $y_r = 0$, all y_i are integers and their sum is even; this allows us to prove that y is in the subgroup generated by $(b_2, b_3, \dots, b_r)$. Consequently $(b_1, \dots, b_r)$ is a $\mathbb{Z}$-basis of P_r. Since the r numbers $2x_i$ are integers of the same parity, it is clear that $\tilde{\varphi}(x, b_j)$ is an integer for $j = 2, 3, \dots, r$; the same assertion is true when $j = 1$ because $\sum_{i=1}^r x_i = 2(x_1 + x_r) - 2\tilde{\varphi}(x, b_1)$. This proves that $\tilde{\varphi}$ induces a $\mathbb{Z}$-bilinear form φ on P_r, which is obviously positive definite. It is nondegenerate because $\det(\varphi) = (\det(\lambda_{i,j}))^2 = 1$. Since $\varphi(b_1, b_1) = r/4$ and $\varphi(b_i, b_i) = 2$ for $i = 2, 3, \dots, r$, we conclude that φ has even type if and only if r is divisible by 8. □

A serious study of the bilinear spaces P_r is proposed in (2.ex.25); except P_4 which is isomorphic to $G_{4,0}$, no other bilinear space P_r admits orthogonal bases. Here we are especially interested in P_8. The bilinear space P_r can also be defined by means of the matrix $(\varphi(b_i, b_j))$ (often called *Milnor's matrix* when $r = 8$); the entries $\varphi(b_i, b_i)$ are calculated just above; and $\varphi(b_i, b_j)$ with $i < j$ is equal to -1 when (i, j) is equal to $(i, i+2)$ with $i = 1, 2$, or equal to $(i, i+1)$ with $i \geq 3$; all other entries with $i < j$ are 0.

Now we calculate the Witt groups of $\mathbb{Z}$.

(2.8.10) **Lemma.** *Let φ be a nondegenerate symmetric bilinear form of odd (resp. even) type on a free $\mathbb{Z}$-module M containing a nonzero x such that $\varphi(x, x) = 0$. This x is contained in a submodule of (M, φ) isomorphic to $G_{1,1}$ (resp. H_2).* According to (2.6.1) this submodule is an orthogonal summand.

Proof. After dividing x by a suitable integer, we can suppose that its components in a basis of M are coprime integers, so that there exists $f \in \mathrm{Hom}(M, \mathbb{Z})$ such that $f(x) = 1$. Since φ is nondegenerate, there exists z in M such that $\varphi(x, z) = 1$. If $\varphi(z, z)$ is even, we set $y = z - \varphi(z, z)x/2$, whence $\varphi(x, x) = \varphi(y, y) = 0$ and $\varphi(x, y) = 1$. Thus x and y generate a submodule isomorphic to H_2. If $\varphi(z, z)$ is odd, we set

$$u = z - \frac{1}{2}(\varphi(z, z) - 1)\, x \quad \text{and} \quad v = z - \frac{1}{2}(\varphi(z, z) + 1)\, x\,;$$

this implies $\varphi(u, u) = -\varphi(v, v) = 1$ and $\varphi(u, v) = 0$. Consequently the submodule generated by u and v (which contains $x = u - v$) is isomorphic to $G_{1,1}$.

When φ has odd type, it may happen that $\varphi(z, z)$ is even, and then we still prove that we can replace z with an element z' such that $\varphi(x, z') = 1$ and $\varphi(z', z')$ is odd. Indeed let w be any element such that $\varphi(w, w)$ is odd; if we set $z' = w + (1 - \varphi(x, w))z$, all the required conditions are fulfilled. □

(2.8.11) **Theorem.** *Every indefinite* $\mathbb{Z}$*-bilinear space* A *of odd type is isomorphic to some* $G_{m,n}$.

Of course m and n are the integers determined by the rank r and the signature s as above.

Proof. The determinant of the bilinear form on A is ± 1, and moreover $r \geq 2$ since A is indefinite; consequently we can apply (2.8.7) or (2.8.8) to A; and in all cases we deduce from (2.8.10) that A contains a submodule isomorphic to $G_{1,1}$. When $r > 2$, we make an induction on r. Indeed A is isomorphic to some orthogonal sum $G_{1,1} \perp B$ which we can view as $G_{1,0} \perp B \perp G_{0,1}$; one of the bilinear spaces $G_{1,0} \perp B$ or $B \perp G_{0,1}$ is still indefinite of odd type, and we can apply to it the induction hypothesis. □

(2.8.12) **Theorem.** *The natural morphism* $\mathrm{WB}(\mathbb{Z}) \rightarrow \mathrm{W}(\mathbb{R})$ *is an isomorphism; in other words, the Witt class of a bilinear* $\mathbb{Z}$*-space is determined by its signature.*

Proof. The existence of the bilinear spaces $G_{m,n}$ proves that this morphism is surjective, and we have just to prove that it is injective. Let A be a bilinear space of signature 0, not reduced to 0, and therefore indefinite of even rank; from (2.8.7), (2.8.8) and (2.8.10), and by induction on the rank, we deduce that A is isomorphic to an orthogonal sum of planes isomorphic to $G_{1,1}$ or H_2; this proves that A is metabolic. □

From now on, $\mathrm{WB}(\mathbb{Z})$ and $\mathrm{W}(\mathbb{R})$ are identified with $\mathbb{Z}$.

(2.8.13) **Theorem.** *The natural morphism* $\mathrm{WQ}(\mathbb{Z}) \rightarrow \mathrm{WB}(\mathbb{Z})$ (*corresponding to* $q \longmapsto \mathrm{b}_q$) *is injective, and its image is* $8\mathbb{Z}$.

Proof. Let (M, q) be a $\mathbb{Z}$-quadratic space of null signature; as explained in the proof of (2.8.12), (M, b_q) is an orthogonal sum of planes isomorphic to H_2 (not to $G_{1,1}$ since b_q has even type); consequently (M, q) is hyperbolic. This proves the injectiveness of the morphism $\mathrm{WQ}(\mathbb{Z}) \rightarrow \mathbb{Z}$. The existence of P_8 (see (2.8.9)) proves that $8\mathbb{Z}$ is contained in its image. Thus it remains to prove that the signature of a $\mathbb{Z}$-quadratic space (M, q) is always divisible by 8.

Indeed from (2.8.11) we derive that $(M, \mathrm{b}_q) \perp G_{1,1}$ is isomorphic to $G_{m+1,n+1}$; we write φ' and φ'' for the symmetric bilinear forms on $G_{1,1}$ and $G_{m+1,n+1}$. From (2.8.5) and the extension $\mathbb{Z} \rightarrow \mathbb{Z}/2\mathbb{Z}$ we derive the existence of an element ξ'' in $G_{m+1,n+1}$ such that

$$\forall x'' \in G_{m+1,n+1} \qquad \varphi''(x'', x'') \equiv \varphi''(\xi'', x'') \quad \text{modulo } 2\mathbb{Z} \ ;$$

moreover all the elements that can play the same role as ξ'' are the elements $\xi'' + 2x''$ with an arbitrary $x'' \in G_{m+1,n+1}$. Now observe that

$$\begin{aligned} \varphi''(\xi'' + 2x'',\ \xi'' + 2x'') &= \varphi''(\xi'', \xi'') + 4\varphi''(\xi'', x'') + 4\varphi''(x'', x'') \\ &\equiv \varphi''(\xi'', \xi'') + 4\varphi''(x'', x'') + 4\varphi''(x'', x'') \quad \text{modulo } 8\mathbb{Z} \\ &\equiv \varphi''(\xi'', \xi'') \quad \text{modulo } 8\mathbb{Z} \ ; \end{aligned}$$

consequently the image of $\varphi''(\xi'',\xi'')$ in $\mathbb{Z}/8\mathbb{Z}$ is well determined. And if we find analogous elements ξ and ξ' in the bilinear spaces (M, b_q) and $G_{1,1}$, we can write

$$\varphi''(\xi'',\xi'') \equiv \mathrm{b}_q(\xi,\xi) + \varphi'(\xi',\xi') \quad \text{modulo } 8\mathbb{Z}.$$

Since b_q has even type, we can choose $\xi = 0$. In $G_{1,1}$ we can choose $\xi' = (1,1)$, whence $\varphi'(\xi',\xi') = 0$. Similarly we can choose $\xi'' = (1,1,\dots,1,1,\dots,1)$, whence $\varphi''(\xi'',\xi'') = m - n = s$. All this proves that s belongs to $8\mathbb{Z}$. □

Now let us prove that an indefinite bilinear form is determined (up to isomorphy) by its rank, its signature and its type (even or odd). For the odd type, this follows from (2.8.11); for the even type, it follows from the next theorem.

(2.8.14) **Theorem.** *Let A be an indefinite $\mathbb{Z}$-bilinear space of even type, r its rank, s its signature, m and n as above; we know that $s = 8k$ for some $k \in \mathbb{Z}$. When $s = 0$, then A is isomorphic to H_r. When $s > 0$ (resp. $s < 0$), then A is isomorphic to*

$$H_{2n} \perp (P_8 \otimes G_{k,0}) \qquad (\textit{resp.} \quad H_{2m} \perp (P_8 \otimes G_{0,-k})).$$

Proof. The case $s = 0$ has been settled in the proof of (2.8.13); therefore we suppose $s \neq 0$. From (2.8.7) and (2.8.10) we deduce that A is isomorphic to some orthogonal sum $H_2 \perp B$, and Theorem (2.8.14) states that it must be isomorphic to $H_2 \perp C$, where C is either $H_{2n-2} \perp (P_8 \otimes G_{k,0})$ or $H_{2m-2} \perp (P_8 \otimes G_{0,-k})$. It is easy to check that B and C have the same rank and the same signature; thus (2.8.11) implies that $G_{1,1} \perp B$ and $G_{1,1} \perp C$ are isomorphic; the proof ends with the following more difficult lemma.

(2.8.15) **Lemma.** *If B and C are two $\mathbb{Z}$-bilinear spaces of even type such that $G_{1,1} \perp B$ and $G_{1,1} \perp C$ are isomorphic, then $H_2 \perp B$ and $H_2 \perp C$ too are isomorphic.*

Proof. Let $\frac{1}{2}\mathbb{Z}$ be the free additive group of integers and half integers, L a free $\mathbb{Z}$-module with basis (e_1', e_2'), and $\beta : L \times L \to \frac{1}{2}\mathbb{Z}$ the symmetric $\mathbb{Z}$-bilinear mapping such that

$$\beta(e_1', e_1') = \beta(e_2', e_2') = 0 \quad \text{and} \quad \beta(e_1', e_2') = 1/2.$$

Let us set $e_1 = e_1' + e_2'$ and $e_2 = e_1' - e_2'$; thus we get:

$$\beta(e_1, e_1) = -\beta(e_2, e_2) = 1 \quad \text{and} \quad \beta(e_1, e_2) = 0\ ;$$

consequently e_1 and e_2 generate a submodule which we can identify with $G_{1,1}$. Secondly it is easy to verify that $2L$ is the subgroup of all $x \in G_{1,1}$ such that $\beta(x,x)$ belongs to $2\mathbb{Z}$. Thirdly $L/2L$ is isomorphic to $(\mathbb{Z}/2\mathbb{Z})^2$, and consequently there are exactly three groups strictly intermediate between L and $2L$; beside $G_{1,1}$ there are the subgroup H_2' generated by $(2e_1', e_2')$ and the subgroup H_2'' generated by $(e_1', 2e_2')$, both isomorphic to H_2. Here is a fourth observation: $\beta(x, x')$ is an integer whenever x belongs to L and x' to $2L$, and thus β determines an isomorphism from L onto the dual module $\mathrm{Hom}(2L, \mathbb{Z})$.

Now we consider B and C, the orthogonal sums $L \perp B$ and $L \perp C$ (in the category of symmetric bilinear mappings with target $\frac{1}{2}\mathbb{Z}$) and the bilinear mappings φ and ψ associated with $L \perp B$ and $L \perp C$. The previous four observations remain valid for $L \perp B$ and $L \perp C$; thus $L \perp B$ contains a subgroup identified with $G_{1,1} \perp B$, whereas $(2L) \perp B$ is the subgroup of all $y \in G_{1,1} \perp B$ such that $\varphi(y, y)$ belongs to $2\mathbb{Z}$ (indeed B has even type); there are three subgroups strictly intermediate between $L \perp B$ and $2L \perp B$, which are all $\mathbb{Z}$-bilinear spaces, and moreover φ determines an isomorphism between $L \perp B$ and $\mathrm{Hom}(2L \perp B,\ \mathbb{Z})$. Let us assume that f is an isomorphism from $G_{1,1} \perp B$ onto $G_{1,1} \perp C$; consequently

$$\forall y \in G_{1,1} \perp B, \quad \forall z \in G_{1,1} \perp C, \quad \varphi(y,\ f^{-1}(z)) = \psi(f(y),\ z).$$

Since $2L$ is the subgroup of all $x \in G_{1,1}$ such that $\beta(x, x) \in 2\mathbb{Z}$, from f we derive by restriction an isomorphism from $2L \perp B$ onto $2L \perp C$, whence an isomorphism from $\mathrm{Hom}(2L \perp C,\ \mathbb{Z})$ onto $\mathrm{Hom}(2L \perp B,\ \mathbb{Z})$, and finally an isomorphism g from $L \perp C$ onto $L \perp B$. From the construction of g we deduce that

$$\forall y \in 2L \perp B, \quad \forall z \in L \perp C, \quad \varphi(y,\ g(z)) = \psi(f(y),\ z) \ ;$$

this proves that g extends f^{-1}. Consequently g^{-1} is an isomorphism intertwining φ and ψ. It must map $H_2' \perp B$ onto a subgroup strictly intermediate between $L \perp C$ and $(2L) \perp C$, therefore either $H_2' \perp C$ or $H_2'' \perp C$. In both cases $H_2 \perp B$ is isomorphic to $H_2 \perp C$. □

Let us assume that the bilinear spaces A and B have the same Witt–Grothendieck class; thus $A \perp C$ and $B \perp C$ are isomorphic for some C, and the ring extension $\mathbb{Z} \to \mathbb{R}$ shows that A and B have the same rank and the same signature. Consequently a pair of integers (r, s) can be associated with every Witt–Grothendieck class.

(2.8.16) **Proposition.** *By mapping every Witt–Grothendieck class to the pair* (r, s) *we get injective ring morphisms* $\mathrm{WGB}(\mathbb{Z}) \to \mathbb{Z} \times \mathbb{Z}$ *and* $\mathrm{WGQ}(\mathbb{Z}) \to \mathbb{Z} \times \mathbb{Z}$. *The former gives all pairs* (r, s) *such that* r *and* s *have the same parity, and the latter gives all* (r, s) *such that* r *is even and* s *divisible by* 8.

Proof. If the bilinear spaces A and B have the same rank and the same signature, then $A \perp G_{1,1}$ and $B \perp G_{1,1}$ are isomorphic because of (2.8.11); consequently A and B have the same class in $\mathrm{WGB}(\mathbb{Z})$. If moreover A and B have even type, we deduce from (2.8.14) that $A \perp H_2$ and $B \perp H_2$ are isomorphic, and that A and B have the same class in $\mathrm{WGQ}(\mathbb{Z})$. The previous results show which couples (r, s) can be obtained in each case. □

It is clear that the canonical mapping $\mathrm{WIB}(\mathbb{Z}) \to \mathrm{WGB}(\mathbb{Z})$ is not injective since $G_{1,1}$ and H_2 have the same image in $\mathrm{WGB}(\mathbb{Z})$; this is banal because $G_{1,1}$ is metabolic but not hyperbolic. It is less banal that $G_{12,0}$ and P_{12} (both positive definite of odd type) have the same class in $\mathrm{WGB}(\mathbb{Z})$ although they are not isomorphic (see (2.ex.25)).

The analogous mapping $\mathrm{WIQ}(\mathbb{Z}) \to \mathrm{WGQ}(\mathbb{Z})$ neither is injective; indeed $P_8 \perp P_8$ and P_{16} are not isomorphic (see (2.ex.25)). For every multiple r of 8, let N_r be the number of isomorphy classes of positive definite $\mathbb{Z}$-quadratic spaces of rank r; all these quadratic spaces have the same class in $\mathrm{WGQ}(\mathbb{Z})$. Witt already knew that $N_8 = 1$ and $N_{16} = 2$, and in [Nimeier 1973] it is stated that $N_{24} = 24$. The subsequent numbers N_r (still unknown) are discouragingly great; for instance $N_{32} > 8 \times 10^7$ (see [Serre]).

Exercises

(2.ex.1) Let M and N be two K-modules; assume that M is a torsion module (for every $x \in M$ there is a nonzero $\lambda \in K$ such that $\lambda x = 0$) and that N is a torsionless module (the equality $\lambda y = 0$ implies $\lambda = 0$ if y is a nonzero element of N). Prove that all linear mappings $M \to N$ and all quadratic mappings $M \to N$ are null.

(2.ex.2) Let M and N be K-modules, and $q : M \to N$ a K-quadratic mapping; prove that

$$\forall x, y \in M \ , \quad q(x+y) + q(x-y) \; = \; 2q(x) + 2q(y).$$

Conversely let M and N be additive groups, and $q : M \to N$ a mapping satisfying the above equality for all x and $y \in M$; prove that the mapping $x \longmapsto 2q(x)$ is $\mathbb{Z}$-quadratic.

Hint. Prove that $2q(0) = 0$ and that $q(nx) = n^2 q(x) + (1-n)q(0)$ for all $x \in M$ and all $n \in \mathbb{Z}$; then set

$$\mathrm{b}(x,y) = q(x+y) - q(x) - q(y) \quad \text{and} \quad \mathrm{t}(x,y,z) = \mathrm{b}(x,\ y+z) - \mathrm{b}(x,y) - \mathrm{b}(x,z) \ ;$$

prove that $\mathrm{b}(2x,y) = \mathrm{b}(x,2y) = 2\mathrm{b}(x,y) + q(0)$, whence

$$\mathrm{t}(2x,y,z) = \mathrm{t}(x,2y,2z) = 2\mathrm{t}(x,y,z) - q(0) \ ;$$

since $\mathrm{t}(x,y,z)$ is invariant by all permutations of $\{x,y,z\}$, the vanishing of $2\mathrm{t}(x,y,z)$ follows.

(2.ex.3) Let φ be a symmetric bilinear form on the K-module M, n an integer ≥ 1, and $(a_1, a_2, \dots, a_n)$ and $(b_1, b_2, \dots, b_n)$ two families of n elements of M. By means of two $(n \times n)$-matrices $A = (\lambda_{i,j})$ and $B = (\mu_{i,j})$ with entries in K we define two other families $(a'_1, \dots, a'_n)$ and $(b'_1, \dots, b'_n)$ of n elements of M :

$$a'_j = \sum_i \lambda_{i,j} a_i \ , \quad \text{and} \quad b'_j = \sum_i \mu_{i,j} b_i \ .$$

Now let Φ (resp. Φ') be the $(n \times n)$-matrix $(\varphi(a_i, b_j))$ (resp.$(\varphi(a'_i, b'_j))$). The notation ${}^tA$ means the matrix derived from A by transposition. Prove that $\Phi' = {}^tA\,\Phi\,B$. Consequently

$$\det(\Phi') \; = \; \det(A)\det(B)\det(\Phi)\,.$$

(2.ex.4) We use the same notation as in (2.ex.3). Let J_n be the ideal of K generated by the determinants of all the matrices Φ defined as in (2.ex.3). Prove that $J_n = 0$ whenever M is finitely generated of rank $< n$ at every prime ideal of K. When M is a finitely generated projective module of constant rank n, prove that φ is nondegenerate if and only if $J_n = K$.

(2.ex.5)* Here is an example of a quadratic module (M, q) that is not a quadratic space (it is not even finitely generated) although all its localizations are quadratic spaces. Let F be a field, and K the ring of all functions $\lambda : \mathbb{N} \to F$ that remain constant when the variable n is large enough (larger than some value depending on λ); the value of λ at the point n is denoted by λ_n, and its constant value for n large enough is denoted by λ_∞. Let $(M_n, q_n)_{n\in\mathbb{N}}$ be a family of quadratic spaces over F such that $M_n \neq 0$ for every n; the direct sum $M = \bigoplus_n M_n$ is a K-module in this way:

$$\lambda\,(x_0, x_1, x_2, \dots) \;=\; (\lambda_0 x_0,\ \lambda_1 x_1,\ \lambda_2 x_2, \dots);$$

we get a quadratic form q on M if we set

$$q(x_0, x_1, x_2, \dots) = \sum_n q_n(x_n);$$

this equality is meaningful because all x_n vanish except a finite number.

(a) Prove that the prime ideals of K are the kernels of the morphisms $\lambda \longmapsto \lambda_n$ from K to F associated with the points n of $\mathbb{N} \cup \{\infty\}$. How are the localizations of K and M?

(b) Prove that all the localizations of (M, q) are quadratic spaces, but that the mapping $\mathrm{d}_q : M \to M^*$ is not surjective.

(2.ex.6)* Let K be a ring in which 2 is a divisor of zero. A quadratic module (M, q) over K is said to be *defective* if $\mathrm{Ker}(\mathrm{b}_q) \neq \mathrm{Ker}(q)$, and its *defect* is represented by the quotient $\mathrm{Def}(M, q) = \mathrm{Ker}(\mathrm{b}_q)/\mathrm{Ker}(q)$. It shall be proved that the defect cannot be arbitrarily large, because there is always an injective linear mapping from $\mathrm{Def}(M, q)$ into some K-module J^{sq} only depending on K.

(a) Let J be first any module over the ring $K/2K$ (and consequently over K too); prove the existence of a K-module J^{sq} satisfying these properties: as an additive group it is isomorphic to J by means of a canonical isomorphism $j \longmapsto j^{sq}$, and the operation in J^{sq} of any $\lambda \in K$ is defined according to this formula: $\lambda\, j^{sq} = (\lambda^2 j)^{sq}$.

From now on, J is the kernel of the mapping $\lambda \longmapsto 2\lambda$ from K into itself; this ideal J is a module over $K/2K$ in a natural way, and gives another K-module J^{sq}.

(b) Prove that $\mathrm{Def}(M, q)$ is a module over $K/2K$ in a natural way, that $q(\mathrm{Ker}(\mathrm{b}_q))$ is contained in the ideal J defined above, and that q induces an injective K-linear mapping $\mathrm{Def}(M, q) \to J^{sq}$.

(c) *Application.* When K is a field of characteristic 2, then $J = K = K/2K$ and the subset K^2 of all λ^2 with $\lambda \in K$ is a subfield of K. Suppose that the dimension of K over K^2 is finite and equal to d; prove that $\mathrm{Def}(M, q)$ has finite dimension over K, not exceeding d. Observe that $d = 1$ when K is finite.

(d) Here is an example proving that the image of the injective mapping defined in (b) may be equal to J^{sq}. Prove that the mapping $q' : J^{sq} \longrightarrow K$ defined by $q'(j^{sq}) = j$ is K-quadratic, and calculate the derived mapping $\mathrm{Def}(J^{sq}, q') \to J^{sq}$.

(2.ex.7) Let P and Q be submodules of a module M provided with a quadratic form or a symmetric bilinear form.

(a) Suppose that P and Q are orthogonally closed (in other words, $P = P^{\perp\perp}$ and Q=$Q^{\perp\perp}$); prove that $P \cap Q$ too is orthogonally closed.

In (b) and (c) just below there are counter-examples against the equalities $P+Q = (P+Q)^{\perp\perp}$ and $P^{\perp} + Q^{\perp} = (P \cap Q)^{\perp}$ even with P and Q orthogonally closed.

(b) Let F be a field, and K the quotient of $F[X, Y]$ by the ideal generated by the polynomial XY; thus $K = F[x, y]$ with $xy = 0$. Let P and Q be the ideals respectively generated by x and y. Consider the nondegenerate K-bilinear form on K defined by $(\lambda, \mu) \longmapsto \lambda\mu$, and prove that $Q = P^{\perp}$, $P = Q^{\perp}$ and $(P + Q)^{\perp} = 0$. Then examine the inclusions $P + Q \subset (P + Q)^{\perp\perp}$ and $P^{\perp} + Q^{\perp} \subset (P \cap Q)^{\perp}$.

(c) Let K be a field, and M an infinite dimensional vector space over K with a basis $(e_n)_{n\in\mathbb{Z}}$. Let P (resp. Q) be the subspace spanned by all e_n such that $n > 0$ (resp. $n < 0$). On M a weakly nondegenerate bilinear form can be defined in this way:

$$\left(\sum\nolimits_{n\in\mathbb{Z}} \lambda_n e_n \,, \sum\nolimits_{n\in\mathbb{Z}} \mu_n e_n\right) \longmapsto \sum_{n\neq 0} (\lambda_n\mu_n - \lambda_0\mu_n - \lambda_n\mu_0) \;;$$

remember that all λ_n and μ_n vanish except a finite number. Prove that $Q = P^{\perp}$, $P = Q^{\perp}$ and $(P + Q)^{\perp} = 0$. Then examine the same inclusions as in (b).

(2.ex.8) Let M and M' be two K-modules provided with alternate bilinear forms $\psi : M \times M \to K$ and $\psi' : M' \times M' \to K$. Prove that there exists a unique quadratic form $q'' : M \otimes M' \to K$ such that

$$\mathrm{b}_{q''} = \psi \otimes \psi' \quad \text{and} \quad q''(x \otimes x') = 0$$

for all $x \in M$ and $x' \in M'$.

(2.ex.9)* Let K be a local ring with maximal ideal $\mathfrak{m}$, and $\omega : \mathfrak{m} \to \mathfrak{m}$ the mapping defined by $\omega(\mu) = \mu - \mu^2$.

(a) Prove that ω is injective.

Prove that the bijectiveness of ω is a necessary and sufficient condition for this assertion to be true: a quadratic space (M, q) of rank 2 over K is hyperbolic whenever M contains an element x such that $q(x) \in \mathfrak{m}$ but $x \notin \mathfrak{m}M$.

(b) Suppose that ω is bijective, and let (M, q) be a quadratic space of even rank $2r$ over K. Prove that (M, q) is hyperbolic if it contains a direct summand P of rank r such that $q(P) \subset \mathfrak{m}$.

(c) Prove that the ring $\mathbb{Z}[[t]]$ of formal series with integer coefficients contains an element $f(t)$ such that $f(t) - f(t)^2 = t$ and $f(0) = 0$.

(d) Consider the ring morphism $\mathbb{Z}[t] \to K$ that maps t to some given element μ of $\mathfrak{m}$, and prove that μ belongs to the image of ω if this ring morphism extends to a ring morphism $\mathbb{Z}[[t]] \to K$. This extension exists whenever μ is nilpotent.

Comment. When $\mathfrak{m}$ is a finitely generated K-module, any equality $\mathfrak{m}^k = \mathfrak{m}^{k+1}$ implies $\mathfrak{m}^k = 0$ (see (1.12.1)); consequently every element of $\mathfrak{m}$ is nilpotent if the sequence $(\mathfrak{m}^k)$ is not strictly decreasing.

(e) Let F be a field, and K the localization of the ring $F[t]$ at the maximal ideal generated by t; prove that K contains no element $f(t)$ such that $f(t) - f(t)^2 = t$.

Properties of $q^{-1}(0)$

(2.ex.10) Let (M, q) be a quadratic module over K. Prove that M is generated by $q^{-1}(0)$ if the following sufficient condition is satisfied: there are two elements a and b in M such that $q(a) = 0$ and $\mathrm{b}_q(a, b)$ is invertible. When K is a field, this shows that either $q^{-1}(0)$ generates M, or $q^{-1}(0) = \mathrm{Ker}(q)$.

Hint. $Ka + Kb$ is a hyperbolic subspace, therefore an orthogonal summand.

(2.ex.11)* Here we assume that K is a field.

(a) Prove the following preliminary lemma. Let a_0, a_1, a_2, a_3, b_0, b_1, b_2, b_3 be elements of K with which we define these two functions on K^2 :

$$f(x_1, x_2) = a_0 + a_1 x_1 + a_2 x_2 + a_3 x_1 x_2 \ ,$$
$$g(x_1, x_2) = b_0 + b_1 x_1 + b_2 x_2 + b_3 x_1 x_2 \ ;$$

if $f^{-1}(0)$ and $g^{-1}(0)$ are equal, there is a nonzero $\lambda \in K$ such that $b_i = \lambda a_i$ for $i = 0, 1, 2, 3$.

Comment. When $K = \mathbb{Z}/2\mathbb{Z}$, we get a bijection between the 16 functions f and the 16 subsets of K^2.

(b) Let M be a vector space over K, and q and q' two quadratic forms on M such that $q^{-1}(0)$ and $q'^{-1}(0)$ are equal and generate M. Prove the existence of a nonzero $\lambda \in K$ such that $q' = \lambda q$.

Hint. Let $(e_j)_{j \in J}$ be a basis of M made of elements of $q^{-1}(0)$, $a_{i,j} = \mathrm{b}_q(e_i, e_j)$ and $b_{i,j} = \mathrm{b}_{q'}(e_i, e_j)$; assume $q \neq 0$ and $a_{u,v} \neq 0$ for some $(u, v) \in J^2$, and set $\lambda = b_{u,v}/a_{u,v}$. For which $(x_u, x_v) \in K^2$ does $e_j + x_u e_u + x_v e_v$ belong to $q^{-1}(0)$? Prove that $b_{u,j} = \lambda a_{u,j}$, ... and finally $b_{i,j} = \lambda a_{i,j}$ for all $(i, j) \in J^2$.

(2.ex.12) Let q be the quadratic form on $\mathbb{Q}^4$ that maps every (x_1, x_2, x_3, x_4) to $x_1^2 + x_2^2 - 3x_3^2 - 3x_4^2$. Prove that $q^{-1}(0)$ is reduced to $\{0\}$.

Hint. If q vanished on a nontrivial element of $\mathbb{Q}^4$, it would vanish on a nontrivial element of $\mathbb{Z}^4$; reduce modulo $9\mathbb{Z}$.

Half determinants

(2.ex.13) Let n be an odd integer ≥ 1, and $\mathbb{Z}[t_{i,j}]$ the ring of polynomials in $n(n+1)/2$ indeterminates $t_{i,j}$ such that $0 \leq i \leq j \leq n$, and with coefficients in $\mathbb{Z}$. We set $t_{i,j} = 0$ when $i > j$ and $x_{i,j} = t_{i,j} + t_{j,i}$ for all $(i,j) \in \{1,2,\ldots,n\}^2$ (whence $x_{i,i} = 2t_{i,i}$). Prove that there exists a polynomial $P(t_{i,j}) \in \mathbb{Z}[t_{i,j}]$ such that the determinant of the matrix $(x_{i,j})$ is equal to $2P(t_{i,j})$. This polynomial is denoted by $\frac{1}{2}\det(x_{i,j})$.

When $(a_1, a_2, \ldots, a_n)$ is a family of elements in a quadratic module (M, q), this allows us to define $\frac{1}{2}\det(\mathrm{b}_q(a_i, a_j))$, provided that n is odd.

Now let $(\lambda_{i,j})$ be an $(n \times n)$-matrix with entries in K, and $a'_j = \sum_i \lambda_{i,j} a_i$ for $j = 1, 2, \ldots, n$; prove that

$$\frac{1}{2}\det(\mathrm{b}_q(a'_i, a'_j)) = \frac{1}{2}\det(\mathrm{b}_q(a_i, a_j))\ (\det(\lambda_{i,j}))^2 .$$

(2.ex.14) If M is a free module of finite odd rank n, and q a quadratic form $M \to K$, we say that (M, q) is *almost nondegenerate* if for some basis $(a_1, a_2, \ldots, a_n)$ of M the half determinant $\frac{1}{2}\det(\mathrm{b}_q(a_i, a_j))$ (see (2.ex.13)) is invertible in K.

(a) Explain why this property does not depend on the choice of the basis $(a_1, \ldots$ $\ldots, a_n)$.

(b) Assume that K is a local ring with maximal ideal $\mathfrak{m}$, that M is free of finite odd rank n, and that q is almost nondegenerate. Prove that (M, q) contains a nondegenerate quadratic submodule (M', q') of even rank $n-1$, and that the submodule orthogonal to (M', q') is generated by some element $e \in M$ such that $q(e)$ is invertible in K.

Hint. If 2 is invertible in K, this is clear because q is nondegenerate; if $2 \in \mathfrak{m}$, prove that b_q induces an alternate bilinear form on $M/\mathfrak{m}M$, the kernel of which has dimension 1 over $K/\mathfrak{m}K$.

(c) If M is a finitely generated projective module of constant odd rank n, a quadratic form q on M is said to be almost nondegenerate if all its localizations are almost nondegenerate. Prove the existence of a faithfully flat extension $K \to L$ such that $L \otimes (M, q)$ is the orthogonal sum of a hyperbolic subspace and a free submodule of rank 1 on which $L \otimes q$ takes invertible values.

For other applications of half determinants, see (4.ex.4), (5.ex.15) and (7.ex.17).

In other books (for instance [Knus, 1991]) the words "regular" and "semiregular" are used instead of "nondegenerate" and "almost nondegenerate".

Witt rings

(2.ex.15) Here L, L',... are finitely generated projective K-modules of constant rank 1, and we are interested in nondegenerate quadratic mappings or nondegenerate symmetric bilinear mappings with target one of these modules.

(a) Let P be a finitely generated projective module, and

$$\mathbf{H}(P,L) = P \oplus \mathrm{Hom}(P,L);$$

prove that $\mathbf{H}(P,L)$ is provided with a natural nondegenerate quadratic mapping $\mathbf{H}(P,L) \to L$; with this quadratic mapping it becomes the *L-hyperbolic space* associated with P. State the essential properties of L-hyperbolic spaces, for instance the proposition analogous to (2.5.5). Let $\mathrm{WIB}(K,L)$ be the additive monoid of all isomorphy classes of nondegenerate quadratic mappings $M \to L$ in which the source M is a finitely generated projective K-module; prove that the isomorphy classes of L-hyperbolic spaces constitute an absorbent submonoid, whence a Witt group $\mathrm{WQ}(K,L)$. Prove that there is a natural $\mathbb{Z}$-bilinear mapping from $\mathrm{WQ}(K,L) \times \mathrm{WQ}(K,L')$ into $\mathrm{WQ}(K, L \otimes L')$; in particular $\mathrm{WQ}(K,L)$ is a module over $\mathrm{WQ}(K)$.

(b) In an analogous way define L-metabolic spaces and Witt rings $\mathrm{WB}(K,L)$; state their essential properties.

(2.ex.16) Let (M,q) be an anisotropic space over the field $\mathbb{Q}$ of rational numbers; let r be its dimension, and s its signature (that is the signature of $\mathbb{R} \otimes (M,q)$). From Meyer's theorem (see (2.8.7)) deduce the following statements:

(a) If $r \geq 5$, then (M,q) is definite positive or negative, whence $s = \pm r$.

(b) If $|s| \geq 3$, then $r = |s|$.

(c) If $|s| \leq 4$, then $|s| \leq r \leq 4$.

Comment. In [Serre] it is proved that the determinant of q is a square when $r = 4$ (see an example in (2.ex.12)), and consequently $|s|$ is 0 or 4 in this case.

(2.ex.17) Let λ and μ be two positive integers; we are interested in the $\mathbb{Q}$-quadratic space $\langle \lambda, -\mu \rangle$ (with quadratic form $(x,y) \longmapsto (\lambda x^2 - \mu y^2)/2$) and in the order of its Witt class w in the additive group $\mathrm{W}(\mathbb{Q})$.

(a) Suppose that both λ and μ are sums of two squares in $\mathbb{Z}$, and prove that w has order 0 or 2.

(b) It is well known that λ and μ are sums of four squares in $\mathbb{Z}$:

$$\lambda = a_1^2 + a_2^2 + a_3^2 + a_4^2 \quad \text{and} \quad \mu = b_1^2 + b_2^2 + b_3^2 + b_4^2;$$

prove that w has order 0, 2 or 4.

Hint. Calculate the quadratic form on $\mathbb{Q}^4$ that maps every (x_1, x_2, x_3, x_4) to

$$(a_1x_1 - a_2x_2 - a_3x_3 - a_4x_4)^2 + (a_2x_1 + a_1x_2 - a_4x_3 + a_3x_4)^2$$
$$+ (a_3x_1 + a_4x_2 + a_1x_3 - a_2x_4)^2 + (a_4x_1 - a_3x_2 + a_2x_3 + a_1x_4)^2.$$

(c) Prove that every element in the kernel of the ring morphism $\mathrm{W}(\mathbb{Q}) \to \mathrm{W}(\mathbb{R})$ (derived from the field extension $\mathbb{Q} \to \mathbb{R}$) has order 0, 2 or 4.

Quadratic spaces over $\mathbb{Z}/2\mathbb{Z}$

(2.ex.18) Let (M, q) be a quadratic space of dimension 4 over the field $\mathbb{Z}/2\mathbb{Z}$, x_0 a nonzero isotropic element of M, and x_1 an element such that $q(x_1) = 1$. Let G be the group of automorphisms of (M, q), A the set of anisotropic planes contained in M, H the set of hyperbolic planes, D the set of planes on which q is degenerate but not null, and T the set of totally isotropic planes. Let g, a, h, d, t be the cardinals of G, A, H, D, T. For $i = 0, 1$, let g_i be the cardinal of the subgroup of all $\gamma \in G$ such that $\gamma(x_i) = x_i$, let a_i be the cardinal of the subset of all $P \in A$ such that $x_i \in P$, and so forth.... Calculate all these cardinals. The answers are given just after (2.ex.26).

(2.ex.19)* Let (M, q) be a quadratic space of dimension $2m$ over the field $\mathbb{Z}/2\mathbb{Z}$. In (M, q), a plane P (that is a subspace of dimension 2) is said to be nondegenerate (resp. degenerate) if the restriction of q to P is nondegenerate (resp. degenerate). Let $X(2m)$ be the number of planes in M, and $Z(2m)$ the number of nondegenerate planes in (M, q); thus $X(2m) - Z(2m)$ is the number of degenerate planes. This exercise intends to prove that

$$\frac{X(2m) - Z(2m)}{Z(2m)} = 1 - \frac{1}{4^{m-1}}.$$

(a) Assume that M is the orthogonal sum of N and N', and prove that the plane P is nondegenerate if and only if its projection in one summand N or N' is a nondegenerate plane, whereas its projection in the other summand is either a degenerate plane, or a line, or 0.
Hint. P is nondegenerate if and only if $\sum_{x \in P} q(x) = 1$.

(b) Calculate $X(2m)$ and $Z(2m)$.
Hint. The calculation of $X(2m)$ is a classical problem: $X(2m) = (2^{2m} - 1)\,(2^{2m} - 2)\,/\,6$; then use (a) to prove this induction formula:

$$Z(2m + 2) = 4\,Z(2m) + 4^{2m}.$$

Bilinear spaces over $\mathbb{Z}$

(2.ex.20) Let (M, φ) and (M', φ') be two positive definite bilinear spaces over $\mathbb{Z}$, the former of even type and the latter of odd type; prove that there is no orthogonal basis in their orthogonal sum. Consequently $P_8 \perp G_{1,0}$ is not isomorphic to $G_{9,0}$.

(2.ex.21) Let (M, φ) be an indefinite bilinear space over $\mathbb{Z}$. Prove that the image of the quadratic form $x \longmapsto \varphi(x, x)$ is $\mathbb{Z}$ or $2\mathbb{Z}$ according to the type of φ, except when (M, φ) is isomorphic to $G_{1,1}$. What happens in this exceptional case?

(2.ex.22) Let q be a quadratic form (not necessarily nondegenerate) on a free $\mathbb{Z}$-module M of finite rank r. We are interested in the image $q(M)$ of q.

(a) Suppose that M contains elements x and y such that $q(x) = 0$ and $\mathrm{b}_q(x, y) \neq 0$, and denote by δ the greatest common divisor of $\mathrm{b}_q(x, y)$ and $q(y)$. Prove that $q(M)$ contains $\delta^{-1}\mathrm{b}_q(x, y)^2\,\mathbb{Z}$.

(b) Suppose that q is anisotropic and takes a positive (resp. negative) value on some point of M, and that $r \geq 4$. Deduce from Meyer's theorem (see (2.8.7)) that for every positive integer k there exist $x \in M$ and a nonzero $\lambda \in \mathbb{N}$ such that $q(x) = k\lambda^2$ (resp. $q(x) = -k\lambda^2$).

(2.ex.23) Consider $\mathbb{R}^8$ with its canonical basis $(e_1, e_2, \ldots, e_8)$ and its usual bilinear form $\tilde{\varphi}$ such that $\tilde{\varphi}(x, x') = \sum_{i=1}^{8} x_i x'_i$. Let f be the automorphism of $\mathbb{R}^8$ that maps every $(x_1, x_2, \ldots, x_8)$ to

$$(\ x_1 + x_2\ ,\ x_1 - x_2\ ,\ x_3 + x_4\ ,\ x_3 - x_4\ ,\ x_5 + x_6\ ,\ x_5 - x_6\ ,\ x_7 + x_8\ ,\ x_7 - x_8\ )\ ;$$

f is even an isomorphism from $(\mathbb{R}^8, \tilde{\varphi})$ onto $(\mathbb{R}^8,\ \tilde{\varphi}/2)$. As in (2.8.9) let P be the subgroup of all $x \in \mathbb{R}^8$ such that the eight numbers $2x_i$ are integers of the same parity, and $\sum_{i=1}^{8} x_i$ is even; we know that the restriction of $\tilde{\varphi}$ to P induces a nondegenerate $\mathbb{Z}$-bilinear form φ of even type. Besides, if z is any element of $\mathbb{Z}^8$ (considered as a subgroup of $\mathbb{R}^8$), the "odd support" of z is by definition the subset of all i such that z_i is odd; let Q be the subset of all elements of $\mathbb{Z}^8$ with "odd support" equal either to $\emptyset$, or to $\{1, 2, \ldots, 8\}$, or to one of these 14 subsets:

$$\{1,2,3,4\},\ \{1,2,5,6\},\ \{1,2,7,8\},\ \{1,3,5,7\},\ \{1,3,6,8\},\ \{1,4,5,8\},\ \{1,4,6,7\},$$
$$\{5,6,7,8\},\ \{3,4,7,8\},\ \{3,4,5,6\},\ \{2,4,6,8\},\ \{2,4,5,7\},\ \{2,3,6,7\},\ \{2,3,5,8\}.$$

(a) Prove that Q is a free group of rank 8, and that $\tilde{\varphi}(z, z)/4$ is an integer for every $z \in Q$. Consequently the restriction of $\tilde{\varphi}/2$ to Q induces a bilinear form $\psi : Q \times Q \to \mathbb{Z}$.

Hint. An element z of $\mathbb{Z}^8$ belongs to Q if and only if $z_1 + z_2$, $z_3 + z_4$, $z_5 + z_6$ and $z_7 + z_8$ have the same parity, and moreover $z_1 + z_3 + z_5 + z_7$ is even.

(b) Prove that f induces an isomorphism from (P, φ) onto (Q, ψ). Therefore ψ is nondegenerate.

Hint. Prove that $f(P) \subset Q$; then $f(P)$ is an orthogonal summand of (Q, ψ).

(2.ex.24)* This exercise intends to prove that all positive definite $\mathbb{Z}$-bilinear spaces of rank 2 are isomorphic to each other, in other words, they all contain an orthogonal basis. To do this, some special knowledge about the ring $\mathbb{Z}[i]$ of Gauss integers is needed: here $i = \sqrt{-1}$ and $\mathbb{Z}[i] = \mathbb{Z} \oplus \mathbb{Z}i$ is a subring of the field $\mathbb{C}$ of complex numbers. It is recalled that every nonzero element of $\mathbb{Z}[i]$ is a product of irreducible (or prime) elements in an essentially unique way, and that every positive prime integer p that is not congruent to -1 modulo 4, admits a factorization $p = (u + iv)(u - iv)$ in $\mathbb{Z}[i]$.

Here is a symmetric bilinear form on $\mathbb{Z}^2$:

$$\varphi((x,y),(x',y')) \;=\; axx' + b(xy'+yx') + cyy' \;;$$

assume $ac - b^2 = 1$ and $a > 0$ so that φ is positive and nondegenerate.

(a) Explain that there exists an orthogonal basis for φ if and only there exist u, v, u', v' in $\mathbb{Z}$ such that

$$\begin{pmatrix} a & b \\ b & c \end{pmatrix} = \begin{pmatrix} u & v \\ u' & v' \end{pmatrix} \begin{pmatrix} u & u' \\ v & v' \end{pmatrix}.$$

(b) Write a and c as products of positive prime numbers in $\mathbb{Z}$:

$$a \;=\; p_1^{\alpha_1} p_2^{\alpha_2} \cdots p_k^{\alpha_k} \;, \quad \text{and} \quad c \;=\; p_1^{\gamma_1} p_2^{\gamma_2} \cdots p_k^{\gamma_k} \;;$$

the exponents α_j and γ_j are ≥ 0 (yet $\alpha_j + \gamma_j > 0$), and the primes numbers p_j are pairwise distinct. Note that $b^2 + 1 \equiv 0$ modulo p_j for $j = 1, 2, \ldots, k$, and prove the existence of a decomposition $p_j = (u_j + iv_j)(u_j - iv_j)$ in $\mathbb{Z}[i]$. Then prove that

$$b + i \;=\; \lambda\,(u_1 + iv_1)^{\beta'_1}(u_1 - iv_1)^{\beta''_1} \cdots (u_k + iv_k)^{\beta'_k}(u_k - iv_k)^{\beta''_k} \;,$$

with some λ equal to ± 1 or $\pm i$, and with exponents β'_j and β''_j such that

$$\beta'_j + \beta''_j \;=\; \alpha_j + \gamma_j \quad \text{for } j = 1, 2, \ldots, k.$$

(c) Choose exponents α'_j and α''_j such that

$$\alpha'_j + \alpha''_j = \alpha_j \;, \qquad 0 \leq \alpha'_j \leq \beta'_j \quad \text{and} \quad 0 \leq \alpha''_j \leq \beta''_j \quad \text{for } j = 1, 2, \ldots, k;$$

then set $\gamma'_j = \beta'_j - \alpha'_j$ and $\gamma''_j = \beta''_j - \alpha''_j$, and prove that the following equalities defined a suitable family of integers u, v, u', v' :

$$\begin{aligned} u + iv \;&=\; (u_1 + iv_1)^{\alpha'_1}(u_1 - iv_1)^{\alpha''_1} \cdots (u_k + iv_k)^{\alpha'_k}(u_k - iv_k)^{\alpha''_k} \;; \\ u' - iv' \;&=\; (u_1 + iv_1)^{\gamma'_1}(u_1 - iv_1)^{\gamma''_1} \cdots (u_k + iv_k)^{\gamma'_k}(u_k - iv_k)^{\gamma''_k}. \end{aligned}$$

(2.ex.25)* Let (P_r, φ) be the bilinear space of rank r (a multiple of 4) presented in (2.8.9).

(a) First suppose that r is not divisible by 8, in other words, $r = 8s + 4$. Prove that $\varphi(x,x) \geq 2s+1$ whenever $\varphi(x,x)$ is odd, and that P_r contains exactly 2^{r-1} elements x such that $\varphi(x,x) = 2s+1$. Prove that P_4 contains orthogonal bases (in other words, it is isomorphic to $G_{4,0}$), whereas P_r never contains orthogonal bases when $s > 0$. Prove that $P_{8(s+t)+4}$ is never isomorphic to $P_{8s} \perp P_{8t+4}$ when $s > 0$.

(b) Now suppose r divisible by 8. Let U_r be the subset of all $x \in P_r$ such that $\varphi(x,x) = 2$, and let N_r be the subgroup generated by U_r. Prove that the cardinal of U_r is $2r(r-1)$ when $r \geq 16$, whereas the cardinal of U_8 is 240. Prove that P_r/N_r is a group of order 2 when $r \geq 16$, whereas $N_8 = P_8$.

Prove that $P_r \perp P_{r'}$ is never isomorphic to $P_{r+r'}$ when r and r' are positive integers divisible by 8.

Hint. When $(r,r') \neq (8,8)$, the cardinal of $U_{r+r'}$ is not the sum of the cardinals of U_r and $U_{r'}$; when $(r,r') = (8,8)$, you must consider N_8 and N_{16}.

(c) According to Theorem (2.8.11), $P_8 \perp G_{0,1}$ contains orthogonal bases because it is isomorphic to $G_{8,1}$; find such an orthogonal basis. A solution is given after (2.ex.26).

Quadratic forms in abelian categories

(2.ex.26) A category $\mathcal{C}$ is said to be abelian (or additive) if all sets of morphisms $\mathrm{Hom}_{\mathcal{C}}(M,N)$ are additive groups, and if some other properties are also fulfilled, for instance the existence of a direct sum $M \oplus_{\mathcal{C}} N$ for each couple of objects (see **1.3**), so that $\mathcal{C}$ becomes a monoidal category with neutral element (see **2.4**). A functor $\mathcal{F}$ between abelian categories is said to be additive if it determines a group morphism $\mathrm{Hom}(M,N) \rightarrow \mathrm{Hom}(\mathcal{F}(M), \mathcal{F}(N))$ for every pair of objects. Here we also assume that there is a contravariant additive functor from $\mathcal{C}$ to $\mathcal{C}$, denoted by $M \longmapsto M^*$ (and $f \longmapsto f^*$), the iteration of which is isomorphic to the identity functor of $\mathcal{C}$; in other words, there are canonical isomorphisms $M \rightarrow M^{**}$ intertwining the couples (f, f^{**}).

By definition a "bilinear form" on an object M is a morphism $b : M \rightarrow M^*$; for this b, the notation b^* means $M \rightarrow M^{**} \rightarrow M^*$ and not merely $M^{**} \rightarrow M^*$. We say that b is symmetric if $b = b^*$, and that it is alternate if there exists $c : M \rightarrow M^*$ such that $b = c - c^*$. The group of "bilinear forms" on M is denoted by $\mathrm{F}(M)$, the subgroups of symmetric and alternate forms are denoted by $\mathrm{Bil}(M)$ and $\mathrm{A}(M)$, and the quotient $\mathrm{Quad}(M) = \mathrm{F}(M)/\mathrm{A}(M)$ is called the group of "quadratic forms" on M; the "quadratic form" derived from any $b \in \mathrm{F}(M)$ is denoted by $[b]$. Finally any morphism $f : M \rightarrow N$ in $\mathcal{C}$ gives a group morphism $\mathrm{F}(f) : \mathrm{F}(N) \rightarrow \mathrm{F}(M)$ defined by $b \longmapsto f^* \circ b \circ f$.

(a) Verify that this $\mathrm{F}(f)$ induces group morphisms $\mathrm{Bil}(N) \rightarrow \mathrm{Bil}(M)$, $\mathrm{A}(N) \rightarrow \mathrm{A}(M)$ and $\mathrm{Quad}(N) \rightarrow \mathrm{Quad}(M)$. Then define a canonical mapping $q \longmapsto \mathrm{b}_q$ from $\mathrm{Quad}(M)$ into $\mathrm{Bil}(M)$, which gives the multiplication by 2 when it is composed (on either side) by the mapping $b \longmapsto [b]$ from $\mathrm{Bil}(M)$ into $\mathrm{Quad}(M)$.

(b) *Example.* What give the previous definitions when $\mathcal{C}$ is the category of finitely generated projective modules over K, and when M^* means $\mathrm{Hom}_K(M,K)$ as usual?

Hint. Remember (2.5.3). The isomorphism $M \longrightarrow M^{**}$ is here $x \longmapsto (h \longmapsto h(x))$.

(c) By definition a quadratic object of $\mathcal{C}$ is a couple (M,q) with $q \in \mathrm{Quad}(M)$, and it is said to be nondegenerate if $\mathrm{b}_q : M \rightarrow M^*$ is bijective. Define the orthogonal sum of two quadratic objects, so that the Propositions (2.4.1), (2.4.2) and (2.4.3) are still valid.

Comment. The usefulness of such developments appears for instance in [Quebbemann et al. 1976].

Here are the answers to (2.ex.18).

According as (M, q) is hyperbolic or not, you must find that

$$\begin{array}{rcccc} (g,a,h,d,t) & = & (72,2,18,9,6) & \text{or} & (120,10,10,15,0)\ , \\ (g_0,a_0,h_0,d_0,t_0) & = & (8,0,4,1,2) & \text{or} & (24,0,4,3,0)\ , \\ (g_1,a_1,h_1,d_1,t_1) & = & (12,1,3,3,0) & \text{or} & (12,3,1,3,0)\ . \end{array}$$

Here is a solution to (2.ex.25)(c).

First you get an orthogonal basis $(\varepsilon_1, \ldots, \varepsilon_4)$ of P_4 if you set

$$\varepsilon_i = e_i \ + \frac{1}{2}(-e_1 - e_2 - e_3 + e_4) \quad \text{for } i = 1,2,3,$$
$$\varepsilon_4 = \frac{1}{2}(e_1 + e_2 + e_3 + e_4).$$

Then you can identify $P_8 \perp G_{0,1}$ with a subgroup of $\mathbb{R}^9$ provided with the bilinear form $\tilde{\psi}$ such that $\tilde{\psi}(y,y') = \sum_{i=1}^{8} y_i y'_i - y_9 y'_9$. You get an orthogonal basis $(\eta_1, \eta_2, \ldots, \eta_8, \eta_9)$ if you set:

$$\eta_i = \varepsilon_i \ + \frac{1}{2}\left(\sum_{j=5}^{8} e_j\right) + e_9 \qquad \text{for } i = 1,2,3,4,$$
$$\eta_i = e_4 + e_i + e_9 \qquad \text{for } i = 5,6,7,8,$$
$$\eta_9 = 2e_4 \ + \sum_{j=5}^{8} e_j \ + 3e_9.$$

Chapter 3

Clifford Algebras

This chapter begins with elementary properties of Clifford algebras, in particular their parity grading (presented in **3.2**). Exterior algebras are treated as Clifford algebras of null quadratic forms. More sophisticated properties are expounded in the next chapter.

The main part of this chapter is devoted to Clifford algebras of quadratic spaces (see Definitions (2.5.1)). In **3.7** it is explained that they belong to a very special kind of algebras, namely the graded Azumaya algebras. The general study of graded Azumaya algebras is undertaken in **3.5**, after a section **3.4** devoted to graded quadratic extensions. Some complements are added in the last section **3.8**, but a great amount of more difficult knowledge (including all properties of separability) is postponed until Chapter **6**.

3.1 Definitions and elementary properties

Let M be a module over the ring K, and $q : M \to K$ a quadratic form; with (M, q) we associate the category $\mathcal{A}_K(M, q)$: its objects are the linear mappings f from M into any (associative) algebra A such that $f(a)^2 = q(a)1_A$ for all $a \in M$; a morphism from $f : M \to A$ to $g : M \to B$ is an algebra morphism $u : A \to B$ such that $g = u \circ f$. If this category $\mathcal{A}_K(M, q)$ contains an initial universal object ρ (which is then unique up to isomorphism, see (1.2.1)), its target is called the *Clifford algebra associated with* (M, q) and denoted by $C\ell_K(M, q)$. This notation, and the abbreviations like $C\ell(M, q)$ or $C\ell(q)$ or $C\ell(M)$, are quite classical, but not the notations ρ for the canonical mapping $M \to C\ell(M, q)$, and 1_q for the unit element of $C\ell(M, q)$; indeed when the canonical algebra morphism $K \to C\ell(M, q)$ and the canonical mapping $\rho : M \to C\ell(M, q)$ are both injective, usually K and M are systematically identified with their images in $C\ell(M, q)$, and the notations 1_q and $\rho(a)$ are replaced with 1 and a. But as long as the injectiveness of these canonical mappings is not sure, notations like 1_q and $\rho(a)$ are necessary.

The universality of ρ means that *for every linear mapping* $f : M \to A$ *such that* $f(a)^2 = q(a)1_A$ *for all* $a \in M$, *there exists a unique algebra morphism* $f' : \mathrm{C}\ell_K(M, q) \to A$ *such that* $f = f' \circ \rho$. Since ρ is an object of $\mathcal{A}_K(M, q)$, the equality $\rho(a)^2 = q(a)1_q$ holds for all $a \in M$; it implies, for all a and $b \in M$,

$$\rho(a)\rho(b) + \rho(b)\rho(a) \;=\; \mathrm{b}_q(a, b)\; 1_q.$$

Consequently $\rho(a)$ and $\rho(b)$ anticommute if a and b are orthogonal. It is clear that $\mathrm{C}\ell(M, q) = K$ when M is reduced to 0.

The following lemma states that $\mathcal{A}_K(M, q)$ always contains an initial universal object, and constructs it by means of the tensor algebra $\mathrm{T}_K(M)$ defined in **1.4**; as usual the components $\mathrm{T}^0(M)$ and $\mathrm{T}^1(M)$ of this graded algebra $\mathrm{T}(M)$ are identified with K and M.

(3.1.1) **Lemma.** *Let* $\mathrm{J}(M, q)$ *be the two-sided ideal of* $\mathrm{T}(M)$ *generated by all elements* $a \otimes a - q(a)$ *where* a *runs through* M, *let* $\mathrm{C}\ell(M, q)$ *be the quotient of* $\mathrm{T}(M)$ *by* $\mathrm{J}(M, q)$; *the natural linear mapping*

$$\rho :\; M \longrightarrow \mathrm{T}(M) \longrightarrow \mathrm{C}\ell(M, q) = \mathrm{T}(M)/\mathrm{J}(M, q)$$

is an initial universal object in $\mathcal{A}_K(M, q)$.

Proof. Let $f : M \to A$ be an object of $\mathcal{A}_K(M, q)$. First it extends to an algebra morphism $f'' : \mathrm{T}(M) \to A$ (see (1.4.1)). The equality $f(a)^2 = q(a)1_A$ implies that f'' vanishes on all elements $a \otimes a - q(a)$, and consequently on $\mathrm{J}(M, q)$. Because of (1.3.1), f'' induces an algebra morphism $f' : \mathrm{C}\ell(M, q) \to A$. Obviously $f = f' \circ \rho$, and f' is the only algebra morphism satisfying this equality because $\mathrm{C}\ell(M, q)$ is generated as an algebra by all elements $\rho(a)$ (and 1_q which is always silently joined with the generators). □

As usual, with the universal property of Clifford algebras is associated a functor $\mathrm{C}\ell$ (more precisely $\mathrm{C}\ell_K$), which is a covariant functor from the category of K-quadratic modules (denoted by $\mathcal{C}_K(K)$ in **2.4**) to the category $\mathcal{A}lg(K)$. If $u : (M, q) \to (M', q')$ is a morphism of quadratic modules, in other words, if $q'(u(x)) = q(x)$ for all $x \in M$, then the algebra morphism $\mathrm{C}\ell(u)$ is defined in this way: the equality $\rho'(u(x))^2 = q'(x)1_{q'}$ and the universal property of $\rho' : M' \to \mathrm{C}\ell(M', q')$ imply the existence of a unique morphism $\mathrm{C}\ell(u) : \mathrm{C}\ell(M, q) \to \mathrm{C}\ell(M', q')$ such that $\rho' \circ u = \mathrm{C}\ell(u) \circ \rho$.

(3.1.2) **Example.** When M is a free module generated by one element e, then $\mathrm{T}(M)$ is isomorphic to the polynomial algebra $K[e]$, and by this isomorphism $\mathrm{J}(M, q)$ becomes the ideal generated by the polynomial $e^2 - q(e)$; thus $\mathrm{C}\ell(M, q)$ is a free module with basis $(1, e)$. If 2 and $q(e)$ are invertible in K, it is a free quadratic extension according to the definition given in **2.6**.

(3.1.3) **Example.** This is a pathological example in which the canonical mappings $K \to \mathrm{C}\ell(M, q)$ and $\rho : M \to \mathrm{C}\ell(M, q)$ are not injective. Although K is here the ring $\mathbb{Z}/4\mathbb{Z}$, we begin with a quadratic module over $\mathbb{Z}$. Let $\hat{q}$ be the quadratic form on $\mathbb{Z} \oplus \mathbb{Z}$ defined by

$$\hat{q}(x,y) = x^2 + y^2 \ , \quad \text{whence} \quad \mathrm{b}_{\hat{q}}((x,y),(x',y')) \ = \ 2xx' + 2yy' \ .$$

Let N be the submodule $2\mathbb{Z}\oplus 4\mathbb{Z}$ in $\mathbb{Z}^2$. For all $(x,y) \in N$, and all $(x',y') \in \mathbb{Z}\oplus\mathbb{Z}$, $\hat{q}(x,y)$ and $\mathrm{b}_{\hat{q}}((x,y),(x',y'))$ belong to $4\mathbb{Z}$; this proves that $\hat{q}$ induces a quadratic form q on the quotient group

$$M = \mathbb{Z}^2/N \cong (\mathbb{Z}/2\mathbb{Z}) \oplus (\mathbb{Z}/4\mathbb{Z})$$

considered as a module over $K = \mathbb{Z}/4\mathbb{Z}$. Let a and b be the images of $(1,0)$ and $(0,1)$ in the quotient M; thus $2a = 4b = 0$. Consequently

$$\begin{aligned} &2 \times 1_q = 2\rho(a)^2 = \rho(2a)\rho(a) = 0 \quad \text{whereas} \quad 2 \neq 0 \text{ in } K \ ; \\ &\rho(2b) = (2 \times 1_q)\rho(b) = 0 \qquad\qquad\quad \text{whereas} \quad 2b \neq 0 \text{ in } M \ . \end{aligned}$$

Thus neither of the canonical mappings is injective. The equality $2 \times 1_q = 0$ shows that $\mathrm{C}\ell(M,q)$ is also an algebra over the field $F = K/2K$, and the orthogonality of a and b implies that $\rho(a)$ and $\rho(b)$ commute. Let A be the quotient of $F[X,Y]$ by the ideal generated by the polynomials $X^2 - 1_F$ and $Y^2 - 1_F$; if x and y are the images of X and Y in A, then $(1_F, x, y, xy)$ is a basis of A over F; but A is also a K-algebra. There is a K-linear mapping $f : M \to A$ that maps a and b to x and y, and since f is an object of $\mathcal{A}(M,q)$, it determines a K-algebra morphism $f' : \mathrm{C}\ell(M,q) \to A$. Conversely there is an F-algebra morphism $f'' : A \to \mathrm{C}\ell(M,q)$ mapping x and y to $\rho(a)$ and $\rho(b)$; obviously f' and f'' are reciprocal isomorphisms.

Grade automorphism, reversion and conjugation

Let (M,q) be a quadratic module. Obviously $-\mathrm{id}_M$ is an involutive automorphism of (M,q), and the functor $\mathrm{C}\ell$ associates with it an involutive automorphism σ of $\mathrm{C}\ell(M,q)$; it is called the *grade automorphism* for reasons explained in **3.2**.

With every algebra A is associated an *opposite algebra* A^o which is isomorphic to A as a K-module by means of a canonical isomorphism $x \longmapsto x^o$, but in which the product of two elements x^o and y^o (canonical images in A^o of two elements of A) is by definition $x^o y^o = (yx)^o$. Obviously every algebra morphism $u : A \to B$ gives an algebra morphism $u^o : A^o \to B^o$ defined by $u^o(x^o) = u(x)^o$.

Now the mapping $a \longmapsto \rho(a)^o$ from M into $\mathrm{C}\ell(M,q)^o$ is also an object of $\mathcal{A}_K(M,q)$; consequently it induces an algebra morphism $\mathrm{C}\ell(M,q) \to \mathrm{C}\ell(M,q)^o$, and there is a linear mapping τ from $\mathrm{C}\ell(M,q)$ into itself such that this algebra morphism is $x \longmapsto \tau(x)^o$. The properties of τ stated in the next proposition explain why it is called the *reversion*.

(3.1.4) **Proposition.** *There is a unique linear mapping τ from $Cl(M,q)$ into itself such that*

$$\tau(1_q) = 1_q\ , \quad \tau(\rho(a)) = \rho(a) \quad \textit{for all } a \in M,$$
$$\textit{and } \tau(xy) = \tau(y)\tau(x) \quad \textit{for all } x \textit{ and } y \in C\ell(M,q).$$

It is an involution of $C\ell(M,q)$ (*see Definition* (1.13.7)).

Proof. The above explanations already prove the properties of τ mentioned in (3.1.4), except the fact that it is involutive; but since τ^2 is an automorphism of $C\ell(M,q)$ that induces the identity on $\rho(M)$, it is the identity everywhere on $C\ell(M,q)$. □

It is clear that $\sigma\tau$ and $\tau\sigma$ are anti-automorphisms of $C\ell(M,q)$ that both induce $-\mathrm{id}$ on $\rho(M)$; therefore they are equal, and $\sigma\tau$ is also an involution of $C\ell(M,q)$; it is often called the *conjugation*.

Let us calculate σ, τ and $\sigma\tau$ when q is the null quadratic form; in this case the Clifford algebra is the *exterior algebra* $\bigwedge(M)$ for which it is customary to use a proper symbol $\wedge$ of multiplication.

According to (3.1.1), $\bigwedge(M)$ is the quotient of $\mathrm{T}(M)$ by the ideal $\mathrm{J}(M)$ generated by all $a\otimes a$ with $a \in M$; it is a graded ideal, in other words, it is the direct sum of the intersections $\mathrm{J}^n(M) = \mathrm{T}^n(M) \cap \mathrm{J}(M)$, and obviously $\mathrm{J}^0(M) = \mathrm{J}^1(M) = 0$; consequently the algebra $\bigwedge(M)$ is graded over the semi-group $\mathbb{N}$, and $\bigwedge^0(M)$ and $\bigwedge^1(M)$ can be identified with K and M. A serious study of exterior algebras shall be undertaken in **4.3**; here we just state that, for every $x \in \bigwedge^n(M)$,

$$(3.1.5) \qquad \sigma(x) = (-1)^n x\ , \quad \tau(x) = (-1)^{n(n-1)/2} x\ , \quad \sigma\tau(x) = (-1)^{n(n+1)/2} x.$$

Indeed it suffices to calculate $\sigma(x)$ and $\tau(x)$ when x is an exterior product $a_1 \wedge a_2 \wedge \cdots \wedge a_n$ of elements of A; since σ is an algebra automorphism, the first conclusion follows from $\sigma(a_i) = -a_i$ for $i = 1, 2, \dots, n$. Then $\tau(x) = a_n \wedge a_{n-1} \wedge \cdots \wedge a_1$, and since the elements of M pairwise anticommute in $\bigwedge(M)$, we must calculate the signature of the permutation $(n, n-1, \dots, 2, 1)$; when n is even (resp. odd), it depends on the parity of $n/2$ (resp. $(n-1)/2$); this accounts for $(-1)^{n(n-1)/2}$. □

The natural filtration of a Clifford algebra

If A is an algebra, a family $(A^{\leq k})_{k\in\mathbb{Z}}$ of submodules of A is called an *increasing filtration* of A if the following conditions are fulfilled: first every $A^{\leq k}$ is contained in $A^{\leq k+1}$, secondly xy belongs to $A^{\leq j+k}$ whenever x belongs to $A^{\leq j}$ and y to $A^{\leq k}$; moreover $1_A \in A^{\leq 0}$. This filtration is said to be *regular* if the intersection of all $A^{\leq k}$ is 0, and their union is A.

In Chapter **5** we shall consider various filtrations of Clifford algebras, but here we only need the natural filtration of $C\ell(M,q)$, which is inherited from the natural increasing filtration of $\mathrm{T}(M)$, but which can also be defined without the help of $\mathrm{T}(M)$. For every negative integer k we set $C\ell^{\leq k}(M,q) = 0$, and for every integer $k \geq 0$ the subset $C\ell^{\leq k}(M,q)$ of elements of degree $\leq k$ is the submodule of $C\ell(M,q)$ generated by all products $\rho(a_1)\rho(a_2)\cdots\rho(a_j)$ such that $0 \leq j \leq k$. When $j = 0$, this product means 1_q; thus the filtration begins with $C\ell^{\leq 0}(M,q) = K1_q$ and $C\ell^{\leq 1}(M,q) = K1_q \oplus \rho(M)$. It is clear that the submodules $C\ell^{\leq k}(M,q)$ constitute a regular increasing filtration of the algebra $C\ell(M,q)$. More details about filtrations will be expounded in **5.2**; here we especially need the following lemma.

(3.1.6) **Lemma.** *Let k be a positive integer, s a permutation of the set $\{1,2,\dots,k\}$, and* $\text{sgn}(s)$ *its signature; for all $a_1,\dots,a_k$ in M,*

$$\rho(a_1)\rho(a_2)\cdots\rho(a_k) \;-\; \text{sgn}(s)\;\rho(a_{s(1)})\rho(a_{s(2)})\cdots\rho(a_{s(k)}) \;\in\; C\ell^{\leq k-2}(M,q).$$

Proof. Since every permutation of factors in a product can be achieved by means of successive transpositions of two consecutive factors, it suffices to verify the announced result when s is the transposition of two consecutive indices j and $j+1$. In this case the conclusion follows from the equality

$$\rho(a_j)\rho(a_{j+1}) + \rho(a_{j+1})\rho(a_j) \;=\; \text{b}_q(a_j,\; a_{j+1})\; 1_q\;. \qquad \square$$

(3.1.7) **Corollary.** *Let $(a_j)_{j\in J}$ be a family of generators of M indexed by a totally ordered set J; the products*

$$\rho(a_{j_1})\rho(a_{j_2})\cdots\rho(a_{j_n}) \quad \textit{with} \quad n \geq 0 \quad \textit{and} \quad j_1 < j_2 < \cdots < j_n$$

constitute a family of generators of the module $C\ell(M,q)$.

When $n = 0$, it must be understood that the above product is 1_q.

Proof. It is clear that we get a family of generators of the module $C\ell(M,q)$ if we do not require the sequence $(j_1, j_2, \dots, j_n)$ to be strictly increasing; consequently it suffices to prove by induction on n that such a product with an arbitrary sequence of indices is a linear combination of similar products with strictly increasing sequences of indices. From (3.1.6) we deduce that such a product is congruent modulo $C\ell^{\leq n-2}(M,q)$ to a similar product with an increasing sequence of indices; when this sequence is not strictly increasing, there is an integer $i \in \{1,2,\dots,n-1\}$ such that $j_i = j_{i+1}$; since $\rho(a)^2 = q(a)1_q$ for all $a \in M$, the product under consideration belongs to $C\ell^{\leq n-2}(M,q)$. By the induction hypothesis every element of $C\ell^{\leq n-2}(M,q)$ is a linear combination of products of factors $\rho(a_j)$ with a strictly increasing sequence of indices. $\square$

The products described in (3.1.7) are in bijection with the finite subsets of J, and when J has a finite cardinal r, their number is 2^r. Let us notice and once for all that $C\ell(M,q)$ *is a finitely generated module when M is finitely generated.* When the module M is free with basis $(a_j)_{j\in J}$, the products described in (3.1.7) constitute a basis of $C\ell(M,q)$; this is proved in this chapter in some easy cases, and in the next chapter in full generality.

Let B be an algebra graded over $\mathbb{Z}$, in other words, $B = \bigoplus_{k\in\mathbb{Z}} B^k$ and $B^j B^k \subset B^{j+k}$ for all $(j,k) \in \mathbb{Z}^2$. Automatically B is provided with a regular increasing filtration if we set $B^{\leq k} = \bigoplus_{i\leq k} B^i$. Conversely if the filtration of the algebra A is regular, it is said that *its filtration comes from a grading* if every submodule $A^{\leq k-1}$ admits a supplementary submodule A^k in $A^{\leq k}$ in such a way that the submodules A^k constitute a grading of the algebra A.

Let us still assume that the filtration of A is regular; in all cases we can derive a graded algebra $\mathrm{Gr}(A)$ from it, in such a way that, in case the filtration would come from a grading, the algebra A with this grading should be canonically isomorphic to $\mathrm{Gr}(A)$. Here is the construction of $\mathrm{Gr}(A)$: it is the direct sum of the quotients $\mathrm{Gr}^k(A) = A^{\leq k}/A^{\leq k-1}$ provided with the multiplication induced by that of A. Indeed let us consider the two mappings

$$\begin{aligned} A^{\leq i} \otimes A^{\leq j} &\longrightarrow \mathrm{Gr}^i(A) \otimes \mathrm{Gr}^j(A) \ , \\ A^{\leq i} \otimes A^{\leq j} &\longrightarrow A^{\leq i+j} \longrightarrow \mathrm{Gr}^{i+j}(A) \ ; \end{aligned}$$

the kernel of the former mapping is generated by the images of $A^{\leq i-1} \otimes A^{\leq j}$ and $A^{\leq i} \otimes A^{\leq j-1}$ (see (1.6.3)); by the latter mapping they are mapped first into $A^{\leq i+j-1}$ and then to 0 in $\mathrm{Gr}^{i+j}(A)$; thus we get a mapping $\mathrm{Gr}^i(A) \otimes \mathrm{Gr}^j(A) \to \mathrm{Gr}^{i+j}(A)$. It is easy to prove that all these mappings together make $\mathrm{Gr}(A)$ become an associative algebra with unit. And when the filtration of A comes from a grading $A = \bigoplus_k A^k$, the natural bijections $A^k \to \mathrm{Gr}^k(A)$ together afford an algebra isomorphism $A \to \mathrm{Gr}(A)$.

If we apply this construction to the natural filtration of a Clifford algebra, we get the following lemma.

(3.1.8) **Lemma.** *The canonical mapping* $M \to C\ell^{\leq 1}(M,q) \to \mathrm{Gr}^1(C\ell(M,q))$ *induces a surjective algebra morphism from* $\bigwedge(M)$ *onto* $\mathrm{Gr}(C\ell(M,q))$.

Proof. Let $\tilde{a}$ be the image of a (any element of M) in $\mathrm{Gr}^1(C\ell(M,q))$. Since $\rho(a)^2 = q(a)1_A$ belongs to $C\ell^{\leq 0}(M,q)$, and since $\tilde{a}^2$ is the image of $\rho(a)^2$ in $\mathrm{Gr}^2(C\ell(M,q))$, we realize that $\tilde{a}^2 = 0$. Consequently there is an algebra morphism $\bigwedge(M) \to \mathrm{Gr}(C\ell(M,q))$ mapping every a to $\tilde{a}$. It is surjective because the target is generated (as an algebra with unit element) by all elements $\tilde{a}$. □

Change of basic ring

Let (M,q) be a quadratic module over K, and $f : K \to K'$ a ring morphism; from Theorem (2.2.3) we derive a quadratic module $K' \otimes (M,q)$ over K', and subsequently a Clifford algebra $C\ell_{K'}(K' \otimes (M,q))$ over K'; let ρ' be the canonical mapping from $K' \otimes M$ into this algebra.

(3.1.9) **Proposition.** *There is a canonical isomorphism from* $C\ell_{K'}(K' \otimes (M,q))$ *onto* $K' \otimes C\ell(M,q)$ *which maps every* $\rho'(\lambda \otimes a)$ *(with* $\lambda \in K'$ *and* $a \in M$*) to* $\lambda \otimes \rho(a)$.

Proof. Let us set $(M',q') = K' \otimes (M,q)$. For every element $\sum_j \lambda_j \otimes a_j$ of $M' = K' \otimes M$,

$$q'\left(\sum_j \lambda_j \otimes a_j\right) = \sum_j \lambda_j^2 \otimes q(a_j) + \sum_{i<j} \lambda_i\lambda_j \otimes \mathfrak{b}_q(a_i, a_j) ,$$

and thus the square of $\sum_j \lambda_j \otimes \rho(a_j)$ in $K' \otimes C\ell(M,q)$ is equal to $q'(\sum_j \lambda_j \otimes a_j) \otimes 1_q$. Because of the universal property of $C\ell_{K'}(M',q')$ there is a K'-algebra morphism from $C\ell_{K'}(M',q')$ into $K' \otimes C\ell(M,q)$ which maps every $\rho'(\lambda \otimes a)$ (with $\lambda \in K'$ and $a \in M$) to $\lambda \otimes \rho(a)$. Conversely, because of the universal property of $C\ell(M,q)$, the K-linear mapping $a \longmapsto \rho'(1_{K'} \otimes a)$ determines a K-algebra morphism $C\ell(M,q) \to C\ell_{K'}(M',q')$, whence a K'-algebra morphism from $K' \otimes C\ell(M,q)$ into $C\ell_{K'}(M',q')$ which maps every $\lambda \otimes \rho(a)$ to $\rho'(\lambda \otimes a)$. Obviously we have two reciprocal isomorphisms of K'-algebras. □

(3.1.10) **Remark.** As a particular case of (3.1.9), we consider the ring extension $K \to S^{-1}K$ determined by a multiplicative subset S of K; it affords an isomorphism of $(S^{-1}K)$-algebras

$$C\ell_{S^{-1}K}(S^{-1}M, S^{-1}q) \cong S^{-1}C\ell(M,q) ;$$

the quadratic form $S^{-1}q$ is defined by $a/s \longmapsto q(a)/s^2$. When S is the subset complementary to a prime ideal $\mathfrak{p}$, we get isomorphisms $C\ell_{K_\mathfrak{p}}(M_\mathfrak{p}, q_\mathfrak{p}) \cong C\ell(M,q)_\mathfrak{p}$.

(3.1.11) **Remark.** If e is an idempotent of K, then $M = eM \oplus (1-e)M$, and we can treat eM as a module over Ke, on which q induces a quadratic form $q_e : eM \to Ke$. Let ρ_e be the canonical mapping $eM \to C\ell_{Ke}(eM, q_e)$. The mapping $\lambda \longmapsto e\lambda$ is a ring extension $K \to Ke$, and there is an isomorphism $Ke \otimes (M,q) \to (eM, q_e)$ that maps every $e \otimes a$ to ea, whence an isomorphism $Ke \otimes C\ell(M,q) \to C\ell_{Ke}(eM,q_e)$. On the other side $C\ell(M,q)$ is the direct sum of the ideals $eC\ell(M,q)$ and $(1-e)C\ell(M,q)$, and in $\mathcal{A}lg(Ke)$ there is an isomorphism $Ke \otimes C\ell(M,q) \to eC\ell(M,q)$ mapping every $e \otimes x$ to ex. All this gives an isomorphism $C\ell_{Ke}(eM,q_e) \to eC\ell(M,q)$ mapping every $\rho_e(ea)$ to $e\rho(a)$; it allows us to identify the ideal $eC\ell(M,q)$ with $C\ell_{Ke}(eM,q_e)$. Such identifications are silently performed whenever we apply (1.12.8) to reduce the case of a projective and finitely generated quadratic module to the case of a module of constant rank.

(3.1.12) **Remark.** The isomorphism mentioned in (3.1.9) results in an isomorphism between two functors defined on $\mathcal{C}_K(K)$; indeed for each morphism $u : (M_1,q_1) \to (M_2,q_2)$ in $\mathcal{C}_K(K)$ there is a commutative diagram

$$\begin{array}{ccc} C\ell_{K'}(K' \otimes (M_1,q_1)) & \longleftrightarrow & K' \otimes C\ell(M_1,q_1) \\ \downarrow{\scriptstyle C\ell(K'\otimes u)} & & \downarrow{\scriptstyle K'\otimes C\ell(u)} \\ C\ell_{K'}(K' \otimes (M_2,q_2)) & \longleftrightarrow & K' \otimes C\ell(M_2,q_2) \end{array}$$

in which the left-hand column corresponds to the functor $K' \otimes \cdots$ followed by the functor $C\ell_{K'}$ whereas the right-hand column corresponds to the functor $C\ell_K$

followed by the functor $K' \otimes \cdots$. The existence of this isomorphism of functors allows us to say that the Clifford functors $C\ell$ commute with the ring extensions up to isomorphy. Besides, in the case of two consecutive ring extensions $K \to K' \to K''$ there is a transitivity property based on canonical isomorphisms of the following kind:

$$K'' \otimes_{K'} (K' \otimes_K M) \cong (K'' \otimes_{K'} K') \otimes_K M \cong K'' \otimes_K M.$$

3.2 The parity grading of Clifford algebras

In (3.1.1) the Clifford algebra $C\ell(M,q)$ has been constructed as a quotient of the tensor algebra $\mathrm{T}(M)$; from $\mathrm{T}(M)$ it inherits an increasing filtration, but not a grading over $\mathbb{N}$ unless q is the null quadratic form. Nevertheless $C\ell(M,q)$ inherits from $\mathrm{T}(M)$ a parity grading, that is a grading over the group $\mathbb{Z}/2\mathbb{Z}$. Indeed let us set

$$\mathrm{T}_0(M) = \bigoplus_m \mathrm{T}^{2m}(M) \quad \text{and} \quad \mathrm{T}_1(M) = \bigoplus_m \mathrm{T}^{2m+1}(M) \ ;$$

the lower indices 0 or 1 are the elements of $\mathbb{Z}/2\mathbb{Z}$ and may be called "even" and "odd". As a submodule, the ideal $\mathrm{J}(M,q)$ defined in (3.1.1) is generated by all products $y \otimes (a \otimes a - q(a)) \otimes z$ with a in M, y in some $\mathrm{T}^j(M)$ and z in some $\mathrm{T}^k(M)$; therefore $\mathrm{J}(M,q)$ is graded in the following sense: it is the direct sum of its intersections $\mathrm{J}_0(M,q)$ and $\mathrm{J}_1(M,q)$ with $\mathrm{T}_0(M)$ and $\mathrm{T}_1(M)$. Consequently $C\ell(M,q)$ is the direct sum of two submodules $C\ell_0(M,q)$ and $C\ell_1(M,q)$ respectively isomorphic to $\mathrm{T}_i(M)/\mathrm{J}_i(M,q)$ with $i = 0,1$. The elements of $C\ell_0(M,q)$ or $C\ell_1(M,q)$ are said to be respectively *even* or *odd*.

From now on, when it is not otherwise specified, every grading will be a parity grading over the group $\mathbb{Z}/2\mathbb{Z}$, and the parities are indicated by lower indices 0 and 1. A parity grading on a module P is merely a decomposition of P into a direct sum of two submodules P_0 and P_1, the elements of which are respectively called even or odd; an element z is said to be *homogeneous* if it is even or odd, and its degree ∂z is 0 or 1 according to its parity; when the notation ∂z appears, it is often silently assumed that z is homogeneous.

Every graded algebra $A = A_0 \oplus A_1$ admits a *grade automorphism* σ such that $\sigma(x) = (-1)^{\partial x} x$ for every homogeneous x. For a Clifford algebra, σ has already been defined in **3.1**. If the mapping $x \longmapsto 2x$ is injective from A into itself, the grading of A is determined by the grade automorphism. When this mapping $x \longmapsto 2x$ is bijective, every involutive automorphism of A determines a parity grading for which it is the grade automorphism. For instance the algebra $K^2 = K \times K$ is provided with an involutive *swap automorphism* $(\lambda, \mu) \longmapsto (\mu, \lambda)$ for which the diagonals of K^2 (the subset of all (λ, λ) and the subset of all $(\lambda, -\lambda)$) are eigenspaces; yet K^2 is the direct sum of its diagonals if and only if the mapping $\lambda \longmapsto 2\lambda$ is bijective from K onto itself; only in this case does the swap automorphism determine a parity grading.

The graded algebras constitute a new category $\mathcal{GA}lg(K)$; a *graded algebra morphism* $f : A \longrightarrow B$ is an algebra morphism such that $f(A_i) \subset B_i$ for $i = 0, 1$. With each object (M, q) of the category $\mathcal{C}_K(K)$ we have associated an object $\mathrm{C}\ell(M, q)$ of $\mathcal{GA}lg(K)$, and with each morphism $u : (M, q) \longrightarrow (M', q')$ we have associated an algebra morphism $\mathrm{C}\ell(u)$ which is graded because it maps the odd generators $\rho(a)$ of $\mathrm{C}\ell(M, q)$ to odd elements of $\mathrm{C}\ell(M', q')$. Thus the *functor* $\mathrm{C}\ell$ *is also a functor from the category* $\mathcal{C}_K(K)$ *of* K*-quadratic modules to the category* $\mathcal{GA}lg(K)$ *of graded* K*-algebras.*

We get a functor $\mathrm{C}\ell_0$ from $\mathcal{C}_K(K)$ to $\mathcal{A}lg(K)$ if we associate with every quadratic module (M, q) the even Clifford subalgebra $\mathrm{C}\ell_0(M, q)$; like $\mathrm{C}\ell$, this functor $\mathrm{C}\ell_0$ commutes with the extensions of the basic ring (see (3.1.12). In (3.ex.7) there is another interesting property of $\mathrm{C}\ell_0$.

Parity grading and natural filtration appear together in this technical lemma.

(3.2.1) **Lemma.** *If* M *is a finitely generated module of rank* $\leq r$ *at every maximal ideal, then for all homogeneous* y, $z \in \mathrm{C}\ell(M, q)$,

$$zy \equiv (-1)^{\partial y \partial z} yz \equiv (-1)^{(r-1)\partial y} yz \equiv (-1)^{(r-1)\partial z} yz \quad \text{modulo } \mathrm{C}\ell^{\leq r-1}(M, q).$$

Proof. By localization (see (3.1.10)) we reduce the problem to the case of a module generated by a family $(a_1, a_2, \dots, a_r)$ of r elements. With every subset F of $E = \{1, 2, \dots, r\}$ we associate the product e_F of all $\rho(a_i)$ with $i \in F$ in the increasing order of the indices; thus $\tau(e_F)$ is the product of the same factors in the decreasing order of the indices. As a module, $\mathrm{C}\ell(M, q)$ is generated by all $\tau(e_F)$ as well as by all e_F ; therefore it suffices to prove (3.2.1) when $y = \tau(e_F)$ and $z = e_G$ for some subsets F and G of E. Let m and n be the cardinals of F and G. From (3.1.6) it follows immediately that $zy - (-1)^{mn} yz$ belongs to $\mathrm{C}\ell^{\leq m+n-2}(M, q)$. If we manage to prove that both yz and zy belong to $\mathrm{C}\ell^{\leq r-1}(M, q)$ unless F and G are complementary subsets in E, the proof is finished, because ∂y and ∂z are the parities of m and n, and moreover $(r-1)m$ and $(r-1)n$ have the same parity as mn when $m + n = r$. It is clear that yz and zy are in $\mathrm{C}\ell^{\leq r-1}(M, q)$ when $F \cup G \neq E$. Let us suppose that $F \cap G$ is not empty, let k be the lowest element of $F \cap G$, and F' the subset of all $i \in F$ such that $i \leq k$; thus $\tau(e_F)e_G$ is the product of $\tau(e_{F'})e_G$ and other factors $\rho(a_i)$ (with $i > k$) on the left side. In general the indices of the factors in $\tau(e_{F'})e_G$ are not arranged in increasing order; yet by applying the formulas $\rho(a_i)^2 = q(a_i)1_q$ and

$$\rho(a_i)\rho(a_j) = -\rho(a_j)\rho(a_i) + \mathrm{b}_q(\rho(a_i), \rho(a_j))\, 1_q$$

as often as we meet two consecutive factors $\rho(a_i)$ and $\rho(a_j)$ such that $i \geq j$, we can transform $\tau(e_{F'})e_G$ into a linear combination of elements e_U (with $U \subset E$). Each subset U is strictly smaller than $F \cup G$, since it has lost at least one index j such that $j \leq k$. When we complete the multiplication of $\tau(e_F)$ and e_G by reinserting the indices that are in F but not in F', no factor $\rho(a_i)$ with $i \leq k$ may appear again since in $\tau(e_F)$ the factors are arranged in the decreasing order of the indices.

Consequently $\tau(e_F)e_G$ is a linear combination of elements e_V with $V \neq E$, since V has lost at least one element $\leq k$. In the same way we can prove that $e_G\tau(e_F)$ too belongs to $C\ell^{\leq r-1}(M,q)$. □

The four algebras A, A^o, A^t, A^{to}

In **3.1** an opposite algebra A^o was associated with every algebra A. When A is graded, then A^o is graded in an obvious way, and besides A^o there are two other graded algebras A^t and A^{to} derived from A. As graded modules, the *twisted algebra* A^t and the *twisted opposite algebra* A^{to} (also called the graded opposite algebra) are isomorphic to A through canonical isomorphisms $x \longmapsto x^t$ and $x \longmapsto x^{to}$, but the product of two homogeneous elements of A^t or A^{to} are defined by these formulas:

$$x^t y^t = (-1)^{\partial x \partial y}(xy)^t \quad \text{and} \quad x^{to}y^{to} = (-1)^{\partial x \partial y}(yx)^{to}.$$

The associativity of A^t and A^{to} follows from the equality $\partial x\partial y + \partial(xy)\partial z = \partial x\partial(yz) + \partial y\partial z$, which is an easy consequence of $\partial(xy) = \partial x + \partial y$ and $\partial(yz) = \partial y + \partial z$.

Obviously every graded algebra morphism f determines graded algebra morphisms f^o, f^t and f^{to}. Moreover $(x^o)^o$ is identified with x, $(x^t)^t$ is identified with x, $(x^t)^o$ and $(x^o)^t$ are both identified with x^{to}, and so forth…; in other words, we consider that a group of order 4 is acting on the set $\{A,\ A^o,\ A^t,\ A^{to}\}$.

Whereas the opposite algebra $C\ell(M,q)^o$ is involved in the definition of the reversion (see (3.1.4)), the other algebras derived from $C\ell(M,q)$ are involved in this proposition.

(3.2.2) **Proposition.** *There is a canonical graded isomorphism from the Clifford algebra $C\ell(M,-q)$ onto the twisted (resp. twisted opposite) algebra $C\ell(M,q)^t$ (resp. $C\ell(M,q)^{to}$) which maps $\rho(a)$ to $\rho(a)^t$ (resp. $\rho(a)^{to}$) for every $a \in M$.*

Proof. Since $(\rho(a)^{to})^2 = -q(a)1_A^{to}$, the mapping $a \longmapsto \rho(a)^{to}$ is an object in the category $\mathcal{A}_K(M,-q)$ (defined like $\mathcal{A}_K(M,q)$ in **3.1**); whence a graded algebra morphism $f : C\ell(M,-q) \to C\ell(M,q)^{to}$. There is a similar morphism $g : C\ell(M,q) \to C\ell(M,-q)^{to}$. Obviously $f^{to} \circ g$ is an algebra morphism from $C\ell(M,q)$ into itself that leaves invariant every element $\rho(a)$; consequently $f^{to} \circ g$ is the identity mapping of $C\ell(M,q)$. Similarly $g^{to} \circ f$ is the identity mapping of $C\ell(M,-q)$; all this implies that f is bijective. With $C\ell(M,q)^t$ the proof is similar. □

The Clifford algebra of an orthogonal sum

When A and B are graded modules (with parity gradings), their tensor product $A \otimes B$ is graded in this way:

$$(A \otimes B)_0 = (A_0 \otimes B_0) \oplus (A_1 \otimes B_1)\ ,$$
$$(A \otimes B)_1 = (A_0 \otimes B_1) \oplus (A_1 \otimes B_0)\ .$$

When A and B are graded algebras, besides their tensor product $A \otimes B$ defined in **1.3** (which obviously is still a graded algebra), there is also the *twisted tensor product* (or graded tensor product) $A \hat{\otimes} B$ defined in this way: as a graded module, it is the same thing as $A \otimes B$, but it is provided with the following multiplication:

$$(x \otimes y)\,(x' \otimes y') \;=\; (-1)^{\partial x' \partial y}\; xx' \otimes yy' \;.$$

It is easy to prove that $A \hat{\otimes} B$ is associative with unit element $1_A \otimes 1_B$. When the equality $2 = 0$ holds in K, then $A \hat{\otimes} B$ coincides with $A \otimes B$; but in general these algebras are different; indeed in $A \hat{\otimes} B$ the following equalities hold, and lead to the universal property stated in the next lemma:

$$x \otimes y \;=\; (x \otimes 1_B)(1_A \otimes y) \;=\; (-1)^{\partial x \partial y}(1_A \otimes y)\,(x \otimes 1_B).$$

(3.2.3) **Lemma.** *Let A and B be graded algebras, and let $f_1 : A \to D$ and $f_2 : B \to D$ be algebra morphisms such that*

$$\forall x \in A,\ \forall y \in B, \quad f_1(x)\, f_2(y) \;=\; (-1)^{\partial x \partial y}\, f_2(y)\, f_1(x) \;;$$

there exists a unique algebra morphism $f' : A \hat{\otimes} B \to D$ such that

$$\forall x \in A,\ \forall y \in B, \quad f_1(x) = f'(x \otimes 1_B) \quad \text{and} \quad f_2(y) = f'(1_A \otimes y).$$

This lemma is similar to (1.3.3) and its proof is omitted. It leads to a theorem analogous to (1.5.1).

(3.2.4) **Theorem.** *The Clifford algebra of an orthogonal sum*

$$(M'', q'') = (M, q) \perp (M', q')$$

is canonically isomorphic to the twisted tensor product $C\ell(M,q) \hat{\otimes} C\ell(M',q')$; if (a,b) is any element of $M'' = M \oplus M'$, the image of $\rho''(a,b)$ in the twisted tensor product is $\rho(a) \otimes 1_{q'} + 1_q \otimes \rho'(b)$.

Proof. It is easy to verify the following equality in the twisted tensor product C'' of $C\ell(M,q)$ and $C\ell(M',q')$:

$$(\rho(a) \otimes 1_{q'} + 1_q \otimes \rho'(b))^2 \;=\; (q(a) + q'(b))\; 1_q \otimes 1_{q'} \;;$$

this proves the existence of an algebra morphism f'' from $C\ell(M'',q'')$ into C'' which maps every $\rho''(a,b)$ to $\rho(a) \otimes 1_{q'} + 1_q \otimes \rho'(b)$. Conversely the canonical injections of (M,q) and (M',q') into (M'',q'') induce algebra morphisms

$$g : C\ell(M,q) \to C\ell(M'',q'') \quad \text{and} \quad g' : C\ell(M',q') \to C\ell(M'',q'') \;;$$

since $(a,0)$ and $(0,b)$ are orthogonal in (M'',q''), $g(\rho(a))$ and $g'(\rho'(b))$ anticommute, and consequently g and g' satisfy the property required in (3.2.3) to define an algebra morphism $g'' : C'' \to C\ell(M'',q'')$. The action of f'' on the generators $\rho''(a,b)$, and the action of g'' on the generators $\rho(a) \otimes 1_{q'}$ and $1_q \otimes \rho'(b)$ show that f'' and g'' are reciprocal morphisms. □

The following corollary applies to free quadratic modules provided with an orthogonal basis, for instance quadratic spaces over a local ring in which 2 is invertible (see (2.6.2)); it also applies to exterior algebras of free modules of finite rank, since every basis is orthogonal for the null quadratic form.

(3.2.5) **Corollary.** *If the quadratic module (M,q) admits an orthogonal basis $(e_1, e_2, \dots, e_r)$, the canonical mappings from K and M into $C\ell(M,q)$ are injective, $C\ell(M,q)$ is a free module of rank 2^r, and the products $e_{j_1}e_{j_2}\cdots e_{j_n}$ with $n \geq 0$ and $j_1 < j_2 < \cdots < j_n$ constitute a basis of $C\ell(M,q)$.*

This can be proved by induction on r; if $r = 1$, then $C\ell(M) = K \oplus Ke_1$ (see (3.1.2)); and when $r > 1$, we apply the induction hypothesis to the submodule M' spanned by $(e_1, e_2, \dots, e_{r-1})$ and draw the conclusion from (3.2.4):

$$C\ell(M) \cong C\ell(M') \hat{\otimes} C\ell(Ke_r) \cong C\ell(M') \oplus C\ell(M')e_r. \qquad \square$$

(3.2.6) **Corollary.** *When M is a finitely generated projective module, the exterior algebra $\bigwedge(M)$ is a finitely generated projective module; its rank at a prime ideal $\mathfrak{p}$ of K is 2^r if $r = \mathrm{rk}(M,\mathfrak{p})$. Moreover the subset of all $x \in \bigwedge(M)$ such that $a \wedge x = 0$ for all $a \in M$ is a direct summand $\bigwedge^{\max}(M)$ of constant rank 1; if M has constant rank r, then $\bigwedge^{\max}(M) = \bigwedge^r(M)$.*

Since all localisations of M are free, this can be proved by localization by means of (3.1.10) and (3.2.5). The study of $\bigwedge^{\max}(M)$ can be reduced to the case of a module M of constant rank r because of (1.12.8). $\square$

(3.2.7) **Remark.** The statement (3.2.5) remains valid when (M,q) has an infinite orthogonal basis $(e_j)_{j \in J}$ indexed by a totally ordered set J. Indeed let $\mathcal{P}$ be the set of all finite subsets of J, and $K^{(\mathcal{P})}$ the free module with basis (e_F) indexed by $F \in \mathcal{P}$; for every $E \in \mathcal{P}$ let M_E be the submodule of M spanned by all e_j with $j \in E$; if $E \supset F$, with e_F we associate the product in $C\ell(M_E)$ of all e_j such that $j \in F$, according to the increasing order of the indices. It is easy to define a K-bilinear multiplication on $K^{(\mathcal{P})}$ in such a way that it becomes the Clifford algebra of (M,q); to define the product $e_F e_G$ it suffices to calculate it in any algebra $C\ell(M_E)$ such that E contains $F \cup G$, since the result does not depend on E :

$$e_F e_G = \pm\Big(\prod_{j \in F \cap G} q(e_j)\Big)\, e_S \quad \text{if } S = (F \cup G) \setminus (F \cap G).$$

To verify the associativity relation $(e_F e_G)e_H = e_F(e_G e_G)$, it suffices to do it in any algebra $C\ell(M_E)$ such that E contains $F \cup G \cup H$; thus $K^{(\mathcal{P})}$ becomes an algebra (with unit element $1 = e_\emptyset$) in which $(e_{\{j\}})^2 = q(e_j)$ for all $j \in J$, and in which all $e_{\{j\}}$ pairwise anticommute. Consequently the linear mapping $M \to K^{(\mathcal{P})}$ defined by $e_j \longmapsto e_{\{j\}}$ induces an algebra morphism $C\ell(M,q) \to K^{(\mathcal{P})}$ which is an isomorphism because of (3.1.7). The leading idea in this argument is that M is the direct limit of the submodules M_E (see (3.ex.8)).

In particular the exterior algebra of a free module is a free module. If M is projective, there exists a module N such that $M \oplus N$ is free; since $\bigwedge(M)$ is isomorphic to a direct summand of the free module $\bigwedge(M \oplus N)$, it is a projective module.

When A and B are graded algebras, both algebras $A \otimes B$ and $A \hat{\otimes} B$ have the same grading, and consequently the same grade automorphism σ, and it is clear that $\sigma(x \otimes y) = \sigma(x) \otimes \sigma(y)$ for all $x \in A$ and all $y \in B$. When A and B are Clifford algebras $C\ell(M, q)$ and $C\ell(M', q')$, the canonical isomorphism $A \hat{\otimes} B \cong C\ell((M, q) \perp (M', q'))$ allows us to define a reversion τ on their twisted tensor product.

(3.2.8) **Lemma.** *If x and y are homogeneous elements of $C\ell(M, q)$ and $C\ell(M', q')$, then*

$$\tau(x \otimes y) \quad = \quad (-1)^{\partial x \partial y}\, \tau(x) \otimes \tau(y).$$

Indeed this is true when $x = 1_q$ or $y = 1_{q'}$; consequently

$$\begin{aligned} \tau(x \otimes y) &= \tau((x \otimes 1_{q'})(1_q \otimes y)) = \tau(1_q \otimes y)\ \tau(x \otimes 1_{q'}) \\ &= (1_q \otimes \tau(y))\ (\tau(x) \otimes 1_{q'}) = (-1)^{\partial x \partial y}\ \tau(x) \otimes \tau(y). \end{aligned}$$ □

3.3 Clifford algebras of free modules of rank 2

Theorem (2.6.2) emphasizes the importance of orthogonal sums of free quadratic modules of rank 1 or 2, and Theorem (3.2.4) encourages the study of their Clifford algebras. The Clifford algebra of a free module of rank 1 has been presented in (3.1.2). Before more powerful theories are expounded in the next chapter, the case of a free module of rank 2 can be treated by means of the following proposition, which presents the Cayley–Dickson extension process; it involves algebras provided with a unit element and a standard involution (see Definition (1.13.7)), but which are not necessarily associative.

(3.3.1) **Proposition.** *Let B be an algebra, faithful as a module over K, provided with a unit element (denoted by 1) and a standard involution φ, let α be any element of K, and let C be the module $B \oplus B$ provided with this multiplication:*

$$(b, c)\ (b', c') \quad = \quad (bb' + \alpha c' \varphi(c)\ ,\ \varphi(b)c' + b'c).$$

In all cases $(1, 0)$ is a unit element in C and the mapping $(b, c) \longmapsto (\varphi(b), -c)$ is a standard involution. When moreover the algebra B is commutative and associative, then C is an associative algebra.

Proof. Obviously $(1, 0)$ is a unit element. The mapping $(b, c) \longmapsto (\varphi(b), -c)$ is a standard involution because

$$\begin{aligned} (\varphi(b'), -c')\ (\varphi(b), -c) &= (\varphi(bb') + \alpha\varphi(c'\varphi(c)),\ -\varphi(b)c' - b'c)\ , \\ (b, c)\ (\varphi(b), -c) &= (b\varphi(b) - \alpha c\varphi(c),\ 0). \end{aligned}$$

With the only hypothesis that φ is an involution of B, let us test the associativity of the product of three factors in C; with the usual short notations $\bar{b} = \varphi(b)$, $\bar{b}' = \varphi(b')$,... a straightforward calculation gives

$$\begin{aligned}
&((b,c)(b',c'))(b'',c'') - (b,c)((b',c')(b'',c'')) \\
&= \big((bb')b'' - b(b'b'') + \alpha c''(\bar{c}'b) - \alpha b(c''\bar{c}') \\
&\qquad + \alpha c''(\bar{c}\bar{b}') - \alpha(\bar{b}'c'')\bar{c} + \alpha(c'\bar{c})b'' - \alpha(b''c')\bar{c}\ , \\
&\quad b''(b'c) - (b'b'')c + b''(\bar{b}c') - \bar{b}(b''c') + (\bar{b}'\bar{b})c'' \\
&\qquad - \bar{b}(\bar{b}'c'') + \alpha(c\bar{c}')c'' - \alpha(c''\bar{c}')c\big)\ ;
\end{aligned}$$

after eight little verifications we realize that the algebra C is associative when B is commutative and associative. □

When the algebra B is still associative but not commutative, in general C is not associative. Nevertheless the equality $(xy)z = x(yz)$ is true whenever $x = y$; indeed when $b = b'$ and $c = c'$, the above calculation shows that $(b,c)^2(b'',c'') = (b,c)((b,c)(b'',c''))$ because $b + \bar{b}$ and $c\bar{c}$ commute with all elements of B, and moreover c and $\bar{c}$ commute. Since C admits an involution, the equality $(xy)z = x(yz)$ also holds when $y = z$. Thus the trilinear mapping $(x,y,z) \longmapsto (xy)z - x(yz)$ is alternate; when this mapping is alternate, the algebra C is said to be *alternative.*

In the algebra C defined in (3.3.1) we can identify $(b,0)$ with b for every $b \in B$, and set $j = (0,1)$ so that $(b,c) = b + jc$ and $j^2 = \alpha \in K$. With these notations,

$$(b + jc)(b' + jc') = bb' + j^2c'\varphi(c) + j(\varphi(b)c' + b'c)\ ;$$

since $b + cj = b + j\varphi(c)$, we can also write

$$(b + cj)(b' + c'j) = bb' + \varphi(c')cj^2 + (c'b + c\varphi(b'))j.$$

We can extend φ by setting $\varphi(b + jc) = \varphi(b) - jc$. For the norm $\mathcal{N}$ derived from φ we can write $\mathcal{N}(b + jc) = \mathcal{N}(b) - j^2\mathcal{N}(c)$.

As it is reported in [Schafer 1966], § III,1, Artin has proved that in an alternative algebra every subalgebra generated by two elements is associative. This fact ensures that $\mathcal{N}(xy) = \mathcal{N}(x)\mathcal{N}(y)$ for all x and y; indeed $\varphi(x)$ belongs to the subalgebra generated by x since $\varphi(x) = \mathrm{tr}(x) - x$ (see **1.13**).

(3.3.2) **Examples.** Here K is the field $\mathbb{R}$ of real numbers. We apply the process described in (3.3.1) first to the algebra $A_0 = \mathbb{R}$ in order to get an algebra A_1 of dimension 2, then we apply it to A_1 to get an algebra A_2 of dimension 4, and finally we derive from A_2 an algebra A_3 of dimension 8. We always set $\alpha = -1$ so that the norm $\mathcal{N}$ remains positive definite; consequently every nonzero $x \in A_3$ has an inverse $x^{-1} = \varphi(x)/\mathcal{N}(x)$, and A_3 is a division algebra; every equality like $x(x^{-1}y) = y$ is true since A_3 is alternative. In the algebra A_1 it is convenient to set $i = (0,1)$; since $i^2 = -1$, it is isomorphic to the field $\mathbb{C}$ of complex numbers. In A_2 we identify 1 and i with $(1,0,0,0)$ and $(0,1,0,0)$ and we set $j = (0,0,1,0)$ whence

$ij = -ji = (0, 0, 0, -1)$; this associative algebra A_2 generated by two elements i and j such that $i^2 = j^2 = -1$ and $ij + ji = 0$ is the division ring $\mathbb{H}$ of real quaternions (or at least is isomorphic to it if you prefer another definition of $\mathbb{H}$). Then A_3 is an alternative algebra generated by three elements i, j, k such that

$$i^2 = j^2 = k^2 = -1 \quad \text{and} \quad ij + ji = jk + kj = ki + ik = (ij)k + k(ij) = 0 \ ,$$

whence

$$(ij)k = (jk)i = (ki)j = k(ji) = j(ki) = i(kj) \ ;$$

the family $(1, i, j, k, ij, ik, jk, (ij)k)$ is a basis of A_3 over $\mathbb{R}$, and $\mathrm{Ker}(\varphi + \mathrm{id})$ contains all elements of this basis except 1 (which spans $\mathrm{Ker}(\varphi - \mathrm{id})$); this algebra A_3 is called the *Cayley algebra of octonions* (or *octaves*).

Obviously A_1 is the Clifford algebra of a negative definite quadratic space of dimension 1 (see (3.1.2)). In A_2 the equality $(\lambda i + \mu j)^2 = -\lambda^2 - \mu^2$ holds for all λ and $\mu \in \mathbb{R}$, and thus $\mathbb{R}i \oplus \mathbb{R}j$ is a negative definite quadratic space of dimension 2; because of (3.1.7), the dimension of its Clifford algebra cannot exceed 4; therefore the natural injection $\mathbb{R}i \oplus \mathbb{R}j \to \mathbb{H}$ extends to an isomorphism from this Clifford algebra onto $\mathbb{H}$.

In 1878 Frobenius proved that every associative division algebra that has finite dimension over $\mathbb{R}$, is isomorphic either to $\mathbb{R}$ or to $\mathbb{C}$ or to $\mathbb{H}$, and in 1933 Zorn proved that every alternative division algebra that has finite dimension over $\mathbb{R}$ and that is not associative, is isomorphic to the algebra of octonions. A very simple proof of these theorems is presented in (3.ex.17).

In (3.3.2) we have found the Clifford algebra of a real quadratic space of dimension 2; let us generalize this process with an arbitrary ring K.

(3.3.3) **Lemma.** *Let $L = K[\alpha, \beta, \gamma]$ be a polynomial ring in three indeterminates, and $(N, \tilde{q})$ a free quadratic module over L with basis (e, e') such that $\tilde{q}(e) = \alpha$, $\mathrm{b}_{\tilde{q}}(e, e') = \beta$ and $\tilde{q}(e') = \gamma$. The Clifford algebra $C\ell_L(N, \tilde{q})$ is a free module over L with basis $(1, e, e', ee')$, and the conjugation $\tilde{\sigma}\tilde{\tau}$ in $C\ell_L(N, \tilde{q})$ is a standard involution.*

Proof. First we prove that the four elements $1_{\tilde{q}}$, $\tilde{\rho}(e)$, $\tilde{\rho}(e')$ and $\tilde{\rho}(e)\tilde{\rho}(e')$ are linearly independent over L in $C\ell_L(N, \tilde{q})$; this allows us to denote them by 1, e, e' and ee', and to claim that they constitute a basis of $C\ell_L(N, \tilde{q})$ because of (3.1.7). It is easy to verify that the equality

$$(\tilde{\rho}(e)\tilde{\rho}(e'))^2 \ = \ \beta \, \tilde{\rho}(e)\tilde{\rho}(e') \ - \alpha\gamma \, 1_{\tilde{q}}$$

holds in $C\ell_L(N, \tilde{q})$. Besides, for reasons that shall appear in the last calculations, we need the extension $L \to L_\alpha$ to the ring of fractions with denominator a power of α ; since α is not a divisor of zero, we can treat L as a subring of L_α and write α^{-1} instead of the fraction $1/\alpha$. Let B be the quotient of the polynomial algebra $L_\alpha[Z]$ by the ideal $(Z^2 - \beta Z + \alpha\gamma)$; thus B admits a basis $(1, z)$ over L_α such that $z^2 = \beta z - \alpha\gamma$, and a standard involution φ such that $\varphi(z) = \beta - z$. Let $C = B \oplus B$ be the algebra derived from B as it is explained in (3.3.1). As above we identify B with a subalgebra of C and we set $j = (0,1)$ (whence $j^2 = \alpha$). Obviously the four elements 1, z, j and $jz\alpha^{-1}$ constitute a basis of C over L_α and therefore are linearly independent over L. Straightforward calculations show that for all λ and $\mu \in L$,

$$\begin{aligned}(\lambda j + \mu zj\alpha^{-1})^2 &= \alpha\ (\lambda + \mu z\alpha^{-1})\ \varphi(\lambda + \mu z\alpha^{-1}) \\ &= \alpha\lambda^2 + \beta\lambda\mu + \gamma\mu^2\ ;\end{aligned}$$

consequently there is an L-algebra morphism from $C\ell_L(N,\tilde{q})$ into C that maps $\tilde{\rho}(e)$ and $\tilde{\rho}(e')$ respectively to j and $jz\alpha^{-1}$; since it maps $\tilde{\rho}(e)\tilde{\rho}(e')$ to $j(jz\alpha^{-1}) = z$, the first conclusion follows from the fact that 1, z, j and $jz\alpha^{-1}$ are linearly independent over L.

From (3.3.1) we know that C is provided with a standard involution φ that maps j and $zj\alpha^{-1}$ respectively to $-j$ and $-zj\alpha^{-1}$; this agrees with the fact that the involution $\tilde{\sigma}\tilde{\tau}$ extends $-\mathrm{id}_N$; consequently this involution is also a standard involution. □

(3.3.4) **Corollary.** *When (M,q) is a free quadratic module of rank 2 over K with basis (b,b'), then $C\ell(M,q)$ is a free module with basis $(1,b,b',bb')$. Moreover the conjugation $\sigma\tau$ is a standard involution of $C\ell(M,q)$.*

Proof. The quadratic module (M,q) can be treated as an extension of the L-quadratic module $(N,\tilde{q})$ mentioned in (3.3.3); indeed there is a K-algebra morphism $L \to K$ that maps α, β, γ respectively to $q(b)$, $\mathrm{b}_q(b,b')$, $q(b')$, and thus (M,q) is isomorphic to $K \otimes_L (N,\tilde{q})$; by this isomorphism, the images of b and b' are $1{\otimes}e$ and $1{\otimes}e'$. Since $C\ell(M,q)$ is isomorphic to $K{\otimes}C\ell_L(N,\tilde{q})$ (see (3.1.9)), from the L-basis $(1,e,e',ee')$ of $C\ell_L(N,\tilde{q})$ we derive a K-basis $(1,b,b',bb')$ of $C\ell(M,q)$. The conjugation $\tilde{\sigma}\tilde{\tau}$ in $C\ell_L(N,\tilde{q})$ gives the conjugation $\sigma\tau$ in $C\ell(M,q)$ by the extension $L \to K$, whence the end of the proof. □

Since Clifford algebras of free modules of rank 1 and 2 are now well described, we can draw more corollaries from Theorem (3.2.4) about orthogonal sums.

(3.3.5) **Corollary.** *When (M,q) is an orthogonal sum of free submodules of rank 1 or 2, then $C\ell(M,q)$ is a free module of rank 2^r if $r = \mathrm{rk}(M)$. Consequently, if $(e_1, e_2, \dots, e_r)$ is any basis of M, the products $e_{j_1}e_{j_2}\cdots e_{j_n}$ with $n \geq 0$ and $j_1 < j_2 < \cdots < j_n$ constitute a basis of $C\ell(M,q)$.*

The first statement in (3.3.5) can be proved by induction on r. The second statement (that gives a basis of $C\ell(M,q)$) means the bijectiveness of some linear mapping $P \to C\ell(M,q)$ with source a free module P of rank 2^r; from (3.2.8) we know that it is surjective, and with (1.13.5) we conclude that it is bijective. □

(3.3.6) **Corollary.** *When M is a vector space of finite dimension r over a field K, the Clifford algebra $C\ell(M,q)$ is a vector space of dimension 2^r on K.*

Proof. It suffices to prove that M is an orthogonal direct sum of lines and planes. Indeed M is the orthogonal sum of $\mathrm{Ker}(\mathrm{b}_q)$ and any supplementary subspace M'; all bases of $\mathrm{Ker}(\mathrm{b}_q)$ are orthogonal; moreover the restriction of q to M' is weakly nondegenerate, and consequently nondegenerate, and because of (2.6.2) M' is an orthogonal sum of lines or planes. □

(3.3.7) **Corollary.** *When (M, q) is a quadratic space, $\mathrm{C}\ell(M, q)$ is a finitely generated projective module and the canonical mappings $K \to \mathrm{C}\ell(M, q)$ and $\rho : M \to \mathrm{C}\ell(M, q)$ are injective. The rank of $\mathrm{C}\ell(M, q)$ at a prime ideal $\mathfrak{p}$ is 2^r if $r = \mathrm{rk}(M, \mathfrak{p})$. Besides, the conjugation $\sigma\tau$ is a standard involution if the rank of M is everywhere ≤ 2.*

Indeed all localizations of (M, q) are orthogonal sums of submodules of rank 1 or 2 (see (2.6.2)), and the conclusions follow from (3.3.5) (already from (3.3.4) when the rank is ≤ 2), and from (1.12.9). □

Historical comments

The discovery of "geometric algebras" in 1878 by Clifford, and independently by Lipschitz in 1880, can be understood as a step in the long investigations of complex and hypercomplex numbers. A premonition of these "geometric algebras" already appeared in a letter that Leibniz sent to Huygens in 1679 to explain that it should be possible to calculate with geometrical objects in an algebraic way (see [Crowe 1985], p. 3). To imagine complex numbers it was first necessary to learn the property of linear independence over $\mathbb{R}$; some mathematicians of the Italian Renaissance already were familiar with it, for instance Bombelli who said that 1 and $\sqrt{-1}$ cannot be added up (see [Bombelli 1579]). Later C. Wessel achieved one of the first sound works in this topic when he presented an "Essay on the analytical representation of direction" to the Royal Academy of Denmark (see [Wessel 1797]). Then came J.R. Argand who published in 1806 an "Essay on a manner of representing imaginary quantities in geometrical constructions". In his Treatise of Algebra published in 1685, Wallis already had a premonition about a bijection between the complex numbers and the points of a plane; this idea became quite clear and effective in the Inaugural Dissertation published by Gauss in 1799. Another milestone was passed when quaternions were discovered by several mathematicians; Hamilton mentioned them in 1843. The algebras imagined by the mathematician and logician A. de Morgan (especially in odd dimensions) were less useful, since he did not worry about their divisors of zero; anyhow his dreams were ruined by Frobenius's theorem (see (3.3.2) above). The field $\mathbb{C}$ and the division algebra $\mathbb{H}$ probably helped Clifford to imagine the geometric algebras that now have his name; in [Clifford 1878] he constructed them by means of generators and relations. Two years later (see [Lipschitz 1880]), Lipschitz sent Hermite a letter explaining his "Principles of an algebraic calculation containing as particular cases the calculation of imaginary quantities and that of quaternions". Some years later (see [Lipschitz 1886]), he declared that in 1880 he did not know Clifford's work. Yet Lipschitz went much farther than Clifford, since he was the first to imagine an application of Clifford algebras to the study of orthogonal groups. Almost at the same time Cayley and Graves (independently of one another) constructed the algebra of octonions; but since it is no longer associative, it shows another direction in this field of research.

Clifford algebras of quadratic spaces of dimension 4 were discovered again by P.A.M. Dirac (see [Dirac]) for his "theory of electron spin". As a matter of fact, at that time spinors were already known by Elie Cartan (see [Cartan 1913]), but the theory of spinors only won acknowledgement after [Brauer, Weyl 1935], [Cartan 1938], and especially [Chevalley 1954]. Meanwhile important researches were led on this subject in mathematical physics, for instance by E. Majorana (see [Amaldi]), who mysteriously disappeared in 1938, and by M. Schenberg (1914–1990) (see [Schenberg 1941 and 1943]), whose works about Quantum mechanics remained scarcely noticed probably because they were written in the Portuguese language; more information about Schenberg can be found in [Rocha Barros] and [Fernandes].

3.4 Graded quadratic extensions

Let us remember that every algebra A that is a finitely generated projective module of constant rank 2 is commutative and admits a unique standard involution φ (see (1.13.8) and (1.13.10)). From φ we derive a norm and a trace defined by $\mathcal{N}(x) = x\varphi(x)$ and $\mathrm{tr}(x) = x + \varphi(x)$.

We can identify K with a direct summand of this algebra A (see (1.13.2)); if we write $A = K \oplus P$, then *A is a free module if and only if P is free.* Indeed from the isomorphism $\bigwedge(K) \hat{\otimes} \bigwedge(P) \cong \bigwedge(A)$ (see (3.2.4)) we deduce an isomorphism $\bigwedge^1(K) \otimes \bigwedge^1(P) \cong \bigwedge^2(A)$ which shows that P is isomorphic to $\bigwedge^2(A)$, therefore free if A is free; and conversely A is free if P is free. Consequently if A is a free module, there is an element $z \in A$ such that $(1, z)$ is a basis of A, and there are elements β and γ of K such that $z^2 = \beta z - \gamma$; in other words, A is isomorphic to the quotient of $K[Z]$ by the ideal $(Z^2 - \beta Z + \gamma)$.

(3.4.1) **Definitions.** An algebra A is called a *quadratic extension of* K if A is a finitely generated projective module of constant rank 2 and if the norm $\mathcal{N}$ derived from its standard involution φ is a nondegenerate quadratic form on A. It is called a *graded quadratic extension* if it is a quadratic extension and if it is provided with a parity grading: $A = A_0 \oplus A_1$. Every nongraded quadratic extension A can be treated as a *trivially graded* quadratic extension such that $A_0 = A$. A quadratic extension A of K is said to be *trivial* if it is trivially graded (or nongraded) and isomorphic to K^2, that is, the direct product $K \times K$.

The algebra K^2 is actually a quadratic extension because its standard involution is the swap automorphism $(\lambda, \mu) \longmapsto (\mu, \lambda)$ and its norm is the hyperbolic quadratic form $(\lambda, \mu) \longmapsto \lambda\mu$. Since K^2 is generated by the idempotent $(1,0)$ as a K-algebra, it is isomorphic to the quotient $K[Z]/(Z^2 - Z)$. In **2.6** it is proved that, when $\beta^2 - 4\gamma$ is invertible, the quotient $K[Z]/(Z^2 - \beta Z + \gamma)$ is isomorphic to K^2 if and only if the polynomial $Z^2 - \beta Z + \gamma$ has a root in K.

Let us verify that (3.4.1) agrees with the definition of free quadratic extensions given in **2.6**.

(3.4.2) **Lemma.** *Let A be an algebra that is a free module with basis $(1, z)$ such that $z^2 = \beta z - \gamma$. It is a quadratic extension if and only if the discriminant $\beta^2 - 4\gamma$ of the polynomial $Z^2 - \beta Z + \gamma$ is invertible in K.*

Indeed $\varphi(z) = \beta - z$ and consequently, for all λ, $\mu \in K$,

$$\mathcal{N}(\lambda + \mu z) \quad = \quad (\lambda + \mu z)(\lambda + \mu(\beta - z)) \quad = \quad \lambda^2 + \beta\lambda\mu + \gamma\mu^2;$$

thus the determinant of the quadratic form $\mathcal{N}$ is $4\gamma - \beta^2$. □

As it is explained in (2.3.3) and (2.3.4), the nondegeneracy of $\mathcal{N}$ can be tested by localization and even by extension to residue fields; by such a process, the finitely generated projective module A of rank 2 becomes a free module to which (3.4.2) can be applied. Now (2.3.2) suggests studying what happens to the extension $K' \otimes A$ of a quadratic extension A; of course $K' \otimes A$ inherits a parity grading from A when A is graded.

(3.4.3) **Proposition.** *Let $K \to K'$ be an extension of K. If A is a graded quadratic extension of K, then $K' \otimes A$ is a graded quadratic extension of K'. The converse statement is also true when the extension is faithfully flat.*

Proof. Since A is a finitely generated projective module of constant rank 2 over K, $K' \otimes A$ is a finitely generated projective module of constant rank 2 over K' (see **1.9** and (1.12.12)). Conversely, when K' is faithfully flat over K, and $K' \otimes A$ is a finitely generated projective module of constant rank 2 over K', then A is a finitely generated projective module of constant rank 2 over K (see (1.9.10) and (1.12.13)). Obviously $K' \otimes \varphi$ is the standard involution of $K' \otimes A$, and $K' \otimes \mathcal{N}$ is its norm; from (2.3.2) we deduce that $K' \otimes \mathcal{N}$ is nondegenerate if $\mathcal{N}$ is nondegenerate, and the converse statement is true if K' is faithfully flat over K. □

In (3.4.4) there is a faithfully flat extension that deserves special attention; a first example of its usefulness is presented in (3.4.5).

(3.4.4) **Proposition.** *When A is a nongraded quadratic extension of K, then $A \otimes A$ is isomorphic to the trivial A-quadratic extension A^2 through an isomorphism mapping every $x \otimes y$ to $(xy,\ x\varphi(y))$.*

Proof. Obviously there is an A-algebra morphism F such that $F(x \otimes y) = (xy,\ x\varphi(y))$ for all x, $y \in A$. The bijectiveness of F can be tested by localization; therefore we may assume that A is a free module with basis $(1, z)$ such that

$z^2 = \beta z - \gamma$. In $A \otimes A$ we consider the A-basis $(1 \otimes 1,\ 1 \otimes z)$ and in A^2 the basis $((1,0),(0,1))$; let us look for the matrix of F with respect to these bases, and calculate its determinant:

$$\det \begin{pmatrix} 1 & z \\ 1 & \beta - z \end{pmatrix} = \beta - 2z \ ;$$

it is invertible because $(\beta - 2z)^2 = \beta^2 - 4\gamma$. □

(3.4.5) **Lemma.** *If A and B are quadratic extensions of K such that $A \subset B$, then $A = B$.*

Proof. The faithfully flat extension $K \to A \otimes B$ makes both quadratic extensions A and B (considered without grading) become trivial; consequently we can assume that they are already trivial over K. By localization we reduce the problem to the case of a local ring K in which the only idempotents are 0 and 1. Thus the only idempotents of K^2 other than $(0,0)$ and $(1,1)$ are $(1,0)$ and $(0,1)$; a trivial quadratic extension contained in K^2 is generated as an algebra by an idempotent, either $(1,0)$ or $(0,1)$; consequently it coincides with K^2. □

Now we consider $\mathrm{Ker}(\varphi - \mathrm{id})$ and $\mathrm{Im}(\varphi - \mathrm{id})$; the former is the subalgebra of elements invariant by φ and the latter is called the *discriminant module of the quadratic extension A* for reasons explained in (3.4.7). We shall compare them with $\mathrm{Im}(\varphi + \mathrm{id})$ and $\mathrm{Ker}(\varphi + \mathrm{id})$, the image and the kernel of the mapping $x \longmapsto \mathrm{tr}(x)$. When A is the trivial quadratic extension K^2, then φ is the swap automorphism and $\mathrm{Ker}(\varphi - \mathrm{id})$ and $\mathrm{Im}(\varphi - \mathrm{id})$ are the diagonals of K^2.

(3.4.6) **Proposition.** *When A is a quadratic extension with standard involution φ, then*

$$\mathrm{Ker}(\varphi - \mathrm{id}) = \mathrm{Im}(\varphi + \mathrm{id}) = K \quad \textit{and} \quad \mathrm{Im}(\varphi - \mathrm{id}) = \mathrm{Ker}(\varphi + \mathrm{id}) \ .$$

Moreover $\mathrm{Im}(\varphi - \mathrm{id})$ is a direct summand D of A of constant rank 1, and by mapping every $x \otimes y \in D \otimes D$ to xy we get an isomorphism $D \otimes D \to K$. When 2 is invertible in K, then $A = K \oplus D$.

Proof. All these statements are evident when A is the trivial quadratic extension K^2, and by means of (3.4.4) we can reduce the problem to this case. The faithfully flat extension $K \to A$ even allows us to prove that D is a direct summand of A, because this property is equivalent to the surjectiveness of the natural mapping $\mathrm{Hom}(A, D) \to \mathrm{Hom}(D, D)$ (see (1.13.1)), and its surjectiveness can be tested with such an extension. □

(3.4.7) **Example.** Let us assume that A is a free module with basis $(1, z)$ such that $z^2 = \beta z - \gamma$, since by localization we can always reduce the problem to this case. Then $D = \mathrm{Im}(\varphi - \mathrm{id})$ is the submodule generated by $\varphi(z) - z = \beta - 2z$. Since $(\beta - 2z)^2 = \beta^2 - 4\gamma$, *the bijectiveness of the mapping $D \otimes D \to K$ is equivalent to* *the invertibility of the discriminant* $\beta^2 - 4\gamma$, and this explains why D is called the discriminant module. Besides, it is worth noticing that $D = K$ when the equality $2 = 0$ holds in K.

In the previous propositions we have ignored that A may have a parity grading; now we must recognize it.

(3.4.8) **Proposition.** *Let* $A = A_0 \oplus A_1$ *be a graded quadratic extension, and* $D = \mathrm{Im}(\varphi - \mathrm{id})$ *its discriminant module. There is an idempotent* e *in* K *such that* $A_1 = eD$ *(whence* $A_1 \subset D$*), and* $2e$ *is invertible in* Ke*. Conversely if* e *is an idempotent such that* $2e$ *is invertible in* Ke*, there is a grading* $A = A_0 \oplus A_1$ *such that* $A_1 = eD$ *and* $A_0 = (1-e)A \oplus Ke$*.*

Proof. Since A_0 contains K, the rank of the direct summand A_1 is either 0 or 1, and from (1.12.8) we deduce the existence of an idempotent e such that eA_1 is a Ke-module of constant rank 1, whereas $(1-e)A_1 = 0$. By localization we reduce the problem to the case of a free module A_1 of rank 0 or 1. If its rank is 1, then $A_0 = K$ and A_1 is generated by an element z^2 such that z^2 is even; therefore we can write $z^2 = \beta z - \gamma$ with $\beta = 0$, and since $\beta = z + \varphi(z)$, we conclude that z belongs to D, and moreover the invertibility of $\beta^2 - 4\gamma$ shows that 2 is invertible. This proves the first part of (3.4.8). The converse part follows from the fact that z^2 belongs to K for all $z \in D$, and that $A = K \oplus D$ when 2 is invertible. □

From (3.4.8) and (3.4.6) it follows that φ is a graded automorphism: $\varphi(A_i) = A_i$ for $i = 0, 1$.

(3.4.9) **Remark.** If A is a graded quadratic extension such that A_1 has constant rank 1, then 2 is invertible in K, the mapping $x \longmapsto x^2$ induces a nondegenerate quadratic form on the discriminant module D and the natural injection $D \to A$ extends to an isomorphism $C\ell(D) \to A$. Conversely if (D, q) is a quadratic space of constant rank 1, there is a bilinear form $\gamma : D \times D \to K$ such that $\gamma(x, x) = q(x)$ for all $x \in D$ (see (2.5.3)); this allows us to define a multiplication on $K \oplus D$:

$$(\lambda, x)(\mu, y) \quad = \quad (\lambda\mu + \gamma(x, y),\ \lambda y + \mu x) \ ;$$

it is easy to prove by localization that this is a graded quadratic extension in which D is the submodule of odd elements, and thus $K \oplus D$ is the Clifford algebra of (D, q).

The group $Q^g(K)$

The graded quadratic extensions of K are classified by a group $Q^g(K)$ in which the operation is derived from the following theorem; it involves the twisted tensor products of graded algebras defined in **3.2**.

(3.4.10) **Theorem.** *Let* A *and* A' *be two graded quadratic extensions,* φ *and* φ' *their standard involutions, and* $\varphi \otimes \varphi'$ *the resulting graded automorphism of* $A \hat{\otimes} A'$*. The* *subset $A \star A'$ of all elements of $A \hat{\otimes} A'$ invariant by $\varphi \otimes \varphi'$ is a graded quadratic extension. The standard involution of $A \star A'$ is the common restriction of $\varphi \otimes A'$ and $A \otimes \varphi'$ to the subalgebra $A \star A'$, and the natural injection $D \otimes D' \to A \hat{\otimes} A'$ induces a bijection from $D \otimes D'$ onto the discriminant module of $A \star A'$.*

Proof. First we notice that $A \star A'$ is a graded subalgebra of $A \hat{\otimes} A'$ because φ and φ' are graded automorphisms of A and A'. Secondly, $A \star A'$ contains $D \otimes D'$ because $D = \text{Ker}(\varphi + \text{id})$ and $D' = \text{Ker}(\varphi' + \text{id})$ (see (3.4.6)). Thirdly, if it is true that $A \star A'$ is a quadratic extension and that its standard involution is the common restriction φ'' of $\varphi \otimes A'$ and $A \otimes \varphi'$, then its determinant module $D'' = \text{Ker}(\varphi'' + \text{id})$ contains $D \otimes D'$, whence $D'' = D \otimes D'$ because D'' and $D \otimes D'$ are direct summands of $A \star A'$ of constant rank 1. Therefore it suffices to prove that $A \star A'$ is a quadratic extension with standard involution φ''.

We reduce the problem to the case of graded algebras A and A' with odd components A_1 and A'_1 of constant ranks, either by localization, or by means of suitable idempotents of K as is suggested by (3.4.8). Thus it suffices to consider these two cases: either 2 is invertible in K (as it happens when A_1 or A'_1 has constant rank 1), or the algebra $A \hat{\otimes} A'$ is equal to $A \otimes A'$ (as it happens when A_1 or A'_1 has constant rank 0). When 2 is invertible, we can write $A = K \oplus D$ and $A' = K \oplus D'$, and it immediately follows that

$$A \star A' = (K \otimes K) \oplus (D \otimes D').$$

For all $c,\ d \in D$, and for all $c',\ d' \in D'$, the product $(c \otimes c')(d \otimes d')$ is equal to $\pm cd \otimes c'd'$ (where $\pm$ means $+$ if $A_1 = 0$ or $A'_1 = 0$, but $-$ if $A_1 = D$ and $A'_1 = D'$); in all cases it belongs to $K \otimes K$. Consequently φ'' is the standard involution of $A \star A'$ and $\text{Im}(\varphi'' - \text{id}) = D \otimes D'$. From the bijectiveness of $D \otimes D \to K$ and $D' \otimes D' \to K$ we deduce the bijectiveness of the mapping

$$(D \otimes D') \otimes (D \otimes D') \longrightarrow K\ , \quad (c \otimes c') \otimes (d \otimes d') \longmapsto \pm cdc'd'\ ;$$

this implies that $A \star A'$ is actually a quadratic extension with discriminant module $D \otimes D'$.

When $A \hat{\otimes} A'$ is equal to $A \otimes A'$, we can forget the gradings of A and A' and by means of the faithfully flat extension $K \to A \otimes A'$ we reduce the problem to the case of two trivial quadratic extensions. Therefore we can assume that A and A' are both equal to K^2; we identify $A \otimes A'$ with K^4 through this isomorphism:

$$K^2 \otimes K^2 \longrightarrow K^4\ , \quad (\lambda, \mu) \otimes (\lambda', \mu') \longmapsto (\lambda\lambda',\ \lambda\mu',\ \mu\lambda',\ \mu\mu').$$

Let ψ and ψ' be the involutions of K^4 corresponding to the involutions $\varphi \otimes A'$ and $A \otimes \varphi'$ of $A \otimes A'$; since φ and φ' are swap automorphisms, we realize that ψ, ψ' and $\psi\psi'$ map every $(\lambda_1, \lambda_2, \lambda_3, \lambda_4) \in K^4$ respectively to $(\lambda_3, \lambda_4, \lambda_1, \lambda_2)$, $(\lambda_2, \lambda_1, \lambda_4, \lambda_3)$ and $(\lambda_4, \lambda_3, \lambda_2, \lambda_1)$. The subalgebra $A \star A'$ corresponds to the subalgebra of elements $(\lambda_1, \lambda_2, \lambda_2, \lambda_1)$ invariant by $\psi\psi'$, and therefore is isomorphic to K^2; moreover $A \otimes \varphi'$ induces the standard involution on $A \star A'$ because ψ' induces the standard involution on $\text{Ker}(\psi\psi' - \text{id})$. □

(3.4.11) **Example.** Let us suppose that A and A' are free with bases $(1,z)$ and $(1,z')$ such that $z^2 = \beta z - \gamma$ and $z'^2 = \beta' z' - \gamma'$. Obviously we get an element z'' of $A \star A'$ if we set

$$z'' \;=\; z \otimes \varphi'(z') + \varphi(z) \otimes z' \;=\; z \otimes \beta' + \beta \otimes z' - 2z \otimes z'.$$

Let us calculate the coefficients β'' and γ'' such that $z''^2 = \beta'' z'' - \gamma''$. When $A \hat{\otimes} A'$ coincides with $A \otimes A'$, after some calculations we find

$$\beta'' = \beta\beta' \quad \text{and} \quad \gamma'' = \beta^2\gamma' + \gamma\beta'^2 - 4\gamma\gamma' \;;$$

since $\beta''^2 - 4\gamma'' = (\beta^2 - 4\gamma)(\beta'^2 - 4\gamma')$, the subalgebra $K \oplus Kz''$ is a quadratic extension because of (3.4.2), and is equal to $A \star A'$ because of (3.4.5). Now let us assume that $A_1 = D$ and $A'_1 = D'$. Since D is generated by $\beta - 2z$, and D' by $\beta' - 2z'$, the discriminant module D'' of $A \star A'$ is generated by $(\beta - 2z) \otimes (\beta' - 2z')$ which is equal to $\beta\beta' - 2z''$; since $A \star A' = K \oplus D''$, it is already sure that $A \star A' = K \oplus Kz''$. From the equality

$$(\beta\beta' - 2z'')^2 = -(\beta - 2z)^2 \otimes (\beta' - 2z')^2 = -(\beta^2 - 4\gamma) \otimes (\beta'^2 - 4\gamma')$$

it is easy to deduce that

$$\beta'' = \beta\beta' \quad \text{and} \quad \gamma'' = \frac{1}{2}\beta^2\beta'^2 - \beta^2\gamma' - \gamma\beta'^2 + 4\gamma\gamma'.$$

In this case $\beta''^2 - 4\gamma'' = -(\beta^2 - 4\gamma)(\beta'^2 - 4\gamma')$.

(3.4.12) **Proposition.** *When A, A', A'' are graded quadratic extensions, then*

(a) *$A \star A'$ and $A' \star A$ are canonically isomorphic;*

(b) *$(A \star A') \star A''$ and $A \star (A' \star A'')$ are the same subalgebra of $A \hat{\otimes} A' \hat{\otimes} A''$;*

(c) *$A \star K^2$ (with K^2 trivially graded) is isomorphic to A;*

(d) *$A \star A$ is isomorphic to K^2 if A is trivially graded;*

(e) *$A \star A \star A \star A$ is always isomorphic to K^2.*

Proof. The statement (a) comes from the canonical isomorphism $A \hat{\otimes} A' \to A' \hat{\otimes} A$ defined by $x \otimes x' \longmapsto (-1)^{\partial x \partial x'} x' \otimes x$. To prove (b), we identify $(A \star A') \star A''$ with a subalgebra of $A \hat{\otimes} A' \hat{\otimes} A''$; first $(A \star A') \otimes A''$ is the subset of all elements of $A \otimes A' \otimes A''$ invariant by $\varphi \otimes \varphi' \otimes A''$; then $(A \star A') \star A''$ is the subset of all elements of $(A \star A') \otimes A''$ invariant by $A \otimes \varphi' \otimes \varphi''$. This proves (b) because similarly $A \star (A' \star A'')$ is the subset of all elements of $A \otimes A' \otimes A''$ invariant both by $A \otimes \varphi' \otimes \varphi''$ and $\varphi \otimes \varphi' \otimes A''$.

When $A' = K^2$ and consequently $\varphi'(\lambda, \mu) = (\mu, \lambda)$, we can identify $A \hat{\otimes} A'$ with the ordinary direct product $A^2 = A \times A$, and thus $\varphi \otimes \varphi'$ becomes the automorphism $(x, y) \longmapsto (\varphi(y), \varphi(x))$. Thus the mapping $x \longmapsto (x, \varphi(x))$ is an isomorphism from A onto $A \star K^2$, and (c) is proved.

When A is trivially graded, by the isomorphism $A \otimes A \rightarrow A \times A$ presented in (3.4.4) the involution $\varphi \otimes \varphi$ of $A \otimes A$ corresponds to the involution (φ, φ) of $A \times A$, and therefore $A \star A$ corresponds to $K \times K$. This proves (d). And (e) follows immediately from (d) because $A \star A$ is always trivially graded; indeed the odd component of $A \star A$ is contained in its discriminant module $D \otimes D$, in which all elements are even. □

Let $\mathrm{Q}^g(K)$ be the set of isomorphy classes of graded quadratic extensions; the proposition (3.4.12) shows that the operation $\star$ on graded quadratic extensions provides $\mathrm{Q}^g(K)$ with a structure of commutative group, in which every element has order 1, 2 or 4. The set $\mathrm{Q}(K)$ of isomorphy classes of trivially graded (or nongraded) quadratic extensions is a subgroup of $\mathrm{Q}^g(K)$, in which every element has order 1 or 2.

Besides, the set $\mathrm{Ip}(K)$ of all idempotents of K is a boolean ring for the "boolean addition" defined by $(e, e') \longmapsto e \tilde{+} e' = e + e' - 2ee'$ and the ordinary multiplication $(e, e') \longmapsto ee'$. This fact can be proved either by a direct verification or by the following geometrical considerations: there is a canonical bijection from $\mathrm{Ip}(K)$ onto the algebra $\mathcal{C}(\mathrm{Spec}(K), \mathbb{Z}/2\mathbb{Z})$ of continuous functions $\mathrm{Spec}(K) \rightarrow \mathbb{Z}/2\mathbb{Z}$. Indeed on one side the mapping $e \longmapsto \mathcal{V}(K(1-e))$ is a bijection from $\mathrm{Ip}(K)$ onto the set of subsets of $\mathrm{Spec}(K)$ that are both open and closed (see (1.11.3)), and on the other side the mapping $f \longmapsto f^{-1}(1)$ is a bijection from $\mathcal{C}(\mathrm{Spec}(K), \mathbb{Z}/2\mathbb{Z})$ onto the same set of open and closed subsets; thus each $e \in \mathrm{Ip}(K)$ is mapped to the function $f_e : \mathrm{Spec}(K) \rightarrow \mathbb{Z}/2\mathbb{Z}$ such that $f_e(\mathfrak{p}) = 1$ if $e \notin \mathfrak{p}$, and $f_e(\mathfrak{p}) = 0$ if $e \in \mathfrak{p}$. This allows us to verify that the "boolean addition" of idempotents (resp. the multiplication of idempotents) corresponds exactly to the ordinary addition (resp. multiplication) of continuous functions $\mathrm{Spec}(K) \rightarrow \mathbb{Z}/2\mathbb{Z}$.

In (3.4.8) an idempotent of K is associated with every graded quadratic extension; let us study the resulting mapping $\mathrm{Q}^g(K) \rightarrow \mathrm{Ip}(K)$.

(3.4.13) **Proposition.** *By associating with every graded quadratic extension A of K the idempotent e such that $A_1 = eD$ we obtain a morphism of commutative groups from $\mathrm{Q}^g(K)$ into $\mathrm{Ip}(K)$; its kernel is the subgroup $\mathrm{Q}(K)$, and its image is the subgroup $\mathrm{Ip}'(K)$ of all $e \in \mathrm{Ip}(K)$ such that $2e$ is invertible in Ke.*

In other words, there is an exact sequence

$$0 \longrightarrow \mathrm{Q}(K) \longrightarrow \mathrm{Q}^g(K) \longrightarrow \mathrm{Ip}'(K) \longrightarrow 0.$$

Proof. Let e and e' be the idempotents associated with the graded quadratic extensions A and A', and let us set $e'' = e + e' - 2ee'$ and $A'' = A \star A'$; we must prove that $(1 - e'')A''_1 = 0$ and A''_1 is faithful over Ke''. Since this can be done by localizations, we can reduce the proof to the case of free modules A_1 and A'_1 of ranks 0 or 1, for which the associated idempotents are 0 or 1. If $(e, e') = (0, 0)$, all gradings are trivial, and the conclusion is trivial. In all other cases, 2 is invertible in K, and (e, e', e'') is either $(1, 0, 1)$ or $(0, 1, 1)$ or $(1, 1, 0)$; consequently we must prove that one (and only one) of the three modules A_1 or A'_1 or A''_1 is reduced to 0. Let z and z' be generators of D and D'; thus $z'' = z \otimes z'$ is a generator of $D'' = D \otimes D'$. When A_1 is not reduced to 0, it is equal to D, and the same for A'_1 and A''_1. Since one (and only one) of the three generators z or z' or z'' is even, it is sure that one (and only one) of the three idempotents is 0. Besides, the kernel of $\mathrm{Q}^g(K) \to \mathrm{Ip}(K)$ is obviously $\mathrm{Q}(K)$, and its image is given by the second statement in (3.4.8). □

When $K \to K'$ is an extension of the basic ring K, with every graded K-extension A is associated a graded K'-quadratic extension $K' \otimes A$ (see (3.4.3)). By mapping the isomorphy class of A to the class of $K' \otimes A$ we get a mapping $\mathrm{Q}^g(K) \to \mathrm{Q}^g(K')$; let us prove that it is a group morphism. If A and A' are graded quadratic extensions of K with standard involutions φ and φ', from (1.9.6) we derive

$$K' \otimes (A \,\hat{\otimes}\, A') \;\cong\; (K' \otimes A) \,\hat{\otimes}_{K'} (K' \otimes A') \;;$$

the elements of the left-hand member that are invariant by $K' \otimes (\varphi \otimes \varphi')$ correspond exactly to the elements of the right-hand member that are invariant by $(K' \otimes \varphi) \otimes (K' \otimes \varphi')$; thus we get isomorphisms of graded quadratic extensions

$$K' \otimes (A \star A') \;\cong\; (K' \otimes A) \star_{K'} (K' \otimes A').$$

When A is a nongraded quadratic extension of K, it is worth observing that the class of A belongs to the kernel of the morphism $\mathrm{Q}^g(K) \to \mathrm{Q}^g(A)$; this is one of the many consequences of (3.4.4).

Inside $\mathrm{Q}^g(K)$ it is sometimes useful to consider the subset $\mathrm{Q}^g_f(K)$ of isomorphy classes of graded quadratic extensions which are free modules provided with a basis $(1, z)$ such that $z^2 = z - \gamma$ for some $\gamma \in K$; similarly inside $\mathrm{Q}(K)$ we consider the subset $\mathrm{Q}_f(K)$ of isomorphy classes of trivially graded quadratic extensions satisfying the same condition. From (3.4.11) it immediately follows that $\mathrm{Q}^g_f(K)$ (resp. $\mathrm{Q}_f(K)$) is a subgroup of $\mathrm{Q}^g(K)$ (resp. $\mathrm{Q}(K)$).

Often the study of $\mathrm{Q}^g_f(K)$ is a useful step in the study of $\mathrm{Q}^g(K)$. Sometimes $\mathrm{Q}^g_f(K)$ is even equal to $\mathrm{Q}^g(K)$; this happens for instance when K is a local ring. Indeed every quadratic extension A over a local ring K admits a basis $(1, z)$; let us set $z^2 = \beta z - \gamma$, and prove that $(1, z)$ can be replaced by a basis $(1, y)$ such that $y^2 - y \in K$. When 2 is invertible in K, this happens with $y = z - (\beta - 1)/2$; and when 2 is not invertible, the invertibility of $\beta^2 - 4\gamma$ implies that β is invertible, and we set $y = z/\beta$.

There is an exact sequence $0 \to \mathrm{Q}_f(K) \to \mathrm{Q}^g_f(K) \to \mathrm{Ip}'(K) \to 0$ (see (3.4.13)) which shows that the knowledge of $\mathrm{Q}_f(K)$ leads to the knowledge of $\mathrm{Q}^g_f(K)$. To get information about $\mathrm{Q}_f(K)$, we can use the exact sequences that are now explained. Let $K\lceil 2\rfloor$ be the subset of all $\kappa \in K$ such that $1 - 2\kappa$ is invertible, and $K\lceil 4\rfloor$ the subset of all $\gamma \in K$ such that $1 - 4\gamma$ is invertible; we provide $K\lceil 2\rfloor$

and $K\lceil 4\rfloor$ with the following operations:

$$\kappa \tilde{+} \kappa' = \kappa + \kappa' - 2\kappa\kappa' \qquad \text{and} \qquad \gamma \star \gamma' = \gamma + \gamma' - 4\gamma\gamma'.$$

It is easy to verify that $K\lceil 2\rfloor$ and $K\lceil 4\rfloor$ are commutative groups for these operations, and that the mapping $\kappa \longmapsto \kappa - \kappa^2$ is a group morphism from $K\lceil 2\rfloor$ into $K\lceil 4\rfloor$. Moreover $K\lceil 2\rfloor$ contains $\mathrm{Ip}(K)$; indeed $1-2e$ is invertible for all $e \in \mathrm{Ip}(K)$ since $(1-2e)^2 = 1$.

(3.4.14) **Lemma.** *There is an exact sequence*

$$0 \longrightarrow \mathrm{Ip}(K) \longrightarrow K\lceil 2\rfloor \longrightarrow K\lceil 4\rfloor \longrightarrow \mathrm{Q}_f(K) \longrightarrow 0$$

in which the arrows are defined in this way; the second arrow is the natural injection, the third arrow maps every $\kappa \in K\lceil 2\rfloor$ to $\kappa - \kappa^2$, and the fourth arrow maps every $\gamma \in K\lceil 4\rfloor$ to the isomorphy class of the quadratic extension $K \oplus Kz$ such that $z^2 = z - \gamma$.

When 2 is invertible in K, there is an equivalent exact sequence of multiplicative groups:

$$1 \longrightarrow \mu_2(K) \longrightarrow K^\times \longrightarrow K^\times \longrightarrow \mathrm{Q}_f(K) \longrightarrow 1\ ;$$

$\mu_2(K)$ is the group of square roots of 1 in K, the second arrow is the natural injection, the third arrow is the morphism $\lambda \longmapsto \lambda^2$, and the fourth arrow maps every $\delta \in K^\times$ to the isomorphy class of the quadratic extension $K \oplus Ky$ such that $y^2 = \delta$.

Proof. The exactness of the first sequence only needs an explanation at the right end: the extension $K \oplus Kz$ (with $z^2 = z - \gamma$) is trivial if and only the polynomial $Z^2 - Z + \gamma$ has a root in K (see **2.6**), and this means precisely that γ is in the image of the morphism $\kappa \longmapsto \kappa - \kappa^2$. When 2 is invertible in K, the mapping $\kappa \longmapsto 1 - 2\kappa$ is an isomorphism from $K\lceil 2\rfloor$ onto $K^\times$ that maps the subgroup $\mathrm{Ip}(K)$ onto $\mu_2(K)$; and the mapping $\gamma \longmapsto 1 - 4\gamma$ is an isomorphism from $K\lceil 4\rfloor$ onto $K^\times$; this allows us to deduce the second exact sequence from the first one. Indeed the equality $\gamma = \kappa - \kappa^2$ is equivalent to $\delta = \lambda^2$ if $\lambda = 1 - 2\kappa$ and $\delta = 1 - 4\gamma$, and the equality $z^2 = z - \gamma$ is equivalent to $y^2 = \delta$ if $\delta = 1 - 4\gamma$ and $y = 1 - 2z$. □

Automorphisms of quadratic extensions

Later in **5.5** we shall need the group $\mathrm{Aut}(A)$ of all automorphisms of the K-algebra A when A is a quadratic extension. When A is graded, we can already predict that every such automorphism f is graded (in other words, $f(A_i) = A_i$ for $i = 0, 1$); indeed the grading of A is determined by φ and some idempotent of K (see (3.4.8)), and f commutes with φ (see (1.13.8)).

(3.4.15) **Proposition.** *If f is an automorphism of the quadratic extension A, there exists a unique idempotent e in K such that the restriction of f to $(1-e)A$ is the identity mapping, whereas its restriction to eA is its standard involution.*

Proof. When K is a local ring, (3.4.15) means that f is either id or φ. Indeed let $(1, z)$ be a basis of A; we can write $z^2 = \beta z - \gamma$ with $\beta^2 - 4\gamma$ invertible in K, and $\varphi(z) = \beta - z$. It is clear that $(1, f(z))$ is also a basis of A, and $f(z)^2 = \beta f(z) - \gamma$. If we prove that z and $\varphi(z)$ are the only elements $u \in A$ such that $(1, u)$ is a basis of A and $u^2 - \beta u + \gamma = 0$, then it follows that f is either id or φ. Since $(1, u)$ is a basis, we can write $u = \lambda + \mu z$ with $\mu \in K^\times$. Now the equality $u^2 - \beta u + \gamma = 0$ is equivalent to $(u - z)(u - \varphi(z)) = 0$, and either $u - z$ or $u - \varphi(z)$ is invertible because

$$\mathcal{N}(u - z) - \mathcal{N}(u - \varphi(z)) \;=\; (u - \varphi(u))(z - \varphi(z)) \;=\; \mu(\beta^2 - 4\gamma)\ ;$$

consequently u is either z or $\varphi(z)$ and the particular case of a local ring K is settled.

When K is an arbitrary ring, by localization we realize that $f(x) - x$ belongs to the discriminant module for all $x \in A$, and consequently $(f(x) - x)^2 \in K$. Let $\mathfrak{a}$ be the ideal of K generated by all $(f(x) - x)^2$, and $\mathfrak{b}$ the annihilator of $\mathfrak{a}$, that is the subset of all $\lambda \in A$ such that $\lambda\mathfrak{a} = 0$. Since $\mathfrak{a}$ is finitely generated, for every multiplicative subset S of K the vanishing of $S^{-1}\mathfrak{a}$ is equivalent to $S \cap \mathfrak{b} \neq \emptyset$. For every prime ideal $\mathfrak{p}$ of K, either $f_\mathfrak{p}$ is the standard involution of $A_\mathfrak{p}$ and then $\mathfrak{p}$ contains $\mathfrak{b}$ but not $\mathfrak{a}$ since $\mathfrak{a}_\mathfrak{p} = K_\mathfrak{p}$; or $f_\mathfrak{p}$ is the identity mapping and then $\mathfrak{p}$ contains $\mathfrak{a}$ but not $\mathfrak{b}$ since $\mathfrak{a}_\mathfrak{p} = 0$. This proves that $\mathrm{Spec}(K)$ is the disjoint union of the closed subsets $\mathcal{V}(\mathfrak{a})$ and $\mathcal{V}(\mathfrak{b})$. From Theorem (1.11.3) we deduce the existence of an idempotent $e \in K$ such that $\mathcal{V}(\mathfrak{a}) = \mathcal{V}(Ke)$ and $\mathcal{V}(\mathfrak{b}) = \mathcal{V}(K(1-e))$. Thus $\mathfrak{a} = Ke$ and f has the same localizations as $e\varphi + (1-e)\mathrm{id}$. □

(3.4.16) **Corollary.** *If we map every* $f \in \mathrm{Aut}(A)$ *to the idempotent* e *mentioned in* (3.4.15) *we get a group isomorphism* $\mathrm{Aut}(A) \to \mathrm{Ip}(K)$.

Indeed this mapping is bijective, and by localization (as in the proof of (3.4.13)) it is easy to prove that it is a group morphism. □

3.5 Graded Azumaya algebras

As in the previous sections, here we are concerned with parity gradings (over the group $\mathbb{Z}/2\mathbb{Z}$). Let us recall that $M \otimes N$ is graded whenever M and N are graded:

$$(M \otimes N)_0 = (M_0 \otimes N_0) \oplus (M_1 \otimes N_1) \quad \text{and} \quad (M \otimes N)_1 = (M_1 \otimes N_0) \oplus (M_0 \otimes N_1)\ ;$$

moreover $\mathrm{Hom}(M, N)$ is also graded; $\mathrm{Hom}_0(M, N)$ is the submodule of all graded morphisms (the morphisms f such that $f(M_i) \subset N_i$ for $i = 0, 1$), and $\mathrm{Hom}_1(M, N)$ is the submodule of all f such that $f(M_i) \subset N_{1-i}$ for $i = 0, 1$. In particular, $\mathrm{End}(M) = \mathrm{Hom}(M, M)$ is a graded algebra.

When A is a graded algebra, there is a canonical morphism

$$A \,\hat{\otimes}\, A^{to} \longrightarrow \mathrm{End}(A)\ , \quad a \otimes b^{to} \;\longmapsto\; (x \longmapsto (-1)^{\partial b \partial x} axb)\ ;$$

it is easy to verify that it is a graded algebra morphism; it plays an essential role in the following definition.

(3.5.1) **Definition.** A graded algebra A is said to be a *graded Azumaya algebra over* K if A is a finitely generated and faithful projective K-module, and if the canonical morphism $A \hat{\otimes} A^{to} \to \mathrm{End}(A)$ is bijective.

To work with this definition, we need some assorted additional definitions.

(3.5.2) **Definitions.** Let $A = A_0 \oplus A_1$ be a graded module; its grading is said to be *trivial* if $A_1 = 0$. Every nongraded module is silently given the trivial grading. When A is a finitely generated projective module, its grading is said to be *balanced* if at every prime ideal of K the ranks of A_0 and A_1 are equal. When this module A is a graded algebra, its grading is said to be *regular* if the multiplication mapping $A_1 \otimes A_1 \longrightarrow A_0$ (defined by $x \otimes y \longmapsto xy$) is surjective. Moreover we define in A the following graded subalgebras:

$$\begin{aligned}
\mathrm{Z}(A) &= \{a \mid a \in A \ ; \ \forall b \in A, \ ab = ba \ \}, \ \text{ the center of } A, \\
\mathrm{Z}_0(A) &= \{a \mid a \in A_0 \ ; \ \forall b \in A, \ ab = ba \ \}, \ \text{ the even center of } A, \\
\mathrm{Z}^g(A) &= \mathrm{Z}_0^g(A) \oplus \mathrm{Z}_1^g(A), \ \text{ the graded center of } A, \text{ with components } \mathrm{Z}_i^g(A) = \\
&\qquad \{a \mid a \in A_i \ ; \ \forall b \in A_0 \cup A_1, \ ab = (-1)^{\partial a \partial b} ba \ \} \ \text{ for } i = 0, 1, \\
\mathrm{Z}(A_0) &= \{a \mid a \in A_0 \ ; \ \forall b \in A_0, \ ab = ba \ \}, \ \text{ the center of } A_0, \\
\mathrm{Z}(A_0, A) &= \{a \mid a \in A \ ; \ \forall b \in A_0, \ ab = ba \ \}, \ \text{ the centralizer of } A_0 \text{ in } A.
\end{aligned}$$

When A is trivially graded, $\mathrm{Z}_0(A)$, $\mathrm{Z}^g(A)$, $\mathrm{Z}(A_0, A)$ and $\mathrm{Z}(A_0)$ are all equal to $\mathrm{Z}(A)$. In all cases $\mathrm{Z}(A)$ and $\mathrm{Z}^g(A)$ have the same even subalgebra $\mathrm{Z}_0(A) = \mathrm{Z}_0^g(A)$, and $\mathrm{Z}(A_0)$ is the even subalgebra of $\mathrm{Z}(A_0, A)$. There are obvious inclusions:

$$\mathrm{Z}_0(A) \subset \mathrm{Z}(A_0), \quad \mathrm{Z}(A) \subset \mathrm{Z}(A_0, A) \quad \text{and} \quad \mathrm{Z}^g(A) \subset \mathrm{Z}(A_0, A).$$

When A is moreover a finitely generated projective module, the presence of invertible elements in A_1 (or in each localization of A_1) implies that the grading of A is balanced and regular.

Here are some consequences of the definition (3.5.1).

(3.5.3) **Theorem.** *Let $K \to K'$ be an extension of the ring K. When A is a graded Azumaya algebra over K, then $K' \otimes A$ is a graded Azumaya algebra over K'. The converse statement is also true if this extension is faithfully flat.*

Proof. When A is a finitely generated and faithful projective K-module, then $K' \otimes A$ is a finitely generated and faithful projective K'-module for every extension $K \to K'$; when faithfulness is involved, remember (1.13.3) and (1.12.12); moreover (1.9.6) and (1.9.7) ensure the bijectiveness of these algebra morphisms:

$$\begin{aligned}
K' \otimes (A \hat{\otimes} A^{to}) &\longrightarrow (K' \otimes A) \hat{\otimes}_{K'} (K' \otimes A^{to}) \ , \\
K' \otimes \mathrm{End}(A) &\longrightarrow \mathrm{End}_{K'}(K' \otimes A) \ ;
\end{aligned}$$

consequently $K'\otimes A$ is a graded Azumaya K'-algebra when A is a graded Azumaya K-algebra. When the extension $K \to K'$ is faithfully flat, this argument also works in the converse way; indeed a K-linear morphism $M \to N$ is injective or surjective if and only if the same is true for $K' \otimes M \to K' \otimes N$; when projectiveness and finiteness are involved, remember (1.9.10); and for faithfulness, use (1.12.13). □

(3.5.4) **Lemma.** *Let P and Q be two graded finitely generated and projective modules; we get a graded algebra isomorphism*

$$\mathrm{End}(P)\,\hat{\otimes}\,\mathrm{End}(Q) \longrightarrow \mathrm{End}(P\otimes Q)$$

if we map every $f\otimes g \in \mathrm{End}(P)\,\hat{\otimes}\,\mathrm{End}(Q)$ to the endomorphism $f\,\hat{\otimes}\,g$ of $P\otimes Q$ defined by

$$(f\,\hat{\otimes}\,g)(x\otimes y) \;=\; (-1)^{\partial g\partial x}\; f(x)\otimes g(y)$$

for all (homogeneous) $x\in P$ and $y \in Q$.

Proof. Obviously the mapping $f\otimes g \longmapsto f\,\hat{\otimes}\,g$ is graded, and it is easy to verify that

$$(f\,\hat{\otimes}\,g)\circ(f'\,\hat{\otimes}\,g') \;=\; (-1)^{\partial g\partial f'}\;(f\circ f')\,\hat{\otimes}\,(g\circ g').$$

To prove the bijectiveness of this mapping, we can use the isomorphism

$$P\otimes P^* \longrightarrow \mathrm{End}(P)\;, \quad x\otimes u \;\longmapsto\; (\,x' \longmapsto u(x')\,x\,)\;,$$

and the similar isomorphisms involving Q and $P\otimes Q$, and also the isomorphism $P^*\otimes Q^* \to (P\otimes Q)^*$ which maps every $u\otimes v$ to the linear form $x\otimes y \longmapsto (-1)^{\partial v\partial x}u(x)v(y)$. The bijectiveness of all these isomorphisms appears when they are localized. They allow us to derive the bijectiveness of the above mapping $f\otimes g \longmapsto f\,\hat{\otimes}\,g$ from the evident bijectiveness of this mapping:

$$\begin{aligned}(P\otimes P^*)\otimes(Q\otimes Q^*) &\longrightarrow (P\otimes Q)\otimes(P^*\otimes Q^*)\;,\\ (x\otimes u)\otimes(y\otimes v) &\longmapsto (-1)^{\partial u\partial y}(x\otimes y)\otimes(u\otimes v)\;.\end{aligned}$$

□

(3.5.5) **Theorem.** *When A and B are graded Azumaya algebras, then $A\,\hat{\otimes}\,B$ too is a graded Azumaya algebra.*

Proof. On one side $A\,\hat{\otimes}\,B$ is also a finitely generated and faithful projective module. On the other side there is a canonical isomorphism of graded algebras

$$\begin{aligned}(A\,\hat{\otimes}\,B)\,\hat{\otimes}\,(A\,\hat{\otimes}\,B)^{to} &\longrightarrow (A\,\hat{\otimes}\,A^{to})\,\hat{\otimes}\,(B\,\hat{\otimes}\,B^{to})\;,\\ (x\otimes y)\otimes(x'\otimes y')^{to} &\longmapsto (-1)^{\partial y\partial x'}(x\otimes x'^{to})\otimes(y\otimes y'^{to})\;;\end{aligned}$$

thus $(A\,\hat{\otimes}\,B)\,\hat{\otimes}\,(A\,\hat{\otimes}\,B)^{to}$ is isomorphic to $\mathrm{End}(A)\,\hat{\otimes}\,\mathrm{End}(B)$, which by the previous lemma is isomorphic to $\mathrm{End}(A\otimes B)$. The conclusion follows after some evident calculations. □

When we wish to know whether a graded algebra is a graded Azumaya algebra, this property can be tested by localization, and even by means of extensions to residue fields, as it is now explained.

(3.5.6) **Theorem.** *When a graded algebra A is a finitely generated projective K-module, the following assertions are equivalent:*

(a) *A is a graded Azumaya algebra over K;*
(b) *for every prime ideal $\mathfrak{p}$ of K, $A_\mathfrak{p}$ is a graded Azumaya algebra over $K_\mathfrak{p}$;*
(c) *for every maximal ideal $\mathfrak{m}$ of K, $A_\mathfrak{m}$ is a graded Azumaya algebra over $K_\mathfrak{m}$;*
(d) *for every maximal ideal $\mathfrak{m}$ of K, $A/\mathfrak{m}A$ is a graded Azumaya algebra over the field $K/\mathfrak{m}$.*

Proof. The equivalences (a)⇔(b)⇔(c) are immediate consequences of (1.11.7), (1.10.5) and (1.10.8), and from (3.5.3) we can derive (a)⇒(d). Conversely, when the assertion (d) is true, by the argument explained in the proof of (3.5.3) we get isomorphisms

$$(K/\mathfrak{m}) \otimes (A \hat{\otimes} A^{to}) \longrightarrow (K/\mathfrak{m}) \otimes \mathrm{End}(A)$$

for all maximal ideals $\mathfrak{m}$; thus the conclusion follows from (1.13.5). □

(3.5.7) **Remark.** Let us suppose that in K there is an equality $1 = e_1 + e_2 + \cdots + e_n$ involving idempotents e_i such that $e_i e_j = 0$ if $i \neq j$. It is easy to prove that *a graded algebra A is a graded Azumaya algebra over K if and only if Ae_i is a graded Azumaya algebra over Ke_i for $i = 1, 2, \dots, n$.* When A_0 and A_1 are finitely generated projective modules, it follows from (1.12.8) that there are idempotents $e_1, e_2, \dots, e_n$ in K such that, on one side $1 = \sum_i e_i$ and $e_i e_j = 0$ if $i \neq j$, and on the other side $e_i A_0$ and $e_i A_1$ have constant rank as modules over Ke_i for $i = 1, 2, \dots, n$. Consequently it suffices to study graded Azumaya algebras A such that A_0 and A_1 have constant ranks.

Examples of graded Azumaya algebras

Important consequences will follow from our first example.

(3.5.8) **Proposition.** *When P is a graded finitely generated and faithful projective module, and $A = \mathrm{End}(P)$, then A is a graded Azumaya algebra, and $\mathrm{Z}(A) = \mathrm{Z}^g(A) = K$. Moreover A_1 is a faithful module if and only if P_0 and P_1 are both faithful modules. And when A_1 is a faithful module, all these statements are true:*

- *the grading of A is regular;*
- *$\mathrm{Z}(A_0) = \mathrm{Z}(A_0, A)$, and $\mathrm{Z}(A_0, A)$ is isomorphic to the trivial quadratic extension K^2;*
- *if φ is the standard involution of $\mathrm{Z}(A_0, A)$, the equality $yz = \varphi(z)y$ is valid for all $y \in A_1$ and all $z \in \mathrm{Z}(A_0, A)$;*
- *when its grading is forgotten, A is a nongraded Azumaya algebra over K;*
- *A_0 is a nongraded Azumaya algebra over its center.*

Proof. Let us consider this graded K-linear mapping:

$$\mathrm{End}(P)^{to} \longrightarrow \mathrm{End}(P^*) \ , \quad f^{to} \longmapsto (u \longmapsto (-1)^{\partial f \partial u} \ u \circ f) \ ;$$

a straightforward verification shows that it is an algebra morphism. To prove that it is bijective, we also consider the bijective mapping $P \otimes P^* \to \mathrm{End}(P)$ already mentioned in the proof of (3.5.4), and the similar bijective mapping $P^* \otimes P^{**} \to \mathrm{End}(P^*)$. Because of all the already introduced twisting signs, we identify P with P^{**} by means of the following bijection:

$$P \longrightarrow P^{**} \ , \quad x \longmapsto (u \longmapsto (-1)^{\partial u \partial x} u(x)).$$

All these bijections allow us to derive the bijectiveness of $\mathrm{End}(P)^{to} \to \mathrm{End}(P^*)$ from the bijectiveness of the mapping $P \otimes P^* \to P^* \otimes P$ defined by $x \otimes u \longmapsto (-1)^{\partial u \partial x} u \otimes x$. Another straightforward calculation shows that the canonical mapping $A \hat{\otimes} A^{to} \to \mathrm{End}(A)$ is equal to the following composition of bijections:

$$\mathrm{End}(P) \hat{\otimes} \mathrm{End}(P)^{to} \longrightarrow \mathrm{End}(P) \hat{\otimes} \mathrm{End}(P^*) \longrightarrow \mathrm{End}(P \otimes P^*) \longrightarrow \mathrm{End}(\mathrm{End}(P)) \ ;$$

consequently A is a graded Azumaya algebra.

Later the equality $\mathrm{Z}(A) = K$ will be an immediate consequence of Morita theory (see (6.4.5)), but an elementary proof can be achieved already now. Indeed by localization we reduce the problem to the case of a free module P with a basis $(e_1, e_2, \dots, e_r)$ such that $r > 0$ (because P is faithful); let g_1 be the parallel projection onto Ke_1 with respect to the submodule spanned by $e_2,\dots,e_r$; if f belongs to $\mathrm{Z}(A)$, the equality $fg_1 = g_1 f$ implies that e_1 is an eigenvector of f, and similarly $e_2,\dots,e_r$ are also eigenvectors; since $(e_1 + e_2, \ e_2, \dots, e_r)$ is also a basis, $e_1 + e_2$ is also an eigenvector, and also every $e_i + e_j$ with $i < j$; consequently all eigenvalues are equal to some $\lambda \in K$, and f is the canonical image of λ in A.

The even subalgebra A_0 and the odd component A_1 are respectively isomorphic to

$$\mathrm{End}(P_0) \oplus \mathrm{End}(P_1) \qquad \text{and} \qquad \mathrm{Hom}(P_0, P_1) \oplus \mathrm{Hom}(P_1, P_0) \ .$$

If m and n are the ranks of P_0 and P_1 at some prime ideal of K, the ranks of A_0 and A_1 are respectively $m^2 + n^2$ and $2mn$. Consequently A_1 is a faithful module if and only if P_0 and P_1 are both faithful. Let p_0 and p_1 be the projections $P \to P_0$ and $P \to P_1$. Every element f of $\mathrm{Z}(A_0, A)$ must commute with these even endomorphisms p_0 and p_1; consequently $f(P_i) \subset P_i$ for $i = 0, 1$, in other words, f is even. Thus we have proved that $\mathrm{Z}(A_0, A) = \mathrm{Z}(A_0)$ and $\mathrm{Z}^g(A) = \mathrm{Z}_0(A) = \mathrm{Z}(A) = K$. The center of A_0 is $Kp_0 \oplus Kp_1$ because it is naturally isomorphic to the direct sum of the centers of $\mathrm{End}(P_0)$ and $\mathrm{End}(P_1)$; therefore it is isomorphic to K^2 when P_0 and P_1 are both faithful.

The image of the multiplication mapping $A_1 \otimes A_1 \to A_0$ is obviously an ideal of A_0; let us prove that it contains id_P when P_0 and P_1 are both faithful; by localization we reduce the problem to the case of free modules P_0 and P_1; let

$(e_1, e_2, \dots, e_r)$ be a basis of P which begins with a basis of P_0 and ends with a basis of P_1; it suffices to prove that the above projection g_1 onto Ke_1 belongs to the image of $A_1 \otimes A_1 \to A_0$ since $\mathrm{id}_P = g_1 + g_2 + \cdots + g_r$; as a matter of fact, it is easy to find two odd endomorphisms f' and f'' such that $g_1 = f'f''$; for instance we can require that $f''(e_1) = e_r$, $f'(e_r) = e_1$, and $f''(e_i) = 0$ if $i > 1$.

When P_0 and P_1 are faithful modules, the standard involution φ of $\mathrm{Z}(A_0, A)$ permutes p_0 and p_1. It is easy to verify that $hp_0 = p_1 h$ and $hp_1 = p_0 h$ for every $h \in A_1$, and with slightly different notations this is the same thing as the equality $yz = \varphi(z)y$.

At last, the grading of A is forgotten if we give P the trivial grading, and this does not invalidate the conclusion that A is an Azumaya algebra over K. Let us suppose that P_0 and P_1 are both faithful, and let us consider A_0 as an algebra over its center $Z = Kp_0 \oplus Kp_1$; since p_0 and p_1 are idempotents, A_0 is an Azumaya algebra over Z if and only if $p_i A_0$ is an Azumaya algebra over Kp_i for $i = 0, 1$, or in other words, if and only if $\mathrm{End}(P_i)$ is an Azumaya algebra over K for $i = 0, 1$; actually $\mathrm{End}(P_i)$ is an Azumaya algebra like $\mathrm{End}(P)$. □

General consequences are immediately derived from (3.5.8).

(3.5.9) **Proposition.** *The equalities $\mathrm{Z}^g(A) = \mathrm{Z}_0(A) = K$ are valid for every graded Azumaya algebra A over K. Moreover the grading of A is regular if and only if A_1 is a faithful K-module. When K is a field, A contains no graded (two-sided) ideal other than 0 and A.*

Proof. Since A is faithfully flat, the mapping $x \longmapsto x \otimes 1^{to}$ is injective from A into $A \hat{\otimes} A^{to}$; if x belongs to $\mathrm{Z}^g(A)$, then $x \otimes 1^{to}$ belongs to $\mathrm{Z}^g(A \hat{\otimes} A^{to})$ which is equal to $K \otimes 1^{to}$ because of the isomorphism $A \hat{\otimes} A^{to} \to \mathrm{End}(A)$; consequently x belongs to K. Let $A_1 A_1$ be the image of $A_1 \otimes A_1$ in A_0, or in other words, the ideal of A_0 generated by the products of two odd elements. Since A_0 is always faithful (indeed $K \subset A_0$), the equality $A_1 A_1 = A_0$ implies that A_1 too is faithful. Conversely if A_1 is faithful, the grading of $\mathrm{End}(A)$ is regular, and also that of $A \hat{\otimes} A^{to}$; the ideal of $(A \hat{\otimes} A^{to})_0$ generated by the products of two odd elements is the direct sum of $A_1 \otimes A_1^{to}$ and the following ideal N of $A_0 \otimes A_0^{to}$:

$$\begin{aligned} N &= (A_1 A_1 \otimes A_0^{to}) + (A_0 \otimes (A_1 A_1)^{to}) \\ &= \mathrm{Ker}\ (A_0 \otimes A_0^{to} \longrightarrow (A_0/A_1 A_1) \otimes (A_0/A_1 A_1)^{to})\ ; \end{aligned}$$

if $A_1 A_1$ were not equal to A_0, the quotient algebra $A_0/A_1 A_1$ would not be reduced to 0; when an algebra B is not reduced to 0, the same is true for $B \hat{\otimes} B^{to}$, since the multiplication in B determines a nonzero mapping $B \otimes B \to B$; all this would imply that the ideal N would not be equal to $A_0 \otimes A_0^{to}$, contrary to the fact that the grading of $A \hat{\otimes} A^{to}$ is regular; consequently the grading of A is also regular.

When K is a field, $\mathrm{End}(A)$ contains no ideal other than 0 or itself; later this will be an immediate consequence of Morita theory (see (6.4.5)), but elementary proofs are also available: see (3.ex.15). If $\mathfrak{a}$ is a graded ideal of A (such that

$\mathfrak{a} = (\mathfrak{a} \cap A_0) \oplus (\mathfrak{a} \cap A_1))$, then $\mathfrak{a} \otimes A^{to}$ is also a graded ideal of $A \hat{\otimes} A^{to}$, which must be equal to 0 or the whole algebra; consequently $\mathfrak{a}$ is 0 or A. □

(3.5.10) **Remark.** When K is a field, a finite-dimensional and graded K-algebra A (not reduced to 0) in which $Z^g(A) = K$ and every graded ideal is equal to 0 or A, is called a *graded central simple algebra.* The study of graded central simple algebras in **6.6** is an important step toward the most difficult theorems about graded Azumaya algebras, that have been postponed up to **6.7**; the importance of this step comes from the assertion (d) in (3.5.6). In Theorem (6.6.4) it is stated that conversely every graded central simple algebra over a field is a graded Azumaya algebra.

Here is another example of a graded Azumaya algebra.

(3.5.11) **Proposition.** *Let A be a graded algebra that is a finitely generated projective module of constant rank 2. It is a graded Azumaya algebra if and only if A is a quadratic extension and A_1 has constant rank 1. These conditions require 2 to be invertible in K.*

Proof. Since A is commutative, $Z_0(A) = A_0$. If A is an Azumaya algebra, then $Z_0(A) = K$ and consequently A_0 has constant rank 1, and A_1 too. By localization we can reduce the problem to the case of an algebra $A = K \oplus Kz$ in which z generates A_1; consequently $z^2 = -\gamma$ for some $\gamma \in K$. The images of $1 \otimes 1^{to}$, $z \otimes z^{to}$, $z \otimes 1^{to}$, $1 \otimes z^{to}$ in $\mathrm{End}(A)$ are described by the following matrices:

$$\begin{pmatrix} 1 & 0 \\ 0 & 1 \end{pmatrix}, \quad \begin{pmatrix} -\gamma & 0 \\ 0 & \gamma \end{pmatrix}, \quad \begin{pmatrix} 0 & -\gamma \\ 1 & 0 \end{pmatrix}, \quad \begin{pmatrix} 0 & \gamma \\ 1 & 0 \end{pmatrix};$$

with these matrices we get a basis of $\mathrm{End}(A)$ if and only if 2γ is invertible, and this means exactly that A is a quadratic extension. □

When 2 is invertible in K, with every nongraded algebra B we associate the graded algebra $(B^2)^g$ which is the algebra $B \times B$ provided with the following grading: $(B^2)^g_0$ is the subalgebra of all (b, b) with $b \in B$, whereas $(B^2)^g_1$ is the submodule of all $(b, -b)$. Obviously $(B^2)^g$ is isomorphic to $(K^2)^g \otimes B$. Its grading is always balanced and regular, because $(B^2)^g_1$ contains the invertible element $(1_B, -1_B)$.

(3.5.12) **Proposition.** *We assume that 2 is invertible in K. Let P be a nongraded finitely generated and faithful projective module, and $A = (\mathrm{End}(P)^2)^g$. Then A is a graded Azumaya algebra, provided with a regular and balanced grading. Moreover $Z(A_0) = K$, $Z(A) = Z(A_0, A)$, and $Z(A_0, A)$ is a quadratic extension isomorphic to $(K^2)^g$. If φ is the standard involution of $Z(A_0, A)$, the equality $yz = (-1)^{\partial z}\varphi(z)y$ is valid for all $y \in A_1$ and all homogeneous $z \in Z(A_0, A)$. At last, A_0 is a nongraded Azumaya algebra over K, and when the grading of A is forgotten, it is a nongraded Azumaya algebra over its center.*

Proof. First A is a graded Azumaya algebra because it is isomorphic to $(K^2)^g \otimes \mathrm{End}(P)$; indeed $(K^2)^g$ is a graded Azumaya algebra (see (3.5.11)), and $\mathrm{End}(P)$ is a trivially graded Azumaya algebra (see (3.5.8)). Obviously A_0 is isomorphic to $\mathrm{End}(P)$, and therefore is a nongraded Azumaya algebra over K; whence $\mathrm{Z}(A_0) = K$. From (3.5.3) we derive that A without its grading is a nongraded Azumaya algebra over K^2; therefore $\mathrm{Z}(A) \cong K^2$, and with its grading $\mathrm{Z}(A)$ is isomorphic to $(K^2)^g$. The natural algebra morphism $\mathrm{Z}(A) \otimes A_0 \to A$ is obviously bijective, and since $\mathrm{Z}(A_0) = K$, we come easily to $\mathrm{Z}(A_0, A) = \mathrm{Z}(A)$. Consequently the equality $yz = zy$ is true for all $y \in A_1$ and all $z \in \mathrm{Z}(A_0, A)$, and since $\varphi(z) = (-1)^{\partial z} z$, it implies $yz = (-1)^{\partial z}\varphi(z)y$. □

Remark. Let A be $\mathrm{End}(P)$ as in (3.5.8), or $(\mathrm{End}(P)^2)^g$ as in (3.5.12), and let z be a homogeneous element of $\mathrm{Z}(A_0, A)$; in both cases the equality $yz = (-1)^{\partial z}\varphi(z)y$ is valid for all $y \in A_1$; remember that $\partial z = 0$ in the case of (3.5.8). From the definition of $\mathrm{Z}(A_0, A)$ it follows that $yz = zy$ if y is even. Both equalities can be combined in the following one, that is valid for all (homogeneous) $y \in A$ and all (homogeneous) $z \in \mathrm{Z}(A_0, A)$:

(3.5.13) $$yz = (-1)^{\partial y \partial z} \varphi^{\partial y}(z)\, y \ ;$$

observe that φ^n is well defined when the exponent n is an element of $\mathbb{Z}/2\mathbb{Z}$, because φ is involutive, and φ^n is equal to id or φ according to the parity of n.

Other general theorems

The previous examples lead us to the following theorem, which here is proved only with an additional hypothesis; anyhow this hypothesis will always be fulfilled when this theorem is used in this chapter, and in **6.7** it is proved that the theorem holds true without this hypothesis.

(3.5.14) **Theorem.** *Let A be a graded Azumaya algebra such that A_0 and A_1 have constant ranks. Then the rank of A is either a square, or the double of a square; in the former case we say that A has* even type, *and in the latter case that it has* odd type. *Now three cases can be distinguished:*

(a) *When $A_1 = 0$, then A has even type and*

$$\mathrm{Z}^g(A) = \mathrm{Z}_0(A) = \mathrm{Z}(A) = \mathrm{Z}(A_0) = \mathrm{Z}(A_0, A) = K \ .$$

(b) *When A has even type and $A_1 \neq 0$, there are two integers m and n such that the ranks of A_0 and A_1 are respectively equal to $m^2 + n^2$ and $2mn$. Moreover*

$$\mathrm{Z}^g(A) = \mathrm{Z}_0(A) = \mathrm{Z}(A) = K \quad \textit{and} \quad \mathrm{Z}(A_0) = \mathrm{Z}(A_0, A) \ .$$

(c) *When A has odd type, 2 must be invertible in K and the grading of A is balanced. Moreover*

$$\mathrm{Z}^g(A) = \mathrm{Z}_0(A) = \mathrm{Z}(A_0) = K \quad \textit{and} \quad \mathrm{Z}(A) = \mathrm{Z}(A_0, A) \ ;$$

in this case, the multiplication mapping $\mathrm{Z}(A) \otimes A_0 \to A$ is bijective.

In both cases (b) *and* (c), *the grading of A is regular and* $\mathrm{Z}(A_0, A)$ *is a graded quadratic extension of K, which is trivially graded in the case* (b), *whereas* $\mathrm{Z}_1(A_0, A)$ *has constant rank* 1 *in the case* (c). *If φ is the standard involution of* $\mathrm{Z}(A_0, A)$, *in both cases* (b) *and* (c) *the equality* (3.5.13) *is valid for all (homogeneous) $y \in A$ and $z \in \mathrm{Z}(A_0, A)$. At last A_0 is always a nongraded Azumaya algebra over its center, and when the grading of A is forgotten, it is a nongraded Azumaya algebra over its center.*

Beginning of the proof. Here we prove (3.5.14) only when this additional hypothesis is fulfilled: there exists a faithfully flat extension $K \to L$ such that $L \otimes A$ is isomorphic either to $\mathrm{End}_L(P)$ or to $(\mathrm{End}_L(P)^2)^g$ with P a finitely generated and faithful projective L-module, graded in the former case, nongraded in the latter case. Obviously A has even type in the former case, and odd type in the latter case; moreover 2 must be invertible in K in the latter case.

When the additional hypothesis is fulfilled, then (3.5.14) is an immediate consequence of (3.5.8) or (3.5.12), and the various theorems stating that some properties are true if they prove to be true after a faithfully flat extension. With the help of (1.12.12) and (1.12.13) we get the statements involving the ranks of A, A_0 and A_1. The case (a) occurs when $L \otimes A$ is isomorphic to $\mathrm{End}(P)$ with P a trivially graded module over L, the case (b) when it is isomorphic to $\mathrm{End}(P)$ with P_0 and P_1 both faithful over L, and the case (c) when it is isomorphic to $(\mathrm{End}(P)^2)^g$. Since we already know that $\mathrm{Z}^g(A) = \mathrm{Z}_0(A) = K$ (see (3.5.9)), we only consider $\mathrm{Z}(A)$, $\mathrm{Z}(A_0)$ and $\mathrm{Z}(A_0, A)$; they can be described as finite intersections of kernels of mappings like $a \longmapsto ab - ba$ because A and A_0 are finitely generated modules; and since L is faithfully flat, we can write

$$\mathrm{Z}(L \otimes A) = L \otimes \mathrm{Z}(A)\ , \quad \mathrm{Z}(L \otimes A_0) = L \otimes \mathrm{Z}(A_0)\ ,$$
$$\mathrm{Z}(L \otimes A_0,\ L \otimes A) = L \otimes \mathrm{Z}(A_0, A).$$

Besides, when M runs through the set of K-submodules of A, the mapping $M \longmapsto L \otimes M$ is injective into the set of L-submodules of $L \otimes A$. Consequently the announced properties of $\mathrm{Z}(A)$, $\mathrm{Z}(A_0)$ and $\mathrm{Z}(A_0, A)$ are immediate consequences of properties of their L-extensions which can be deduced from (3.5.8) and (3.5.12); in particular $\mathrm{Z}(A_0, A)$ is a quadratic extension in both cases (b) and (c) because $L \otimes \mathrm{Z}(A_0, A)$ is isomorphic either to L^2 in the case (b), or to $(L^2)^g$ in the case (c).

The faithful flatness of L also allows us to carry the equality (3.5.13) from $L \otimes A$ back to A. The mapping $\mathrm{Z}(A) \otimes A_0 \to A$ is an isomorphism when A has odd type, because $\mathrm{Z}_0(A_0, A)$ (that is K) and $\mathrm{Z}_1(A_0, A)$ (the discriminant module) have constant rank 1, and every localization of $\mathrm{Z}_1(A_0, A)$ is generated by an invertible element. At last, A_0 and A without its grading are nongraded Azumaya algebras over their centers if some mappings are bijective (see (3.5.1)), and their bijectiveness can be tested by means of a faithfully flat extension.

When K is a field, later in (6.6.5) it is proved that the above additional hypothesis is fulfilled for every graded central simple algebra over K, and thus

(3.5.14) becomes valid over a field without more hypotheses. Consequently when K is not a field, (3.5.14) becomes valid for all the extensions $(K/\mathfrak{m}) \otimes A$ with $\mathfrak{m}$ a maximal ideal of K; but more work is required until (3.5.14) is completely proved in **6.7**, since the separability theory presented in **6.5** plays an important role at this moment. □

The following theorem is only stated for information, since we do not really need it here; we only need it in (3.6.6), (3.6.7) and (3.6.8) below, and later in (6.ex.11).

(3.5.15) **Theorem.** *Let A be a graded Azumaya algebra such that A_0 and A_1 have constant ranks. If A has even type, there exists a faithfully flat extension $K \to L$ and a graded L-module $P = P_0 \oplus P_1$ such that P_0 and P_1 are free modules and $L \otimes A$ is isomorphic to* $\mathrm{End}_L(P)$. *If A has odd type, there exists a faithfully flat extension $K \to L$ and a free L-module P such that $L \otimes A$ is isomorphic to* $(\mathrm{End}_L(P)^2)^g$.

In other words, the additional hypothesis with which (3.5.14) has been proved, is always fulfilled. The fact that (3.5.15) mentions free modules instead of projective modules is not meaningful, since every finitely generated projective module of constant rank gives a free module after a suitable Zariski extension (see (1.ex.21)); free modules are anyhow indispensable in the proof of (3.6.6). Nevertheless it would be an illusion to believe that (3.5.15) might help to prove (3.5.14), because the proof of (3.5.14) is rather a step toward the proof of (3.5.15). In [Knus 1991] there is a proof of (3.5.15) for (trivially graded) Azumaya algebras which uses almost all the arguments involved in the proof of (3.5.14), and other arguments still more difficult. Then (3.5.14) is needed to complete the proof of (3.5.15) with nontrivial gradings.

(3.5.16) **Remarks.** When A_0 and A_1 do not have constant rank, A is said to have even type (resp. odd type) if all its localizations have even type (resp. odd type). When A and B have constant types, by examining the rank of $A \hat{\otimes} B$ at each prime ideal we find that it has also a constant type, which is even (resp. odd) if and only if A and B have the same type (resp. different types); in other words, the type of $A \hat{\otimes} B$ is the sum of the types of A and B.

When the type of A is not constant, there is an idempotent e in K such that $(1-e)A$ is a graded Azumaya algebra of even type over $K(1-e)$, whereas eA is a graded Azumaya algebra of odd type over Ke. Thus $e\mathrm{Z}_1(A_0, A)$ has constant rank 1 over Ke, whereas $(1-e)\mathrm{Z}_1(A_0, A) = 0$. Besides, $\mathrm{Z}(A_0, A)$ is a quadratic extension of K whenever A_1 is a faithful module.

(3.5.17) **Theorem.** *Let A and B be graded Azumaya algebras such that A_1 and B_1 are faithful modules, and let us identify $\mathrm{Z}(A_0, A) \hat{\otimes} \mathrm{Z}(B_0, B)$ with a subalgebra of $A \hat{\otimes} B$. Then*

$$\mathrm{Z}((A \hat{\otimes} B)_0\,,\, A \hat{\otimes} B) \;=\; \mathrm{Z}(A_0, A) \star \mathrm{Z}(B_0, B)\ .$$

Proof. The identification advised in (3.5.17) is sensible because $\mathrm{Z}(A_0, A)$ and B are faithfully flat modules (indeed they are faithful and projective), and afford injective morphisms

$$\mathrm{Z}(A_0, A) \hat{\otimes} \mathrm{Z}(B_0, B) \longrightarrow \mathrm{Z}(A_0, A) \hat{\otimes} B \longrightarrow A \hat{\otimes} B.$$

Let us set $C = A \hat{\otimes} B$. We first prove that $\mathrm{Z}(C_0, C)$ is contained in

$$\mathrm{Z}(A_0, A) \hat{\otimes} \mathrm{Z}(B_0, B).$$

By localization we reduce the problem to the case of a free module B with a basis $(b'_1, \dots, b'_n)$; then every $z'' \in \mathrm{Z}(C_0, C)$ can be written (in a unique way) $z'' = \sum_i a_i \otimes b'_i$ for some $a_1, \dots, a_n$ in A; from the fact that z'' commutes with $x \otimes 1$ for every $x \in A_0$, we derive the equality

$$\sum_{i=1}^{n} (xa_i - a_i x) \otimes b'_i \;=\; 0 \;;$$

this proves that each a_i commutes with all $x \in A_0$, and consequently z'' belongs to $\mathrm{Z}(A_0, A) \hat{\otimes} B$. Then a similar argument using a basis of $\mathrm{Z}(A_0, A)$ shows that z'' belongs to $\mathrm{Z}(A_0, A) \hat{\otimes} \mathrm{Z}(B_0, B)$.

Now let φ and φ' be the standard involutions of $\mathrm{Z}(A_0, A)$ and $\mathrm{Z}(B_0, B)$, and let y, z, y', z' be homogeneous elements respectively in A, $\mathrm{Z}(A_0, A)$, B and $\mathrm{Z}(B_0, B)$. From the equality (3.5.13) (and the similar equality involving φ') we deduce

$$(y \otimes y')\;(z \otimes z') \;=\; (-1)^{(\partial y + \partial y')(\partial z + \partial z')}\;(\varphi^{\partial y} \otimes \varphi'^{\partial y'})(z \otimes z')\;(y \otimes y');$$

this shows that an element z'' of $\mathrm{Z}(A_0, A) \hat{\otimes} \mathrm{Z}(B_0, B)$ belongs to $\mathrm{Z}(C_0, C)$ if it is invariant by $\varphi \otimes \varphi'$; indeed it commutes with every $y \otimes y'$ in $A_0 \otimes B_0$ or in $A_1 \otimes B_1$; consequently $\mathrm{Z}(A_0, A) \star \mathrm{Z}(B_0, B)$ is contained in $\mathrm{Z}(C_0, C)$. This inclusion is an equality because of (3.4.5). □

In some contexts it is useful to associate a graded quadratic extension $\mathrm{QZ}(A)$ with each graded Azumaya algebra A, even when A_1 is not a faithful module; of course when A_1 is faithful, $\mathrm{QZ}(A)$ is merely $\mathrm{Z}(A_0, A)$. When A_1 is not faithful, there is a unique idempotent e in K such that eA_1 is faithful over Ke whereas $(1-e)A_1 = 0$; with the trivially graded algebra $(1-e)A$ we associate the trivial quadratic extension $(1-e)K^2$ of $(1-e)K$ and consequently, by definition,

$$\mathrm{QZ}(A) \;=\; (1-e)K^2 \;\oplus\; e\mathrm{Z}(A_0, A)\;.$$

(3.5.18) **Corollary.** *When A and B are graded Azumaya algebras, then $\mathrm{QZ}(A \hat{\otimes} B)$ is isomorphic to $\mathrm{QZ}(A) \star \mathrm{QZ}(B)$.*

Proof. As explained in (3.5.7), we can assume that A_0, A_1, B_0, B_1 have constant ranks. When A_1 and B_1 are faithful modules, (3.5.18) follows from (3.5.17). When A_1 and B_1 are both zero, (3.5.18) is a triviality. It remains to consider this case: A_1 is zero whereas B_1 is faithful; in this case, (3.5.18) means that $\mathrm{Z}(C_0, C)$ (with $C = A \hat{\otimes} B$ as above) is isomorphic to $\mathrm{Z}(B_0, B)$. As explained in the proof of (3.5.17), $\mathrm{Z}(C_0, C)$ is contained in $\mathrm{Z}(A_0, A) \hat{\otimes} \mathrm{Z}(B_0, B)$, which is now the quadratic extension $K \otimes \mathrm{Z}(B_0, B)$. Thus the conclusion follows from (3.4.5). □

The Brauer–Wall group $\mathrm{Br}^g(K)$

Proposition (3.5.5) shows that the isomorphy classes of graded Azumaya algebras over K constitute a monoid, in which the class of K is the unit element. Lemma (3.5.4) shows that the isomorphy classes of all graded algebras $\mathrm{End}(P)$, with P a graded finitely generated and faithful projective module, constitute a submonoid; this submonoid is absorbent (see the definition in the first part of **2.7**) because $A \hat{\otimes} A^{to}$ is isomorphic to $\mathrm{End}(A)$ when A is a graded Azumaya algebra. Despite the multiplicative notations involved in this monoid, from Proposition (2.7.1) we deduce that the quotient of this monoid by this absorbent submonoid is a group; this group is called the *Brauer–Wall group* (or graded Brauer group) of K, and denoted by $\mathrm{Br}^g(K)$. If A and B are graded Azumaya algebras over K, their classes $[A]$ and $[B]$ in $\mathrm{Br}^g(K)$ are equal if there exist graded finitely generated and faithful projective modules P and Q such that $A \hat{\otimes} \mathrm{End}(P)$ is isomorphic to $B \hat{\otimes} \mathrm{End}(Q)$. Observe that the inverse class $[A]^{-1}$ is the class of A^{to}.

When A and B are trivially graded, the above condition equivalent to $[A] = [B]$ now means that the ordinary tensor products $A \otimes \mathrm{End}(P)$ and $B \otimes \mathrm{End}(Q)$ are isomorphic, and if they are isomorphic as graded algebras, they are still isomorphic when the gradings are forgotten; consequently $\mathrm{Br}^g(K)$ contains a subgroup $\mathrm{Br}(K)$, the ordinary *Brauer group*, which can be defined by means of isomorphy classes of trivially graded Azumaya algebras. Often $\mathrm{Br}(K)$ and $\mathrm{Br}^g(K)$ are treated as multiplicative groups, but there are also sensible reasons to treat them rather as additive groups (see (3.7.9)).

Because of (3.5.3), every extension $K \longrightarrow K'$ induces a group morphism $\mathrm{Br}^g(K) \to \mathrm{Br}^g(K')$. Thus we get a functor Br^g from the category of commutative rings to the category of commutative groups.

With each graded Azumaya algebra A is associated a quadratic extension $\mathrm{QZ}(A)$ defined just before (3.5.18); when A is isomorphic to some $\mathrm{End}(P)$, then $\mathrm{QZ}(A)$ is trivial (see (3.5.8)); because of (3.5.18) there is a group morphism $\mathrm{Br}^g(K) \to \mathrm{Q}^g(K)$ defined by $[A] \longmapsto [\mathrm{QZ}(A)]$. In **3.8**, the exactness of the following sequence shall be proved:

$$1 \longrightarrow \mathrm{Br}(K) \longrightarrow \mathrm{Br}^g(K) \longrightarrow \mathrm{Q}^g(K) \longrightarrow 1. \tag{3.5.19}$$

This exact sequence allows us to calculate $\mathrm{Br}^g(K)$ when $\mathrm{Br}(K)$ is known, provided that $\mathrm{Q}^g(K)$ has been already calculated by means of (3.4.13) and (3.4.14).

(3.5.20) **Examples.** When K is a field, a graded algebra B of finite dimension over K is called a *graded division algebra* if every nonzero homogeneous element is invertible; if moreover $\mathrm{Z}^g(B) = K$, we say that B is a *graded central division algebra*. When the grading of B is trivial, it is a division algebra if every nonzero element is invertible, a central division algebra if moreover $\mathrm{Z}(B) = K$. It is worth noticing that $(K^2)^g$ is a graded central division algebra over K (when 2 is invertible in K), although it is neither a division algebra nor a central algebra when its grading is forgotten. Obviously every graded central division algebra B is a graded central simple algebra, consequently a graded Azumaya algebra because of Theorem (6.6.4). The Brauer–Wall group of a field K classifies the graded central division algebras over K, because Theorem (6.6.2) implies that every graded Azumaya algebra A has the same class as a graded central division algebra B, which is uniquely determined by A up to isomorphy.

When K is a field, the group $\mathrm{Ip}(K)$ has order 2, but the subgroup $\mathrm{Ip}'(K)$ appearing in (3.4.13) has order 2 or 1 according as 2 is invertible or not in K. Thus the knowledge of $\mathrm{Q}(K)$ and $\mathrm{Br}(K)$ gives information about $\mathrm{Br}^g(K)$ with the help of (3.4.13) and (3.5.19). For instance if K is an algebraically closed field, then $\mathrm{Br}(K) = 1$, because in each division algebra of finite dimension over K the subalgebra generated by any nonzero element is a (commutative) field, therefore equal to K. Then (3.4.14) shows that $\mathrm{Q}(K) = \mathrm{Q}_f(K) = 1$ for this algebraically closed field K, and consequently $\mathrm{Br}^g(K)$ is isomorphic to $\mathrm{Q}^g(K)$ and to $\mathrm{Ip}'(K)$.

Wedderburn's theorem states that every finite division ring is a (commutative) field; consequently $\mathrm{Br}(K) = 1$ when K is a finite field, and $\mathrm{Br}^g(K)$ is isomorphic to $\mathrm{Q}^g(K)$. Let n be the cardinal of K. When n is odd, $\mathrm{Q}(K)$ is isomorphic to the quotient of $K^\times$ (a cyclic group of order $n-1$) by the subgroup of all squares; therefore $\mathrm{Q}(K)$ has order 2 and $\mathrm{Q}^g(K)$ has order 4. If A is a nontrivially graded quadratic extension over this field, then $A \star A$ is isomorphic to $K[i]$ with $i^2 = -1$ (because of the twisting in $A \hat{\otimes} A$); if $n-1$ is divisible by 4, then -1 has a square root in K and the class $[A]$ has order 2 in $\mathrm{Q}^g(K)$; consequently $\mathrm{Q}^g(K)$ is isomorphic to $(\mathbb{Z}/2\mathbb{Z})^2$; but if $n-1$ is not divisible by 4, then -1 has no square root in K and $\mathrm{Q}^g(K)$ is a cyclic group generated by $[A]$. When n is even, $\mathrm{Q}^g(K)$ and $\mathrm{Br}^g(K)$ are isomorphic to $\mathrm{Q}(K) = \mathrm{Q}_f(K)$ which has order 2 because of (3.4.14).

When K is the field $\mathbb{R}$ of real numbers, it is clear that $\mathrm{Q}(\mathbb{R})$ is a group of order 2. Since the class of $(\mathbb{R}^2)^g$ has order 4 in $\mathrm{Q}^g(\mathbb{R})$, this group is cyclic of order 4. In (3.ex.17) it is proved that $\mathrm{Br}(\mathbb{R})$ is a group of order 2. Consequently $\mathrm{Br}^g(\mathbb{R})$ contains eight elements, and in (3.ex.22) Clifford algebras are used to prove that this group is cyclic and generated by the Brauer class of $(\mathbb{R}^2)^g$.

When K is the ring $\mathbb{Z}$ of integers, it is easy to prove that $\mathrm{Q}^g(\mathbb{Z})$ is a group of order 1 (see (3.ex.10)) but much more work is necessary to prove that $\mathrm{Br}(\mathbb{Z}) = 1$, and consequently $\mathrm{Br}^g(\mathbb{Z}) = 1$.

3.6 Traces and determinants

This section presents more specialized information about Azumaya algebras, which hurried readers may omit until they really need it.

When P is a finitely generated projective module, we can define the *trace* (resp. the *determinant*) of every endomorphism f of P. When P is a free module with finite bases, there is no problem in defining it as the trace (resp. the determinant) of its matrix in any basis of P, since changes of bases involve invertible matrices. When P is not free, there is a module P' such that $P \oplus P'$ is free with finite bases; we extend f with zero on P' (resp. with the identity mapping of P') in order to get an endomorphism f' of $P \oplus P'$, and by definition the trace of f (resp. its determinant) is that of f'. Of course we must check that the extension f'' of f to another free module $P \oplus P''$ gives the same trace (resp. the same determinant); indeed the module $P \oplus (P \oplus P' \oplus P'')$ is also free with finite bases, and gives another extension f''' of f by means of zero (resp. the identity mapping) on $P \oplus P' \oplus P''$; it is clear that

$$\mathrm{tr}(f') = \mathrm{tr}(f''') = \mathrm{tr}(f'') \quad (\text{resp. } \det(f') = \det(f''') = \det(f'')).$$

Traces and determinants can be calculated by localization. For instance let us calculate $\mathrm{tr}(\mathrm{id}_P)$ when the rank of P takes the values $r_1, r_2, \dots, r_k$ and $e_1, e_2, \dots, e_k$ are the corresponding idempotents of K as in (1.12.8):

$$\mathrm{tr}(\mathrm{id}_P) \quad = \quad r_1 e_1 + r_2 e_2 + \cdots + r_k e_k \ ;$$

consequently $\mathrm{tr}(\mathrm{id}_P)$ is invertible in K if and only if r_i is invertible in Ke_i for $i = 1, 2, \dots, k$.

The usual properties of traces and determinants remain valid in this context; for instance the following three lemmas can be proved by localization with ordinary matrix calculus.

(3.6.1) **Lemma.** *Let P be a finitely generated projective module and $A = \mathrm{End}(P)$. The linear form $\mathrm{tr} : A \to K$ is surjective and its kernel (which is consequently a direct summand of A) is the submodule $[A, A]$ generated by all Lie brackets $[f, g] = fg - gf$. Moreover the bilinear form $(f, g) \longmapsto \mathrm{tr}(fg)$ is symmetric and nondegenerate.*

(3.6.2) **Lemma.** *Let M be a finitely generated projective module of constant rank m that contains a direct summand N of constant rank n, and let $p : M \to N$ be a projector onto N with respect to a supplementary submodule. Then the left (or right) ideal generated by p is a direct summand of $\mathrm{End}(M)$ of constant rank mn, and $\mathrm{tr}(p)$ is the image of n in K.*

(3.6.3) **Lemma.** *If P is a finitely generated projective module of constant rank 2, the mapping $f \longmapsto \mathrm{tr}(f)\mathrm{id}_P - f$ is a standard involution of the algebra $\mathrm{End}(P)$, and the associated norm is the quadratic form $f \longmapsto \det(f)$.*

The next lemma is more difficult.

(3.6.4) **Lemma.** *If P is a finitely generated projective module, and w an automorphism of the algebra* $\mathrm{End}(P)$, *the equalities* $\mathrm{tr}(w(f)) = \mathrm{tr}(f)$ *and* $\det(w(f)) = \det(f)$ *hold for every* $f \in \mathrm{End}(P)$.

Proof. Both linear forms tr and $\mathrm{tr} \circ w$ on $\mathrm{End}(P)$ are surjective and have the same kernel; consequently the latter is the product of the former by some $\kappa \in K^\times$. By localization we reduce the calculation of κ to the case of a free module P; then $\mathrm{End}(P)$ contains projectors p such that $\mathrm{Im}(p)$ has constant rank 1, and $w(p)$ too is a projector since it is an idempotent like p; since the left ideals generated by p and $w(p)$ have the same rank, we conclude that $\mathrm{tr}(p) = \mathrm{tr}(w(p)) = 1$, whence $\kappa = 1$. For the determinant we need a stronger argument: in **6.6** it is proved that w is an inner automorphism when K is a local ring, in other words, there is an automorphism u of P such that $w(f) = ufu^{-1}$ for all $f \in \mathrm{End}(P)$; the conclusion follows from the multiplicative property of determinants. □

The existence of traces and determinants is a general property of all Azumaya algebras, not only a property of the algebras $\mathrm{End}(P)$ with P as above. On every graded Azumaya algebra A there is a *reduced trace* tr and a reduced determinant, rather called *reduced norm* and denoted by $\mathcal{N}$, satisfying analogous properties. The word "reduced" probably recalls this property, in which A is assumed to have constant rank r^2 or $2r^2$: if x is any element of A, and if L_x and R_x are the endomorphisms of A defined by $L_x(y) = xy$ and $R_x(y) = yx$, then

$$\mathrm{tr}(L_x) = \mathrm{tr}(R_x) = r\ \mathrm{tr}(x) \quad \text{and} \quad \det(L_x) = \det(R_x) = \mathcal{N}(x)^r.$$

Reduced characteristic polynomials can be derived from reduced norms in the usual way, and give the reduced trace as a particular coefficient. The existence of reduced traces or norms can be proved by Descent Theory; the aim of this theory is to determine, in case of a faithfully flat extension $K \to L$, whether an object defined over L is an L-extension of an object defined over K. The next lemma is the basic trick in this theory; in the subsequent two propositions only (reduced) traces are mentioned since here we do not need more.

(3.6.5) **Lemma.** *Every faithfully flat extension $K \to L$ allows us to identify K with a subalgebra of L, and an element $\lambda \in L$ belongs to K if and only if the equality $\lambda \otimes 1 = 1 \otimes \lambda$ holds in $L \otimes L$.*

Proof. We must prove the exactness of the sequence

$$0 \longrightarrow K \longrightarrow L \longrightarrow L \otimes L$$

in which the last arrow is $\lambda \longmapsto \lambda \otimes 1 - 1 \otimes \lambda$. Since the extension is faithfully flat, it suffices to prove the exactness of the sequence derived from the previous one by the functor $L \otimes \cdots$:

$$0 \longrightarrow L \longrightarrow L \otimes L \longrightarrow L \otimes L \otimes L.$$

The second arrow $\lambda \longmapsto \lambda \otimes 1$ is injective because we obtain id_L if we compose it with $\lambda\otimes\mu \longmapsto \lambda\mu$. If we compose the third arrow $\lambda\otimes\mu \longmapsto \lambda\otimes\mu\otimes 1 - \lambda\otimes 1\otimes\mu$ with the mapping $\lambda\otimes\mu\otimes\nu \longmapsto \lambda\mu\otimes\nu$, we obtain $\lambda\otimes\mu \longmapsto \lambda\mu\otimes 1 - \lambda\otimes\mu$. Consequently every element $\sum_j \lambda_j\otimes\mu_j$ in the kernel of the third arrow belongs to the image $L\otimes 1$ of the second arrow because

$$\sum_j \lambda_j\mu_j\otimes 1 - \sum_j \lambda_j\otimes\mu_j = 0. \qquad \square$$

(3.6.6) **Proposition.** *Let us assume that A is a graded Azumaya algebra of constant rank r^2, and that there is an isomorphism $L\otimes A \to \mathrm{End}_L(P)$ involving a faithfully flat extension $K \to L$ and a graded L-module P with free even and odd components. There is a linear form $\mathrm{tr} : A \to K$ such that $\mathrm{tr}(x)$ is equal for all $x \in A$ to the trace of the image of $1\otimes x$ in $\mathrm{End}_L(P)$. This linear form does not depend on the choice of the isomorphism $L\otimes A \to \mathrm{End}_L(P)$. It is surjective onto K, its kernel is $[A,A]$ and contains A_1. Moreover the bilinear form $(x,y) \longmapsto \mathrm{tr}(xy)$ is symmetric and nondegenerate. And if $r = 2$, the mapping $x \longmapsto \mathrm{tr}(x) - x$ is a standard involution of A.*

Proof. The grading of A has no importance here; in the conclusions it only appears when it is stated that $\mathrm{tr}(A_1) = 0$, and (when the other conclusions are proved) this detail immediately follows from the vanishing of $\mathrm{tr}(f)$ whenever f is an endomorphism of P such that $f(P_0) \subset P_1$ and $f(P_1) \subset P_0$. Consequently we forget the parity gradings. First we must prove that $\mathrm{tr}(f)$ belongs to K when f is the L-endomorphism of P associated with some element x of A; since P is free, we can replace $\mathrm{End}_L(P)$ with the matrix algebra $\mathcal{M}(r,L)$ of square matrices of order r. Secondly, when $L'\otimes A \to \mathcal{M}(r,L')$ is another isomorphism of the same kind, we must prove that $\mathrm{tr}(f) = \mathrm{tr}(f')$ is f' is the image of x in $\mathcal{M}(r,L')$. Both statements can be proved at the same time. Indeed let w be the automorphism of the $(L\otimes L')$-algebra $\mathcal{M}(r,\ L\otimes L')$ defined in this way:

$$\begin{array}{ccccc} \mathcal{M}(r,\ L\otimes L') & \longrightarrow & \mathcal{M}(r,L)\otimes L' & \longrightarrow & (L\otimes A)\otimes L' \\ & & & & \downarrow \\ \mathcal{M}(r,\ L\otimes L') & \longleftarrow & L\otimes\mathcal{M}(r,L') & \longleftarrow & L\otimes(L'\otimes A) \end{array}$$

This automorphism w maps the image g of $f\otimes 1$ in $\mathcal{M}(r,\ L\otimes L')$ to the image g' of $1\otimes f'$, and it is clear that $\mathrm{tr}(g) = \mathrm{tr}(f)\otimes 1$ and $\mathrm{tr}(g') = 1\otimes\mathrm{tr}(f')$. Since $\mathrm{tr}(g') = \mathrm{tr}(g)$ (see (3.6.4)), we realize that $\mathrm{tr}(f)\otimes 1 = 1\otimes\mathrm{tr}(f')$. When the isomorphism $L'\otimes A \to \mathcal{M}(r,L')$ is equal to $L\otimes A \to \mathcal{M}(r,L)$, we get the equality $\mathrm{tr}(f)\otimes 1 = 1\otimes\mathrm{tr}(f)$ which proves that $\mathrm{tr}(f)$ belongs to K (see (3.6.5)). Similarly $\mathrm{tr}(f')$ belongs to K, and the above equality involving both $\mathrm{tr}(f)$ and $\mathrm{tr}(f')$ now proves their equality. The remainder of the proof is evident, since it merely means that some properties are true when they prove to be true after a faithfully flat extension. $\square$

(3.6.7) **Proposition.** *Let us assume that A is a graded Azumaya algebra of constant rank $2r^2$, and that there is an isomorphism $L \otimes A \to (\mathrm{End}_L(P)^2)^g$ involving a faithfully flat extension $K \to L$ and a free L-module P. There is a linear form $\mathrm{tr} : A \to K$ such that $\mathrm{tr}(x)$ is equal for all $x \in A$ to $\mathrm{tr}(f) + \mathrm{tr}(g)$ if (f, g) is the image of $1 \otimes x$ in $(\mathrm{End}_L(P)^2)^g$. This linear form does not depend on the choice of the isomorphism $L \otimes A \to (\mathrm{End}_L(P)^2)^g$. It is surjective onto K, its kernel is $[A_0, A_0] \oplus A_1$ and contains $[A, A]$. Moreover the bilinear form $(x, y) \longmapsto \mathrm{tr}(xy)$ is symmetric and nondegenerate.*

Proof. In this case A is isomorphic to $\mathrm{Z}(A) \otimes A_0$ (see (3.5.14)); thus the isomorphism $L \otimes A \to (\mathrm{End}_L(P)^2)^g$ gives the same information as the two isomorphisms $L \otimes \mathrm{Z}(A) \to (L^2)^g$ and $L \otimes A_0 \to \mathrm{End}_L(P)$, and (3.6.7) is an easy consequence of (3.6.6). It is worth observing that $\mathrm{Z}(A)$ and A_0 are both provided with traces, and that $\mathrm{tr}(zx) = \mathrm{tr}(z)\mathrm{tr}(x)$ for all $z \in \mathrm{Z}(A)$ and all $x \in A_0$; indeed $\mathrm{tr}(z) = 2$ if z is the unit element of $\mathrm{Z}(A)$, and $\mathrm{tr}(z) = 0$ for all z in the discriminant module $\mathrm{Z}_1(A)$. □

If we admit that Theorem (3.5.15) is true for all graded Azumaya algebras of constant rank, and remember what Remark (3.5.7) says about nonconstant ranks, we realize that every graded Azumaya algebra A is provided with a trace; it is a surjective linear form $A \to K$, and its kernel is $[A, A]+A_1$. Consequently $[A, A]+A_1$ is a direct summand of A and every supplementary submodule is isomorphic to K. The canonical image of K in A is supplementary to $[A, A] + A_1$ if and only if $\mathrm{tr}(1)$ is invertible, and $\mathrm{tr}(1)$ is invertible if and only if $\mathrm{tr}(\mathrm{id}_A)$ is invertible.

Remark. When the faithfully flat extension $K \to L$ is a Zariski extension, it is an easy exercise to deduce (1.13.9) from (3.6.5), provided that it is known that for every pair (s, t) of elements of K the natural morphism $K_s \otimes K_t \to K_{st}$ is an isomorphism; thus the proof of (1.13.10) was our first application of Descent Theory. Interested readers may find more information about this important but difficult theory in [Knus, Ojanguren 1974] or [Knus 1991].

Quaternion algebras

A *quaternion algebra* is a (nongraded) Azumaya algebra of constant rank 4. Because of (3.5.15) an equivalent definition may be this one: an algebra A is a quaternion algebra if there exists a faithfully flat extension $K \to L$ such that $L \otimes A$ is isomorphic to the matrix algebra $\mathcal{M}(2, L)$. Because of (3.6.6), every quaternion algebra is provided with a standard involution (also called the *quaternionic conjugation*).

(3.6.8) **Proposition.** *If A is a K-algebra, these assertions are equivalent:*

(a) *A is a quaternion algebra;*
(b) *A is a finitely generated projective module of constant rank 4, the algebra A is provided with a standard involution φ, and the derived norm $\mathcal{N}$ is a nondegenerate quadratic form.*

Proof. The Azumaya property can be tested by means of extensions to residue fields (see (3.5.6)), and the nondegeneracy of the quadratic form $\mathcal{N}$ too (see (2.3.4)); consequently we can assume that K is a field. A quaternion algebra A over a field contains no ideal other than 0 and A (see (3.5.9)). But *if* $\mathcal{N}$ *is the norm derived from a standard involution* φ, *then* $\mathrm{Ker}(\mathrm{b}_\mathcal{N})$ *is an ideal.* Indeed, for all x, $y \in A$,

$$\mathrm{b}_\mathcal{N}(x,y) = x\varphi(y) + y\varphi(x) = \mathrm{tr}(x\varphi(y)) \ ;$$

therefore x belongs to $\mathrm{Ker}(\mathrm{b}_\mathcal{N})$ if and only if $\mathrm{tr}(xy) = \mathrm{tr}(yx) = 0$ for all $y \in A$, and from this property it is easy to deduce that $\mathrm{Ker}(\mathrm{b}_\mathcal{N})$ is an ideal. Moreover $\mathrm{Ker}(\mathrm{b}_\mathcal{N}) \neq A$ if $\mathrm{b}_\mathcal{N}(1,x) = \mathrm{tr}(x) \neq 0$ for some $x \in A$. When A is a quaternion algebra over a field, all this implies that $\mathrm{Ker}(\mathrm{b}_\mathcal{N}) = 0$ and that $\mathcal{N}$ is nondegenerate.

Conversely let us assume that $\mathcal{N}$ is nondegenerate; we can assume that $\mathcal{N}$ is a hyperbolic quadratic form because a faithfully flat extension reduces the problem to this case (see (2.6.6)); and then, as above, we can assume that K is a field. Consequently an element x of A is invertible if and only if $\mathcal{N}(x) \neq 0$. Since $\mathcal{N}$ is hyperbolic, A contains noninvertible elements other than 0, and consequently contains left and right ideals other than 0 and A. Such a left or right ideal is totally isotropic for $\mathcal{N}$, and its dimension must be 1 or 2; it is always 2 for the following reasons. Let x be a nonzero element in a left ideal P other than 0 or A, and let R_x be the multiplication $y \longmapsto yx$. Since $\mathrm{Im}(R_x)$ and $\mathrm{Ker}(R_x)$ are left ideals different from 0 and A, and since the sum of their dimensions is 4, both $\mathrm{Im}(R_x)$ and $\mathrm{Ker}(R_x)$ have dimension 2, and $P = \mathrm{Im}(R_x)$. Moreover if P and Q are left ideals of dimension 2, then either $P = Q$ or $A = P \oplus Q$; indeed if $P \cap Q$ contains a nonzero element x, then P and Q are both equal to $\mathrm{Im}(R_x)$. Let u be the algebra morphism $A \to \mathrm{End}(P)$ that maps every z to the restriction of the left multiplication L_z ($y \longmapsto zy$) to P; if we manage to prove that A contains no ideal other 0 and A, it follows that u is injective, therefore bijective, and that A is an Azumaya algebra, therefore a quaternion algebra. To prove that A contains no ideal of dimension 2, it suffices to prove that the above left ideal P cannot be a two-sided ideal; if we prove that the dimension of $P\cap\varphi(P)$ is 1, then the right ideal $\varphi(P)$ cannot be a left ideal like P, and the conclusion follows. First $P\cap\varphi(P) \neq 0$, since the intersection of P and $\mathrm{Ker}(\mathrm{tr})$ (subspaces of dimension respectively 2 and 3) is not reduced to 0, and $\varphi(x) = -x$ for every x in this intersection. Secondly $P \neq \varphi(P)$; indeed, because of (2.5.4), there are noninvertible elements y outside P, and the left ideal $Q = \mathrm{Im}(R_y)$ is supplementary to P since $P \neq Q$; the projections ε and ε' of 1 in P and Q are idempotents, and the general formula $x^2 = \mathrm{tr}(x)x - \mathcal{N}(x)$ shows that $\mathrm{tr}(\varepsilon) = 1$, whence $\varphi(\varepsilon) = \varepsilon'$ and $\varphi(P) \neq P$. □

(3.6.9) **Example.** Let Z be a quadratic extension, α an invertible element of K and A the algebra $Z \oplus Z$ derived from Z and α by the Cayley–Dickson process explained in **3.3**; we set $j = (0,1)$ and $A = Z \oplus jZ$ as in **3.3**. The standard involution of Z extends to a standard involution φ of A, with a norm $\mathcal{N}$ such that $\mathcal{N}(b + jc) = \mathcal{N}(b) - \alpha\mathcal{N}(c)$ for all b, $c \in Z$. Since $\mathcal{N}$ is nondegenerate on A, this algebra A is a quaternion algebra. The equality $(jc)^2 = -\alpha\mathcal{N}(c)$ holds for all $c \in Z$ because $\mathrm{tr}(jc) = 0$. Therefore if we provide jZ with the quadratic form $jc \longmapsto -\alpha\mathcal{N}(c)$, the natural injection $jZ \to A$ extends to an algebra morphism $\mathrm{C}\ell(jZ) \to A$, the bijectiveness of which easily follows from (3.3.4) by localization.

3.7 Clifford algebras of quadratic spaces

Remember that a quadratic space over the ring K is a finitely generated projective K-module provided with a nondegenerate quadratic form. In this section we intend to prove that the Clifford algebra of a quadratic space is always a graded Azumaya algebra. When (M, q) is a quadratic space, from (3.3.7) we know that the canonical mappings $K \to \mathrm{C}\ell(M,q)$ and $M \to \mathrm{C}\ell(M,q)$ are injective; this allows us to identify 1_q with 1, and $\rho(x)$ with x, so as to get easily readable calculations. We begin with a preliminary lemma.

(3.7.1) **Lemma.** *Let P be a K-module, and h a linear form on P; there exists a unique linear mapping D_h from $\bigwedge(P)$ into itself satisfying the following two conditions:*

$$\forall a \in P, \quad D_h(a) = h(a) \ ;$$
$$\forall x,\ y \in \textstyle\bigwedge(P), \quad D_h(x \wedge y) \ = \ D_h(x) \wedge y + (-1)^{\partial x} x \wedge D_h(y).$$

Besides, $D_h \circ D_h = 0$.

The second condition means that D_h is a twisted derivation, and implies $D_h(1) = 0$ when $x = y = 1$. Both conditions imply, for all $a_1,\dots,a_k$ in P,

$$D_h(a_1 \wedge a_2 \wedge \cdots \wedge a_k) \ = \ \sum_{i=1}^{k} (-1)^{i-1}\ a_1 \wedge \cdots \wedge h(a_i) \wedge \cdots \wedge a_k \ .$$

Proof. The unicity of D_h is obvious. Let D'_h be the linear mapping $\mathrm{T}(P) \to \mathrm{T}(P)$ defined by $D'_h(1) = 0$ and, for every $k \geq 1$,

$$D'_h(a_1 \otimes a_2 \otimes \cdots \otimes a_k) \ = \ \sum_{i=1}^{k} (-1)^{i-1}\ a_1 \otimes \cdots \otimes h(a_i) \otimes \cdots \otimes a_k \ .$$

Remember that $\bigwedge(P)$ is the quotient of $\mathrm{T}(P)$ by the ideal $\mathrm{J}(P)$ generated by all $a \otimes a$. It is easy to check that $D'_h(\mathrm{J}(P)) = 0$ and $(D'_h)^2 = 0$. This implies the existence of D_h and the equality $(D_h)^2 = 0$. Anyhow (3.7.1) is also an immediate consequence of the interior multiplications presented later in **4.4**. □

(3.7.2) **Theorem.** *Let P be a finitely generated projective module, and $\mathbf{H}[P]$ the derived hyperbolic space; the graded algebras $\mathrm{C}\ell(\mathbf{H}[P])$ and $\mathrm{End}(\bigwedge(P))$ are isomorphic.*

The grading of $\mathrm{End}(\bigwedge(P))$ comes from the grading of $\bigwedge(P) = \bigwedge_0(P) \oplus \bigwedge_1(P)$; all gradings are balanced when P is a faithful module.

Proof in three steps. The identity mapping of $\bigwedge(P)$ is denoted by $\mathrm{id}_\wedge$.

First step. We construct an algebra morphism $\Phi : C\ell(\mathbf{H}[P]) \to \mathrm{End}(\bigwedge(P))$. Because of the universal property of $C\ell(\mathbf{H}[P])$, the construction of Φ merely requires a linear mapping $\varphi : P^* \oplus P \to \mathrm{End}(\bigwedge(P))$ such that

$$\forall h \in P^*,\ \forall a \in P, \quad (\varphi(h,a))^2 \;=\; h(a)\ \mathrm{id}_\wedge\ .$$

This condition is fulfilled if we set $L_a(y) = a \wedge y$ for all $y \in \bigwedge(P)$, and then

$$\varphi(h,a) \;=\; D_h + L_a\ ;$$

indeed we know that $(D_h)^2 = 0$ and $(L_a)^2 = 0$, and the equality

$$D_h L_a + L_a D_h \;=\; h(a)\ \mathrm{id}_\wedge$$

follows from the fact that D_h is a twisted derivation. Since D_h and L_a both permute $\bigwedge_0(P)$ and $\bigwedge_1(P)$, they are odd elements in $\mathrm{End}(\bigwedge(P))$, and consequently Φ is a graded algebra morphism.

Second step. We prove that Φ is surjective when P is free. Let $(e_1, \dots, e_m)$ be a basis of P; we derive from it a basis $\mathcal{B}$ of $\bigwedge(P)$, the elements of which are all the products

$$\varepsilon \;=\; e_{j_1} \wedge e_{j_2} \wedge \cdots \wedge e_{j_k} \quad \text{with} \;\; k \geq 0 \;\; \text{and} \;\; j_1 < j_2 < \cdots < j_k\ ;$$

we write $\mathcal{B}_k$ for the set of all elements of $\mathcal{B}$ contained in $\bigwedge^k(P)$, so that $\mathcal{B}$ is the union of $\mathcal{B}_0, \mathcal{B}_1, \dots, \mathcal{B}_m$. Let $(e_1^*, \dots, e_m^*)$ be the dual basis of P^*; with each element ε of $\mathcal{B}$ (written as above), we associate

$$\varepsilon^* \;=\; e_{j_k}^* \wedge \cdots \wedge e_{j_2}^* \wedge e_{j_1}^*$$

(take notice of the reversion of the indices).

The functor $C\ell$ associates two algebra morphisms u and v with the natural injections from $P^* \oplus 0$ and $0 \oplus P$ into $\mathbf{H}[P]$:

$$u : \bigwedge(P^*) \longrightarrow C\ell(\mathbf{H}[P]) \quad \text{and} \quad v : \bigwedge(P) \longrightarrow C\ell(\mathbf{H}[P])\ ;$$

thus with each ε in $\mathcal{B}$ are associated two elements $u(\varepsilon^*)$ and $v(\varepsilon)$ in $C\ell(\mathbf{H}[P])$. The following properties are easy consequences of the definitions:

$$\forall \varepsilon,\ \theta \in \mathcal{B}\ , \quad \Phi(v(\theta))(\varepsilon) \;=\; \theta \wedge \varepsilon\ ;$$

$$\forall \eta \in \mathcal{B}_k\ , \quad \Phi(u(\eta^*))(\textstyle\bigwedge^j(P)) \;\subset\; \bigwedge^{j-k}(P)\ , \text{ whence } \Phi(u(\eta^*))(\bigwedge^{<k}(P)) = 0;$$

$$\forall \varepsilon,\ \eta \in \mathcal{B}_k\ , \quad \Phi(u(\eta^*))(\varepsilon) = 0 \;\text{ if } \eta \neq \varepsilon, \;\text{ but } \Phi(u(\eta^*))(\eta) = 1.$$

There is a basis of $\mathrm{End}(\bigwedge(P))$ that is constituted of all endomorphisms $E_{\theta,\eta}$ defined in this way: for every $(\eta, \theta) \in \mathcal{B} \times \mathcal{B}$, $E_{\theta,\eta}$ maps all $\varepsilon \in \mathcal{B}$ to 0, except η which is mapped to θ. By means of a *decreasing* induction on the degree k

of η, we prove that every $E_{\theta,\eta}$ belongs to the image of Φ. The induction begins with the greatest value $k = m$; if η is the unique element in $\mathcal{B}_m$, it is clear that $E_{\theta,\eta} = \Phi(v(\theta)u(\eta^*))$. Now let us take η in some $\mathcal{B}_k$ and assume that $\mathrm{Im}(\Phi)$ contains $E_{\theta',\eta'}$ whenever η' has degree $> k$; in other words, $\mathrm{Im}(\Phi)$ contains all endomorphisms of $\bigwedge(P)$ vanishing on $\bigwedge^{\leq k}(P)$. We observe that $\Phi(v(\theta)u(\eta^*))$ vanishes on all ε of degree $\leq k$, except η which is mapped to θ; the difference between $\Phi(v(\theta)u(\eta^*))$ and $E_{\theta,\eta}$ is an endomorphism vanishing on $\bigwedge^{\leq k}(P)$, and by the induction hypothesis it belongs to $\mathrm{Im}(\Phi)$. Consequently $E_{\theta,\eta} \in \mathrm{Im}(\Phi)$.

Third step. We complete the proof without the assumption that P is free. We already know that Φ is surjective because all its localisations are surjective. If m is the rank of P at some prime ideal $\mathfrak{p}$, the ranks of $C\ell(\mathbf{H}[P])$ and $\mathrm{End}(\bigwedge(P))$ at $\mathfrak{p}$ are both equal to 2^{2m}; now the bijectiveness of Φ follows from (1.13.5). □

(3.7.3) **Theorem.** *We suppose that 2 is invertible in K, and that the quadratic space (M,q) is the orthogonal sum of the hyperbolic space $\mathbf{H}[P]$ (associated with some finitely generated projective module P) and a free submodule W of rank 1 generated by an element w such that $q(w) = 1$. The graded algebras $C\ell(M,q)$ and $(\mathrm{End}(\bigwedge(P))^2)^g$ are isomorphic.*
Here the algebra $\mathrm{End}(\bigwedge(P))$ is trivially graded; the even (resp. odd) elements of $(\mathrm{End}(\bigwedge(P))^2)^g$ are the elements (f,f) (resp. $(f,-f)$).

Proof. To construct an algebra morphism Ψ from $C\ell(M,q)$ into

$$A = (\mathrm{End}(\bigwedge(P))^2)^g,$$

it suffices to construct a linear mapping $\psi : P^* \oplus P \oplus W \to A$ such that

$$\forall h \in P^*,\ \forall a \in P,\ \forall \lambda \in K, \quad (\psi(h,a,\lambda w))^2 \ = \ (h(a) + \lambda^2)\ (\mathrm{id}_\wedge\,,\ \mathrm{id}_\wedge)\ .$$

This condition is fulfilled if we set $L_a(y) = a \wedge y$ and $\sigma(y) = (-1)^{\partial y} y$ for all y in $\bigwedge(P)$, and then

$$\psi(h,a,\lambda w) \ = \ (D_h + L_a + \lambda\sigma\ ,\ -D_h - L_a - \lambda\sigma)\ ;$$

indeed to the explanations given in the proof of (3.7.2) we have just to add the following one: D_h and L_a anticommute with σ because they permute $\bigwedge_0(P)$ and $\bigwedge_1(P)$. The resulting algebra morphism Ψ is graded because all $\psi(h,a,\lambda w)$ are odd elements in A.

Now $C\ell(M,q)$ is the twisted tensor product of $C\ell(\mathbf{H}[P])$ and $C\ell(W) = K \oplus Kw$, and this twisted tensor product is the direct sum of these four submodules:

$$C_0 = C\ell_0(\mathbf{H}[P]) \otimes 1\ , \qquad C_1 = C\ell_1(\mathbf{H}[P]) \otimes 1\ ,$$
$$C_2 = C\ell_1(\mathbf{H}[P]) \otimes w\ , \qquad C_3 = C\ell_0(\mathbf{H}[P]) \otimes w\ .$$

In the following explanations, the notations $\mathrm{End}_0(\bigwedge(P))$ and $\mathrm{End}_1(\bigwedge(P))$ refer to the nontrivial grading of $\mathrm{End}(\bigwedge(P))$ used in the proof of (3.7.2), and g_0 and

g_1 are variables running respectively through $\mathrm{End}_0(\bigwedge(P))$ and $\mathrm{End}_1(\bigwedge(P))$. From the proof of (3.7.2) we deduce that Ψ induces a bijection from C_0 onto the set of all elements (g_0, g_0), a bijection from C_1 onto the set of all $(g_1, -g_1)$, a bijection from C_2 onto the set of all $(g_1\sigma,\ g_1\sigma)$, which is the same thing as the set of all (g_1, g_1), and finally a bijection from C_4 onto the set of all $(g_0\sigma,\ -g_0\sigma)$, which is also the set of all $(g_0, -g_0)$. Since 2 is invertible in K, it is now clear that Ψ is bijective. □

Now remember that in (2.6.6) (resp. (2.6.7)) the following result has been proved: if (M, q) is a quadratic space of even constant rank (resp. odd constant rank), there exists a faithfully flat extension $K \to L$ such that $L \otimes (M, q)$ is a hyperbolic quadratic space over L (resp. the orthogonal sum of a hyperbolic space and a free space of rank 1 on which the quadratic form takes the value 1); moreover this hyperbolic space is associated with a free L-module. From (3.7.2) (resp. (3.7.3)) we derive this immediate corollary.

(3.7.4) **Corollary.** *If (M, q) is a quadratic space of even constant rank r (resp. odd constant rank r) over K, there exists a faithfully flat extension $K \to L$ and a graded (resp. nongraded) free module P of finite rank over L such that $L \otimes C\ell(M, q)$ is isomorphic to $\mathrm{End}_L(P)$ (resp. $(\mathrm{End}_L(P)^2)^g$).*

The rank of P is $2^{r/2}$ or $2^{(r-1)/2}$ according to the parity of r, and this would enable us to prove that $C\ell(M, q)$ has constant rank 2^r, if it were not yet proved in (3.3.7).

By means of (3.5.8) and (3.5.12) we deduce from this corollary that $L \otimes C\ell(M, q)$ is a graded Azumaya algebra over L. Because of (3.5.3), $C\ell(M, q)$ is a graded Azumaya algebra over K, with a regular and balanced grading when $r > 0$ (whereas $C\ell(M, q) = K$ if $r = 0$). When M does not have constant rank, we remember (1.12.8) and (3.1.11), and thus we come to the next statement.

(3.7.5) **Corollary.** *The Clifford algebra of a quadratic space (M, q) is a graded Azumaya algebra; when M has everywhere an even rank (resp. an odd rank), it is a graded Azumaya algebra of even (resp. odd) type. Moreover the following assertions are equivalent:*

- *M is a faithful module;*
- *the odd component $C\ell_1(M, q)$ is a faithful module;*
- *the grading of $C\ell(M, q)$ is balanced;*
- *the grading of $C\ell(M, q)$ is regular.*

It is worth noticing that the regularity of the grading of $C\ell(M, q)$ already follows from a much weaker hypothesis: if the ideal of K generated by $q(M)$ is K, this grading is regular because there exists a finite sequence of elements $x_j \in \rho(M)$ and a finite sequence of $\lambda_j \in K$ such that $\sum_j (\lambda_j x_j) x_j = 1$.

Because of (3.7.4), here we are not concerned with a general proof of Theorem (3.5.15), since anyhow this statement is true for all Clifford algebras of quadratic spaces of constant rank. Remember that Theorem (3.5.14) is already proved for all Azumaya algebras satisfying the property stated in (3.5.15), therefore it can be applied to the Clifford algebra of any quadratic space, whence the following corollary.

(3.7.6) **Corollary.** *Let (M, q) be a quadratic space. The graded center* $\mathrm{Z}^g(\mathrm{C}\ell(M,q))$ *is always reduced to K. When M is a faithful module, the centralizer of* $\mathrm{C}\ell_0(M,q)$ *in* $\mathrm{C}\ell(M,q)$ *is a quadratic extension denoted by* $\mathrm{QZ}(M,q)$. *When the rank of M is always even and nonzero,* $\mathrm{QZ}(M,q)$ *is the center of* $\mathrm{C}\ell_0(M,q)$, *whereas the center of* $\mathrm{C}\ell(M,q)$ *is reduced to K. But when the rank of M is always odd, then* $\mathrm{QZ}(M,q)$ *is the center of* $\mathrm{C}\ell(M,q)$, *its odd component* $\mathrm{QZ}_1(M,q)$ *has constant rank* 1, *and the center of* $\mathrm{C}\ell_0(M,q)$ *is reduced to K.*

The formula (3.5.13) can be used when y and z belong respectively to $\mathrm{C}\ell(M,q)$ and $\mathrm{QZ}(M,q)$. The notation $\mathrm{QZ}(M,q)$ is an abbreviation for $\mathrm{QZ}(\mathrm{C}\ell(M,q))$ and can be used even when M is not a faithful module, according to the explanations presented just before (3.5.18). This quadratic extension is often called the *Arf subalgebra*, and its class in $\mathrm{Q}^g(K)$ is called the *Arf invariant*. When K is a field of characteristic $\neq 2$, the Arf invariant is described by an element of $K^\times$ modulo the subgroup of squares (see (3.4.14)), and has been observed long before Arf; but Arf considered the case of a field of characteristic 2 which did not allow so easy a description (see (3.ex.25)).

Let (M,q) and (M',q') be two quadratic spaces such that M and M' are faithful modules; let us set $(M'',q'') = (M,q) \perp (M',q')$. Because of (3.2.3) we can identify $\mathrm{C}\ell((M'',q'')$ with $\mathrm{C}\ell(M',q') \hat{\otimes} \mathrm{C}\ell(M'',q'')$, and from (3.5.17) we derive

$$\mathrm{QZ}((M,q) \perp (M',q')) \;\cong\; \mathrm{QZ}(M,q) \star \mathrm{QZ}(M',q') \;. \tag{3.7.7}$$

Remember that the Witt class of a quadratic space is trivial whenever it is hyperbolic, whereas the Brauer class of a graded Azumaya algebra is trivial whenever it is isomorphic to some $\mathrm{End}(P)$, with P a graded finitely generated and faithful projective module. Theorem (3.7.2) implies that the Brauer class of the Clifford algebra of a hyperbolic space is trivial, and thus (3.2.4) and (2.7.3) lead to the following consequence.

(3.7.8) **Corollary.** *The Brauer class of the Clifford algebra of a quadratic space only depends on the Witt class of this quadratic space, and by mapping this Witt class to this Brauer class we get a group morphism from the additive group* $\mathrm{WQ}(K)$ *into* $\mathrm{Br}^g(K)$.

(3.7.9) **Remarks.** Often $\mathrm{Br}^g(K)$ is treated as a multiplicative group; nonetheless there are many reasons to prefer additive notation for it. Indeed it is conjectured, and in some cases even proved, that the image $\mathcal{H}(K)$ of $\mathrm{WQ}(K)$ in $\mathrm{Br}^g(K)$ inherits a structure of ring from $\mathrm{WQ}(K)$. This image $\mathcal{H}(K)$ shall be considered again in **8.6**; it may be much smaller than $\mathrm{Br}^g(K)$; indeed Clifford algebras are provided with a reversion (see (3.1.4)), and later (see (3.8.15)) this reversion will imply that every element in the group $\mathcal{H}(K)$ has an order dividing 8; although it is true (yet not evident) that every element of $\mathrm{Br}^g(K)$ has a finite order, its order may be any positive integer.

3.8 Discriminant modules, quadratic extensions and quaternion algebras

The results of this section will be especially useful in **8.6**. Here we show how Clifford algebras can help the study of various objects mentioned in the previous sections. When it is written that A is a graded Azumaya algebra such that $\mathrm{Z}(A_0)$ is a quadratic extension, you must understand that A has even type and that A_1 is a faithful module; we shall be interested in the nongraded (or rather trivially graded) algebra A^{ng} obtained by forgetting the grading, and in the comparison of the Brauer classes $[A]$ and $[A^{ng}]$. The inverse classes $[A]^{-1}$ and $[A^{ng}]^{-1}$ are the classes of A^{to} and $(A^{ng})^o$; when A is provided with an involution, A^{ng} is isomorphic to $(A^{ng})^o$, and its class in $\mathrm{Br}(K)$ has order 1 or 2.

All quaternion algebras involved here are obtained as algebras $C\ell(M,q)^{ng}$ derived from a quadratic space (M,q) of constant rank 2; in (3.3.7) it is stated that the conjugation $\sigma\tau$ is a standard involution of $C\ell(M,q)$; besides, the derived quadratic extension $\mathrm{QZ}(M,q)$ is merely $C\ell_0(M,q)$.

(3.8.1) **Proposition.** *When the quadratic space (M,q) of constant rank 2 contains an element e such that $q(e)=1$, there is an isomorphism $C\ell(M,q)^{ng} \to \mathrm{End}(M)$.*

Proof. Let us map every $a\in M$ to the endomorphism $x\longmapsto axe$ of M; observe that axe belongs to M because here $C\ell_1(M,q)$ is equal to M. The equality $a(axe)e = q(a)x$ shows that this mapping $M\to\mathrm{End}(M)$ extends to an algebra morphism $C\ell(M,q)\to\mathrm{End}(M)$. To prove that it is an isomorphism, we can assume that M is free and contains an element b such that (e,b) is a basis, since localization allows us to reduce the problem to this case. When (e,b) is a basis, we have to verify the invertibility of some square matrix of order 4; easy calculations show that its determinant is $\pm(\mathrm{b}_q(e,b)^2-4q(b))$, and it is invertible since q is nondegenerate. □

(3.8.2) **Proposition.** *When (M,q) is a quadratic space of constant rank 2, the following three assertions are equivalent:*

(a) *(M,q) is hyperbolic;*
(b) *the class of $C\ell(M,q)$ in $\mathrm{Br}^g(K)$ is neutral;*
(c) *$C\ell_0(M,q)$ is a trivial quadratic extension.*

Proof. The implications (a) ⇒ (b) ⇒ (c) follow from (3.7.2) and (3.5.8). When $C\ell_0(M,q)$ is trivial, it contains an idempotent ε such that $C\ell_0(M,q)=K\varepsilon\oplus K(1-\varepsilon)$. Since $C\ell_1(M,q)=M$, this implies $M=\varepsilon M+(1-\varepsilon)M$. The standard involution

φ of $C\ell_0(M,q)$ maps ε to $1-\varepsilon$; thus from (3.5.13) we deduce $\varepsilon a = a(1-\varepsilon)$ for all $a \in M$. All this (together with the evident equality $\varepsilon(1-\varepsilon)=0$) allows us to prove that εM and $(1-\varepsilon)M$ are totally isotropic submodules, and that M is their direct sum. Now it is easy to prove that (M,q) is hyperbolic (see (2.5.5)). □

Discriminant modules

A *discriminant module* D is a finitely generated projective module of constant rank 1, provided with an isomorphism $D \otimes D \longrightarrow K$. This name comes from the fact that every quadratic extension contains such a module D (see (3.4.6)); but when 2 is not invertible in K, it often happens that a discriminant module cannot be the discriminant module of a quadratic extension. Here the isomorphism $D\otimes D \to K$ is always treated as a multiplication $d\otimes d' \longmapsto dd'$; because of (1.12.11) the following equalities hold for all d, d', $d'' \in D$:

$$(3.8.3) \qquad dd' = d'd \quad \text{and} \quad (dd')d'' = (d'd'')d = (d''d)d'.$$

Consequently a discriminant module is the same thing as a bilinear space of constant rank 1.

The tensor product of two discriminant modules D_1 and D_2 is still a discriminant module for the evident multiplication

$$(d_1 \otimes d_2)\ (d'_1 \otimes d'_2) \quad = \quad (d_1 d'_1)\ (d_2 d'_2)\ .$$

The ring K itself is a discriminant module for the natural multiplication mapping $K \otimes K \to K$, and $K \otimes D$ is isomorphic to D as a discriminant module. It is clear that $(D_1 \otimes D_2) \otimes D_3$ and $D_1 \otimes (D_2 \otimes D_3)$ are isomorphic discriminant modules, and (3.8.3) implies that $D \otimes D$ is isomorphic to the discriminant module K. Thus the isomorphy classes of discriminant modules constitute a group $\mathrm{Disc}(K)$ in which every element has order 1 or 2.

If D is a discriminant module over K, and $K \to L$ a ring extension, then $L\otimes D$ is a discriminant module over L. This leads to a group morphism $\mathrm{Disc}(K) \to \mathrm{Disc}(L)$. A functor Disc from the category $\mathcal{C}om(\mathbb{Z})$ to the category of commutative groups has been defined in this way.

If D and D' are the discriminant modules of the quadratic extensions A and A', the discriminant module of $A \star A'$ is $D \otimes D'$ (see (3.4.10)), whence a group morphism $\mathrm{Q}(K) \to \mathrm{Disc}(K)$, and a morphism of functors $\mathrm{Q} \to \mathrm{Disc}$. When 2 is invertible in K, the morphism $\mathrm{Q}(K) \to \mathrm{Disc}(K)$ is bijective; indeed on one side the equality $Z = K \oplus D$ holds for every quadratic extension Z with discriminant module D, and conversely if D is a discriminant module, the following multiplication on $Z = K \oplus D$ turns Z into a quadratic extension with discriminant module D:

$$(\lambda, d)\ (\lambda', d') \quad = \quad (\lambda\lambda' + dd',\ \lambda d' + \lambda' d).$$

In an evident way we can define graded discriminant modules, and associate a graded discriminant module with every graded quadratic extension: see (3.ex.27).

Let (M, q) be a quadratic space. Since a discriminant module D' is a bilinear module of constant rank 1, the notation $D' \otimes (M, q)$ represents a quadratic space of the same rank (see (2.4.5)). If D' is free and generated by an element d such that $d^2 = \lambda$, then $D' \otimes (M, q)$ is isomorphic to $(M, \lambda q)$.

(3.8.4) **Lemma.** *Let (M, q) be a quadratic space of constant rank 2, and D the discriminant module of $C\ell_0(M, q)$. The multiplication mapping $d \otimes a \longmapsto da$ is an isomorphism from $D \otimes (M, q)$ onto $(M, -q)$.*

Proof. This mapping $d \otimes a \longmapsto da$ takes its values in M because $M = C\ell_1(M, q)$; its bijectiveness is ensured by the reciprocal mapping $M \to K \otimes M \to (D \otimes D) \otimes M \to D \otimes (D \otimes M) \to D \otimes M$; and the equality $q(da) = -d^2 q(a)$ follows from $ad = \varphi(d)a$ (see (3.5.13)) and $\varphi(d) = -d$. □

Very often the quadratic space under consideration is a quadratic extension Z provided with its norm $\mathcal{N}$, and the notation Z may be an abbreviation for $(Z, \mathcal{N})$. Remember that $\mathcal{N}(\lambda + d) = \lambda^2 - d^2$ for all $\lambda \in K$ and all $d \in D$. For reasons that shall soon appear, we are especially interested in the quadratic spaces $D' \otimes (Z, \mathcal{N})$ involved in the next lemma.

(3.8.5) **Lemma.** *When D' is a discriminant module, and Z a quadratic extension, the quadratic extension $C\ell_0(D' \otimes Z)$ is isomorphic to Z.*

Proof. We denote the unit element of Z by 1_Z to distinguish it from the unit element 1 of $C\ell(D' \otimes Z)$. Let $(d_1, d'_1, d_2, d'_2, \dots)$ be a finite sequence of elements of D' such that $\sum_i d_i d'_i = 1$, and let us prove that the mapping $x \longmapsto \sum_i (d_i \otimes 1_Z)(d'_i \otimes x)$ is an isomorphism from Z onto $C\ell_0(D' \otimes Z)$. Since $\mathcal{N}(1_Z) = 1$, it is already clear that it maps 1_Z to 1. By localization we can reduce the problem to this case: Z admits a basis $(1, z)$ such that $z^2 = \beta z - \gamma$ and D' is generated by an element d. Let d' be the generator of D' such that $dd' = 1$, so that $\sum_i (d_i \otimes 1_Z)(d'_i \otimes x) = (d \otimes 1_Z)(d' \otimes x)$. Since $d \otimes 1_Z$ is invertible in $C\ell(D' \otimes Z)$, we get a bijection $Z \to C\ell_0(D' \otimes Z)$, and it remains to verify that

$$((d \otimes 1_Z)(d' \otimes z))^2 = \beta \, (d \otimes 1_Z)(d' \otimes z) \; - \gamma \; ;$$

this is a particular example of an equality $(ab)^2 = \mathrm{b}_q(a, b)ab - q(a)q(b)$ which is valid for any pair (a, b) of elements of a quadratic space; indeed $\mathcal{N}(z) = \gamma$ and $\mathrm{b}_{\mathcal{N}}(z, 1_Z) = \mathrm{tr}(z) = \beta$. □

The theorem $A \hat{\otimes} B \cong A \otimes B_D$

Let $B = B_0 \oplus B_1$ be a graded algebra, and D a discriminant module; on the direct sum $B_D = B_0 \oplus (D \otimes B_1)$ we define a multiplication in this way (for all b and $b' \in B_0$, all c and $c' \in B_1$, and all d and $d' \in D$):

$$(b \; + (d \otimes c)) \; (b' \; + (d' \otimes c')) \;\; = \;\; bb' + (dd')cc' \; + \; d \otimes cb' \; + \; d' \otimes bc' \; ;$$

obviously B_D is also a graded algebra; the associativity of its multiplication results from (3.8.3). It is clear that $(B_D)_{D'}$ is canonically isomorphic to $B_{D\otimes D'}$; in particular $(B_D)_D$ is isomorphic to B.

(3.8.6) **Theorem.** *Let A be a graded Azumaya algebra such that $\mathrm{Z}(A_0)$ is a quadratic extension Z, let D be the discriminant module of Z, and B any graded algebra. There is a graded algebra isomorphism $A \otimes B_D \to A \hat{\otimes} B$ that maps every $a \otimes b$ to itself for all $b \in B_0$, and $a \otimes d \otimes c$ to $ad \otimes c$ for all $c \in B_1$. Similarly there is an isomorphism $A \hat{\otimes} B_D \to A \otimes B$.*

Proof. The bijectiveness of the graded mapping $A \otimes B_D \to A \hat{\otimes} B$ defined in (3.8.6) follows from the bijectiveness of $D \otimes D \to K$; indeed we can write $\sum_i d_i d_i' = 1$ for some finite sequence $(d_1, d_1', d_2, d_2', \dots)$ of elements of D, and thus the reciprocal mapping from $A \hat{\otimes} B_1$ into $A \otimes D \otimes B_1$ maps every $a \otimes c$ to $\sum_i ad_i \otimes d_i' \otimes c$. Now remember that $\varphi(d) = -d$ for all $d \in D$; consequently from (3.5.13) we deduce that $ad = (-1)^{\partial a} da$ for all homogeneous $a \in A$ and all $d \in D$. This equality implies (after some straightforward calculations) that the above mapping $A \otimes B_D \longrightarrow A \hat{\otimes} B$ is an algebra morphism. □

Algebras like B_D are very useful in the treatment of Clifford algebras for the following reason.

(3.8.7) **Theorem.** *Let (M', q') be a quadratic module, and D a discriminant module, and let ρ' and ρ'' be the canonical mappings from (M', q') and $D \otimes (M', q')$ into their Clifford algebras. There is an isomorphism from $\mathrm{C}\ell(D \otimes (M', q'))$ into $\mathrm{C}\ell(M', q')_D$ that maps $\rho''(d \otimes a')$ to $d \otimes \rho'(a')$ for all $a' \in M'$ and all $d \in D$.*

Proof. Let q'' be the quadratic form on $D \otimes (M', q')$, let $d_1, d_2, \dots, d_n$ be elements of D, and $a_1, a_2, \dots, a_n$ elements of M'. A direct calculation shows that the square of $\sum_i d_i \otimes \rho'(a_i')$ in the algebra $\mathrm{C}\ell(M', q')_D$ is equal to $q''(\sum_i d_i \otimes a_i')\, 1_{q'}$. Consequently the mapping $d \otimes a' \longmapsto d \otimes \rho'(a')$ determines a graded algebra morphism from $\mathrm{C}\ell(D \otimes (M', q'))$ into $\mathrm{C}\ell(M', q')_D$. To construct a reciprocal morphism, we proceed in this way: there is an analogous algebra morphism from $\mathrm{C}\ell(D \otimes (D \otimes (M', q')))$ into $\mathrm{C}\ell(D \otimes (M', q'))_D$; by means of the isomorphisms $D \otimes D \longleftrightarrow K$ we get the algebra morphism

$$\begin{aligned}\mathrm{C}\ell(M', q')_D &\longrightarrow \mathrm{C}\ell(D \otimes D \otimes (M', q'))_D \\ &\longrightarrow (\mathrm{C}\ell(D \otimes (M', q'))_D)_D \longrightarrow \mathrm{C}\ell(D \otimes (M', q'))\ ;\end{aligned}$$

this is the desired reciprocal algebra morphism, because it maps $d \otimes \rho'(a')$ to $\rho''(d \otimes a')$. □

As an immediate corollary of (3.8.7) we can state that *the even subalgebra $\mathrm{C}\ell_0(D \otimes (M', q'))$ is always isomorphic to $\mathrm{C}\ell_0(M', q')$*. The next statement (the most usual version of the theorem $A \hat{\otimes} B \cong A \otimes B_D$) is an immediate consequence of (3.8.6) and (3.8.7).

(3.8.8) **Corollary.** *Let (M',q') be a quadratic module, and A a graded Azumaya algebra such that $\mathrm{Z}(A_0)$ is a quadratic extension Z with discriminant module D. There are graded algebra isomorphisms*

$$A \otimes C\ell(D \otimes (M',q')) \longrightarrow A \hat{\otimes} C\ell(M',q') ,$$
$$A \hat{\otimes} C\ell(D \otimes (M',q')) \longrightarrow A \otimes C\ell(M',q') .$$

Since the Clifford algebra of an orthogonal sum is a twisted tensor product of Clifford algebras (see (3.2.4)), Corollary (3.8.8) will be repeatedly used when orthogonal sums appear.

(3.8.9) **Proposition.** *Let (M,q) and (M',q') be quadratic spaces with even nonzero ranks at every prime ideal, let Z and Z' be the centers of their even Clifford subalgebras, and D and D' their discriminant modules. There are graded algebra isomorphisms*

$$C\ell((M,q) \perp (M',q')) \otimes C\ell(Z) \cong C\ell(M,q) \otimes C\ell(M',q') \otimes C\ell(D' \otimes Z) ,$$
$$C\ell((M,q) \perp (M',q')) \otimes C\ell(Z') \cong C\ell(M,q) \otimes C\ell(M',q') \otimes C\ell(D \otimes Z') ,$$
$$C\ell((M,q) \perp (M',q')) \otimes C\ell(Z) \otimes C\ell(Z') \cong C\ell(M,q) \otimes C\ell(M',q') \otimes C\ell(Z \perp Z') .$$

Indeed there are isomorphisms

$$\begin{aligned} C\ell((M,q) \perp (M',q')) &\longleftrightarrow C\ell(M,q) \hat{\otimes} C\ell(M',q') \\ &\longleftrightarrow C\ell(M,q) \otimes C\ell(D \otimes (M',q')) , \\ C\ell(D \otimes (M',q')) \otimes C\ell(Z) &\longleftrightarrow C\ell(M',q') \hat{\otimes} C\ell(Z) \\ &\longleftrightarrow C\ell(M',q') \otimes C\ell(D' \otimes Z) , \end{aligned}$$

and so forth.... □

The importance of the following example shall appear later in (3.8.15).

(3.8.10) **Lemma.** *If (M,q) is a quadratic space of constant rank 2, the graded algebra*

$$C\ell(M,q) \hat{\otimes} C\ell(M,q) \hat{\otimes} C\ell(M,q) \hat{\otimes} C\ell(M,q)$$

is isomorphic to $\mathrm{End}(\bigwedge(M \oplus M))$.

Proof. Let D be the discriminant module of $C\ell_0(M,q)$, and X and Y short notations for (M,q) and $D \otimes (M,q)$; because of (3.8.4), Y is isomorphic to $(M,-q)$. With three applications of (3.8.8) we get

$$\begin{aligned} C\ell(X) \hat{\otimes} C\ell(X) \hat{\otimes} C\ell(X) \hat{\otimes} C\ell(X) &\cong C\ell(X \perp X \perp X \perp X) \\ &\cong C\ell(X) \otimes C\ell(Y \perp Y \perp Y) \\ &\cong C\ell(X) \otimes C\ell(Y) \otimes C\ell(X \perp X) \\ &\cong C\ell(X) \otimes C\ell(Y) \otimes C\ell(X) \otimes C\ell(Y) . \end{aligned}$$

In the same way we get

$$\mathrm{C}\ell(X \perp X \perp Y \perp Y) \;\cong\; \mathrm{C}\ell(X) \otimes \mathrm{C}\ell(Y) \otimes \mathrm{C}\ell(Y) \otimes \mathrm{C}\ell(X)\ ,$$

and therefore all the previous algebras are isomorphic to one another. Because of (2.5.8), $X \perp X \perp Y \perp Y$ is a hyperbolic space isomorphic to $\mathbf{H}(M \oplus M)$, and because of (3.7.2) its Clifford algebra is isomorphic to $\mathrm{End}(\bigwedge(M \oplus M))$. □

The Brauer classes of A, A^{ng} and $A^{ng}_{D'}$

When Z is a quadratic extension and D' a discriminant module, on the direct sum $Z[D'] = Z \oplus (D' \otimes Z)$ we define the following multiplication:

$$(y\ + (d \otimes z))\ (y'\ + (d' \otimes z'))\ =\ yy' + (dd')\varphi(z)z'\ +\ d \otimes zy'\ + d' \otimes \varphi(y)z'\ ;$$

in this definition, the standard involution φ of Z gets involved in every reversion of an element of D' and one of Z; this simple rule allows us easily to verify that $Z[D']$ is a (noncommutative) associative algebra of constant rank 4. When D' is a free module, it is easy to verify that $Z[D']$ is isomorphic to a Cayley–Dickson extension of Z (see (3.3.1)). In $Z[D']$ the elements of Z are always even, but we may treat the elements of D' either as even ones, or as odd ones, and accordingly $Z[D']$ receives either a trivial grading or a nontrivial one. If M is a module over Z, we also consider $M[D'] = M \oplus (D' \otimes M)$ and give it a structure of left module over $Z[D']$ in an evident way.

(3.8.11) **Lemma.** *When the elements of D' are given the odd parity, the graded algebra $Z[D']$ is isomorphic to $\mathrm{C}\ell(D' \otimes Z)$.*

Proof. In $Z[D']$ the square of every element $d \otimes z$ is $d^2\mathcal{N}(z)$; it soon appears that the natural injection $D' \otimes Z \rightarrow Z[D']$ extends to an algebra morphism $\mathrm{C}\ell(D' \otimes Z) \rightarrow Z[D']$. This morphism is obviously surjective, therefore bijective (see (1.13.5)). □

Another technical lemma is still necessary; it is an easy consequence of the isomorphism $Z \otimes Z \rightarrow Z \times Z$ described in (3.4.4).

(3.8.12) **Lemma.** *If M is a module over Z, the mapping $z \otimes v \longmapsto (zv,\ \varphi(z)v)$ is a bijection from $Z \otimes_K M$ onto $M \times M$.*

Proof. All the arrows in this sequence are bijective:

$$Z \otimes_K M \longrightarrow Z \otimes_K (Z \otimes_Z M) \longrightarrow (Z \otimes_K Z) \otimes_Z M \longrightarrow (Z \times Z) \otimes_Z M \longrightarrow M \times M\ ;$$

the second arrow is $z \otimes (y \otimes v) \longmapsto (y \otimes z) \otimes v$, and the third arrow comes from (3.4.4). □

Now we are ready to compare the algebras A, A^{ng} and $A^{ng}_{D'}$.

(3.8.13) **Theorem.** *Let A be a graded Azumaya algebra such that $\mathrm{Z}(A_0)$ is a quadratic extension Z, and let D' be any discriminant module. In* $\mathrm{Br}(K)$ *the class of* $A_{D'}^{ng}$ *is the product of the classes of* A^{ng} *and* $\mathrm{C}\ell(D' \otimes Z)^{ng}$.

Proof. Here we give the elements of D' the even parity, so that $Z[D']$ is isomorphic to $\mathrm{C}\ell(D' \otimes Z)^{ng}$; this algebra has a Brauer class of order 1 or 2 because it is provided with a reversion. We begin with the construction of an algebra morphism

$$G \quad \text{from} \quad B = Z[D'] \otimes A_{D'} \otimes A^o \quad \text{into} \quad \mathrm{End}(A[D']) \ ;$$

if we manage to prove that it is bijective, the proof is finished. We treat A as a Z-module, on which Z acts by multiplication on the left side; consequently $Z[D']$ acts on $A[D']$:

$$(y \ + (d \otimes z)) \ (a' \ + (d' \otimes b')) \ = \ ya' + (dd')\varphi(z)b' \ + \ d \otimes za' \ + d' \otimes \varphi(y)b'$$

for all y and $z \in Z$, all d and $d' \in D$, and all a' and $b' \in A$. Then we make $A_{D'}$ act on $A[D']$ in this evident way (for all $a \in A_0$ and all $b \in A_1$):

$$(a \ + (d \otimes b)) \ (a' \ + (d' \otimes b')) \ = \ aa' + (dd')bb' \ + \ d \otimes ba' \ + \ d' \otimes ab' \ ;$$

some calculations are needed to verify that the operation in $A[D']$ of every element of $Z[D']$ commutes with the operation of every element of $A_{D'}$; in these calculations, a (element of A_0) commutes with y and z, but not b (element of A_1), since $bz = \varphi(z)b$ (see (3.5.13)). Of course A^o acts by multiplications on the right side, and the operations of its elements commute with all the previous ones:

$$c^o \ (a' \ + (d' \otimes b')) \ = \ a'c \ + (d' \otimes b'c) \quad \text{for all } c \in A.$$

The algebra morphism G is now well defined. We must deduce its bijectiveness from the bijectiveness of the canonical morphism $F : A \otimes A^o \to \mathrm{End}(A)$ defined by $F(a' \otimes c^o)(b') = a'b'c$. For this purpose we can assume that D' is a free module generated by some element d; let d' be the generator such that $dd' = 1$; in the following calculations d and d' are always these generators. There is a bijection from $\mathrm{End}(A)^4$ onto $\mathrm{End}(A[D'])$, which maps every quartet (f, f', g, g') to this endomorphism of $A[D']$:

$$(f, f', g, g') \ (a' \ + (d' \otimes b')) \ = \ f(a') + f'(b') \ + d' \otimes g(b') \ + d' \otimes g'(a') \ .$$

Consequently the morphism G allows us to associate an element of $\mathrm{End}(A)^4$ with every element of B, and thus to get a mapping $H : B \to \mathrm{End}(A)^4$. Moreover $\mathrm{End}(A)$ is a Z-module: for instance $(zf)(a') = z(f(a'))$.

In B we consider the subalgebra B_0 spanned by all elements like $z \otimes a \otimes c^o$ or like $(d \otimes z) \otimes (d' \otimes b) \otimes c^o$. Easy calculations show that

$$H(z \otimes a \otimes c^o) = (zF(a \otimes c^o), \ 0, \ \varphi(z)F(a \otimes c^o), \ 0) \ ,$$
$$H((d \otimes \varphi(z)) \otimes (d' \otimes b) \otimes c^o) = (zF(b \otimes c^o), \ 0, \ \varphi(z)F(b \otimes c^o), \ 0) \ ;$$

because of (3.8.12), the elements of B_0 are mapped bijectively to the elements $(f,0,g,0)$ of $\mathrm{End}(A)^4$. The submodule B_1 spanned by all elements $z\otimes(d\otimes b)\otimes c^o$ and $(d\otimes z)\otimes a\otimes c^o$ is supplementary to B_0; other easy calculations show that

$$H(z\otimes(d\otimes b)\otimes c^o) = (0,\ zF(b\otimes c^o),\ 0,\ \varphi(z)F(b\otimes c^o))\ ,$$
$$H((d\otimes\varphi(z))\otimes a\otimes c^o) = (0,\ zF(a\otimes c^o),\ 0,\ \varphi(z)F(a\otimes c^o))\ ;$$

consequently the elements of B_1 are mapped bijectively to the elements $(0,f',0,g')$. □

(3.8.14) **Theorem.** *Let A be a graded Azumaya algebra such that $\mathrm{Z}(A_0)$ is a quadratic extension Z. In $\mathrm{Br}^g(K)$, the class of A is the product of the classes of A^{ng} and $C\ell(Z)$ (with its usual grading of Clifford algebra).*

Proof. The proof is quite similar to that of (3.8.13); but now D' is a discriminant module that is isomorphic to K when its grading is forgotten, and that only differs from K because all its elements are given the *odd* parity; consequently $Z[D']$ is now isomorphic to $C\ell(Z)$ as a graded algebra, and $A_{D'}$ is isomorphic to A^{ng}, because all elements of $D'\otimes A_1$ are now even. Since twisted tensor products are here involved, twisting signs shall be necessary, but they will never come from any element of $A_{D'}$. The multiplications in $A_{D'}$ and in $Z[D']$, and their actions on $A[D']$, are the same as in the proof of (3.8.13). Now we construct a graded algebra morphism

$$G' \quad \text{from} \quad B' = Z[D']\,\hat{\otimes}\,A_{D'}\,\hat{\otimes}\,A^{to} \quad \text{into} \quad \mathrm{End}(A[D'])\ ;$$

we have just to make precise the twisting signs involved in the action of A^{to} on $A[D']$:

$$c^{to}\ (a' + (d'\otimes b')) \ = \ (-1)^{\partial c\partial a'}\ a'c \ + \ (-1)^{\partial c(1+\partial b')}\ (d'\otimes b'c)\ ;$$

this operation of c^{to} in $A[D']$ commutes with the operation of all elements of $A_{D'}$; it commutes or anticommutes with the operation of an element of $Z[D']$ according to the awaited twisting sign. We must deduce the bijectiveness of G' from the bijectiveness of $F' : A\,\hat{\otimes}\,A^{to} \to \mathrm{End}(A)$; this is achieved as in the proof of (3.8.13).

Only one detail deserves a mention: because of many twisting signs $(-1)^{\partial c}$ (see for instance $(-1)^{\partial c(1+\partial b')}$ just above), we must slightly modify the bijection $\mathrm{End}(A)^4 \to \mathrm{End}(A[D'])$. We need the automorphism σ^{to} of A^{to} defined by $\sigma^{to}(c^{to}) = (-1)^{\partial c}c^{to}$ and the corresponding automorphism σ_{E} of $\mathrm{End}(A)$: $\sigma_{\mathrm{E}} = F'\circ(A\otimes\sigma^{to})\circ F'^{-1}$. Now an element of $\mathrm{End}(A)^4$ gives an endomorphism of $\mathrm{End}(A[D'])$ in this way (when $dd'=1$):

$$(f,f',g,g')(a' + (d'\otimes b')) = f(a') + \sigma_{\mathrm{E}}(f')(b') + d'\otimes\sigma_{\mathrm{E}}(g)(b') + d'\otimes g'(a')\ . \ \square$$

Here are two direct applications of (3.8.14).

Proof of of the exactness of (3.5.19). $1 \to \mathrm{Br}(K) \to \mathrm{Br}^g(K) \to \mathrm{Q}^g(K) \to 1$. We already know that the second arrow is injective. If the class of A belongs to the kernel of the third arrow, A has even type, and according to the remark (3.5.7) about nonconstant ranks, we can assume that A is either trivially graded (and there is nothing to prove), or that $\mathrm{Z}(A_0)$ is a quadratic extension, therefore a trivial quadratic extension; in this case Theorem (3.8.14) says that A and A^{ng} have the same class in $\mathrm{Br}^g(K)$. It remains to prove the surjectiveness of the third arrow. It suffices to consider the class of a quadratic extension Z such that Z_0 and Z_1 have constant ranks; if Z_1 has constant rank 1, then Z is already a graded Azumaya algebra (see (3.5.11)); and when Z is trivially graded, it is isomorphic to $C\ell_0(Z)$ (see (3.8.5)). □

(3.8.15) **Theorem.** *Let* (M,q) *be a quadratic space; the order of the Brauer class of* $C\ell(M,q)$ *is a divisor of* 8. *If the rank of* M *is even, it is a divisor of* 4.

Proof. The former statement is an immediate consequence of the latter; thus we can assume that the rank of M is even and never vanishes. Because of the reversion in $C\ell(M,q)$, the Brauer class of $C\ell(M,q)^{ng}$ has order 1 or 2. Because of (3.8.14) it suffices to prove that the order of the class of $C\ell(Z)$ (if $Z = \mathrm{QZ}(M,q)$) is a divisor of 4, and this has been done in (3.8.10). □

The bilinear mapping $\mathcal{Q} : \mathrm{Disc}(K) \times \mathrm{Q}(K) \to \mathrm{Br}(K)$

From (3.8.9), and from the triviality of the Brauer class of $C\ell(Z)^{ng}$ (see (3.8.1)), we deduce that the Brauer class of $(C\ell((M,q) \perp (M',q')))^{ng}$ is the product of the Brauer classes of $C\ell(M,q)^{ng}$, $C\ell(M',q')^{ng}$ and $C\ell(D' \otimes Z)^{ng}$. This leads us to consider the mapping $\mathcal{Q}$ which maps the isomorphy class of D' (any discriminant module) and the isomorphy class of Z (any quadratic extension) to the Brauer class of the quaternion algebra $C\ell(D' \otimes Z)^{ng}$. The next theorem shows four properties of this mapping; the first one is a property of symmetry, and the second and third properties mean that $\mathcal{Q}$ is bilinear; indeed this is ordinary $\mathbb{Z}$-bilinearity if $\mathrm{Disc}(K)$, $\mathrm{Q}(K)$ and $\mathrm{Br}(K)$ are treated as additive groups, in other words, objects of $\mathcal{M}od(\mathbb{Z})$.

(3.8.16) **Theorem.**

(a) *If* D *and* D' *are the discriminant modules of the quadratic extensions* Z *and* Z', *then* $C\ell(D' \otimes Z)^{ng}$ *and* $C\ell(D \otimes Z')^{ng}$ *have the same Brauer class.*

(b) *If* D *and* D' *are discriminant modules, and if* Z''' *is a quadratic extension, the Brauer class of* $C\ell(D \otimes Z''')^{ng} \otimes C\ell(D' \otimes Z''')^{ng}$ *is the Brauer class of* $C\ell((D \otimes D') \otimes Z''')^{ng}$.

(c) *If* D''' *is a discriminant module, and if* Z *and* Z' *are quadratic extensions, the Brauer class of* $C\ell(D''' \otimes Z)^{ng} \otimes C\ell(D''' \otimes Z')^{ng}$ *is the Brauer class of* $C\ell(D''' \otimes (Z \star Z'))^{ng}$.

(d) *If Z is a quadratic extension with discriminant module D, and if J is the free discriminant module generated by an element j such that $j^2 = -1$, then $C\ell(D \otimes Z)^{ng} \cong C\ell(J \otimes Z)^{ng}$.*

Proof. From (3.8.8) we deduce that

$$C\ell(Z) \otimes C\ell(D \otimes Z') \;\cong\; C\ell(Z \perp Z') \;\cong\; C\ell(D' \otimes Z) \otimes C\ell(Z') \;;$$

since the Brauer classes of $C\ell(Z)^{ng}$ and $C\ell(Z')^{ng}$ are trivial (see (3.8.1)), we have proved (a). From (3.8.7) we deduce that $C\ell((D \otimes D') \otimes Z''')$ is isomorphic to $C\ell(D' \otimes Z''')_D$; since $C\ell_0(D' \otimes Z''')$ is isomorphic to Z''' (see (3.8.5)), from (3.8.13) we deduce that the class of $C\ell(D' \otimes Z''')_D^{ng}$ is the product of the classes of $C\ell(D' \otimes Z''')^{ng}$ and $C\ell(D \otimes Z''')^{ng}$, as stated in (b). Then (d) is an obvious consequence of (3.8.4) and (3.8.5), and it remains to prove (c). Let D, D', φ, φ' be the discriminant modules and standard involutions of Z and Z', and let us set $Z'' = Z \star Z'$. Since the discriminant module of Z'' is $D \otimes D'$, from (3.8.4) we deduce that the quadratic space $D \otimes D' \otimes Z''$ is isomorphic to $J \otimes Z''$; now two applications of (3.8.8) show that

$$C\ell(D''' \otimes Z) \otimes C\ell(D''' \otimes Z') \otimes C\ell(D''' \otimes Z'') \;\cong\; C\ell(D''' \otimes (M,q))$$

if (M,q) is the quadratic space

$$(M,q) \;=\; Z \perp (D \otimes Z') \perp (J \otimes Z'').$$

If we manage to prove that (M,q) is hyperbolic, the conclusion (c) follows immediately. Let $\Psi : Z \otimes Z' \to M$ be the linear mapping defined in this way:

$$\Psi(z \otimes z') \;=\; (\mathrm{tr}(z')z,\ (z - \varphi(z)) \otimes z',\ j \otimes (\varphi(z) \otimes z' + z \otimes \varphi'(z')) \;;$$

if we manage to prove that $\mathrm{Im}(\Psi)$ is a direct summand of M of constant rank 3, and that it is totally isotropic, from (2.5.5) we deduce that (M,q) is hyperbolic. Because of (1.12.9) and (1.13.1) these properties of $\mathrm{Im}(\Psi)$ can be tested by localization; thus we can assume that Z and Z' have bases $(1,z)$ and $(1,z')$ such that $z^2 = \beta z - \gamma$ and $z'^2 = \beta' z' - \gamma'$. Let us set $z'' = \beta \otimes z' + z \otimes \beta' - 2z \otimes z'$; in (3.4.11) it is proved that $(1, z'')$ is a basis of Z'', and the coefficients β'' and γ'' such that $z''^2 = \beta'' z' - \gamma''$ are calculated there. Now we get:

$$\begin{aligned}
\Psi(1 \otimes 1) &= 2\ (1,\ 0,\ j \otimes 1 \otimes 1)\ , \\
\Psi(1 \otimes z') &= \beta'\ (1,\ 0,\ j \otimes 1 \otimes 1)\ , \\
\Psi(z \otimes 1) &= (2z,\ (2z - \beta) \otimes 1,\ j \otimes \beta \otimes 1)\ , \\
\Psi(z \otimes z') &= (\beta' z,\ (2z - \beta) \otimes z',\ j \otimes z'')\ .
\end{aligned}$$

Since β' and 2 generate K as an ideal (indeed $\beta'^2 - 4\gamma'$ is invertible), we realize that $\mathrm{Im}(\Psi)$ is the free module generated by

$$(1,\ 0,\ j \otimes 1 \otimes 1)\ , \quad (2z - \beta,\ (2z - \beta) \otimes 1,\ 0)\ , \quad (\beta' z,\ (2z - \beta) \otimes z',\ j \otimes z'')\ ,$$

and that $\mathrm{Im}(\Psi)$ is supplementary to $Z \perp (D' \otimes z') \perp 0$. At last the values of β'' and γ'' calculated in (3.4.11) allow us to verify that $\mathrm{Im}(\Psi)$ is totally isotropic. □

Exercises

(3.ex.1) Let a, b, c be three elements in a quadratic module (M, q); prove the following equalities:

$$\rho(a)\rho(b)\rho(c) - \rho(b)\rho(c)\rho(a) = \mathrm{b}_q(a,b)\rho(c) - \mathrm{b}_q(a,c)\rho(b) \ ;$$
$$\rho(a)\rho(b)\rho(c) + \rho(c)\rho(b)\rho(a) = \mathrm{b}_q(a,b)\rho(c) - \mathrm{b}_q(a,c)\rho(b) + \mathrm{b}_q(b,c)\rho(a).$$

(3.ex.2) Let a, b, c, d be elements in a quadratic module (M, q); prove the equality

$$\rho(a)\rho(b)\rho(c)\rho(d) + \rho(d)\rho(c)\rho(b)\rho(a) \ = \ \rho(b)\rho(a)\rho(d)\rho(c) + \rho(c)\rho(d)\rho(a)\rho(b).$$

Hint. $\rho(a)\rho(b)\rho(c)\rho(d) - \rho(b)\rho(a)\rho(d)\rho(c)$ belongs to $C\ell^{\leq 2}(M, q)$.

(3.ex.3)

(a) Let A and B be graded algebras; justify the existence of the following four isomorphisms:

$$\begin{aligned}
A \hat{\otimes} B &\longrightarrow B \hat{\otimes} A \ , & x \otimes y &\longmapsto (-1)^{\partial x \partial y} \ y \otimes x \ ; \\
A \hat{\otimes} B &\longrightarrow (A^o \hat{\otimes} B^o)^o \ , & x \otimes y &\longmapsto (-1)^{\partial x \partial y} \ (x^o \otimes y^o)^o \ ; \\
A \hat{\otimes} B &\longrightarrow (A^t \hat{\otimes} B^t)^t \ , & x \otimes y &\longmapsto (-1)^{\partial x \partial y} \ (x^t \otimes y^t)^t \ ; \\
A \hat{\otimes} B &\longrightarrow (A^{to} \hat{\otimes} B^{to})^{to}, & x \otimes y &\longmapsto \ (x^{to} \otimes y^{to})^{to}.
\end{aligned}$$

(b) Suppose that f and g are graded anti-morphisms respectively from A to B and from B to A; in other words, the mappings $x \longmapsto f(x)^o$ and $y \longmapsto g(y)^o$ are graded algebra morphisms; prove that the mapping

$$A \hat{\otimes} B^{to} \longrightarrow (A \hat{\otimes} B^{to})^t \ , \qquad x \otimes y^{to} \longmapsto (g(y) \otimes f(x)^{to})^t$$

is an algebra morphism.

(3.ex.4) An increasing filtration of a module M is any increasing family of submodules $(M^{\leq j})_{j \in \mathbb{Z}}$. With such a filtered module M is associated a graded module $\mathrm{Gr}(M)$, which is the direct sum of all quotients $\mathrm{Gr}^j(M) = M^{\leq j}/M^{\leq j-1}$. Let $f : M \to N$ be a morphism of filtered modules; this means that $f(M^{\leq j}) \subset N^{\leq j}$ for all $j \in \mathbb{Z}$.

- Prove that f induces a graded mapping $\mathrm{Gr}(f) : \mathrm{Gr}(M) \to \mathrm{Gr}(N)$.
- Suppose that $\mathrm{Gr}(f)$ is injective, that $M^{\leq -1} = 0$ and $\bigcup_j M^{\leq j} = M$. Prove that f is injective.
- Suppose that $\mathrm{Gr}(f)$ is surjective, that $N^{\leq -1} = 0$ and $\bigcup_j N^{\leq j} = N$. Prove that f is surjective.

(3.ex.5) Let M be a finitely generated module; thus there exists an integer r such that the rank of M at every prime ideal is $\leq r$. Prove that $C\ell^{\leq r}(M, q) = C\ell(M, q)$ for every quadratic form q on M.

(3.ex.6) Suppose that 2 is invertible in K, and that the quadratic module (M,q) contains an element e such that $q(e) = -1$. Let M' be the submodule orthogonal to e, and q' the restriction of q to M'. Prove the existence of a surjective algebra morphism $f : C\ell(M',q') \to C\ell_0(M,q)$ such that $f(\rho'(x)) = \rho(e)\rho(x)$ for all $x \in M'$. Prove that it is an isomorphism when M is a quadratic space.

(3.ex.7) Let (M,q) and (M',q') be quadratic modules, ρ and ρ' the canonical mappings from M and M' into the corresponding Clifford algebras, and (u,λ) an element of $\mathrm{Hom}(M,M') \times K$ such that

$$\forall a \in M, \qquad q(a) \;=\; \lambda\, q'(u(a)) \;;$$

when λ is invertible, u is called a *similitude of ratio* λ^{-1}.

(a) When λ admits a square root κ in K, prove the existence of an algebra morphism $C\ell(M,q) \to C\ell(M',q')$ that maps every $\rho(a)$ to $\kappa\rho'(u(a))$.

(b) Prove the existence of an algebra morphism $C\ell_0(u,\lambda)$ from $C\ell_0(M,q)$ into $C\ell_0(M',q')$ that maps every product $\rho(a)\rho(b)$ to $\lambda\rho'(u(a))\rho'(u(b))$.

(c) You can even prove this stronger result: there are two linear mappings

$$C\ell_i(u,\lambda) \;:\; C\ell_i(M,q) \longrightarrow C\ell_i(M',q') \quad \text{for } i = 0,1,$$

satisfying these properties: first $C\ell_1(u,\lambda)$ maps every $\rho(a)$ to $\rho'(u(a))$; secondly for all $(i,j) \in (\mathbb{Z}/2\mathbb{Z})^2$, for all $x \in C\ell_i(M,q)$ and all $y \in C\ell_j(M,q)$,

$$\begin{aligned} C\ell_i(u,\lambda)(x)\; C\ell_j(u,\lambda)(y) &= C\ell_{i+j}(u,\lambda)(xy) && \text{if } ij = 0, \\ C\ell_i(u,\lambda)(x)\; C\ell_j(u,\lambda)(y) &= \lambda\; C\ell_{i+j}(u,\lambda)(xy) && \text{if } ij = 1. \end{aligned}$$

(3.ex.8)* We use the notations of (1.ex.27) and (1.ex.28): J is an ordered set, D the set of all (i,j) such that $i \le j$, and (J,D) satisfies the condition required in the definition of direct limits. Let $((M_j),\ (f_{j,i}))$ be a family of modules and morphisms over (J,D), in which every module M_j is provided with a quadratic form q_j and every $f_{j,i}$ is a morphism of quadratic modules.

(a) Prove the existence of a unique quadratic form $\varinjlim(q_j)$ on $\varinjlim(M_j)$ such that all canonical morphisms $M_i \to \varinjlim(M_j)$ are morphisms of quadratic modules.

(b) Prove that the canonical algebra morphism

$$\varinjlim(C\ell(M_j,q_j))) \longrightarrow C\ell(\varinjlim(M_j), \varinjlim(q_j))$$

is an isomorphism.

(c) Prove that the Clifford algebra of a flat module is flat, by means of the following three results:

- a direct limit of flat modules is flat (see (1.ex.28));
- the Clifford algebra of a free module is a free module (see (4.8.11));
- every flat module is a direct limit of free modules (see for instance [Lazard 1964]).

Graded quadratic extensions and graded Azumaya algebras

(3.ex.9) Let n be an integer ≥ 1, and K the local ring $\mathbb{Z}/2^n\mathbb{Z}$. It is clear that the K-algebra $A = K \oplus Kz$ with $z^2 = 1$ is not a quadratic extension. Let φ be the standard involution of A. Prove that the subalgebra of all elements of $A \otimes A$ invariant by $\varphi \otimes \varphi$ is not a free module, and that its rank (the minimal number of generators) is 4.

Prove that the order of the group $\mathrm{Aut}(A)$ is never 2 (that is the order of $\mathrm{Aut}(Z)$ when Z is a quadratic extension over a local ring); it is 1 (resp. 4, resp. 8) if $n = 1$ (resp. $n = 2$, resp. $n \geq 3$).
Hint. If $n \geq 3$, the group $K^\times$ is the direct product of two cyclic subgroups which are respectively generated by -1 and 5.

(3.ex.10) Let $K \oplus Kz$ be a free and trivially graded quadratic extension: thus $z^2 = \beta z - \gamma$ with $\beta^2 - 4\gamma$ invertible in K. Prove that its isomorphy class belongs to the classifying group $\mathrm{Q}_f(K)$ if and only if there exists $\lambda \in K$ such that $\beta + 2\lambda$ is invertible.

When $K = \mathbb{Z}$, prove that $\mathrm{Q}^g(\mathbb{Z})$ is reduced to one element.
Hint. (2.8.6) and (3.4.14).

(3.ex.11) Suppose that K is the direct product $K' \times K''$ of two rings $\neq 0$. Prove that $\mathrm{Q}^g(K)$ is isomorphic to $\mathrm{Q}^g(K') \times \mathrm{Q}^g(K'')$, and $\mathrm{Br}^g(K)$ to $\mathrm{Br}^g(K') \times \mathrm{Br}^g(K'')$.

(3.ex.12) Let A and A' be (nongraded) quadratic extensions of K. Prove that $A \otimes A'$ is a quadratic extension of $A \star A'$. Which is its standard involution? When z, z' and z'' are defined as in Example (3.4.11), compare $\mathrm{tr}(z \otimes z')$ and $\mathrm{tr}(\varphi(z) \otimes z')$ with z'' and $(\varphi \otimes A')(z'')$.

Compare $A \otimes A'$, $(A \star A') \otimes A$ and $(A \star A') \otimes A'$ as quadratic extensions of $A \star A'$.

(3.ex.13) Let P be a finitely generated projective module of constant rank r over the ring K.

(a) Let f be an endomorphism of P; prove that, for all $a_1, a_2, \ldots, a_r \in P$,

$$\det(f)\; a_1 \wedge a_2 \wedge \cdots \wedge a_r = f(a_1) \wedge f(a_2) \wedge \cdots \wedge f(a_r)\;,$$

$$\mathrm{tr}(f)\; a_1 \wedge a_2 \wedge \cdots \wedge a_r = \sum_{i=1}^{r} a_1 \wedge \cdots \wedge f(a_i) \wedge \cdots \wedge a_r.$$

(b) Now $r = 2$ and we set $\varphi(f) = \mathrm{tr}(f)\mathrm{id}_P - f$ for all $f \in \mathrm{End}(P)$. Prove that

$$\forall a,\ b \in P, \qquad f(a) \wedge b \;=\; a \wedge \varphi(f)(b).$$

Then prove that φ is a standard involution of $\mathrm{End}(P)$ and that the associated norm is the quadratic form $f \longmapsto \det(f)$ (as it is stated in (3.6.3)) by means of this trick: an element $c \in P$ vanishes if and only if $a \wedge c$ vanishes for all $a \in P$.

(3.ex.14) Let A and B be graded Azumaya algebras for which traces $A \to K$ and $B \to K$ have been defined in accordance with (3.6.6) and (3.6.7); there is also a trace $A \hat{\otimes} B \to K$. Prove that $\mathrm{tr}(x \otimes y) = \mathrm{tr}(x)\ \mathrm{tr}(y)$ for all $x \in A$ and all $y \in B$.

(3.ex.15) Let P be a finite-dimensional vector space over a *field* K, and $A = \mathrm{End}(P)$.

(a) For every $f \in A$, prove that the left ideal Af is the subset of all $g \in A$ such that $\mathrm{Ker}(g) \supset \mathrm{Ker}(f)$, whereas the right ideal fA is the subset of all g such that $\mathrm{Im}(g) \subset \mathrm{Im}(f)$.

(b) Let f and g be two elements of A. Prove that the left ideal $Af + Ag$ contains an element the kernel of which is $\mathrm{Ker}(f) \cap \mathrm{Ker}(g)$, whereas the right ideal $fA + gA$ contains an element the image of which is $\mathrm{Im}(f) + \mathrm{Im}(g)$.

(c) Let J be a left ideal (resp. a right ideal) of A, and f an element of J of maximal rank among all elements of J. Prove that $J = Af$ (resp. $J = fA$).

(d) Prove that every two-sided ideal of A is equal to 0 or A.

Comment. If V is an infinite-dimensional vector space, the elements of $\mathrm{End}(V)$ of finite rank constitute a two-sided ideal different from 0 and $\mathrm{End}(V)$.

(3.ex.16) Prove that the group $\mathrm{Q}^g(\mathbb{R})$ derived from the field $\mathbb{R}$ of real numbers is a cyclic group of order 4, generated by the class of $(\mathbb{R}^2)^g$ (the quadratic extension $\mathbb{R} \times \mathbb{R}$ provided with its nontrivial grading).

(3.ex.17) We assume that A is a finite-dimensional algebra over the field $\mathbb{R}$ of real numbers, and that A is also a division ring; we will prove that A is isomorphic either to $\mathbb{R}$ or to $\mathbb{C}$ (the field of complex numbers) or to $\mathbb{H}$ (the division ring of real quaternions); this implies that $\mathrm{Br}(\mathbb{R})$ is a group of order 2 (see (3.5.20)). Let V be the subset of all $v \in A$ such that $v^2 \in \mathbb{R}$ and $v^2 \leq 0$.

(a) Let x be a nonzero element of A, and $f : \mathbb{R}[X] \to A$ the algebra morphism that maps every polynomial $P(X)$ to its value $P(x)$ on x. Prove that the ideal $\mathrm{Ker}(f)$ is generated either by a polynomial of degree 1, or by a polynomial of degree 2 without real roots.

(b) Prove that the mapping $\mathbb{R} \times V \to A$ defined by $(r, v) \longmapsto r + v$ is bijective. Consequently if u and v belong to V, and $u + v$ to $\mathbb{R}$, then $u + v = 0$.

(c) Prove that V is a vector subspace of A.
Hint. For any $(u, v) \in V^2$, set $u + v = s + x$ and $u - v = t + y$ with s and t in $\mathbb{R}$, and x and y in V; observe that $2sx + 2ty \in \mathbb{R}$, and consequently $sx + ty = 0$; this implies $(s + t)u + (s - t)v = s^2 + t^2$, and finally $s = t = 0$. This trick stems from a personal communication of J. Commeau in 1960.

(d) What happens when $\dim(V) \leq 1$?

(e) Suppose that $\dim(V) > 1$, and consider the quadratic form $q : V \longrightarrow \mathbb{R}$ defined by $q(v) = -v^2$, whence $\mathrm{b}_q(u, v) = -uv - vu$. Prove that V contains two elements i and j such that $i^2 = j^2 = -1$ and $ij = -ji$. Then prove that (i, j, ij) is an orthonormal basis of V, and that A is isomorphic to $\mathbb{H}$.

Hint. If an element k of V is orthogonal to i and j (in other words, $ik+ki = jk+kj = 0$), the associativity of A implies that k commutes with ij, and consequently belongs to $\mathbb{R}ij$.

Comment. Everything before (e) is valid for a nonassociative $\mathbb{R}$-algebra A (with unit element) in which every element generates an associative and finite-dimensional subalgebra without divisors of zero. Consequently it is valid when A is a finite-dimensional alternative algebra without divisors of zero; indeed in this case every subalgebra generated by two elements is associative. To prove the theorem of Zorn (see (3.3.2)), it remains to prove that A is isomorphic to the Cayley algebra of octonions if $\dim(V) > 3$. Since A is alternative, every equality $xy + yx = 0$ implies $(wx)y+(wy)x = x(yz)+y(xz) = 0$ for all w, $z \in A$. Let i, j, k be elements of V such that (i,j,ij,k) is an orthonormal family in (V,q); then on one side $(ij)k = (jk)i = (ki)j = k(ji) = i(kj) = j(ik)$ and $(i,j,k,ij,ik,jk,(ij)k)$ is still an orthonormal family; on the other side every element $x \in V$ that is orthogonal to (i,j,k,ij,ik,jk), commutes with $(ij)k$ and consequently belongs to $\mathbb{R}(ij)k$; indeed $x((ij)k) = k((ix)j)$ and $((ij)k)x = (k(ix))j$, and by alternativity

$$(k(ix))j - k((ix)j) = (kj)(ix) - k(j(ix)) = x((kj)i) + x(k(ji)) = 0.$$

All this proves that $\dim(V) = 7$ and that A is isomorphic to the Cayley algebra of octonions.

(3.ex.18)* The following considerations are motivated by the first exact sequence in (3.4.14). For every integer n let $K\lceil n\rfloor$ be the subset of all $\lambda \in K$ such that $1-n\lambda$ is invertible; it is an abelian group for the modified addition $(\lambda,\mu) \longmapsto \lambda+\mu-n\lambda\mu$. We are looking for group morphisms $K\lceil m\rfloor \to K\lceil n\rfloor$ when m and n are nonzero integers. Obviously $\lambda \longmapsto -\lambda$ is an isomorphism from $K\lceil n\rfloor$ onto $K\lceil -n\rfloor$. Moreover in the first exact sequence of (3.4.14) there is a morphism $\kappa \longmapsto \kappa - \kappa^2$ from $K\lceil 2\rfloor$ into $K\lceil 4\rfloor$.

Let $K_0[[t]]$ be the subset of all formal series $x(t) \in K[[t]]$ such that $x(0) = 0$ (so that $1 - nx$ is invertible for all n); first we look for formal series $f \in K_0[[t]]$ such that the following equality is true in the ring of formal series $K[[x,y]]$:

$$\text{(E)} \qquad f(x+y-mxy) \;=\; f(x)+f(y)-nf(x)f(y)\ .$$

(a) Let h be another indeterminate; if (x,y) is replaced with $(t,\ h(1-mt)^{-1})$, the equation (E) becomes

$$\frac{f(t+h)-f(t)}{h} = \frac{1-nf(t)}{1-mt}\ \frac{f(h(1-mt)^{-1})}{h(1-mt)^{-1}}.$$

Then, by replacing h with 0, derive from (E) the new equation

$$\text{(E}'\text{)} \qquad \frac{f'(t)}{1-nf(t)} = \frac{f'(0)}{1-mt}\ .$$

(b) Solve the differential equation (E′) when K is the field $\mathbb{Q}$ of rational numbers. You must find a unique solution f satisfying the conditions $f(0) = 0$ and $f'(0) = a$ (an arbitrary value), namely

$$f(t) = \frac{1}{n}\left(1 - (1-mt)^{an/m}\right) = at - \sum_{k\geq 2} \omega_k t^k$$

with

$$\omega_k = (-1)^k \frac{a}{k!}(an-m)(an-2m)(an-3m)\cdots(an-(k-1)m) .$$

(c) Verify that the above solution of (E′) is also a solution of (E). Let R be the subring of $\mathbb{Q}$ generated by a and the coefficients ω_k; conclude that you get a solution of (E) over the ring K if the canonical morphism $\mathbb{Z} \to K$ extends to a morphism $R \to K$, and that you even get a morphism $K\lceil m \rfloor \longrightarrow K\lceil n \rfloor$ if all the coefficients ω_k have a zero image in K except a finite number.

(d) For every integer $m \geq 2$ prove that the following equality defines a homomorphism f_m from $K\lceil m \rfloor$ into $K\lceil m^2 \rfloor$:

$$f_m(\lambda) = \lambda - \sum_{k=2}^{m} (-1)^k \frac{m!}{k!\,(m-k)!}\, m^{k-2}\lambda^k .$$

Comments. The kernel $\mathrm{Ip}_m(K)$ of f_m has been called the group of generalized m-idempotents of K, and its cokernel $G_m(K)$ the *Villamayor group* of K. When m is not a divisor of zero in K, the equality $1 - m^2 f_m(\lambda) = (1-m\lambda)^m$ yields an easier definition of f_m.

Clifford algebras of quadratic spaces

(3.ex.19) Let (M, q) be a quadratic space of constant rank 4.

(a) Prove that M is the submodule of all elements of $C\ell_1(M,q)$ invariant by the reversion τ.

(b) Let $Z = \mathrm{QZ}(M,q)$ be the center of $C\ell_0(M,q)$. Prove that $\tau(z) = z$ for all $z \in Z$. And for every $x \in C\ell_0(M,q)$ prove that $x\tau(x)$ belongs to Z; in other words, τ induces a standard involution on the Z-algebra $C\ell_0(M,q)$. Since $C\ell_0(M,q)$ is an Azumaya algebra over Z (see (3.5.14)), it is a quaternion algebra over Z; and when Z is isomorphic to K^2, it is isomorphic to the direct product of two quaternion algebras over K.

(c) Prove the bijectiveness of the mapping $Z \otimes M \to C\ell_1(M,q)$ defined by $z \otimes a \longmapsto za$.

(3.ex.20) Let (M, q) be a quadratic space of constant rank 4, such that the center Z of $C\ell_0(M,q)$ is isomorphic to the algebra K^2; thus there is an idempotent ε such that $Z = K\varepsilon \oplus K(1-\varepsilon)$.

(a) Prove that we obtain a trilinear mapping T from M^3 into M if we set, for all a, b, c in M,

$$T(a,b,c) \;=\; \varepsilon abc + cba\varepsilon \ ,$$

and that it satisfies these wonderful properties:

$$T(a,a,b) \;=\; T(b,a,a) \;=\; q(a)b \ , \qquad q(T(a,b,c)) \;=\; q(a)\,q(b)\,q(c) \ ,$$

$$T(T(a,b,c),d,e) \;=\; T(a,T(d,c,b),e) \;=\; T(a,b,T(c,d,e)).$$

Hint. Remember that $x\varepsilon = (1-\varepsilon)x$ for all $x \in C\ell_1(M,q)$ (see (3.5.13)); deduce from (3.ex.19)(b) that $\tau(\varepsilon) = \varepsilon$, and then use (3.ex.19)(a) to prove that $T(a,b,c)$ belongs to M.

(b) Prove that the mapping $a \longmapsto \varepsilon a$ induces a bijection from M onto $\varepsilon C\ell_1(M,q)$. Consequently $T(a,b,c)$ is the *only* element $x \in M$ such that $\varepsilon x = \varepsilon abc$.

Hint. The mapping $(a,b) \longmapsto \varepsilon a + (1-\varepsilon)b$ induces a bijection from M^2 onto $C\ell_1(M,q)$; indeed it is surjective because $a = \varepsilon a + (1-\varepsilon)a$ and $abc = \varepsilon T(a,b,c) + (1-\varepsilon)T(c,b,a)$; its bijectiveness follows from (1.13.5) .

(c) For all a, b, c, d in M we set $\Omega(a,b,c,d) \;=\; \mathrm{b}_q\big(a\,,\ T(b,c,d) - T(d,c,b)\big)$. Prove that Ω is an alternate quadrilinear form.

Comment. Ω gives a generator of the module $\bigwedge^{*4}(M) = \mathrm{Hom}(\bigwedge^4(M),K)$; to prove it, you can assume that K is a field (see (1.13.5)) and that (M,q) is hyperbolic (see (2.6.6)). This property of Ω implies the existence of an isomorphism $\bigwedge^3(M) \to M$ that maps every $b\wedge c\wedge d$ to $T(b,c,d) - T(d,c,b)$.

(d) Let G be the subset of all $a \in M$ such that $q(a)$ is invertible in K; we suppose that G is not empty, and that e is an element of G. For every $(a,b) \in M^2$ we set

$$a * b = q(e)^{-1}T(a,e,b) \quad \text{and} \quad \bar{a} = q(e)^{-1}T(e,a,e) \ ;$$

prove that this multiplication makes M become an associative algebra with unit element e, and that the mapping $a \longmapsto \bar{a}$ is a standard involution of M. Moreover G is a $*$-multiplicative group, and the subset of all a such that $q(a) = q(e)$ is a subgroup.

Conversely $T(b,c,d) \;=\; q(e)\ b * \bar{c} * d$.

(e) The even subalgebra $C\ell_0(M,q)$ can be considered as the direct product of the algebras $C' = \varepsilon C\ell_0(M,q)$ and $C'' = (1-\varepsilon)C\ell_0(M,q)$. Verify that the mapping $a \longmapsto q(e)^{-1}\varepsilon ae$ is an isomorphism from the algebra M defined in (d) onto C', and that the mapping $a \longmapsto q(e)^{-1}(1-\varepsilon)ea$ is an algebra isomorphism from M onto C''. Therefore C', C'' and M are isomorphic quaternion algebras.

(3.ex.21) Let (M,q) be a quadratic space of dimension 4 over a *field* K. Assume that the center Z of $C\ell_0(M,q)$ is isomorphic to K^2, and that M contains nonzero elements on which q vanishes. Prove that (M,q) is hyperbolic.

Hint. Let ε be an idempotent of Z as in (3.ex.20), and e an element of M such that $q(e) \neq 0$; the mapping $a \longmapsto \varepsilon ea$ is a bijection from M onto $\varepsilon C\ell_0(M,q)$; this is a quaternion algebra over K, provided with a norm $\mathcal{N}$; prove that $\mathcal{N}(\varepsilon ea) =$

$q(e)q(a)$; then the argument developed at the end of the proof of (3.6.8) shows that $\varepsilon C\ell_0(M,q)$ is isomorphic to $\mathcal{M}(2,K)$, and consequently is hyperbolic for the quadratic form $\mathcal{N}$.

(3.ex.22) Since the groups $Q^g(\mathbb{R})$ and $\mathrm{Br}(\mathbb{R})$ have respectively order 4 and 2 (see (3.ex.16) and (3.ex.17)), the group $\mathrm{Br}^g(\mathbb{R})$ has order 8. Prove that $\mathrm{Br}^g(\mathbb{R})$ is a cyclic group generated by the class of $(\mathbb{R}^2)^g$.
Hint. Let $[(\mathbb{R}^2)^g]_Q$ and $[(\mathbb{R}^2)^g]_B$ be the classes of $(\mathbb{R}^2)^g$ in $Q^g(\mathbb{R})$ and $\mathrm{Br}^g(\mathbb{R})$; since $[(\mathbb{R}^2)^g]_Q$ has order 4, the order of $[(\mathbb{R}^2)^g]_B$ is a multiple of 4. Now the fourth power of $[(\mathbb{R}^2)^g]_B$ is the class of $C\ell(M,q)$ if (M,q) is a positive definite quadratic space of dimension 4 over $\mathbb{R}$; if (e_1,e_2,e_3,e_4) is an orthonormal basis of M, then $(1 - e_1e_2e_3e_4)/2$ is an idempotent ε in $\mathrm{QZ}(M,q)$; verify that $\varepsilon C\ell_0(M,q)$ is a division ring (isomorphic to $\mathbb{H}$ and not to $\mathcal{M}(2,\mathbb{R})$).

(3.ex.23) Let (M,q) be a quadratic space provided with an orthogonal basis $(e_1,e_2,\dots,e_n)$. Since 2 must be invertible in K, the quadratic extension $\mathrm{QZ}(M,q)$ is the direct sum of K and its discriminant module D. Prove that D is the free module generated by $e_1e_2\cdots e_n$, and that

$$(e_1e_2\cdots e_n)^2 \;=\; (-1)^{n(n-1)/2}\, q(e_1)q(e_2)\cdots q(e_n)\ .$$

(3.ex.24) Let (M,q) and (M',q') be quadratic spaces of constant ranks r and r', and (M'',q'') their orthogonal sum with Clifford algebra $C\ell(M'',q'') = C\ell(M,q)\,\hat{\otimes}\,C\ell(M',q')$. Suppose that $\mathrm{QZ}(M,q)$ is free with basis $(1,z)$ such that $z^2 = \beta z - \gamma$, and that $\mathrm{QZ}(M',q')$ is free with basis $(1,z')$ such that $z'^2 = \beta' z' - \gamma'$. Prove that $\mathrm{QZ}(M'',q'')$ is free with basis $(1\otimes 1,\ z'')$ such that $z'' = z\otimes\beta' + \beta\otimes z' - 2z\otimes z'$ and

$$z''^2 \;=\; \beta\beta' z'' \;-(-1)^{rr'}(\beta^2\gamma' + \gamma\beta'^2 - 4\gamma\gamma') \;-\; \frac{1-(-1)^{rr'}}{4}\,\beta^2\beta'^2.$$

(3.ex.25) Let (M,q) be a quadratic space of nonzero dimension over a field K of characteristic 2. From (3.4.14) we deduce that the class of $\mathrm{QZ}(M,q)$ in $Q(K)$ is given by an element of $K\lceil 4\rfloor$ modulo the image of $K\lceil 2\rfloor$ by the group morphism $\kappa \longmapsto \kappa - \kappa^2$; here $K\lceil 2\rfloor$ and $K\lceil 4\rfloor$ both coincide with the additive group K, and the image of $\kappa \longmapsto \kappa - \kappa^2$ is an additive subgroup denoted by $\wp(K)$. The element of $K/\wp(K)$ representing the Arf subalgebra $\mathrm{QZ}(M,q)$ is called the *Arf invariant* of (M,q); if z is an element of $\mathrm{QZ}(M,q)$ such that $z\notin K$ and $z - z^2 \in K$, the Arf invariant $\mathrm{Arf}(M,q)$ is the class of $z - z^2$ modulo $\wp(K)$.

(a) Prove that $\mathrm{Arf}((M,q)\perp(M',q')) = \mathrm{Arf}(M,q) + \mathrm{Arf}(M',q')$.

(b) Let $2m$ be the dimension of M. Prove that M contains a basis $(e_1,\ f_1,\ e_2,\ f_2,\ \dots,\ e_m,\ f_m)$ such that $\mathrm{b}_q(e_i,f_i) = 1$ for $i = 1,2,\dots,m$, but $\mathrm{b}_q(e_i,f_j) = 0$ if $i \neq j$, and $\mathrm{b}_q(e_i,e_j) = \mathrm{b}_q(f_i,f_j) = 0$ for all (i,j). Calculate $z - z^2$ when $z = \sum_i e_if_i$ and prove that

$$\mathrm{Arf}(M,q) \;=\; \sum_{i=1}^{m} q(e_i)q(f_i) \quad \text{modulo } \wp(K).$$

Discriminant modules and other topics

(3.ex.26) Prove the exactness of the sequence

$$1 \to \mu_2(K) \to K^\times \to K^\times \to \mathrm{Disc}(K) \to \mathrm{Pic}(K) \to \mathrm{Pic}(K) \ ;$$

here $\mu_2(K)$ is the subgroup of square roots of 1 in the group $K^\times$ of invertible elements, the third arrow is the morphism $\lambda \longmapsto \lambda^2$, the fourth arrow maps $\mu \in K^\times$ to the class of K provided with the multiplication $(x, y) \longmapsto \mu xy$, the fifth arrow is the forgetting morphism that maps the discriminant class of D to the class of D in the Picard group of K (defined at the end of **1.12**), and the sixth arrow maps the class of D to the class of $D \otimes D$.

When we consider different basic rings, we get functors μ_2, U (defined by $\mathrm{U}(K) = K^\times$), Q, Disc and Pic from the category $\mathcal{C}om(\mathbb{Z})$ of commutative rings to the category of abelian groups, and the arrows in the above exact sequence can be associated with morphisms between these functors.

(3.ex.27) Work out a parallel theory for *graded discriminant modules*. You must find a group $\mathrm{Disc}^g(K)$ in which every element has order 1, 2 or 4, a group morphism $\mathrm{Q}^g(K) \to \mathrm{Disc}^g(K)$, and a group morphism $\mathrm{Disc}^g(K) \to \mathrm{Pic}^g(K)$ with target isomorphic to $\mathrm{Pic}(K) \times \mathrm{Ip}(K)$. There is also an exact sequence

$$1 \longrightarrow \mathrm{Disc}(K) \longrightarrow \mathrm{Disc}^g(K) \longrightarrow \mathrm{Ip}(K) \longrightarrow 1$$

which splits when -1 has a square root in K.

(3.ex.28)* By localization we get group morphisms $\mathrm{Disc}(K) \to \mathrm{Disc}(K_\mathfrak{m})$ and $\mathrm{Q}(K) \to \mathrm{Q}(K_\mathfrak{m})$ for every maximal ideal $\mathfrak{m}$ of K, whence two group morphisms

$$\mathrm{Disc}(K) \longrightarrow \prod_\mathfrak{m} \mathrm{Disc}(K_\mathfrak{m}) \quad \text{and} \quad \mathrm{Q}(K) \longrightarrow \prod_\mathfrak{m} \mathrm{Q}(K_\mathfrak{m}) \ ;$$

the targets are direct products over the set of all maximal ideals. When K is an integral domain, it is proved in [Bass, 1974] (see Proposition (2.6.2)) that the former morphism is injective; and in [Knus, Paques 1985] (see Theorem (2.10)) it is proved that the latter is injective too. Here is a counterexample with a ring K containing divisors of zero; since 2 is invertible in this ring, the canonical group morphisms $\mathrm{Q}(K) \to \mathrm{Disc}(K)$ and $\mathrm{Q}(K_\mathfrak{m}) \to \mathrm{Disc}(K_\mathfrak{m})$ are bijective, and thus this counterexample proves that neither of these group morphisms is injective, although only the former is considered.

Let F be a field in which 2 is invertible, and K the subring of $F[t]^2$ containing all pairs of polynomials (f, g) such that $f(1) = g(1)$ and $f(-1) = g(-1)$. Let D be the subset of $F[t]^2$ containing all pairs (u, v) such that $u(1) = v(1)$ and $u(-1) = -v(-1)$. There is an evident multiplication $K \times D \to D$ that makes D become a K-module, and there is an evident multiplication $D \times D \to K$. Prove that D is a discriminant module, and that D is *not* a free module over K. Nevertheless

for every prime ideal $\mathfrak{q}$ of K, the discriminant module $D_{\mathfrak{q}}$ is isomorphic to $K_{\mathfrak{q}}$ (provided with its natural multiplication).
Hint. With every prime ideal $\mathfrak{p}$ of $F[t]$ not containing t^2-1 are associated two prime ideals $\mathfrak{p}'$ and $\mathfrak{p}''$ of K, one containing $(t^2-1,\ 0)$, the other one containing $(0,\ t^2-1)$, and both localized rings $K_{\mathfrak{p}'}$ and $K_{\mathfrak{p}''}$ are isomorphic to $F[t]_{\mathfrak{p}}$. But there are still two other prime ideals in K, both containing $(t^2-1,\ 0)$ and $(0,\ t^2-1)$; the localized ring contains all pairs (f,g) of rational functions such that (for one prime ideal) $f(1)$ and $g(1)$ exist and are equal, or (for the other one) $f(-1)$ and $g(-1)$ exist and are equal.

(3.ex.29) Let $G_{m,n}$ be the quadratic space over $\mathbb{R}$ defined in **2.8**; we suppose $(m,n) \neq (0,0)$. Prove that $C\ell(G_{m,n})^{ng}$ (Clifford algebra without grading) and $C\ell_0(G_{m,n})$ are respectively isomorphic to matrix algebras $\mathcal{M}(k,B)$ and $\mathcal{M}(k_0,B_0)$ over rings B and B_0 both belonging to the set $\{\mathbb{R}^2,\ \mathbb{R},\ \mathbb{C},\ \mathbb{H},\ \mathbb{H}^2\}$, and that B and B_0 only depend on the signature $s = m-n$ modulo 8 according to the following table:

s mod 8	0	1	2	3	4	5	6	7
B	$\mathbb{R}$	$\mathbb{R}^2$	$\mathbb{R}$	$\mathbb{C}$	$\mathbb{H}$	$\mathbb{H}^2$	$\mathbb{H}$	$\mathbb{C}$
B_0	$\mathbb{R}^2$	$\mathbb{R}$	$\mathbb{C}$	$\mathbb{H}$	$\mathbb{H}^2$	$\mathbb{H}$	$\mathbb{C}$	$\mathbb{R}$

Hint. Use (3.7.2) and (3.8.8) to prove that B only depends on s (in accordance with (3.7.8)); calculate $QZ(G_{m,n})$ by means of (3.7.7); calculate B by means of (3.8.8) when $G_{m,n}$ is definite (positive or negative) of even dimension; from (3.ex.6) deduce that the B for (m,n) is the B_0 for $(m,n{+}1)$; the isomorphism $Z(A)\otimes A_0 \rightarrow A$, valid for every graded Azumaya algebra of odd type, allows you to complete the table.

(3.ex.30) Let A be a graded Azumaya algebra such that $Z(A_0)$ is a quadratic extension Z, let D be the discriminant module of Z, and J the free discriminant module generated by an element j such that $j^2=-1$. Prove that the algebras A_D and A_J are isomorphic.

(3.ex.31) We assume that 2 is invertible in K. With every discriminant module D is associated the quadratic space D_2 that is the module D with the quadratic form $d \longmapsto d^2$; here all lower indices 2 must be understood in this way. As above, J is the free discriminant module in which $j^2=-1$.

(a) Let D and D' be discriminant modules. Prove that the discriminant module of $C\ell_0(D_2 \perp D'_2)$ is isomorphic to $J\otimes D\otimes D'$. Prove that the quaternion algebras $C\ell(D_2\perp D'_2)^{ng}$ and $C\ell(D_2 \perp (J\otimes D\otimes D')_2)^{ng}$ are isomorphic.
(b) Deduce from (3.8.8), (3.8.4) and (3.7.2), that

$$C\ell(D_2\perp D'_2)\otimes C\ell(D_2\perp D''_2) \cong C\ell(D_2\perp (J\otimes D\otimes D'\otimes D'')_2)\otimes \mathrm{End}(K\oplus D''),$$

and conclude that the Brauer class of $C\ell(D_2\perp(D'\otimes D'')_2)^{ng}$ is the product of the classes of $C\ell(D_2\perp D'_2)^{ng}$ and $C\ell(D_2\perp D''_2)^{ng}$.

(c) Suppose that D is the discriminant module of the quadratic extension Z, and prove that the quaternion algebras $C\ell(D' \otimes Z)^{ng}$ and $C\ell(D_2 \perp D'_2)^{ng}$ are isomorphic. From all the previous results, deduce a new (and much easier) proof of the properties of the mapping $\mathcal{Q}$ stated in (3.8.16).

(d) Let A be a graded Azumaya algebra such that $Z(A_0)$ is a quadratic extension Z with discriminant module D, and let D' be any discriminant module. Prove that

$$A_{D'} \otimes C\ell(K_2 \perp (J \otimes D')_2) \;\cong\; A \otimes C\ell(D_2 \perp (J \otimes D \otimes D')_2) \ ,$$

and give another proof of (3.8.13).

(3.ex.32) Let A be a graded Azumaya algebra over K such that $Z(A_0)$ is a quadratic extension Z. We wish to compare the Z-algebras A_0 and $Z \otimes A$. This comparison is motivated by the fact that the multiplication mapping $Z(B) \otimes B_0 \to B$ is an isomorphism of graded $Z(B)$-algebras when B is a graded Azumaya algebra of odd type.

(a) Justify the existence of a graded algebra morphism

$$A_0 \otimes_Z (Z \otimes_K A^{to}) \longrightarrow \mathrm{End}_Z(A),$$
$$a \otimes z \otimes b^{to} \longmapsto (\ x \ \longmapsto (-1)^{\partial b \partial x}\ azxb\ ) \ ,$$

and prove that it is bijective. Consequently A_0 and $Z \otimes A$ have the same class in $\mathrm{Br}^g(Z)$.

(b) Suppose that A_1 contains an invertible element w and define a graded algebra morphism in this way (for all $z \in Z$, all $a \in A_0$ and all $b \in A_1$):

$$Z \otimes A \longrightarrow \mathcal{M}(1,1;A_0) \ , \qquad z \otimes (a+b) \longmapsto \begin{pmatrix} za & zbw^{-1} \\ zwb & zwaw^{-1} \end{pmatrix} \ ;$$

the notation $\mathcal{M}(1,1;A_0)$ means the matrix algebra $\mathcal{M}(2,A_0)$ with the grading that lets the elements of the diagonal be even, and the others odd. Prove the bijectiveness of this morphism.

Hint. Remember that $zw = w\varphi(z)$ and use (3.8.12).

Remark. The exercise (4.ex.18) about *Weyl algebras* is now feasible.

Chapter 4

Comultiplications. Exponentials. Deformations

Completely different notions are now expounded: first comultiplications and interior multiplications; then exponentials (defined without exponential series); finally deformations of Clifford algebras, which need both exponentials and interior products. Exterior algebras play an important role, because with a weak additional hypothesis (the existence of scalar products) we shall prove that Clifford algebras are isomorphic to them as K-modules (and even as comodules). The first two sections of Chapter **3** are sufficient prerequisites for almost all this chapter.

4.1 Coalgebras and comodules

Coalgebras

The category $\mathcal{A}lg(K)$ is a subcategory of $\mathcal{M}od(K)$; let us find out which particular properties an object A of $\mathcal{A}lg(K)$ does possess, with the requirement that these properties must be understandable inside the category $\mathcal{M}od(K)$. For this purpose, instead of defining the multiplication by a bilinear mapping $A \times A \to A$, we define it by the corresponding linear mapping $\pi_A : A \otimes A \to A$, and instead of mentioning the unit element 1_A, we mention the linear mapping $\varepsilon_A : K \to A$ such that $\varepsilon_A(1) = 1_A$. The associativity of the algebra A and the properties of its unit element are equivalent to the following properties, which only involve objects and morphisms of $\mathcal{M}od(K)$:

$$\begin{aligned}
\pi_A\ (\pi_A \otimes \mathrm{id}_A) &= \pi_A\ (\mathrm{id}_A \otimes \pi_A)\ , \\
\pi_A\ (\mathrm{id}_A \otimes \varepsilon_A) &= \text{canonical isomorphism } A \otimes K \longrightarrow A\ , \\
\pi_A\ (\varepsilon_A \otimes \mathrm{id}_A) &= \text{canonical isomorphism } K \otimes A \longrightarrow A.
\end{aligned}$$

By means of the automorphism $\top$ of $A\otimes A$ such that $\top(a\otimes b) = b\otimes a$, we can write the condition that means that the algebra A is commutative: $\pi_A \top = \pi_A$.

An object A of $\mathcal{M}od(K)$ is called a *coalgebra* if it satisfies the previous three conditions in the dual category of $\mathcal{M}od(K)$; this means the existence of a linear mapping $\pi'_A : A \to A \otimes A$ and a linear mapping $\varepsilon'_A : A \to K$ such that

$$(\pi'_A \otimes \mathrm{id}_A)\ \pi'_A = (\mathrm{id}_A \otimes \pi'_A)\ \pi'_A\ ,$$
$$(\mathrm{id}_A \otimes \varepsilon'_A)\ \pi'_A = \text{canonical isomorphism } A \longrightarrow A \otimes K\ ,$$
$$(\varepsilon'_A \otimes \mathrm{id}_A)\ \pi'_A = \text{canonical isomorphism } A \longrightarrow K \otimes A.$$

The mapping π'_A is called the *comultiplication* (or *coproduct*) of the coalgebra A, and ε'_A is called its *counit*. The coalgebra A is said to be *cocommutative* if moreover $\top \pi'_A = \pi'_A$.

The first motivation of these definitions is the following theorem.

(4.1.1) **Theorem.** *If A is a coalgebra and B an algebra, then* $\mathrm{Hom}_K(A, B)$ *is an algebra when it is provided with the following multiplication:*

$$(u, v) \longmapsto u * v = \pi_B\ (u \otimes v)\ \pi'_A \quad : \quad A \to A \otimes A \to B \otimes B \to B$$

for all u and v in $\mathrm{Hom}(A, B)$*; the unit element of this algebra* $\mathrm{Hom}(A, B)$ *is* $\varepsilon_B \varepsilon'_A$. *It is commutative whenever A is cocommutative and B commutative.*

Proof. A straightforward calculation shows that

$$(u * v) * w \ = \ \pi_B\ (\pi_B \otimes \mathrm{id}_B)\ (u \otimes v \otimes w)\ (\pi'_A \otimes \mathrm{id}_A)\ \pi'_A\ ,$$

and the analogous expression of $u*(v*w)$ shows that the associativity of $\mathrm{Hom}(A, B)$ is a consequence of the associativity of B and the coassociativity of A. Another calculation shows that

$$u * (\varepsilon_B \varepsilon'_A) \ = \ \pi_B\ (\mathrm{id}_B \otimes \varepsilon_B)\ (u \otimes \mathrm{id}_K)(\mathrm{id}_A \otimes \varepsilon'_A)\pi'_A \ = \ u\ ,$$

and in the same way $(\varepsilon_B \varepsilon'_A) * v = v$. The proof of the statement about commutativity is still easier. □

The basic ring K is both an algebra and a coalgebra; π_K is the canonical isomorphism $K \otimes K \to K$, and π'_K is the reciprocal isomorphism $K \to K \otimes K$; both ε_K and ε'_K are equal to id_K. The above theorem will often be used to state that the dual module $A^* = \mathrm{Hom}(A, K)$ is an algebra whenever A is a coalgebra.

Additional information. For interested readers we present the category of coalgebras (but hurried readers may go directly to comodules). When A and B are algebras, a linear mapping $f : A \to B$ is an algebra morphism if (and only if) the two following equalities are true:

$$f \pi_A \ = \ \pi_B\ (f \otimes f) \quad \text{and} \quad f\ \varepsilon_A \ = \ \varepsilon_B\ ;$$

therefore when A and B are coalgebras, we say that $f : A \to B$ is a *coalgebra morphism* if (by definition)

$$\pi'_B f = (f \otimes f)\,\pi'_A \quad \text{and} \quad \varepsilon'_B f = \varepsilon'_A .$$

If $f_1 : A' \to A$ is a coalgebra morphism and $f_2 : B \to B'$ an algebra morphism, then the mapping $\mathrm{Hom}(f_1, f_2)$ (defined in **1.5**) is an algebra morphism from $\mathrm{Hom}(A, B)$ into $\mathrm{Hom}(A', B')$.

When A and B are algebras, the algebra structure put on $C = A \otimes B$ (see **1.3**) corresponds to

$$\pi_C = (\pi_A \otimes \pi_B)\,\top_{2,3} \quad \text{and} \quad \varepsilon_C = (\varepsilon_A \otimes \varepsilon_B)\,\pi'_K \;;$$

here $\top_{2,3}$ means the reversion of the second and third factors, whatever the modules which they belong to may be: $\top_{2,3}(a \otimes b \otimes a' \otimes b') = a \otimes a' \otimes b \otimes b'$. Therefore when A and B are coalgebras, we make $C = A \otimes B$ become a coalgebra by setting

$$\pi'_C = \top_{2,3}\,(\pi'_A \otimes \pi'_B) \quad \text{and} \quad \varepsilon'_C = \pi_K\,(\varepsilon'_A \otimes \varepsilon'_B).$$

When A is both an algebra and a coalgebra, it is called a *bialgebra* when the four mappings π_A, ε_A, π'_A, ε'_A are related together by the four equalities

$$\begin{aligned}
&\text{(a)} \quad \pi'_A \pi_A = (\pi_A \otimes \pi_A)\,\top_{2,3}\,(\pi'_A \otimes \pi'_A)\,, \\
&\text{(b)} \quad \pi'_A \varepsilon_A = (\varepsilon_A \otimes \varepsilon_A)\,\pi'_K\,, \\
&\text{(c)} \quad \varepsilon'_A \pi_A = \pi_K\,(\varepsilon'_A \otimes \varepsilon'_A)\,, \\
&\text{(d)} \quad \varepsilon'_A \varepsilon_A = \mathrm{id}_K \;;
\end{aligned}$$

these four conditions can be interpreted in two different ways; first we can observe that the conditions (a) and (b) mean that $\pi'_A : A \to A \otimes A$ is an algebra morphism, and that (c) and (d) mean that $\varepsilon'_A : A \to K$ is also an algebra morphism; but in a dual way we can also observe that (a) and (c) mean that $\pi_A : A \otimes A \to A$ is a coalgebra morphism, and that (b) and (d) mean that $\varepsilon_A : K \to A$ is also a coalgebra morphism. The ring K is a trivial example of a bialgebra. Later the symmetric algebra $\mathrm{S}(M)$ of a module M and its exterior algebra $\bigwedge(M)$ will receive comultiplications and counits that are algebra morphisms; consequently they will become bialgebras.

Comodules

The notion of *comodule* is derived from the notion of module by duality in an analogous way. First let A be a K-algebra, that is a K-module provided with two mappings π_A and ε_A as above; instead of describing the properties of a left A-module M by means of a bilinear mapping $A \times M \to M$, we will use the associated linear mapping; thus when M is a K-module, we can say that a linear mapping

$\pi_M : A \otimes M \to M$ makes it become a left A-module if these two conditions are satisfied:

$$\pi_M\ (\pi_A \otimes \mathrm{id}_M) = \pi_M\ (\mathrm{id}_A \otimes \pi_M)\ ,$$
$$\pi_M\ (\varepsilon_A \otimes \mathrm{id}_M) = \text{canonical isomorphism } K \otimes M \longrightarrow M.$$

When M is a right A-module, then π_M is a linear mapping from $M \otimes A$ into M satisfying the evident analogous conditions. Later we shall need right comodules, whence the following definition: when A is a K-coalgebra and M a K-module, we say that a linear mapping $\pi'_M : M \to M \otimes A$ makes M become a *right comodule over* A if the two following conditions are satisfied:

$$(\mathrm{id}_M \otimes \pi'_A)\ \pi'_M = (\pi'_M \otimes \mathrm{id}_A)\ \pi'_M\ ,$$
$$(\mathrm{id}_M \otimes \varepsilon'_A)\ \pi'_M = \text{canonical isomorphism } M \longrightarrow M \otimes K.$$

Comodules are interesting because they naturally become modules over suitable algebras. Indeed it is sensible to wonder whether M would be a module over the algebra $\mathrm{Hom}(A, B)$ defined in (4.1.1) when it is a comodule over the coalgebra A and a module over the algebra B. This statement is actually true provided that we use together a structure of comodule on one side and a structure of module on the other side, and require some compatibility between both structures; for instance we can suppose that M is a right A-comodule and a left B-module, and then we must require that the comultiplication π'_M and the multiplication π_M are compatible in the following sense:

$$\pi'_M\ \pi_M \ = \ (\pi_M \otimes \mathrm{id}_A)\ (\mathrm{id}_B \otimes \pi'_M) \qquad \begin{array}{ccc} B \otimes M & \longrightarrow & B \otimes M \otimes A \\ \downarrow & & \downarrow \\ M & \longrightarrow & M \otimes A \end{array} \ ;$$

this requirement may be interpreted in this way: the comultiplication π'_M must be B-linear. Before stating the announced theorem, let us recall other statements in which similar features appear. For instance a change of side also appears in the following statement: $\mathrm{Hom}(M, N)$ is a left B-module when M is a right B-module (and N merely a K-module); this change of side is easily explained by the contravariance of the functor $\mathrm{Hom}(\ldots, N)$. Besides, when we wish to make M become a left module over an algebra $B \otimes C$ (assuming that it is already a left module over B and C), we must also require that the structures of module over B and C are compatible: the operation in M of any element of B must commute with the operation of any element of C (see (1.3.3)); this means that the multiplication $C \otimes M \to M$ must be B-linear. These explanations should make the following theorem look quite natural.

(4.1.2) **Theorem.** *If M is a right comodule over the coalgebra A and a left module over the algebra B, and if the comultiplication π'_M is B-linear, then M is a* *left module over the algebra* Hom(A, B); *the operation in* M *of an element* u *of* Hom(A, B) *is this endomorphism of* M :

$$\pi_M\ (u \otimes \mathrm{id}_M) \top \pi'_M \quad : \quad M \to M \otimes A \to A \otimes M \to B \otimes M \to M.$$

Here $\top$ is the canonical isomorphism $M \otimes A \to A \otimes M$. Yet $\pi_M(u \otimes \mathrm{id}_M)\top\pi'_M$ is the same thing as $\pi_M\top(\mathrm{id}_M \otimes u)\pi'_M$ if $\top$ now means the canonical isomorphism $M \otimes B \to B \otimes M$.

Proof. We must prove that the mapping $u \longmapsto \pi_M(u \otimes \mathrm{id}_M)\top\pi'_M$ is an algebra morphism from Hom(A, B) into End(M). First the unit element $\varepsilon_B\varepsilon'_A$ of Hom(A, B) is mapped to the endomorphism

$$M \longrightarrow M \otimes A \longrightarrow M \otimes K \longrightarrow K \otimes M \longrightarrow B \otimes M \longrightarrow M$$

which is id_M because $(\mathrm{id}_M \otimes \varepsilon'_A)\pi'_M$ is the canonical isomorphism $M \to M \otimes K$, and $\pi_M(\varepsilon_B \otimes \mathrm{id}_M)$ is the canonical isomorphism $K \otimes M \to M$. Now let u and v be two elements of Hom(A, B); the following diagram contains the proof of the equality $u * (v * x) = (u * v) * x$ (for all $x \in M$); the endomorphism $x \longmapsto u * (v * x)$ appears if you go from M to M through the first column, whereas the endomorphism $x \longmapsto (u * v) * x$ appears if you go from M to M through the third column. The places where u and v are involved, are all indicated; the double arrows $\longleftrightarrow$ indicate canonical isomorphisms, that represent either a "commutativity" property of a tensor product, or an "associativity" property, according to the parentheses that are displayed; for instance the mapping $M \otimes (A \otimes A) \to (A \otimes A) \otimes M$ that appears in the third column is the canonical isomorphism $x \otimes a \otimes a' \longmapsto a \otimes a' \otimes x$; in all other arrows a multiplication or a comultiplication is involved:

$$\begin{array}{ccccccc}
M & \longrightarrow & M \otimes A & \longrightarrow & (M \otimes A) \otimes A & \longleftrightarrow & M \otimes (A \otimes A) \\
 & & \updownarrow & & \updownarrow & & \updownarrow \\
 & & A \otimes M & \longrightarrow & A \otimes (M \otimes A) & & \\
 & & \downarrow{\scriptstyle v\otimes\mathrm{id}} & & \downarrow{\scriptstyle v\otimes\mathrm{id}\otimes\mathrm{id}} & & \\
 & & B \otimes M & \longrightarrow & B \otimes (M \otimes A) & & \\
 & & \downarrow & & \updownarrow & & \\
 & & M & & & & (A \otimes A) \otimes M \\
 & & \downarrow & & & & \\
 & & M \otimes A & \longleftarrow & (B \otimes M) \otimes A & & \downarrow{\scriptstyle u\otimes v\otimes\mathrm{id}} \\
 & & \updownarrow & & \updownarrow & & \\
 & & A \otimes M & \longleftarrow & A \otimes (B \otimes M) & & \\
 & & \downarrow{\scriptstyle u\otimes\mathrm{id}} & & \downarrow{\scriptstyle u\otimes\mathrm{id}\otimes\mathrm{id}} & & \\
M & \longleftarrow & B \otimes M & \longleftarrow & B \otimes (B \otimes M) & \longleftrightarrow & (B \otimes B) \otimes M
\end{array}$$

The coassociativity hypothesis $(\mathrm{id}_M \otimes \pi'_A)\pi'_M = (\pi'_M \otimes \mathrm{id}_A)\pi'_M$ is involved in the first line, the associativity hypothesis $\pi_M(\pi_B \otimes \mathrm{id}_M) = \pi_M(\mathrm{id}_B \otimes \pi_M)$ is involved in the last line, and the compatibility hypothesis relating π'_M and π_M is involved in the middle of the first two columns; all other places of this diagram only require trivial verifications. □

Of course the compatibility hypothesis is always fulfilled when $B = K$; thus every right A-comodule is a left A^*-module, and A itself is a A^*-module.

An example: the coalgebra $\mathrm{S}(M)$

The symmetric algebra $\mathrm{S}(M)$ of a K-module M is a cocommutative coalgebra, and the definition of its comultiplication is now explained because later we shall meet a similar but slightly more difficult comultiplication that makes every Clifford algebra become a comodule. The comultiplication $\pi' : \mathrm{S}(M) \to \mathrm{S}(M) \otimes \mathrm{S}(M)$ is the unique algebra morphism extending the linear mapping $M \to \mathrm{S}(M) \otimes \mathrm{S}(M)$ defined by $a \longmapsto a \otimes 1 + 1 \otimes a$; and the counit $\varepsilon' : \mathrm{S}(M) \to K$ is the unique algebra morphism extending the linear mapping $M \to K$ defined by $a \longmapsto 0$. Let us prove that $\mathrm{S}(M)$ is now a coalgebra.

Indeed, if we identify $\mathrm{S}(M) \otimes \mathrm{S}(M)$ with the algebra $\mathrm{S}(M \oplus M)$ (see (1.5.1)), then $a \otimes 1 + 1 \otimes a$ is identified with $(a, a) \in M \oplus M$, and thus π' becomes the mapping $\mathrm{S}(\delta)$ associated by the functor S with the linear mapping $\delta : M \to M \oplus M$ defined by $\delta(a) = (a, a)$. And if we identify K with the symmetric algebra $\mathrm{S}(0)$ of a zero module, then ε' becomes the mapping $\mathrm{S}(\zeta)$ associated by the functor S with the zero mapping $\zeta : M \to 0$. In the equality $(\delta \oplus \mathrm{id}_M)\,\delta = (\mathrm{id}_M \oplus \delta)\,\delta$ both members are equal to the mapping $a \longmapsto (a, a, a)$ from M into $M \oplus M \oplus M$, and it is not more difficult to verify that

$$(\mathrm{id}_M \oplus \zeta)\ \delta = \text{canonical isomorphism } M \to M \oplus 0\ ;$$
$$(\zeta \oplus \mathrm{id}_M)\ \delta = \text{canonical isomorphism } M \to 0 \oplus M\ ;$$

if we transform the previous three equalities by means of the functor S, we get the equalities that mean that $\mathrm{S}(M)$ is a coalgebra. Since δ is invariant by the automorphism $(a, b) \longmapsto (b, a)$ of $M \oplus M$, we can add that $\mathrm{S}(M)$ is a cocommutative coalgebra.

Therefore the dual module $\mathrm{S}^*(M) = \mathrm{Hom}(\mathrm{S}(M), K)$ is a commutative algebra. Let $\mathrm{S}^{*n}(M)$ be the set of all linear forms on $\mathrm{S}(M)$ that vanish on all $\mathrm{S}^j(M)$ such that $j \neq n$. Thus $\mathrm{S}^{*n}(M)$ is naturally isomorphic to $\mathrm{S}^n(M)^* = \mathrm{Hom}(\mathrm{S}^n(M), K)$, and $\mathrm{S}^*(M)$ is isomorphic to the direct product of its submodules $\mathrm{S}^{*n}(M)$. Nevertheless the direct sum of the submodules $\mathrm{S}^{*n}(M)$ is a subalgebra of $\mathrm{S}^*(M)$, because $f \vee g$ belongs to $\mathrm{S}^{*(i+j)}(M)$ for all $f \in \mathrm{S}^{*i}(M)$ and $g \in \mathrm{S}^{*j}(M)$.

The comultiplication of $\mathrm{S}(M)$ is involved in the Leibniz formula which gives the successive derivatives of a product; it is worth explaining this, because analogous Leibniz formulas will appear in the context of exterior and Clifford algebras. Let us assume that M is a vector space of finite dimension over $\mathbb{R}$, and let $\mathcal{C}^\infty(M)$ be the algebra of indefinitely differentiable real functions on M. With each vector

$a \in M$ is associated a derivation ∂_a; the value of a derivative $\partial_a f$ at any point $x \in M$ is

$$\partial_a f(x) \;=\; \lim_{t\to 0}\,(f(x+ta) - f(x))\,t^{-1}\;;$$

it is known that the derivations ∂_a are pairwise commuting, and because of the universal property of $\mathrm{S}(M)$ the mapping $a \longmapsto \partial_a$ extends to an algebra morphism from $\mathrm{S}(M)$ into $\mathrm{End}(\mathcal{C}^\infty(M))$; it maps every $w \in \mathrm{S}(M)$ to an operator ∂_w on $\mathcal{C}^\infty(M)$ which is called a partial differential operator with constant coefficients. The *Leibniz formula* tells how such an operator ∂_w operates on a product fg of two functions:

$$\partial_w(fg) \;=\; \pi_{\mathcal{C}}\,(\partial_{\pi'(w)}(f\otimes g)). \tag{4.1.3}$$

There are two interpretations of this formula: we can consider that each element of $\mathrm{S}(M)\otimes\mathrm{S}(M)$ operates in $\mathcal{C}^\infty(M)\otimes\mathcal{C}^\infty(M)$, so that $\pi'(w)$ operates on $f\otimes g$, and then $\pi_{\mathcal{C}}$ is the multiplication mapping associated with the algebra $\mathcal{C}^\infty(M)$; but we can also identify $\mathrm{S}(M)\otimes\mathrm{S}(M)$ with $\mathrm{S}(M\oplus M)$, and $f\otimes g$ with the function on $M\oplus M$ defined by $(x,y) \longmapsto f(x)g(y)$ (as it is usually done in functional analysis), and then $\pi_{\mathcal{C}}$ is the morphism from $\mathcal{C}^\infty(M\oplus M)$ into $\mathcal{C}^\infty(M)$ which maps each function $(x,y)\mapsto h(x,y)$ to the function $x \mapsto h(x,x)$; both interpretations are legitimate. To show that the above formula is the same thing as the ordinary Leibniz formula, it suffices to replace w with a symmetric power a^n of a vector $a \in M$; then

$$\pi'(a^n) = (\pi'(a))^n = (a\otimes 1 + 1\otimes a)^n \;=\; \sum_{k=0}^{n} \frac{n!}{k!\;(n-k)!}\; a^k\otimes a^{n-k}\;,$$

and thus we get the well-known formula

$$\partial_a^n(fg) \;=\; \sum_{k=0}^{n} \frac{n!}{k!\;(n-k)!}\;(\partial_a^k f)\;(\partial_a^{n-k} g).$$

4.2 Algebras and coalgebras graded by parities

Let G be an additive monoid with zero element (see **2.7**); a K-module M is said to be *graded over* G if it is the direct sum of submodules M_j indexed by the elements j of G. Such a decomposition into a direct sum is called a *grading* (or *gradation*) *over* G. An element $x \in M$ is said to be *homogeneous* if it belongs to some M_j, and j (well defined whenever $x \neq 0$) is called the *degree* of x and denoted by ∂x. Whenever ∂x is written, it is silently assumed that x is homogeneous. When $\gamma : G \to G'$ is a morphism between monoids with zero elements, every module M graded over G is also graded over G' : for all $j' \in G'$, $M_{j'}$ is the direct sum of all M_j such that $\gamma(j) = j'$. Here we shall especially use gradings over $\mathbb{Z}/2\mathbb{Z}$, which are called *parity gradings*. Often they come from gradings over $\mathbb{N}$ or $\mathbb{Z}$ by means of the evident monoid morphisms $\mathbb{N} \to \mathbb{Z} \to \mathbb{Z}/2\mathbb{Z}$. Nevertheless later in **4.5** we shall also use quite different gradings: when an algebra A is the direct sum of a subalgebra A^0 and an ideal A^+, such a decomposition means a grading over a monoid containing two elements, namely "zero" and "positive"; such a grading results from any grading over $\mathbb{N}$ by means of an evident morphism monoid.

When M and N are graded over G, $M \otimes N$ is also graded over it: $(M \otimes N)_j$ is the direct sum of all $M_{j'} \otimes N_{j''}$ such that $j' + j'' = j$. An element f of $\mathrm{Hom}(M, N)$ is said to be *homogeneous of degree* j if $f(x)$ is homogeneous of degree $\partial x + j$ whenever x is homogeneous in M; we write $j = \partial f$. A *graded morphism* is a homogeneous morphism of degree 0. In the category of G-graded K-modules we only accept morphisms which are finite sums of homogeneous morphisms, so that the module of morphisms between two G-graded modules M and N inherits a G-grading; when G is a finite monoid, this module coincides with $\mathrm{Hom}(M, N)$ if its grading is forgotten.

When G-gradings are involved, every module P that has not been given a particular grading, automatically receives the *trivial grading* such that $M_0 = M$ and $M_j = 0$ for all $j \neq 0$. The ring K is always trivially graded.

Let A be an algebra (resp. a coalgebra), the structure of which is defined by the linear mappings π and ε (resp. π' and ε'); A is said to be an *algebra graded over* G (resp. a *coalgebra graded over* G) when the module A is graded over G and when π and ε (resp. π' and ε') are graded morphisms. When A is an algebra, this means that the degree of 1_A is null, and that the degree of a product is the sum of the degrees of the factors. Many definitions impose this rule on other kinds of products too; for instance $\partial f(x) = \partial f + \partial x$ if f is a homogeneous morphism as above, and consequently $\partial(f' \circ f) = \partial f' + \partial f$. When A is a graded coalgebra, the coproduct $\pi'(a)$ of every homogeneous $a \in A$ can be written as a sum $\sum_i b_i \otimes c_i$ such that $\partial b_i + \partial c_i = \partial a$ for each term of this sum, and moreover $\varepsilon'(a) = 0$ if $\partial a \neq 0$.

Of course *graded modules or comodules* are defined in the same way by requiring that the corresponding products or coproducts are determined by graded morphisms.

For algebras and coalgebras graded over $\mathbb{Z}/2\mathbb{Z}$, besides many usual constructions, there are also twisted analogous ones; the twisted tensor products (defined in **3.2**) are typical examples. When we deal with exterior and Clifford algebras, twisting factors ± 1 appear very often, and in order to avoid any trouble with them, it is more convenient to introduce them everywhere according to the following rule.

(4.2.1) **Twisting rule.** Whenever in a product (of any kind) of homogeneous factors the order of two letters is reversed, this reversion must be compensated by a twisting factor that changes the sign if and only if both letters represent odd factors.

In general the letters represent factors with arbitrary parities; if there are n letters, there are 2^n possible distributions of parities; let us assume that for instance the letters $\ell_1, \ell_2, \dots, \ell_k$ are odd factors, and that $\ell_{k+1}, \dots, \ell_n$ are even; if the factors $\ell_1, \dots, \ell_k$ first appear in this order, and after some calculations in a different order, the product of all the twisting factors is the signature of the permutation inflicted on these k letters; since the signature of a product of permutations is the product of their signatures, we can forget which reversions of factors have been committed, and in which order, because at any moment the relative places of these k letters enable us to determine the exact value of the product of all twisting factors. Thus we can behave with twisting factors in an unconcerned way, and replace them with $\pm$ all along the calculations, since we are sure easily to find the value of their product at the end.

The uncompromising observance of the twisting rule is the wonderful remedy that delivers us from wasting an awful lot of energy in the calculation of a tremendous number of factors all belonging to the set $\{+1, -1\}$. We shall observe this rule even when it causes discrepancies with common use. Anyhow, it is hard to find a set of rules that accounts for all the fancies of common use, because it has conceded twisting factors without planning, always under constraint; the usefulness of systematic rules has been acknowledged only recently. The discrepancies with common use will be mentioned here at each occurrence.

When we systematically write a conventional sign $\pm$ which we calculate only at the end according to the permutation inflicted on the letters, we must carefully write every sign that is not automatically implied by the twisting rule. For instance the equality $\tau(xy) = \tau(y)\tau(x)$ involving the reversion τ (see (3.1.4)) may now be written $\tau(xy) = \pm(-1)^{\partial x \partial y}\tau(y)\tau(x)$, because the conventional sign $\pm$ automatically involves a twisting sign $(-1)^{\partial x \partial y}$ that here must be compensated.

When f and g belong respectively to $\mathrm{Hom}(M, M')$ and $\mathrm{Hom}(N, N')$, the ambivalent notation $f \otimes g$ means either an element of $\mathrm{Hom}(M, M') \otimes \mathrm{Hom}(N, N')$ or an element of $\mathrm{Hom}(M \otimes N,\ M' \otimes N')$; when all the involved modules are graded by parities, the latter $f \otimes g$ is replaced with the element $f \hat{\otimes} g$ of $\mathrm{Hom}(M \otimes N,\ M' \otimes N')$ defined in this way:

$$(4.2.2) \qquad (f \hat{\otimes} g)(x \otimes y) \;=\; (-1)^{\partial g \partial x}\, f(x) \otimes g(y) \;;$$

when g is even, $f \hat{\otimes} g$ coincides with $f \otimes g$. This definition implies, for all $f' \in \mathrm{Hom}(M', M'')$ and all $g' \in \mathrm{Hom}(N', N'')$,

$$(4.2.3) \qquad (f' \hat{\otimes} g') \circ (f \hat{\otimes} g) \;=\; (-1)^{\partial g' \partial f}\, f'f \hat{\otimes} g'g \;.$$

A lot of formulas of the same kind might be added, for instance the definition of the twisted tensor product of three morphisms:

$$(f \hat{\otimes} g \hat{\otimes} h)(x \otimes y \otimes z) \;=\; (-1)^{\partial g \partial x + \partial h \partial x + \partial h \partial y}\, f(x) \otimes g(y) \otimes h(z) \;.$$

When A is a coalgebra and B an algebra, both graded over $\mathbb{Z}/2\mathbb{Z}$, besides the algebra $\mathrm{Hom}(A, B)$ defined in (4.1.1), there is the *twisted algebra of morphisms*

$\mathrm{Hom}^\wedge(A,B)$ in which the product of two elements is defined in the following way:

$$u * v \;=\; \pi_B\,(u \,\hat{\otimes}\, v)\;\pi'_A\,. \tag{4.2.4}$$

Theorem (4.1.1) remains valid for this new multiplication (provided that commutativity is replaced with twisted commutativity).

When A is a graded coalgebra, $\mathrm{Hom}^\wedge(A,K)$ is a graded algebra which often is still denoted by A^*. Nevertheless if the parity grading of A comes from a grading $A = \bigoplus A^n$ over $\mathbb{Z}$, in general A^* is not graded over $\mathbb{Z}$, because the $\mathbb{Z}$-grading is only available on the direct sum of all submodules like A^{*n} (the natural image of $(A^n)^*$ in A^*); this direct sum is a subalgebra. Moreover the elements of A^{*n} must be given the degree $-n$.

When M is a graded right comodule over A and a graded left module over B, there is a graded version of Theorem (4.1.2) stating that M is a graded left module over $\mathrm{Hom}^\wedge(A,B)$, provided that the operation of $u \in \mathrm{Hom}^\wedge(A,B)$ on $x \in M$ is defined by means of the twisted reversion $\top^\wedge$:

$$u * x \;=\; \pi_M\,(u \otimes \mathrm{id}_M)\;\top^\wedge\;\pi'_M\,(x) \quad \text{with} \quad \top^\wedge(y \otimes a) \;=\; (-1)^{\partial y \partial a}\, a \otimes y\,. \tag{4.2.5}$$

When C is the twisted tensor product of the graded algebras A and B, then

$$\pi_C \;=\; (\pi_A \otimes \pi_B)\;\top^\wedge_{2,3} \quad \text{with} \quad \top^\wedge_{2,3}(a \otimes b \otimes a' \otimes b') \;=\; (-1)^{\partial b \partial a'}\, a \otimes a' \otimes b \otimes b'.$$

The twisted tensor product C' of two graded coalgebras A' and B' is defined in an analogous way:

$$\pi'_{C'} \;=\; \top^\wedge_{2,3}\;(\pi'_{A'} \otimes \pi'_{B'}) \quad \text{if} \quad C' = A' \,\hat{\otimes}\, B'. \tag{4.2.6}$$

All these definitions are involved in the following proposition.

(4.2.7) **Proposition.** *Let A and A' be graded coalgebras, and B and B' graded algebras; there is a graded algebra morphism (called canonical morphism) from*

$$\mathrm{Hom}^\wedge(A,B) \,\hat{\otimes}\, \mathrm{Hom}^\wedge(A',B') \quad \textit{into} \quad \mathrm{Hom}^\wedge(A \,\hat{\otimes}\, A',\; B \,\hat{\otimes}\, B')$$

that maps every $f \otimes f'$ to $f \,\hat{\otimes}\, f'$.

Proof. Let us consider homogeneous elements f and g in $\mathrm{Hom}(A,B)$, f' and g' in $\mathrm{Hom}(A',B')$, a in A and a' in A'. Let $\sum_i b_i \otimes c_i$ and $\sum_j b'_j \otimes c'_j$ be the coproducts of a and a'. We must prove that

$$(f * g) \,\hat{\otimes}\, (f' * g') \quad \text{and} \quad (-1)^{\partial f' \partial g}\,(f \,\hat{\otimes}\, f') * (g \,\hat{\otimes}\, g')$$

both map $a \otimes a'$ to the same element of $B \otimes B'$. Straightforward applications of the definitions show that they both map it to the element

$$\sum_{i,j} \pm\, f(b_i)g(c_i) \otimes f'(b'_j)g'(c'_j)\;;$$

as explained above, we need not worry about the sign $\pm$; here it is determined by the parity of

$$(\partial f' + \partial g')\partial a \ + \partial g \partial b_i \ + \partial g' \partial b'_j . \qquad \square$$

Remarks.

(a) The nongraded version of (4.2.7) has not been stated; as a matter of fact, it is included in the above graded version when A, A', B, B' are all trivially graded.

(b) When A, B, A', B' are finitely generated projective modules, all canonical morphisms like $B \otimes A^* \to \mathrm{Hom}(A,B)$ or $A^* \otimes A'^* \to (A \otimes A')^*$ are bijective; consequently the canonical morphism described in (4.2.7) is also an isomorphism.

(c) The canonical morphism in (4.2.7) explains why the notation $f \hat{\otimes} f'$ is often replaced with $f \otimes f'$, since the former is the canonical image of the latter. Anyhow, when the twisting rule (4.2.1) is strictly observed, no ambiguity may occur.

4.3 Exterior algebras

The study of exterior algebras has already begun in **3.1** and **3.2** where they are treated as Clifford algebras of null quadratic forms. The exterior algebra of a module M is provided with an $\mathbb{N}$-grading: $\bigwedge(M) = \bigoplus_n \bigwedge^n(M)$. Moreover $\bigwedge^0(M) = K$ and $\bigwedge^1(M) = M$. The even subalgebra $\bigwedge_0(M)$ is the direct sum of all $\bigwedge^{2k}(M)$, and $\bigwedge_1(M)$ the direct sum of all $\bigwedge^{2k+1}(M)$.

Here is the universal property of the algebra $\bigwedge(M)$: every linear mapping f from M into any algebra P such that $f(a)^2 = 0$ for all $a \in M$, extends in a unique way to an algebra morphism $\bigwedge(M) \to P$. Each subspace $\bigwedge^n(M)$ has its own universal property: every alternate n-linear mapping g from M^n into any K-module P determines a unique linear mapping $g'' : \bigwedge^k(M) \to P$ such that

$$g(a_1, a_2, \dots, a_n) \ = \ g''(a_1 \wedge a_2 \wedge \cdots \wedge a_n) \quad \text{for all } a_1, a_2, \dots, a_n \in M \ ;$$

the proof of this statement is analogous to that of (1.4.3).

These universal properties lead to functors $\bigwedge$ and $\bigwedge^n$ (or $\bigwedge_K$ and $\bigwedge_K^n$ when the basic ring must be specified). Indeed any K-linear mapping $f : M \to N$ extends to an algebra morphism $\bigwedge(f)$ from $\bigwedge(M)$ into $\bigwedge(N)$, and determines linear mappings $\bigwedge^n(f) : \bigwedge^n(M) \to \bigwedge^n(N)$ for all $n \in \mathbb{N}$.

If M and N are K-modules, the algebra $\bigwedge(M \oplus N)$ is canonically isomorphic to the twisted tensor product $\bigwedge(M) \hat{\otimes} \bigwedge(N)$; this is a particular case of (3.2.4).

If $K \to L$ is an extension of the basic ring, then $\bigwedge_L(L \otimes M)$ is canonically isomorphic to $L \otimes \bigwedge_K(M)$; this is a particular case of (3.1.9).

Like every Clifford algebra, $\bigwedge(M)$ admits a grade automorphism σ and a reversion τ, for which (3.1.5) and (3.2.8) give precise information.

When M is a free module, then $\bigwedge(M)$ too is a free module; this is stated in (3.2.5) when the rank is finite, in (3.2.7) when it is infinite, and moreover we know how to derive a basis of $\bigwedge(M)$ from every basis of M. The case of a finitely generated projective module M is treated in (3.2.6). When M is merely projective, there exists a module M' such that $M \oplus M'$ is free, consequently $\bigwedge(M) \hat{\otimes} \bigwedge(M')$ is a free module, and $\bigwedge(M)$ is projective because it is a direct summand of this free module.

In **4.5** we shall need the following lemma.

(4.3.1) **Lemma.** *If N is a submodule of M, the algebra $\bigwedge(M/N)$ is canonically isomorphic to the quotient of $\bigwedge(M)$ by the ideal $N \wedge \bigwedge(M)$ generated by N.*

Proof. With the quotient mapping $M \to M/N$ the functor $\bigwedge$ associates an algebra morphism vanishing on the ideal J generated by N in $\bigwedge(M)$, whence an algebra morphism $\bigwedge(M)/J \to \bigwedge(M/N)$. Conversely the mapping $M \to \bigwedge(M) \to \bigwedge(M)/J$ vanishes on N, and gives a linear mapping defined on M/N, which extends to an algebra morphism $\bigwedge(M/N) \to \bigwedge(M)/J$. Thus we have got two reciprocal morphisms. □

The exterior algebra $\bigwedge(M)$ becomes a coalgebra in the same way as $\mathrm{S}(M)$. The comultiplication $\pi' : \bigwedge(M) \to \bigwedge(M) \otimes \bigwedge(M)$ is the algebra morphism from $\bigwedge(M)$ into $\bigwedge(M) \hat{\otimes} \bigwedge(M)$ such that $\pi'(a) = a \otimes 1 + 1 \otimes a$ for all $a \in M$, and the counit $\varepsilon' : \bigwedge(M) \to K$ is the algebra morphism such that $\varepsilon'(a) = 0$ for all $a \in M$. If we identify $\bigwedge(M) \hat{\otimes} \bigwedge(M)$ with $\bigwedge(M \oplus M)$ and K with $\bigwedge(0)$, we recognize that π' and ε' are the algebra morphisms associated by the functor $\bigwedge$ with the linear mappings $\delta : a \longmapsto (a,a)$ and $\zeta : a \longmapsto 0$. This allows us to prove that π' and ε' actually give $\bigwedge(M)$ a structure of coalgebra.

Let us calculate $\pi'(x)$ when x is the exterior product of n elements a_1, $a_2, \dots, a_n$ of M :

$$\pi'(x) \;=\; \sum_{j=0}^{k} \sum_{s} \operatorname{sgn}(s)\ (a_{s(1)} \wedge \cdots \wedge a_{s(j)}) \ \otimes\ (a_{s(j+1)} \wedge \cdots \wedge a_{s(n)})\ ;$$

the second summation runs over the subset of all permutations s such that

$$s(1) < s(2) < \cdots < s(j) \quad \text{and} \quad s(j+1) < s(j+2) < \cdots < s(n)\ ,$$

and $\operatorname{sgn}(s)$ is the signature of s.

The dual space $\bigwedge^*(M) = \mathrm{Hom}^\wedge(\bigwedge(M), K)$ is an algebra for the multiplication defined by (4.2.4), and the specific symbol $\wedge$ is still used for this multiplication. Let us observe that $\bigwedge^*(M)$ is naturally isomorphic to the direct product of the modules $\bigwedge^n(M)^* = \mathrm{Hom}(\bigwedge^n(M), K)$, and even to their direct sum when M is finitely generated. The image of $\bigwedge^n(M)^*$ in $\bigwedge^*(M)$ is denoted by $\bigwedge^{*n}(M)$; its elements are the linear forms vanishing on all $\bigwedge^j(M)$ such that $j \neq n$, and as $\mathbb{Z}$-homogeneous elements, they have the degree $-n$. In an analogous way, $\bigwedge^{*\leq n}(M)$

(resp. $\bigwedge^{*\geq n}(M)$) is the set of all linear forms vanishing on all $\bigwedge^j(M)$ such that $j > n$ (resp. $j < n$).

Let f and g be elements of $\bigwedge^{*j}(M)$ and $\bigwedge^{*k}(M)$ respectively; since $\pi' = \bigwedge(\delta)$ is a morphism of $\mathbb{N}$-graded algebras, $f \wedge g$ belongs to $\bigwedge^{*(j+k)}(M)$; here is its value on the product of $j + k$ elements of M:

$$(f \wedge g)(a_1 \wedge a_2 \wedge \cdots \wedge a_{j+k})$$
$$= \sum_s \operatorname{sgn}(s)\ (-1)^{jk}\ f(a_{s(1)} \wedge \cdots \wedge a_{s(j)})\ g(a_{s(j+1)} \wedge \cdots \wedge a_{s(j+k)})\ ;$$

the summation runs on all permutations s satisfying the conditions required above.

Because of the universal property of $\bigwedge^n(M)$, $\bigwedge^n(M)^*$ can be identified with the set of all alternate n-linear forms on M; thus the above definition of $f \wedge g$ allows us to define the exterior product of an alternate j-linear form and an alternate k-linear form; the definition of the exterior product of two alternate multilinear forms has been classical long before comultiplications were used to explain it; nevertheless the twisting factor $(-1)^{jk}$ which appears above as a consequence of (4.2.2), has not been introduced in this classical definition; consequently a discrepancy with common use appears here: the product here denoted by $f \wedge g$ is understood elsewhere as the exterior product of g and f in this order.

When the twisting rule (4.2.1) is strictly observed, for all $h_1, \dots, h_n$ in $\bigwedge^{*1}(M)$ and all $a_1, \dots, a_n$ in M, the determinant of the matrix $\big(h_j(a_k)\big)$ (in which j, $k = 1, 2, \dots, n$) is equal to the value of $h_n \wedge h_{n-1} \wedge \cdots \wedge h_1$ on $a_1 \wedge a_2 \wedge \cdots \wedge a_n$.

Since $h \wedge h = 0$ for all $h \in \bigwedge^{*1}(M)$, the natural bijection $M^* \to \bigwedge^{*1}(M)$ extends to a canonical algebra morphism from $\bigwedge(M^*)$ into $\bigwedge^*(M)$. It is obviously an isomorphism when M is a free module of finite rank; consequently it is still an isomorphism when M is a finitely generated projective module.

Since $\bigwedge(M)$ is obviously a module over $\bigwedge(M)$ on the left (resp. right) side, $\bigwedge^*(M)$ is a module over $\bigwedge(M)$ on the right (resp. left) side; thus there are interior products $f \lfloor x$ and $x \rfloor f$ for all $f \in \bigwedge^*(M)$ and all $x \in \bigwedge(M)$. Because of the relation $f \lfloor x = (-1)^{\partial f \partial x} x \rfloor f$, both multiplications are equally useful; but the interior multiplication by x on the left side involves the canonical mapping $M \times M^* \to K$ defined by $(a, h) \longmapsto -h(a)$ according to the twisting rule (4.2.1); to avoid unpleasant twisting signs, we prefer the interior multiplication by x on the right side. By definition of $f \lfloor x$ the following identity holds for all $y \in \bigwedge(M)$:

$$(4.3.2) \qquad (f \lfloor x)(y) = f(x \wedge y)\ ;$$

thus $\bigwedge^*(M)$ becomes a right $\bigwedge(M)$-module:

$$(4.3.3) \qquad (f \lfloor x) \lfloor y = f \lfloor (x \wedge y).$$

When f and x belong respectively to $\bigwedge^{*j}(M)$ and $\bigwedge^k(M)$, then $f \lfloor x$ belongs to $\bigwedge^{*(j-k)}(M)$; thus the equality $\partial(f \lfloor x) = \partial f + \partial x$ is valid for $\mathbb{Z}$-degrees since the degrees of f, x and $f \lfloor x$ are respectively $-j$, k and $-j + k$.

The interior multiplication by an element a of M is a twisted derivation of degree $+1$, but since the multiplication in $\bigwedge^*(M)$ has been defined in agreement with the twisting rule (4.2.1), we get a formula slightly different from the usual one:

$$(4.3.4) \qquad (f \wedge g) \lfloor a \;=\; f \wedge (g \lfloor a) + (f \lfloor a) \wedge \sigma(g) \;;$$

indeed, if we write $\pi'(y) = \sum_i y_i' \otimes y_i''$ for some $y \in \bigwedge(M)$, then

$$\begin{aligned} ((f \wedge g) \lfloor a)(y) &= (f \mathbin{\hat{\otimes}} g)\big((a \otimes 1 + 1 \otimes a) \wedge \pi'(y)\big) \\ &= \sum_i (-1)^{\partial g(1+\partial y_i')} f(a \wedge y_i') g(y_i'') + \sum_i (-1)^{(1+\partial g)\partial y_i'} f(y_i') g(a \wedge y_i'') \;; \end{aligned}$$

the former (resp. latter) summation is the value of $(f \lfloor a) \wedge \sigma(g)$ (resp. $f \wedge (g \lfloor a)$) on y. □

When f vanishes on $\bigwedge^{>n}(M)$ and x has no component of degree $< n$, then $f \lfloor x$ belongs to K :

$$(4.3.5) \qquad f \lfloor x \;=\; f(x) \quad \text{for all} \quad f \in \bigwedge^{*\leq n}(M) \quad \text{and all} \quad x \in \bigwedge^{\geq n}(M) \;;$$

for instance $h \lfloor a = h(a)$ for all $a \in M$ and all $h \in \bigwedge^{*1}(M)$.

We can interpret $\bigwedge^*$ as a contravariant functor, namely $\mathrm{Hom}(\bigwedge(\dots), K)$; any linear mapping $w : M \to N$ determines an algebra morphism $\bigwedge^*(w) : \bigwedge^*(N) \to \bigwedge^*(M)$. Let g be an element of $\bigwedge^*(N)$ and x an element of $\bigwedge(M)$; straightforward calculations show that

$$(4.3.6) \qquad \bigwedge^*(w)(g) \;\lfloor\; x \;=\; \bigwedge^*(w)\big(\, g \;\lfloor\; \bigwedge(w)(x)\big).$$

It is natural to define the interior product of the elements

$$f \otimes g \in \bigwedge^*(M) \mathbin{\hat{\otimes}} \bigwedge^*(N) \quad \text{and} \quad x \otimes y \in \bigwedge(M) \mathbin{\hat{\otimes}} \bigwedge(N)$$

by the following formula:

$$(4.3.7) \qquad (f \otimes g) \;\lfloor\; (x \otimes y) \;=\; (1)^{\partial g \partial x}\, (f \lfloor x) \otimes (g \lfloor y).$$

This definition is so much the more sensible as it is compatible with the canonical morphism $f \otimes g \longmapsto f \mathbin{\hat{\otimes}} g$ that is defined according to (4.2.7), and that maps $f \otimes g$ to a linear form on $\bigwedge(M) \mathbin{\hat{\otimes}} \bigwedge(N)$. Indeed, because of the canonical isomorphism $\bigwedge(M) \mathbin{\hat{\otimes}} \bigwedge(N) \cong \bigwedge(M \oplus N)$, we can identify $f \mathbin{\hat{\otimes}} g$ with an element of $\bigwedge^*(M \oplus N)$ and $x \otimes y$ with an element of $\bigwedge(M \oplus N)$; therefore the interior product $(f \mathbin{\hat{\otimes}} g) \lfloor (x \otimes y)$ is meaningful, and after some calculations it becomes clear that

$$(f \otimes g) \;\lfloor\; (x \otimes y) \;\longmapsto\; (f \mathbin{\hat{\otimes}} g) \;\lfloor\; (x \otimes y).$$

These considerations lead us to the Leibniz formula in exterior algebras, which implies the derivation formula (4.3.4) as a particular case. There are two versions of this formula in (4.3.8) below, which correspond to the two possible interpretations of (4.1.3). With the diagonal mapping δ (that is $a \longmapsto (a,a)$) the contravariant functor $\bigwedge^*$ associates an algebra morphism $\bigwedge^*(\delta) : \bigwedge^*(M \oplus M) \to \bigwedge^*(M)$, and from the definition of the exterior product of two elements f and g of $\bigwedge(M)$ it immediately follows that $f \wedge g = \bigwedge^*(\delta)(f \hat{\otimes} g)$. This shows a close relation between $\bigwedge^*(\delta)$ and the morphism $\pi_* : \bigwedge^*(M) \hat{\otimes} \bigwedge^*(M) \to \bigwedge^*(M)$ that represents the multiplication in $\bigwedge^*(M)$. Now we claim that for all $f,\, g \in \bigwedge^*(M)$ and all $x \in \bigwedge(M)$,

$$(4.3.8) \qquad \begin{aligned} (f \wedge g) \rfloor x &= \bigwedge^*(\delta)\big((f \hat{\otimes} g) \ \rfloor\ \pi'(x)\big) \\ &= \ \pi_*\big((f \otimes g) \ \rfloor\ \pi'(x)\big). \end{aligned}$$

Indeed the former right-hand member comes from a direct application of (4.3.6), since $\pi' = \bigwedge(\delta)$ and $f \wedge g = \bigwedge^*(\delta)(f \hat{\otimes} g)$. The latter right-hand member involves the definition (4.3.7) and its compatibility with the canonical morphism $f \otimes g \longmapsto f \hat{\otimes} g$. □

Because of its parity grading, $\bigwedge^*(M)$ has a grade automorphism σ, and the equalities $(\sigma(f))(x) = f(\sigma(x))$ and $\sigma(f \rfloor x) = \sigma(f) \rfloor \sigma(x)$ are obviously true for all $f \in \bigwedge^*(M)$ and $x \in \bigwedge(M)$.

Let us define the reversion τ in $\bigwedge^*(M)$ by the formula $(\tau(f))(x) = f(\tau(x))$. This definition immediately implies that there are formulas analogous to (3.1.5) in $\bigwedge^*(M)$, but more work is necessary to verify that we have got an involution of $\bigwedge^*(M)$, in other words, $\tau(f \wedge g) = \tau(g) \wedge \tau(f) = (-1)^{\partial f \partial g}\tau(f) \wedge \tau(g)$. After some calculations this follows from

$$\begin{aligned} \pi' \circ \tau(x) &= (\tau \otimes \tau) \circ \pi'(x) \qquad \text{for all } x \in \textstyle\bigwedge_1(M), \\ &= (\tau \otimes \sigma\tau) \circ \pi'(x) \quad \text{for all } x \in \textstyle\bigwedge_0(M). \end{aligned}$$

Besides, for all $f \in \bigwedge^*(M)$ and all $x \in \bigwedge(M)$,

$$(4.3.9) \qquad \tau(f \rfloor x) \ = \ (-1)^{\partial x(\partial f + \partial x)}\ \tau(f) \ \rfloor\ \tau(x) \ ;$$

indeed from the definitions (in particular (4.3.2)) it follows that

$$\tau(f \rfloor x)(y) \ = \ (-1)^{\partial x \partial y}\ (\tau(f) \rfloor \tau(x))(y) \ ,$$

and we can suppose that $\partial y = \partial f + \partial x$ since both members of this equality vanish if y has another parity.

This classical interior multiplication $\bigwedge^*(M) \times \bigwedge(M) \to \bigwedge^*(M)$ will serve as a model for the interior multiplication presented in the next section.

4.4 Interior products in Clifford algebras

Let M be a K-module provided with a quadratic form $q : M \to K$, $C\ell(M,q)$ the associated Clifford algebra, and $\rho : M \to C\ell(M,q)$ the canonical morphism. Since the canonical algebra morphism $K \to C\ell(M,q)$ is not always injective, we must distinguish the unit elements 1 in K and 1_q in $C\ell(M,q)$. To get convenient notation, we denote the identity mappings of $\bigwedge(M)$, $\bigwedge^*(M)$ and $C\ell(M,q)$ by $\mathrm{id}_\wedge$, id_* and id_q, and we denote the linear mappings that determine the algebra and coalgebra structures of $\bigwedge(M)$ by π, ε, π', ε', whereas π_* and π_q correspond to the multiplications in $\bigwedge^*(M)$ and $C\ell(M,q)$.

(4.4.1) **Theorem.** *There exists a unique algebra morphism*

$$\pi'_q : C\ell(M,q) \to C\ell(M,q) \hat{\otimes} \bigwedge(M)$$

such that $\pi'_q(\rho(a)) = \rho(a) \otimes 1 + 1_q \otimes a$ *for all* $a \in M$*; it makes* $C\ell(M,q)$ *become a right comodule over the coalgebra* $\bigwedge(M)$*.*

Proof. The unicity of π'_q is evident. Let $\delta : M \to M \oplus M$ be defined as in **4.1** and **4.3**: $\delta(a) = (a,a)$. It is clear that δ is a morphism from the quadratic module (M,q) into the orthogonal sum $(M,q) \perp (M,0)$; consequently it induces an algebra morphism $C\ell(\delta)$ between the associated Clifford algebras; we can identify the Clifford algebra of this orthogonal sum with the twisted tensor product of $C\ell(M,q)$ and $\bigwedge(M)$, and thus $C\ell(\delta)$ maps $\rho(a)$ to $\rho(a) \otimes 1 + 1_q \otimes a$; this proves the existence of π'_q. The comultiplication π' of $\bigwedge(M)$ can also be identified with the algebra morphism derived from δ when δ is understood as a morphism from the trivial quadratic module $(M,0)$ into $(M,0) \perp (M,0)$, and its counit ε' is the algebra morphism associated with the zero morphism $\zeta : (M,0) \to (0,0)$. Consequently the required equalities

$$(\pi'_q \otimes \mathrm{id}_\wedge)\, \pi'_q = (\mathrm{id}_q \otimes \pi')\, \pi'_q \ ,$$
$$(\mathrm{id}_q \otimes \varepsilon')\, \pi'_q = \text{canonical isomorphism } C\ell(M,q) \to C\ell(M,q) \otimes K \ ,$$

are consequences of these equalities:

$$(\delta \oplus \mathrm{id}_M)\, \delta = (\mathrm{id}_M \oplus \delta)\, \delta = \text{mapping } a \longmapsto (a,a,a) \ ,$$
$$(\mathrm{id}_M \oplus \zeta)\, \delta = \text{canonical isomorphism } M \to M \oplus 0. \qquad \square$$

Since $C\ell(M,q)$ is a right comodule over $\bigwedge(M)$ and a module over K, it is a left module over the algebra $\bigwedge^*(M) = \mathrm{Hom}^\wedge(\bigwedge(M), K)$ according to the graded version of Theorem (4.1.2). Consequently there is a multiplication $\bigwedge^*(M) \times C\ell(M,q) \to C\ell(M,q)$ that we shall call an interior multiplication and denote by $(f,x) \longmapsto f \rfloor x$. By definition,

$$(4.4.2) \qquad f \rfloor x = \sum_i (-1)^{\partial x'_i \partial x''_i} f(x''_i)\, x'_i \quad \text{if } \pi'_q(x) = \sum_i x'_i \otimes x''_i \ ;$$

the reversion of x'_i and x''_i with the subsequent twisting sign is due to the presence of the twisted reversion $\top^\wedge$ in (4.2.5). We already know that

$$f\rfloor(g\rfloor x) \;=\; (f\wedge g)\rfloor x. \tag{4.4.3}$$

Let us consider an element $x=\rho(a_1)\rho(a_2)\cdots\rho(a_n)$ which is the product in $C\ell(M,q)$ of n elements of M; the calculation of $\pi'_q(x)$ is exactly similar to that of $\pi'(a_1\wedge a_2\wedge\cdots\wedge a_n)$ in **4.3**, but we write the result in a slightly different way to compensate the reversion of x'_i and x''_i in (4.4.2):

$$\pi'_q(x) \;=\; \sum_{j=0}^{n}\sum_{s}(-1)^{j(n-j)}\,\mathrm{sgn}(s)\,\big(\rho(a_{s(j+1)})\cdots\rho(a_{s(n)})\big)\otimes\big(a_{s(1)}\wedge\cdots\wedge a_{s(j)}\big)\ ;$$

the second summation still runs on the permutations s such that

$$s(1)<s(2)<\cdots<s(j)\quad\text{and}\quad s(j+1)<s(j+2)<\cdots<s(n)\ ;$$

a straightforward application of the definition (4.4.2) shows that

$$\begin{aligned} &f\ \rfloor\ \rho(a_1)\rho(a_2)\cdots\rho(a_n)\\ &\quad=\sum_{j=0}^{n}\sum_{s}\mathrm{sgn}(s)\ f\big(a_{s(1)}\wedge\cdots\wedge a_{s(j)}\big)\ \rho(a_{s(j+1)})\cdots\rho(a_{s(n)})\ .\end{aligned}$$

From this calculation we derive:

$$h\rfloor(xy) \;=\; (h\rfloor x)y\ +\sigma(x)(h\rfloor y)\qquad\text{for all } h\in{\textstyle\bigwedge}^{*1}(M)\ , \tag{4.4.4}$$

$$\begin{aligned} &f\ \rfloor\ \big(\rho(a_1)\rho(a_2)\cdots\rho(a_n)\big)\\ &\quad= f\big(a_1\wedge a_2\wedge\cdots\wedge a_n\big)\ 1_q\qquad\text{for all } f\in{\textstyle\bigwedge}^{*\geq n}(M)\ ;\end{aligned} \tag{4.4.5}$$

in Formula (4.4.5), f must vanish on all $\bigwedge^j(M)$ with $j<n$. As for (4.4.4), it means that interior multiplications by elements of $\bigwedge^{*1}(M)$ are twisted derivations of odd degree. It is also clear that $f\rfloor x$ belongs to $C\ell^{\leq k-j}(M,q)$ when f belongs to $\bigwedge^{*j}(M)$ and x to $C\ell^{\leq k}(M,q)$.

Let w be a morphism from (M,q) into $(N,\tilde q)$, that is a linear mapping such that $\tilde q(w(a))=q(a)$ for all $a\in M$. The functors $C\ell$ and $\bigwedge^*$ associate with w two algebra morphisms $C\ell(w):C\ell(M,q)\to C\ell(N,\tilde q)$ and $\bigwedge^*(w):\bigwedge^*(N)\to\bigwedge^*(M)$. This situation leads to a formula analogous to (4.3.6); for all $x\in M$ and all $g\in\bigwedge^*(N)$,

$$g\ \rfloor\ C\ell(w)(x) \;=\; C\ell(w)\ \Big({\textstyle\bigwedge}^*(w)(g)\ \rfloor\ x\ \Big). \tag{4.4.6}$$

For the reversion τ there is a formula analogous to (4.3.9):

$$\tau(f\rfloor x) \;=\; (-1)^{\partial f(\partial f+\partial x)}\ \tau(f)\ \rfloor\ \tau(x). \tag{4.4.7}$$

Now we come to the Leibniz formulas. Although $\bigwedge^*(M)$ is not always a coalgebra, we can define a linear mapping $\pi^*:\bigwedge^*(M)\longrightarrow(\bigwedge(M)\hat\otimes\bigwedge(M))^*$

that looks like a comultiplication; it is the morphism associated with

$$\pi : \bigwedge(M) \hat{\otimes} \bigwedge(M) \longrightarrow \bigwedge(M)$$

by the contravariant functor $\mathrm{Hom}(\ldots, K)$; consequently

$$\pi^*(f)\ (x \otimes y) \ = \ f(x \wedge y) \quad \text{for all } x,\ y \in \bigwedge(M). \tag{4.4.8}$$

It is worth noticing that π is the algebra morphism $\bigwedge(M) \hat{\otimes} \bigwedge(M) \to \bigwedge(M)$ associated by the functor $\bigwedge$ with the morphism $(a,b) \longmapsto a+b$ from $M \oplus M$ onto M; consequently π^* is the algebra morphism associated by the functor $\bigwedge^*$ with this mapping $(a,b) \longmapsto a+b$.

Besides, from (4.2.7) we derive a canonical morphism from $\bigwedge^*(M) \hat{\otimes} \bigwedge^*(M)$ into $(\bigwedge(M) \hat{\otimes} \bigwedge(M))^*$; when it is an isomorphism (for instance when M is a finitely generated projective module), π^* determines a morphism π'_* from $\bigwedge^*(M)$ into $\bigwedge^*(M) \hat{\otimes} \bigwedge^*(M)$ which makes $\bigwedge^*(M)$ actually become a coalgebra.

Here is the first *Leibniz formula* that shows the effect of the interior multiplication by $f \in \bigwedge^*(M)$ on the product xy of two elements of $C\ell(M,q)$; it is meaningful because $x \otimes y$ and $\pi^*(f)$ can be understood as elements of $C\ell((M,q) \perp (M,q))$ and $\bigwedge^*(M \oplus M)$:

$$f \rfloor (xy) \ = \ \pi_q\big(\pi^*(f)\ \rfloor\ (x \otimes y)\big). \tag{4.4.9}$$

Proof of (4.4.9). In the diagram just below the morphism that goes from

$$C\ell(M,q) \hat{\otimes} C\ell(M,q) \quad \text{to} \quad C\ell(M,q)$$

through the left-hand column, is the mapping $(x \otimes y) \longmapsto f \rfloor (xy)$; but you get the mapping $(x \otimes y) \longmapsto \pi_q(\pi^*(f) \rfloor (x \otimes y))$ if you follow the longer path through the right-hand column. To prove that both paths give the same result, an oblique arrow has been added, which divides the diagram into two parts:

$$\begin{array}{ccc}
C\ell(M,q) \hat{\otimes} C\ell(M,q) & \longrightarrow & (C\ell(M,q) \hat{\otimes} \bigwedge(M)) \hat{\otimes} (C\ell(M,q) \hat{\otimes} \bigwedge(M)) \\
\downarrow & & \updownarrow \\
C\ell(M,q) & & (C\ell(M,q) \hat{\otimes} C\ell(M,q)) \hat{\otimes} (\bigwedge(M) \hat{\otimes} \bigwedge(M)) \\
\downarrow & \swarrow & \updownarrow \\
C\ell(M,q) \hat{\otimes} \bigwedge(M) & & (\bigwedge(M) \hat{\otimes} \bigwedge(M)) \hat{\otimes} (C\ell(M,q) \hat{\otimes} C\ell(M,q)) \\
\updownarrow & & \downarrow \\
\bigwedge(M) \hat{\otimes} C\ell(M,q) & & \bigwedge(M) \hat{\otimes} (C\ell(M,q) \hat{\otimes} C\ell(M,q)) \\
\downarrow & & \downarrow \\
K \otimes C\ell(M,q) & & K \otimes (C\ell(M,q) \hat{\otimes} C\ell(M,q)) \\
\updownarrow & & \updownarrow \\
C\ell(M,q) & \longleftarrow & C\ell(M,q) \hat{\otimes} C\ell(M,q))
\end{array}$$

The upper part of this diagram shows two paths from $C\ell(M,q) \hat{\otimes} C\ell(M,q)$ into $C\ell(M,q) \hat{\otimes} \bigwedge(M)$; they give equal morphisms because π'_q is an algebra morphism: $\pi'_q(xy) = \pi'_q(x)\pi'_q(y)$. The lower part of this diagram only requires a trivial verification. $\square$

Unlike the Leibniz formula (4.3.8) which is an immediate consequence of (4.3.6), here (4.4.9) is not a consequence of (4.4.6); indeed π_q is not an algebra morphism, it is not associated by the functor $C\ell$ with the mapping $(a,b) \longmapsto a+b$ from $M \oplus M$ to M, unless $q = 0$.

Besides, there is a definition analogous to (4.3.7):

$$(f \otimes g) \rfloor (x \otimes y) = (-1)^{\partial g \partial x} (f \rfloor x) \otimes (g \rfloor y)$$

for $f \in \bigwedge^*(M)$, $g \in \bigwedge^*(N)$, $x \in C\ell(M,q)$ and $y \in C\ell(N,\tilde{q})$. This definition too is compatible with the canonical morphism $\bigwedge^*(M) \hat{\otimes} \bigwedge^*(N) \to (\bigwedge(M) \hat{\otimes} \bigwedge(N))^*$. It can be used in case of an application of (4.4.9) when $\pi^*(f)$ is the image of some element $\sum_i f'_i \otimes f''_i \in \bigwedge^*(M) \hat{\otimes} \bigwedge^*(M)$:

$$f \rfloor (xy) = \sum_i (-1)^{\partial x \partial f''_i} (f'_i \rfloor x)\, (f''_i \rfloor y).$$

For instance every $\pi^*(h)$ with $h \in \bigwedge^{*1}(M)$ is the image of $h \otimes 1 + 1 \otimes h$, and this allows us to deduce the derivation formula (4.4.4) from the Leibniz formula.

Besides the Leibniz formula (4.4.9), the product $f \rfloor (xy)$ gives rise to two other formulas which involve the coproduct of x or y, and which are called *composite Leibniz formulas* because they need both interior multiplications $\lfloor$ and $\rfloor$. If $\pi'_q(x) = \sum_i x'_i \otimes x''_i$ and $\pi'_q(y) = \sum_j y'_j \otimes y''_j$ (with homogeneous x'_i and y'_j in $C\ell(M,q)$, and homogeneous x''_i and y''_j in $\bigwedge(M)$), then

(4.4.10)
$$f \rfloor (xy) = \sum_i (-1)^{\partial f \partial x'_i}\, x'_i\, ((f \lfloor x''_i) \rfloor y)\ ,$$
$$f \rfloor (xy) = \sum_j (-1)^{(\partial x + \partial y'_j)\partial y''_j}\, ((f \lfloor y''_j) \rfloor x)\, y'_j.$$

When x or y is an element $\rho(a)$ with $a \in M$, we get the *composite derivation formulas*

(4.4.11)
$$f \rfloor (\rho(a)y) = (f \lfloor a) \rfloor y + \rho(a)(\sigma(f) \rfloor y)\ ,$$
$$f \rfloor (x\rho(a)) = (f \rfloor x)\rho(a) + (f \lfloor a) \rfloor \sigma(x).$$

Proof of (4.4.10). These two formulas are immediate consequences of these easy calculations:

$$f \rfloor (xy) = \sum_i \sum_j \pm f(x''_i \wedge y''_j)\, x'_i y'_j\ ,$$
$$(f \lfloor x''_i) \rfloor y = \sum_j \pm f(x''_i \wedge y''_j)\, y'_i\ ,$$
$$(f \lfloor y''_i) \rfloor x = \sum_i \pm f(y''_i \wedge x''_j)\, x'_j.$$

The three signs $\pm$ are those resulting from the twisting rule (4.2.1); if you wish to calculate them, remember that in the summations you must pay attention only to terms such that $\partial f = \partial x_i'' + \partial y_j''$. $\square$

When $q = 0$, we get an interior multiplication $\bigwedge^*(M) \times \bigwedge(M) \to \bigwedge(M)$ satisfying all the properties stated here, with the consequent little changes of notation. The grading of $\bigwedge(M)$ over $\mathbb{Z}$ is involved in the following assertion: when f belongs to $\bigwedge^{*j}(M)$ and x to $\bigwedge^k(M)$, then $f \rfloor x$ belongs to $\bigwedge^{k-j}(M)$. Remember that this f has degree $-j$.

This interior multiplication appears in the composite Leibniz formula that deals with the same product $(f \wedge g) \lfloor x$ as (4.3.8), but does not involve the coproduct of x. If $\pi^*(f)$ is the image of $\sum_i f_i' \otimes f_i''$ (an element of $\bigwedge^*(M) \hat{\otimes} \bigwedge^*(M)$), then

$$(f \wedge g) \lfloor x \;=\; \sum_i (-1)^{\partial f_i'' \partial g}\, f_i' \wedge (g \lfloor (f_i'' \rfloor x)).$$

Here this composite Leibniz formula is never needed, and its proof is proposed as an exercise. When f belongs to $\bigwedge^{*1}(M)$, it gives a composite derivation formula.

Interior multiplications involving two factors respectively in $\bigwedge^*(M)$ (or $\bigwedge(M^*)$) and $\bigwedge(M)$ appear very often in the literature, yet with systematic discrepancies in the treatment of the twisting signs, since the twisting rule (4.2.1) is not always uncompromisingly enforced as it is here. More fundamental discrepancies appear when the factor undergoing the operation belongs to a Clifford algebra $C\ell(M,q)$, because the assailing factor does not always belong to $\bigwedge^*(M)$ nor to $\bigwedge(M^*)$; sometimes it belongs to $\bigwedge(M)$ or even to $C\ell(M,q)$ (as in (4.ex.8)). All these versions can be derived from the present one because the operation of an assailing factor belonging to $\bigwedge(M^*)$ or to $\bigwedge(M)$ or to $C\ell(M,q)$ is always the operation of its natural image in $\bigwedge^*(M)$ by these natural morphisms:

$$C\ell(M,q) \longrightarrow \bigwedge(M) \longrightarrow \bigwedge(M^*) \longrightarrow \bigwedge^*(M).$$

The first arrow $\Phi_{-\beta} : C\ell(M,q) \to \bigwedge(M)$ is not an algebra morphism but a comodule isomorphism; it is associated with the "canonical scalar product" $\beta = \mathrm{b}_q/2$ as it is later explained at the end of **4.8**. The second arrow is the algebra morphism associated by the functor $\bigwedge$ with $\mathrm{d}_q : M \to M^*$; it is an isomorphism when q is nondegenerate. The third arrow is the algebra morphism that extends the natural injection $M^* \to \bigwedge^*(M)$; it is an isomorphism when M is projective and finitely generated.

This section ends with an easy yet very important result. Remember that for all $a \in M$, $\mathrm{d}_q(a)$ is the element of M^* such that $\mathrm{d}_q(a)(b) = \mathrm{b}_q(a,b)$; here this $\mathrm{d}_q(a)$ is silently identified with its canonical image in $\bigwedge^{*1}(M)$. For all $a \in M$ and all $x \in C\ell(M,q)$, it is stated that

$$\text{(4.4.12)} \qquad \rho(a)\, x \;-\; \sigma(x)\, \rho(a) \;=\; \mathrm{d}_q(a) \rfloor x.$$

Proof. We consider a as fixed. Let D_1 and D_2 be the mappings $x \longmapsto \rho(a)x - \sigma(x)\rho(a)$ and $x \longmapsto \mathrm{d}_q(a) \rfloor x$. On one side, for all $b \in M$,

$$D_1(\rho(b)) = \rho(a)\rho(b) + \rho(b)\rho(a) = \mathrm{b}_q(a,b)\ 1_q = D_2(\rho(b)) .$$

On the other side, for all x, $y \in C\ell(M,q)$, and for $i = 1, 2$,

$$D_i(xy) = D_i(x)\ y + (-1)^{\partial x} x\ D_i(y) ;$$

indeed, when $i = 2$, this is a consequence of (4.4.4); and when $i = 1$, it is easy to verify that every odd element z in a graded algebra determines a twisted derivation $x \longmapsto zx - \sigma(x)z$. Since the algebra $C\ell(M,q)$ is generated by $\rho(M)$, these common properties of D_1 and D_2 imply their equality. □

4.5 Exponentials in even exterior subalgebras

Let M be a K-module and $\bigwedge(M)$ its exterior algebra; the even subalgebra $\bigwedge_0(M)$ is commutative; it is the direct sum of $K = \bigwedge^0(M)$ and the ideal $\bigwedge_0^+(M)$ that is the direct sum of all $\bigwedge^{2i}(M)$ with $i > 0$; all elements in this ideal are nilpotent.

An element of $\bigwedge(M)$ is said to be *decomposable* if it is an element of $K = \bigwedge^0(M)$ or an element of $M = \bigwedge^1(M)$ or an exterior product of elements of M.

(4.5.1) **Theorem.** *There is a unique mapping* Exp *from* $\bigwedge_0^+(M)$ *into* $\bigwedge_0(M)$ *such that*

$$\mathrm{Exp}(x+y) = \mathrm{Exp}(x) \wedge \mathrm{Exp}(y) \quad \textit{for all } x \textit{ and } y \textit{ in } \textstyle\bigwedge_0^+(M),$$

$$\mathrm{Exp}(x) = 1 + x \quad \textit{whenever } x \textit{ is decomposable with even positive degree.}$$

Proof. The unicity of the mapping Exp is evident, since every element of $\bigwedge_0^+(M)$ is a sum of decomposable elements; every decomposition of x as a sum of decomposable elements allows us to calculate $\mathrm{Exp}(x)$, and we must prove that they all give the same value to $\mathrm{Exp}(x)$. If x is decomposable, $1-x$ is the $\wedge$-inverse of $1+x$ because $x \wedge x = 0$; thus it is easy to realize that the existence of the mapping Exp is equivalent to the following statement:

(4.5.2) *If* x_1, x_2, $\dots$, x_r *are decomposable elements in* $\bigwedge_0^+(M)$ *and if their sum vanishes, then*

$$(1+x_1) \wedge (1+x_2) \wedge \cdots \wedge (1+x_r) = 1 .$$

The existence of the mapping Exp is evident when K contains a subring isomorphic to the field $\mathbb{Q}$ of rational numbers, because every $x \in \bigwedge_0^+(M)$ is nilpotent, and the exponential series gives the value of $\mathrm{Exp}(x)$ in such a way that the statement (4.5.1) is true; this fact suggests a proof in three steps.

First step. If (4.5.2) is true for every module over $\mathbb{Z}$, then it is true for every module over K. Indeed the K-module M is also a $\mathbb{Z}$-module, and its identity mapping extends to a ring morphism from $\bigwedge_{\mathbb{Z}}(M)$ into $\bigwedge(M) = \bigwedge_K(M)$; the restricted mapping $\bigwedge_{\mathbb{Z}}^+(M) \to \bigwedge^+(M)$ is surjective, and its kernel is the ideal of $\bigwedge_{\mathbb{Z}}(M)$ generated by all elements $\lambda a \wedge b - a \wedge \lambda b$ with $\lambda \in K$ and a and $b \in M$. The decomposable elements $x_1, \dots, x_n$ in $\bigwedge_0^+(M)$ are images of decomposable elements $y_1, \dots, y_n$ in $\bigwedge_{\mathbb{Z}}(M)$, and since the sum $\sum_i x_i$ vanishes, the sum $\sum_i y_i$ is equal to a sum of several terms like $u \wedge (\lambda a \wedge b - a \wedge \lambda b) \wedge v$ with arbitrary decomposable factors u and v in $\bigwedge_{\mathbb{Z}}(M)$, both even or odd. If we have proved that (4.5.2) is valid in $\bigwedge_{\mathbb{Z}}(M)$, we can assert that the product of the r factors $1 + y_i$ and several other factors like

$$(1 - u \wedge \lambda a \wedge b \wedge v) \wedge (1 + u \wedge a \wedge \lambda b \wedge v)$$

is equal to 1. Therefore the product of the images of all these factors in $\bigwedge(M)$ is also equal to 1. The n factors $1 + y_i$ give the n factors $1 + x_i$ in $\bigwedge(M)$, but the above two factors give a product in $\bigwedge(M)$ equal to 1. It follows that (4.5.2) is also valid in $\bigwedge(M)$.

Second step. (4.5.2) is valid for free additive groups. Indeed if M is a free additive group, it can be considered as a subgroup of $\mathbb{Q} \otimes_{\mathbb{Z}} M$ whereas $\bigwedge_{\mathbb{Z}}(M)$ is a subring of $\mathbb{Q} \otimes_{\mathbb{Z}} \bigwedge_{\mathbb{Z}}(M)$, itself canonically isomorphic to $\bigwedge_{\mathbb{Q}}(\mathbb{Q} \otimes_{\mathbb{Z}} M)$. The statement (4.5.2) is true for the $\mathbb{Q}$-module $\mathbb{Q} \otimes_{\mathbb{Z}} M$, consequently it is also true for the free group M.

Third step. (4.5.2) is true for all additive groups. Indeed if M is not free, there exists a surjective group morphism $N \to M$ defined on a free additive group N. If N_0 is the kernel of this morphism, the kernel of the ring morphism $\bigwedge_{\mathbb{Z}}(N) \to \bigwedge_{\mathbb{Z}}(M)$ is the ideal generated by N_0 (see (4.3.1)). Let $x_1, \dots, x_n$ be decomposable elements in $\bigwedge_{\mathbb{Z},0}^+(M)$, the sum of which is 0; they are the images of decomposable elements $y_1, \dots, y_n$ in $\bigwedge_{\mathbb{Z},0}^+(N)$, the sum of which belongs to the ideal generated by N_0; consequently their sum is also a sum of decomposable elements $z_1, \dots, z_r$ all belonging to this ideal. Since the property (4.5.2) is valid for the free group N, the product of all factors $(1+y_1), \dots, (1+y_n)$, $(1-z_1), \dots, (1-z_r)$ is equal to 1. If we transport this result in $\bigwedge(M)$ by means of the morphism $\bigwedge(N) \to \bigwedge(M)$, we get the awaited equality: the product of the factors $(1+x_1), \dots, (1+x_n)$ is 1. □

(4.5.3) **Example.** Let M be a free K-module of rank 4 with basis (a, b, c, d), and $x = a \wedge b + c \wedge d$; obviously $x \wedge x = 2\, a \wedge b \wedge c \wedge d$ and

$$\text{Exp}(x) = (1 + a \wedge b) \wedge (1 + c \wedge d) = 1 + a \wedge b + c \wedge d + a \wedge b \wedge c \wedge d.$$

When K is the field $\mathbb{Z}/2\mathbb{Z}$, then $x \wedge x = 0$, but $\text{Exp}(x) \neq 1 + x$.

(4.5.4) **Corollary.** *For all $x \in \bigwedge_0^+(M)$ and all $h \in \bigwedge^{*1}(M)$,*

$$h \rfloor \text{Exp}(x) = (h \rfloor x) \wedge \text{Exp}(x).$$

Proof. Formula (4.4.4) shows that $(h \rfloor x) \wedge x = 0$ when x is decomposable with degree ≥ 2; consequently the equality in (4.5.4) is true when x is decomposable. When this equality is true for x and y, it is still true for $x + y$ because (4.4.4) allows us to write

$$\begin{aligned} h \rfloor \mathrm{Exp}(x+y) &= (h \rfloor \mathrm{Exp}(x)) \wedge \mathrm{Exp}(y) \ + \ \mathrm{Exp}(x) \wedge (h \rfloor \mathrm{Exp}(y)) \\ &= (h \rfloor x) \wedge \mathrm{Exp}(x) \wedge \mathrm{Exp}(y) \ + \ \mathrm{Exp}(x) \wedge (h \rfloor y) \wedge \mathrm{Exp}(y) \\ &= (h \rfloor (x+y)) \wedge \mathrm{Exp}(x+y) \ ; \end{aligned}$$

the conclusion follows. □

(4.5.5) **Corollary.** *If $w : M \to N$ is a linear mapping, for all $x \in \bigwedge_0^+(M)$,*

$$\mathrm{Exp}(\textstyle\bigwedge(w)(x)) \ = \ \bigwedge(w)(\mathrm{Exp}(x)).$$

This corollary is evident, and also the following identity involving the reversion in $\bigwedge(M)$:

$$\mathrm{Exp}(\tau(x)) \ = \ \tau(\mathrm{Exp}(x)).$$

Now in the algebra $\bigwedge^*(M)$ we consider the subalgebra $\bigwedge_0^*(M)$ of all elements vanishing on $\bigwedge_1(M)$, and in this subalgebra, the ideal $\bigwedge_0^{*+}(M)$ of all elements also vanishing on $K = \bigwedge^0(M)$. The elements of this ideal are all nilpotent when M is finitely generated, but nonnilpotent elements exist in $\bigwedge^{*2}(M)$ when M is free with infinite bases.

(4.5.6) **Theorem.** *There exists a unique mapping* Exp *from $\bigwedge_0^{*+}(M)$ into $\bigwedge_0^*(M)$ such that the following equalities hold for all $f \in \bigwedge_0^{*+}(M)$ and all $a \in M$:*

$$(\mathrm{Exp}(f))(1) = 1 \quad \textit{and} \quad \mathrm{Exp}(f) \lfloor a \ = \ \mathrm{Exp}(f) \wedge (f \lfloor a) \ .$$

Proof. Let g_k be the restriction of $\mathrm{Exp}(f)$ to the sum $\bigwedge^{\leq k}(M)$ of all components of degree $\leq k$. First g_0 must be the identity mapping of K, and the other g_k are determined by induction on k by the following requirement, for all $a \in M$ and all $x \in \bigwedge^k(M)$ (see (4.3.2)):

$$g_{k+1}(a \wedge x) \ = \ (\mathrm{Exp}(f) \lfloor a)(x) \ = \ (g_k \wedge (f \lfloor a))(x) \ ;$$

this proves the unicity of $\mathrm{Exp}(f)$. Moreover since f vanishes on $\bigwedge_1(M)$, $f \lfloor a$ vanishes on $\bigwedge_0(M)$, and consequently, by induction on k, g_k vanishes on $\bigwedge_1(M) \cap \bigwedge^{\leq k}(M)$; thus the induction that determines the restrictions g_k, implies that $\mathrm{Exp}(f)$ vanishes on $\bigwedge_1(M)$ as required.

We prove the existence of g_k by induction on k. The existence of g_0 and g_1 is evident, and g_1 vanishes on $M = \bigwedge^1(M)$. We assume the existence of g_k for some $k \geq 1$, and we consider the following $(k+1)$-linear form g' on M^{k+1} :

$$g'(a_1, a_2, \ldots, a_{k+1}) \ = \ (g_k \wedge (f \lfloor a_1))(a_2 \wedge a_3 \wedge \cdots \wedge a_{k+1}) \ ;$$

if g' is alternate in all variables, it determines a linear form on $\bigwedge^{k+1}(M)$ that allows us to extend g_k to a linear form g_{k+1} on $\bigwedge^{\leq k+1}(M)$. Obviously g' is alternate in $a_2, a_3, \dots, a_{k+1}$; thus it suffices to prove that it vanishes if $a_1 = a_2$; in other words it suffices to prove the following equality for all $a \in M$ and all $x \in \bigwedge^{\leq k-1}(M)$:

$$(g_k \wedge (f \lfloor a))(a \wedge x) \;=\; 0\;.$$

By means of (4.3.2), (4.3.4) and (4.3.3) we obtain

$$(g_k \wedge (f \lfloor a))(a \wedge x) \;=\; \big((g_k \wedge (f \lfloor a)) \lfloor a\big)(x) \;=\; -\big((g_k \lfloor a) \wedge (f \lfloor a)\big)(x)\;;$$

because of the induction hypothesis, $g_k \lfloor a$ is equal to $g_{k-1} \wedge (f \lfloor a)$; and since $f \lfloor a$ is odd, its exterior square vanishes; all this proves the desired equality and completes the proof. □

(4.5.7) **Corollary.** *For all f and g in $\bigwedge_0^{*+}(M)$,*

$$\mathrm{Exp}(f+g) \;=\; \mathrm{Exp}(f) \wedge \mathrm{Exp}(g)\;.$$

Proof. Indeed by (4.3.4) we can write (for all $a \in M$)

$$\begin{aligned}(\mathrm{Exp}(f) \wedge \mathrm{Exp}(g)) \lfloor a &= \mathrm{Exp}(f) \wedge \mathrm{Exp}(g) \wedge (g \lfloor a) \;+\; \mathrm{Exp}(f) \wedge (f \lfloor a) \wedge \mathrm{Exp}(g)\\ &= \mathrm{Exp}(f) \wedge \mathrm{Exp}(g) \wedge ((f+g) \lfloor a).\end{aligned}$$

□

(4.5.8) **Corollary.** *If $w : M \longrightarrow N$ is a linear mapping, for all $g \in \bigwedge_0^{*+}(N)$*

$$\bigwedge^*(w)(\mathrm{Exp}(g)) \;=\; \mathrm{Exp}(\bigwedge^*(w)(g))\;.$$

This is an easy consequence of (4.3.6); and the following equality involving the reversion in $\bigwedge^*(M)$ is also evident: $\mathrm{Exp}(\tau(g)) = \tau(\mathrm{Exp}(g))$.

Here is a last technical lemma involving a Clifford algebra.

(4.5.9) **Lemma.** *The equality* $\mathrm{Exp}(f) \rfloor x = x$ *holds if x belongs to $C\ell^{\leq k}(M,q)$ and $f(\bigwedge^{\leq k}(M)) = 0$. The same equality holds if x belongs to the subalgebra of $C\ell(M,q)$ generated by a direct summand N of M and $f(\bigwedge(N)) = 0$.*

Proof. From (4.5.6) it is easy to deduce that $\mathrm{Exp}(f)(\bigwedge^j(M)) = 0$ if $f(\bigwedge^{\leq k}(M)) = 0$ and $1 \leq j \leq k$, or that $\mathrm{Exp}(f)(\bigwedge^+(N)) = 0$ if $f(\bigwedge(N)) = 0$. Then $\pi'_q(x)$ can be written as a sum $\sum_i x'_i \otimes x''_i$ in which $(x'_1, x''_1) = (x,1)$ whereas $x''_i \in \bigwedge^+(M)$ for all $i \neq 1$; moreover each x''_i belongs to $\bigwedge^{\leq k}(M)$ (resp. $\bigwedge(N)$) if x belongs to $C\ell^{\leq k}(M,q)$ (resp. to the subalgebra generated by N). Now the conclusion $\mathrm{Exp}(f) \rfloor x = x$ follows from the definition (4.4.2). □

4.6 Systems of divided powers

Section **4.6** is just an appendix to **4.5** and hurried readers are advised to skip it. Although the system of divided powers in $\bigwedge_0(M)$ becomes superfluous if we use the exponentials presented in **4.5**, it is worth knowing that divided powers also account for the existence of these exponentials. Besides, systems of divided powers appear in many other places and lead to the universal algebras $\Gamma(M)$ mentioned below.

Let A be a commutative K-algebra such that the canonical morphism $K \to A$ is injective, and A is the direct sum of the image of K and some ideal A^+; we identify K with its image in A, and write $A = K \oplus A^+$. This situation may also be described in this manner: there are two algebra morphisms $\varepsilon_A : K \to A$ and $\varepsilon'_A : A \to K$ such that $\varepsilon'_A \varepsilon_A = \mathrm{id}_K$; the ideal A^+ is then the kernel of ε'_A. A *system of divided powers* on A is a sequence of mappings $x \longmapsto x^{[n]}$ from A^+ into A, such that the following six conditions are satisfied whenever x and y are in A^+, m and n in $\mathbb{N}$, and λ in K :

(4.6.1) $$x^{[0]} = 1,\ x^{[1]} = x \quad \text{and} \quad x^{[n]} \in A^+ \text{ for all } n > 0\ ;$$

(4.6.2) $$(\lambda x)^{[n]} = \lambda^n x^{[n]}\ ;$$

(4.6.3) $$(x+y)^{[n]} = \sum_{k=0}^{n} x^{[k]}\, y^{[n-k]}\ ;$$

(4.6.4) $$(xy)^{[n]} = x^n y^{[n]} = x^{[n]} y^n\ ;$$

(4.6.5) $$x^{[m]} x^{[n]} = \frac{(m+n)!}{m!\, n!}\, x^{[m+n]}\ ;$$

(4.6.6) $$\text{if } n > 0, \text{ then } (x^{[n]})^{[m]} = \frac{(mn)!}{m!\,(n!)^m}\, x^{[mn]}.$$

It is known that the rational number that appears in (4.6.5) is an integer. The rational number that appears in (4.6.6) is also an integer (provided that $n > 0$); this can be proved by induction on m with the help of the equality

$$\frac{(mn)!}{m!\,(n!)^m} = \frac{((m-1)n)!}{(m-1)!\,(n!)^{m-1}}\ \frac{(mn-1)!}{(mn-n)!\,(n-1)!}.$$

It is possible to prove that (4.6.4) is a consequence of the five other conditions; the proof begins with the identity $xy = (x+y)^{[2]} - x^{[2]} - y^{[2]}$ which follows from (4.6.3) and (4.6.1).

By using (4.6.1) and (4.6.5) and by induction on n it is easy to prove, for all $x \in A^+$ and $n \in \mathbb{N}$,

(4.6.7) $$x^n = n!\, x^{[n]}\ ;$$

this explains why $x^{[n]}$ is called the nth divided power of x (divided by $n!$).

Let us suppose that the canonical morphism $\mathbb{Z} \to K$ extends to a ring morphism $\mathbb{Q} \to K$; then (4.6.7) proves the existence and unicity of a system of divided powers on every algebra A that is decomposable as $A = K \oplus A^+$. Indeed it is easy to prove that the six conditions (4.6.1) to (4.6.6) are consequences of (4.6.7) when all integers $n!$ are invertible in K. Therefore a system of divided powers is interesting only when some integers are *not* invertible in K.

The even exterior algebra $\bigwedge_0(M)$ is provided with a system of divided powers that can be deduced from Theorem (4.5.1) in this way: by means of the polynomial extension $K \to K[t]$ it is possible to define $\mathrm{Exp}(tx)$ in $K[t] \otimes \bigwedge_0(M)$, and then $x^{[m]}$ is the factor multiplied by t^m in the development of $\mathrm{Exp}(tx)$. The ideas underlying the proof of (4.5.1) can also show that a system of divided powers has actually been defined in this way.

The extension $K \to K[t]$ can also serve to define divided powers in $\bigwedge_0^*(M)$. When M is finitely generated, $\bigwedge_{K[t]}^*(K[t] \otimes M)$ can be identified with $K[t] \otimes \bigwedge^*(M)$ which is the direct sum of the submodules $t^n \otimes \bigwedge^*(M)$; in all cases $\bigwedge_{K[t]}^*(K[t] \otimes M)$ can be identified with a subalgebra of the direct product of the submodules $t^n \otimes \bigwedge^*(M)$; this fact gives sense to this definition: $f^{[m]}$ is the factor multiplied by t^m in the development of $\mathrm{Exp}(tf)$. Consequently

$$f^{[m+1]} \lfloor a \;=\; f^{[m]} \wedge (f \lfloor a) \quad \text{for all } a \in M;$$

this allows us to prove that a system of divided powers has been obtained.

Here is another nontrivial example of a system of divided powers. Let M be a K-module. The group $\mathcal{S}_n$ of all permutations of $\{1, 2, 3, \dots, n\}$ acts in the nth tensor power $\mathrm{T}^n(M)$ of M; for all $s \in \mathcal{S}_n$, the action of s^{-1} in $\mathrm{T}^n(M)$ is the following one:

$$s^{-1}(a_1 \otimes a_2 \otimes \cdots \otimes a_n) \;=\; a_{s(1)} \otimes a_{s(2)} \otimes \cdots \otimes a_{s(n)}.$$

The elements of $\mathrm{T}^n(M)$ that are invariant under the action of $\mathcal{S}_n$ make up the submodule $\mathrm{ST}^n(M)$ of all symmetric n-tensors. Of course $\mathrm{ST}^0(M) = K$ and $\mathrm{ST}^1(M) = M$. The direct sum $\mathrm{ST}(M)$ of all $\mathrm{ST}^n(M)$ becomes a commutative algebra when it is provided with the following multiplication; if $y \in \mathrm{ST}^j(M)$ and $z \in \mathrm{ST}^k(M)$, their symmetric product is the symmetrized tensor

$$y \vee z \;=\; \sum_s s^{-1}(y \otimes z) \,,$$

where the summation runs only on those $s \in \mathcal{S}_{j+k}$ such that

$$s(1) < s(2) < s(3) < \cdots < s(j) \quad \text{and} \quad s(j+1) < s(j+2) < \cdots < s(j+k).$$

By induction on n it is easy to prove that the symmetric product of n elements of M is given by

$$a_1 \vee a_2 \vee \cdots \vee a_n \;=\; \sum_s s^{-1}(a_1 \otimes a_2 \otimes \cdots \otimes a_n) \,,$$

with a summation running over all $s \in \mathcal{S}_n$. In particular the nth symmetric power of a (element of M) and its nth tensor power are related by the equality

$$a \vee a \vee \cdots \vee a \;=\; n!\; a \otimes a \otimes \cdots \otimes a.$$

Of course much more work is necessary to prove that $\mathrm{ST}(M)$ is a commutative and associative algebra provided with a system of divided powers such that $a^{[n]}$ is the nth tensor power of a for all $a \in M$ and all $n \in \mathbb{N}$. Yet the unicity of this system of divided power is obvious.

It is known that the algebra $\mathrm{S}^*(M)$ dual to the coalgebra $\mathrm{S}(M)$ is also provided with a system of divided powers (see (4.ex.2)). When M is a finitely generated projective module, each dual space $(\mathrm{S}^n(M))^*$ is canonically isomorphic to $\mathrm{ST}^n(M^*)$.

The algebras provided with a system of divided powers constitute a subcategory $\mathcal{D}iv(K)$ of $\mathcal{C}om(K)$; a morphism in this category is an algebra morphism $f : K \oplus A^+ \longrightarrow K \oplus B^+$ such that $f(A^+) \subset B^+$ and $f(x^{[n]}) = f(x)^{[n]}$ for all $x \in A^+$ and all $n \in \mathbb{N}$. It is sensible to ask wether a module M may freely generate an algebra in this category, in the same way as it generates the algebras $\mathrm{T}(M)$ and $\mathrm{S}(M)$ in the categories $\mathcal{A}lg(K)$ and $\mathcal{C}om(K)$. The answer is positive: with M is associated a universal algebra $\Gamma(M)$ provided with a system of divided powers; it is an $\mathbb{N}$-graded algebra such that $\Gamma^0(M) = K$ and $\Gamma^1(M) = M$. Its universal property says that every linear mapping f from M into an object $K \oplus A^+$ of $\mathcal{D}iv(K)$ such that $f(M) \subset A^+$, extends in a unique way to a morphism $f' : \Gamma(M) \to K \oplus A^+$ in the category $\mathcal{D}iv(K)$.

In particular there is a unique morphism from $\Gamma(M)$ into the previous algebra $\mathrm{ST}(M)$ that maps every $a \in M$ to itself; it is known that it is an isomorphism whenever M is a projective module. Besides, in the category $\mathcal{C}om(K)$ there is a unique algebra morphism from $\mathrm{S}(M)$ into $\Gamma(M)$ that maps every $a \in M$ to itself; it is an isomorphism whenever there is a ring morphism $\mathbb{Q} \to K$.

It remains to report that the submodule $\Gamma^2(M)$ of this algebra $\Gamma(M)$ is canonically isomorphic to the module defined in **2.1** and already denoted by $\Gamma^2(M)$. In other words, for every quadratic mapping $q : M \to N$ there exists a unique linear mapping $\tilde{q} : \Gamma^2(M) \to N$ such that $q(a) = \tilde{q}(a^{[2]})$ for all $a \in M$.

4.7 Deformations of Clifford algebras

This is the main section in Chapter **4**. The notations are those of **4.3** and **4.4**; we still consider the same three algebras, and all the notations referring to each one are here recalled without comment:

$$C\ell(M,q) \;:\; \rho\,,\; 1_q\,,\; \mathrm{id}_q\,,\; \pi_q\,,\; \pi'_q\;;$$

$$\textstyle\bigwedge(M) \;:\; \mathrm{id}_\wedge\,,\; \pi\,,\; \pi'\,,\; \varepsilon\,,\; \varepsilon'\;;$$

$$\textstyle\bigwedge^*(M) \;:\; \mathrm{id}_*\,,\; \pi_*\,,\; \pi^*\;.$$

Although the unit element of $\bigwedge^*(M)$ is ε', it is rather denoted by 1 wherever this notation causes no ambiguity. Each of these three algebras is provided with a grade automorphism σ and a reversion τ.

Now a new figure β appears; it is any bilinear form $\beta : M \times M \to K$. All the notations referring to β are presented here together. With q and β we associate the quadratic form q' such that $q'(a) = q(a) + \beta(a,a)$ for all $a \in M$, and ρ' is the canonical mapping $M \to C\ell(M, q')$. According to the twisting rule (4.2.1), with β we associate the *twisted opposite bilinear form* β^{to} defined by $\beta^{to}(a,b) = -\beta(b,a)$. The linear mappings $M \to M^*$ determined by β and β^{to} are denoted by d_β and d_β^{to} :

$$\mathrm{d}_\beta(a)(b) = \beta(a,b) \quad \text{and} \quad \mathrm{d}_\beta^{to}(a)(b) = -\beta(b,a) \ .$$

The *opposite bilinear form* $(a,b) \longmapsto \beta(b,a)$ appears only later in (4.7.15) when the reversion τ gets involved; it is denoted by $-\beta^{to}$ (rather than β^o). In $\bigwedge^{*2}(M)$ there is an element $[\beta]$ such that

$$[\beta](a \wedge b) \;=\; \beta(a,b) - \beta(b,a) \quad \text{for all } a,\ b \in M.$$

Moreover, since $\bigwedge^2(M \oplus M)$ is canonically isomorphic to the direct sum of $\bigwedge^2(M) \otimes 1$, $1 \otimes \bigwedge^2(M)$ and $M \otimes M$, in $\bigwedge^{*2}(M \oplus M)$ there is a submodule canonically isomorphic to $(M \otimes M)^*$; and since the bilinear forms on M are in bijection with the elements of $(M \otimes M)^*$, β has a canonical image $\beta_{\prime\prime}$ in $\bigwedge^{*2}(M \oplus M)$. Thus $\beta_{\prime\prime}$ is the element of $\bigwedge^{*2}(M \oplus M)$ such that

$$\beta_{\prime\prime}((a_1,b_1) \wedge (a_2,b_2)) \;=\; \beta(a_1,b_2) - \beta(a_2,b_1) \ .$$

Since we can identify $(a_1,b_1) \wedge (a_2,b_2) \in \bigwedge(M \oplus M)$ with

$$(a_1 \wedge a_2) \otimes 1 \ + 1 \otimes (b_1 \wedge b_2) \ + (a_1 \otimes b_2) - (a_2 \otimes b_1) \ \in \bigwedge(M) \,\hat{\otimes}\, \bigwedge(M),$$

we can also say that $\beta_{\prime\prime}$ is the linear form on $\bigwedge(M) \,\hat{\otimes}\, \bigwedge(M)$ that vanishes on $\bigwedge^i(M) \otimes \bigwedge^j(M)$ whenever $(i,j) \neq (1,1)$, and such that $\beta_{\prime\prime}(a \otimes b) = \beta(a,b)$ for all $a,\ b \in M$.

This is not yet sufficient, since we shall also use the three images of β in $\bigwedge^{*2}(M \oplus M \oplus M)$. A notation like $\beta_{\prime\prime .}$ or $\beta_{\prime . \prime}$ or $\beta_{. \prime\prime}$ should clearly enough indicate which of these three images we consider: $\beta_{\prime\prime .}$ (resp. $\beta_{\prime . \prime}$) (resp. $\beta_{. \prime\prime}$) is the linear form on $\bigwedge(M) \,\hat{\otimes}\, \bigwedge(M) \,\hat{\otimes}\, \bigwedge(M)$ that vanishes on $\bigwedge^i(M) \otimes \bigwedge^j(M) \otimes \bigwedge^k(M)$ whenever (i,j,k) is not equal to $(1,1,0)$ (resp. $(1,0,1)$) (resp. $(0,1,1)$) and such that

$$\beta_{\prime\prime .}(a \otimes b \otimes 1) = \beta(a,b) \ , \quad \text{resp. } \ \beta_{\prime . \prime}(a \otimes 1 \otimes b) = \beta(a,b) \ ,$$
$$\text{resp. } \ \beta_{. \prime\prime}(1 \otimes a \otimes b) = \beta(a,b).$$

Later we shall even use $\beta_{\prime . \prime .}$ which is one of the six images of β in $\bigwedge^{*2}(M \oplus M \oplus M \oplus M)$, and we rely on the reader to guess its definition.

(4.7.1) **Definition.** The *deformation* of the algebra $C\ell(M,q)$ by the bilinear form β is the K-module $C\ell(M,q)$ provided with the following multiplication:

$$(x,y) \longmapsto x \star y = \pi_q\big(\mathrm{Exp}(\beta_{//}) \rfloor (x \otimes y)\big) \ ;$$

this deformation is denoted by $C\ell(M,q;\beta)$.

Usually the word "deformation" (when it does not mean an infinitesimal deformation) refers to a family of multiplications depending on a parameter t in such a way that the initial multiplication is obtained for $t = 0$. Nonetheless there is no impassable gap between this usual concept of deformation and Definition (4.7.1); indeed if we consider the polynomial extension $K \to K[t]$, the deformation of $K[t] \otimes C\ell(M,q)$ by the bilinear form $t \otimes \beta$ gives the initial multiplication when t is replaced with 0, and the new one when t is replaced with 1.

The notation $\bigwedge(M;\beta)$ means $C\ell(M,0;\beta)$; nonetheless the modified multiplication of $\bigwedge(M;\beta)$ is simply denoted by $(x,y) \longmapsto xy$ (instead of $x \star y$), since the initial multiplication in $\bigwedge(M)$ is already denoted by the proper symbol $\wedge$.

Since $\mathrm{Exp}(\beta_{//})$ is even, it is clear that the deformed algebra $C\ell(M,q;\beta)$ is graded by the same subspaces $C\ell_0(M,q)$ and $C\ell_1(M,q)$ as $C\ell(M,q)$.

More than the half of this section is devoted to the proof of the following five theorems, and near the end, a sixth theorem shall be added.

(4.7.2) **Theorem.** *The deformation $C\ell(M,q;\beta)$ is an associative algebra with the same unit element 1_q.*

(4.7.3) **Theorem.** *These two equalities are true for all $a \in M$ and all $x \in C\ell(M,q)$:*

(a) $\rho(a) \star x = \rho(a)\, x + \mathrm{d}_\beta(a) \rfloor x$;

(b) $x \star \rho(a) = x\, \rho(a) + \mathrm{d}_\beta^{to}(a) \rfloor \sigma(x)$.

Here $\mathrm{d}_\beta(a)$ and $\mathrm{d}_\beta^{to}(a)$ must be understood as elements of $\bigwedge^{*1}(M)$, that are linear forms on $\bigwedge(M)$ vanishing on all $\bigwedge^j(M)$ such that $j \neq 1$; such identifications of elements of M^* with their image in $\bigwedge^{*1}(M)$ will be silently committed when the context obviously requires them.

(4.7.4) **Theorem.** *Among all the associative multiplications on the K-module $C\ell(M,q)$ that admit 1_q as a unit element, the multiplication defined by* (4.7.1) *is the only one satisfying the equality* (a) *in* (4.7.3) *(for all $a \in M$ and all $x \in C\ell(M,q)$). It is also the only one satisfying the equality* (b).

(4.7.5) **Theorem.** *Let us set $q'(a) = q(a) + \beta(a,a)$. There is a unique algebra morphism Φ_β from $C\ell(M,q')$ into $C\ell(M,q;\beta)$ such that $\Phi_\beta(\rho'(a)) = \rho(a)$ for all $a \in M$; it is a morphism of graded algebras. It is also a morphism of right comodules over $\bigwedge(M)$, and consequently a morphism of left modules over $\bigwedge^*(M)$.*

(4.7.6) **Theorem.** *The algebra morphisms $\Phi_\beta : C\ell(M,q') \to C\ell(M,q;\beta)$ and $\Phi_{-\beta} : C\ell(M,q) \to C\ell(M,q';-\beta)$ are reciprocal bijections.*

Proof of the five theorems, and corollaries

Proof of (4.7.2). Since $\beta_{\prime\prime}$ vanishes on $\bigwedge(M)\otimes 1$ and $1\otimes\bigwedge(M)$, the equalities $\mathrm{Exp}(\beta_{\prime\prime})\rfloor(x\otimes 1_q) = x\otimes 1_q$ and $\mathrm{Exp}(\beta_{\prime\prime})\rfloor(1_q\otimes x) = 1_q\otimes x$ follow from (4.5.9) and imply $x\star 1_q = 1_q\star x = x$ for all $x\in C\ell(M,q)$. The associativity of the $\star$-multiplication is the most difficult stage in this section because it involves two Leibniz formulas slightly more sophisticated than the simple formula (4.4.9). Let C be a graded right comodule over a graded coalgebra A; later C and A will be $C\ell(M,q)$ and $\bigwedge(M)$. Thus $C\otimes C\ell(M,q)$ (resp. $C\ell(M,q)\otimes C$) is a left module over the algebra $A^*\hat{\otimes}\bigwedge^*(M)$ (resp. $\bigwedge^*(M)\hat{\otimes}A^*$). If f (resp. g) is an element of this algebra, if x, y, z are elements of $C\ell(M,q)$, and ξ, ζ elements of C, then

$$f\rfloor(\xi\otimes yz) = (\mathrm{id}_C\otimes\pi_q)\left((\mathrm{id}_A\otimes\pi)^*(f)\ \rfloor\ (\xi\otimes y\otimes z)\right),$$
$$g\rfloor(xy\otimes\zeta) = (\pi_q\otimes\mathrm{id}_C)\left((\pi\otimes\mathrm{id}_A)^*(g)\ \rfloor\ (x\otimes y\otimes\zeta)\right);$$

these formulas are proved exactly like (4.4.9); the mappings $(\mathrm{id}_A\otimes\pi)^*$ and $(\pi\otimes\mathrm{id}_A)^*$ are associated by the functor $\mathrm{Hom}(\dots,K)$ with $\mathrm{id}_A\otimes\pi$ and $\pi\otimes\mathrm{id}_A$.

Now let us calculate $x\star(y\star z)$. By means of Definition (4.7.1), the Leibniz formula devoted to $f\rfloor(\xi\otimes yz)$, and also (4.4.3), (4.5.7) and (4.5.8), we obtain

$$\begin{aligned} x\star(y\star z) &= \pi_q\ \big(\mathrm{Exp}(\beta_{\prime\prime})\ \rfloor\ \big((\mathrm{id}_q\otimes\pi_q)\ ((1\otimes\mathrm{Exp}(\beta_{\prime\prime}))\rfloor(x\otimes y\otimes z))\big)\big)\\ &= \pi_q\ (\mathrm{id}_q\otimes\pi_q)\ \big(\mathrm{Exp}\big((\mathrm{id}_\wedge\otimes\pi)^*(\beta_{\prime\prime})\ + 1\otimes\beta_{\prime\prime}\big)\ \rfloor\ (x\otimes y\otimes z)\big).\end{aligned}$$

Obviously $1\otimes\beta_{\prime\prime} = \beta_{\cdot\prime\prime}$. Let us verify that $(\mathrm{id}_\wedge\otimes\pi)^*(\beta_{\prime\prime}) = \beta_{\prime\prime\cdot} + \beta_{\prime\cdot\prime}$; indeed

$$\begin{aligned}&(\mathrm{id}_\wedge\otimes\pi)^*(\beta_{\prime\prime})\ (a_1\otimes b_1\otimes 1 + a_2\otimes 1\otimes b_2 + 1\otimes a_3\otimes b_3)\\ &\quad = \beta_{\prime\prime}(a_1\otimes b_1 + a_2\otimes b_2 + 1\otimes(a_3\wedge b_3)) = \beta(a_1,b_1)+\beta(a_2,b_2).\end{aligned}$$

All this shows that

$$x\star(y\star z) = \pi_q\ (\mathrm{id}_q\otimes\pi_q)\ \big(\mathrm{Exp}(\beta_{\prime\prime\cdot}+\beta_{\prime\cdot\prime}+\beta_{\cdot\prime\prime})\ \rfloor\ (x\otimes y\otimes z)\big).$$

In the same way we can calculate that

$$(x\star y)\star z = \pi_q\ (\pi_q\otimes\mathrm{id}_q)\ \big(\mathrm{Exp}(\beta_{\prime\prime\cdot}+\beta_{\prime\cdot\prime}+\beta_{\cdot\prime\prime})\ \rfloor\ (x\otimes y\otimes z)\big).$$

We remember that $\pi_q(\pi_q\otimes\mathrm{id}_q) = \pi_q(\mathrm{id}_q\otimes\pi_q)$ because the algebra $C\ell(M,q)$ is associative, and the proof is complete. □

If we calculated the product of four factors in the algebra $C\ell(M,q;\beta)$, we should find a similar result involving the six images of β in $\bigwedge^{*2}(M\oplus M\oplus M\oplus M)$, and so forth....

Proof of (4.7.3). By means of (4.3.2) we get (for all $a\in M$)

$$\beta_{\prime\prime}\lfloor(a\otimes 1) = 1\otimes\mathrm{d}_\beta(a)\quad\text{and}\quad\beta_{\prime\prime}\lfloor(1\otimes a) = \mathrm{d}_\beta^{to}(a)\otimes 1\ ;$$

here $\mathrm{d}_\beta(a)$ and $\mathrm{d}_\beta^{to}(a)$ must be understood as elements of $\bigwedge^{*1}(M)$. Now the formulas (a) and (b) are immediate consequences of the equalities

$$\begin{aligned} \operatorname{Exp}(\beta_{\prime\prime}) \rfloor (\rho(a) \otimes x) &= \rho(a) \otimes x \; + \; 1_q \otimes (\mathrm{d}_\beta(a) \rfloor x) \ , \\ \operatorname{Exp}(\beta_{\prime\prime}) \rfloor (x \otimes \rho(a)) &= x \otimes \rho(a) \; + \; (\mathrm{d}_\beta^{to}(a) \rfloor \sigma(x)) \otimes 1_q \ , \end{aligned}$$

which themselves can be easily proved by means of the composite derivations formulas (4.4.11). Indeed $\rho(a) \otimes x$ (for instance) is the product of $\rho(a) \otimes 1_q$ and $1_q \otimes x$, and in the proof of (4.7.2) we have already noticed that the second factor is invariant by the interior multiplication by $\operatorname{Exp}(\beta_{\prime\prime})$; thus (4.4.11) implies

$$\operatorname{Exp}(\beta_{\prime\prime}) \rfloor (\rho(a) \otimes x) \;=\; \rho(a) \otimes x \; + \big(\operatorname{Exp}(\beta_{\prime\prime}) \lfloor (a \otimes 1)\big) \rfloor (1_q \otimes x) \ ;$$

then (4.5.6) implies

$$\operatorname{Exp}(\beta_{\prime\prime}) \lfloor (a \otimes 1) \;=\; \operatorname{Exp}(\beta_{\prime\prime}) \wedge (1 \otimes \mathrm{d}_\beta(a)) \ ;$$

this allows us to complete the proof of (a) with the help of (4.3.7) and (4.5.9):

$$\begin{aligned} \big(\operatorname{Exp}(\beta_{\prime\prime}) \lfloor (\rho(a) \otimes 1)\big) \rfloor (1_q \otimes x) &= \operatorname{Exp}(\beta_{\prime\prime}) \rfloor \big((1 \otimes \mathrm{d}_\beta(a)) \rfloor (1_q \otimes x)\big) \\ &= \; 1_q \otimes (\mathrm{d}_\beta(a) \rfloor x). \end{aligned}$$

The proof of the formula (b) is similar. □

(4.7.7) **Examples** of applications of (4.7.3). For all a, b, $c \in M$ we can write

$$\begin{aligned} \rho(a) \star \rho(b) &= \rho(a)\, \rho(b) \; + \beta(a,b)\, 1_q \ , \\ \rho(a) \star \rho(a) &= (q(a) + \beta(a,a))\, 1_q \ , \\ \rho(a) \star \rho(b) \star \rho(c) &= \rho(a)\, \rho(b)\, \rho(c) \; + \beta(b,c)\, \rho(a) - \beta(a,c)\, \rho(b) + \beta(a,b)\, \rho(c) \ . \end{aligned}$$

In the proof of (4.7.4) we shall use the filtration of the algebra $C\ell(M,q)$ by the submodules $C\ell^{\leq k}(M,q)$ defined in **3.1**; when x belongs to $C\ell^{\leq k}(M,q)$, then $f \rfloor x$ also belongs to it for all $f \in \bigwedge^*(M)$.

Proof of (4.7.4). Let us forget the algebra $C\ell(M,q;\beta)$ and assume that there is an associative multiplication on the K-module $C\ell(M,q)$ admitting 1_q as a unit element and satisfying (4.7.3)(b) (for instance) for all $a \in M$ and all $x \in C\ell(M,q)$; we denote it by $(x,y) \longmapsto x \star y$. Obviously the product $x \star y$ is uniquely determined for all y in $C\ell^{\leq 1}(M,q)$. Let us assume that it is uniquely determined for all y in $C\ell^{\leq k}(M,q)$, and let us prove that it is still uniquely determined for all y in $C\ell^{\leq k+1}(M,q)$. We can suppose that $y = z\,\rho(a)$ for some z in $C\ell^{\leq k}(M,q)$. The condition (b) still determines the value of $x \star y$:

$$x \star (z\,\rho(a)) \;=\; x \star (z \star \rho(a) - \mathrm{d}_\beta^{to}(a) \rfloor \sigma(z)) \;=\; (x \star z) \star \rho(a) \; - x \star (\mathrm{d}_\beta^{to}(a) \rfloor \sigma(z)) \ ;$$

it suffices to remember that $f \rfloor \sigma(z)$ belongs to $C\ell^{\leq k}(M,q)$ for all $f \in \bigwedge^*(M)$. With the condition (4.7.3)(a) the proof is similar. □

Here is a corollary of (4.7.4); it involves the twisted opposite algebra $C\ell(M,q)^{to}$ provided with the multiplication $x^{to}y^{to} = (-1)^{\partial x \partial y}(yx)^{to}$ (see **3.2**).

(4.7.8) **Corollary.** *The mapping* $x \longmapsto x^{to}$ *is an algebra isomorphism from*

$$C\ell(M,q;-\mathrm{b}_q) \quad onto \quad C\ell(M,q)^{to}.$$

Proof. For all $a \in M$ and all $x \in C\ell(M,q)$, the product of $\rho(a)^{to}$ and x^{to} in $C\ell(M,q)^{to}$ is equal to $(\sigma(x)\rho(a))^{to}$; from (4.4.12) we deduce that

$$\sigma(x)\rho(a) \;=\; \rho(a)x - \mathrm{d}_q(a) \rfloor x \;;$$

the right-hand member is the product of $\rho(a)$ and x in $C\ell(M,q;-\mathrm{b}_q)$, and this suffices to conclude. □

Proof of (4.7.5). The equality $\rho(a) \star \rho(a) = q'(a,a)1_q$ (see (4.7.7)) proves the existence of the algebra morphism Φ_β from $C\ell(M,q')$ into $C\ell(M,q;\beta)$. Since the algebra $C\ell(M,q;\beta)$ admits the same parity grading as $C\ell(M,q)$, Φ_β is a graded morphism. The main assertion in (4.7.5) is that Φ_β is a morphism of right comodules over $\bigwedge(M)$, in other words,

$$\pi'_q \circ \Phi_\beta \;=\; (\Phi_\beta \otimes \mathrm{id}_\wedge) \circ \pi'_{q'}.$$

The right-hand member of this equality is the algebra morphism from $C\ell(M,q')$ into $C\ell(M,q;\beta) \hat{\otimes} \bigwedge(M)$ that maps every $\rho'(a)$ to $\rho(a) \otimes 1 + 1_q \otimes a$; therefore it suffices to prove that π'_q is also an algebra morphism from $C\ell(M,q;\beta)$ into $C\ell(M,q;\beta) \hat{\otimes} \bigwedge(M)$. Let us denote by Π the linear mapping representing the multiplication in $C\ell(M,q) \hat{\otimes} \bigwedge(M)$; the product of two elements ξ and ζ in $C\ell(M,q;\beta) \hat{\otimes} \bigwedge(M)$ is

$$\xi \star \zeta \;=\; \Pi\,(\mathrm{Exp}(\beta_{\prime.\prime.}) \rfloor (\xi \otimes \zeta)) \;;$$

consequently it suffices to prove the following equality for $x,\, y \in C\ell(M,q)$:

$$\Pi\big(\,\mathrm{Exp}(\beta_{\prime.\prime.}) \rfloor (\pi'_q(x) \otimes \pi'_q(y))\big) \;=\; \pi'_q \circ \pi_q\big(\,\mathrm{Exp}(\beta_{\prime\prime}) \rfloor (x \otimes y)\big).$$

Since π'_q is an algebra morphism, $\pi'_q \circ \pi_q = \Pi \circ (\pi'_q \otimes \pi'_q)$, and thus it suffices to prove that

$$\mathrm{Exp}(\beta_{\prime.\prime.}) \rfloor (\pi'_q(x) \otimes \pi'_q(y)) \;=\; (\pi'_q \otimes \pi'_q)\big(\mathrm{Exp}(\beta_{\prime\prime}) \rfloor (x \otimes y)\big) \;.$$

This last equality is an immediate consequence of (4.4.6), when the morphism $w : (M,q) \to (N,\tilde{q})$ appearing there is here replaced with the morphism

$$\Delta : (M,q) \perp (M,q) \longrightarrow (M,q) \perp (M,0) \perp (M,q) \perp (M,0)$$

defined by $\Delta(a,b) = (a,a,b,b)$; indeed all this implies that

$$C\ell(\Delta) = \pi'_q \otimes \pi'_q \quad \text{and} \quad \bigwedge^*(\Delta)(\beta_{\prime.\prime.}) = \beta_{\prime\prime} \;.$$ □

Since Φ_β is a morphism of comodules, the equality $\Phi_\beta(f \rfloor x) = f \rfloor \Phi_\beta(x)$ holds for all $x \in C\ell(M,q')$ and all $f \in \bigwedge^*(M)$. Besides, it is clear that the objects presented here behave nicely in the case of a direct sum; if (M,q) is the direct sum of two orthogonal submodules (M_1,q_1) and (M_2,q_2), and if β is the direct sum of two bilinear forms β_1 and β_2 respectively on M_1 and M_2, then $C\ell(M,q;\beta)$ is canonically isomorphic to the twisted tensor product of $C\ell(M_1,q_1;\beta_1)$ and $C\ell(M_2,q_2;\beta_2)$, and by this isomorphism Φ_β becomes $\Phi_{\beta_1} \otimes \Phi_{\beta_2}$. Whence the following consequence of (4.7.5).

(4.7.9) **Corollary.** *For every bilinear form $\beta' : M \times M \to K$, the algebra morphism Φ_β from $C\ell(M,q')$ into $C\ell(M,q;\beta)$ is also an algebra morphism from $C\ell(M,q';\beta')$ into $C\ell(M,q;\ \beta+\beta')$.*

Proof. We must prove this equality for all $\xi \in C\ell(M,q') \otimes C\ell(M,q')$:

$$\Phi_\beta \circ \pi_{q'}(\ \mathrm{Exp}(\beta'_{/\!/})\ \rfloor\ \xi\ ) \ = \ \pi_q \circ (\ \mathrm{Exp}(\beta_{/\!/} + \beta'_{/\!/})\ \rfloor\ (\Phi_\beta \otimes \Phi_\beta)(\xi)\ )\ ;$$

since Φ_β is a morphism from $C\ell(M,q')$ into $C\ell(M,q;\beta)$, we know that

$$\Phi_\beta \circ \pi_{q'}(\ \mathrm{Exp}(\beta'_{/\!/})\ \rfloor\ \xi\ ) \ = \ \pi_q \circ (\ \mathrm{Exp}(\beta_{/\!/})\ \rfloor\ (\Phi_\beta \otimes \Phi_\beta)(\mathrm{Exp}(\beta'_{/\!/})\ \rfloor\ \xi)\ )\ ;$$

since the interior multiplication by $\mathrm{Exp}(\beta_{/\!/}+\beta'_{/\!/})$ is equivalent to successive interior multiplications by $\mathrm{Exp}(\beta'_{/\!/})$ and $\mathrm{Exp}(\beta_{/\!/})$, it suffices to verify that

$$(\Phi_\beta \otimes \Phi_\beta)(\mathrm{Exp}(\beta'_{/\!/})\ \rfloor\ \xi) \ = \ \mathrm{Exp}(\beta'_{/\!/})\ \rfloor\ (\Phi_\beta \otimes \Phi_\beta)(\xi)\ ;$$

since $\Phi_\beta \otimes \Phi_\beta$ can be identified with the algebra morphism

$$\Phi_{\beta\perp\beta}\ :\ C\ell((M,q') \perp (M,q')) \longrightarrow C\ell((M,q) \perp (M,q)\ ;\ \beta \perp \beta)\ ,$$

the conclusion follows from the fact that $\Phi_{\beta\perp\beta}$ is a morphism of comodules. □

Proof of (4.7.6). Because of (4.7.9), Φ_β is also an algebra morphism from

$$C\ell(M,q';-\beta) \quad \text{into} \quad C\ell(M,q).$$

Consequently $\Phi_\beta \circ \Phi_{-\beta}$ is an algebra morphism from $C\ell(M,q)$ into itself which maps every $\rho(a)$ to itself; this proves that $\Phi_\beta \circ \Phi_{-\beta}$ is the identity mapping. And the same for $\Phi_{-\beta} \circ \Phi_\beta$. □

More generally, if we set $q''(a) = q'(a)+\beta'(a,a)$ and consider the isomorphism $\Phi_{\beta'}$ from $C\ell(M,q'')$ onto $C\ell(M,q';\beta')$, we can deduce from (4.7.9) that

(4.7.10) $$\Phi_\beta \circ \Phi_{\beta'} \ = \ \Phi_{\beta+\beta'}.$$

Here are other corollaries of the previous results.

(4.7.11) **Corollary.** *For all $k \in \mathbb{N}$, $\Phi_\beta(C\ell^{\leq k}(M,q')) = C\ell^{\leq k}(M,q)$; moreover Φ_β induces an algebra isomorphism between the graded algebras $\mathrm{Gr}(C\ell(M,q'))$ and $\mathrm{Gr}(C\ell(M,q))$ defined in* **3.1**.

Proof. Since the interior multiplication by $\mathrm{d}_\beta(a)$ maps $C\ell^{\leq k}(M,q)$ into itself, the equality (4.7.3)(a) shows (by induction on k) that $\Phi_\beta(C\ell^{\leq k}(M,q')) \subset C\ell^{\leq k}(M,q)$. The opposite inclusion is proved by means of $\Phi_{-\beta}$. Thus it is clear that Φ_β induces bijections

$$C\ell^{\leq k}(M,q') \,/\, C\ell^{\leq k-1}(M,q') \longrightarrow C\ell^{\leq k}(M,q) \,/\, C\ell^{\leq k-1}(M,q) \ ,$$

resulting in a bijection $\mathrm{Gr}(\Phi_\beta) : \mathrm{Gr}(C\ell(M,q')) \to \mathrm{Gr}(C\ell(M,q))$.

For every $x \in C\ell^{\leq k}(M,q)$, $\rho(a) \star x$ and $\rho(a)x$ are congruent modulo $C\ell^{\leq k}(M,q)$; consequently, for all $b \in \mathrm{Gr}^1(C\ell(M,q'))$ and all $y \in \mathrm{Gr}^k(C\ell(M,q'))$,

$$\mathrm{Gr}(\Phi_\beta)(by) = \mathrm{Gr}(\Phi_\beta)(b)\ \mathrm{Gr}(\Phi_\beta)(y) \quad \text{in} \quad \mathrm{Gr}^{k+1}(C\ell(M,q'))\ ;$$

since the algebra $\mathrm{Gr}(C\ell(M,q'))$ is generated by $\mathrm{Gr}^1(C\ell(M,q'))$, this suffices to conclude that $\mathrm{Gr}(\Phi_\beta)$ is an algebra morphism. □

(4.7.12) **Corollary.** *The mapping $x \longmapsto x^{to}$ is an algebra isomorphism from*

$$C\ell(M,q;\ \beta^{to} - \mathrm{b}_q) \quad \textit{onto} \quad C\ell(M,q;\beta)^{to}.$$

Proof. Let F be this mapping, and F' the analogous mapping $C\ell(M,q') \to C\ell(M,q')^{to}$. Since Φ_β and $\Phi_{-\beta}$ are reciprocal bijections, we can write $F = \Phi_\beta^{to} \circ F' \circ \Phi_{-\beta}$. Moreover an easy calculation shows that $\beta^{to} - \mathrm{b}_q$ and $\beta - \mathrm{b}_{q'}$ are the same thing. Now $\Phi_{-\beta}$ is an isomorphism from $C\ell(M,q;\ \beta - \mathrm{b}_{q'})$ onto $C\ell(M,q';-\mathrm{b}_{q'})$ because of (4.7.9); then F' is an isomorphism from $C\ell(M,q';-\mathrm{b}_{q'})$ onto $C\ell(M;q')^{to}$ because of (4.7.8); and finally Φ_β^{to} is an isomorphism $C\ell(M;q')^{to} \to C\ell(M,q;\beta)^{to}$. □

Additional information

A sixth theorem is now added to the five previous ones.

(4.7.13) **Theorem.** *Let β and β' be two bilinear forms $M \times M \to K$ such that $\beta(a,a) = \beta'(a,a)$ for all $a \in M$, and let f be the element of $\bigwedge^{*2}(M)$ such that $f(a \wedge b) = \beta'(a,b) - \beta(a,b)$ for all $a, b \in M$. Then $\Phi_{\beta'-\beta}$ (that is the unique algebra isomorphism $C\ell(M,q;\beta) \to C\ell(M,q;\beta')$ leaving all elements of $\rho(M)$ invariant) is the mapping $x \longmapsto \mathrm{Exp}(f) \rfloor x$.*

Proof. The bilinear form $(a,b) \longmapsto \beta'(a,b) - \beta(a,b)$ is alternate and defines an element $f \in \bigwedge^{*2}(M)$. From (4.7.9) we deduce that $\Phi_{\beta'-\beta}$ is an algebra isomorphism $C\ell(M,q;\beta) \to C\ell(M,q;\beta')$. As a linear endomorphism of $C\ell(M,q)$ it is characterized by these two properties: it leaves invariant all elements of $C\ell^{\leq 1}(M,q)$, and for every $a \in M$ and $x \in C\ell(M,q)$ it maps $\rho(a) \star x$ (product in $C\ell(M,q;\beta)$) to the product of $\rho(a)$ and $\Phi_{\beta'-\beta}(x)$ in $C\ell(M,q;\beta')$. Let us verify that the mapping $x \longmapsto \mathrm{Exp}(f) \rfloor x$ satisfies these two properties. First it leaves invariant all elements of $C\ell^{\leq 1}(M,q)$ because of (4.5.9). Secondly we must verify that, for every $a \in M$ and $x \in C\ell(M,q)$,

$$\mathrm{Exp}(f) \rfloor (\rho(a)x) + (\mathrm{Exp}(f) \wedge \mathrm{d}_\beta(a)) \rfloor x = \rho(a)(\mathrm{Exp}(f) \rfloor x) + (\mathrm{d}_{\beta'}(a) \wedge \mathrm{Exp}(f)) \rfloor x$$

From the definitions (in particular (4.3.2)) it follows immediately that $\mathrm{d}_{\beta'}(a) - \mathrm{d}_\beta(a) = f \lfloor a$; consequently the previous equality is equivalent to

$$\mathrm{Exp}(f) \rfloor (\rho(a)x) \;=\; (\mathrm{Exp}(f) \wedge (f \lfloor a)) \rfloor x + \rho(a)(\mathrm{Exp}(f) \rfloor x) \;;$$

since $\mathrm{Exp}(f) \wedge (f \lfloor a)$ is the same thing as $\mathrm{Exp}(f) \lfloor a$ (see (4.5.6)), this is an example of a composite derivation formula like (4.4.11). □

Like $C\ell(M,q')$, the algebra $C\ell(M,q;\beta)$ admits a reversion that we shall now calculate as a corollary of (4.7.13); the definition of $[\beta] \in \bigwedge^{*2}(M)$ has been given at the beginning.

(4.7.14) **Proposition.** *The reversion τ_β in $C\ell(M,q;\beta)$ maps every $x \in C\ell(M,q;\beta)$ to*

$$\tau_\beta(x) \;=\; \mathrm{Exp}([\beta]) \;\rfloor\; \tau(x) \;=\; \tau(\;\mathrm{Exp}(-[\beta]) \;\rfloor\; x) \;.$$

Proof. Because of (4.7.13), the mapping $x \longmapsto \mathrm{Exp}([\beta]) \rfloor x$ is the isomorphism of $C\ell(M,q;-\beta^{to})$ onto $C\ell(M,q;\beta)$ that leaves invariant all elements of $\rho(M)$. Thus the proof of (4.7.14) is completed by the following lemma.

(4.7.15) **Lemma.** *The mapping $x \longmapsto \tau(x)^o$ is an isomorphism from $C\ell(M,q;\beta)$ onto the opposite algebra $C\ell(M,q;-\beta^{to})^o$.*

Proof. We must prove that

$$\tau\big(\;\mathrm{Exp}(\beta_{\prime\prime}) \;\rfloor\; (\tau(x) \otimes \tau(y))\;\big) \;=\; \mathrm{Exp}(-\beta_{\prime\prime}^{to}) \;\rfloor\; (y \otimes x) \;;$$

we observe that $\tau(\beta_{\prime\prime}) = -\beta_{\prime\prime}$ (see (3.1.5)), and because of (3.2.8) and (4.4.7) the previous equality is equivalent to

$$(-1)^{\partial x \partial y}\, \mathrm{Exp}(-\beta_{\prime\prime}) \;\rfloor\; (x \otimes y) \;=\; \mathrm{Exp}(-\beta_{\prime\prime}^{to}) \;\rfloor\; (y \otimes x) \;;$$

this is an immediate consequence of (4.4.6) when w is the reversion mapping $M \oplus M \to M \oplus M$ defined by $w(a,b) = (b,a)$; indeed

$$C\ell(w)(x \otimes y) = (-1)^{\partial x \partial y} y \otimes x \;, \quad \text{and} \quad \bigwedge^{*}(w)(\beta_{\prime\prime}) = \beta_{\prime\prime}^{to} \;. \qquad \square$$

Examples. It is clear that $[\beta] = 0$ if and only if β is symmetric. When 2 is invertible in K, for every pair (q,q') of quadratic forms on M, there exists a unique symmetric bilinear form β such that $q'(a) = q(a) + \beta(a,a)$ for all $a \in M$. On the contrary, when the equality $2 = 0$ holds in K, then $[\beta]$ is strictly determined by $q' - q$ because

$$[\beta](a \wedge b) \;=\; \beta(a,b) + \beta(b,a) \;=\; \mathrm{b}_{q'}(a,b) - \mathrm{b}_q(a,b).$$

This sections ends with some routine information.

(4.7.16) **Proposition.** *Let* (M, q) *and* (M', q') *be two quadratic modules,* β *and* β' *bilinear forms respectively on* M *and* M'*, and* $f : M \to M'$ *a linear mapping such that*

$$\forall a, b \in M, \quad q(a) = q'(f(a)) \quad \text{and} \quad \beta(a,b) = \beta'(f(a), f(b)) \ ;$$

then the algebra morphism $\mathrm{C}\ell(f) : \mathrm{C}\ell(M, q) \to \mathrm{C}\ell(M', q')$ *is also an algebra morphism from* $\mathrm{C}\ell(M, q; \beta)$ *into* $\mathrm{C}\ell(M', q'; \beta')$*.*

Proof. For every x, $y \in \mathrm{C}\ell(M, q)$ we must verify that

$$\mathrm{C}\ell(f) \circ \pi_q\big(\,\mathrm{Exp}(\beta_{\prime\prime}) \;\rfloor\; (x \otimes y)\big) \;=\; \pi_{q'}\big(\,\mathrm{Exp}(\beta'_{\prime\prime}) \;\rfloor\; (\mathrm{C}\ell(f)(x) \otimes \mathrm{C}\ell(f)(y))\,\big).$$

If f_2 is the morphism from $(M, q) \perp (M, q)$ into $(M', q') \perp (M', q')$ such that $f_2(a, b) = (f(a), f(b))$, it is obvious that

$$\mathrm{C}\ell(f)(x) \otimes \mathrm{C}\ell(f)(y) = \mathrm{C}\ell(f_2)(x \otimes y) \quad \text{and} \quad \mathrm{C}\ell(f) \circ \pi_q = \pi_{q'} \circ \mathrm{C}\ell(f_2) \ ,$$

and it is easy to verify that $\beta_{\prime\prime} = \bigwedge^*(f_2)(\beta'_{\prime\prime})$. Thus the conclusion follows from (4.4.6). □

Interior products and exponentials have been presented in **4.4** and **4.5** without mentioning their behaviour in case of an extension $K \to K'$ of the basic ring; indeed it is clear that they behave as expected. Since later we shall again use localizations in a systematic way, we just add the following evident statement.

(4.7.17) **Lemma.** *Let* $K \to K'$ *be a ring morphism,* q' *and* β' *the quadratic form and the bilinear form on* $K' \otimes M$ *derived from* q *and* β*. The algebra* $\mathrm{C}\ell_{K'}(K' \otimes M,\ q'; \beta')$ *is canonically isomorphic to* $K' \otimes \mathrm{C}\ell(M, q; \beta)$*.*

4.8 Applications of deformations

A Clifford algebra $\mathrm{C}\ell(M, q)$ is not at all convenient when the canonical mappings $K \to \mathrm{C}\ell_0(M, q)$ and $\rho : M \to \mathrm{C}\ell_1(M, q)$ are not both injective; whence the following definition.

(4.8.1) **Definition.** A quadratic module (M, q), or the quadratic form q itself, is said to be *cliffordian* if the canonical morphisms from K and M into $\mathrm{C}\ell(M, q)$ are both injective, and allow us to identify K and M with submodules of $\mathrm{C}\ell(M, q)$. It is said to be *strongly cliffordian* if moreover K is a direct summand of $\mathrm{C}\ell_0(M, q)$, and M a direct summand of $\mathrm{C}\ell_1(M, q)$.

When (M, q) is cliffordian, K and M are silently identified with their canonical images in $\mathrm{C}\ell(M, q)$ unless it is otherwise specified (for instance when M itself is an algebra already containing K as a subalgebra). Almost everywhere in the literature additional hypotheses ensure the quadratic modules to be strongly cliffordian, and let the definition (4.8.1) become useless; nevertheless an example of a non-cliffordian quadratic form has been given in (3.1.3).

(4.8.2) **Proposition.** *Let us suppose that (M, q) is not a cliffordian quadratic module; let K' be the image of K in $C\ell(M, q)$, and $M' = \rho(M)$ the image of M, considered as a module over K'. We get a cliffordian quadratic form q' on M' if we set $q'(a') = a'^2$ for all $a' \in M'$. Moreover the identity mapping of M' extends to an isomorphism $C\ell_{K'}(M', q') \to C\ell_K(M, q)$ of algebras over K or K'.*

Proof. It is clear that q' is a K'-quadratic form $M' \to K'$; since $C\ell(M, q)$ is also a K'-algebra, the universal property of $C\ell_{K'}(M', q')$ associates with $\mathrm{id}_{M'}$ an algebra morphism from $C\ell_{K'}(M', q')$ into $C\ell_K(M, q)$. This already proves that K' and M' are mapped injectively into $C\ell_{K'}(M', q')$. Conversely for all $a' = \rho(a) \in M'$ we can write $q'(a')1_{q'} = (q(a)1_q)1_{q'} = q(a)1_{q'}$ and consequently with $\rho : M \to M'$ the universal property of $C\ell_K(M, q)$ associates an algebra morphism from $C\ell_K(M, q)$ into $C\ell_{K'}(M', q')$. Obviously these algebra morphisms are reciprocal isomorphisms. □

Proposition (4.8.2) shows that the Clifford algebra of a non-cliffordian quadratic module (M, q) is also the Clifford algebra of a cliffordian one (M', q') canonically derived from it, and that the properties of (M, q) are observable in its Clifford algebra only as far as they are inherited by (M', q'). Up to now, nothing seems to be known about what is inherited and what is lost.

Here is an immediate consequence of the definition (4.8.1).

(4.8.3) **Proposition.** *Let $q : M \to K$ be a cliffordian quadratic form; for every $\lambda \in K$, λq is also a cliffordian quadratic form. When λ is invertible, and q strongly cliffordian, then λq too is strongly cliffordian.*

Proof. On the module $C\ell(M, q)$ we define the following multiplication:

$$\begin{aligned}(x, y) \longmapsto x * y &= xy \quad \text{if } x \text{ or } y \text{ is even,}\\ &= \lambda xy \text{ if } x \text{ and } y \text{ are odd.}\end{aligned}$$

It is easy to prove that this new multiplication is still associative, with the same unit element 1_q. Since $\rho(a) * \rho(a) = \lambda q(a)$ for all $a \in M$, the mapping ρ induces an algebra morphism g from $C\ell(M, \lambda q)$ into the new algebra $C\ell(M, q)$. Since K and M are mapped injectively into the module $C\ell(M, q)$, they are already mapped injectively into $C\ell(M, \lambda q)$.

When λ is invertible, the stronger conclusion follows from the bijectiveness of g. It is bijective because similarly there is an algebra morphism g' from $C\ell(M, q)$ into $C\ell(M, \lambda q)$ provided with a new multiplication such that $x * y = \lambda^{-1}xy$ when x and y are odd; thus gg' and $g'g$ are algebra endomorphisms of respectively $C\ell(M, q)$ and $C\ell(M, \lambda q)$. Since gg' and $g'g$ leave invariant the elements of M, they are the identity automorphisms. There is another proof using (3.8.7), because $(M, \lambda q)$

is isomorphic to the tensor product of (M,q) and the free discriminant module generated by an element d such that $d^2 = \lambda$. $\square$

The following statement is an immediate corollary of the five theorems at the beginning of **4.7**.

(4.8.4) **Proposition.** *Let q and q' be two quadratic forms on M; if there exists a bilinear form $\beta : M \times M \to K$ such that $q'(a) = q(a) + \beta(a,a)$ for all $a \in M$, then q' is cliffordian (resp. strongly cliffordian) if and only if q is cliffordian (resp. strongly cliffordian).*

Here is a less trivial application of the results of **4.7**.

(4.8.5) **Theorem.** *Let us suppose that M is the direct sum of the submodules M' and M''; let q be a quadratic form on M, and q' and q'' its restrictions to M' and M''; and let $f' : C\ell(M',q') \to C\ell(M,q)$ and $f'' : C\ell(M'',q'') \to C\ell(M,q)$ be the algebra morphisms derived from the canonical injections $M' \to M$ and $M'' \to M$.*

(a) *The following multiplication mapping is bijective:*

$$C\ell(M',q') \otimes C\ell(M'',q'') \longrightarrow C\ell(M,q)\ , \qquad x' \otimes x'' \longmapsto f'(x')\ f''(x'').$$

(b) *When q'' is strongly cliffordian, then f' is injective and allows us to identify $C\ell(M',q')$ with a subalgebra of $C\ell(M,q)$.*

(c) *When q' and q'' are both strongly cliffordian, then q too is strongly cliffordian.*

Proof. Let β be the bilinear form on M such that $\beta(a,b)$ vanishes whenever a belongs to M', and also whenever b belongs to M'', but is equal to $-\mathrm{b}_q(a,b)$ when a and b belong respectively to M'' and M'; in other words, β is the bilinear form such that $\beta(M',M) = \beta(M,M'') = 0$ and such that M' and M'' are orthogonal for the quadratic form $a \longmapsto q(a) + \beta(a,a)$. Therefore the isomorphism Φ_β can be identified with an isomorphism from $C\ell(M',q') \hat{\otimes} C\ell(M'',q'')$ onto $C\ell(M,q;\beta)$. Since the restrictions of β to M' and M'' vanish, f' and f'' are also algebra morphisms into $C\ell(M,q;\beta)$. Since $\beta(M',M'') = 0$, for all $x' \in M'$ and all $x'' \in M''$ the product of $f'(x')$ and $f''(x'')$ in $Cl(M,q;\beta)$ is equal to their product in $C\ell(M,q)$; indeed this can be proved with the help of the formula (4.7.3)(a) and by induction on k for every $x' \in C\ell^{\leq k}(M',q')$. Consequently Φ_β maps every $x' \otimes x''$ to $f'(x')f''(x'')$. This proves the statement (a).

When q'' is strongly cliffordian, K is a direct summand of $C\ell(M'',q'')$, and $C\ell(M',q')$ is isomorphic to a direct summand of $C\ell(M',q') \hat{\otimes} C\ell(M'',q'')$ by the mapping $x' \longmapsto x' \otimes 1_{q''}$; thus the injectiveness of f' follows from that of Φ_β. When q' and q'' are both strongly cliffordian, then $K \otimes K$ is a direct summand of $C\ell(M',q') \hat{\otimes} C\ell(M'',q'')$, and $M' \otimes K$ and $K \otimes M''$ too; consequently K and $M' \oplus M''$ are direct summands of $C\ell(M,q)$. $\square$

Admissible scalar products

The null quadratic form on M is obviously strongly cliffordian because $C\ell(M,0) = \bigwedge(M)$; this observation leads to the following definition.

(4.8.6) **Definition.** A bilinear form $\beta : M \times M \to K$ is called an *admissible scalar product* (or simply a *scalar product*) for the quadratic form $q : M \to K$ if $q(a) = \beta(a,a)$ for all $a \in M$.

When β is an admissible scalar product for q, then $\mathrm{b}_q = \beta - \beta^{to}$.

Here is an immediate consequence of several results of **4.7**, especially (4.7.11).

(4.8.7) **Theorem.** *When q admits a scalar product β, then q is strongly cliffordian and there is a comodule isomorphism $\Phi_{-\beta} : \bigwedge(M) \to C\ell(M,q)$ such that*

$$C\ell^{\leq k}(M,q) \;=\; C\ell^{\leq k-1}(M,q) \oplus \Phi_{-\beta}\left(\bigwedge^k(M)\right) \quad \textit{for all } k > 0\ .$$

Besides, the canonical morphism $\bigwedge(M) \to \mathrm{Gr}(C\ell(M,q))$ defined in (3.1.8) *is an isomorphism.*

Now we state sufficient conditions ensuring the existence of scalar products.

(4.8.8) **Theorem.** *Let (M,q) be a quadratic module.*

(a) *When M is a projective module, there are always admissible scalar products for q, and $C\ell(M,q)$ too is a projective module.*
(b) *When the mapping $a \longmapsto 2a$ is bijective from M onto M, there is a unique* symmetric *scalar product β admissible for q, which is defined by the equality $\beta(a,b) = 2\,\mathrm{b}_q(a/2,\, b/2)$; this symmetric scalar product is called the* canonical scalar product *derived from q.*

Proof. When M is projective, the existence of admissible scalar products has already been proved for a quite different purpose: see Lemma (2.5.3). It is already known that $\bigwedge(M)$ too is a projective module (see (3.2.6), and (3.2.7) if M is not finitely generated); because of the comodule isomorphism $\Phi_{-\beta} : \bigwedge(M) \to C\ell(M,q)$, the same is true for $C\ell(M,q)$.

When the mapping $a \longmapsto 2a$ is bijective, there is a reciprocal mapping $a \longmapsto a/2$. If β is an admissible *symmetric* scalar product, then $\mathrm{b}_q = 2\beta$ and consequently $2\,\mathrm{b}_q(a/2,\, b/2) = \beta(a,b)$. Conversely if β is defined by this equality, it is an admissible scalar product because $\mathrm{b}_q(a,a) = 2q(a)$ for all $a \in M$. □

When an admissible scalar product β has been chosen, the algebra $C\ell(M,q)$ is often replaced with $\bigwedge(M;\beta)$ which is then treated as a module provided with two multiplications: the Clifford multiplication $(x,y) \longmapsto xy$ and the exterior multiplication $(x,y) \longmapsto x \wedge y$. The equalities (a) and (b) in (4.7.3) are now written in this way (for all $a \in M$ and $x \in \bigwedge(M)$):

(4.8.9) $$ax \;=\; a \wedge x + \mathrm{d}_\beta(a) \rfloor x \quad \text{and} \quad xa \;=\; x \wedge a + \mathrm{d}_\beta^{to}(a) \rfloor \sigma(x)\ ;$$

here is more information about the relations between both multiplications.

(4.8.10) **Proposition.** *Let x and y be elements of $\bigwedge^j(M)$ and $\bigwedge^k(M)$ respectively, and xy their product in the Clifford algebra $\bigwedge(M;\beta)$. If $j \leq k$ (resp. $j \geq k$), xy belongs to*

$$\bigwedge^{j+k}(M)\oplus\bigwedge^{j+k-2}(M)\oplus\bigwedge^{j+k-4}(M)\oplus\cdots\oplus\bigwedge^{k-j}(M)\ \Big(\textit{resp.}\cdots\oplus\bigwedge^{j-k}(M)\Big).$$

Its component of degree $j+k$ is always $x\wedge y$. Its component of degree $|j-k|$ is

$$\bigwedge(\mathrm{d}_\beta)(x)\ \rfloor\, y \quad (\textit{resp.}\ (-1)^{\partial x \partial y}\ \bigwedge(\mathrm{d}_\beta^{to})(y)\ \rfloor\, x\ ).$$

Here the notation $\bigwedge(\mathrm{d}_\beta)$ means the algebra morphism $\bigwedge(M) \to \bigwedge(M^*)$ associated by the functor $\bigwedge$ with the mapping d_β; because of the canonical morphism $\bigwedge(M^*) \to \bigwedge^*(M)$, the interior multiplication by $\bigwedge(\mathrm{d}_\beta)(x)$ is meaningful. The strict observance of the twisting rule (4.2.1) gives immediately the correct sign for the component in $\bigwedge^{j-k}(M)$ when $j \geq k$; nevertheless we must remember that this rule has given $\bigwedge^*(M)$ a multiplication that for some people is that of $(\bigwedge^*(M))^o$.

Proof of (4.8.10). Everything is trivial when $j=0$ or $k=0$; and when $j=1$ or $k=1$, we have just to use the equalities (4.8.9), that here can be written in this way (for a and b in M):

$$ay \;=\; a\wedge y \;+\; \mathrm{d}_\beta(a)\rfloor y\,, \quad \text{and} \quad xb \;=\; x\wedge b \;+\; (-1)^{\partial x}\ \mathrm{d}_\beta^{to}(b)\rfloor x\,.$$

Then we proceed by induction. Let us treat for instance the case $j \geq k$ which requires an induction on k. Our induction hypothesis is that (4.8.10) is true for (j,k) and $(j,\ k-1)$, and we want to prove that it is true for $(j,\ k+1)$ if $j > k$. Consequently we replace (x,y) with $(x,\ y\wedge b)$:

$$\begin{aligned} x\ (y\wedge b) &= xyb \;-\; x\ (\mathrm{d}_\beta^{to}(b)\rfloor\sigma(y)) \\ &= (xy)\wedge b \;+\; \mathrm{d}_\beta^{to}(b)\rfloor\sigma(xy) \;-\; x\ (\mathrm{d}_\beta^{to}(b)\rfloor\sigma(y)) \\ &= (xy)\wedge b \;+\; (\mathrm{d}_\beta^{to}(b)\rfloor\sigma(x))\ \sigma(y); \end{aligned}$$

the component of $x(y\wedge b)$ in $\bigwedge^{j+k+1}(M)$ comes from the first term $(xy)\wedge b$ and is equal to $x\wedge y\wedge b$; its component in $\bigwedge^{j-k-1}(M)$ comes from the second term and is equal to

$$(-1)^{(1+\partial x)\partial y}\bigwedge(\mathrm{d}_\beta^{to})(\sigma(y))\rfloor\ (\mathrm{d}_\beta^{to}(b)\rfloor\sigma(x)) = (-1)^{\partial x(1+\partial y)}\bigwedge(\mathrm{d}_\beta^{to})(y\wedge b)\rfloor x.$$

□

Now we prove that the Clifford algebra of a free module is a free module.

(4.8.11) **Proposition.** *Let M be a free module with a basis $(e_j)_{j\in J}$ indexed by a totally ordered set J. The linear mapping $\Phi : \bigwedge(M) \to C\ell(M,q)$ such that*

$$\Phi(e_{j_1}\wedge e_{j_2}\wedge\cdots\wedge e_{j_k}) = e_{j_1}e_{j_2}\cdots e_{j_k} \quad \text{whenever} \quad j_1<j_2<\cdots<j_k ,$$

is an isomorphism of K-modules, and even an isomorphism of comodules over $\bigwedge(M)$.

Proof. Let β be the bilinear form on M defined in this way: $\beta(e_i,e_j)$ is equal to 0 when $i<j$, equal to $q(e_j)$ when $i=j$, and equal to $\mathrm{b}_q(e_i,e_j)$ when $i>j$. In $\bigwedge(M;\beta)$ the following equality can be proved by induction on k:

$$e_{j_1}e_{j_2}\cdots e_{j_k} = e_{j_1}\wedge e_{j_2}\wedge\cdots\wedge e_{j_k} \quad \text{whenever} \quad j_1<j_2<\cdots<j_k;$$

this proves that $\Phi=\Phi_\beta$. □

In the situation of (4.8.5) the identification of $C\ell(M',q')$ with a subalgebra of $C\ell(M,q)$ is legitimate whenever M is a projective module (and M' a direct summand of M); this result must now be improved.

(4.8.12) **Lemma.** *Let M be a finitely generated projective module, M' a direct summand of M, q a quadratic form on M, and q' its restriction to M'. Thus $C\ell(M',q')$ can be identified with a subalgebra of $C\ell(M,q)$. An element $x\in C\ell(M,q)$ belongs to $C\ell(M',q')$ if and only if $h\rfloor x=0$ for every linear form $h\in M^*$ such that $h(M')=0$. If q is nondegenerate, or more generally if d_q induces a surjective mapping $M'^\perp\to(M/M')^*$, this condition is equivalent to $\mathrm{d}_q(a)\rfloor x=0$ for every $a\in M'^\perp$.*

Proof. It is clear that $h\rfloor x=0$ if $x\in C\ell(M',q')$ and $h(M')=0$. Conversely let us suppose that $h\rfloor x=0$ whenever $h(M')=0$. By means of localizations we can reduce the problem to the case of free modules M and M'; let $(e_1,e_2,\dots,e_m)$ be a basis of M such that $(e_1,e_2,\dots,e_n)$ is a basis of M'; thus the products $e_{j_1}e_{j_2}\cdots e_{j_k}$ with $j_1<j_2<\cdots<j_k$ constitute a basis of $C\ell(M,q)$, and if we moreover require $j_k\le n$, we get a basis of $C\ell(M',q')$. Suppose that x does not belong to $C\ell(M',q')$; by writing x in the above basis of $C\ell(M,q)$, we would find some $k\in\{n{+}1,\ n{+}2,\dots,m\}$ such that $x=ye_k+z$ with some y and z both in the subalgebra generated by $(e_1,e_2,\dots,e_{k-1})$, and with $y\neq 0$; let h be the linear form such that $h(e_j)=0$ whenever $j\neq k$, but $h(e_k)=1$; now $h\rfloor x=(-1)^{\partial y}y\neq 0$; this contradicts the assumption that $h\rfloor x=0$ whenever $h(M')=0$.

If the mapping $M'^\perp\to(M/M')^*$ induced by d_q is surjective, the submodule of all $h\in M^*$ such that $h(M')=0$ is equal to the submodule of all $\mathrm{d}_q(a)$ with $a\in M'^\perp$. And from (2.3.7) we know that this mapping is bijective if q is nondegenerate. □

Canonical scalar products

Canonical scalar products, which exist when the mapping $a \longmapsto 2a$ is bijective (see (4.8.8)), have special properties that deserve a separate exposition. The bijectiveness of this mapping does not require 2 to be invertible in K; but when M is finitely generated, it implies that $M_\mathfrak{p} = 0$ for every prime ideal $\mathfrak{p}$ containing the image of 2 in K; indeed, because of Nakayama's lemma (1.12.1), the equality $M_\mathfrak{p} = 2M_\mathfrak{p}$ implies $M_\mathfrak{p} = 0$ if the image of 2 falls in $\mathfrak{p}$. The notation $\mathrm{b}_q/2$ means the canonical scalar product even if 2 is not invertible in K.

When β is a symmetric scalar product, from (4.7.14) we deduce that the Clifford algebra $\bigwedge(M;\beta)$ and the exterior algebra $\bigwedge(M)$ have the same reversion τ, which is described by (3.1.5). The equalities (4.8.9) now look like this:

$$ax = a \wedge x + \mathrm{d}_\beta(a) \rfloor x \quad \text{and} \quad xa = a \wedge \sigma(x) - \mathrm{d}_\beta(a) \rfloor \sigma(x) ;$$

they are equivalent to the following equalities (with $a \in M$), sometimes attributed to Riesz:

$$(4.8.13) \qquad 2\, a \wedge x = ax + \sigma(x)a \quad \text{and} \quad \mathrm{d}_q(a) \rfloor x = ax - \sigma(x)a.$$

Whereas the latter equality in (4.8.13) is the same thing as (4.4.12), the former equality is a new result; it leads to paying some attention to formulas of the following kind, in which E is some subset of the group $\mathcal{S}_n$ of permutations of $\{1, 2, \dots, n\}$:

$$(4.8.14) \qquad \mathrm{card}(E)\; a_1 \wedge a_2 \wedge \cdots \wedge a_n = \sum_{s \in E} \mathrm{sgn}(s)\; a_{s(1)} a_{s(2)} \cdots a_{s(n)} ;$$

indeed from (4.8.13) we can derive by induction on n the existence of a subset E of cardinal 2^{n-1} for which the equality (4.8.14) holds. Nevertheless when $n > 2$, it is well known that this equality already holds with a smaller subset, because of the following equality which is an easy consequence of (4.7.7):

$$2\, a_1 \wedge a_2 \wedge a_3 = a_1 a_2 a_3 - a_3 a_2 a_1 .$$

It is still an open question to know whether (4.8.14) holds with a subset E of cardinal 2^k if k is the greatest integer such that $2k \leq n$. For $n = 5$ the answer is already known since

$$4\, a_1 \wedge a_2 \wedge a_3 \wedge a_4 \wedge a_5$$
$$= a_1 a_2 a_3 a_4 a_5 + a_1 a_5 a_4 a_3 a_2 - a_2 a_5 a_4 a_3 a_1 - a_3 a_4 a_5 a_2 a_1 .$$

It is worth adding that (4.8.14) also holds when E is the whole group $\mathcal{S}_n$; often this assertion has been proved with the assumption of an orthogonal basis in M; for a quite general proof see (4.ex.14).

The bijection $\Phi_{-\beta}$ (with $\beta = \mathrm{b}_q/2$) allows us to carry onto $C\ell(M,q)$ the $\mathbb{N}$-grading of $\bigwedge(M)$; therefore we set $C\ell^n(M,q) = \Phi_{-\beta}(\bigwedge^n(M))$ for every $n \in \mathbb{N}$. The notation $C\ell^n(M,q)$ can also be understood as an abbreviation of $C\ell^n(M,q;-\beta)$ since the algebra $C\ell(M,q;-\beta)$ is canonically isomorphic to the $\mathbb{N}$-graded algebra $\bigwedge(M)$. The subspaces $C\ell^n(M,q)$ determine a grading of the module $C\ell(M,q)$ that is compatible with the natural filtration of the algebra $C\ell(M,q)$ (see (4.8.7)). Besides, every automorphism g of (M,q) determines an automorphism $C\ell(g)$ of the algebra $C\ell(M,q)$ which leaves every subspace $C\ell^n(M,q)$ invariant; indeed (4.8.13) implies that $C\ell(g)$ is also an automorphism of the exterior algebra $C\ell(M,q;-\beta)$.

The usefulness of these subspaces $C\ell^n(M,q)$ appears for instance in the research of the quadratic extension $\mathrm{QZ}(M,q)$ mentioned in (3.7.6).

(4.8.15) **Proposition.** *When 2 is invertible in K and (M,q) is a quadratic space of constant rank r, then $\mathrm{QZ}(M,q)$ is equal to $C\ell^0(M,q)\oplus C\ell^r(M,q)$, and its discriminant module is $C\ell^r(M,q)$. Besides, for $k=0,1,2,\dots,r$, the multiplication mapping π_q induces an isomorphism*

$$C\ell^r(M,q)\otimes C\ell^k(M,q)\longrightarrow C\ell^{r-k}(M,q)\ ,\quad z\otimes x\longmapsto zx\ =(-1)^{k(r-1)}xz.$$

Proof. We know that $\mathrm{QZ}(M,q)$ is the direct sum of $K=C\ell^0(M,q)$ and its discriminant module, which is the submodule of all $z\in\mathrm{QZ}(M,q)$ mapped to $-z$ by its standard involution φ. Because of (3.5.13), this means that $az=-\sigma(z)a$ for all $a\in M$; because of (4.8.13), this is equivalent to the equality $a\wedge z=0$ which characterizes the elements z of $\bigwedge^r(M)$ in $\bigwedge(M)$ (see (3.2.6)), and the elements of $C\ell^r(M,q)$ in $C\ell(M,q)$.

To prove the last assertion of (4.8.15), we can suppose that M admits an orthogonal basis $(e_1,e_2,\dots,e_r)$ such that $q(e_1)$, $q(e_2)$,…,$q(e_r)$ are all invertible, since localizations allow us to reduce the general case to that one. For every subset F of $B=\{1,2,\dots,r\}$, we denote by e_F the product of all elements e_j such that $j\in F$, when the order of the factors is the order of their indices; their product in $C\ell(M,q)$ is also their exterior product in $C\ell(M,q;-\beta)$ since every $\beta(e_i,e_j)$ vanishes if $i\neq j$. Moreover $C\ell^r(M,q)$ is the submodule generated by e_B. Now it suffices to observe that

$$e_Be_F\ =\ \pm\,(\prod_{j\in F}\ q(e_j))\ e_{B\setminus F}\ ,$$

and that the products e_F (resp. $e_{B\setminus F}$), with F a subset of cardinal k, constitute a basis of $C\ell^k(M,q)$ (resp. $C\ell^{r-k}(M,q)$). □

For all $x\in C\ell(M,q)$, the parallel projection of x in $C\ell^0(M,q)=K$ with respect to $C\ell^{>0}(M,q)$ is called the *scalar component* of x and denoted by $\mathrm{Scal}(x)$.

(4.8.16) **Proposition.** *The bilinear form $(x,y)\longmapsto\mathrm{Scal}(xy)$ is symmetric and the submodules $C\ell^n(M,q)$ are pairwise orthogonal for it. It is nondegenerate whenever (M,q) is a quadratic space.*

Proof. Let x and y be elements of $C\ell^j(M,q)$ and $C\ell^k(M,q)$ respectively; from (4.8.10) we deduce that $\mathrm{Scal}(xy)=0$ whenever $j\neq k$. Consequently we only have to compare Scal(xy) and Scal(yx) when $j = k$. Since the formulas (3.1.5) are also valid for the reversion τ of $C\ell(M,q)$, we can write (when $j = k$)

$$\mathrm{Scal}(xy) = \mathrm{Scal}(\tau(xy)) = \mathrm{Scal}(\tau(y)\tau(x)) = \mathrm{Scal}(yx).$$

When (M,q) is a quadratic space, we can suppose that (M,q) admits an orthogonal basis $(e_1, e_2, \dots, e_r)$ as in the last part of the proof of (4.8.15); then the products e_F defined above constitute a basis of $C\ell(M,q)$. Since $e_1,\dots,e_r$ are pairwise anticommuting, it is easy to prove that it is an orthogonal basis of $(C\ell(M,q),\ \mathrm{Scal})$. The nondegeneracy of Scal follows from the fact that $(e_F)^2$ is always an invertible element of K, since it is $\pm\prod_{j\in F} q(e_j)$. □

When M is a finitely generated projective module, more precise properties of the linear form $\mathrm{Scal} : C\ell(M,q) \to K$ are stated in (4.8.17); they imply that it is invariant by all automorphisms of $C\ell(M,q)$, and not only by the automorphisms $C\ell(g)$ derived from an automorphism g of (M,q). The trace of an endomorphism of a finitely generated projective module has been defined by elementary means at the beginning of **3.6**; for instance if M is a finitely generated projective module, if its rank takes the values $r_1, r_2, \dots, r_k$ and if $e_1, e_2, \dots, e_k$ are the corresponding idempotents of K (see (1.12.8)), than the trace of the identity mapping of $C\ell(M,q)$ is

$$\mathrm{tr}(\mathrm{id}_q) = 2^{r_1}e_1 + 2^{r_2}e_2 + \dots + 2^{r_n}e_n \ ;$$

the bijectiveness of the mapping $a \longmapsto 2a$ implies that $r_i = 0$ whenever $2e_i$ is not invertible in Ke_i; therefore $\mathrm{tr}(\mathrm{id}_q)$ is invertible in K.

(4.8.17) **Proposition.** *When M is a finitely generated projective module (such that the mapping $a \longmapsto 2a$ is bijective), for all $x \in C\ell(M,q)$ the traces of the multiplications $y \longmapsto xy$ and $y \longmapsto yx$ are both equal to* $\mathrm{tr}(\mathrm{id}_q)\,\mathrm{Scal}(x)$.

Proof. This is obviously true when x belongs to $K = C\ell^0(M,q)$; therefore it suffices to prove that the traces of the multiplications by x both vanish when x belongs to $C\ell^{>0}(M,q)$. This is clear when x is odd, because these multiplications permute $C\ell_0(M,q)$ and $C\ell_1(M,q)$. When x is even, it is a sum of elements like $2a \wedge z$ with $a \in M$ and $z \in C\ell_1(M,q)$; because of (4.8.13), $2a \wedge z$ is a Lie bracket $az - za$ (in the ordinary nongraded sense); the multiplication by a Lie bracket (on either side) is a Lie bracket of multiplications, and the trace of a Lie bracket of endomorphisms is always 0. □

Comment. The invertibility of $\mathrm{tr}(\mathrm{id}_q)$ ensures the interest of the property of Scal(x) stated in (4.8.17), and allows us to compare it with the reduced trace tr(x) when $C\ell(M,q)$ is a graded Azumaya algebra. Reduced traces are defined in (3.6.6) and (3.6.7); when A is a graded Azumaya algebra of constant rank n^2 or $2n^2$, the traces of the multiplications by an element $x \in A$ are both equal to $n\,\mathrm{tr}(x)$. Therefore if (M,q) is a quadratic space of constant rank $2k$ or $2k-1$, the equality $\mathrm{tr}(x) = 2^k\mathrm{Scal}(x)$ holds for all $x \in C\ell(M,q)$.

Exercises

(4.ex.1) Let A be an algebra (associative with unit 1_A); suppose that the underlying K-module A is graded over an additive group G (therefore $A = \bigoplus_{j\in G} A_j$) in such a way that $A_iA_j \subset A_{i+j}$ for all $(i,j) \in G^2$. Prove that 1_A belongs to A_0 (and consequently A is a graded algebra).

Give a counterexample when G is merely an additive monoid (for instance $G = \{\text{"zero"}, \text{"positive"}\}$).

(4.ex.2)* Let M be a module. Since $\mathrm{S}(M)$ is a coalgebra, the dual module $\mathrm{S}^*(M)$ is an algebra; it is the direct sum of $\mathrm{S}^{*0}(M)$ which is isomorphic to K, and the ideal $\mathrm{S}^{*+}(M)$ of all linear forms vanishing on $\mathrm{S}^0(M) = K$.

(a) Define an interior multiplication $\mathrm{S}^*(M) \times \mathrm{S}(M) \to \mathrm{S}^*(M)$ that makes $\mathrm{S}^*(M)$ become a module over $\mathrm{S}(M)$, and give its elementary properties. The notation $f \rfloor x$ is still suitable, but the twisting rule (4.2.1) is not relevant for symmetric algebras.

(b) Prove that there exists a unique mapping Exp from $\mathrm{S}^{*+}(M)$ into $\mathrm{S}^*(M)$ such that the following equalities hold for all $f \in \mathrm{S}^{*+}(M)$ and all $a \in M$:

$$\mathrm{Exp}(f)(1) = 1 \quad \text{and} \quad \mathrm{Exp}(f) \rfloor a \;=\; \mathrm{Exp}(f) \vee (f \rfloor a) \ .$$

(c) Let $\mathrm{ST}^k(M^*)$ be the submodule of symmetric tensors in $\mathrm{T}^k(M^*)$. Define a canonical mapping $\mathrm{ST}^k(M^*) \to \mathrm{S}^{*k}(M)$ and prove that it is bijective when M is a finitely generated projective module.

(4.ex.3) Consider the exterior algebra $\bigwedge(M)$, and prove that the four mappings π, ε, π', ε' (defined in **4.3**) let $\bigwedge(M)$ become a bialgebra according to the definition in **4.1**, provided that this definition is adapted to the twisting rule (4.2.1). Prove that the automorphism σ (that is $x \longmapsto (-1)^{\partial x}x$) is the inverse of $\mathrm{id}_\wedge$ in the algebra $\mathrm{Hom}^\wedge(\bigwedge(M), \bigwedge(M))$ defined by (4.2.4):

$$\pi \circ (\mathrm{id}_\wedge \otimes \sigma) \circ \pi' \;=\; \varepsilon \circ \varepsilon' \;=\; \pi \circ (\sigma \otimes \mathrm{id}_\wedge) \circ \pi' \ .$$

Such an inverse of the identity mapping is called an *antipode*, and a bialgebra with an antipode is called a *Hopf algebra*.

(4.ex.4) Let (M,q) be a quadratic module; is it possible to make the exterior powers $\bigwedge^k(M)$ become quadratic modules in a natural way?

(a) First suppose that 2 is invertible in K. By means of the algebra morphism $\bigwedge(\mathrm{d}_q) : \bigwedge(M) \longrightarrow \bigwedge(M^*)$, prove the existence of a unique quadratic form $\hat{q}$ on $\bigwedge(M)$ such that the submodules $\bigwedge^k(M)$ are pairwise orthogonal, and such that this equality holds for every sequence $(a_1, b_1, a_2, b_2, \dots, a_k, b_k)$ of elements of M :

$$\mathrm{b}_{\hat{q}}(a_1 \wedge a_2 \wedge \cdots \wedge a_k \ , \ b_k \wedge \cdots b_2 \wedge b_1) \;=\; \det(\mathrm{b}_q(a_i, b_j))_{1\le i,j\le k}.$$

Prove that $(\bigwedge(M),\ \hat{q})$ is a quadratic space whenever (M,q) is one.

(b)* Without any hypothesis on K, deduce from the concept of "half-determinant" (see (2.ex.13)) that $\bigwedge^k(M)$ is still a quadratic module at least for every *odd* exponent k.

Hint. First consider free modules.

(4.ex.5) Assume that q and q' are strongly cliffordian quadratic forms on the module M, according to Definition (4.8.1); prove that $q+q'$ is still strongly cliffordian.

Hint. Use the morphism $a \longmapsto (a,a)$ from $(M,\, q+q')$ into $(M,q) \perp (M,q')$; observe that $M\oplus M$ is the direct sum of $M\oplus 0$ and the image Δ of this morphism; deduce from (4.8.5) that the subalgebra generated by Δ in $C\ell((M,q)\perp(M,q'))$ is isomorphic to $C\ell(M,\ q+q')$.

(4.ex.6) Let (M,q) be a quadratic module such that M is a finitely generated projective module of nonzero constant rank r. Because of (4.8.7) there is a surjective mapping $p : C\ell(M,q) \to \bigwedge^r(M)$ with kernel $C\ell^{<r}(M,q)$. We also consider the dual module $C\ell^*(M,q) = \mathrm{Hom}(C\ell(M,q),K)$ and its parity grading: for $i=0,1$, $C\ell_i^*(M,q)$ is the submodule of all linear forms vanishing on $C\ell_{1-i}(M,q)$.

(a) For every $\omega^* \in \bigwedge^{*r}(M)$, let F_{ω^*} be the linear form on $C\ell(M,q)$ defined by $F_{\omega^*}(x) = \omega^*(p(x))$. Prove that the linear forms F_{ω^*} make up a direct summand of $C\ell^*(M,q)$ of constant rank 1, that they have the same parity as r, and deduce from (3.2.1) the equality

$$\forall x,\ y\in C\ell(M,q),\quad F_{\omega^*}(xy) \;=\; (-1)^{\partial x \partial y}\, F_{\omega^*}(yx)\ .$$

(b) Suppose that (M,q) is a quadratic space, and consider the discriminant module D of $\mathrm{QZ}(M,q)$ (the centralizer of $C\ell_0(M,q)$ in $C\ell(M,q)$). With every $w \in \bigwedge^{*r}(M)\otimes D$ we associate a linear form G_w on $C\ell(M,q)$ in this way:

$$\forall \omega^* \in \bigwedge^{*r}(M)\ ,\ \forall d\in D\ ,\ \forall x\in C\ell(M,q)\ ,\quad G_{\omega^*\otimes d}(x) \;=\; \omega^*(p(dx)).$$

Prove that the linear forms G_w make up a direct summand of $C\ell^*(M,q)$ of constant rank 1, that they are all even, and satisfy the equality

$$\forall x,\ y\in C\ell(M,q),\quad G_w(xy) \;=\; G_w(yx)\ .$$

Comment. When A is a graded Azumaya algebra, in (6.ex.11) it is proved that the submodule of all $h\in A_0^*$ such that $h(xy)=h(yx)$ for all $x,\ y\in A$, is a free direct summand of constant rank 1, and that it is generated by the reduced trace $x\longmapsto \mathrm{tr}(x)$ defined in **3.6**; consequently when $A = C\ell(M,q)$, it is the submodule just found above; since it is free, $D\otimes\bigwedge^r(M)$ too is free, whence $D\cong\bigwedge^r(M)$.

(c) Suppose that (M,q) is a quadratic space such that the mapping $a\longmapsto 2a$ is bijective from M onto M; in this case D is equal to $C\ell^r(M,q)$ (see (4.8.15)), which p maps bijectively onto $\bigwedge^r(M)$. Verify that

$$G_{\omega^*\otimes d}(x) \;=\; \omega^*(p(d))\ \mathrm{Scal}(x)\ .$$

Scalar products β and algebras $\bigwedge(M;\beta)$

(4.ex.7) Let (M,q) be a quadratic module with M a finitely generated projective module, and β an admissible scalar product for q (see (4.8.6)). Here we shall prove the existence of an associative multiplication on $\bigwedge(M)$ admitting 1 as a unit element and satisfying the conditions (4.8.9), without the help of (4.7.1) and the subsequent theorems.

(a) From the universal properties of $C\ell(M,q)$ and $C\ell(M,-q)$ deduce the existence of an algebra morphism Ψ from $C\ell(M,q)\,\hat{\otimes}\,C\ell(M,q)^{to}$ into $\mathrm{End}(\bigwedge(M))$ such that, for all $a\in M$ and all $x\in\bigwedge(M)$,

$$\Psi(\rho(a)\otimes 1_q^{to})(x) = a\wedge x + \mathrm{d}_\beta(a)\rfloor x$$
$$\text{and}\qquad \Psi(1_q\otimes\rho(a)^{to})(x) = a\wedge x + \mathrm{d}_\beta^{to}(a)\rfloor x.$$

(b) For every $z\in C\ell(M,q)$ we set $f(z)=\Psi(z\otimes 1_q^{to})(1)$ and $g(z)=\Psi(1_q\otimes z^{to})(1)$; prove that f and g are bijections from $C\ell(M,q)$ onto $\bigwedge(M)$ such that $f(1_q)=g(1_q)=1$ and $f(\rho(a))=g(\rho(a))=a$ for all $a\in M$.
Hint. Localizations, (3.1.7) and perhaps (3.ex.4).

(c) Let $\bigwedge(M;\beta)$ be the module $\bigwedge(M)$ provided with the following multiplication:

$$\begin{aligned}(x,y)\ \longmapsto\ xy &= \Psi(f^{-1}(x)\otimes g^{-1}(y)^{to})(1)\\ &= \Psi(f^{-1}(x)\otimes 1_q^{to})(y) \ = \ (-1)^{\partial x\partial y}\ \Psi(1_q\otimes g^{-1}(y)^{to})(x)\ .\end{aligned}$$

Verify that $xy=f(f^{-1}(x)\,f^{-1}(y))=g(g^{-1}(x)\,g^{-1}(y))$. Prove that $\bigwedge(M;\beta)$ is an associative algebra with unit element 1, in which both equalities (4.8.9) are valid. Moreover $f=g$.
Comment. This construction of $\bigwedge(M;\beta)$ comes from [Chevalley 1954]; there is no doubt that Chevalley knew both equalities (4.8.9); but he thought (in accordance with (4.7.4)) that the first one was sufficient to characterize the multiplication in $\bigwedge(M;\beta)$, and consequently he only defined the algebra morphism $z\longmapsto\Psi(z\otimes 1_q^{to})$ from $C\ell(M,q)$ into $\mathrm{End}(\bigwedge(M))$.

(d) Verify (without (4.7.5)) that the interior multiplication by any $h\in M^*$ is also a twisted derivation of $\bigwedge(M;\beta)$; it suffices to verify that $h\rfloor(ax)=h(a)x-a(h\rfloor x)$ for all $a\in M$.

(4.ex.8) Let (M,q) be a quadratic module, β an admissible scalar product for q, and $\bigwedge(M;\beta)$ the derived algebra. When x and y are elements of respectively $\bigwedge^j(M)$ and $\bigwedge^k(M)$, by definition the interior product $x\rfloor y$ (resp. $x\lfloor y$) is the component of the Clifford product xy in $\bigwedge^{k-j}(M)$ (resp. $\bigwedge^{j-k}(M)$). Consequently $x\rfloor y$ vanishes whenever $j>k$, whereas $x\lfloor y$ vanishes whenever $j<k$ (see (4.8.10)). Moreover $x\rfloor y$ and $x\lfloor y$ are the same element of K when $j=k$; for instance $a\rfloor b=a\lfloor b=\beta(a,b)$ for all a and $b\in M$. By bilinearity the interior products $x\rfloor y$ and $x\lfloor y$ are defined for all x and y in $\bigwedge(M)$.

(a) Prove these equalities for all x, y, z in $\bigwedge(M)$:

$$(x \wedge y) \rfloor z \;=\; x \rfloor (y \rfloor z) \quad \text{and} \quad (x \lfloor y) \lfloor z \;=\; x \lfloor (y \wedge z)\ .$$

(b) Prove these equalities for all $x,\ y \in \bigwedge(M)$ and all $a \in M$:

$$\begin{aligned} a \rfloor (x \wedge y) &= (a \rfloor x) \wedge y + \sigma(x) \wedge (a \rfloor y)\ , \\ a \rfloor (xy) &= (a \rfloor x)\ y + \sigma(x)\ (a \rfloor y)\ , \\ (x \wedge y) \lfloor a &= \ (x \lfloor a)\sigma(y) + x \wedge (y \lfloor a)\ , \\ (xy) \lfloor a &= \ (x \lfloor a)\ \sigma(y) + x\ (y \lfloor a)\ . \end{aligned}$$

(c) Suppose that β is symmetric and that x and y are homogeneous for the parity grading of $\bigwedge(M)$; prove that $y \lfloor x \;=\; (-1)^{\partial x(1+\partial y)}\ x \rfloor y$.

Comment. Such a concept of interior multiplication is only advisable in elementary presentations of Clifford algebras, when K is a field of characteristic $\neq 2$, and β is nondegenerate and symmetric; thus the algebra $\bigwedge(M;\beta)$ can be constructed in an elementary way (without quotient of $\mathrm{T}(M)$) by means of an orthogonal basis of M. For many applications of Clifford algebras, this may be sufficient.

(4.ex.9) Let (M,q) be a quadratic module, β an admissible scalar product for q, and $\bigwedge(M;\beta)$ the derived algebra; besides, let g be an automorphism of (M,q). The functors $\bigwedge$ and $C\ell$ associate with g an automorphism $\bigwedge(g)$ of $\bigwedge(M)$ and an automorphism $C\ell(g)$ of $C\ell(M,q)$. If we replace $C\ell(M,q)$ with $\bigwedge(M;\beta)$, we get an automorphism of $\bigwedge(M;\beta)$ also denoted by $C\ell(g)$; here we are interested in a comparison between $\bigwedge(g)$ and $C\ell(g)$, both considered as linear automorphisms of the module $\bigwedge(M)$.

(a) Prove the existence of $\delta \in \bigwedge^{*2}(M)$ such that

$$\forall a,\ b \in M, \qquad \delta(a \wedge b) \;=\; \beta(g(a), g(b)) - \beta(a,b).$$

(b) Prove the following equalities, for all $x \in \bigwedge(M)$:

$$C\ell(g)(x) \;=\; \bigwedge(g)(\,\mathrm{Exp}(\delta)\ \rfloor\ x\,) \;=\; \mathrm{Exp}\left(\bigwedge^{*}(g^{-1})(\delta)\right)\ \rfloor\ \bigwedge(g)(x).$$

Hint. Let us set $\theta(x) = \bigwedge(g)(\mathrm{Exp}(\delta) \rfloor x)$; the main difficulty is to prove that $\theta(xy) = \theta(x)\theta(y)$; here is the beginning of the calculations:

$$\begin{aligned} \theta(xy) &= \bigwedge(g)(\ \mathrm{Exp}(\delta)\ \rfloor\ \pi(\mathrm{Exp}(\beta_{\prime\prime}) \rfloor (x \otimes y))) \\ &= \bigwedge(g) \circ \pi(\ \mathrm{Exp}(\pi^*(\delta) + \beta_{\prime\prime})\ \rfloor\ (x \otimes y))\ ; \\ \theta(x)\ \theta(y) &= \pi\left(\ \mathrm{Exp}(\beta_{\prime\prime})\ \rfloor\ \bigwedge(g,g)((\mathrm{Exp}(\delta) \rfloor x) \otimes (\mathrm{Exp}(\delta) \rfloor y))\right) \\ &= \bigwedge(g) \circ \pi\left(\ \mathrm{Exp}\left(\bigwedge^{*}(g,g)(\beta_{\prime\prime}) + \delta \otimes 1 + 1 \otimes \delta\right)\ \rfloor\ (x \otimes y)\right)\ . \end{aligned}$$

(c) Suppose that 2 is invertible in K, and prove that the above equalities are equivalent to this one, in which $[\beta]$ is defined as in **4.7**:

$$\mathrm{C}\ell(g)(x) \;=\; \mathrm{Exp}\left(\frac{1}{2}[\beta]\right) \rfloor \bigwedge(g)\left(\mathrm{Exp}\left(\frac{-1}{2}[\beta]\right) \rfloor x\right).$$

(4.ex.10)* With every bilinear form β on the module M is associated an algebra $\bigwedge(M;\beta)$ that is isomorphic to the Clifford algebra of the quadratic form $a \longmapsto \beta(a,a)$. In a dual way some people have associated a "Clifford coalgebra" with any element γ of $M \otimes M$. Let $\gamma_{//}$ be the natural image of γ in $\bigwedge(M) \hat{\otimes} \bigwedge(M)$; the comultiplication $\pi'_\gamma : \bigwedge(M) \to \bigwedge(M) \otimes \bigwedge(M)$ is defined by

$$\pi'_\gamma(x) \;=\; \mathrm{Exp}(\gamma_{//}) \wedge \pi'(x)\ .$$

Prove that π'_γ and ε' (that is the projection $\bigwedge(M) \longrightarrow \bigwedge^0(M) = K$) make $\bigwedge(M)$ become a coalgebra.

Now suppose that M is a finitely generated projective module, so that the algebras $\bigwedge(M^*)$ and $\bigwedge^*(M)$ can be identified; since γ induces a bilinear form on M^*, a deformation $\bigwedge(M^*;\gamma)$ can be defined; let $\pi_\gamma : \bigwedge(M^*) \otimes \bigwedge(M^*) \to \bigwedge(M^*)$ be the corresponding multiplication mapping. Prove the following equality for all $x \in \bigwedge(M)$ and all f and $g \in \bigwedge(M^*)$:

$$(\pi_\gamma(f \otimes g))(x) \;=\; (f \hat{\otimes} g)(\pi'_\gamma(x))\ .$$

Canonical scalar products

(4.ex.11) Let (M,q) be a quadratic module with M a finitely generated module of rank ≤ 4 at every prime ideal, and such that the mapping $a \longmapsto 2a$ is bijective from M onto M.

(a) Prove that $\mathrm{C}\ell^0(M,q) \oplus \mathrm{C}\ell^4(M,q)$ is a subalgebra contained in the center of $\mathrm{C}\ell_0(M,q)$, and that it is the submodule of all $x \in \mathrm{C}\ell_0(M,q)$ such that $\tau(x) = x$.
(b) Prove that M is the submodule of all $x \in \mathrm{C}\ell_1(M,q)$ such that $\tau(x) = x$. This has been proved in (3.ex.19) with less hypotheses but with more difficulty.
(c) When (M,q) is a quadratic space of constant rank 4, prove that $\mathrm{C}\ell^2(M,q)$ is a projective module of constant rank 3 over $\mathrm{QZ}(M,q) = \mathrm{C}\ell^0(M,q) \oplus \mathrm{C}\ell^4(M,q)$ (see (4.8.15)).

(4.ex.12) Let (M,q) be again a quadratic module with M a finitely generated module of rank ≤ 4, and such that the mapping $a \longmapsto 2a$ is bijective from M onto M. The main purpose of this exercise is to prove that the square of every element of $\mathrm{C}\ell^3(M,q)$ or $\mathrm{C}\ell^4(M,q)$ belongs to $K = \mathrm{C}\ell^0(M,q)$.

(a) Prove this when (M,q) is a quadratic space of constant rank 4, by means of (4.8.15) or (2.6.2).

In the following parts, where (M,q) is merely a quadratic module of rank ≤ 4, the proofs are not so simple, and $C\ell(M,q)$ is replaced with the algebra $\bigwedge(M;\beta)$ (where $\beta = \mathrm{b}_q/2$) provided with a Clifford multiplication and an exterior one.

(b) Take y and y' in $\bigwedge^3(M)$ and prove that the Clifford product yy' has no component in $\bigwedge^4(M)$; consequently $yy' + y'y$ belongs to K. Besides, when h is an element of M^*, verify that $(h\rfloor y)\,(h\rfloor y')$ has no component in $\bigwedge^4(M)$.
Hint. After localization, you can suppose that M is generated by (a,b,c,d), and that $y = a\wedge b\wedge c$ and $y' = a\wedge b\wedge d$; deduce from (4.7.7) that

$$a\wedge b\wedge c = -cba + \beta(b,c)a - \beta(a,c)b + \beta(a,b)c\ ,$$
$$a\wedge b\wedge d = abd - \beta(b,d)a + \beta(a,d)b - \beta(a,b)d\ ,$$

and remember (4.8.10); finally observe that $yy' + y'y$ is invariant by τ. A direct calculation proves that

$$(h\rfloor(a\wedge b\wedge c))\wedge(h\rfloor(a\wedge b\wedge d)) \;=\; 0.$$

(c) Prove that zz' belongs to K for all z and z' in $\bigwedge^4(M)$.
Hint. After localization, you can suppose that $z = z' = a\wedge y$ for some $a\in M$ and some $y\in\bigwedge^3(M)$; from (b) above you know that $y^2\in K$; set $h = \mathrm{d}_q(a)$ and use (4.8.13) in this way:

$$4\,(a\wedge y)^2 \;=\; (ay - ya)^2 \;=\; h\rfloor(yay)\; - ay^2a - ya^2y$$
$$\text{with}\quad yay \;=\; (h\rfloor y)\,y\; - ay^2\ .$$

For another point of view, see (4.ex.16).

(4.ex.13) Let (M,q) be a quadratic space of constant rank r, such that the mapping $a\longmapsto 2a$ is bijective from M onto M, and let N be a direct summand of M of constant rank s. The restrictions of q to N and $N^\perp$ are denoted by q' and q'', and $C\ell(N,q')$ and $C\ell(N^\perp,q'')$ are treated as subalgebras of $C\ell(M,q)$. Prove that the multiplication

$$C\ell^r(M,q)\otimes C\ell^s(M,q)\longrightarrow C\ell^{r-s}(M,q) \qquad \text{(see (4.8.15))}$$

induces a bijection

$$C\ell^r(M,q)\otimes C\ell^s(N,q')\longrightarrow C\ell^{r-s}(N^\perp,q'').$$

Hint. By means of (4.8.13) calculate $2b\wedge(zx)$ when b, z and x belong respectively to $N^\perp$, $C\ell^r(M,q)$ and $C\ell^s(N,q')$; you must discover that $b\wedge(zx) = 0$; this implies that zx lies in $C\ell^{r-s}(N^\perp,q'')$.

(4.ex.14)* We suppose that 2 is invertible in K and we look for a method allowing us to prove formulas like (4.8.14) for suitable subsets E of the group $\mathcal{S}_n$ of permutations of $\{1,2,\dots,n\}$. Such formulas involve algebras $\bigwedge(M;\beta)$ with $\beta = \mathrm{b}_q/2$, which are provided with exterior and Clifford multiplications. First, for any subset E, we define the following n-multilinear mapping P_E from M^n into $\bigwedge(M;\beta)$:

$$P_E(a_1, a_2, \dots, a_n) = \sum_{s\in E} \operatorname{sgn}(s)\ a_{s(1)}a_{s(2)}\cdots a_{s(n)}\ .$$

Besides let Π_n be the set of all sets $\{B_1, B_2, \dots, B_k\}$ such that $0 \le 2k \le n$ and B_1, B_2, ..., B_k are pairwise disjoint subsets of $\{1, 2, \dots, n\}$ all of cardinal 2. When $k > 0$, they constitute a partition of a subset of even cardinal $2k$. With every $\varpi \in \Pi_n$ we associate another n-multilinear mapping P_ϖ defined in this way: $P_\varpi(a_1, a_2, \dots, a_n)$ is the exterior product of all elements a_i such that i does not belong to $B_1 \cup B_2 \cup \dots \cup B_k$, still multiplied by all $\beta(a_i, a_j)$ such that $\{i, j\}$ is one of the sets B_1, B_2, ..., B_k; of course the order of the factors in the exterior product is the order of their indices.

(a) Explain that there exist integers N_ϖ independent of K, M and q, which allow you to write

$$P_E(a_1, a_2, \dots, a_n) = \sum_{\varpi} N_\varpi\ P_\varpi(a_1, a_2, \dots, a_n).$$

Moreover $N_\varpi = \operatorname{card}(E)$ when ϖ is the element of Π_n such that $k = 0$.

(b) Suppose that for all quadratic modules (M, q) over the field $\mathbb{Q}$ the n-linear mapping P_E vanishes whenever the variables a_i and a_j (with $i \neq j$) are equal. Prove that $N_\varpi = 0$ if i or j or both belong to $B_1 \cup B_2 \cup \dots \cup B_k$. What happens when P_E is always an alternate n-multilinear mapping?

(c) *Example.* Prove that the 5-linear mapping

$$(a_1, a_2, a_3, a_4, a_5) \longmapsto a_1a_2a_3a_4a_5 + a_1a_5a_4a_3a_2 - a_2a_5a_4a_3a_1 - a_3a_4a_5a_2a_1$$

is always alternate.

(4.ex.15) We suppose that (M, q) is a quadratic module such that the mapping $a \longmapsto 2a$ is bijective from M onto M. Let J be the kernel of the symmetric bilinear form $(x, y) \longmapsto \operatorname{Scal}(xy)$ mentioned in (4.8.16). Prove that J is an ideal of $C\ell(M, q)$, that it is the direct sum of all the intersections $J \cap C\ell^k(M, q)$, and that $J \cap C\ell^0(M, q) = 0$.

Comment. When (M, q) is a quadratic space, $C\ell(M, q)$ is a graded Azumaya algebra, and Proposition (6.7.4) implies $J = 0$ when J is a graded ideal of $C\ell(M, q)$ such that $J \cap K = 0$; the equality $J = 0$ only means that the bilinear form $(x, y) \longmapsto \operatorname{Scal}(xy)$ is weakly nondegenerate; compare with (4.8.16).

(4.ex.16) Let (M, q) be a quadratic module such that the mapping $a \longmapsto 2a$ is bijective from M onto M. We suppose that M is a finitely generated module, and consequently there exists an integer r such that the rank of M at every prime ideal is $\le r$. Let x and y be elements of $C\ell^j(M, q)$ and $C\ell^k(M, q)$ respectively; for every integer m between 0 and r, let $\Gamma_m(x, y)$ be the component of xy in $C\ell^m(M, q)$.

Prove that $\Gamma_m(y,x)$ vanishes when these two conditions are not both satisfied: first $|j-k| \leq m \leq \inf(j+k,\ 2r-j-k)$, secondly the parity of m must be that of $j+k$. Moreover, the number of values of m satisfying both conditions is $1+\inf(j,\ k,\ r-j,\ r-k)$.

Hint. The first result follows from $yx = \tau(\tau(x)\tau(y))$; then the inequalities $|j-k| \leq m \leq j+k$ follow from (4.8.10); but the inequality $m \leq 2r-j-k$ (only valid for a symmetric β) requires more work. By localizations reduce the problem to the case of a module M generated by r elements $a_1, a_2, \dots, a_r$; you can suppose that x and y are exterior products of some of these r elements; thus xy is a sum of terms of this kind: exterior products of elements a_i multiplied by some factors $\beta(a_h, a_i)$ and some *universal* integers (independent of M, q and K); to prove that these integers vanish when $m > 2r-j-k$, you can assume that $K = \mathbb{Q}$, and use an orthogonal basis $(b_1, b_2, \dots, b_r)$ of M besides the basis $(a_1, a_2, \dots, a_r)$.

A characterization of Clifford algebras

(4.ex.17) According to Theorem (4.8.7) the canonical morphism $\bigwedge(M) \to \mathrm{Gr}(C\ell(M,q))$ is often an isomorphism. Conversely, if the graded algebra $\mathrm{Gr}(A)$ derived from some filtered algebra A is isomorphic to an exterior algebra, in some cases A must be isomorphic to a Clifford algebra. Observe that a graded algebra isomorphism $\bigwedge(N) \to \mathrm{Gr}(A)$ already implies $A^{\leq -1} = 0$ and $A^{\leq 0} = K$. Following [Roy 1964], we will prove the following statement: if 2 is invertible in K, and if $\mathrm{Gr}(A)$ is isomorphic (as a graded algebra) to the exterior algebra of a free module N, then there exists a unique submodule M of A such that $A^{\leq 1} = K \oplus M$ and a^2 belongs to K for all $a \in M$; moreover the mapping $a \longmapsto a^2$ is a quadratic form q on M, and id_M extends to an algebra isomorphism from $C\ell(M,q)$ onto A.

For every $k \in \mathbb{N}$ the notation g_k means the canonical mapping $A^{\leq k} \to \mathrm{Gr}^k(A)$. The proof will be achieved in four steps.

(a) Since N is free, there is a family $(b_j)_{j\in J}$ constituting a basis of a submodule of $A^{\leq 1}$ supplementary to K. Since $g_1(b_j)^2$ vanishes, b_j^2 belongs to $A^{\leq 1}$; and since b_j and b_j^2 commute, there are scalars λ_j and μ_j such that $b_j^2 = \lambda_j b_j + \mu_j$. Replace b_j with $e_j = b_j - \lambda_j/2$, so that $e_j^2 \in K$. Let M be the submodule generated by all the e_j.

(b) If i and j are distinct elements of J, for the same reasons $(e_i+e_j)^2$ can be written $\lambda_{i,j}(e_i+e_j) + \mu_{i,j}$ for some scalars $\lambda_{i,j}$ and $\mu_{i,j}$. Observe that $(e_i+e_j)^2$ commutes with e_j, whence $\lambda_{i,j} = 0$. Conclude that $a^2 \in K$ for every $a \in M$.

(c) Prove that M is the only submodule satisfying the above stated properties.

(d) Prove the bijectiveness of $C\ell(M,q) \to A$.

(e) The assumption about the invertibility of 2 has been used several times above; the following counter-example shows that it is probably indispensable. Let K be the field $\mathbb{Z}/2\mathbb{Z}$, and A the quotient of the polynomial ring $K[x]$ by the ideal generated by $x^2 - x - 1$, which inherits the natural increasing filtration of $K[x]$. Let b be the image of x in A; since $g_1(b)^2 = g_2(b^2) = 0$, $\mathrm{Gr}(A)$ is isomorphic to an exterior algebra. Prove that A is not isomorphic to a Clifford algebra.

Weyl algebras (for interested readers)

(4.ex.18) Let M be a K-module, and ψ an alternate bilinear form on M; the Weyl algebra $\mathrm{W}(M,\psi)$ (or $\mathrm{W}_K(M,\psi)$) is the quotient of the tensor algebra $\mathrm{T}(M)$ by the ideal generated by all elements

$$a \otimes b - b \otimes a - \psi(a,b) \quad \text{with} \quad a, b \in M.$$

The natural morphism $M \to \mathrm{T}(M) \to \mathrm{W}(M,\psi)$ is denoted by ρ, and 1_ψ is the unit element of $\mathrm{W}(M,\psi)$. This first exercise about $\mathrm{W}(M,\psi)$ only requires the knowledge expounded in **3.1** and **3.2**.

State the universal property directly derived from this definition. Develop an elementary theory for this algebra $\mathrm{W}(M,\psi)$ by following the ideas presented in **3.1** and **3.2**. In particular, you must explain that $\mathrm{W}(M,\psi)$ is provided with a *twisted reversion* τ, such that $\tau(\rho(a)) = \rho(a)$ for all $a \in M$, and $\tau(xy) = (-1)^{\partial x \partial y}\tau(y)\tau(x)$ for all (homogeneous) x and $y \in \mathrm{W}(M,\psi)$.

Prove the theorem analogous to (3.2.4): the Weyl algebra of $(M,\psi) \perp (M',\psi')$ is isomorphic to the ordinary tensor product $\mathrm{W}(M,\psi) \otimes \mathrm{W}(M',\psi')$.

(4.ex.19) The notation is the same as in (4.ex.18). Explain why $\mathrm{W}(M,\psi)$ is a comodule over the coalgebra $\mathrm{S}(M)$. Define the interior product $f \rfloor x$ of an element f of $\mathrm{S}^*(M) = \mathrm{Hom}(\mathrm{S}(M), K)$ and an element x of $\mathrm{W}(M,\psi)$, and state the elementary properties of this operation.

(4.ex.20) The notation is the same as in (4.ex.18); we also consider the symmetric algebra $\mathrm{S}(M)$. An admissible scalar product for ψ is a bilinear form β on M such that $\psi(a,b) = \beta(a,b) - \beta(b,a)$ for all a and $b \in M$. Of course, when 2 is invertible in K, there is a canonical scalar product $\beta = \psi/2$. Assuming that M is a finitely generated projective module, prove the existence of scalar products β for any alternate bilinear form ψ on M. Then, following (4.ex.7), define an algebra morphism Ψ from $\mathrm{W}(M,\psi) \otimes \mathrm{W}(M,\psi)^o$ into $\mathrm{End}(\mathrm{S}(M))$, and bijections f and g from $\mathrm{W}(M,\psi)$ onto $\mathrm{S}(M)$ that enable you to define an algebra $\mathrm{S}(M;\beta)$ isomorphic to $\mathrm{W}(M,\psi)$. Besides, K and M can be identified with their images in $\mathrm{W}(M,\psi)$, and the notations 1_ψ and $\rho(a)$ can be replaced with 1 and a.

(4.ex.21) The notation is the same as in (4.ex.18) and (4.ex.19). Let β be any bilinear form on M, and ψ' the alternate bilinear form defined by $\psi'(a,b) = \psi(a,b) + \beta(a,b) - \beta(b,a)$. Here $\beta_{//}$ is the element of $S^{*2}(M \oplus M)$ naturally derived from β. To define $\mathrm{Exp}(\beta_{//})$ in $S^*(M \oplus M)$, you may either assume that the natural ring morphism $\mathbb{Z} \to K$ extends to a ring morphism $\mathbb{Q} \to K$, or use the results of (4.ex.2) if you are more courageous. On $\mathrm{W}(M,\psi)$ a new multiplication is defined in this way:

$$(x,y) \longmapsto x \star y = \pi_\psi\big(\mathrm{Exp}(\beta_{//}) \rfloor (x \otimes y)\big) .$$

Prove that this multiplication admits 1_ψ as a unit element, that it satisfies equalities analogous to (a) and (b) in (4.7.3), that it is associative, and that the resulting algebra $\mathrm{W}(M,\psi;\beta)$ is isomorphic to $\mathrm{W}(M,\psi')$ through a bijection Φ_β that is also an isomorphism of comodules over $\mathrm{S}(M)$.

(4.ex.22)* Let β be a bilinear form on M, and ψ the alternate bilinear form defined by $\psi(a,b) = \beta(a,b) - \beta(b,a)$. Moreover let $L = K[[t]]$ be the ring of formal series with coefficients in K. We identify $\mathrm{S}_L(L \otimes M)$ with $L \otimes \mathrm{S}(M)$, and we embed it into the algebra $\bar{\mathrm{S}}_L(L \otimes M)$ which is by definition the direct product of all submodules $t^j \otimes S^k(M)$. For every $n \in \mathbb{N}$ let $F^{\geq n}$ be the direct product of all submodules $t^j \otimes S^k(M)$ such that $2j + k \geq n$; these ideals $F^{\geq n}$ determine a decreasing filtration of $\bar{\mathrm{S}}_L(L \otimes M)$: $F^{\geq m} \vee F^{\geq n} \subset F^{\geq m+n}$; mind that $t \otimes 1$ has degree 2. We provide $\bar{\mathrm{S}}_L(L \otimes M)$ with the topology for which these ideals $F^{\geq n}$ constitute a basic family of neighbourhoods of 0; thus the subalgebra $K[t] \otimes \mathrm{S}(M)$ is dense in $\bar{S}_L(L \otimes M)$.

According to (4.ex.21), or to (4.ex.20) (when M is finitely generated and projective), we can define an algebra $\mathrm{S}_L(L \otimes M;\ t \otimes \beta)$, that is the module $L \otimes \mathrm{S}(M)$ provided with a new multiplication which lets it become isomorphic to $\mathrm{W}_L(L \otimes M,\ t \otimes \psi)$:

$$\forall a,\ b \in M, \quad (1 \otimes a)(1 \otimes b) - (1 \otimes b)(1 \otimes a) = t\ \psi(a,b).$$

Prove that this new multiplication extends by continuity to $\bar{\mathrm{S}}_L(L \otimes M)$. Moreover the submodules $F^{\geq n}$ also determine a filtration for this new multiplication: $F^{\geq m}\ F^{\geq n} \subset F^{\geq m+n}$.

Comment. Thus we get a "formal enlargement" $\bar{\mathrm{S}}_L(L \otimes M;\ t \otimes \beta)$ of the algebra $\mathrm{W}_L(L \otimes M,\ t \otimes \psi)$. When x is any element of $L \otimes \mathrm{S}^{\geq 1}(M)$, any power series in x is formally convergent in this enlargement.

(4.ex.23)* Let M be a vector space of finite dimension r over $\mathbb{R}$, β an element of $M \otimes M$, and ψ the element of $M \otimes M$ derived from β by the skew symmetrization $a \otimes b \longmapsto a \otimes b - b \otimes a$. We treat ψ as an alternate bilinear form on the dual space M^*, and β as an admissible scalar product. From ψ we can derive a Weyl algebra $\mathrm{W}(M^*,\psi)$, but here we are rather concerned with the Weyl algebra $\mathrm{W}_{\mathbb{C}}(\mathbb{C} \otimes M^*,\ i \otimes \psi)$ (where $i = \sqrt{-1}$), and we are going to construct an "enlargement" of this complex Weyl algebra. We use Fourier transformation according to this definition: the letters x and y represent variables running respectively through M and M^*, and the Fourier transform of a regular enough function f on M is defined in this way:

$$\mathcal{F}(f)(y) = (2\pi)^{-r/2} \int_M \exp(iy(x))\ f(x)\ dx\ ;$$

let $\mathrm{A}(M)$ be the space of functions $f : M \to \mathbb{C}$ such that $\mathcal{F}(f)$ is a distribution on M^* with compact support; $\mathrm{A}(M)$ contains the algebra $\mathrm{S}(M^*)$ identified with the algebra of polynomial functions on M. If f and g are elements of $\mathrm{A}(M)$, it is known that their ordinary product fg satisfies this equality, in which φ is any "test function" on M^* (that is an infinitely derivable function, the derivatives of which are all "rapidly vanishing" at infinity):

$$\int_{M^*} \varphi(y)\ \mathcal{F}(fg)(y)\ dy \ = \ (2\pi)^{-r/2} \int_{M^*\oplus M^*} \varphi(y_1+y_2)\ \mathcal{F}(f)(y_1)\mathcal{F}(g)(y_2)\ dy_1 dy_2 .$$

Their $\star$-product is the element $f \star g$ of $\mathrm{A}(M)$ defined by this equality, which, according to the principles of Fourier analysis, is the natural translation of the definition proposed in (4.ex.21):

$$\int_{M^*} \varphi(y)\ \mathcal{F}(f\star g)(y)\ dy$$
$$= (2\pi)^{-r/2} \int_{M^*\oplus M^*} \varphi(y_1+y_2)\ \exp(-i\beta(y_1,y_2))\ \mathcal{F}(f)(y_1)\mathcal{F}(g)(y_2)\ dy_1 dy_2 .$$

Prove that this multiplication on $\mathrm{A}(M)$ is associative, admits the constant function 1 as a unit element, and satisfies this equality:

$$\forall f,\ g \in M^*, \quad f\star g - g\star f \ = \ i\ \psi(f,g)\ .$$

You can even write formulas analogous to (4.8.9) for a $\star$-product $f \star g$ in which f or g belongs to M^*; for instance if f belongs to M^*, and if ∂_a is the partial derivation along the vector $a \in M$ such that $h(a) = \beta(f,h)$ for all $h \in M^*$, then $f \star g = fg + i\partial_a(g)$.

Comment. Unfortunately when neither $\mathcal{F}(f)$ nor $\mathcal{F}(g)$ has a compact support in M^*, the above definition in general fails to define a $\star$-product $f\star g$; existence theorems for this $\star$-product (with various additional hypotheses) require sophisticated functional analysis, and are outside the scope of this book. Of course when the supports of $\mathcal{F}(f)$ and $\mathcal{F}(g)$ are not compact, it becomes important that φ and all its derivatives rapidly vanish at infinity; and above all, the factor i always present beside β plays a capital role, because the function $\exp(-i\beta(y_1,y_2))$ is bounded on $M^* \oplus M^*$.

Chapter 5

Orthogonal Groups and Lipschitz Groups

In this chapter, M is a K-module provided with a *cliffordian* quadratic form q; according to Definition (4.8.1), this means that the canonical mappings $K \to \mathrm{C}\ell(M,q)$ and $M \to \mathrm{C}\ell(M,q)$ are injective; thus *the unit element* 1_q *of* $\mathrm{C}\ell(M,q)$ *is identified with the unit element* 1 *of* K, *and every* $a \in M$ *is identified with its image* $\rho(a)$ *in* $\mathrm{C}\ell(M,q)$; from now on, these identifications are done without warning. The existence of an admissible scalar product β (see (4.8.6)) is still the only available general criterion allowing us to recognize whether a quadratic form is cliffordian (see (4.8.7)).

Since Chapter **5** is long and eventful, a short summary might be helpful. In Section **5.1** it is explained that some automorphisms of (M,q) have been privileged and called "orthogonal transformations"; an automorphism g of (M,q) is an orthogonal transformation if (by definition) its extension $\mathrm{C}\ell(g)$ to an automorphism of $\mathrm{C}\ell(M,q)$ is a "generalized twisted inner automorphism".

Moreover it is conjectured that these orthogonal transformations are somehow related to a "Lipschitz monoid" $\mathrm{Lip}(M,q)$, which is some multiplicative subset of $\mathrm{C}\ell(M,q)$, and that the Lie algebra naturally derived from $\mathrm{Lip}(M,q)$ should be $\mathrm{C}\ell_0^{\leq 2}(M,q)$. The definition of the Lipschitz monoid requires particular filtrations of Clifford algebras that are explained in **5.2**. Their first properties and the derived Lipschitz groups are presented in **5.3**, and the especially important "invariance property" in **5.4**. Then in **5.5** the relations between $\mathrm{Lip}(M,q)$ and the Lie algebra $\mathrm{C}\ell_0^{\leq 2}(M,q)$ are explored.

The following three sections are devoted to the conjecture according to which all orthogonal transformations can be derived from $\mathrm{Lip}(M,q)$. This proves to be true for all nondegenerate or tamely degenerate quadratic forms, and even for some quadratic forms on modules that are not finitely generated (in particular for all quadratic modules over fields).

In **5.10** precise results involving Lipschitz monoids *over fields* are presented; they require some additional knowledge about exterior algebras which is expounded in **5.9**. They show that Lipschitz monoids over fields coincide with the objects that Lipschitz's own works actually suggest.

5.1 Twisted inner automorphisms and orthogonal groups

The automorphisms of the quadratic module (M, q) are of course the isomorphisms $(M, q) \to (M, q)$ in the category $\mathcal{C}_K(K)$ defined in **2.4**; in other words, they are the K-linear bijections $g : M \to M$ such that $q(g(a)) = q(a)$ for all $a \in M$. These automorphisms are sometimes called "linear isometries". They constitute a group $\mathrm{Aut}(M, q)$. Since the functor $C\ell$ (see **3.1**) transforms every such automorphism g into a graded automorphism $C\ell(g)$ of $C\ell(M, q)$, the group $\mathrm{Aut}(M, q)$ is naturally isomorphic to a subgroup of graded automorphisms of $C\ell(M, q)$, and the following lemma shows that it is exactly the subgroup of all automorphisms θ leaving invariant the subspace M. If $\theta(M) = M$, it is clear that the automorphism θ is graded: it leaves $C\ell_0(M, q)$ and $C\ell_1(M, q)$ invariant.

(5.1.1) **Lemma.** *If θ is an automorphism of $C\ell(M, q)$ such that $\theta(M) = M$, the restriction of θ to M is an automorphism of (M, q).*

Proof. For every $a \in M$, $q(\theta(a)) = \theta(a)^2 = \theta(a^2) = \theta(q(a)) = q(a)$. □

Whereas this group $\mathrm{Aut}(M, q)$ is quite satisfactory for a quadratic space (M, q) (in other words, when M is projective and finitely generated, and $\mathrm{d}_q : M \to M^*$ is bijective), in other cases it has been noticed that some automorphisms of (M, q), called "orthogonal transformations", should be privileged; they constitute a privileged subgroup $\mathrm{GO}(M, q)$ called the "orthogonal group". The names "orthogonal group" and "orthogonal transformation" have been imposed by common use, although their meanings do not fit the usual meaning of the word "orthogonal". The distinction between $\mathrm{Aut}(M, q)$ and $\mathrm{GO}(M, q)$ corresponds to the following idea: an automorphism g of (M, q) is called an *orthogonal transformation* if $C\ell(g)$ is a "generalized twisted inner automorphism" of $C\ell(M, q)$ according to Definition (5.1.5) beneath. When (M, q) is a quadratic space, the groups $\mathrm{Aut}(M, q)$ and $\mathrm{GO}(M, q)$ will prove to be equal (see (5.8.1)), and in all other cases in which it has been possible to calculate both groups, the eventual discrepancy between them seems sensible. For instance when K is a field, it shall be proved (in **5.8**) that $\mathrm{GO}(M, q)$ is the subgroup of all $g \in \mathrm{Aut}(M, q)$ such that $\mathrm{Ker}(g - \mathrm{id}) = P^\perp$ for some subspace P of finite dimension in M; when $\mathrm{Ker}(q) = 0$, because of (5.6.1) this is equivalent to $\mathrm{Im}(g - \mathrm{id})$ having a finite dimension, a condition generally admitted by people studying weakly nondegenerate quadratic forms on infinite dimensional vector spaces (see for instance [Zassenhaus 1962]); when

$\mathrm{Ker}(\mathrm{b}_q)$ has finite codimension in M, this is equivalent to $\mathrm{Ker}(g - \mathrm{id}) \supset \mathrm{Ker}(\mathrm{b}_q)$, and the relevance of this condition is corroborated by the following example.

(5.1.2) **Example.** Let E be an affine euclidean space; this means that there is a vector space E_v of finite dimension m over $\mathbb{R}$ that operates on E in a simply transitive way (by "translations"), and that is provided with a positive definite quadratic form q. For more than 2000 years, mathematicians have studied the group $\mathrm{Aut}(E, E_v, q)$ of isometries of E; they are the affine transformations $g : E \to E$ with a vectorial part g_v in $\mathrm{Aut}(E_v, q)$. To reduce its study to the study of a group of linear transformations, we introduce the space E' of all affine forms $E \to \mathbb{R}$; it is a vector space of dimension $m + 1$, containing the space N of all constant functions $E \to \mathbb{R}$. Since the quotient E'/N is naturally isomorphic to $E_v^* = \mathrm{Hom}(E_v, \mathbb{R})$, and since q induces on E_v^* a dual positive definite quadratic form, the space E' is naturally provided with a degenerate quadratic form q', the kernel of which is N. Every affine transformation g of E induces a linear transformation g' of E' in a contravariant way, and g' obviously leaves invariant every element of N. It is easy to verify that the group morphism $g \longmapsto g'^{-1}$ induces by restriction an isomorphism from $\mathrm{Aut}(E, E_v, q)$ onto the group of all automorphisms g' of (E', q') such that $\mathrm{Ker}(g' - \mathrm{id}') \supset N$.

Generalized twisted inner automorphisms

Let $A = A_0 \oplus A_1$ be a graded algebra (associative with unit) containing K as a central subalgebra, and let $\mathrm{Aut}^g(A)$ be the group of its graded automorphisms. If x is an invertible homogeneous element of A, its inverse has the same parity as x, and with x and x^{-1} is associated a *twisted inner automorphism* Θ_x defined in this way for all homogeneous $a \in A$:

$$(5.1.3) \qquad \Theta_x(a) = (-1)^{\partial x \partial a}\, x a x^{-1} ;$$

it is easy to verify that Θ_x is a graded automorphism of A such that $\Theta_x(x) = \sigma(x)$. Instead of a homogeneous factor x we can also use a *locally homogeneous* one.

(5.1.4) **Lemma.** *Let x be an element of the graded algebra A; the following three assertions are equivalent, and when they are true, x is said to be locally homogeneous:*

(a) *there exists $\lambda \in K$ such that $(1 - \lambda)x$ is even and λx is odd;*
(b) *for every prime ideal $\mathfrak{p}$ of K, $x/1$ is homogeneous in the localized algebra $A_{\mathfrak{p}}$;*
(c) *for every maximal ideal $\mathfrak{m}$ of K, $x/1$ is homogeneous in $A_{\mathfrak{m}}$.*

Moreover $(1 - \lambda)\lambda x = 0$ *when the assertion* (a) *is true.*

Proof. If $(1-\lambda)x$ and λx are respectively even and odd, $(1-\lambda)\lambda x$ vanishes because it is both even and odd. Moreover if $\mathfrak{p}$ is a prime ideal, $(1 - \lambda)/1$ and $\lambda/1$ cannot be both non-invertible in $K_{\mathfrak{p}}$; if $(1 - \lambda)/1$ is invertible, $x/1$ is even like $(1 - \lambda)x/1$, and if $\lambda/1$ is invertible, $x/1$ is odd like $\lambda x/1$. Since obviously (b)$\Rightarrow$(c), it remains to prove (c)⇒(a). Let x_0 and x_1 be the even and odd components of x, let $\mathfrak{a}$ (resp. $\mathfrak{b}$) be the ideal of all $\lambda \in K$ such that $\lambda x_1 = 0$ (resp. $\lambda x_0 = 0$); if $\mathfrak{m}$ is any maximal ideal, the image of x in $A_\mathfrak{m}$ is even (resp. odd) if and only if $\mathfrak{a}$ (resp. $\mathfrak{b}$) is not contained in $\mathfrak{m}$; the assumption that this image is homogeneous implies that $\mathfrak{a}+\mathfrak{b}$ is never contained in $\mathfrak{m}$; consequently $\mathfrak{a}+\mathfrak{b} = K$ and there is an element λ in $\mathfrak{b}$ such that $1-\lambda$ belongs to $\mathfrak{a}$. □

When x is invertible and locally homogeneous, the vanishing of $(1-\lambda)\lambda x$ implies that λ is an idempotent, and that A is the direct sum of the ideals $(1-\lambda)A$ and λA; moreover $(1-\lambda)x^{-1}$ and λx^{-1} are respectively even and odd. It is natural to apply (5.1.3) in each ideal $(1-\lambda)A$ and λA and to define the twisted inner automorphism Θ_x in this way:

$$\Theta_x(a) \quad = \quad (1-\lambda)xax^{-1} + \lambda x\sigma(a)x^{-1}.$$

This definition still implies $\Theta_x(x) = \sigma(x)$.

The product of two locally homogeneous elements is still locally homogeneous, since all its localizations are homogeneous. Therefore the invertible locally homogeneous elements of A constitute a group, and the mapping $x \longmapsto \Theta_x$ is a group morphism from this group into $\mathrm{Aut}^g(A)$.

Nevertheless this concept of twisted inner automorphism is much too restrictive. The choice of a suitable generalization of this concept is still a subject of discussions; yet the following definition is manageable enough, and well suitable to the quadratic modules later under consideration.

(5.1.5) **Definition.** If θ is a graded automorphism of A, we denote by $\mathrm{Z}^g(\theta)$ the graded submodule generated by all homogeneous $x \in A$ such that $\theta(a)x = (-1)^{\partial x \partial a}xa$ for all homogeneous $a \in A$. Thus $\mathrm{Z}^g(\theta^{-1})$ is the submodule generated by all homogeneous $x' \in A$ such that $x'\theta(a) = (-1)^{\partial x \partial a}ax'$ for all homogeneous $a \in A$. We say that θ is a *generalized twisted inner automorphism* of A if there is a finite sequence $(x_1, x_2, \dots, x_n)$ of homogeneous elements of $\mathrm{Z}^g(\theta)$ (of any length $n > 0$) and a sequence $(x'_1, x'_2, \dots, x'_n)$ of homogeneous elements of $\mathrm{Z}^g(\theta^{-1})$ such that $x_i x'_i \in K$ for $i = 1, 2, \dots, n$, and K is generated as an ideal by the n elements $x_i x'_i$. Below in (5.1.6) it shall appear that we get an equivalent definition if we replace $\mathrm{Z}^g(\theta)$ (resp. $\mathrm{Z}^g(\theta^{-1})$) with the submodule $\mathrm{Z}^r(\theta)$ (resp. $\mathrm{Z}^r(\theta^{-1})$) of all $x \in \mathrm{Z}^g(\theta)$ (resp. $x' \in \mathrm{Z}^g(\theta^{-1})$) such that $\theta(x) = \sigma(x)$ (resp. $\theta(x') = \sigma(x')$).

It is clear that the graded center $\mathrm{Z}^g(A)$ defined in (3.5.2) is equal to $\mathrm{Z}^g(\mathrm{id}_A)$. By definition the *reduced center* $\mathrm{Z}^r(A)$ is the same thing as $\mathrm{Z}^r(\mathrm{id}_A)$. Its even component is $\mathrm{Z}^r_0(A) = \mathrm{Z}^g_0(A) = \mathrm{Z}_0(A)$, and its odd component $\mathrm{Z}^r_1(A)$ is the subset of all $x \in \mathrm{Z}^g_1(A)$ such that $2x = 0$, and also the subset of all $x \in \mathrm{Z}_1(A)$ such that $2x = 0$. It is an easy exercise to prove that $\mathrm{Z}^r(A)$ is the intersection of $\mathrm{Z}(A)$ and $\mathrm{Z}^g(A)$ when the grading of A is regular (see Definitions (3.5.2)).

The following properties of the submodules $\mathrm{Z}^g(\theta)$ and $\mathrm{Z}^r(\theta)$ should be evident. If $\theta = \Theta_x$ for some invertible locally homogeneous x, it is clear that $x \in \mathrm{Z}^r(\theta)$

and $x^{-1} \in \mathrm{Z}^r(\theta^{-1})$. Conversely if $\mathrm{Z}^g(\theta)$ contains an invertible locally homogeneous element x, then $\Theta_x = \theta$. If x and y belong respectively to $\mathrm{Z}^g(\theta_1)$ and $\mathrm{Z}^g(\theta_2)$, it is easy to verify that xy belongs to $\mathrm{Z}^g(\theta_1\theta_2)$. If x and y belong respectively to $\mathrm{Z}^r(\theta_1)$ and $\mathrm{Z}^r(\theta_2)$, then xy belongs to $\mathrm{Z}^r(\theta_1\theta_2)$; indeed the equality $\theta_1\theta_2(xy) = \sigma(xy)$ follows from this calculation:

$$\begin{aligned}\theta_1\theta_2(xy) &= \theta_1(\theta_2(x)\theta_2(y)) = (-1)^{\partial y}\theta_1(\theta_2(x)\,y) = (-1)^{\partial y+\partial x\partial y}\theta_1(yx)\\ &= (-1)^{\partial y+\partial x\partial y}\theta_1(y)\theta_1(x) = (-1)^{\partial x+\partial y+\partial x\partial y}\theta_1(y)\,x = (-1)^{\partial x+\partial y}xy\ .\end{aligned}$$

Consequently every submodule $\mathrm{Z}^g(\theta)$ (resp. $\mathrm{Z}^r(\theta)$) is a module (on the left or right side) over the algebra $\mathrm{Z}^g(A)$ (resp. $\mathrm{Z}^r(A)$). And the products xx' and $x'x$ belong to $\mathrm{Z}^g(A)$ (resp. $\mathrm{Z}^r(A)$) whenever x belongs to $\mathrm{Z}^g(\theta)$ (resp. $\mathrm{Z}^r(\theta)$) and x' to $\mathrm{Z}^g(\theta^{-1})$ (resp. $\mathrm{Z}^r(\theta^{-1})$). In particular all products $x_ix'_j$ and x'_jx_i belong to $\mathrm{Z}^g(A)$ when $(x_1,\dots,x_n)$ and $(x'_1,\dots,x'_n)$ are the sequences mentioned in Definition (5.1.5).

If K' is an extension of K, and $K'\otimes\theta$ the derived automorphism of $K'\otimes A$, there is a natural mapping $K'\otimes\mathrm{Z}^g(\theta)\to\mathrm{Z}^g(K'\otimes\theta)$; therefore if θ is a generalized twisted inner automorphism, the same is true for $K'\otimes\theta$. When A is finitely generated as an algebra, $\mathrm{Z}^g(\theta)$ can be presented as a finite intersection of kernels of mappings $x\longmapsto\theta(a)x-(-1)^{\partial x\partial a}xa$, and thus for a flat extension $K\to K'$ there is an equality $K'\otimes\mathrm{Z}^g(\theta)=\mathrm{Z}^g(K'\otimes\theta)$. For the same reasons $K'\otimes\mathrm{Z}^r(\theta)=\mathrm{Z}^r(K'\otimes\theta)$. When K' is a ring of fractions of K, this equality means that every element of $\mathrm{Z}^r(K'\otimes\theta)$ is a fraction with numerator in $\mathrm{Z}^r(\theta)$.

The next proposition explains why $\mathrm{Z}^r(\theta)$ is more convenient than $\mathrm{Z}^g(\theta)$, and proves that θ^{-1} too is a generalized twisted inner automorphism if θ is such an automorphism.

(5.1.6) **Proposition.** *If θ is a generalized twisted inner automorphism, every element of $\mathrm{Z}^r(\theta)$ commutes with every element of $\mathrm{Z}^r(\theta^{-1})$. Moreover there is a sequence $(u_1,u_2,\dots,u_n)$ of homogeneous elements of $\mathrm{Z}^r(\theta)$ and a sequence $(u'_1,u'_2,\dots,u'_n)$ of homogeneous elements of $\mathrm{Z}^r(\theta^{-1})$ such that $u_iu'_i=u'_iu_i\in K$ for $i=1,2,\dots,n$, and K is generated by the n products $u_iu'_i$ as an ideal.*

Proof. If x is a homogeneous element of $\mathrm{Z}^g(\theta)$, and x' a homogeneous element of $\mathrm{Z}^g(\theta^{-1})$, then $\theta(x)x=(-1)^{\partial x}x^2$ and $x'\theta(x')=(-1)^{\partial x'}x'^2$. If their product $s=xx'$ belongs to K, it follows that $sx\in\mathrm{Z}^r(\theta)$ and $sx'\in\mathrm{Z}^r(\theta^{-1})$; indeed:

$$\begin{aligned}\theta(sx) &= \theta(x)xx' = (-1)^{\partial x}x^2x' = \sigma(sx)\ ,\\ \theta(sx') &= xx'\theta(x') = (-1)^{\partial x'}xx'^2 = \sigma(sx')\ .\end{aligned}$$

If $s=0$, it is clear that sx' commutes with x. If $s\neq0$, then $\partial x=\partial x'$, thus $x'x$ belongs to $\mathrm{Z}^g_0(A)=\mathrm{Z}_0(A)$, and it is still true that sx' commutes with x :

$$(sx')x = xx'(x'x) = x(x'x)x' = sxx' = x(sx').$$

Consequently sx and sx' commute and their product is s^3. We can derive the sequences (u_i) and (u'_i) mentioned in (5.1.6) from the sequences (x_i) and (x'_i)

given in (5.1.5) in this way: we set $s_i = x_i x'_i$, $u_i = s_i x_i$ and $u'_i = s_i x'_i$ for $i = 1$, $2, \dots, n$. If K is generated by the elements s_i as an ideal, it is generated by the elements s_i^3 too. Of course we can remove all pairs (x_i, x'_i) such that $\partial x_i \neq \partial x'_i$ (whence $s_i = 0$).

Let us prove that every homogeneous $x \in \mathrm{Z}^r(\theta)$ commutes with every homogeneous $x' \in \mathrm{Z}^r(\theta^{-1})$. There are elements $\lambda_i \in K$ such that $1 = \sum_{i=1}^n \lambda_i u_i u'_i$; since $u'_j x$ and xx' belong to $\mathrm{Z}^r(A)$, therefore to $\mathrm{Z}(A)$, we can write

$$x'x = \sum_i \sum_j x' \lambda_i u_i u'_i \lambda_j u_j (u'_j x) = \sum_i \sum_j \lambda_i \lambda_j (u'_j x) x' u_i u'_i u_j$$
$$= \sum_i \sum_j \lambda_i \lambda_j u'_j (xx') u_i u'_i u_j = \sum_i \sum_j (xx') \lambda_j u'_j (\lambda_i u_i u'_i) u_j \;;$$

since $u'_j u_j = u_j u'_j$, the final result is $x'x = xx'$. □

It is not true that every $x \in \mathrm{Z}^g(\theta)$ commutes with every $x' \in \mathrm{Z}^g(\theta^{-1})$ for every generalized twisted inner automorphism θ. Indeed there are graded algebras (for instance exterior algebras) such that $\mathrm{Z}^g(A)$ is not a commutative algebra. Now it is explained how to construct generalized twisted inner automorphisms.

(5.1.7) **Theorem.** *Let $(x_1, x_2, \dots, x_n)$ and $(x'_1, x'_2, \dots, x'_n)$ be two sequences of homogeneous elements of A (of the same length $n > 0$) such that*

- *$x_i x'_i \in K$ for $i = 1, 2, \dots, n$, and K is generated as an ideal by the n elements $x_i x'_i$,*
- *$x_i x'_j$ and $x'_j x_i$ belong to $\mathrm{Z}^g(A)$ for all i and j.*

There exists a generalized twisted inner automorphism θ satisfying all these properties:

(a) *θ is the only graded automorphism of A such that $\mathrm{Z}^g(\theta)$ contains x_i for $i = 1, 2, \dots, n$, and also the only one such that $\mathrm{Z}^g(\theta^{-1})$ contains all x'_i.*
(b) *$\mathrm{Z}^g(\theta)$ (resp. $\mathrm{Z}^g(\theta^{-1})$) is the submodule over $\mathrm{Z}^g(A)$ generated by the n elements x_i (resp. x'_i).*
(c) *an element x (resp. x') of A belongs to $\mathrm{Z}^g(\theta)$ (resp. $\mathrm{Z}^g(\theta^{-1})$) if and only if xx'_i (resp. $x'x_i$) belongs to $\mathrm{Z}^g(A)$ for $i = 1, 2, \dots, n$, or equivalently, if and only if $x'_i x$ (resp. $x_i x'$) belongs to $\mathrm{Z}^g(A)$ for all i.*

Besides, this theorem still holds if we replace $\mathrm{Z}^g(A)$, $\mathrm{Z}^g(\theta)$ and $\mathrm{Z}^g(\theta^{-1})$ with $\mathrm{Z}^r(A)$, $\mathrm{Z}^r(\theta)$ and $\mathrm{Z}^r(\theta^{-1})$.

Proof. We can assume that $\partial x_i = \partial x'_i$ for $i = 1, 2, \dots, n$, and we can write $1 = \sum_i \lambda_i x_i x'_i$ for some family $(\lambda_1, \lambda_2, \dots, \lambda_n)$ of elements of K. If θ is a graded automorphism such that $\mathrm{Z}^g(\theta)$ contains x_i for $i = 1, 2, \dots, n$, then for every homogeneous $a \in A$,

$$\theta(a) = \sum_i \lambda_i \theta(a) x_i x'_i = \sum_i (-1)^{\partial a \partial x_i} \lambda_i x_i a x'_i \;;$$

this proves the unicity of θ. The same conclusion holds if $\mathrm{Z}^g(\theta^{-1})$ contains all x'_i :

$$\theta(a) = \sum_i \lambda_i x_i x'_i \theta(a) = \sum_i (-1)^{\partial a \partial x_i} \lambda_i x_i a x'_i.$$

Conversely let us set

$$\theta(a) = \sum_i (-1)^{\partial a \partial x_i} \lambda_i x_i a x'_i \quad \text{and} \quad \theta'(a) = \sum_i (-1)^{\partial a \partial x_i} \lambda_i x'_i a x_i \ ,$$

and let us prove that $\theta' = \theta^{-1}$ and that θ satisfies all the announced properties. We first achieve the proof with the supplementary hypothesis $x_i x'_i = x'_i x_i$ for $i = 1, 2, \dots, n$. Since all $x_j x'_i$ belong to $\mathrm{Z}^g(A)$, we get for all homogeneous $a \in A$,

$$\begin{aligned}\theta\theta'(a) &= \sum_i \sum_j (-1)^{\partial a(\partial x_i + \partial x_j)} \lambda_i x_i (\lambda_j x'_j a x_j) x'_i \\ &= \sum_i \sum_j (\lambda_i x_i (\lambda_j x'_j x_j) x'_i) a \ = \ a \ ,\end{aligned}$$

whence $\theta\theta' = \mathrm{id}_A$. And similarly $\theta'\theta = \mathrm{id}_A$ because all $x'_j x_i$ are in $\mathrm{Z}^g(A)$. Let us prove that θ is an algebra morphism. It is clear that $\theta(1) = 1$. Moreover, since all $x'_i x_j$ belong to $\mathrm{Z}^g(A)$,

$$\begin{aligned}\theta(a)\theta(b) &= \sum_i \sum_j (-1)^{\partial a \partial x_i} (\lambda_i x_i a x'_i)(-1)^{\partial b \partial x_j} (\lambda_j x_j b x'_j) \\ &= \sum_i \sum_j (-1)^{(\partial a + \partial b)\partial x_i} (\lambda_i x_i a b x'_i)(\lambda_j x_j x'_j) = \theta(ab).\end{aligned}$$

Let x be a homogeneous element such that $x'_i x \in \mathrm{Z}^g(A)$ for $i = 1, 2, \dots, n$, and let us prove that $x \in \mathrm{Z}^g(\theta)$; indeed for all homogeneous $a \in A$ we get

$$\theta(a)x = \sum_i (-1)^{\partial a \partial x_i} (\lambda_i x_i a x'_i) x = \sum_i (-1)^{\partial a \partial x} (\lambda_i x_i x'_i) x a = (-1)^{\partial a \partial x} x a \ ;$$

if all $x'_i x$ belong to $\mathrm{Z}^r(A)$, then $x \in \mathrm{Z}^r(\theta)$ because

$$\theta^{-1}(x) = \sum_i (-1)^{\partial x_i \partial x} \lambda_i x'_i x x_i = \sum_i (-1)^{\partial x_i} \lambda_i x_i x'_i x = \sum_i \lambda_i \sigma(x_i x'_i x) = \sigma(x).$$

If we suppose that all $x x'_i$ are in $\mathrm{Z}^g(A)$, then we write

$$\theta(a)x = \sum_i \theta(a) x (\lambda_i x'_i x_i) = \sum_i (-1)^{\partial a(\partial x + \partial x_i)} x (\lambda_i x'_i \theta(a) x_i) = (-1)^{\partial a \partial x} x a \ ,$$

whence $x \in \mathrm{Z}^g(\theta)$; and if all $x x'_i$ belong to $\mathrm{Z}^r(A)$, we verify the equality $\theta(x) = \sigma(x)$ as above and still conclude that $x \in \mathrm{Z}^r(\theta)$.

In the remainder of the proof only submodules $\mathrm{Z}^g(\dots)$ are mentioned, but the argument is still valid if we replace them with $\mathrm{Z}^r(\dots)$. From the previous results it follows that the n elements x_i belong to $\mathrm{Z}^g(\theta)$ (indeed all $x_i x'_j$ are in $\mathrm{Z}^g(A)$); therefore $\mathrm{Z}^g(\theta)$ contains the submodule over $\mathrm{Z}^g(A)$ that they generate. Similarly $\mathrm{Z}^g(\theta^{-1})$ contains the submodule over $\mathrm{Z}^g(A)$ generated by the n elements x'_i. Now it is clear that all xx'_i and $x'_i x$ belong to $\mathrm{Z}^g(A)$ for every $x \in \mathrm{Z}^g(\theta)$. This allows us to prove that every $x \in \mathrm{Z}^g(\theta)$ is in the $\mathrm{Z}^g(A)$-module generated by all x_i :

$$x \;=\; \sum_i x(\lambda_i x'_i x_i) \;=\; \sum_i (\lambda_i x x'_i) x_i .$$

Since the equalities $x_i x'_i = x'_i x_i$ let the sequences (x_i) and (x'_i) play symmetrical roles, we can claim that the three statements (a), (b), (c) have been proved.

Now we must get rid of the supplementary hypothesis $x_i x'_i = x'_i x_i$. Let us set $x''_i = (x_i x'_i) x'_i$ for all i. From the calculations at the beginning of the proof of (5.1.6) we know that $x_i x''_i = x''_i x_i = (xx'_i)^2$ for all i. Since K is also generated as an ideal by all $(x_i x'_i)^2$, we can apply the previous results to the pair of sequences (x_i) and (x''_i), and thus we prove again the three statements (a), (b), (c). In particular each x_i belongs to $\mathrm{Z}^g(\theta)$ because $x_i x''_j$ belongs to $\mathrm{Z}^g(A)$ for $j = 1, 2, \dots, n$, and each x'_i belongs to $\mathrm{Z}^g(\theta^{-1})$ because $x'_i x_j$ belongs to $\mathrm{Z}^g(A)$ for $j = 1, 2, \dots, n$. □

The first corollary is evident. In the subsequent two corollaries there are tensor products over the commutative ring $\mathrm{Z}^r(A)$; in such tensor products the parity gradings must be forgotten.

(5.1.8) **Corollary.** *If $\mathrm{Z}^g(A) = \mathrm{Z}^r(A)$, then $\mathrm{Z}^g(\theta) = \mathrm{Z}^r(\theta)$ whenever θ is a generalized twisted inner automorphism.*

(5.1.9) **Corollary.** *If θ is a generalized twisted inner automorphism, the mapping $x \otimes x' \longmapsto xx' = x'x$ induces an isomorphism from $\mathrm{Z}^r(\theta) \otimes_{\mathrm{Z}^r(A)} \mathrm{Z}^r(\theta^{-1})$ onto $\mathrm{Z}^r(A)$.*

Proof. The bijectiveness of this mapping can be proved by localization at every prime ideal $\mathfrak{p}$ of K; because of (1.10.5) it is equivalent to prove the bijectiveness of

$$\mathrm{Z}^r(\theta)_{\mathfrak{p}} \otimes_{\mathrm{Z}^r(A)_{\mathfrak{p}}} \mathrm{Z}^r(\theta^{-1})_{\mathfrak{p}} \longrightarrow \mathrm{Z}^r(A)_{\mathfrak{p}} .$$

From (5.1.5) we deduce the existence of $x \in \mathrm{Z}^r(\theta)$ and $x' \in \mathrm{Z}^r(\theta^{-1})$ such that xx' is an element $s \in K$ outside $\mathfrak{p}$, and we first observe that $(x/s) \otimes (x'/1)$ is mapped to $1/1$ in $\mathrm{Z}^r(A)_{\mathfrak{p}}$. If we prove that $\mathrm{Z}^r(\theta)_{\mathfrak{p}}$ and $\mathrm{Z}^r(\theta^{-1})_{\mathfrak{p}}$ are free modules over $\mathrm{Z}^r(A)_{\mathfrak{p}}$ respectively generated by $x/1$ and $x'/1$, the conclusion follows. Since $x/1$ and $x'/1$ are invertible, they generate free modules. Moreover every y/t with $y \in \mathrm{Z}^r(\theta)$ is in the submodule generated by $x/1$ because

$$\frac{y}{t} \;=\; \frac{yx'}{st}\,\frac{x}{1} \quad \text{with} \quad yx' \in \mathrm{Z}^r(A).$$

Similarly $\mathrm{Z}^r(\theta^{-1})_{\mathfrak{p}}$ is generated by $x'/1$. □

(5.1.10) **Corollary.** *If θ_1 and θ_2 are generalized twisted inner automorphisms, then $\theta_1\theta_2$ too is such an automorphism, and the mapping $x \otimes y \longmapsto xy$ induces an isomorphism from $\mathrm{Z}^r(\theta_1) \otimes_{\mathrm{Z}^r(A)} \mathrm{Z}^r(\theta_2)$ onto $\mathrm{Z}^r(\theta_1\theta_2)$.*

Proof. If θ_1 is determined by two sequences of n elements (x_i) and (x'_i) as in (5.1.7), and θ_2 by two sequences of p elements (y_j) and (y'_j), then the np products x_iy_j (resp. $y'_jx'_i$) belong to $\mathrm{Z}^r(\theta_1\theta_2)$ (resp. $\mathrm{Z}^r((\theta_1\theta_2)^{-1})$). Moreover from two equalities $1 = \sum_i \lambda_i x_i x'_i = \sum_j \mu_j y_j y'_j$ we easily deduce $1 = \sum_i \sum_j (\lambda_i\mu_j)(x_iy_j)(y'_jx'_i)$. Consequently $\theta_1\theta_2$ is the generalized twisted inner automorphism determined by the two families (x_iy_j) and $(y'_jx'_i)$, and $\mathrm{Z}^r(\theta_1\theta_2)$ is the $\mathrm{Z}^r(A)$-module generated by all x_iy_j. This shows the surjectiveness of the mapping $\mathrm{Z}^r(\theta_1) \otimes_{\mathrm{Z}^r(A)} \mathrm{Z}^r(\theta_2) \rightarrow \mathrm{Z}^r(\theta_1\theta_2)$. It is even bijective because of (1.13.5); indeed (5.1.9) shows that the source and the target are invertible modules over $\mathrm{Z}^r(A)$. $\square$

The previous corollary shows that the generalized twisted inner automorphisms constitute a group. Now we consider again rings of fractions of K; the Zariski extensions mentioned beneath in (5.1.11) are defined in (1.10.6), and you may understand the localizations of K as localizations at prime or maximal ideals at your convenience.

(5.1.11) **Theorem.** *The following three assertions are equivalent for every $\theta \in \mathrm{Aut}^g(A)$:*

(a) *θ is a generalized twisted inner automorphism;*
(b) *there is a Zariski extension $L = \prod_{i=1}^n K_{s_i}$ such that $L \otimes \theta = \Theta_z$ for some invertible $z = (z_1, z_2, \dots, z_n)$ in which the numerators of all fractions z_i (resp. z_i^{-1}) are homogeneous elements of $\mathrm{Z}^r(\theta)$ (resp. $\mathrm{Z}^r(\theta^{-1})$);*
(c) *every localization of θ is the twisted inner automorphism determined by an invertible homogeneous fraction ξ such that the numerators of ξ and ξ^{-1} are respectively in $\mathrm{Z}^r(\theta)$ and $\mathrm{Z}^r(\theta^{-1})$.*

If A is finitely generated as an algebra, or more generally if $\mathrm{Z}^r(K'\otimes\theta) = K'\otimes\mathrm{Z}^r(\theta)$ (and the same with θ^{-1}) for every ring of fractions K', these assertions are also equivalent to the following ones:

(d) *there is a Zariski extension $L = \prod_{i=1}^n K_{s_i}$ such that $L \otimes \theta = \Theta_z$ for some invertible $z = (z_1, z_2, \dots, z_n)$ with n homogeneous components z_i;*
(e) *every localization of θ is a twisted inner automorphism derived from some invertible homogeneous element.*

Proof. It is clear that (a)$\Rightarrow$(b) and (a)$\Rightarrow$(c). Conversely let $K' = S^{-1}K$ be a ring of fractions of K such that $K' \otimes \theta = \Theta_\xi$ for some invertible homogeneous fraction ξ such that $\xi = x/t$ with $x \in \mathrm{Z}^r(\theta)$ and $\xi^{-1} = x'/t'$ with $x' \in \mathrm{Z}^r(\theta^{-1})$; of course t and t' are in S. Since these fractions are inverse to each other, $uxx' = utt'$ for some $u \in S$, and thus we have elements $x \in \mathrm{Z}^r(\theta)$ and $x'' = ux' \in \mathrm{Z}^r(\theta^{-1})$ with a product xx'' in S. When K' is a component K_{s_i} of a Zariski extension L, then xx'' is a power of s_i; now if K is generated as an ideal by some elements s_1,

$s_2, \dots, s_n$, it is also generated by any sequence of powers of these elements, and this proves (b)⇒(a). To prove (c)⇒(a), it suffices to remember that, from any subset of K that is not contained in any maximal ideal, we can extract a *finite* sequence that generates K as an ideal. At last it is evident that (b)⇔(d) and (c)⇔(e) when the additional hypothesis is true. □

(5.1.12) **Remarks.**

(a) When $\mathrm{Z}_1^g(A)$ contains an invertible element x, then $1 = xx^{-1} = -x^{-1}x = -1$; thus the equality $2 = 0$ holds in K, and $\mathrm{Z}^r(A) = \mathrm{Z}^g(A) = \mathrm{Z}(A)$.

(b) The reduced odd component $\mathrm{Z}_1^r(A)$ plays a capital role when we try to assign a parity to a generalized twisted inner automorphism θ at some prime ideal $\mathfrak{p}$ of K. Indeed $\mathrm{Z}^r(\theta)_\mathfrak{p}$ is a free module over $\mathrm{Z}^r(A)_\mathfrak{p}$ generated by some invertible homogeneous element ξ (look at the proof of (5.1.9)), and each element of $\mathrm{Z}^r(\theta)_\mathfrak{p}$ is equal to $\zeta\xi$ for some $\zeta \in \mathrm{Z}^r(A)_\mathfrak{p}$. If the odd component $\mathrm{Z}_1^r(A)_\mathfrak{p}$ contains no invertible elements, all homogeneous invertible elements of $\mathrm{Z}^r(\theta)_\mathfrak{p}$ have the same parity as ξ, and thus θ has a parity (that is the parity of ξ) at the point $\mathfrak{p}$. Because of the previous remark, it has a parity at this point whenever the image of 2 in $K_\mathfrak{p}$ does not vanish.

Some of the generalized twisted inner automorphisms are especially convenient and manageable, and shall now be presented.

A graded submodule X of A is called a *graded invertible submodule of* A (and is said to be *invertible inside* A) if there exists a graded submodule X' of A such that the multiplication mapping $\pi_A : A \otimes A \to A$ induces two isomorphisms $X \otimes X' \to K$ and $X' \otimes X \to K$. This implies that X and X' are finitely generated projective modules of constant rank 1 (see (1.12.10)). It is clear that every localization $X_\mathfrak{p}$ at a prime ideal of K is generated by an invertible homogeneous element $\xi \in A_\mathfrak{p}$, and that $X'_\mathfrak{p}$ is generated by ξ^{-1}; consequently X' is determined by X, it is called the *inverse submodule* of X and denoted by X^{-1}. It is clear that every $x \in X$ commutes with every $x' \in X^{-1}$.

If X and Y are graded invertible submodules of A, it is clear that XY (the submodule generated by all products xy with $x \in X$ and $y \in Y$) is still invertible inside A. Thus the graded invertible submodules of A constitute a group. Among them the free submodules are exactly the submodules generated by an invertible locally homogeneous element of A.

If K is a direct summand of A (therefore of A_0), every graded invertible submodule of A is a graded direct summand; indeed the multiplication mapping π_A induces a bijection $X \otimes A \to A$ which maps the graded direct summand $X \otimes K$ onto X. This property is always true when A is a finitely generated projective module (see (1.13.2)).

If X is a graded invertible submodule of A, there are homogeneous $x_1, \dots, x_n \in X$ and homogeneous $x'_1, \dots, x'_n \in X^{-1}$ such that $\sum_i x_i x'_i = 1$; since moreover all products $x_i x'_j = x'_j x_i$ belong to K, from (5.1.7) it immediately follows that there exists a unique graded automorphism Θ_X such that $\mathrm{Z}^r(\Theta_X)$ contains X; it is called the (generalized) twisted inner automorphism derived from X. It is clear that the mapping $X \longmapsto \Theta_X$ is a group morphism.

Let θ be a graded automorphism of A such that $\mathrm{Z}^g(\theta)$ contains a graded invertible submodule X of A. Then from (5.1.7) we deduce that Θ_X coincides with θ, that $X \subset \mathrm{Z}^r(\theta)$ and $X^{-1} \subset \mathrm{Z}^r(\theta^{-1})$.

In $\mathrm{Aut}^g(A)$ we can distinguish the subgroup G of all automorphisms Θ_x derived from an invertible locally homogeneous x, the larger subgroup G' of all Θ_X derived from a graded invertible submodule X, and the still larger subgroup G'' of all generalized twisted inner automorphisms. These subgroups of $\mathrm{Aut}^g(A)$ are normal subgroups because $\mathrm{Z}^r(fgf^{-1}) = f(\mathrm{Z}^r(g))$ for all $f, g \in \mathrm{Aut}^g(A)$. When $\mathrm{Z}^r(A) = K$, then $G' = G''$ (see (5.1.9)); when $\mathrm{Pic}(K)$ is trivial, then $G = G'$.

Orthogonal transformations

The orthogonal transformations of (M,q) are by definition the automorphisms g of (M,q) such that $C\ell(g)$ is a generalized twisted inner automorphism of $C\ell(M,q)$ according to Definition (5.1.5). All localizations of orthogonal transformations are orthogonal transformations, and when M is finitely generated, an automorphism of (M,q) is an orthogonal transformation if and only if its localizations are orthogonal transformations (see (5.1.11)). In the orthogonal group $\mathrm{GO}(M,q)$ we can distinguish the subgroup of all g such that $C\ell(g) = \Theta_X$ for some graded invertible submodule X of $C\ell(M,q)$, and the subgroup of all g such that $C\ell(g) = \Theta_x$ for some invertible locally homogeneous $x \in C\ell(M,q)$; the group $\mathrm{GO}(M,q)$ and these two subgroups are normal subgroups of $\mathrm{Aut}(M,q)$.

The reversion τ (see (3.1.4)) plays an important role in the study of an automorphism g of (M,q), because

$$\mathrm{Z}^r(C\ell(g)^{-1}) \;=\; \tau(\,\mathrm{Z}^r(C\ell(g))\,). \tag{5.1.13}$$

Indeed, since the algebra $C\ell(M,q)$ is generated by M, x belongs to $\mathrm{Z}^g(C\ell(g))$ if and only if $g(a)x = \sigma(x)a$ for all $a \in M$, and this is equivalent to $\tau(x)\,g(a) = a\,\sigma\tau(x)$ which means that $\tau(x)$ belongs to $\mathrm{Z}^g(C\ell(g)^{-1})$. And since $C\ell(g)$ commutes with τ, the equality $C\ell(g)(x) = \sigma(x)$ is equivalent to $C\ell(g)(\tau(x)) = \sigma\tau(x)$. □

Since we distinguish a subgroup $\mathrm{GO}(M,q)$ in $\mathrm{Aut}(M,q)$, we must face this important question: which geometrical properties (not involving $C\ell(M,q)$) characterize the elements of $\mathrm{GO}(M,q)$ inside $\mathrm{Aut}(M,q)$? The next proposition reveals an important property of every orthogonal transformation g : there exists a *finitely generated* submodule P of M such that $\mathrm{Ker}(g-\mathrm{id})$ contains $P^\perp$. For all quadratic modules under consideration in **5.8**, this property suffices to distinguish orthogonal transformations from other automorphisms of (M,q).

(5.1.14) **Proposition.** *Let g be an orthogonal transformation of (M,q) and let $(x_1,\dots,x_n)$ and $(x'_1,\dots,x'_n)$ be sequences of homogeneous elements of $\mathrm{Z}^r(C\ell(g))$ and $\mathrm{Z}^r(C\ell(g)^{-1})$ satisfying the properties required in* (5.1.5). *If all these x_i and x'_i*

belong to the subalgebra generated by some submodule P of M, then $g(a) = a$ for all $a \in P^\perp$, and $g(a) - a \in P$ for all $a \in M$.

Proof. This is a consequence of (4.4.12): $ax - \sigma(x)a = \mathrm{d}_q(a) \rfloor x$. If we write $1 = \sum_i \lambda_i x_i x'_i$ as in the proof of (5.1.7), then, for all $a \in M$,

$$g(a) - a \;=\; \sum_i \lambda_i(\sigma(x_i)a - ax_i)\; x'_i \;=\; -\sum_i \lambda_i(\mathrm{d}_q(a) \rfloor x_i)\; x'_i.$$

By induction on k it is easy to prove that $\mathrm{d}_q(a) \rfloor x$ belongs to the subalgebra generated by P if x is the product of k elements of P, and even that it vanishes if moreover a belongs to $P^\perp$. The same can be said about all factors $\mathrm{d}_q(a) \rfloor x_i$ above, whence the conclusions. □

Now it seems natural to consider the group $\mathrm{GC}\ell(M,q)$ of all invertible locally homogeneous $x \in \mathrm{C}\ell(M,q)$ such that $\Theta_x(M) = M$, the group $\mathrm{G'C}\ell(M,q)$ of all graded invertible submodules $X \subset \mathrm{C}\ell(M,q)$ such that $\Theta_X(M) = M$, and the group $\mathrm{G''C}\ell(M,q)$ of all submodules $\mathrm{Z}^r(\mathrm{C}\ell(g))$ derived from orthogonal transformations. Therefore $\mathrm{G''C}\ell(M,q)$ is canonically isomorphic to $\mathrm{GO}(M,q)$. Elements x such that $\Theta_x(M) = M$ were still called "Clifford–Lipschitz numbers" by E. Cartan and M. Schenberg, but after 1950, for obscure reasons, $\mathrm{GC}\ell(M,q)$ became known as the "Clifford group". Here we accept this name although it is historically incorrect; indeed Clifford died too soon to concern himself with automorphisms of quadratic spaces, and Lipschitz studied automorphisms of real positive definite quadratic spaces before he became aware that he used the algebras that Clifford had just discovered two years before him.

For quadratic spaces over fields or local rings, $\mathrm{GC}\ell(M,q)$ is perhaps sufficient. For quadratic spaces over arbitrary rings, $\mathrm{G'C}\ell(M,q)$ may still help us effectively. But for more general quadratic modules (even over fields), $\mathrm{GC}\ell(M,q)$ and $\mathrm{G'C}\ell(M,q)$ are misleading. All invertible locally homogeneous $x \in \mathrm{Z}^r(\mathrm{C}\ell(M,q))$ belong to $\mathrm{GC}\ell(M,q)$, all graded invertible $X \subset \mathrm{Z}^r(\mathrm{C}\ell(M,q))$ belong to $\mathrm{G'C}\ell(M,q)$, but since Θ_x or Θ_X is the identity automorphism of $\mathrm{C}\ell(M,q)$, they are superfluous elements with a trivial image id_M in $\mathrm{GO}(M,q)$. When $\mathrm{Z}^r(\mathrm{C}\ell(M,q))$ is very large, they are not only superfluous, but most of them are even harmful because they raise impassable obstructions to the generalization of traditional theorems beyond the traditional nondegenerate case. The disorder becomes obvious when we examine the derived Lie algebras (see (5.5.5) farther). Indeed this examination suggests that all orthogonal transformations should be derived from groups associated with the Lie algebra $\mathrm{C}\ell_0^{\leq 2}(M,q)$.

We get groups associated with $\mathrm{C}\ell_0^{\leq 2}(M,q)$ if we derive them from some multiplicative subset $\mathrm{Lip}(M,q)$ of $\mathrm{C}\ell(M,q)$ which we propose to call the "Lipschitz monoid". A historically correct name is better when no name has yet become usual; moreover Lipschitz's original ideas and methods are more recognizable in Lipschitz monoids than in Clifford groups. Whereas Lipschitz only studied real positive definite quadratic forms, more general Lipschitz monoids appeared in [Sato,

Miwa, Jimbo 1978,...] (yet without reference to Lipschitz), and also in [Helmstetter 1977,...] (although this author formerly called them "Clifford monoids" before being gently rebuked by P. Lounesto for his ignorance of Lipschitz's contribution). The relevance of Lipschitz monoids is confirmed by several stability properties, in particular the stability by interior multiplications (see (5.3.13)), their functorial property (see (5.3.8)) and the invariance property (see (5.4.1)), and by their important role in the study of hyperbolic spaces (see Chapter **7**). The Japanese team Sato-Miwa-Jimbo also proved their relevance in applications to remote topics.

Instead of the three Clifford groups $\mathrm{GC}\ell(M,q)$, $\mathrm{G'C}\ell(M,q)$ and $\mathrm{G''C}\ell(M,q)$, we shall use three subgroups $\mathrm{GLip}(M,q)$, $\mathrm{G'Lip}(M,q)$ and $\mathrm{G''Lip}(M,q)$ derived from the Lipschitz monoid $\mathrm{Lip}(M,q)$. For all quadratic modules here under consideration we shall realize that $\mathrm{G''Lip}(M,q) = \mathrm{G''C}\ell(M,q)$; for the other quadratic modules this equality can be proposed as a main conjecture. Another less important conjecture has been confirmed as far as the study has reached: when $\mathrm{GC}\ell(M,q)$ (resp. $\mathrm{G'C}\ell(M,q)$) is strictly larger than $\mathrm{GLip}(M,q)$ (resp. $\mathrm{G'Lip}(M,q)$), its non-lipschitzian elements are actually superfluous.

5.2 Filtrations of Clifford algebras

Here we only consider increasing filtrations; therefore the word "increasing" will be omitted. A *filtration* on a K-module A is a family of submodules $(A^{\leq k})_{k\in\mathbb{Z}}$ such that $A^{\leq k} \subset A^{\leq k+1}$ for all $k \in \mathbb{Z}$. It is said that the filtration begins with the degree j if $A^{\leq j-1} = 0$ and $A^{\leq j} \neq 0$; it is said that it ends with the degree k if $A^{\leq k-1} \neq A$ and $A^{\leq k} = A$. When no precise filtration has been given to a K-module B, it is automatically provided with the *trivial filtration* that begins and ends with the degree 0; in particular, this convention holds for K itself. We set

$$A^{-\infty} = \bigcap_{k\in\mathbb{Z}} A^{\leq k} \quad \text{and} \quad A^{<+\infty} = \bigcup_{k\in\mathbb{Z}} A^{\leq k}$$

and we say that the filtration is *regular* at $-\infty$ (resp. at $+\infty$) if $A^{-\infty} = 0$ (resp. $A^{<+\infty} = A$). There are two *ludicrous filtrations* which should be avoided (when $A \neq 0$): for the former one all $A^{\leq k}$ are 0 (this is the maximal irregularity at $+\infty$); for the latter one, all $A^{\leq k}$ are A itself (this is the maximal irregularity at $-\infty$).

The filtered modules constitute a category; a *filtered morphism* $f : A \to B$ is a linear mapping f such that $f(A^{\leq k}) \subset B^{\leq k}$ for all k. Moreover $A \otimes B$ and $\mathrm{Hom}(A,B)$ are filtered in this way: $(A\otimes B)^{\leq k}$ is the sum of the images in $A\otimes B$ of all modules $A^{\leq j}\otimes B^{\leq k-j}$ (with an arbitrary j), and $\mathrm{Hom}^{\leq k}(A,B)$ is the submodule of all $f \in \mathrm{Hom}(A,B)$ such that $f(A^{\leq j}) \subset B^{\leq j+k}$ for all j.

An algebra A (associative with unit 1) is a *filtered algebra* if it is a filtered module and if the linear mappings $\pi : A \otimes A \to A$ and $\varepsilon : K \to A$ (see **4.1**) are filtered; in other words, if 1 belongs to $A^{\leq 0}$ and $A^{\leq j}A^{\leq k} \subset A^{\leq j+k}$ for all j and k; this implies that $A^{\leq 0}$ is a subalgebra. Of course the filtered algebras (together with the filtered algebra morphisms) constitute a new category. If a filtered K-module M is a left module over a filtered algebra A, it is said to be a *filtered left A-module* if the mapping $A \otimes M \to M$ is filtered, in other words if $A^{\leq j} M^{\leq k} \subset M^{\leq j+k}$ for all j and k. *Filtered coalgebras* and *filtered comodules* are defined in an analogous way.

The *filtering degree* (or degree) of an element x of A is the smallest k such that x belongs to $A^{\leq k}$. When the notation ∂x refers to a filtration, we only impose the condition $\partial xy \leq \partial x + \partial y$, and in all new definitions we shall require this condition to hold for all kinds of products.

It is clear that the tensor product of two filtered algebras (resp. coalgebras) is a filtered algebra (resp. coalgebra); the following statement deserves more attention.

(5.2.1) **Lemma.** *Let A be a filtered coalgebra, and B a filtered algebra.*

- (a) *The algebra* $\mathrm{Hom}(A, B)$ *defined in* (4.1.1) *is a filtered algebra.*
- (b) *When B is provided with the trivial filtration,* (*for instance when $B = K$ and* $\mathrm{Hom}(A, B) = A^*$), *then* $\mathrm{Hom}^{\leq k}(A, B)$ *is the subset of all $f \in \mathrm{Hom}(A, B)$ vanishing on $A^{\leq -k-1}$.*
- (c) *When P is a filtered right comodule over A and a filtered left module over B, and when the comultiplication $P \to P \otimes A$ is B-linear* (*see* (4.1.2)), *then P is a filtered left module over* $\mathrm{Hom}(A, B)$.

Proof. These statements are direct consequences of the definitions; only the first one will be justified here. When f belongs to $\mathrm{Hom}^{\leq j}(A, B)$, g to $\mathrm{Hom}^{\leq k}(A, B)$ and x to $A^{\leq i}$, then $\pi'_A(x)$ is a sum of tensor products $x'_n \otimes x''_n$ satisfying this condition: there exists h (depending on n) such that $x'_n \in A^{\leq h}$ and $x''_n \in A^{\leq i-h}$. Now $(f*g)(x)$ is the sum of all $f(x'_n)g(x''_n)$, and we know that $f(x'_n) \in B^{\leq h+j}$ and $g(x''_n) \in B^{\leq i-h+k}$. Consequently $(f * g)(x) \in B^{\leq i+j+k}$ and $f * g \in \mathrm{Hom}^{\leq j+k}(A, B)$. □

When we consider modules provided both with a filtration and a grading, it is often necessary to require some compatibility between them. The grading of $M = \bigoplus_{j \in G} M_j$ and its filtration are said to be *compatible* if each submodule $M^{\leq k}$ is graded, in other words, if it is the direct sum of its intersections $M_j^{\leq k}$ with the grading submodules M_j. When we use algebras, coalgebras, modules or comodules which are all provided with a parity grading and a compatible filtration, there is no difficulty in generalizing the previous statements to the algebra $A \hat{\otimes} B$ when A and B are algebras, or to the coalgebra $A \hat{\otimes} B$ when A and B are coalgebras, or to the algebra $\mathrm{Hom}^\wedge(A, B)$ (see (4.2.4)) when A is a coalgebra and B an algebra; and there is a graded version of Lemma (5.2.1).

The filtering submodules $C\ell(M,q;V)^{\leq k}$

The definition of Lipschitz monoids requires other filtrations of Clifford algebras than the natural one presented in **3.1**; we need filtrations of $C\ell(M,q)$ for which the elements of M have a degree ≤ 1, yet exceptionally ≤ -1 if they belong to some submodule V.

(5.2.2) **Theorem.** *Let V be a totally isotropic submodule of M (in other words, $q(V)=0$). For all $k \in \mathbb{Z}$ let $C\ell(M,q;V)^{\leq k}$ be the submodule of $C\ell(M,q)$ generated by all products $a_1a_2\cdots a_{i+j}b_1b_2\cdots b_i$ in which all factors belong to M, the last ones $b_1,\dots,b_i$ to V, and $j \leq k$. The submodules $C\ell(M,q;V)^{\leq k}$ constitute a filtration of the algebra $C\ell(M,q)$ for which all elements of V have a degree ≤ -1. This filtration is always regular at $+\infty$; it is regular at $-\infty$ whenever V is a direct summand of M. It begins with a finite degree ≤ 0 whenever V is finitely generated, and it ends with a finite degree ≥ 0 whenever M/V is finitely generated. Besides, it is compatible with the parity grading of $C\ell(M,q)$, and it is invariant by the reversion:*

$$\tau(C\ell(M,q;V)^{\leq k}) = C\ell(M,q;V)^{\leq k} \quad \textit{for all } k.$$

Proof. It is clear that 1 belongs to $C\ell(M,q;V)^{\leq 0}$ since a product of zero factor is 1 by definition. Now we must prove that $x'x$ belongs to $C\ell(M,q;V)^{\leq k'+k}$ when x' and x belong respectively to $C\ell(M,q;V)^{\leq k'}$ and $C\ell(M,q;V)^{\leq k}$. Since we can assume x' to be some product $a'_1a'_2\cdots a'_{i'+j'}b'_1b'_2\cdots b'_{i'}$ (with $b'_1,\dots,b'_{i'}$ in V, and $j' \leq k'$), it suffices to prove that $a'x$ belongs to $C\ell(M,q;V)^{\leq k+1}$ for all $a' \in M$, and that $b'x$ belongs to $C\ell(M,q;V)^{\leq k-1}$ for all $b' \in V$; since the former statement is evident, we focus our attention on the latter. We can assume that $x = yb_1b_2\cdots b_i$ with a first factor y in $C\ell^{\leq i+k}(M,q)$, and all the following ones in V; from (4.4.12) we deduce that

$$b'y \;=\; \sigma(y)b' \;+ \mathrm{d}_q(b') \rfloor y \;;$$

it is clear that $\sigma(y)b'b_1b_2\cdots b_i$ falls into $C\ell(M,q;V)^{\leq k-1}$; moreover the interior multiplication by $\mathrm{d}_q(b')$ maps $C\ell^{\leq i+k}(M,q)$ into $C\ell^{\leq i+k-1}(M,q)$, and consequently the same conclusion holds for $(\mathrm{d}_q(b') \rfloor y)b_1b_2\cdots b_i$. Thus we have proved that the submodules $C\ell(M,q;V)^{\leq k}$ constitute an algebra filtration.

Since the submodules $C\ell(M,q;V)^{\leq k}$ are generated by homogeneous elements, the filtration is compatible with the parity grading. Besides, it is clear that the reversion τ is a filtered mapping; since it is involutive, it leaves the filtration invariant.

The filtration by the submodules $C\ell(M,q;V)^{\leq k}$ is regular at $+\infty$ because $C\ell(M,q;V)^{\leq k}$ contains $C\ell^{\leq k}(M,q)$. Let us assume that M is the direct sum of V and a supplementary module U, and let u and v be the evident algebra morphisms from $C\ell(U,q_{|U})$ and $\bigwedge(V)$ into $C\ell(M,q)$. From (4.8.5) we deduce the bijectiveness of this mapping:

$$C\ell(U,q_{|U}) \hat{\otimes} \bigwedge(V) \longrightarrow C\ell(M,q)\,, \qquad y \otimes z \longmapsto u(y)\,v(z).$$

Let C_i be the submodule generated by all $u(y)v(z)$ with $z \in \bigwedge^i(V)$; thus $\mathrm{C}\ell(M,q)$ is the direct sum of all submodules C_i. If a nonzero x belongs to $\mathrm{C}\ell(M,q;V)^{\leq k}$ (with $k<0$), it is easy to prove that x may have a nonzero component in some C_i only if $k \geq -i$; this inequality shows that x cannot belong to all $\mathrm{C}\ell(M,q;V)^{\leq k}$, and that the filtration is regular at $-\infty$.

If V is generated by n elements, then $\bigwedge^{n+1}(V) = 0$; therefore in $\mathrm{C}\ell(M,q)$ every product of $n+1$ or more elements of V vanishes, and $\mathrm{C}\ell(M,q;V)^{\leq -n-1} = 0$. Now let us assume that M/V is finitely generated, and that $a_1,\dots,a_m$ are elements of M, the images of which generate M/V; consequently M is generated by these m elements and some family of generators of V; from (3.1.7) we derive that every element of $\mathrm{C}\ell(M,q)$ is a sum of products in which the first factors are distinct elements in the set $\{a_1, a_2, \dots, a_m\}$ and the following factors all belong to the set of generators of V; this proves that $\mathrm{C}\ell(M,q)$ is equal to $\mathrm{C}\ell(M,q;V)^{\leq m}$. □

Comments. Perhaps you have been astonished because the hypothesis $q(V)=0$ has not been used to prove that the submodules $\mathrm{C}\ell(M,q;V)^{\leq k}$ constitute an algebra filtration; we do not need this hypothesis to prove the existence of the filtration, but to prove that it is sensible. Indeed let us assume on the contrary that the ideal of K generated by $q(V)$ is K; this would imply that 1 belongs to $\mathrm{C}\ell(M,q;V)^{\leq -2}$, and since $x = 1^k x$ for all x in $\mathrm{C}\ell(M,q)$ and all exponents k, we would have got the ludicrous filtration which assigns the degree $-\infty$ to all elements.

When $V=0$, then $\mathrm{C}\ell(M,q;0)^{\leq k} = \mathrm{C}\ell^{\leq k}(M,q)$. And when $V=M$, then $q=0$ and $\mathrm{C}\ell(M,q;M)^{\leq k} = \bigwedge^{\geq -k}(M)$; in other words, the exterior algebra admits a *decreasing* filtration by submodules $\bigwedge^{\geq k}(M)$, and the decreasing group morphism $k \longmapsto -k$ transforms this decreasing filtration into an increasing one.

The proof of the following lemmas is evident and will be omitted.

(5.2.3) **Lemma.** *Let V and V' be totally isotropic submodules of (M,q) and (M',q') respectively, and $g : M \to M'$ a morphism of quadratic modules such that $g(V) \subset V'$. The resulting algebra morphism $\mathrm{C}\ell(g) : \mathrm{C}\ell(M,q) \to \mathrm{C}\ell(M',q')$ is filtered for the filtrations determined by V and V'.*

(5.2.4) **Lemma.** *Let V and V' be totally isotropic submodules of (M,q) and (M',q') respectively, and let us filter $\mathrm{C}\ell((M,q) \perp (M',q'))$ by means of $V \oplus V'$. This filtered algebra is canonically isomorphic to $\mathrm{C}\ell(M,q) \hat{\otimes} \mathrm{C}\ell(M',q')$ provided with the tensor product filtration.*

(5.2.5) **Corollary.** *For the filtrations of $\mathrm{C}\ell(M,q)$ and $\bigwedge(M)$ determined by the totally isotropic submodule V, the comultiplication*

$$\pi'_q : \mathrm{C}\ell(M,q) \to \mathrm{C}\ell(M,q) \hat{\otimes} \bigwedge(M)$$

is a filtered morphism, and thus $\mathrm{C}\ell(M,q)$ is a filtered comodule over $\bigwedge(M)$.

(5.2.6) **Lemma.** *Let $K \to K'$ be a ring extension. With the totally isotropic subspace V of (M,q) is associated a totally isotropic subspace V', image of $K' \otimes V$ in $(M',q') = K' \otimes (M,q)$; and for all $k \in \mathbb{Z}$, $C\ell_{K'}(M',q';V')^{\leq k}$ is the image of $K' \otimes C\ell(M,q;V)^{\leq k}$.*

Consequently the filtrations presented here have a nice behaviour with respect to localizations.

The grading submodules $C\ell(M,q;U,V)^k$

Sometimes the filtration under consideration comes from a grading.

(5.2.7) **Proposition.** *Let us suppose that (M,q) is the direct sum of two totally isotropic submodules U and V. For every $k \in \mathbb{Z}$ let $C\ell(M,q;U,V)^k$ be the submodule of $C\ell(M,q)$ generated by all products $a_1a_2\cdots a_{i+k}b_1b_2\cdots b_i$ such that $a_1,\dots,a_{i+k}$ belong to U, and $b_1,\dots,b_i$ to V. Then $C\ell(M,q)$ is the direct sum of the submodules $C\ell(M,q;U,V)^k$ which provide it with an algebra grading over $\mathbb{Z}$. The elements of U (resp. V) have degree 1 (resp. -1). Moreover the filtration determined by V comes from this grading, whereas the filtration determined by U comes from the grading by the submodules $C\ell(M,q;V,U)^k = C\ell(M,q;U,V)^{-k}$. Conversely $C\ell(M,q;U,V)^k$ is the intersection of $C\ell(M,q;V)^{\leq k}$ and $C\ell(M,q;U)^{\leq -k}$.*

Proof. We use the same notation as in the proof of (5.2.2); since U and V are both totally isotropic, now we get a bijection

$$\bigwedge(U) \hat{\otimes} \bigwedge(V) \longrightarrow C\ell(M,q)\ , \qquad y \otimes z \longmapsto u(y)\, v(z).$$

It proves that $C\ell(M,q)$ is the direct sum of the images of all submodules $\bigwedge^{i+k}(U)\otimes \bigwedge^{i}(V)$, and consequently it is the direct sum of the submodules $C\ell(M,q;U,V)^k$. A slight modification of the proof of (5.2.2) allows us here to prove that we get an algebra grading. Here $x = yb_1b_2\cdots b_i$ with y in $u(\bigwedge^{i+k}(U))$ and $b_1,\dots,b_i$ in V, so that x belongs to $C\ell(M,q;U,V)^k$; we use the equality $b'y = \sigma(y)b' + \mathrm{d}_q(b') \rfloor y$ when b' belongs to V; since $\mathrm{d}_q(b') \rfloor y$ belongs to $u(\bigwedge^{i+k-1}(U))$, we realize that $b'x \in C\ell(M,q;U,V)^{k-1}$. The last statements in (5.2.7) still follow from the fact that $C\ell(M,q)$ is the direct sum of the images of all $\bigwedge^{i+k}(U) \otimes \bigwedge^{i}(V)$. □

Remark. Hyperbolic quadratic spaces are particular examples of direct sums of two totally isotropic submodules. In Theorem (3.7.2) it is stated that $C\ell(\mathbf{H}[P])$ is isomorphic to $\mathrm{End}(\bigwedge(P))$, which inherits from $\bigwedge(P)$ a grading over $\mathbb{Z}$; this grading corresponds exactly to the one defined in (5.2.7): the elements of P^* have degree -1 whereas the elements of P have degree 1.

Interior products and exponentials

Let us consider the algebra $\bigwedge^*(M)$ dual to the coalgebra $\bigwedge(M)$. If V is any submodule of M, the filtration of the coalgebra $\bigwedge(M)$ by the submodules $\bigwedge(M;V)^{\leq k}$ determines a filtration of the algebra $\bigwedge^*(M)$ by submodules denoted by $\bigwedge^*(M;V)^{\leq k}$; as explained in (5.2.1), $\bigwedge^*(M;V)^{\leq k}$ is the set of all $f \in \bigwedge^*(M)$ such that $f(\bigwedge(M;V)^{\leq -k-1}) = 0$.

When $V = 0$, then $\bigwedge^*(M;0)^{\leq k}$ is the submodule $\bigwedge^{*\geq -k}(M)$ canonically isomorphic to $(\bigwedge^{\geq -k}(M))^*$. When $V = M$, then $\bigwedge^*(M;M)^{\leq k}$ is the submodule $\bigwedge^{*\leq k}(M)$ canonically isomorphic to $(\bigwedge^{\leq k}(M))^*$.

Since $\bigwedge(M)$ is a filtered coalgebra, and $C\ell(M,q)$ a filtered comodule over it (see (5.2.5)), the following statement is evident.

(5.2.8) **Lemma.** *When V is a totally isotropic submodule in the quadratic module (M,q), then*

$$f \lfloor x \in \bigwedge^*(M;V)^{\leq i+k} \quad \text{and} \quad f \rfloor y \in C\ell(M,q;V)^{\leq j+k}$$

for every $x \in \bigwedge(M;V)^{\leq i}$, every $y \in C\ell(M,q;V)^{\leq j}$ and every $f \in \bigwedge^*(M;V)^{\leq k}$.

Of course when M is a direct sum of two totally isotropic subspaces, there is an analogous lemma involving the gradings determined by them.

We need a last lemma about exponentials.

(5.2.9) **Lemma.** *Let x and f be elements of $\bigwedge_0^+(M)$ and $\bigwedge_0^{*+}(M)$ respectively, so that their exponentials exist. When x belongs to $\bigwedge(M;V)^{\leq 0}$, then* $\mathrm{Exp}(x)$ *too belongs to it. And when f belongs to $\bigwedge^*(M;V)^{\leq 0}$, then* $\mathrm{Exp}(f)$ *too belongs to it.*

Proof. The assertion about $\mathrm{Exp}(x)$ is an immediate consequence of the definitions, but there is more work with $\mathrm{Exp}(f)$. It suffices to prove by induction on k that $\mathrm{Exp}(f)(x)$ vanishes for all $x \in \bigwedge^{\leq k}(M) \cap \bigwedge(M;V)^{\leq -1}$. Since $\mathrm{Exp}(f)$ is even, the induction begins trivially when $k = 1$. In order to go from k to $k+1$, we replace x with $b \wedge x$ such that b is in V and x in $\bigwedge^k(M) \cap \bigwedge(M;V)^{\leq 0}$. Let $\sum_n x'_n \otimes x''_n$ be the coproduct $\pi'(x)$; since the comultiplication is filtered, there exists i (depending on n) such that x'_n and x''_n belong respectively to $\bigwedge(M;V)^{\leq i}$ and $\bigwedge(M;V)^{\leq -i}$. Let us write

$$\mathrm{Exp}(f)(b \wedge x) \;=\; (\mathrm{Exp}(f) \wedge (f \lfloor b))(x) \;=\; \sum_n \mathrm{Exp}(f)(x'_n)\; f(b \wedge x''_n)\;;$$

when $i \geq 0$, then $f(b \wedge x''_n) = 0$, and when $i < 0$, then $\mathrm{Exp}(f)(x'_n) = 0$ because of the induction hypothesis which assumes that $\mathrm{Exp}(f)$ vanishes on $\bigwedge^{\leq k}(M) \cap \bigwedge(M;V)^{\leq -1}$. $\square$

5.3 Lipschitz monoids and derived groups

Let (M, q) be a quadratic module. We consider the orthogonal sum $(M, q) \perp (M, -q)$ in which there are two conspicuous totally isotropic submodules, namely the first diagonal Δ which is the subset of all (a, a) with $a \in M$, and the second diagonal Δ' which is the subset of all $(a, -a)$. Both are direct summands of $M \oplus M$. Each one leads to a filtration of the Clifford algebra under consideration, and determines a subalgebra of elements of degree ≤ 0. From (3.2.4) and (3.2.2) we deduce that the mapping $(a, b) \longmapsto a \otimes 1^{to} + 1 \otimes b^{to}$ extends to an isomorphism

$$\mathrm{C}\ell((M, q) \perp (M, -q)) \longrightarrow \mathrm{C}\ell(M, q) \hat{\otimes} \mathrm{C}\ell(M, q)^{to} ;$$

consequently the filtration determined by Δ or Δ' can be carried onto

$$\mathrm{C}\ell(M, q) \hat{\otimes} \mathrm{C}\ell(M, q)^{to},$$

and thus we get submodules

$$(\mathrm{C}\ell(M, q) \hat{\otimes} \mathrm{C}\ell(M, q)^{to};\ \Delta)^{\leq k} \quad \text{or} \quad (\mathrm{C}\ell(M, q) \hat{\otimes} \mathrm{C}\ell(M, q)^{to};\ \Delta')^{\leq k}.$$

The next proposition contains the definition of the *Lipschitz monoid*, the elements of which are called *lipschitzian elements*, and the following two propositions give its elementary properties; in particular $\mathrm{Lip}(M, q)$ is actually a monoid (or multiplicative subset) in $\mathrm{C}\ell(M, q)$.

(5.3.1) **Proposition and definition.** *For every locally homogeneous element $x \in \mathrm{C}\ell(M, q)$ these two assertions are equivalent:*

$x \otimes \tau(x)^{to}$ *belongs to* $(\mathrm{C}\ell(M, q) \hat{\otimes} \mathrm{C}\ell(M, q)^{to};\ \Delta)^{\leq 0}$ *;*

$x \otimes \tau(x)^{to}$ *belongs to* $(\mathrm{C}\ell(M, q) \hat{\otimes} \mathrm{C}\ell(M, q)^{to};\ \Delta')^{\leq 0}$ *;*

moreover x satisfies these properties if and only if its homogeneous components satisfy them.

By definition the Lipschitz monoid $\mathrm{Lip}(M, q)$ is the set of all locally homogeneous elements satisfying these properties.

Proof. Let us write $x = (1 - \lambda)x + \lambda x$ as in (5.1.4); the following equalities show that each assertion in (5.3.1) is true for x if and only if it is true for its homogeneous components $(1 - \lambda)x$ and λx :

$$(1 - \lambda)x \otimes \tau((1 - \lambda)x)^{to} = (1 - \lambda)\ (x \otimes \tau(x)^{to})\ ,$$

$$\lambda x \otimes \tau(\lambda x)^{to} = \lambda\ (x \otimes \tau(x)^{to})\ ,$$

$$x \otimes \tau(x)^{to} = (1 - \lambda)x \otimes \tau((1 - \lambda)x)^{to} + \lambda x \otimes \tau(\lambda x)^{to}.$$

Therefore it suffices to consider a homogeneous x. The mapping $(a, b) \longmapsto (a, -b)$ is an automorphism of $(M, q) \perp (M, -q)$ which permutes Δ and Δ'; it induces an automorphism of $\mathrm{C}\ell(M, q) \hat{\otimes} \mathrm{C}\ell(M, q)^{to}$, exactly this one: $x \otimes y^{to} \longmapsto x \otimes \sigma(y)^{to}$. It suffices to observe that $x \otimes \tau(x)^{to}$ is invariant by this automorphism when x is even, and mapped to the opposite element when x is odd. $\square$

(5.3.2) **Proposition.** *When x and y belong to* $\mathrm{Lip}(M,q)$, *then xy and $\tau(x)$ too belong to it. Moreover* $\mathrm{Lip}(M,q)$ *contains all elements of M and all elements $\lambda + ab$ with λ in K and a and b in M.*

Proof. When x and y are homogeneous, then

$$(x \otimes \tau(x)^{to})\ (y \otimes \tau(y)^{to}) \ = \ xy \otimes \tau(xy)^{to} \ ;$$

the validity of this equality obviously extends to the case of locally homogeneous x and y. Since the submodule of all elements of filtering degree ≤ 0 is a subalgebra, it proves that xy belongs to $\mathrm{Lip}(M,q)$ whenever x and y belong to it. All filtrations defined in (5.2.2) are invariant by the reversion; from (3.2.8) we derive that $\tau(x \otimes \tau(x)^{to})$ is equal to $(-1)^{\partial x}\tau(x) \otimes x^{to}$; consequently $\tau(x)$ belongs to $\mathrm{Lip}(M,q)$ if (and only if) x belongs to it. The following equality proves that $\mathrm{Lip}(M,q)$ contains every element a of M :

$$a \otimes a^{to} \ = \ (a \otimes 1^{to})(a \otimes 1^{to} \ + 1 \otimes a^{to}) - q(a) \otimes 1^{to} \ .$$

Let us notice that the filtering degree of $ab \otimes 1^{to} + 1 \otimes (ba)^{to}$ is ≤ 0 :

$$\begin{aligned} & ab \otimes 1^{to} + 1 \otimes (ba)^{to} \\ & \quad = (a \otimes 1^{to})(b \otimes 1^{to} + 1 \otimes b^{to}) \ - (a \otimes 1^{to} + 1 \otimes a^{to})(1 \otimes b^{to}) \ ; \end{aligned}$$

it follows that $\lambda + ab$ belongs to $\mathrm{Lip}(M,q)$:

$$\begin{aligned} & (\lambda + ab) \otimes (\lambda + ba)^{to} \\ & \quad = \lambda^2 \otimes 1^{to} \ + \lambda(ab \otimes 1^{to} + 1 \otimes (ba)^{to}) \ + (a \otimes a^{to})(b \otimes b^{to}). \end{aligned}$$

□

(5.3.3) **Proposition.** *When x belongs to* $\mathrm{Lip}(M,q)$ *and y to* $C\ell^{\leq k}(M,q)$ *for some k, then $xy\tau(x)$ too belongs to* $C\ell^{\leq k}(M,q)$. *In particular $x\tau(x)$ belongs to K, and $xa\tau(x)$ belongs to M for every $a \in M$. Moreover the equality $x\tau(x) = \tau(x)x$ is true for every lipschitzian x such that $x\tau(x)$ or $\tau(x)x$ is not a divisor of zero in K; and if the mapping $a \longmapsto 2a$ is bijective from M onto M, it is true for every lipschitzian x.*

Proof. We can suppose that x is homogeneous (because it is locally homogeneous) and that y is homogeneous (because $xy\tau(x)$ depends linearly on it). We consider again the linear mapping π_q from $C\ell(M,q) \otimes C\ell(M,q)$ onto $C\ell(M,q)$ such that $\pi_q(x_1 \otimes x_2) = x_1x_2$, and we observe that $xy\tau(x) = \pi_q(xy \otimes \tau(x))$. This equality is interesting because $xy \otimes \tau(x)^{to}$ has a filtering degree $\leq k$ for the filtration determined by Δ'; indeed

$$xy \otimes \tau(x)^{to} \ = \ \pm(x \otimes \tau(x)^{to})\ (y \otimes 1^{to}).$$

Consequently we can decompose it into a sum of terms of the following two kinds. First there are terms $x_1 \otimes x_2^{to}$ which have a filtering degree $\leq k$ for the natural filtration that ignores Δ'; their contributions $\pi_q(x_1 \otimes x_2)$ to the calculation of $xy\tau(x)$ obviously belong to $C\ell^{\leq k}(M,q)$. And there are terms that can be written as the product of some factor $x_1 \otimes x_2^{to}$ by some $a \otimes 1^{to} - 1 \otimes a^{to}$ that has degree -1 for the filtration determined by Δ'. The following equalities show that their contributions to $xy\tau(x)$ all vanish:

$$(x_1 \otimes x_2^{to})\,(a \otimes 1^{to} - 1 \otimes a^{to}) = x_1 a \otimes \sigma(x_2)^{to} - x_1 \otimes (a\sigma(x_2))^{to},$$
$$\pi_q(\,x_1 a \otimes \sigma(x_2) - x_1 \otimes a\sigma(x_2)) = 0.$$

When $k = 0$, it follows that $x\tau(x)$ belongs to K, and also $\tau(x)x$, since $\tau(x)$ too is lipschitzian. When $k = 1$, we observe that $xa\tau(x)$ belongs to M because it is an odd element of $C\ell^{\leq 1}(M,q)$.

Since $x\tau(x)$ and $\tau(x)x$ both belong to the even center $Z_0(C\ell(M,q))$, it is easy to verify that

$$(x\tau(x))^2 \;=\; (x\tau(x))\,(\tau(x)x) \;=\; (\tau(x)x)^2\ ;$$

the equality $x\tau(x) = \tau(x)x$ follows when either member is not a divisor of zero in K. When the mapping $a \longmapsto 2a$ is bijective, this equality is a consequence of (4.8.16):

$$x\tau(x) = \mathrm{Scal}(x\tau(x)) = \mathrm{Scal}(\tau(x)x) = \tau(x)x. \qquad \square$$

Remarks.

(a) When $q = 0$, the Lipschitz monoid lies in $\bigwedge(M)$; it is denoted by $\mathrm{Lip}(M)$ and called the *neutral Lipschitz monoid.* From (5.3.2) and (4.5.1) we deduce that $\mathrm{Lip}(M)$ contains $\mathrm{Exp}(u)$ for every $u \in \bigwedge^2(M)$.

(b) The general validity of the equality $x\tau(x) = \tau(x)x$ in (5.3.3) is still an open question; although the failure of all attempts to answer it is a vexation, it does not raise any serious hindrance. From (5.3.3) it follows that this equality is valid for all lipschitzian elements if K is an integral domain. Besides, if x belongs to the submodule $Z^r(C\ell(g))$ determined by an orthogonal transformation g, then $\tau(x)$ belongs to $Z^r(C\ell(g)^{-1})$ (see (5.1.13)), therefore commutes with x (see (5.1.6)).

(c) Instead of the isomorphism $C\ell(M,-q) \to C\ell(M,q)^{to}$ we could use the isomorphism $C\ell(M,-q) \to C\ell(M,q)^t$ also mentioned in (3.2.2); a locally homogeneous x is lipschitzian if and only if $x \otimes x^t$ has filtering degree ≤ 0 in $C\ell(M,q) \hat{\otimes} C\ell(M,q)^t$ for the filtration determined by Δ or Δ'.

(d) When $K \to L$ is an extension of K, every lipschitzian element x in $C\ell(M,q)$ gives a lipschitzian element in $L\otimes C\ell(M,q)$, the Clifford algebra of $L\otimes(M,q)$. If this ring extension is faithfully flat, the converse statement is also true. The next lemmas give more precise results for rings of fractions of K.

(5.3.4) **Lemma.** *If S is a multiplicative subset of K, every homogeneous lipschitzian element of $S^{-1}C\ell(M,q)$ (the Clifford algebra of $S^{-1}(M,q)$) can be written as a fraction in which the numerator is a lipschitzian homogeneous element of $C\ell(M,q)$.*

Proof. A homogeneous $x \in C\ell(M,q)$ is lipschitzian if (by definition) $x \otimes \tau(x)^{to}$ can be written as a sum of terms like $(u \otimes v^{to}) \prod_{i=1}^{m} (a_i \otimes 1^{to} + 1 \otimes a_i^{to})$, with $u \in C\ell^{\leq j}(M,q)$ and $v \in C\ell^{\leq k}(M,q)$ such that $j+k \leq m$. A homogeneous fraction x/s in $S^{-1}C\ell(M,q)$ is lipschitzian if and only if $(x \otimes \tau(x)^{to})/s^2$ can be written as a fraction in which the numerator is a sum of terms like the previous ones. This means that tx is lipschitzian in $C\ell(M,q)$ for some $t \in S$. It suffices to replace the fraction x/s with $(tx)/(st)$. □

(5.3.5) **Lemma.** *For every locally homogeneous $x \in C\ell(M,q)$ these three assertions are equivalent:*

x is lipschitzian;

for every prime ideal $\mathfrak{p}$ its image in $C\ell(M_\mathfrak{p}, q_\mathfrak{p})$ is lipschitzian;

for every maximal ideal $\mathfrak{m}$ its image in $C\ell(M_\mathfrak{m}, q_\mathfrak{m})$ is lipschitzian.

Proof. Because of (5.2.6), there is no problem in localizing the filtrations of $C\ell(M,q) \hat{\otimes} C\ell(M,q)^{to}$. The fact that $x \otimes \tau(x)^{to}$ belongs to some submodule of $C\ell(M,q) \hat{\otimes} C\ell(M,q)^{to}$ is a local property because it is equivalent to an inclusion between two submodules (see (1.11.6)); consequently for a locally homogeneous element to be lipschitzian it is also a local property. □

Derived Lipschitz groups

From (5.3.3) we deduce that a lipschitzian element x is invertible in $C\ell(M,q)$ if and only if $x\tau(x)$ is invertible in K; indeed $\tau(x)$ is invertible whenever x is invertible, and conversely the invertibility of $x\tau(x)$ implies $x\tau(x) = \tau(x)x \in K$ (see (5.3.3)), whence $x^{-1} = \mu^{-1}\tau(x)$ if $\mu = x\tau(x)$. It follows from (5.3.3) that the twisted inner automorphism Θ_x leaves M invariant; indeed for every $a \in M$,

$$\Theta_x(a) = \mu^{-1} x a \sigma\tau(x) \in M \quad \text{and} \quad \Theta_x^{-1}(a) = \mu^{-1}\tau(x) a \sigma(x) \in M \ ;$$

the presence of σ in these calculations is not a hindrance since the homogeneous components of x are lipschitzian. Consequently x determines an orthogonal transformation G_x of (M,q). The following proposition is now evident; it leads to the *Lipschitz group* $\mathrm{GLip}(M,q)$ *of invertible elements.*

(5.3.6) **Proposition.** *The subset $\mathrm{GLip}(M,q)$ of all invertible lipschitzian elements is a group; there is a canonical morphism from $\mathrm{GLip}(M,q)$ into the orthogonal group $\mathrm{GO}(M,q)$ which maps every x to the orthogonal transformation G_x defined by $G_x(a) = x a\, \sigma(x)^{-1}$. The kernel of this morphism is the group of invertible lipschitzian elements in $\mathrm{Z}^r(C\ell(M,q))$. Besides, the mapping $x \longmapsto x\tau(x)$ is a group morphism from $\mathrm{GLip}(M,q)$ into $K^\times$.*

An *invertible lipschitzian submodule* of $C\ell(M,q)$ is a graded invertible submodule contained in $\mathrm{Lip}(M,q)$. Every localization of such a submodule X is generated by an invertible lipschitzian element, and consequently the inverse module is

$\tau(X)$. Conversely if X is a submodule such that every localization at any maximal ideal is generated by an invertible lipschitzian element, then by localization we can prove the bijectiveness of the multiplication mappings $X \otimes \tau(X) \to K$ and $\tau(X) \otimes X \to K$; from (5.3.5) we deduce that all elements of X are lipschitzian, and thus X is an invertible lipschitzian submodule. Since $\mathrm{Lip}(M,q)$ is stable by multiplication, the invertible lipschitzian submodules constitute a group $\mathrm{G'Lip}(M,q)$, in which the unit element is the submodule K; it is called the *Lipschitz group of invertible submodules.*

An element of $\mathrm{G'Lip}(M,q)$ is a free module if and only if it is generated by an invertible lipschitzian element, whence the exact sequence

$$(5.3.7) \qquad 1 \longrightarrow K^\times \longrightarrow \mathrm{GLip}(M,q) \longrightarrow \mathrm{G'Lip}(M,q) \longrightarrow \mathrm{Pic}(K) \ ;$$

the Picard group $\mathrm{Pic}(K)$ is defined in **1.12**. When $\mathrm{Pic}(K) = \{1\}$ (for instance when K is a local ring), then $\mathrm{G'Lip}(M,q)$ is isomorphic to the quotient $\mathrm{GLip}(M,q)/K^\times$.

By localization it is easy to verify that the twisted inner automorphism Θ_X associated with each $X \in \mathrm{G'Lip}(M,q)$ leaves M invariant; whence an orthogonal transformation G_X of (M,q) and a group morphism $\mathrm{G'Lip}(M,q) \to \mathrm{GO}(M,q)$. Its kernel is the subgroup of all invertible submodules of $\mathrm{Z}^r(\mathrm{C}\ell(M,q))$; it is injective when $\mathrm{Z}^r(\mathrm{C}\ell(M,q)) = K$. Later we shall prove that the canonical morphisms $\mathrm{G'Lip}(M,q) \to \mathrm{GO}(M,q) \to \mathrm{Aut}(M,q)$ are bijective when (M,q) is a quadratic space.

Unfortunately this morphism $\mathrm{G'Lip}(M,q) \to \mathrm{GO}(M,q)$ is not always surjective (see (5.ex.14)); therefore we must still contemplate a group $\mathrm{G''Lip}(M,q)$ called the *Lipschitz group of* Z^r*-submodules.* Its elements are the submodules over $\mathrm{Z}^r(\mathrm{C}\ell(M,q))$ generated by a finite sequence of homogeneous lipschitzian elements $(x_1,\dots,x_n)$ satisfying these properties: K is generated as an ideal by the n elements $x_i\tau(x_i)$, and the n^2 products $x_i\tau(x_j)$ and the n^2 products $\tau(x_j)x_i$ all belong to $\mathrm{Z}^r(\mathrm{C}\ell(M,q))$. Because of Theorem (5.1.7), such a sequence $(x_1,\dots,x_n)$ determines a graded automorphism θ of $\mathrm{C}\ell(M,q)$, and even an orthogonal transformation of (M,q) if we manage to prove that $\theta(M) = M$. This can be proved either by localization with the help of (5.3.6), or by a direct calculation: $\theta(a) = \sum_i \lambda_i x_i a\, \sigma\tau(x_i)$ for every $a \in M$ (and with suitable $\lambda_i \in K$), whence $\theta(M) \subset M$ because of (5.3.3), and similarly $\theta^{-1}(M) \subset M$. Conversely from (5.1.7) we know that $\mathrm{Z}^r(\theta)$ is the $\mathrm{Z}^r(\mathrm{C}\ell(M,q))$-submodule under consideration.

Since $\mathrm{Lip}(M,q)$ is stable by multiplication and invariant by the reversion τ, from (5.1.10) we deduce that $\mathrm{G''Lip}(M,q)$ is a group with unit element $\mathrm{Z}^r(\mathrm{C}\ell(M,q))$, and that there is a group morphism $\mathrm{G''Lip}(M,q) \to \mathrm{GO}(M,q)$. This morphism is obviously injective, and its surjectiveness is equivalent to the equality $\mathrm{G''Lip}(M,q) = \mathrm{G''C}\ell(M,q)$ which is the main conjecture in this chapter. As long as it is not proved in all cases, the image of $\mathrm{G''Lip}(M,q)$ in $\mathrm{GO}(M,q)$ is denoted by $\mathrm{GO}_{\mathrm{Lip}}(M,q)$.

(5.3.8) **Proposition.** *Let w be a morphism from (M,q) into (M',q'); the resulting algebra morphisms* $C\ell(w)$ *maps* $\mathrm{Lip}(M,q)$ *into* $\mathrm{Lip}(M',q')$ *and determines group morphisms* $\mathrm{GLip}(M,q) \to \mathrm{GLip}(M',q')$ *and* $\mathrm{G'Lip}(M,q) \to \mathrm{G'Lip}(M',q')$. *If* $C\ell(w)$ *maps* $Z^r(C\ell(M,q))$ *into* $Z^r(C\ell(M',q'))$, *it determines group morphisms* $\mathrm{G''Lip}(M,q) \to \mathrm{G''Lip}(M',q')$ *and* $\mathrm{GO_{Lip}}(M,q) \to \mathrm{GO_{Lip}}(M',q')$.

Proof. It suffices to consider this morphism of quadratic modules:

$$(M,q) \perp (M,-q) \longrightarrow (M',q') \perp (M',-q') \ , \quad (a,b) \longmapsto (w(a), w(b)) \ ;$$

since it maps Δ into the diagonal of $M' \oplus M'$, it follows from (5.2.3) that the algebra morphism $C\ell(w) \otimes C\ell(w)^{to}$ is filtered for the filtrations determined by the diagonals; consequently $C\ell(w)$ maps $\mathrm{Lip}(M,q)$ into $\mathrm{Lip}(M',q')$. The subsequent statements in (5.3.8) are now evident. □

It sometimes occurs that a morphism $w : (M,q) \to (M',q')$ induces a morphism $\mathrm{GO}(M,q) \to \mathrm{GO}(M',q')$ in a canonical way; this occurs for instance if (M,q) is a quadratic space, because in this case w is injective and M' is the direct sum of $w(M)$ and $w(M)^\perp$ (see (2.3.8)), and thus every automorphism of (M,q) gives an automorphism of (M',q') that leaves invariant all elements of $w(M)^\perp$. The existence of the morphism $\mathrm{GO_{Lip}}(M,q) \to \mathrm{GO_{Lip}}(M',q')$ is ensured by a weaker hypothesis involving the reduced centers.

It remains to find methods allowing us to recognize whether an automorphism g of (M,q) belongs to $\mathrm{GO_{Lip}}(M,q)$. If we read again the proof of (5.1.11) and take (5.3.4) and (5.1.13) into account, we come to the following conclusions.

(5.3.9) **Theorem.** *The following three assertions are equivalent for every* $g \in \mathrm{Aut}(M,q)$:

- (a) g *is in* $\mathrm{GO_{Lip}}(M,q)$ *(the image of* $\mathrm{G''Lip}(M,q) \to \mathrm{GO}(M,q)$*);*
- (b) *there is a Zariski extension* $L = \prod_{i=1}^n K_{s_i}$ *and a family of homogeneous* $x_i \in \mathrm{Lip}(M,q) \cap Z^r(C\ell(g))$ *(with* $i = 1,2,\dots,n$*) such that* $z = (x_1/1, \dots, x_n/1)$ *is invertible in* L, *and* $L \otimes g = G_z$;
- (c) *every localization of* g *is the orthogonal transformation determined by an invertible lipschitzian element that belongs to the localization of* $Z^r(C\ell(g))$.

If M *is finitely generated, or more generally if* $Z^r(K' \otimes C\ell(g)) = K' \otimes Z^r(C\ell(g))$ *for every ring of fractions* K', *these assertions are also equivalent to the following ones:*

- (d) *there is a Zariski extension* $L = \prod_{i=1}^n K_{s_i}$ *and a family of homogeneous* $x_i \in \mathrm{Lip}(M,q)$ *(with* $i = 1,2,\dots,n$*) such that* $z = (x_1/1, \dots, x_n/1)$ *is invertible in* L, *and* $L \otimes g = G_z$;
- (e) *every localization of* g *is the orthogonal transformation determined by some invertible lipschitzian element.*

Improvements when 2 is invertible

When the mapping $a \longmapsto 2a$ is bijective from M onto M, then $M \oplus M$ is the direct sum of the diagonals Δ and Δ', and instead of the filtration determined by either diagonal we can use the grading determined by both diagonals (see (5.2.7)).

(5.3.10) **Proposition.** *Let us consider the grading of $\mathrm{C}\ell(M,q) \hat{\otimes} \mathrm{C}\ell(M,q)^{to}$ determined by the diagonals Δ and Δ'. For every locally homogeneous element x of $\mathrm{C}\ell(M,q)$ these four assertions are equivalent:*

- *x is lipschitzian;*
- *$x \otimes \tau(x)^{to}$ belongs to $(\mathrm{C}\ell(M,q) \hat{\otimes} \mathrm{C}\ell(M,q)^{to};\ \Delta,\Delta')^0$;*
- *for every $j > 0$ the component of $x \otimes \tau(x)^{to}$ of degree $4j$ vanishes;*
- *for every $j > 0$ the component of $x \otimes \tau(x)^{to}$ of degree $-4j$ vanishes.*

Proof. The equivalence of the first two assertions follows from the fact that the subalgebra of elements of degree 0 (for the grading determined by Δ and Δ') is the intersection of the subalgebras

$$(\mathrm{C}\ell(M,q) \hat{\otimes} \mathrm{C}\ell(M,q)^{to};\ \Delta)^{\leq 0} \quad \text{and} \quad (\mathrm{C}\ell(M,q) \hat{\otimes} \mathrm{C}\ell(M,q)^{to};\ \Delta')^{\leq 0}.$$

By means of the automorphism $y \otimes z \longmapsto y \otimes \sigma(z)$ (already used in the proof of (5.3.1)) we can prove that the component of $x \otimes \tau(x)^{to}$ of degree k vanishes if and only if its component of degree $-k$ vanishes; consequently the last two assertions are also equivalent. The proof ends with the next lemma.

(5.3.11) **Lemma.** *For every locally homogeneous element x we can write*

$$x \otimes \tau(x)^{to} \ \in \ \bigoplus_{j\in\mathbb{Z}} (\mathrm{C}\ell(M,q) \hat{\otimes} \mathrm{C}\ell(M,q)^{to};\ \Delta,\Delta')^{4j}.$$

Proof. Since $x \otimes \tau(x)^{to}$ is even, all its components of odd degree vanish; we must prove that moreover its component of degree k vanishes whenever k is even but not divisible by 4. It is easy to verify (see (3.ex.3)) that we get an algebra isomorphism

$$\Omega \ :\ \mathrm{C}\ell(M,q) \hat{\otimes} \mathrm{C}\ell(M,q)^{to} \longrightarrow (\mathrm{C}\ell(M,q) \hat{\otimes} \mathrm{C}\ell(M,q)^{to})^t$$

if we set $\Omega(y \otimes z^{to}) = (\tau(z) \otimes \tau(y)^{to})^t$; the target of this isomorphism is a twisted algebra according to the definition in **3.2** (twisted multiplication without reversion). The following equalities are evident:

$$\begin{aligned} \Omega(x \otimes \tau(x)^{to}) &= (x \otimes \tau(x)^{to})^t \ , \\ \forall a \in M, \quad \Omega(a \otimes 1^{to} + 1 \otimes a^{to}) &= (a \otimes 1^{to} + 1 \otimes a^{to})^t \ , \\ \forall b \in M, \quad \Omega(b \otimes 1^{to} - 1 \otimes b^{to}) &= -(b \otimes 1^{to} - 1 \otimes b^{to})^t \ ; \end{aligned}$$

from the second and third equalities we shall soon deduce that

$$\forall\, \zeta \in\ (\mathrm{C}\ell(M,q) \mathbin{\hat{\otimes}} \mathrm{C}\ell(M,q)^{to};\ \Delta, \Delta')^k\ , \qquad \Omega(\zeta)\ =\ (-1)^{k(k-1)/2}\ \zeta^t\ ;$$

remember that $k(k-1)/2$ is even if and only if k or $k-1$ is divisible by 4; consequently, when the above equality is proved, we will claim that every ξ such that $\Omega(\xi) = \xi^t$ has a zero component in every degree k that is even but not divisible by 4, and this will be valid when $\xi = x \otimes \tau(x)^{to}$.

To prove it, we need a little piece of information about twisted algebras: when $a_1,\dots,a_p$ are odd elements in a graded algebra A, then (by induction on p)

$$a_1^t a_2^t \cdots a_p^t\ =\ (-1)^{p(p-1)/2}\ (a_1 a_2 \cdots a_p)^t\ .$$

Now let us suppose that the previous ζ is the product of m elements $u_i = a_i \otimes 1^{to} + 1 \otimes a_i^{to}$ and n elements $v_i = b_i \otimes 1^{to} - 1 \otimes b_i^{to}$; for the grading determined by Δ and Δ' its degree is $k = m - n$. Since $\Omega(u_i) = u_i^t$ for $i = 1, 2, \dots, m$ and $\Omega(v_i) = -v_i^t$ for $i = 1, 2, \dots, n$, we can write

$$\Omega(u_1 \cdots u_m v_1 \cdots v_n)\ =\ (-1)^{(m+n)(m+n-1)/2}\ (-1)^n\ (u_1 \cdots u_m v_1 \cdots v_n)^t\ ;$$

the proof ends with this banal equality:

$$\frac{1}{2}(m+n)(m+n-1)\ +n\ =\ \frac{1}{2}(m-n)(m-n-1)\ +2mn. \qquad \square$$

When M is finitely generated, there exists r such that the rank of M is $\leq r$ at every prime ideal, and then for the grading determined by Δ and Δ' all nonzero homogeneous elements of $\mathrm{C}\ell(M,q) \mathbin{\hat{\otimes}} \mathrm{C}\ell(M,q)^{to}$ have a degree between $-r$ and $+r$ (see (5.ex.2)); therefore in the last two assertions of (5.3.10) we may require $0 < 4j \leq r$. This proves that every locally homogeneous element is lipschitzian when $r \leq 3$; in the next section this fact is proved with weaker hypotheses: see (5.4.3).

Dual Lipschitz monoids

The *dual Lipschitz monoid* $\mathrm{Lip}^*(M)$ still deserves a mention; it is the subset of all locally homogeneous elements f of $\bigwedge^*(M)$ satisfying the following equivalent properties:

$$f \otimes \tau(f)\ \text{ belongs to } \textstyle\bigwedge^*(M \oplus M;\ \Delta)^{\leq 0}\ ;$$

$$f \otimes \tau(f)\ \text{ belongs to } \textstyle\bigwedge^*(M \oplus M;\ \Delta')^{\leq 0}.$$

Remember that $f \otimes \tau(f)$ has filtering degree ≤ 0 if it vanishes on all elements of filtering degree ≤ -1 (see (5.2.1)), and since it is even, here this condition is already fulfilled when it vanishes on all elements of filtering degree ≤ -2.

(5.3.12) **Proposition.** *When* $\mathrm{Lip}^*(M)$ *contains* f_1 *and* f_2*, it contains* $f_1 \wedge f_2$ *too. When it contains* f*, it contains* $\tau(f)$ *too, and moreover* $f \wedge \tau(f)$ *belongs to* $K = \bigwedge^{*0}(M)$*. Besides,* $\mathrm{Lip}^*(M)$ *contains all elements of* $\bigwedge^{*1}(M)$*, all elements like* $\lambda + h_1 \wedge h_2$ *with* $\lambda \in K$ *and* $h_1, h_2 \in \bigwedge^{*1}(M)$*, and also all exponentials of elements of* $\bigwedge^{*2}(M)$ *.*

Proof. The beginning of (5.3.12) is proved like (5.3.2). Let us consider $f \wedge \tau(f)$; by definition, for all x in $\bigwedge(M)$ and all $a \in M$,

$$(f \wedge \tau(f))(x) = (f \hat{\otimes} \tau(f)) \circ \pi'(x) \quad \text{and} \quad \pi'(a) = a \otimes 1 + 1 \otimes a \quad \text{for all} \quad a \in M;$$

all $\pi'(a)$ have degree ≤ -1 for the filtration determined by Δ; consequently when f is lipschitzian, $f \wedge \tau(f)$ vanishes on every x in $\bigwedge^+(M)$, in other words, it belongs to $\bigwedge^{*0}(M)$. To prove that $\mathrm{Lip}^*(M)$ contains all elements h and $\lambda + h_1 \wedge h_2$, the calculations presented in the proof of (5.3.2) are still valid, because the elements of the first (resp. second) diagonal of $\bigwedge^{*1}(M \oplus M)$ have degree ≤ -1 for the filtration determined by the second diagonal Δ' (resp. the first one Δ). The last assertion deserves more attention; if w belongs to $\bigwedge^{*2}(M)$, then

$$\mathrm{Exp}(w) \otimes \tau(\mathrm{Exp}(w)) \;=\; \mathrm{Exp}(w \otimes 1 - 1 \otimes w) \qquad \text{because } \tau(w) = -w\ ;$$

because of (5.2.9), it suffices to prove that $(w \otimes 1 - 1 \otimes w)$ has a filtering degree ≤ 0, in other words, that it vanishes on all products

$$(a \otimes 1 + 1 \otimes a)\ (b \otimes 1 + 1 \otimes b) \;=\; (a \wedge b) \otimes 1\ + 1 \otimes (a \wedge b)\ + a \otimes b - b \otimes a\ ;$$

but this is obvious. □

Remarks. When f is even, the assertion $f \wedge \tau(f) \in K$ means that $f \wedge \tau(f) = \lambda^2$ if λ is the component of f in $K = \bigwedge^{*0}(M)$; when f is odd, it means that $f \wedge \tau(f) = 0$. All generalized twisted inner automorphisms of $\bigwedge^*(M)$ are trivial because $f \wedge g = (-1)^{\partial f \partial g} g \wedge f$ for all homogeneous $f, g \in \bigwedge^*(M)$; and the same remark is valid for $\bigwedge(M)$; consequently the usefulness of $\mathrm{Lip}^*(M)$ and $\mathrm{Lip}(M)$ does not come from automorphisms that might be derived from their elements. The usefulness of $\mathrm{Lip}(M)$ comes only from the invariance property stated in the next section, and the usefulness of $\mathrm{Lip}^*(M)$ from the following proposition.

(5.3.13) **Proposition.** *If* x*,* y *and* f *are elements of* $\mathrm{Lip}(M)$, $\mathrm{Lip}(M,q)$ *and* $\mathrm{Lip}^*(M)$ *respectively, then* $f \rfloor x \in \mathrm{Lip}^*(M)$ *and* $f \lrcorner y \in \mathrm{Lip}(M,q)$*.*

Proof. By means of (4.3.7), (4.3.9) and (4.4.7) we get

$$(f \otimes \tau(f))\ \rfloor\ (x \otimes \tau(x)) = (-1)^{\partial x}\ (f \rfloor x) \otimes \tau(f \rfloor x)\ ,$$
$$(f \otimes \tau(f))\ \lrcorner\ (y \otimes \tau(y)) = (-1)^{\partial f}\ (f \lrcorner y) \otimes \tau(f \lrcorner y)\ ;$$

the conclusions follow from (5.2.8) since the even and odd components of x or f are lipschitzian. □

The following evident property too deserves to be stated.

(5.3.14) **Lemma.** *For every linear mapping $w : M \to M'$, the algebra morphism $\bigwedge(w)$ maps $\mathrm{Lip}(M)$ into $\mathrm{Lip}(M')$, whereas $\bigwedge^*(w)$ maps $\mathrm{Lip}^*(M')$ into $\mathrm{Lip}^*(M)$.*

5.4 The invariance property

The invariance property is a quite superb property of Lipschitz monoids, and a very convincing motivation to use them rather than the traditional Clifford groups.

We use the notation of **4.7**: β is any bilinear form on M, q' is defined by $q'(a) = q(a) + \beta(a,a)$, and Φ_β is the resulting isomorphism $C\ell(M,q') \to C\ell(M,q;\beta)$.

(5.4.1) **Invariance theorem.** $\Phi_\beta(\mathrm{Lip}(M,q')) = \mathrm{Lip}(M,q)$.

The invariance theorem states that Φ_β induces a bijection between the sets $\mathrm{Lip}(M,q')$ and $\mathrm{Lip}(M,q)$; of course it is not in general a monoid morphism. The proof requires a preliminary lemma. When V is a totally isotropic submodule for the quadratic form q', for every $n \in \mathbb{Z}$ we set

$$C\ell(M,q;\beta;V)^{\leq n} = \Phi_\beta(C\ell(M,q';V)^{\leq n}).$$

(5.4.2) **Lemma.** *Let β and β' be two bilinear forms on M, and V a submodule of M that is totally isotropic for both quadratic forms $a \longmapsto q(a) + \beta(a,a)$ and $a \longmapsto q(a) + \beta'(a,a)$. We even suppose that $\beta' - \beta$ vanishes on $V \times V$. Then for every $n \in \mathbb{Z}$,*

$$C\ell(M,q;\beta;V)^{\leq n} = C\ell(M,q;\beta';V)^{\leq n}.$$

Proof. Let us denote the multiplications in $C\ell(M,q;\beta)$ and $C\ell(M,q;\beta')$ respectively by $\star$ and $\star'$. It follows immediately from (4.7.11) that the natural filtration of $C\ell(M,q;\beta)$ or $C\ell(M,q;\beta')$ coincides with that of $C\ell(M,q)$. Let us prove that $C\ell(M,q;\beta';V)^{\leq n}$ is contained in $C\ell(M,q;\beta;V)^{\leq n}$ for every n. We shall prove precisely that for every $(j,k) \in \mathbb{N}\times\mathbb{N}$, for every $x \in C\ell^{\leq k}(M,q)$ and for every sequence $(b_1, b_2, \dots, b_j)$ of elements of V, the product of x, b_1, b_2, ..., b_j in $C\ell(M,q;\beta')$ is a sum of terms like $y \star c_1 \star \cdots \star c_{j-i}$ with $i \leq j$, y in $C\ell^{\leq k-i}(M,q)$ and c_1, ..., c_{j-i} all in $\{b_1, b_2, \dots, b_j\}$. This is trivial if $j = 0$. Then, proceeding by induction on j, we consider $(y \star c_1 \star c_2 \star \cdots \star c_{j-i}) \star' b_{j+1}$; the following equality shows that it is a sum of two terms of the predicted type:

$$\begin{aligned}(y \star c_1 \star \cdots \star c_{j-i}) &\star' b_{j+1} \\ &= y \star c_1 \star \cdots \star b_{j+1} + (-1)^{j-i} (\mathrm{d}^{to}_{\beta'-\beta}(b_{j+1}) \rfloor \sigma(y)) \star c_1 \star \cdots \star c_{j-i};\end{aligned}$$

the justification of this equality involves the formula (b) in (4.7.3) (applied both to β and β'), the formula (4.4.4) which explains how the interior multiplication by $\mathrm{d}^{to}_{\beta'-\beta}(b_{j+1})$ operates on a product, the theorem (4.7.5) which shows that (4.4.4) is also valid for $\star$-products, and finally the vanishing of $(\beta' - \beta)(c_m, b_{j+1})$ for $m = 1, 2, \ldots, j - i$. □

Proof of (5.4.1). Let x be a locally homogeneous element of $C\ell(M, q)$; we treat $x \otimes \tau(x)$ as an element of $C\ell(M \oplus M,\ q \perp q)$. In (4.7.8) it is stated that $C\ell(M, q; -b_q)$ is isomorphic to $C\ell(M, q)^{to}$ by the mapping $y \longmapsto y^{to}$; consequently x is in $\mathrm{Lip}(M, q)$ if and only if

$$x \otimes \tau(x)\ \in\ C\ell(M \oplus M,\ q \perp q;\ 0 \perp (-b_q);\ \Delta)^{\leq 0}\ .$$

Now let us consider $\mathrm{Lip}(M, q; \beta) = \Phi_\beta(\mathrm{Lip}(M, q'))$, and the reversion τ_β in $C\ell(M, q; \beta)$; in (4.7.12) it is stated that $C\ell(M, q; \beta^{to} - b_q)$ is isomorphic to $C\ell(M, q; \beta)^{to}$ by the mapping $y \longmapsto y^{to}$; consequently x belongs to $\mathrm{Lip}(M, q; \beta)$ if and only if

$$x \otimes \tau_\beta(x)\ \in\ C\ell(M \oplus M,\ q \perp q;\ \beta \perp (\beta^{to} - b_q);\ \Delta)^{\leq 0}\ .$$

We must prove that the latter condition is equivalent to the former; the local homogeneousness of x is not here involved. From (4.7.14) we deduce that

$$x \otimes \tau_\beta(x)\ =\ \mathrm{Exp}(1 \otimes [\beta])\ \rfloor\ (x \otimes \tau(x))\ .$$

From (4.7.13) we deduce that the interior multiplication by $\mathrm{Exp}(1 \otimes [\beta])$ is the isomorphism

$$C\ell(M \oplus M,\ q \perp q;\ \beta \perp (-\beta - b_q)) \longrightarrow C\ell(M \oplus M,\ q \perp q;\ \beta \perp (\beta^{to} - b_q))$$

that leaves invariant all the elements of $M \oplus M$. Consequently the latter condition is equivalent to this one:

$$x \otimes \tau(x)\ \in\ C\ell(M \oplus M,\ q \perp q;\ \beta \perp (-\beta - b_q);\ \Delta)^{\leq 0}\ .$$

It suffices to observe that the bilinear form $\beta \perp (-\beta)$ vanishes on $\Delta \times \Delta$, and to apply (5.4.2); thus we come to the conclusion that the latter condition is equivalent to the former. □

The invariance property allows the involvement of the neutral Lipschitz monoid $\mathrm{Lip}(M)$ whenever the quadratic form under consideration admits scalar products. The proof of the following proposition explains the important role it may play, and how it is advisable to use it.

(5.4.3) **Proposition.** *When M is a finitely generated module of rank ≤ 3 at each prime ideal of K, and q a quadratic form on M admitting a scalar product, then* $\mathrm{Lip}(M, q)$ *is the set of all locally homogeneous elements of* $C\ell(M, q)$.

Proof. Because of the invariance property, we can suppose $q = 0$. Then we replace $\bigwedge(M)^{to}$ with $\bigwedge(M)$, since $x \longmapsto x^{to}$ is an isomorphism from the latter onto the former. Besides, we may use an automorphism of $M \oplus M$ which maps Δ'

to $M \oplus 0$, for instance $(a,b) \longmapsto (a,\ a+b)$; the corresponding automorphism of $\bigwedge(M) \hat{\otimes} \bigwedge(M)$ maps every $x \otimes y$ to $\pi'(x) \wedge (1 \otimes y)$ where π' is the comultiplication of $\bigwedge(M)$. Consequently a locally homogeneous element x belongs to Lip(M) if and only if

$$\pi'(x) \wedge (1 \otimes \tau(x)) \ \in \ (\bigwedge(M) \hat{\otimes} \bigwedge(M) \ ; \ M \oplus 0)^{\leq 0} \ .$$

The subalgebra of $\bigwedge(M) \hat{\otimes} \bigwedge(M)$ appearing just above is the direct sum of all submodules $\bigwedge^i(M) \otimes \bigwedge^j(M)$ with $i \geq j$. Consequently this condition is satisfied if and only if the component of $\pi'(x) \wedge (1 \otimes \tau(x))$ in every $\bigwedge^i(M) \otimes \bigwedge^j(M)$ with $i < j$ vanishes. Besides, we can suppose that x is homogeneous and that M is generated by three elements a, b, c, since by localization we can reduce the problem to this case (remember (5.3.5)). The detailed calculations (only using elementary properties of $\bigwedge(M)$) do not deserve to be written up here; we consider an even or odd element like

$$x \ = \ \kappa \ + \lambda b \wedge c \ + \mu c \wedge a \ + \nu a \wedge b \quad \text{or} \quad x \ = \ \kappa a \wedge b \wedge c \ + \lambda a + \mu b + \nu c \ .$$

Truly $\pi'(x) \wedge (1 \otimes \tau(x))$ is an enormous sum of 52 or 56 terms, since for instance $\pi'(a \wedge b)$ alone brings four terms into the first factor $\pi'(x)$:

$$\pi'(a \wedge b) \ = \ (a \wedge b) \otimes 1 \ + 1 \otimes (a \wedge b) \ + a \otimes b - b \otimes a \ ;$$

nevertheless, since a, b, c play symmetrical roles, and since all 52 or 56 terms are even, it suffices to verify that among them there are exactly two terms containing $a \otimes (a \wedge b \wedge c)$ (multiplied by an element of K) which are opposite to one another, and two terms containing $1 \otimes (b \wedge c)$ which are also opposite to one another. It does take a long time to find out these four terms in this big sum. □

Here is another consequence of the invariance property; in the traditional theory of Clifford groups nothing similar can be obtained without much stronger hypotheses.

(5.4.4) **Proposition.** *Let (M,q) be a quadratic module, M' and M'' two supplementary submodules of M, and q' and q'' the restrictions of q to M' and M''. Let us suppose that q'' admits scalar products, so that $C\ell(M',q')$ can be identified with a subalgebra of $C\ell(M,q)$ (see (4.8.5)). Then*

$$\mathrm{Lip}(M',q') \ = \ C\ell(M',q') \cap \mathrm{Lip}(M,q).$$

Proof. To prove that every element of $\mathrm{Lip}(M',q')$ belongs to $\mathrm{Lip}(M,q)$ it suffices to consider the natural injection $(M',q') \rightarrow (M,q)$ and to apply (5.3.8). Conversely let us prove that every element of $C\ell(M',q')$ that is lipschitzian in $C\ell(M,q)$ is already lipschitzian in $C\ell(M',q')$. When q'' vanishes, and M' and M'' are orthogonal, the projection $M \rightarrow M'$ is a morphism of quadratic modules, and the resulting morphism $C\ell(M,q) \rightarrow C\ell(M',q')$ leaves invariant all elements of the subalgebra $C\ell(M',q')$; thus the conclusion still follows from (5.3.8). Now we reduce the general case to this very particular case by means of a suitable deformation of $C\ell(M,q)$. Let β be a bilinear form $M \times M \to K$ satisfying these properties: its restriction to $M'' \times M''$ is an admissible scalar product for q'', its restriction to $M' \times M''$ coincides with that of b_q, and its restrictions to $M'' \times M'$ and $M' \times M'$ both vanish. Thus the quadratic form $a \longmapsto q(a) - \beta(a,a)$ vanishes on M'', and M' and M'' are orthogonal to each other for it; moreover $C\ell(M',q')$ is a subalgebra of $C\ell(M,q;-\beta)$ since $\beta(M',M') = 0$. Because of the invariance property, every element of $C\ell(M',q')$ that is lipschitzian in $C\ell(M,q)$, is also lipschitzian in $C\ell(M,q;-\beta)$, and consequently is already lipschitzian in $C\ell(M',q')$. □

The conclusion of (5.4.4) is valid whenever M is a projective module, and M' a direct summand of M. In some cases we can still improve it.

(5.4.5) **Proposition.** *Let M' be a direct summand of a finitely generated projective module M provided with a quadratic form q. We suppose that q is nondegenerate, or at least that d_q induces a surjective mapping $M'^{\perp} \to (M/M')^*$. For every $X \in G'\mathrm{Lip}(M,q)$ these two assertions are equivalent:*

- *X belongs to the subgroup $G'\mathrm{Lip}(M',q')$;*
- *$G_X(a) = a$ for every $a \in M'^{\perp}$.*

Proof. According to (4.8.12), X is contained in $C\ell(M',q')$ if and only if $d_q(a) \rfloor x = 0$ for all $x \in X$ and all $a \in M'^{\perp}$. Now let a be any element of M; because of (4.4.12) the equality $d_q(a) \rfloor x = 0$ is equivalent to $ax = \sigma(x)a$. This last equality is true for all $x \in X$ if and only if $G_X(a) = a$. □

5.5 Associated Lie algebras

A Lie bracket on a module A is an alternate bilinear mapping $A \times A \to A$ usually denoted by $(x,y) \longmapsto [x,y]$ that satisfies the Jacobi equality

$$(5.5.1) \qquad [x,[y,z]] + [y,[z,x]] + [z,[x,y]] \;=\; 0\;.$$

When A is an associative algebra, it is also a Lie algebra for the bracket $[x,y] = xy - yx$. When A is an associative algebra with unit, and G a multiplicative group in A, it often occurs that a Lie subalgebra is associated with G. Here is an example of *sufficient* conditions ensuring the existence of this Lie algebra. Suppose that there is a polynomial mapping Ψ from A into some module such that G is the subset of all invertible $x \in A$ satisfying the equation $\Psi(x) = 0$. Like all polynomial mappings, Ψ has a differential $d\Psi$ in the algebraical sense: $d\Psi$ is a mapping defined on $A \times A$, it is linear with respect to the second variable, and it satisfies the following equality, which shall be explained at once:

$$\Psi(x+ty) \;\equiv\; \Psi(x) \;+\; t\; d\Psi(x;y) \quad \text{modulo } (t^2)\;;$$

the words "modulo (t^2)" mean that we have inflicted the ring extension $K \to K[t]/t^2K[t]$ to all modules under consideration; in other words we use "developments limited to the order 1". Let $\mathfrak{g}$ be the subset of all $y \in A$ such that $\Psi(1+ty) \equiv 0$ modulo (t^2), or equivalently, $d\Psi(1;y) = 0$. It often occurs that $\mathfrak{g}$ is a Lie subalgebra; it occurs if for every $y \in \mathfrak{g}$ there exists z such that $\Psi(1+ty+t^2z) \equiv 0$ modulo (t^3), and if the equation $\Psi(x+ty+t^2z) \equiv 0$ modulo (t^3) also determines a multiplicative group in $(K[t]/t^3K[t]) \otimes A$. Such hypotheses are quite natural, especially in the theory of algebraic groups. If $1+ty+t^2z$ and $1+ty'+t^2z'$ belong to this group, their inverses $1-ty+t^2(y^2-z)$ and $1-ty'+t^2(y'^2-z')$ also belong to it, and also the product

$$\begin{aligned}(1+ty+t^2z)(1+ty'+t^2z')(1-ty+t^2(y^2-z))(1-ty'+t^2(y'^2-z'))\\ \equiv\ 1+t^2(yy'-y'y) \quad \text{modulo } (t^3)\ ;\end{aligned}$$

consequently $d\Psi(1;\,[y,y']) = 0$, and it follows that $\mathfrak{g}$ is a Lie subalgebra.

Let us come back to the quadratic module (M,q). We have defined two groups in $\mathrm{End}(M)$, namely $\mathrm{Aut}(M,q)$ and $\mathrm{GO}(M,q)$ (see **5.1**); we have also defined two groups in $C\ell(M,q)$, namely $\mathrm{GC}\ell(M,q)$ and $\mathrm{GLip}(M,q)$, which are relevant at least when K is a local ring (whereas with more general rings we need groups like $\mathrm{G'Lip}(M,q)$ and even $\mathrm{G''Lip}(M,q)$). With each of this four groups is associated a Lie algebra, which is a submodule of $\mathrm{End}(M)$ or $C\ell(M)$ defined in accordance with the above ideas; yet a direct and easy verification (independent of the above speculative argument) shall show that it is actually a Lie subalgebra (stable by bracket).

An endomorphism f of M is called an *infinitesimal automorphism* of (M,q) if $\mathrm{id}_M + tf$ is an automorphism modulo (t^2), in other words if $\mathrm{b}_q(a, f(a)) = 0$ for all $a \in M$. It is easy to verify that these infinitesimal automorphisms constitute a Lie subalgebra of $\mathrm{End}(M)$. When the mapping $\lambda \longmapsto 2\lambda$ is injective from K into K, they are also called *skew symmetric operators*, because the above condition means that the bilinear mapping $(a,b) \longmapsto \mathrm{b}_q(a, f(b))$ is skew symmetric.

If y belongs to $C\ell(M,q)$, let us first observe that $1+ty$ is locally homogeneous if and only if y is even. Consequently we say that an even element y of $C\ell(M,q)$ determines an infinitesimal orthogonal transformation if the inner automorphism $x \longmapsto (1+ty)x(1-ty)$ induces a bijection from $(K[t]/t^2K[t]) \otimes M$ onto itself. This occurs if and only if $[y,a]$ belongs to M for every $a \in M$. If it does, the mapping F_y defined by $a \longmapsto [y,a]$ is an infinitesimal automorphism of (M,q); anyhow the equality $\mathrm{b}_q(a,\,[y,a]) = 0$ can be corroborated by a direct verification. We call F_y the *infinitesimal orthogonal transformation* derived from y. By means of (5.5.1) it is easy to verify these two facts: first if y and z determine infinitesimal orthogonal transformations (in other words, $[y,M] \subset M$ and $[z,M] \subset M$), then so does $[y,z]$; secondly the mapping $y \longmapsto F_y$ is a morphism of Lie algebras (in other words, $F_{[y,z]} = F_yF_z - F_zF_y$). This morphism $y \longmapsto F_y$ is the infinitesimal counterpart of the group morphism $\mathrm{GC}\ell(M,q) \to \mathrm{GO}(M,q)$. Since the group $\mathrm{GC}\ell(M,q)$ is here considered as less important than its subgroup $\mathrm{GLip}(M,q)$, nothing more will be said about its Lie algebra.

An even element y of $C\ell(M,q)$ is said to be an *infinitesimal lipschitzian element* if $1+ty$ is lipschitzian modulo (t^2), or equivalently,

$$y \otimes 1^{to} + 1 \otimes \tau(y)^{to} \;\in\; (C\ell(M,q) \,\hat{\otimes}\, C\ell(M,q]^{to};\ \Delta)^{\leq 0}\ ;$$

of course Δ may be replaced with Δ'. The infinitesimal lipschitzian elements y make up a Lie algebra because

$$[\, y \otimes 1^{to} + 1 \otimes \tau(y)^{to},\ z \otimes 1^{to} + 1 \otimes \tau(z)^{to}] \;=\; [y,z] \otimes 1^{to} + 1 \otimes \tau([y,z])^{to}.$$

From (5.3.3) we deduce that each infinitesimal lipschitzian element determines an infinitesimal orthogonal transformation $a \longmapsto [y,a]$. Indeed, since $(1+ty)(1+t\tau(y))$ and $(1+ty)a(1+t\tau(y))$ belong respectively to $K \oplus tK$ and $M \oplus tM$ modulo (t^2), it follows that $y+\tau(y)$ and $ya + a\tau(y)$ belong respectively to K and M, and this implies that $[y,a]$ belongs to M. Besides, from (5.4.1) we deduce the following invariance property of the set of infinitesimal lipschitzian elements.

(5.5.2) **Proposition.** *With the same hypotheses as in* (5.4.1), *an even element y of $C\ell(M,q')$ is an infinitesimal lipschitzian element if and only if $\Phi_\beta(y)$ is an infinitesimal lipschitzian element in $C\ell(M,q)$.*

Although the Lie algebra of infinitesimal lipschitzian elements is here considered as an essential thing, no special notation will be proposed for it, because, at least in all somewhat regular cases, it coincides with the following Lie subalgebra of $C\ell(M,q)$, to which we shall at once pay much attention:

$$C\ell_0^{\leq 2}(M,q) \;=\; C\ell_0(M,q) \cap C\ell^{\leq 2}(M,q)\ .$$

(5.5.3) **Theorem.** *The submodule $C\ell_0^{\leq 2}(M,q)$ is a Lie subalgebra of $C\ell(M,q)$, in which each element y is an infinitesimal lipschitzian element and determines an infinitesimal automorphism F_y as explained above: $F_y(a) = [y,a] = ya - ay$. When (M,q) is a quadratic space, we get in this way a surjective morphism from $C\ell_0^{\leq 2}(M,q)$ onto the Lie algebra of all infinitesimal automorphisms of (M,q); its kernel is K.*

Proof. Let y be an element of $C\ell_0^{\leq 2}(M,q)$. From (4.4.12) we deduce that, for all $a \in M$,

$$[y,a] \;=\; -ay + ya \;=\; -\mathrm{d}_q(a) \rfloor y \;\in\; C\ell_1^{\leq 1}(M,q) = M\ .$$

Since $[y,M] \subset M$, y determines an infinitesimal orthogonal transformation F_y. Now remember that the mapping $x \longmapsto [y,x]$ is a derivation of $C\ell(M,q)$, in other words, $[y,\,xx'] = [y,x]\,x' + x\,[y,x']$; since it leaves M stable, it leaves stable all submodules $C\ell^{\leq k}(M,q)$; since moreover it respects the parity grading, it leaves $C\ell_0^{\leq 2}(M,q)$ stable; this proves that $C\ell_0^{\leq 2}(M,q)$ is a Lie subalgebra.

At the end of the proof of (5.3.2) it is proved that $ab \otimes 1^{to} + 1 \otimes (ba)^{to}$ has degree ≤ 0 for the filtration determined by Δ. This means that ab is an infinitesimal lipschitzian element, and by linearity this is true for all elements of $\mathrm{C}\ell_0^{\leq 2}(M,q)$.

When (M,q) is a quadratic space, we can use an admissible scalar product β and replace $\mathrm{C}\ell(M,q)$ with $\bigwedge(M;\beta)$. Thus $\mathrm{C}\ell_0^{\leq 2}(M,q)$ is replaced with $K \oplus \bigwedge^2(M)$. It is clear that $F_y = 0$ for all $y \in K$. Consequently we must prove that the mapping $y \longmapsto F_y$ is a bijection from $\bigwedge^2(M)$ onto the module of infinitesimal automorphisms. Let us calculate F_y when $y = b \wedge c$:

$$F_{b\wedge c}(a) \;=\; [b \wedge c,\; a] \;=\; -\mathrm{d}_q(a) \rfloor (b \wedge c) \;=\; \mathrm{b}_q(a,c)b - \mathrm{b}_q(a,b)c\ ;$$

this result suggests the following diagram, in which all arrows are isomorphisms, except two which are injections:

$$\begin{array}{ccccccc}
\bigwedge^2(M) & \longrightarrow & M \otimes M & \longleftrightarrow & M \otimes M^* & \longleftrightarrow & \mathrm{End}(M) \\
\updownarrow & & & & \uparrow & & \\
\bigwedge^2(M^*) & & & & | & & \\
\updownarrow & & & & \downarrow & & \\
\bigwedge^{*2}(M) & \longrightarrow & (M \otimes M)^* & \longleftrightarrow & M^* \otimes M^* & &
\end{array}$$

In this diagram three arrows come from the isomorphism $\mathrm{d}_q : M \to M^*$; they are:

$$\bigwedge^2(M) \longleftrightarrow \bigwedge^2(M^*) \quad \text{and} \quad M \otimes M \longleftrightarrow M \otimes M^* \longleftrightarrow M^* \otimes M^*.$$

The morphism $\bigwedge^2(M) \to M \otimes M$ is defined by $b \wedge c \longmapsto b \otimes c - c \otimes b$; it is injective because all its localizations are injective. The other injection $\bigwedge^{*2}(M) \to (M \otimes M)^*$ merely means that every alternate bilinear form is a bilinear form. All the other arrows come from canonical morphisms which here are bijective because M is projective and finitely generated; let us just remember that the canonical morphism $M \otimes M^* \to \mathrm{End}(M)$ is defined in this way: every $b \otimes h$ is mapped to the endomorphism $a \longmapsto h(a)b$. The calculation of $F_{b\wedge c}$ shows that the mapping $y \longmapsto F_y$ is the mapping $\bigwedge^2(M) \to \mathrm{End}(M)$ appearing in the first line of the diagram; it is injective. Besides, if f is an endomorphism of M, its image in $(M \otimes M)^*$ is the mapping $a \otimes b \longmapsto \mathrm{b}_q(a, f(b))$. Consequently f is an infinitesimal automorphism of (M,q) if and only if its image in $(M \otimes M)^*$ is an alternate bilinear form; this means that it comes from an element of $\bigwedge^{*2}(M)$ and consequently $f = F_y$ for some $y \in \bigwedge^2(M)$. □

(5.5.4) **Remark.** In the previous proof, $\bigwedge^2(M)$ is not in general a Lie subalgebra. Nevertheless it is actually a Lie subalgebra when the mapping $a \longmapsto 2a$ is bijective, and β is the canonical scalar product. Indeed (4.8.13) shows that every derivation of the Clifford algebra $\bigwedge(M; \mathrm{b}_q/2)$ that leaves M stable, is also a derivation of the exterior algebra $\bigwedge(M)$; consequently, for every degree k,

$$[\,C\ell^2(M,q),\ C\ell^k(M,q)\,] \subset C\ell^k(M,q).$$

Among all the reasons that lead to the conclusion that $GC\ell(M,q)$ is not an interesting group, the following theorem is an important one. It means that its associated Lie algebra is the direct sum of $C\ell_0^{\leq 2}(M,q)$ and some superfluous Lie subalgebra contained in the kernel of the morphism $y \longmapsto F_y$.

(5.5.5) **Theorem.** *When q admits a scalar product β, the Lie subalgebra of all $y \in C\ell_0(M,q)$ such that $[y,M] \subset M$, is the direct sum of $C\ell_0^{\leq 2}(M,q)$ and some submodule of $\mathrm{Z}_0(C\ell(M,q))$.*

Proof. We can replace $C\ell(M,q)$ with $\bigwedge(M;\beta)$. At the beginning of the proof of (5.5.3) there is the equality $[y,a] = -\mathrm{d}_q(a) \rfloor y$. It shows that the Lie subalgebra mentioned in (5.5.5) is the subset of all $y \in \bigwedge(M;\beta)$ such that $\mathrm{d}_q(a) \rfloor y$ belongs to M for all $a \in M$. Since Φ_β is an isomorphism of comodules (see (4.7.5)), this interior product is the same in $\bigwedge(M;\beta)$ as in $\bigwedge(M)$. But in $\bigwedge(M)$ there is a grading over $\mathbb{N}$, and the interior multiplication by $\mathrm{d}_q(a)$ maps each $\bigwedge^k(M)$ into $\bigwedge^{k-1}(M)$. The condition $\mathrm{d}_q(a) \rfloor y \in M$ is satisfied by y if and only if it is satisfied by all its homogeneous components $y_0, y_2, y_4, \ldots$ for the $\mathbb{N}$-grading. It is always satisfied by y_0 and y_2. But for an even degree $k \geq 4$, it is satisfied if and only if $\mathrm{d}_q(a) \rfloor (y_k) = 0$, or equivalently $[a, y_k] = 0$. When this condition is required from y_k for all $a \in M$, it means that y_k belongs to the center of $C\ell(M,q)$. □

Theorem (5.5.5) implies that the Lie algebra $C\ell_0^{\leq 2}(M,q)$ gives as many infinitesimal orthogonal transformations as the Lie algebra associated with the group $GC\ell(M,q)$. When the ring K allows an effective Lie theory showing close relations between algebraic groups and their Lie algebras, this suggests that we should forsake the group $GC\ell(M,q)$ and prefer a smaller group with Lie algebra $C\ell_0^{\leq 2}(M,q)$. The fact that every morphism $(M,q) \to (M',q')$ of quadratic modules induces a morphism $C\ell_0^{\leq 2}(M,q) \to C\ell_0^{\leq 2}(M',q')$ of Lie algebras also pleads for these Lie algebras. This functorial property of these Lie algebras recalls the functorial property of Lipschitz monoids mentioned in (5.3.8). We already know that the Lie algebra of all infinitesimal lipschitzian elements contains $C\ell_0^{\leq 2}(M,q)$ (see (5.5.3); if these Lie algebras prove to be equal, at least in the most useful cases, this fact will still support the preference for both the Lie algebra $C\ell_0^{\leq 2}(M,q)$ and the monoid $\mathrm{Lip}(M,q)$. The following theorem cannot yet be applied to all quadratic forms admitting scalar products, but already to many of them.

(5.5.6) **Theorem.** *The Lie algebra of all infinitesimal lipschitzian elements is equal to $C\ell_0^{\leq 2}(M,q)$ whenever M is a projective module. The same conclusion holds when all these hypotheses are satisfied: q admits a scalar product, the multiplication by* 2 *is injective on* $\bigwedge^{2j}(M)$ *for every* $j \geq 2$, *and the multiplication by* j *is injective on* $\bigwedge^{2j}(M)$ *for every odd integer* $j \geq 3$.

The proof of (5.5.6) begins with the following lemma.

(5.5.7) **Lemma.** *For every* $j \geq 2$ *let* W_{2j} *be the submodule of all* $y \in \bigwedge^{2j}(M)$ *such that*

$$\pi'(y) + 1 \otimes \tau(y) \in \bigoplus_{i=j}^{2j} \bigwedge^{i}(M) \otimes \bigwedge^{2j-i}(M).$$

If the quadratic form $q : M \to K$ *admits scalar products, the following assertions are equivalent:*

(a) *every infinitesimal lipschitzian element belongs to* $\mathrm{C}\ell_0^{\leq 2}(M, q)$ *;*
(b) $W_{2j} = 0$ *for all* $j \geq 2$.

Proof. Because of the invariance property (5.5.2), it suffices to prove (5.5.7) when $q = 0$. As in the proof of (5.4.3), we use the automorphism $(a, b) \longmapsto (a,\ a + b)$ of $M \oplus M$ that maps Δ' to $M \oplus 0$; thus an element $y \in \bigwedge_0(M)$ is an infinitesimal lipschitzian element if and only if

$$\pi'(y) + 1 \otimes \tau(y) \in (\bigwedge(M) \hat{\otimes} \bigwedge(M);\ M \oplus 0)^{\leq 0}.$$

The comultiplication π' and the reversion τ are graded mappings for the $\mathbb{N}$- grading of $\bigwedge(M)$; consequently y satisfies this condition if and only if all the components of y in the submodules $\bigwedge^{2j}(M)$ satisfy it. When $j \geq 2$ and y belongs to $\bigwedge^{2j}(M)$, the above condition is satisfied if and only if y belongs to W_{2j}; consequently the vanishing of W_{2j} means that no infinitesimal lipschitzian element may have a nonzero component of degree $2j$. □

Proof of (5.5.6). We already know that every $y \in \mathrm{C}\ell_0^{\leq 2}(M, q)$ is an infinitesimal lipschitzian element, and we must prove the converse statement, at least in the two cases mentioned in (5.5.6). For every $y \in \bigwedge(M)$ and every $i \geq 0$, let $\pi_i'(y)$ be the component of $\pi'(y)$ in $\bigwedge^{i}(M) \otimes \bigwedge(M)$. When y has degree $2j$ and $0 \leq i < 2j$, the component of $\pi'(y) + 1 \otimes \tau(y)$ in $\bigwedge^{i}(M) \otimes \bigwedge^{2j-i}(M)$ is equal to $\pi_i'(y)$; but since $\tau(y) = (-1)^j y$, its component in $1 \otimes \bigwedge^{2j}(M)$ is $1 \otimes 2y$ when j is even, and always 0 when j is odd. This already shows that $W_{2j} = 0$ when j is even ≥ 2 and the multiplication by 2 is injective from $\bigwedge^{2j}(M)$ into itself. When j is odd ≥ 3, we consider the component in $M \otimes \bigwedge^{2j-1}(M)$ and observe that $W_{2j} = 0$ if π_1' is injective on $\bigwedge^{2j}(M)$. Now let us consider this mapping:

$$\bigwedge^{2j}(M) \longrightarrow M \otimes \bigwedge^{2j-1}(M) \longrightarrow \bigwedge^{2j}(M), \qquad y \longmapsto \pi(\pi_1'(y)) ;$$

it is easy to verify that it is the multiplication by $2j$ in $\bigwedge^{2j}(M)$. Consequently $W_{2j} = 0$ if both multiplications by 2 and j in $\bigwedge^{2j}(M)$ are injective.

Admissible scalar products always exist for quadratic forms on projective modules (see (2.5.3)). The case of a projective module M can be reduced to the case of a free module, either by localization if it is finitely generated, or by using another module M' such that $M \oplus M'$ is free. Let us prove that the restriction of π_1' to $\bigwedge^{2j}(M)$ is injective for all $j \geq 2$; this implies $W_{2j} = 0$ as above. Let $(e_s)_{s \in S}$ be a basis of M indexed by a totally ordered set S. From it we derive a basis of $\bigwedge^{2j}(M)$ and a basis of $M \otimes \bigwedge^{2j-1}(M)$ in the usual way; both bases are totally ordered if we use the lexicographic order on the set of all sequences of $2j$ elements of S. An element of the basis of $\bigwedge^{2j}(M)$ looks like

$$e_{s_1} \wedge e_{s_2} \wedge \cdots \wedge e_{s_{2j}} \quad \text{with} \quad s_1 < s_2 < \cdots < s_{2j};$$

π_1' maps it to the sum of $e_{s_1} \otimes (e_{s_2} \wedge \cdots \wedge e_{s_{2j}})$ and something in the submodule generated by other elements of the basis of $M \otimes \bigwedge^{2j-1}(M)$ standing above this one in the lexicographic order; this is sufficient to conclude that π_1' is injective. □

In the proof of (5.5.6) it is clear that the best way of using the lemma (5.5.7) has not yet been found. If somebody finds more effective methods, perhaps the conclusion of (5.5.6) will follow from the only hypothesis that q admits a scalar product.

5.6 First results about orthogonal transformations

Let us come back to the three Lipschitz groups defined in **5.3**, and to these group morphisms:

$$\mathrm{GLip}(M,q) \longrightarrow \mathrm{G'Lip}(M,q) \longrightarrow \mathrm{G''Lip}(M,q) \longrightarrow \mathrm{GO}(M,q) \longrightarrow \mathrm{Aut}(M,q).$$

The first morphism is surjective when $\mathrm{Pic}(K)$ is a trivial group (see (5.3.7)), in particular when K is a local ring. The second morphism maps every $X \in \mathrm{G'Lip}(M,q)$ to the submodule it generates over $\mathrm{Z}^r(\mathrm{C}\ell(M,q))$; it is bijective when $\mathrm{Z}^r(\mathrm{C}\ell(M,q)) = K$ (see (5.1.9)). The third morphism is injective; its bijectiveness is equivalent to the equality $\mathrm{G''Lip}(M,q) = \mathrm{G''C}\ell(M,q)$ which here is the main conjecture. The fourth morphism is a natural injection, and the determination of its image is also a main problem.

When g is an automorphism of (M,q) and N a submodule of M, it is well known that g maps $N^\perp$ onto $g(N)^\perp$; in particular g leaves $N^\perp$ invariant if it leaves N invariant. Here is a less evident property.

(5.6.1) **Proposition.** *When g is an automorphism of (M,q), the kernel and the image of $g - \mathrm{id}$ are orthogonal submodules. When moreover* $\mathrm{Ker}(q) = 0$, *then* $\mathrm{Ker}(g - \mathrm{id})$ *is the submodule of* all *elements orthogonal to* $\mathrm{Im}(g - \mathrm{id})$.

Proof. For all a and b in M we can write

$$\mathrm{b}_q(g(a) - a,\ b) \ = \ -\mathrm{b}_q(g(a),\ g(b) - b)\ ;$$

when $g(b) = b$, then b is orthogonal to all $g(a) - a$. Conversely if b is orthogonal to all $g(a) - a$, then $g(b) - b$ obviously belongs to $\mathrm{Ker}(\mathrm{b}_q)$; the following equalities show that it even belongs to $\mathrm{Ker}(q)$:

$$\begin{aligned} q(g(b) - b) &= q(g(b)) + q(b) - \mathrm{b}_q(g(b), b) \\ &= \mathrm{b}_q(b, b) - \mathrm{b}_q(g(b), b) = -\mathrm{b}_q(g(b) - b,\ b) = 0\ ; \end{aligned}$$

therefore $g(b) - b$ must vanish if $\mathrm{Ker}(q) = 0$. □

Proposition (5.1.14) states other properties of $\mathrm{Ker}(g-\mathrm{id})$ and $\mathrm{Im}(g-\mathrm{id})$ when g is an orthogonal transformation; they imply that $\mathrm{Ker}(g - \mathrm{id})$ contains $\mathrm{Ker}(\mathrm{b}_q)$, and that id_M is the only orthogonal transformation when b_q is the null bilinear form. Unless otherwise specified, it is always assumed that $\mathrm{Ker}(\mathrm{b}_q) \neq M$. Here are other properties which are not true for all automorphisms of quadratic modules; they involve determinants which are defined in the very beginning of **3.6**.

(5.6.2) **Proposition.** *Let q be a quadratic form on a finitely generated projective module M, and g an orthogonal transformation of (M, q). If $\mathrm{Z}^g(\mathrm{C}\ell(g))$ contains an invertible even (resp. odd) element, then* $\det(g) = 1$ *(resp.* $\det(g) = -1$*).*

Proof. Because of (1.12.8) we can assume that M has a constant rank r. By localization it is easy to verify that $\bigwedge(g)(z) = \det(g)z$ for every $z \in \bigwedge^r(M)$; since $\bigwedge^r(M)$ is a projective module of constant rank 1, this property determines $\det(g)$. Since the canonical morphism $\bigwedge(M) \to \mathrm{Gr}(\mathrm{C}\ell(M, q))$ is bijective (see (4.8.7)), we get an isomorphism from $\mathrm{C}\ell(M, q)/\mathrm{C}\ell^{<r}(M, q)$ onto $\bigwedge^r(M)$ if we map every product $b_1 b_2 \cdots b_r$ (modulo $\mathrm{C}\ell^{<r}(M, q)$) to $b_1 \wedge b_2 \wedge \cdots \wedge b_r$. If x is an invertible homogeneous element of $\mathrm{Z}^g(\mathrm{C}\ell(g))$, then $g(b_i) = (-1)^{\partial x} x b_i x^{-1}$ for $i = 1, 2, \ldots, r$, whence

$$x b_1 b_2 \cdots b_r x^{-1} \equiv (-1)^{r\partial x} \det(g)\, b_1 b_2 \cdots b_r \quad \text{modulo } \mathrm{C}\ell^{<r}(M, q).$$

In (3.2.1) it is stated that yz and $(-1)^{(r-1)\partial y} zy$ are congruent modulo $\mathrm{C}\ell^{<r}(M, q)$ for all homogeneous y, $z \in \mathrm{C}\ell(M, q)$; when $y = x b_1 \cdots b_r$ and $z = x^{-1}$, we obtain

$$x b_1 b_2 \cdots b_r x^{-1} \equiv (-1)^{(r-1)\partial x}\, b_1 b_2 \cdots b_r \quad \text{modulo } \mathrm{C}\ell^{<r}(M, q).$$

The conclusion follows. □

(5.6.3) **Corollary.** *If g is an orthogonal transformation of (M, q), and if M is finitely generated and projective, then* $\det(g)$ *is a square root of* 1*, and the fraction* $\det(g)/1$ *is equal to* $1/1$ *or* $-1/1$ *in every localization $K_{\mathfrak{p}}$ of K.*

Indeed since $\mathrm{C}\ell(g)$ is a generalized twisted inner automorphism, from (5.1.5) we know that every localization of $\mathrm{Z}^g(\mathrm{C}\ell(g))$ contains an invertible homogeneous element. □

A great part of the results shall be proved by localizations with the help of (5.3.9), especially the assertion (e). Several times we shall have to prove that an automorphism of (M,q) can be derived from some invertible lipschitzian element. This problem will be reduced by successive steps to a simpler question involving two elements of M: if $q(a) = q(a')$, is there an invertible lipschitzian x such that $G_x(a) = a'$? Two kinds of transformations G_x will help us, and the simplest ones are the reflections. A *reflection* in (M,q) is the transformation G_d derived from an invertible element d of M:

$$G_d(b) \;=\; -dbd^{-1} \;=\; b \;-\, \mathrm{b}_q(d,b)\; d^{-1}\;;$$

G_d maps d to $-d$, and leaves invariant every b orthogonal to d. When 2 is invertible in K, then M is the direct sum of Kd and the orthogonal submodule, and thus G_d may remind us of the reflection of a ray of light on a mirror; but when 2 is not invertible, G_d may look quite differently. In all cases a reflection is involutive.

(5.6.4) **Proposition.** *Let a and a' be elements of M such that $q(a) = q(a')$. If $a' - a$ is invertible, the reflection $G_{a'-a}$ maps a to a'.*

Proof. Indeed $a'(a'-a) = -(a'-a)a\,.$ □

When $a' - a$ is not invertible as in (5.6.4), we may try an element $x = a'd + da$ with some suitable $d \in M$; such an element is always lipschitzian, since

$$x = a'd + da \;=\; \mathrm{b}_q(a,d) + (a'-a)d \;=\; \mathrm{b}_q(a',d) + d(a-a')\;;$$

its invertibility depends on the invertibility of

$$x\tau(x) \;=\; (a'd+da)(ad+da') \;=\; q(a'-a)q(d) + \mathrm{b}_q(a,d)\mathrm{b}_q(a',d).$$

(5.6.5) **Proposition.** *Let a, a' and d be elements of M such that $q(a) = q(a')$, and let us set $x = a'd + da$. If x is invertible, then $G_x(a) = a'$. Besides, if d is invertible, then x is a product of two elements of M.*

Proof. Indeed $a'(a'd+da) = (a'd+da)a\,$. Besides, since $x = \mathrm{b}_q(a,d) + (a'-a)d$, it is easy to verify that xd belongs to M; consequently if d is invertible, x is the product of two elements of M. Of course the same conclusion holds when $a' - a$ is invertible, but in this case it is wiser to apply (5.6.4). □

Let us add this more technical lemma.

(5.6.6) **Lemma.** *The projective quadratic module (M,q) is assumed to be the orthogonal sum of the submodules N and N'', and $\tilde{q}$ is the restriction of q to N. If g is an automorphism of (M,q) such that $g(N) = N$ and* $\mathrm{Ker}(g - \mathrm{id}) \supset N''$, *and if $\tilde{g}$ is its restriction to N, then $\mathrm{Z}^r(\tilde{g}) = \mathrm{Z}^r(g) \cap C\ell(N,\tilde{q})$.*

Indeed $\mathrm{Z}^r(\tilde{g}) \supset \mathrm{Z}^r(g) \cap C\ell(N,\tilde{q})$, and conversely every $x \in \mathrm{Z}^r(\tilde{g})$ belongs to $\mathrm{Z}^r(g)$ because (4.4.12) implies $ax = \sigma(x)a$ for all $a \in N''$ and all $x \in C\ell(N,\tilde{q})$. □

Tamely degenerate quadratic forms

It seems very hard to obtain effective results for *all* quadratic modules. Here we are especially interested in quadratic spaces (over any ring) and in quadratic modules (of any kind) over fields. The methods that are suitable for them, will also enable us to treat many other quadratic modules without getting any more tired. It looks sensible to define a type of quadratic modules for which these methods are effective, and to forsake the other modules until we meet some good reasons to tackle them too. Quadratic forms that are either nondegenerate or tamely degenerate, are convenient objects for the methods that shall be elaborated here. In **5.8** some other quadratic modules will still be mentioned, especially some projective modules that are not finitely generated.

When M is a finitely generated projective module, a quadratic form q on M (and the quadratic module (M,q) too) is said to be *tamely degenerate* if $\mathrm{Ker}(\mathrm{b}_q)$ is a direct summand of M other than 0 and M, and if b_q induces a nondegenerate symmetric bilinear form on the quotient $M/\mathrm{Ker}(\mathrm{b}_q)$. Other authors characterize them by the equivalent property stated hereafter.

(5.6.7) **Lemma.** *Let q be a quadratic form on a finitely generated projective module M. It is tamely degenerate if and only if the image of $\mathrm{d}_q : M \to M^*$ is a direct summand of M^* other than 0 and M^*.*

Proof. If q is tamely degenerate, any submodule M' supplementary to $\mathrm{Ker}(\mathrm{b}_q)$ is a quadratic space, and the restriction of d_q to M' induces a bijection $M' \to M'^*$. Moreover M^* can be identified with $\mathrm{Ker}(\mathrm{b}_q)^* \oplus M'^*$, and thus we realize that $\mathrm{Im}(\mathrm{d}_q)$ is the direct summand M'^*. Conversely let us suppose that $\mathrm{Im}(\mathrm{d}_q)$ is a direct summand of M^*. Since the canonical mapping $M \to (M^*)^*$ is bijective, M is the direct sum of two submodules M' and N such that M'^* and N^* can be identified respectively with $\mathrm{Im}(\mathrm{d}_q)$ and a supplementary submodule in M^*. Now N is the submodule of all $a \in M$ such that $h(a) = 0$ for all $h \in \mathrm{Im}(\mathrm{d}_q)$; therefore $N = \mathrm{Ker}(\mathrm{d}_q) = \mathrm{Ker}(\mathrm{b}_q)$. Consequently d_q induces a bijection $M' \to M'^*$. □

The next lemma might be an easy consequence of (6.7.7), but a more elementary proof can be given already now.

(5.6.8) **Lemma.** *If q is a tamely degenerate quadratic form on a finitely generated projective module M, the graded center of $C\ell(M,q)$ is the subalgebra generated by $\mathrm{Ker}(\mathrm{b}_q)$), and the natural injection $\mathrm{Ker}(\mathrm{b}_q) \to M$ extends to an isomorphism $C\ell(\mathrm{Ker}(\mathrm{b}_q)) \to \mathrm{Z}^g(C\ell(M,q))$.*

Proof. Since every $a \in \mathrm{Ker}(\mathrm{b}_q)$ anticommutes with every $b \in M$, it is clear that $\mathrm{Z}^g(C\ell(M,q))$ contains $\mathrm{Ker}(\mathrm{b}_q)$ and the subalgebra it generates. The injectiveness of the algebra morphism $C\ell(\mathrm{Ker}(\mathrm{b}_q)) \to \mathrm{Z}^g(C\ell(M,q))$ follows from (4.8.5), and it remains to prove its surjectiveness. Let M' be a submodule supplementary to $\mathrm{Ker}(\mathrm{b}_q)$; thus $C\ell(M,q)$ (or shortly $C\ell(M)$) is canonically isomorphic to $C\ell(\mathrm{Ker}(\mathrm{b}_q)) \hat{\otimes} C\ell(M')$. If we prove that $\mathrm{Z}^g(C\ell(M))$ is contained in $C\ell(\mathrm{Ker}(\mathrm{b}_q)) \hat{\otimes}$

$\mathrm{Z}^g(\mathrm{C}\ell(M'))$, we have finished, because $\mathrm{C}\ell(M')$ is a graded Azumaya algebra and $\mathrm{Z}^g(\mathrm{C}\ell(M')) = K$. We can suppose that $\mathrm{C}\ell(\mathrm{Ker}(\mathrm{b}_q))$ is a free module, since by localization we can reduce the problem to this case; let $(x_1, x_2, \dots)$ be a basis of $\mathrm{C}\ell(\mathrm{Ker}(\mathrm{b}_q))$. Every $z \in \mathrm{Z}^g(\mathrm{C}\ell(M))$ can be written *in a unique way* as a sum $\sum_i x_i \otimes y_i$ for suitable $y_i \in \mathrm{C}\ell(M')$, and from the equalities $a'z = \sigma(z)a'$ (with $a' \in M'$) it is easy to deduce that all $y_1, y_2, \dots$ must belong to $\mathrm{Z}^g(\mathrm{C}\ell(M'))$. □

Local parities

If g is an orthogonal transformation of (M,q), every localization $g_\mathfrak{p}$ is equal to G_ξ for some invertible homogeneous $\xi \in \mathrm{Z}^r(\mathrm{C}\ell(g))_\mathfrak{p}$ (see (5.1.5)), and $\mathrm{Z}^r(\mathrm{C}\ell(g))_\mathfrak{p}$ is the submodule over $\mathrm{Z}^r(\mathrm{C}\ell(M,q))_\mathfrak{p}$ generated by ξ (see (5.1.7)). If $\mathrm{Z}^r_1(\mathrm{C}\ell(M,q))_\mathfrak{p}$ contains no invertible elements, then every homogeneous invertible element of $\mathrm{Z}^r(\mathrm{C}\ell(g))_\mathfrak{p}$ has the same parity as ξ, and a parity can be assigned to g at the prime ideal $\mathfrak{p}$. As it has been observed in (5.1.12), there is a parity at $\mathfrak{p}$ whenever the image of 2 in $K_\mathfrak{p}$ does not vanish.

For each idempotent $e \in \mathrm{Ip}(K)$ let $\mathrm{GO}_e(M,q)$ be the subset of all $g \in \mathrm{GO}(M,q)$ such that $(1-e)\mathrm{Z}^r(\mathrm{C}\ell(g))$ is contained in $\mathrm{C}\ell_0(M,q)$, and $e\mathrm{Z}^r(\mathrm{C}\ell(g))$ in $\mathrm{C}\ell_1(M,q)$; this implies that g has an even parity at every prime ideal $\mathfrak{p}$ that contains e, yet an odd parity at every $\mathfrak{p}$ that does not contain e. Because of (5.6.2) it also implies $\det(g) = 1 - 2e$. There is a canonical morphism $e \longmapsto 1 - 2e$ from $\mathrm{Ip}(K)$ into the group $\mu_2(K)$ of square roots of 1, which is bijective when 2 is invertible in K; it already appeared in (3.4.14). Nevertheless when 2 is not invertible, there are orthogonal transformations g such that $\det(g)$ does not belong to the image of this morphism; there is a counterexample in (5.ex.14).

The subset $\mathrm{GO}_0(M,q)$ is a subgroup often denoted by $\mathrm{SO}(M,q)$, and called the *special orthogonal group*. In the notation $\mathrm{GO}_0(M,q)$ or $\mathrm{GO}_1(M,q)$ the lower index 0 or 1 can be read as an idempotent of K, or as an element of $\mathbb{Z}/2\mathbb{Z}$ indicating that $\mathrm{Z}^r(\mathrm{C}\ell(g))$ is entirely contained in $\mathrm{C}\ell_0(M,q)$ or $\mathrm{C}\ell_1(M,q)$. If g and g' belong respectively to $\mathrm{GO}_e(M,q)$ and $\mathrm{GO}_{e'}(M,q)$, it is easy to prove that gg' belongs to $\mathrm{GO}_{e''}(M,q)$ with $e'' = e \tilde{+} e' = e + e' - 2ee'$.

It may occur that $\mathrm{GO}(M,q)$ is the union of all subsets $\mathrm{GO}_e(M,q)$ with $e \in \mathrm{Ip}(K)$; this occurs in the two cases described in the following propositions.

(5.6.9) **Proposition.** *If g is an orthogonal transformation of a quadratic space (M,q), there exists $e \in \mathrm{Ip}(K)$ such that g belongs to $\mathrm{GO}_e(M,q)$; moreover $\mathrm{C}\ell(g)$ induces the identity automorphism on $(1-e)\mathrm{QZ}(M,q)$ and the standard involution on $e\mathrm{QZ}(M,q)$.*
When M is a faithful module, $\mathrm{QZ}(M,q)$ is the centralizer of $\mathrm{C}\ell_0(M,q)$ in $\mathrm{C}\ell(M,q)$ (see (3.7.6)), and every automorphism of this quadratic extension is determined by an idempotent e (see (3.4.15)).

Proof. Because of (5.1.9), $\mathrm{Z}^r(\mathrm{C}\ell(g))$ is a graded invertible module; consequently there exists $e \in \mathrm{Ip}(K)$ such that $(1-e)\mathrm{Z}^r(\mathrm{C}\ell(g))$ and $e\mathrm{Z}^r(\mathrm{C}\ell(g))$ are contained respectively in $C\ell_0(M,q)$ and $C\ell_1(M,q)$. If z is an element of $\mathrm{QZ}(M,q)$, then from (3.5.13) we deduce that $xz = zx$ for all $x \in C\ell_0(M,q)$ (in particular for all $x \in (1-e)\mathrm{Z}^r(C\ell(g))$), and that $xz = (-1)^{\partial z}\varphi(z)x$ for all $x \in C\ell_1(M,q)$ (in particular for all $x \in e\mathrm{Z}^r(C\ell(g))$) if φ is the standard involution of $\mathrm{QZ}(M,q)$. The conclusion follows immediately. □

(5.6.10) **Proposition.** *If M is a finitely generated projective module such that the mapping $a \longmapsto 2a$ is bijective from M onto M, the group $\mathrm{GO}(M,q)$ is the union of the subsets $\mathrm{GO}_e(M,q)$ with $e \in \mathrm{Ip}(K)$.*

Proof. There is an idempotent e_0 such that e_0M is a faithful module over Ke_0 whereas $(1-e_0)M = 0$; the bijectiveness of $a \longmapsto 2a$ implies that $2e_0$ is invertible in Ke_0. Consequently we can reduce the proof to the case of a ring K in which 2 is invertible; thus $\mathrm{Z}_1^r(C\ell(M,q)) = 0$. Let g be an orthogonal transformation of (M,q); since $\mathrm{Z}^r(C\ell(g))$ is an invertible module over $\mathrm{Z}^r(C\ell(M,q))$ (see (5.1.9)), every localization of $\mathrm{Z}^r(C\ell(g))$ is entirely even or entirely odd. Since $\det(g)$ is a square root of 1 (see (5.6.3)), and since 2 is invertible in K, there is a unique idempotent $e \in \mathrm{Ip}(K)$ such that $\det(g) = 1-2e$. Because of (5.6.2) this means that $\mathrm{Z}^r(C\ell(p))$ has an even (resp. odd) localization at every prime ideal that contains e (resp. $(1-e)$); therefore $(1-e)\mathrm{Z}^r(C\ell(g))$ is even, and $e\mathrm{Z}^r(C\ell(g))$ odd. □

5.7 Products of reflections when K is a local ring

When g is an automorphism of a quadratic space (M,q) over a field, the problem of decomposing g into a product of reflections was tackled first by E. Cartan, and later by J. Dieudonné who accepted even fields of characteristic 2, and thus discovered that such decompositions did not always exist when (M,q) was a hyperbolic space of dimension 4 over $\mathbb{Z}/2\mathbb{Z}$. In all other cases g is a product of reflections with a number of factors $\leq \dim(M)$. Still later R. Baeza considered quadratic spaces over semilocal rings (with a finite number of maximal ideals).

Here we treat this problem when M is a free module of finite rank r over a local ring K with maximal ideal $\mathfrak{m}$, and q a nondegenerate or tamely degenerate quadratic form on M. Hereafter are the main results of this section.

(5.7.1) **Theorem.** *With the above hypotheses, $\mathrm{GO}(M,q)$ is the subgroup of all $g \in \mathrm{Aut}(M,q)$ such that $\mathrm{Ker}(g-\mathrm{id}) \supset \mathrm{Ker}(\mathrm{b}_q)$, and the group morphism $\mathrm{GLip}(M,q) \to \mathrm{GO}(M,q)$ is surjective.*

(5.7.2) **Theorem.** *If the residue field $K/\mathfrak{m}$ contains more than 2 elements, every $g \in \mathrm{GO}(M,q)$ is a product of reflections with a number of factors $\leq 2r'$ if r' is the rank of $M/\mathrm{Ker}(\mathrm{b}_q)$.*

The upper bound $2r'$ in (5.7.2) is probably not the best one, but we are not interested in improving it. Let M' be a submodule supplementary to $\mathrm{Ker}(\mathrm{b}_q)$ in M, and q' the nondegenerate restriction of q to M'. Besides, let $q/\mathfrak{m}$ be the quadratic form on the extension $(K/\mathfrak{m})\otimes(M,q)$ over the residue field. When $q(\mathrm{Ker}(\mathrm{b}_q)) \subset \mathfrak{m}$, then $\mathrm{Ker}(q/\mathfrak{m}) = \mathrm{Ker}(\mathrm{b}_{q/\mathfrak{m}})$, and the quadratic space $(K/\mathfrak{m}) \otimes (M', q')$ over the residue field can be identified with the quotient of $(K/\mathfrak{m}) \otimes (M,q)$ by $\mathrm{Ker}(q/\mathfrak{m})$.

(5.7.3) **Theorem.** *If the residue field $K/\mathfrak{m}$ has only two elements, every $g \in \mathrm{GO}(M,q)$ is a product of reflections (with a number of factors $\leq 3r'$) except when these three conditions are all satisfied: first the rank r of M is ≥ 3, secondly $q(\mathrm{Ker}(\mathrm{b}_q)) \subset \mathfrak{m}$, and thirdly $(K/\mathfrak{m}) \otimes (M', q')$ is a hyperbolic space of dimension r' equal to 2 or 4.*

It is sensible to distinguish two exceptional cases in (5.7.3). The exceptional case with $r' = 4$ is called Dieudonné's exceptional case, since its existence is directly predictable from Dieudonné's contribution. The other exceptional case with $r' = 2$ and $r \geq 3$ does not allow q to be nondegenerate.

It is worth looking at $\mathrm{Z}^r(\mathrm{C}\ell(M,q))$ since the kernel of the morphism $\mathrm{GLip}(M,q) \to \mathrm{GO}(M,q)$ is the group of invertible lipschitzian elements in this subalgebra. When q is nondegenerate, then $\mathrm{Z}^g(\mathrm{C}\ell(M,q)) = K$. When q is tamely degenerate, $\mathrm{Z}^g(\mathrm{C}\ell(M,q))$ is the subalgebra generated by $\mathrm{Ker}(\mathrm{b}_q)$ (see (5.6.8)), and we know that $\mathrm{Z}^r_0(\mathrm{C}\ell(M,q)) = \mathrm{Z}^g_0(\mathrm{C}\ell(M,q))$, whereas $\mathrm{Z}^r_1(\mathrm{C}\ell(M,q))$ is the subset of all $x \in \mathrm{Z}^g_1(\mathrm{C}\ell(M,q))$ such that $2x = 0$. When 2 is invertible in K, then $\mathrm{Ker}(\mathrm{b}_q) = \mathrm{Ker}(q)$ and we can identify $\mathrm{Z}^g(\mathrm{C}\ell(M,q))$ with $\bigwedge(\mathrm{Ker}(q))$, and $\mathrm{Z}^r(\mathrm{C}\ell(M,q))$ with $\bigwedge_0(\mathrm{Ker}(q))$. When the equality $2 = 0$ holds in K, then $\mathrm{Z}^r(\mathrm{C}\ell(M,q)) = \mathrm{Z}^g(\mathrm{C}\ell(M,q))$. When $\mathrm{Ker}(\mathrm{b}_q)$ contains an invertible element a, the equality $2 = 0$ must hold in K since $2q(a) = \mathrm{b}_q(a,a) = 0$; this a is an invertible element of $\mathrm{Z}^r_1(\mathrm{C}\ell(M,q))$ which prevents us from assigning a parity to any element of $\mathrm{GO}(M,q)$. When $\mathrm{Ker}(\mathrm{b}_q)$ contains no invertible element (in other words, when $q(\mathrm{Ker}(\mathrm{b}_q)) \subset \mathfrak{m}$), it is sure that $\mathrm{Z}^r_1(\mathrm{C}\ell(M,q))$ cannot contain invertible elements; indeed we can replace $\mathrm{C}\ell(M,q)$ with an isomorphic algebra $\bigwedge(M;\beta)$ as in **4.8**, we can require that $\beta(a,b) \in \mathfrak{m}$ for all a, $b \in \mathrm{Ker}(\mathrm{b}_q)$, and by means of (4.8.9) we can easily prove, for all x, $y \in \mathrm{Z}^r_1(\mathrm{C}\ell(M,q))$, that the component of the product xy in $\bigwedge^0(M) = K$ belongs to $\mathfrak{m}$.

Proof of (5.7.1) and (5.7.2)

We prove them by induction on r'. When $r' = 0$, then $M = \mathrm{Ker}(\mathrm{b}_q)$ and there is nothing to prove since $\mathrm{GO}(M,q)$ is reduced to one element. Now we assume that $r' > 0$, and that g is an automorphism of (M,q) such that $\mathrm{Ker}(\mathrm{b}_q) \subset \mathrm{Ker}(g - \mathrm{id})$.

When 2 is invertible in K, there is an orthogonal basis $(a_1, a_2, \dots, a_{r'})$ in M' (see (2.6.2)). If there exists an invertible lipschitzian x such that $G_x(a_{r'}) = g(a_{r'})$, then $G_x^{-1}g$ leaves $a_{r'}$ invariant, therefore leaves invariant the submodule N orthogonal to $a_{r'}$. Because of the induction hypothesis, the restriction of $G_x^{-1}g$ to N is equal to G_y for some product y of elements of N with a number of factors $\leq 2(r'-1)$. As an element of $\mathrm{GLip}(M,q)$, y gives an orthogonal transformation G_y that leaves invariant every element of $N^\perp$, in particular $a_{r'}$ (see (5.1.14)); thus the equality $g = G_x G_y$ can be extended from N to M. If x is a product of one or two elements of M, we have actually proved both (5.7.1) and (5.7.2) by induction on r'. Thus we reduce the demonstration of these theorems to the following lemma.

(5.7.4) **Lemma.** *Let (M, q) be a quadratic module over a local ring K in which 2 is invertible, and let a and a' be elements of M such that $q(a) = q(a') \in K^\times$. Then $G_x(a) = a'$ for some $x \in \mathrm{GLip}(M, q)$ that is an element of M or a product of two elements of M.*

Proof. If $q(a' - a)$ is invertible, we can choose $x = a' - a$ (see (5.6.4)). If $q(a' + a)$ is invertible, we can choose $x = (a' + a)a$, because G_a maps a to $-a$, and $G_{a'+a}$ maps $-a$ to a'. Either $q(a' - a)$ or $q(a' + a)$ is invertible since $q(a' - a) + q(a' + a) = 4q(a) \in K^\times$. □

When 2 is not invertible in K, then r' is even, and in (2.6.2) it is stated that M' is an orthogonal sum of submodules of rank 2. Each of these submodules contains a basis (a_i, b_i) (with $i = 1, 2, \dots, r'/2$) such that $q(a_i)$ and $\mathrm{b}_q(a_i, b_i)$ are invertible. Thus the induction that allows us to prove (5.7.1) and (5.7.2) is based on the following lemma.

(5.7.5) **Lemma.** *Let (M, q) be a quadratic module over a local ring K in which 2 is not invertible, and let a, a', b, b' be elements of M such that*

$$q(a) = q(a') \in K^\times, \quad \mathrm{b}_q(a, b) = \mathrm{b}_q(a', b') \in K^\times, \quad q(b) = q(b').$$

There exists an invertible lipschitzian element x such that $G_x(a) = a'$ and $G_x(b) = b'$. Moreover this x is a product of elements of M with at most four factors if at least one of these hypotheses is true:

(a) *if $a' - a$ is either null or invertible;*
(b) *if $q(b)$ too is invertible, and $\mathrm{b}_q(a, a')$, $\mathrm{b}_q(a, b')$, $\mathrm{b}_q(b, a')$, $\mathrm{b}_q(b, b')$ are not all in $\mathfrak{m}$;*
(c) *if $K/\mathfrak{m}$ contains more than two elements.*

Proof. For the first part of (5.7.5) it suffices to prove these two statements: first there exists an invertible lipschitzian x such that $G_x(a) = a'$; secondly when $a' = a$, there exists an invertible lipschitzian y such that $G_y(a) = a$ and $G_y(b) = b'$. For the second part of (5.7.5) it suffices to prove that x is a product of one or two elements of M, and y too. Since this is true for y without any additional hypothesis (a) or (b) or (c), we begin with the construction of y.

Let us suppose that $a = a'$, and find a suitable y. When $b' - b$ is invertible, we take $y = b' - b$ because $b' - b$ is orthogonal to a (remember $\mathrm{b}_q(a, b) = \mathrm{b}_q(a, b')$), whence $G_y(a) = a$. When $b' - b$ is not invertible, we choose an invertible $e \in M$ that is orthogonal to a, for instance $e = \mathrm{b}_q(a, b)a - 2q(a)b$, and we set $y = b'e + eb$. The invertibility of y depends on the invertibility of

$$(b'e + eb)(be + eb') \;=\; q(b' - b)q(e) + \mathrm{b}_q(b, e)\mathrm{b}_q(b', e) \;;$$

y is invertible because $\mathrm{b}_q(b,e)$ and $\mathrm{b}_q(b',e)$ are both invertible; moreover y is a product of two elements of M because e is invertible (see (5.6.5)). At last G_y maps b to b' (again (5.6.5)), and leaves a invariant because a is orthogonal to $b'-b$ and c, and $y = \mathrm{b}_q(b,e) + (b'-b)e$.

Now we prove the existence of x such that $G_x(a) = a'$ when $a \neq a'$. When $a'-a$ is invertible, we take $x = a'-a$. By the way, we have already proved the second part of (5.7.5) with the hypothesis (a). When $a'-a$ is not invertible, we take $x = a'd + da$ with some $d \in M$ such that $\mathrm{b}_q(a,d)$ and $\mathrm{b}_q(a',d)$ are both invertible, because this is the condition for x to be invertible. When $\mathrm{b}_q(a',b)$ is invertible, we can choose $d = b$. When $\mathrm{b}_q(a,b')$ is invertible, we can choose $d = b'$. When neither $\mathrm{b}_q(a',b)$ nor $\mathrm{b}_q(a,b')$ is invertible, we can choose $d = b'-b$.

Now let us prove the second part of (5.7.5) with the hypothesis (b); this hypothesis has nothing to do with the proof of (5.7.2), but anticipates later discussions. A problem may occur only when $a'-a$ is neither null nor invertible; then $\mathrm{b}_q(a,a')$ belongs to $\mathfrak{m}$; consequently $\mathrm{b}_q(a,b')$, $\mathrm{b}_q(b,a')$, $\mathrm{b}_q(b,b')$ are not all in $\mathfrak{m}$. If $\mathrm{b}_q(a,b')$ or $\mathrm{b}_q(b,a')$ is invertible, then $x = a'd + da$ with d equal to b or b', consequently invertible; thus x is a product of two elements of M. When $\mathrm{b}_q(a,a')$, $\mathrm{b}_q(a,b')$, $\mathrm{b}_q(b,a')$ are all in $\mathfrak{m}$, then $\mathrm{b}_q(b,b')$ is invertible, and $q(b'-b)$ too; therefore the problem is settled since we have chosen $d = b'-b$ in this case.

Now let us consider the hypothesis (c), the only one that is involved in the proof of (5.7.2). Since $K/\mathfrak{m}$ has more than two elements, K contains an element κ such that $\kappa(\kappa-1)$ is invertible. If $q(b)$ is not invertible, we replace b with $b-a$ or $\kappa b - a$, and similarly b' with $b'-a'$ or $\kappa b' - a'$. Since $q(b-a)$ or $q(\kappa b - a)$ is invertible, it suffices to settle the problem with the extra hypothesis $q(b) \in K^\times$. As with the hypothesis (b), it remains to examine what happens when $\mathrm{b}_q(a,a')$, $\mathrm{b}_q(a,b')$, $\mathrm{b}_q(b,a')$ are all in $\mathfrak{m}$. If $\mathrm{b}_q(b',b)$ is invertible, we still choose $d = b'-b$; but if it is not, we choose $d = \kappa b' - b$. □

This corollary of (5.7.1) shows that there cannot be exceptional cases with $r < 3$ in (5.7.3).

(5.7.6) **Corollary.** *If (M,q) is a quadratic space of rank 2 over a local ring, every automorphism g of (M,q) is a reflection or a product of two reflections.*

Indeed on one side $g = G_x$ for some homogeneous $x \in \mathrm{GLip}(M,q)$, on the other side $C\ell_1(M,q) = M$, and M contains an invertible a. Therefore every $x \in C\ell_0(M,q)$ is equal to ab for some $b \in M$. □

Proof of (5.7.3)

After the proof of (5.7.1) it appears that every $g \in \mathrm{GO}(M,q)$ (with the hypotheses mentioned just before (5.7.1)) is equal to G_x for some x that is a product of factors that belong to M, or can be written $\lambda + ab$ with $\lambda \in K$ and a, $b \in M$. When is $\lambda + ab$ itself a product of elements of M? When the submodule $Ka + Kb$ contains an invertible c, then it is easy to verify that $c(\lambda + ab)$ belongs to M, and that

$\lambda + ab = cd$ for some $d \in M$. This argument has been used to prove that $a'd + da$ in (5.6.5) is a product of two elements of M when d is invertible. Consequently it remains to look for a factorization of $\lambda + ab$ into a product of elements of M when $q(Ka + Kb) \subset \mathfrak{m}$. The upper bound $3r'$ in (5.7.3) comes from the fact that its factorization needs four factors in M. Besides, this $\lambda + ab$ is invertible if and only if λ is invertible.

(5.7.7) **Factorization lemma.** *Let (M, q) be a quadratic module over a local ring K, λ an invertible element of K, and a, b two elements of M such that $q(a)$, $q(b)$, $\mathrm{b}_q(a, b)$ all belong to $\mathfrak{m}$. Then $\lambda + ab$ is a product of four elements of M if there exists an invertible $c \in M$ such that both $\mathrm{b}_q(a, c)$ and $\mathrm{b}_q(b, c)$ belong to $\mathfrak{m}$. When $K/\mathfrak{m}$ contains more than two elements, the same conclusion holds if there exists an invertible $c \in M$ such that only $\mathrm{b}_q(a, c)$ belongs to $\mathfrak{m}$.*

Proof. Let c be an invertible element of M such that $\mathrm{b}_q(a, c) \in \mathfrak{m}$, and let μ and ν be two still unknown elements of K. A routine calculation shows that

$$\begin{aligned}(\lambda + ab)(c + \mu a)(c + \nu b) &= \big(\lambda q(c) - \lambda\nu \mathrm{b}_q(b, c) - \mu\nu q(a)q(b)\big) \; + \kappa ab + dc \\ \text{with} \quad \kappa &= q(c) + \lambda\mu\nu + \mu\nu \mathrm{b}_q(a, b) + \nu \mathrm{b}_q(b, c) \\ \text{and} \quad d &= (\lambda\mu + \mu \mathrm{b}_q(a, b) - \nu q(b))\, a \; - (\lambda\nu + \mu q(a))\, b.\end{aligned}$$

The equality $\kappa = 0$ is equivalent to $\mu\nu(\lambda + \mathrm{b}_q(a, b)) = -q(c) - \nu \mathrm{b}_q(b, c)$; it always allows us to calculate μ when ν is an invertible element. Besides, $(c + \mu a)$ is always invertible, and $(c + \nu b)$ is invertible if and only if $q(c) + \nu \mathrm{b}_q(b, c)$ is not in $\mathfrak{m}$. When $\kappa = 0$, it is clear that $(\lambda + ab)(c + \mu a)(c + \nu b)c$ belongs to M, and if ν has been chosen so that $(c + \nu b)$ is invertible, it follows that $\lambda + ab$ is a product of four elements of M.

When $\mathrm{b}_q(b, c)$ too belongs to $\mathfrak{m}$, then $(c + \nu b)$ is invertible for all $\nu \in K^\times$, and the calculation of μ is always possible. Now let us suppose that $\mathrm{b}_q(b, c)$ is invertible and that $K/\mathfrak{m}$ contains more than two elements, although this case is not involved in the proof of (5.7.3). When $(c + b)$ is invertible, we choose $\nu = 1$ and the calculation of μ follows. When $(c + b)$ is not invertible, we look for an element $\nu \in K$ such that $\nu(\nu - 1)$ is invertible, and then $(c + \nu b)$ is invertible. □

Proof of (5.7.3). We already know that there is no problem when $r < 3$. If $\mathrm{Ker}(\mathrm{b}_q)$ contains an invertible element c, with c we can factorize all elements $\lambda + ab$ such that $q(Ka + Kb) \subset \mathfrak{m}$. Consequently we suppose $r \geq 3$ and $q(\mathrm{Ker}(\mathrm{b}_q)) \subset \mathfrak{m}$. We also suppose that $(K/\mathfrak{m}) \otimes (M', q')$ is not a hyperbolic space of dimension 2 or 4, and we prove this statement: if a and b are elements of M such that $q(Ka + Kb) \subset \mathfrak{m}$, there is an invertible $c \in M$ such that $\mathrm{b}_q(a, c)$ and $\mathrm{b}_q(b, c)$ are in $\mathfrak{m}$. If $\tilde{a}$ and $\tilde{b}$ are the images of a and b in $M/\mathfrak{m}M$, it is equivalent to say that the submodule N of $M/\mathfrak{m}M$ orthogonal to $\tilde{a}$ and $\tilde{b}$ is not totally isotropic; this submodule of $M/\mathfrak{m}M$ has codimension ≤ 2. A totally isotropic submodule of $M'/\mathfrak{m}M'$ has dimension $\leq r'/2$; therefore a totally isotropic submodule of $M/\mathfrak{m}M$ has codimension $\geq r'/2$; this prove that the above N cannot be totally isotropic when $r' \geq 6$. When $r' = 4$

and M' is not hyperbolic, a totally isotropic submodule of $M'/\mathfrak{m}M'$ has dimension ≤ 1, a totally isotropic submodule of $M/\mathfrak{m}M$ has codimension ≥ 3, and the same conclusion follows. When $r' = 2$ and $M'/\mathfrak{m}M'$ is not hyperbolic, then $M'/\mathfrak{m}M'$ is anisotropic; therefore $\tilde{a}$ and $\tilde{b}$ are orthogonal to *all* elements of $M/\mathfrak{m}M$. □

Dieudonné's exceptional case

In (2.8.3) it is stated that a hyperbolic space of dimension 4 over the field $\mathbb{Z}/2\mathbb{Z}$ is in a unique way the orthogonal sum $P_1 \perp P_2$ of two anisotropic planes. Let (a_j, b_j) be a basis of P_j for $j = 1, 2$. There are six invertible elements in $P_1 \perp P_2$, the elements of the basis (a_1, b_1, a_2, b_2), and the other nonzero elements $a_1 + b_1$ and a_2+b_2 in P_1 and P_2. The reflection determined by a_1 (for instance) permutes b_1 and $a_1 + b_1$ and leaves invariant a_1 and all elements of P_2. Obviously every reflection leaves P_1 and P_2 invariant, and every product of reflections too. Nevertheless they are orthogonal transformations permuting P_1 and P_2. For instance if we set

$$\begin{aligned} z = a_2(b_1 + b_2) + (b_1 + b_2)a_1 &= a_1(b_1 + b_2) + (b_1 + b_2)a_2 \\ = \ b_2(a_1 + a_2) + (a_1 + a_2)b_1 &= b_1(a_1 + a_2) + (a_1 + a_2)b_2 \ , \end{aligned}$$

we get an invertible lipschitzian z such that G_z maps

$$(a_1, b_1, a_2, b_2) \quad \text{to} \quad (a_2, b_2, a_1, b_1) \qquad \text{(see (5.6.5))}.$$

Now let (M, q) be a quadratic module belonging to *Dieudonné's exceptional case.* This means first that it is a nondegenerate or tamely degenerate quadratic module of finite rank over a local ring K such that $K/\mathfrak{m}$ is isomorphic to $\mathbb{Z}/2\mathbb{Z}$, secondly that $q(\mathrm{Ker}(\mathrm{b}_q)) \subset \mathfrak{m}$, and thirdly that $(K/\mathfrak{m}) \otimes (M', q')$ is hyperbolic of rank 4 over $K/\mathfrak{m}$. Every automorphism g of (M, q) induces an automorphism $\tilde{g}$ of $(K/\mathfrak{m}) \otimes (M', q')$ (because it is canonically isomorphic to the quotient of $(K/\mathfrak{m}) \otimes (M, q)$ by $\mathrm{Ker}(q/\mathfrak{m})$). If g is a product of reflections, then $\tilde{g}$ too is a product of reflections, but this is impossible if $\tilde{g}$ permutes the two anisotropic planes in $(K/\mathfrak{m}) \otimes (M', q')$. Yet it is easy to find an invertible lipschitzian z in $C\ell(M, q)$ such that G_z permutes the anisotropic planes of $(K/\mathfrak{m}) \otimes (M', q')$; we can even require $z^2 \in K$, so that G_z is involutive. Indeed let (a_1, b_1, a_2, b_2) be a basis of M' such that $Ka_1 + Kb_1$ and $Ka_2 + Kb_2$ are mapped onto the anisotropic planes of $(K/\mathfrak{m}) \otimes (M', q')$, and λ a still unknown element of $K^\times$; if we set $z = \lambda a_2(b_1 + b_2) + (b_1 + b_2)a_1$, it is sure that z is an invertible lipschitzian element such that G_z permutes the anisotropic planes in $(K/\mathfrak{m}) \otimes (M', q')$. Moreover a routine calculation shows that

$$z^2 \ - \mathrm{b}_q(\lambda a_2 + a_1,\ b_1 + b_2)\ (\lambda a_2 - a_1)(b_1 + b_2) \ \in \ K \ ;$$

therefore z^2 belongs to K if and only if $\mathrm{b}_q(\lambda a_2 + a_1,\ b_1 + b_2) = 0$; since $\mathrm{b}_q(a_j,\ b_1 + b_2)$ is invertible for $j = 1, 2$, this equation has a unique and invertible solution λ.

(5.7.8) **Proposition.** *Let (M,q) be a quadratic module that belongs to Dieudonné's exceptional case. An orthogonal transformation of (M,q) is a product of reflections if and only if it leaves invariant the two anisotropic planes in $(K/\mathfrak{m})\otimes(M',q')$. Besides, there are elements $c,\ d \in M$ such that $q(Kc+Kd)\subset \mathfrak{m}$, and G_{1+cd} is an involutive orthogonal transformation that permutes these anisotropic planes.*

Proof. The proof of the second part is almost complete; it suffices to observe that the above z can be written $\mathrm{b}_q(a_1,\ b_1+b_2)+(\lambda a_2-a_1)(b_1+b_2)$; if we set $c=\lambda a_2-a_1$ and $d=\mathrm{b}_q(a_1,\ b_1+b_2)^{-1}(b_1+b_2)$, we get a suitable element $1+cd$. In the first part of (5.7.8) it remains to prove that an orthogonal transformation g that does not permute the anisotropic planes of $(K/\mathfrak{m})\otimes(M',q')$, is a product of reflections. Let (a_1',b_1',a_2',b_2') be the image of (a_1,b_1,a_2,b_2) by g. It suffices to prove the existence of a product x of elements of M (at most four factors) such that G_x maps a_1 and b_1 to a_1' and b_1', because after this first step we meet a subsequent problem in the quadratic module $Ka_2\oplus Kb_2\oplus\mathrm{Ker}(\mathrm{b}_q)$ which is not exceptional (see (5.7.3)). Since g does not permute the anisotropic planes, a_1' is congruent either to a_1 or to b_1 or to a_1+b_1 modulo $\mathfrak{m}M+\mathrm{Ker}(\mathrm{b}_q)$; consequently either $\mathrm{b}_q(a_1,a_1')$ or $\mathrm{b}_q(b_1,a_1')$ is invertible, and the existence of x follows from the second part of (5.7.5) with the hypothesis (b). □

The subgroup $\Gamma(M,q)$ generated by the reflections is a normal subgroup of $\mathrm{Aut}(M,q)$ because $gG_dg^{-1}=G_{g(d)}$ for all invertible $d\in M$ and all $g\in\mathrm{Aut}(M,q)$. When $\mathrm{GO}(M,q)\neq\Gamma(M,q)$, it may be interesting to know a subgroup supplementary to $\Gamma(M,q)$ in $\mathrm{GO}(M,q)$, in other words, a subgroup that is mapped bijectively onto the quotient $\mathrm{GO}(M,q)/\Gamma(M,q)$. When (M,q) is the quadratic module mentioned in (5.7.8), then $\{\mathrm{id},\ G_{1+cd}\}$ is a supplementary subgroup.

The other exceptional case

Now let us suppose that (M,q) belongs to *the other exceptional case.* This means first that it is a tamely degenerate quadratic module of finite rank ≥ 3 over a local ring K such that $K/\mathfrak{m}$ is isomorphic to $\mathbb{Z}/2\mathbb{Z}$, secondly that $q(\mathrm{Ker}(\mathrm{b}_q))\subset\mathfrak{m}$, and thirdly that $(K/\mathfrak{m})\otimes(M',q')$ is hyperbolic of rank 2 over $K/\mathfrak{m}$.

There is a basis (a,b) of M' such that both $q(a)$ and $\mathrm{b}_q(a,b)$ are invertible, whereas $q(b)$ and $q(a+b)$ fall into $\mathfrak{m}$. Every invertible element of M is congruent to a modulo $\mathfrak{m}M+\mathrm{Ker}(\mathrm{b}_q)$. Consequently $g(a)-a$ is in $\mathfrak{m}M+\mathrm{Ker}(\mathrm{b}_q)$ for all $g\in\mathrm{Aut}(M,q)$. If g is an orthogonal transformation, in other words if g leaves invariant every element of $\mathrm{Ker}(\mathrm{b}_q)$, a short calculation shows that $g(a)-a$ and $g(a'')-a''$ are congruent modulo $\mathfrak{m}M$ if a'' is another invertible element of M. This suggests to consider the mapping $D:\mathrm{GO}(M,q)\to\mathrm{Ker}(\mathrm{b}_q)/\mathfrak{m}\mathrm{Ker}(\mathrm{b}_q)$ which maps every $g\in\mathrm{GO}(M,q)$ to the class of $g(a)-a$ modulo $\mathfrak{m}M$. It is a group morphism because, for all $g,\ g'\in\mathrm{GO}(M,q)$,

$$gg'(a)-a \;=\; g(g'(a)-a)+g(a)-a \;\equiv\; (g'(a)-a)+(g(a)-a)$$

modulo $\mathfrak{m}M$. This group morphism maps every reflection G_d to 0; indeed $G_d(a) - a = -\mathrm{b}_q(a,d)d^{-1}$ and $\mathrm{b}_q(a,d)$ belongs to $\mathfrak{m}$.

(5.7.9) **Proposition.** *When (M,q) belongs to the other exceptional case, there is a canonical group morphism D from the multiplicative group $\mathrm{GO}(M,q)$ onto the additive group $\mathrm{Ker}(\mathrm{b}_q)/\mathfrak{m}\mathrm{Ker}(\mathrm{b}_q)$. It is surjective and all products of reflections belong to its kernel. Conversely every orthogonal transformation in $\mathrm{Ker}(D)$ is a product of reflections if at least one of these extra hypotheses is true:*

- *if $\mathrm{Ker}(q) = \mathrm{Ker}(\mathrm{b}_q)$;*
- *if there is a hyperbolic plane M' among the submodules supplementary to $\mathrm{Ker}(\mathrm{b}_q)$.*

Proof. For every $c \in \mathrm{Ker}(\mathrm{b}_q)$ the lipschitzian element $1 + bc$ is invertible and

$$G_{1+bc}(a) - a \;=\; -(1 + q(b)q(c))^{-1}\mathrm{b}_q(a,b)\ (c + q(c)b)\ ,$$

whence $D(G_{1+bc}) \equiv c$ modulo $\mathfrak{m}\mathrm{Ker}(\mathrm{b}_q)$. This proves the surjectiveness of D, and implies that some orthogonal transformations are not products of reflections, since we already know that all reflections are mapped to 0.

Conversely let g be an element of $\mathrm{Ker}(D)$. If we manage to prove that $g_1g(a)$ belongs to M' for some $g_1 \in \mathrm{GO}(M,q)$ that is a product of reflections, then it is easy to prove that g is a product of reflections. Indeed let us set $x = ab + bg_1g(a)$; since $g_1g(a) - a \in \mathfrak{m}M'$, we are sure that both $\mathrm{b}_q(a,b)$ and $\mathrm{b}_q(g_1g(a),b)$ are invertible, whereas $q(g_1g(a)-a) \in \mathfrak{m}$, whence the invertibility of x, and $G_xg_1g(a) = a$. Moreover x is a product of two elements of M' since it belongs to $C\ell_0(M',q')$ (see (5.7.6)). Then (5.7.5) (with the hypothesis (a)) proves the existence of a product g_2 of reflections such that $g_2G_xg_1g$ maps a to a, and b to b, whence $g_2G_xg_1g = \mathrm{id}$ since all orthogonal transformations leave invariant all elements of $\mathrm{Ker}(\mathrm{b}_q)$. Let us prove the existence of g_1 with each of the extra hypotheses. When there exists a hyperbolic plane M' supplementary to $\mathrm{Ker}(\mathrm{b}_q)$ we can assume that $q(b) = 0$. When $\mathrm{Ker}(q) = \mathrm{Ker}(\mathrm{b}_q)$, then $q(c) = 0$ for all $c \in \mathrm{Ker}(\mathrm{b}_q)$. Let $(c_1, c_2, \dots, c_s)$ be a basis of $\mathrm{Ker}(\mathrm{b}_q)$, and let us write

$$g(a) = (1+\kappa)a + \lambda b + \sum_{i=1}^{s} \mu_i c_i \quad \text{with} \quad \kappa, \lambda, \mu_1, \dots, \mu_s \in \mathfrak{m}.$$

The hypothesis $D(g) = 0$ means precisely that all μ_i belong to $\mathfrak{m}$. Let us prove the existence of some invertible y that is a product of elements of M, and that allows us to write

$$G_yg(a) = (1+\kappa)a + \lambda_s b + \sum_{i=1}^{s-1} \mu_i c_i \quad \text{for some } \lambda_s \in \mathfrak{m}\ ;$$

if we manage to prove it, the existence of g_1 follows from an induction on s. Let us set $y = 1 + \nu_s bc_s$ for some still unknown $\nu_s \in \mathfrak{m}$. This y is invertible because

$$(1 + \nu_s bc_s)(1 + \nu_s c_s b) = 1 + \nu_s^2 q(b)q(c_s) = 1\ ;$$

indeed either $q(b)$ or $q(c_s)$ vanishes. Then easy calculations show that

$$(1+\nu_s bc_s)\ a\ (1+\nu_s c_s b) = a\ - \nu_s \mathrm{b}_q(a,b)\ (c_s + \nu_s q(c_s)b)\ ,$$
$$(1+\nu_s bc_s)\ b\ (1+\nu_s c_s b) = b\ - 2\nu_s q(b)\ (c_s + \nu_s q(c_s)b).$$

Consequently we must choose

$$\nu_s\ =\ \mu_s\,((1+\kappa)\mathrm{b}_q(a,b) + 2\lambda q(b))^{-1}$$

to get the above equality with

$$\lambda_s\ =\ \lambda\ - \nu_s^2\, q(c_s)\,((1+\kappa)\mathrm{b}_q(a,b) + 2\lambda q(b))\ .$$

It remains to realize that y is a product of four elements of M because the factorization Lemma (5.7.7) can be applied to the invertible a such that $\mathrm{b}_q(a, c_s) = 0$ and $\mathrm{b}_q(a,\ \nu_s b) \in \mathfrak{m}$. Consequently the wanted g_1 is the orthogonal transformation derived from a lipschitzian element looking like

$$(1+\nu_1 bc_1)\ (1+\nu_2 bc_2)\cdots(1+\nu_s bc_s)\ .$$

When $q(b) = 0$, it is worth observing that the s factors $(1+\nu_i bc_i)$ pairwise commute, and that their product is $1 + bc$ with $c = \sum_i \nu_i c_i$. When $\mathrm{Ker}(q) = \mathrm{Ker}(\mathrm{b}_q)$, it is worth observing that $\lambda = \lambda_s = \lambda_{s-1} = \cdots$; but it is still more interesting to notice that, instead of the above y, it is preferable to use $y' = 1+(a+\nu_s' b)c_s$ with a suitable $\nu_s' \in \mathfrak{m}$, because the invertibility of $(a+\nu_s' b)$ implies that y' is a product of only two elements of M. The calculations with y' lead to

$$\nu_s'\ =\ (\mu_s - 2(1+\kappa)q(a) - \lambda \mathrm{b}_q(a,b))\ ((1+\kappa)\mathrm{b}_q(a,b) + 2\lambda q(b))^{-1}. \qquad \square$$

When the subgroup $\Gamma(M,q)$ generated by the reflections is equal to $\mathrm{Ker}(D)$, a subgroup of $\mathrm{GO}(M,q)$ is supplementary to $\Gamma(M,q)$ if and only if D maps it bijectively onto $\mathrm{Ker}(\mathrm{b}_q)/\mathfrak{m}\mathrm{Ker}(\mathrm{b}_q)$. Here it is more difficult to find such a supplementary subgroup than in Dieudonné's exceptional case. When $q(b) = 0$, the mapping $c \longmapsto G_{1+bc}$ is a group morphism from $\mathrm{Ker}(\mathrm{b}_q)$ into $\mathrm{GO}(M,q)$, and moreover $D(G_{1+bc}) \equiv c$ modulo $\mathfrak{m}\mathrm{Ker}(\mathrm{b}_q)$. Consequently the transformations G_{1+bc} constitute a supplementary subgroup when $K = \mathbb{Z}/2\mathbb{Z}$. When K contains a subgroup supplementary to $\mathfrak{m}$ (in the category $\mathcal{M}od(\mathbb{Z})$ of additive groups), then $\mathrm{Ker}(\mathrm{b}_q)$ contains a subgroup supplementary to $\mathfrak{m}\mathrm{Ker}(\mathrm{b}_q)$, which gives in $\mathrm{GO}(M,q)$ a subgroup supplementary to $\Gamma(M,q)$. Such a subgroup of K supplementary to $\mathfrak{m}$ exists for instance when K is the ring $(\mathbb{Z}/2\mathbb{Z})[[t]]$ of formal series, but does not exist when $K = \mathbb{Z}/2^n\mathbb{Z}$ with $n \geq 2$.

5.8 Further results about orthogonal transformations

When (M,q) is a finitely generated quadratic module, Theorem (5.3.9) allows us to recognize by localization whether an automorphism of (M,q) belongs to the image of the morphism $\mathrm{G''Lip}(M,q) \to \mathrm{GO}(M,q)$, and in the previous section we have obtained precise information about a great amount of quadratic modules over local rings. Here are the immediate consequences.

(5.8.1) **Theorem.** *When (M,q) is a quadratic space, all these canonical group morphisms are bijective:*

$$\mathrm{G'Lip}(M,q) \longrightarrow \mathrm{G''Lip}(M,q) \longrightarrow \mathrm{GO}(M,q) \longrightarrow \mathrm{Aut}(M,q).$$

(5.8.2) **Theorem.** *When M is a finitely generated projective module provided with a tamely degenerate quadratic form q, then $\mathrm{GO}(M,q)$ is the subgroup of all $g \in \mathrm{Aut}(M,q)$ such that $\mathrm{Ker}(g - \mathrm{id}) \supset \mathrm{Ker}(\mathrm{b}_q)$ and the group morphism $\mathrm{G''Lip}(M,q) \to \mathrm{GO}(M,q)$ is bijective.*

The group morphism $\mathrm{GLip}(M,q) \to \mathrm{G'Lip}(M,q)$ is already well described by the exact sequence (5.3.7), but it is a pity that (5.8.2) says nothing about $\mathrm{G'Lip}(M,q)$. There is more information about $\mathrm{G'Lip}(M,q)$ only in the nondefective case that shall be soon presented. Although nondegenerate and tamely degenerate quadratic forms already represent a great amount of various cases, their study does not yet enable us to guess what might occur with orthogonal groups of other interesting quadratic forms.

For instance look at the quadratic form $q : K \to K$ defined by $\lambda \longmapsto \lambda^2$ when $K = \mathbb{Z}/8\mathbb{Z}$. All the four elements μ of $K^\times$ are square roots of 1, and consequently every linear transformation $\lambda \longmapsto \mu\lambda$ is an automorphism of (K,q); nevertheless $\mathrm{GO}(K,q)$ contains only two elements, id and the reflection $-\mathrm{id}$. This quadratic form q is neither nondegenerate nor tamely degenerate, but it is "almost nondegenerate": see (2.ex.14); in the "almost nondegenerate" case an automorphism is an orthogonal transformation if and only if its determinant satisfies the property stated in (5.6.3): see (5.ex.15).

The surjectiveness of $\mathrm{G''Lip}(M,q) \to \mathrm{GO}(M,q)$ in (5.8.1) and (5.8.2) proves (for the quadratic modules here under consideration) the equality $\mathrm{G''Lip}(M,q) = \mathrm{G''C\ell}(M,q)$ which has been considered as a main conjecture. Besides, the elements of $\mathrm{G''C\ell}(M,q)$ are modules over the subalgebra $\mathrm{Z}^r(\mathrm{C\ell}(M,q))$; this is a subalgebra of $\mathrm{Z}^g(\mathrm{C\ell}(M,q))$, which here is the subalgebra generated by $\mathrm{Ker}(\mathrm{b}_q)$ (see (5.6.8)). If the rank of $\mathrm{Ker}(\mathrm{b}_q)$ is everywhere ≤ 3, then all elements of $\mathrm{Z}^g(\mathrm{Ker}(\mathrm{b}_q))$ are lipschitzian (see (5.4.3)), and the equality $\mathrm{G''Lip}(M,q) = \mathrm{G''C\ell}(M,q)$ implies $\mathrm{G'Lip}(M,q) = \mathrm{G'C\ell}(M,q)$ and $\mathrm{GLip}(M,q) = \mathrm{GC\ell}(M,q)$. But it is easy to construct examples for which these last two equalities are false: see (5.8.5) below.

The nondefective case

Let q be a tamely degenerate quadratic form on a finitely generated projective module; we say that q is *nondefective* if $\mathrm{Ker}(q) = \mathrm{Ker}(\mathrm{b}_q)$. In (2.2.1) there are two sufficient conditions for this equality to be true. When q is nondefective, $\mathrm{GO}(M,q)$ is in a natural way a semidirect product of two subgroups; in other words, it contains a remarkable normal subgroup and easily noticeable supplementary subgroups.

(5.8.3) **Proposition.** *Let (M,q) be a tamely degenerate quadratic space such that $\mathrm{Ker}(q) = \mathrm{Ker}(\mathrm{b}_q)$, and (M',q') a submodule supplementary to $\mathrm{Ker}(\mathrm{b}_q)$ as above. Let us map every $h \in \mathrm{Hom}(M',\mathrm{Ker}(q))$ to the endomorphism $F_0(h)$ of M defined in this way:*

$$\forall c \in \mathrm{Ker}(q),\ \forall a' \in M', \quad F_0(h)(c+a') = c + h(a') + a'.$$

Thus we get an injective morphism F_0 from the additive group $\mathrm{Hom}(M',\mathrm{Ker}(q))$ into the multiplicative group $\mathrm{GO}(M,q)$; its image $\mathrm{GO}^n(M,q)$ is the subgroup of all $g \in \mathrm{GO}(M,q)$ such that $\mathrm{Im}(g-\mathrm{id}) \subset \mathrm{Ker}(q)$; it is a normal subgroup of $\mathrm{GO}(M,q)$.

Moreover let us map every $g' \in \mathrm{GO}(M',q')$ to the endomorphism $F_1(g')$ of M defined in this way:

$$\forall c \in \mathrm{Ker}(q),\ \forall a' \in M', \quad F_1(g')(c+a') = c + g'(a').$$

Thus we get an injective group morphism $F_1 : \mathrm{GO}(M',q') \to \mathrm{GO}(M,q)$; its image $\mathrm{GO}'(M,q)$ is the subgroup of all $g \in \mathrm{GO}(M,q)$ such that $g(M') = M'$; it is a subgroup of $\mathrm{GO}(M,q)$ supplementary to the normal subgroup $\mathrm{GO}^n(M,q)$.

Proof. It is easy to verify that $F_0(h)$ and $F_1(g')$ are automorphisms of (M,q) leaving invariant all $c \in \mathrm{Ker}(q)$, therefore orthogonal transformations. The elements of $\mathrm{GO}^n(M,q)$ are *all* the linear transformations g of M such that $\mathrm{Ker}(q) \subset \mathrm{Ker}(g-\mathrm{id})$ and $\mathrm{Im}(g-\mathrm{id}) \subset \mathrm{Ker}(q)$; consequently $\mathrm{GO}^n(M,q)$ is actually a normal subgroup in the group of all linear transformations leaving $\mathrm{Ker}(q)$ invariant. The description of $\mathrm{GO}'(M,q)$ in (5.8.3) is an evidence because $\mathrm{GO}(M',q') = \mathrm{Aut}(M',q')$, and $\mathrm{GO}(M,q)$ is the group of all $g \in \mathrm{Aut}(M,q)$ leaving invariant all $c \in \mathrm{Ker}(q)$. It remains to prove that every $g \in \mathrm{GO}(M,q)$ can be written as a product $F_1(g')F_0(h)$ for a unique couple (h,g'). On one side $F_1(g')F_0(h)(a') = h(a') + g'(a')$ for all $a' \in M'$. On the other side, for every $g \in \mathrm{GO}(M,q)$ there is a unique $h \in \mathrm{Hom}(M',\mathrm{Ker}(q))$ and a unique $g' \in \mathrm{End}(M')$ such that $g(a') = h(a') + g'(a')$ for all $a' \in M'$, and it is clear that g' must belong to $\mathrm{GO}(M',q')$. □

(5.8.4) **Corollary.** *If (M,q) is a tamely degenerate and nondefective quadratic module, the group morphism $\mathrm{G'Lip}(M,q) \to \mathrm{GO}(M,q)$ is surjective.*

Proof. Because of (5.8.1) the morphism $\mathrm{G'Lip}(M',q') \to \mathrm{GO}(M',q')$ is bijective, and from F_1 in (5.8.3) we deduce a bijective morphism $\mathrm{GO}(M',q') \to \mathrm{GO}'(M,q)$. Consequently it suffices to prove that the normal subgroup $\mathrm{GO}^n(M,q)$ is contained in the image of $\mathrm{G'Lip}(M,q) \to \mathrm{GO}(M,q)$. As a matter of fact it lies even in the image of $\mathrm{GLip}(M,q) \to \mathrm{GO}(M,q)$. Indeed there is a bijection $\mathrm{Ker}(q) \otimes M'^* \to \mathrm{Hom}(M', \mathrm{Ker}(q))$ because M' is finitely generated and projective; and there is a bijection $M' \to M'^*$ because q' is nondegenerate; whence a bijection $\mathrm{Ker}(q) \otimes M' \to \mathrm{Hom}(M', \mathrm{Ker}(q))$, which is exactly this one:

$$c \otimes b' \longmapsto (\ a' \longmapsto \mathrm{b}_q(b', a')c\ ) \ .$$

It is easy to verify that

$$(1 + cb')(1 + b'c) = 1 \quad \text{and} \quad (1 + cb')a'(1 + b'c) = a' + \mathrm{b}_q(b', a')c \ .$$

This means that, when h is the mapping $a' \longmapsto \mathrm{b}_q(b', a')c$, then $G_{1+cb'} = F_1(h)$. It follows that for every $h \in \mathrm{Hom}(M', \mathrm{Ker}(q))$ there exists $x \in \mathrm{GLip}(M,q)$ such that $F_1(h) = G_x$. □

(5.8.5) **Remark.** At least in the nondefective case we know that the group $\mathrm{G'Lip}(M,q)$ gives all orthogonal transformations, and consequently the non-lipschitzian elements of $\mathrm{G'C\ell}(M,q)$ are superfluous. But might the Clifford group $\mathrm{GC\ell}(M,q)$ give more orthogonal transformations than $\mathrm{GLip}(M,q)$? In other words, if $G_X = G_Y$ for some $X \in \mathrm{G'Lip}(M,q)$ and some $Y \in \mathrm{G'C\ell}(M,q)$, is it possible for Y to be free and X not free? The equality $G_X = G_Y$ implies that $X\tau(Y)$ is contained in $\mathrm{Z}^r(\mathrm{C\ell}(M,q))$, a subalgebra of $\bigwedge(\mathrm{Ker}(q))$ since (M,q) is assumed to be nondefective. By localization it is possible to prove that the projection $\bigwedge(\mathrm{Ker}(q)) \to \bigwedge^0(\mathrm{Ker}(q)) = K$ induces a bijection $X\tau(Y) \to K$ because $X\tau(Y)$ is invertible inside $\bigwedge(\mathrm{Ker}(q))$; consequently $X\tau(Y)$ is free and X and Y have the same image in $\mathrm{Pic}(K)$.

Other investigations have confirmed that no advantages may be awaited from $\mathrm{GC\ell}(M,q)$ or $\mathrm{G'C\ell}(M,q)$. But they may bring disadvantages. For instance in the traditional nondegenerate case the equality $X^{-1} = \tau(X)$ holds for every $X \in \mathrm{G'C\ell}(M,q)$; this is always true for all $X \in \mathrm{G'Lip}(M,q)$ (see **5.3**); but with a tamely degenerate and nondefective (M,q) over a local ring in which 2 is invertible, it is easy to construct a counterexample with some $X \in \mathrm{G'C\ell}(M,q)$. Indeed let us suppose that the rank of $\mathrm{Ker}(q)$ is ≥ 4 (since it is explained above that $\mathrm{G'C\ell}(M,q) = \mathrm{G'Lip}(M,q)$ when this rank is ≤ 3), and let $(e_1, e_2, \dots)$ be a basis of $\mathrm{Ker}(q)$; every invertible $x \in \mathrm{Z}_0(\mathrm{C\ell}(M,q))$ belongs to $\mathrm{GC\ell}(M,q)$, for instance $x = 1 + e_1e_2e_3e_4$; on one side $\tau(x) = x$, on the other side x^2 is not in K; consequently this x is not lipschitzian and the equality $X^{-1} = \tau(X)$ is false when $X = Kx$.

The infinite case

It is not difficult to study quadratic modules of infinite dimension over a field with the same methods. Several facts help us to reduce the infinite case to the finite case, for instance the fact that every element of $C\ell(M,q)$ belongs to the subalgebra generated by a finite subset of M, and also the following fact.

(5.8.6) **Lemma.** *Let q be a weakly nondegenerate quadratic form on a infinite dimensional vector space M over a field K. If P is any finite dimensional subspace of M, there exists a finite dimensional subspace N containing P on which the restriction of q is nondegenerate.*

Proof. Let us set $P_0 = P \cap P^{\perp}$ and let P_1 be a subspace supplementary to P_0 in P; the restriction of q to P_1 is weakly nondegenerate, therefore nondegenerate, and $M = P_1 \oplus P_1^{\perp}$. Let us map every $a \in P_1^{\perp}$ to the restriction of $\mathrm{d}_q(a)$ to P_0; thus we obtain a mapping $P_1^{\perp} \to P_0^*$. If it were not surjective, there would be an element $b \in P_0$ such that $\mathrm{d}_q(a)(b) = 0$ for all $a \in P_1^{\perp}$, and consequently for all $a \in M$, contrary to the hypothesis that q is weakly nondegenerate. Therefore this mapping is surjective, and there is a subspace P_0' that is mapped bijectively onto P_0^*. Obviously $P_0' \cap P = 0$ and the restriction of q to $P_0 \oplus P_0'$ is nondegenerate, even hyperbolic. Now we can set $N = P \oplus P_0' = (P_0 \oplus P_0') \perp P_1$. □

In the remainder of this section M is a projective module over some ring K, M is *not* finitely generated, and M is provided with a quadratic form q for which these two hypotheses are true:

(i) $\mathrm{Ker}(\mathrm{b}_q)$ is a direct summand of M, and every finitely generated submodule of $\mathrm{Ker}(\mathrm{b}_q)$ is contained in a finitely generated direct summand.
(ii) if M' is any submodule supplementary to $\mathrm{Ker}(\mathrm{b}_q)$, and q' the restriction of q to M', then every finitely generated submodule of M' is contained in a finitely generated submodule on which the restriction of q' is nondegenerate.

It may happen that $\mathrm{Ker}(\mathrm{b}_q)$ or M' (but not both) is finitely generated. If $\mathrm{Ker}(\mathrm{b}_q)$ is finitely generated, the hypothesis (i) only means that $\mathrm{Ker}(\mathrm{b}_q)$ is a direct summand. When M' is finitely generated, the hypothesis (ii) means that (M', q') is a quadratic space. In all cases every submodule of M' on which the restriction of q' is nondegenerate, is an orthogonal summand (see (2.3.8)).

(5.8.7) **Lemma.** *If the hypotheses* (i) *and* (ii) *are true,* $\mathrm{Z}^g(C\ell(M,q))$ *is the subalgebra generated by* $\mathrm{Ker}(\mathrm{b}_q)$.

Proof. Obviously every element of $\mathrm{Ker}(\mathrm{b}_q)$ is in $\mathrm{Z}^g(C\ell(M,q))$. Conversely every $x \in \mathrm{Z}^g(C\ell(M,q))$ belongs to the subalgebra generated by a finitely generated submodule N, and we can suppose that N is the direct sum of a submodule N_0 of $\mathrm{Ker}(\mathrm{b}_q)$ and a submodule N_1 of M'; because of the hypotheses (i) and (ii) we can suppose that N_0 is a direct summand of $\mathrm{Ker}(\mathrm{b}_q)$, and that the restriction of q' to N_1 is nondegenerate; thus the Clifford algebra $C\ell(N)$ can be identified with the subalgebra generated by N in $C\ell(M,q)$ (see (4.8.5)). The restriction of q to N is tamely degenerate (even nondegenerate if $N_0 = 0$), and since x belongs to $\mathrm{Z}^g(C\ell(N))$, from (5.6.8) we deduce that x belongs to the subalgebra generated by N_0. □

If g is an orthogonal transformation of (M,q), we know that $\mathrm{Ker}(g-\mathrm{id}) \supset P^\perp$ for some finitely generated submodule P (see (5.1.14)). Let us prove that this property characterizes the orthogonal transformations inside $\mathrm{Aut}(M,q)$.

(5.8.8) **Lemma.** *Let (M,q) be a quadratic space satisfying the above hypotheses* (i) *and* (ii), *and g an automorphism of (M,q) such that* $\mathrm{Ker}(g-\mathrm{id}) \supset P^\perp$ *for some finitely generated submodule P. In M there are* orthogonal *supplementary submodules N and N'' such that*

- $P \subset N$, $g(N) = N$ *and* $\mathrm{Ker}(g-\mathrm{id}) \supset N''$,
- *N is finitely generated and* $N = (N \cap \mathrm{Ker}(\mathrm{b}_q)) \oplus (N \cap M')$,
- *the restriction of q' to $(N \cap M')$ is nondegenerate.*

Proof. Let P_0 and P_1 be finitely generated submodules of respectively $\mathrm{Ker}(\mathrm{b}_q)$ and M' such that $P \subset P_0 \oplus P_1$. Thus $\mathrm{Ker}(g-\mathrm{id}) \supset P^\perp \supset P_1^\perp$. Because of the hypothesis (ii) we can suppose that the restriction of q' to P_1 is nondegenerate, so that $M = P_1 \oplus P_1^\perp$. It follows that $\mathrm{Im}(g-\mathrm{id}) = (g-\mathrm{id})(P_1)$ and that $\mathrm{Im}(g-\mathrm{id})$ is finitely generated. Let N_0 and N_1 be finitely generated submodules of $\mathrm{Ker}(\mathrm{b}_q)$ and M' such that $P_0 + P_1 + \mathrm{Im}(g-\mathrm{id}) \subset N_0 \oplus N_1$. Because of the hypotheses (i) and (ii) we can suppose that N_0 is a direct summand of $\mathrm{Ker}(\mathrm{b}_q)$ (whence $\mathrm{Ker}(\mathrm{b}_q) = N_0 \oplus N_0''$ for a suitable submodule N_0''), and that the restriction of q' to N_1 is nondegenerate. Now we can set $N = N_0 \oplus N_1$ and $N'' = N_0'' \oplus (M' \cap N_1^\perp)$. □

(5.8.9) **Corollary.** *Let (M,q) and g be the same as in* (5.8.8). *If $K \to K'$ is a flat extension of K, then* $\mathrm{Z}^r(K' \otimes C\ell(g)) = K' \otimes \mathrm{Z}^r(C\ell(g))$.

Proof. It suffices to prove that every $x \in \mathrm{Z}^r(K' \otimes C\ell(g))$ belongs to $K' \otimes \mathrm{Z}^r(C\ell(g))$. Since the inclusion $\mathrm{Ker}(g-\mathrm{id}) \supset P^\perp$ remains true when we increase P, we can suppose that x belongs to the subalgebra generated by $K' \otimes P$. Let N and N'' be the submodules constructed in (5.8.8), and let $\tilde{g}$ be the restriction of g to N. It is clear that x is in $\mathrm{Z}^r(K' \otimes C\ell(\tilde{g}))$. Since N is finitely generated, x is in $K' \otimes \mathrm{Z}^r(C\ell(\tilde{g}))$, and it suffices to remember $\mathrm{Z}^r(C\ell(\tilde{g})) \subset \mathrm{Z}^r(C\ell(g))$ (see (5.6.6)). □

Corollary (5.8.9) shows that we can apply to g all the conclusions of Theorem (5.3.9). When K is a local ring we can also apply all the results of the previous section. Indeed let g, N and N'' satisfy the properties stated in (5.8.8), and $\tilde{g}$ the restriction of g to N. If there is some $x \in \mathrm{GLip}(N,\tilde{q})$ such that g and G_x coincide on N, then they coincide everywhere on M because G_x leaves invariant every element of N'' (see (5.1.14)); therefore every decomposition of $\tilde{g}$ into a product of reflections gives a decomposition of g into a product of reflections. Thus we have proved these two theorems.

(5.8.10) **Theorem.** *If (M,q) is a quadratic module satisfying the hypotheses* (i) *and* (ii), *the orthogonal transformations of (M,q) are the automorphisms q of (M,q) such that* $\mathrm{Ker}(g-\mathrm{id}) \supset P^\perp$ *for some finitely generated submodule P of M. The group morphism* $\mathrm{G''Lip}(M,q) \to \mathrm{GO}(M,q)$ *is bijective. When moreover* $\mathrm{Ker}(\mathrm{b}_q) =$

$\mathrm{Ker}(q)$ (*resp.* $\mathrm{Ker}(\mathrm{b}_q) = 0$), *the group morphism* $\mathrm{G'Lip}(M,q) \to \mathrm{GO}(M,q)$ *is surjective* (*resp. bijective*).

(5.8.11) **Theorem.** *If* (M,q) *is a quadratic module satisfying the hypotheses* (i) *and* (ii) *over a local ring* K *with maximal ideal* $\mathfrak{m}$, *the group morphism* $\mathrm{GLip}(M,q) \to \mathrm{GO}(M,q)$ *is surjective and every orthogonal transformation is a product of reflections, except when the residue field has only two elements and these two conditions are satisfied:* $q(\mathrm{Ker}(\mathrm{b}_q)) \subset \mathfrak{m}$ *and* $(K/\mathfrak{m}) \otimes (M',q')$ *is a hyperbolic space of dimension* 2 *or* 4.

The exceptional cases in (5.8.11) can be explained as in (5.7.8) and (5.7.9).

(5.8.12) **Corollary.** *Let* (M,q) *be as in* (5.8.10). *If* $\mathrm{Ker}(q) = 0$, $\mathrm{GO}(M,q)$ *is the subgroup of all* $g \in \mathrm{Aut}(M,q)$ *such that* $\mathrm{Im}(g - \mathrm{id})$ *is a finitely generated module.*

Indeed during the proof of (5.8.8) we realized that $\mathrm{Im}(g - \mathrm{id})$ is finitely generated for every $g \in \mathrm{GO}(M,q)$. Conversely in (5.6.1) it is stated that $\mathrm{Ker}(g - \mathrm{id}) = (\mathrm{Im}(g - \mathrm{id}))^\perp$ when $\mathrm{Ker}(q) = 0$. □

(5.8.13) **Corollary.** *Let* (M,q) *be as in* (5.8.10). *If* M' *is finitely generated,* $\mathrm{GO}(M,q)$ *is the subgroup of all* $g \in \mathrm{Aut}(M,q)$ *such that* $\mathrm{Ker}(g - \mathrm{id}) \supset \mathrm{Ker}(\mathrm{b}_q)$.

Indeed $\mathrm{Ker}(\mathrm{b}_q) = M'^\perp$. □

When K is a field, the next lemma improves the characterization of orthogonal transformations inside $\mathrm{Aut}(M,q)$, and leads to an example of an automorphism g that is not an orthogonal transformation although it satisfies all conditions required both in (5.8.12) and (5.8.13).

(5.8.14) **Lemma.** *Let* (M,q) *be an infinite dimensional quadratic module over a field, and let* V *be any submodule of* M. *The following assertions are equivalent:*

(a) *there is a finite dimensional submodule* P *such that* $V \supset P^\perp$;
(b) *there is a finite dimensional submodule* P *such that* $V = P^\perp$;
(c) *the codimension of* V *is finite and* $V = V^{\perp\perp}$.

Proof. When $\dim(P)$ is finite, it is clear that the codimension of $P^\perp$ is $\leq \dim(P)$. When $P \cap \mathrm{Ker}(\mathrm{b}_q) \neq 0$, we do not change $P^\perp$ if we replace P with a submodule supplementary to $P \cap \mathrm{Ker}(\mathrm{b}_q)$ in P; therefore we can require $P \cap \mathrm{Ker}(\mathrm{b}_q) = 0$. With this assumption, $\mathrm{d}_q : M \to M^*$ induces an injective mapping $P \to (M/P^\perp)^*$, and since $\dim(M/P^\perp) \leq \dim(P)$, this injective mapping is bijective. It determines a duality between P and $M/P^\perp$, and consequently a bijection between the subspaces of P and the subspaces of $M/P^\perp$. In particular if $V \supset P^\perp$, with the subspace $V/P^\perp$ of $M/P^\perp$ this bijection associates a subspace Q of P such that $V = Q^\perp$. Thus we have proved (a)⇒(b).

Now let us suppose that $V = P^\perp$ as in (b); it is clear that V has finite codimension $\leq \dim(P)$, and the equality $V = V^{\perp\perp}$ announced in (c) follows from

$P^{\perp\perp\perp} = P^{\perp}$ (see (2.3.6)). Conversely if $V = V^{\perp\perp}$, then $V = P^{\perp}$ for every submodule P supplementary to $\mathrm{Ker}(\mathrm{b}_q)$ in $V^{\perp}$; and since the mapping $P \to (M/P^{\perp})^*$ is injective, P has finite dimension if V has finite codimension. $\square$

(5.8.15) **Example.** Let K be a field of characteristic other than 2, and M a vector space over K with an infinite basis $(e_0, e_1, e_2, \cdots)$. Let us set

$$q(\sum_{i\geq 0} \lambda_i e_i) = \sum_{i\geq 1} \lambda_i^2 \quad \text{and} \quad h(\sum_{i\geq 0} \lambda_i e_i) = \sum_{i\geq 1} \lambda_i \ ;$$

these definitions are meaningful since all $\lambda_i \in K$ vanish except a finite number. Thus q is a quadratic form such that $\mathrm{Ker}(q) = Ke_0$, and h is a linear form such that $e_0 \in \mathrm{Ker}(h)$. But h is not in the image of $\mathrm{d}_q : M \longrightarrow M^*$. Since $\mathrm{Ker}(h)$ contains all $e_j - e_k$ with $0 < j < k$, the equalities $\lambda_j = \lambda_k$ hold for every element $\sum_i \lambda_i e_i$ orthogonal to $\mathrm{Ker}(h)$, whence $\lambda_i = 0$ for all $i > 0$, since anyhow all λ_i vanish except a finite number; consequently $\mathrm{Ker}(h)^{\perp} = Ke_0$ and $\mathrm{Ker}(h)^{\perp\perp} = M$. We get an automorphism g of (M, q) if we set $g(a) = a + h(a)e_0$ for all $a \in M$. Obviously $\mathrm{Ker}(g - \mathrm{id}) = \mathrm{Ker}(h)$; thus the equality $V = V^{\perp\perp}$ (see (5.8.14)(c)) is not true when $V = \mathrm{Ker}(g - \mathrm{id})$. Therefore g is not an orthogonal transformation, although $\mathrm{Im}(g - \mathrm{id})$ has finite dimension and $\mathrm{Ker}(g - \mathrm{id})$ contains $\mathrm{Ker}(\mathrm{b}_q)$. The calculation of $\mathrm{Z}^g(C\ell(g))$ shows that it is the submodule generated by e_0, which contains no invertible element.

5.9 More information about exterior algebras

The invariance property (5.4.1) confers a great importance on the Lipschitz monoid $\mathrm{Lip}(M)$ in the exterior algebra $\bigwedge(M)$, although this neutral Lipschitz monoid does not give any transformation of M other than id_M. In all cases $\mathrm{Lip}(M)$ contains all elements of K and M and all exponentials of elements of $\bigwedge^2(M)$, in accordance with Lipschitz's works about 1880. When K is a field, then $\mathrm{Lip}(M)$ is the monoid generated in $\bigwedge(M)$ by these elements; before this is proved in the next section, some additional information about exterior algebras must be expounded.

Let K be a field, and M a vector space of finite nonzero dimension r over K. The dual space M^* can be identified with the subspace $\bigwedge^{*1}(M)$ of $\bigwedge^*(M)$, and the injection $M^* \longrightarrow \bigwedge^*(M)$ extends to an isomorphism $\bigwedge(M^*) \to \bigwedge^*(M)$ which allows us to identify $\bigwedge^*(M)$ with $\bigwedge(M^*)$; the purpose of this identification is to let M and M^* play symmetric roles. The symmetry that opposes M^* and M, also opposes the left side (the side of M^*) to the right side (the side of M). The same letter σ is used for the grade automorphism $x \longmapsto (-1)^{\partial x} x$ in $\bigwedge(M)$ and the grade automorphism in $\bigwedge(M^*)$.

With every subspace P of M is associated its annihilator P^{an} in M^*, which is the subspace of all $h \in M^*$ such that $h(P) = 0$, and it is known that $\mathrm{codim}(P^{an}) = \dim(P)$. Conversely with every subspace Q of M^* is associated its annihilator Q^{an}

in M, which the subspace of all $a \in M$ such that $h(a) = 0$ for all $h \in Q$. It is known that $P = (P^{an})^{an}$ and $Q = (Q^{an})^{an}$.

Let ω and ω^* be nonzero elements of $\bigwedge^r(M)$ and $\bigwedge^r(M^*)$ such that $\omega^*(\omega) = 1$. There is a basis $(e_1, e_2, \dots, e_r)$ of M such that $\omega = e_1 \wedge e_2 \wedge \cdots \wedge e_r$; if $(e_1^*, e_2^*, \dots, e_r^*)$ is the dual basis of M^*, then $\omega^* = e_r^* \wedge e_{r-1}^* \wedge \cdots \wedge e_1^*$. We define two linear mappings

$$\mathcal{F} : \bigwedge(M) \longrightarrow \bigwedge(M^*) , \quad x \longmapsto \omega^* \lfloor x ,$$
$$\text{and} \quad \mathcal{F}_* : \bigwedge(M^*) \longrightarrow \bigwedge(M) , \quad z \longmapsto z \rfloor \omega .$$

Here are immediate consequences of these definitions:

$$\mathcal{F}(1) = \omega^* , \qquad \mathcal{F}(\omega) = 1 , \qquad \mathcal{F} \circ \sigma = (-1)^r \sigma \circ \mathcal{F} ,$$
$$\mathcal{F}_*(1) = \omega , \qquad \mathcal{F}_*(\omega^*) = 1 , \qquad \mathcal{F}_* \circ \sigma = (-1)^r \sigma \circ \mathcal{F}_* .$$

Although $\mathcal{F}$ and $\mathcal{F}_*$ somewhat recall the star-Hodge operators, they were used in algebra long before Hodge elaborated his great machinery for Kählerian manifolds. Here they interest us because they turn exterior multiplications into interior ones, and conversely; this property is stated at the end of the next theorem, and is closely related to a well-known property of Fourier transformation.

(5.9.1) **Theorem.** *The transformations $\mathcal{F}$ and $\mathcal{F}_*$ are bijections such that*

$$\mathcal{F}_* \circ \mathcal{F} \;=\; \sigma^{r-1} \quad \textit{and} \quad \mathcal{F} \circ \mathcal{F}_* \;=\; \sigma^{r-1}.$$

Moreover, for every $k \in \{0, 1, 2, \dots, r\}$ we can write

$$\mathcal{F}(\bigwedge^k(M)) = \bigwedge^{r-k}(M^*) \quad \textit{and} \quad \mathcal{F}_*(\bigwedge^k(M^*)) = \bigwedge^{r-k}(M).$$

At last, for all x and x' in $\bigwedge(M)$, and all z and z' in $\bigwedge(M^*)$, we can write

$$\mathcal{F}(x \wedge x') = \mathcal{F}(x) \;\lfloor\; x' , \qquad \mathcal{F}(z \rfloor x) = \sigma^{r-1}(z) \wedge \mathcal{F}(x) ,$$
$$\mathcal{F}_*(z \wedge z') = \; z \;\rfloor\; \mathcal{F}_*(z') , \qquad \mathcal{F}_*(z \lfloor x) = \mathcal{F}_*(z) \wedge \sigma^{r-1}(x) .$$

Proof. We must prove eight equalities appearing in four pairs; since M and M^* play symmetric roles, it suffices to prove one equality in each pair. Moreover the restrictions of $\mathcal{F}$ to $\bigwedge^0(M)$ and $\bigwedge^r(M)$ are already known, and the same for $\mathcal{F}_*$; this allows us to begin with an element $x \in \bigwedge^k(M)$ of degree k such that $0 < k < r$. We suppose that x is an exterior product $e_1 \wedge e_2 \wedge \cdots \wedge e_k$ of k linearly independent vectors; there are vectors $e_{k+1}, \dots, e_r$ such that $\omega = e_1 \wedge e_2 \wedge \cdots \wedge e_r$. From (4.3.3) we deduce

$$\mathcal{F}(x) \;=\; (\cdots(((e_r^* \wedge \cdots \wedge e_{k+1}^* \wedge e_k^* \wedge \cdots \wedge e_1^*) \lfloor e_1) \lfloor e_2) \lfloor \cdots) \lfloor e_k ;$$

we must remember that $e_i^* \lfloor e_i = 1$ for $i = 1, 2, \dots, r$, whereas $e_i^* \lfloor e_j = 0$ if $i \neq j$; the successive interior multiplications by $e_1, e_2, \dots, e_k$ are calculated by means of (4.3.4), and lead to this first result:

$$\mathcal{F}(x) = e_r^* \wedge e_{r-1}^* \wedge \cdots \wedge e_{k+1}^* ;$$

this already proves that $\mathcal{F}(\bigwedge^k(M)) = \bigwedge^{r-k}(M^*)$. We continue in this way:

$$\begin{aligned}\mathcal{F}_*(\mathcal{F}(x)) &= (-1)^{(r-k)k}(e_r^* \wedge \cdots \wedge e_{k+1}^*) \rfloor (e_{k+1} \wedge e_{k+2} \wedge \cdots \wedge e_r) \wedge (e_1 \wedge \cdots \wedge e_k) \\ &= (-1)^{(r-k)k}\ e_1 \wedge e_2 \wedge \cdots \wedge e_k ;\end{aligned}$$

since $(r-k)k$ has the same parity as $(r-1)k$, the first equality in (5.9.1) is now proved.

The equality applying to $\mathcal{F}(x \wedge x')$ is a trivial consequence of (4.3.3):

$$\mathcal{F}(x \wedge x') = \omega^* \lfloor (x \wedge x') = (\omega^* \lfloor x) \lfloor x' = \mathcal{F}(x) \lfloor x'.$$

The same argument proves the symmetric equality with $\mathcal{F}_*(z \wedge z')$. To prove the equality applying to $\mathcal{F}(z \rfloor x)$, we verify that $\mathcal{F}_*$ transforms both members into the same element of $\bigwedge(M)$; on the left side

$$\mathcal{F}_*(\mathcal{F}(z \rfloor x)) = \sigma^{r-1}(z \rfloor x) = \sigma^{r-1}(z) \rfloor \sigma^{r-1}(x) ;$$

on the right side we use the previous result about $\mathcal{F}_*(z \wedge z')$:

$$\mathcal{F}_*(\sigma^{r-1}(z) \wedge \mathcal{F}(x)) = \sigma^{r-1}(z) \rfloor \mathcal{F}_*(\mathcal{F}(x)) = \sigma^{r-1}(z) \rfloor \sigma^{r-1}(x).$$

The proof is now complete. □

We also need more information about the elements u of $\bigwedge^2(M)$ and the elements w of $\bigwedge^2(M^*)$. Since w is determined by the alternate bilinear form $(a, b) \longmapsto w(a \wedge b)$, it is natural to associate with w the linear mapping $\mathrm{d}_w : M \to M^*$ such that $\mathrm{d}_w(a)(b) = w(a \wedge b)$ for all $a, b \in M$. This implies that $\mathrm{d}_w(a) = w \lfloor a$ (see (4.3.2)). Similarly with u we associate the mapping $\mathrm{d}_u : M^* \to M$ defined by $\mathrm{d}_u(h) = h \rfloor u$ and the alternate bilinear form $M^* \times M^* \to K$ defined by $(h, h') \longmapsto (h \wedge h')(u) = h(\mathrm{d}_u(h'))$. The image of d_u (resp. d_w) is called the *support* of u (resp. w) for reasons appearing in the next lemma, and the dimension of the support is called the *rank* of u (resp. w).

(5.9.2) **Lemma.** *For each $u \in \bigwedge^2(M)$, the image of d_u is the annihilator of $\mathrm{Ker}(\mathrm{d}_u)$, the dimension of $\mathrm{Im}(\mathrm{d}_u)$ is always even, and there is a basis $(e_1, e_2, \dots, e_{2k})$ of $\mathrm{Im}(\mathrm{d}_u)$ such that*

$$u = e_1 \wedge e_2 + e_3 \wedge e_4 + \cdots + e_{2k-1} \wedge e_{2k}.$$

There is an analogous statement for each $w \in \bigwedge^2(M^*)$.

Proof. It is more convenient to prove (5.9.2) for $w \in \bigwedge^2(M^*)$. The bilinear form $(a,b) \longmapsto w(a \wedge b)$ is alternate, whence the inclusion $\mathrm{Im}(\mathrm{d}_w) \subset (\mathrm{Ker}(\mathrm{d}_w))^{an}$ which is an equality because the subspaces on both sides have the same dimension. Consequently $\mathrm{Im}(\mathrm{d}_w)$ can be identified with the dual space of any subspace M' supplementary to $\mathrm{Ker}(\mathrm{d}_w)$ in M. Since d_w induces a bijection $M' \rightarrow \mathrm{Im}(\mathrm{d}_w)$, the restriction to M' of the bilinear form $(a,b) \longmapsto w(a \wedge b)$ is nondegenerate; this allows us to prove that M' is an orthogonal sum of planes, with almost the same argument as in the proof of (2.6.2)(b). Let $(e_1, e_2, \dots, e_r)$ be a basis of M such that $(e_1, e_2, \dots, e_{2k})$ is a basis of M' and the k planes $Ke_{2i-1} \oplus Ke_{2i}$ (with $i = 1, 2, \dots, k$) are orthogonal for the bilinear form $(a,b) \longmapsto w(a \wedge b)$. We can even require $w(e_{2i-1} \wedge e_{2i}) = 1$ for $i = 1, 2, \dots, k$. All this determines w, and if $(e_1^*, e_2^*, \dots, e_r^*)$ is the dual basis in M^*, it is easy to verify that

$$w \;=\; e_2^* \wedge e_1^* \,+ e_4^* \wedge e_3^* \,+ \cdots + e_{2k}^* \wedge e_{2k-1}^* \;,$$

whence $\mathrm{d}_w(e_{2i-1}) = e_{2i}^*$ and $\mathrm{d}_w(e_{2i}) = -e_{2i-1}^*$ for $i = 1, 2, \dots, k$. The conclusions follow. □

By definition $\ell ip(M)$ is the multiplicative monoid generated in $\bigwedge(M)$ by the elements of K and M and the exponentials of elements of $\bigwedge^2(M)$; we define $\ell ip(M^*)$ in an analogous way. From now on, we shall search useful properties of $\ell ip(M)$ and $\ell ip(M^*)$, until they enable us to prove that $\ell ip(M) = \mathrm{Lip}(M)$ and to make effective calculations with elements of $\mathrm{Lip}(M)$. If w is any linear mapping $M \rightarrow N$, it is already clear that its extension $\bigwedge(w)$ to the exterior algebras maps $\ell ip(M)$ into $\ell ip(N)$, and that $\bigwedge^*(w)$ maps $\ell ip(N^*)$ into $\ell ip(M^*)$; this statement agrees with (5.3.14). It is also clear that $\ell ip(M)$ is invariant by the grade automorphism σ and by the reversion τ, and the same for $\ell ip(M^*)$.

Let us precisely describe a nonzero element $x \in \ell ip(M)$. We can directly apply (5.9.2) when $x = \lambda\,\mathrm{Exp}(u)$ with $\lambda \in K^\times$ and $u \in \bigwedge^2(M)$. But when x is the exterior product of $\mathrm{Exp}(u)$ and some vectors $d_1, d_2, \dots, d_n$ (linearly independent in M), we must consider the subspace N spanned by these n vectors, and remember that any equality like $d_1 \wedge \cdots \wedge d_n \wedge y = d_1 \wedge \cdots \wedge d_n \wedge y'$ means that y and y' have the same image in $\bigwedge(M/N)$; consequently we can replace u with any element of $\bigwedge^2(M)$ that has the same image as u in $\bigwedge(M/N)$; thus we can assume that the support of u is a subspace P of M such that $N \cap P = 0$. Let $(e_1, \dots, e_{2k})$ be a basis of P that enables us to write u as in (5.9.2); the vectors $d_1, d_2, \dots, d_n$, $e_1, e_2, \dots, e_{2k}$ are linearly independent and

$$x = d_1 \wedge d_2 \wedge \cdots \wedge d_n \wedge (1 + e_1 \wedge e_2) \wedge \cdots \wedge (1 + e_{2k-1} \wedge e_{2k}).$$

It is worth observing that the nonzero components of x of lowest and highest degrees are always *decomposable elements* (either scalars or vectors or exterior products of vectors).

The property similar to (5.4.3) is also evident: when $r \leq 3$, every homogeneous element of $\bigwedge(M)$ belongs to $\ell ip(M)$, and the same for $\bigwedge(M^*)$. Indeed every even element $\lambda + a \wedge b$ is equal to $\lambda \operatorname{Exp}(\lambda^{-1} a \wedge b)$ when $\lambda \neq 0$, and every odd element of $\bigwedge(M)$ is the product of an element of M and an even element.

(5.9.3) **Theorem.** $\mathcal{F}(\ell ip(M)) = \ell ip(M^*)$ and $\mathcal{F}_*(\ell ip(M^*)) = \ell ip(M)$.

Proof. When $r \leq 3$, this is an immediate consequence of the fact that all homogeneous elements fall into $\ell ip(M)$ or $\ell ip(M^*)$. Let us assume that $r > 3$ and that x is a nonzero element $x \in \ell ip(M)$. We can still assume that the component of x in $K = \bigwedge^0(M)$ is either 1 or 0. From the above description of x we deduce the existence of a linearly independent family of vectors $(d_1, \dots, d_n, e_1, \dots, e_{2k})$ (with $n \geq 0$ and $k \geq 0$) such that x is the exterior product of the n factors $x_i = d_i$ and the k factors $x_{n+i} = 1 + e_{2i-1} \wedge e_{2i}$. Let us complete $(d_1, \dots, d_n, e_1, \dots, e_{2k})$ into a basis $(d_1, \dots, d_n, e_1, \dots, e_{2k}, c_1, \dots, c_m)$ of M, and set $x_{n+k+i} = 1$ for $i = 1, 2, \dots, m$. Thus we come to this first result: M is the direct sum of some subspaces M_1, M_2,…,M_s of dimension 1 or 2 (the number of which is $s = n + k + m = r - k$) in such a way that x is the exterior product of even or odd factors x_1, x_2,…,x_s each of which belongs to the corresponding subalgebra $\bigwedge(M_1)$, $\bigwedge(M_2)$, …, $\bigwedge(M_s)$. The dual space M^* can be identified with the direct sum of the dual spaces M_1^*, …, M_s^*. We also identify $\bigwedge(M)$ with the twisted tensor product of the subalgebras $\bigwedge(M_1)$, …, $\bigwedge(M_s)$, and the same for $\bigwedge(M^*)$. With these identifications we can write $x = x_1 \otimes x_2 \otimes \cdots \otimes x_s$. Similarly $\omega = \omega_1 \otimes \omega_2 \otimes \cdots \otimes \omega_s$ for suitable $\omega_i \in \bigwedge^{\max}(M_i)$, and with evident notation, $\omega^* = \pm \omega_1^* \otimes \omega_2^* \otimes \cdots \otimes \omega_s^*$. Because of (4.3.7) we can also write

$$\mathcal{F}(x) = \pm\ (\omega_1^* \lfloor x_1) \otimes (\omega_2^* \lfloor x_2) \otimes \cdots \otimes (\omega_s^* \lfloor x_s).$$

In other words, by means of partial $\mathcal{F}$-transformations, we have reduced the problem to the case of spaces of dimension 1 or 2, for which the theorem is a triviality. □

(5.9.4) **Corollary.** *If x and z belong respectively to $\ell ip(M)$ and $\ell ip(M^*)$, then $z \rfloor x$ and $z \lfloor x$ belong respectively to $\ell ip(M)$ and $\ell ip(M^*)$.*

This statement corresponds exactly to (5.3.13).

Proof. Because of (5.9.3), it suffices to prove that $\mathcal{F}(z \rfloor x)$ and $\mathcal{F}_*(z \lfloor x)$ belong respectively to $\ell ip(M^*)$ and $\ell ip(M)$. From (5.9.1) we derive

$$\mathcal{F}(z \rfloor x) \;=\; \sigma^{r-1}(z) \wedge \mathcal{F}(x) \quad \text{and} \quad \mathcal{F}_*(z \lfloor x) \;=\; \mathcal{F}_*(z) \wedge \sigma^{r-1}(x)\ ;$$

and the conclusion follows from another application of (5.9.3). □

With (5.9.4) it is already possible to prove (in the next section) that $\ell ip(M) = \operatorname{Lip}(M)$. The subsequent propositions only aim to give effective tools for precise calculations.

(5.9.5) **Proposition.** *When the dimension of M is even, and the support of $u \in \bigwedge^2(M)$ is exactly M, then there exist $\kappa \in K$ and $w \in \bigwedge^2(M^*)$ such that*

$$\mathcal{F}(\mathrm{Exp}(u)) \;=\; \kappa\,\mathrm{Exp}(w)\,, \quad \kappa^2 = \det(\mathrm{d}_u) \quad \textit{and} \quad \mathrm{d}_w = -\mathrm{d}_u^{-1}\;;$$

the second equality is valid if the determinant of d_u is calculated with a basis $(e_1, e_2, \dots, e_r)$ of M such that $\omega = e_1 \wedge e_2 \wedge \cdots \wedge e_r$, and with the dual basis in M^*.

Proof. There exists a basis $(e_1, \dots, e_r)$ of M and a nonzero $\lambda \in K$ such that simultaneously

$$\omega = e_1 \wedge e_2 \wedge \cdots \wedge e_r \quad \text{and} \quad u = \lambda e_1 \wedge e_2 \;+ e_3 \wedge e_4 \;+ \cdots + e_{r-1} \wedge e_r\;;$$

as explained in the proof of (5.9.3), we decompose M into the direct sum of $r/2$ subspaces M_i with basis (e_{2i-1}, e_{2i}), and we set $\omega_i = e_{2i-1} \wedge e_{2i}$ and $\omega_i^* = e_{2i}^* \wedge e_{2i-1}^*$. Since all factors ω_i^* are even, there is no problem about twisting signs and $\mathcal{F}(\mathrm{Exp}(u))$ is obtained as a product of $r/2$ factors, among which we consider the first one, the only one involving λ :

$$(e_2^* \wedge e_1^*) \;\lfloor\; \mathrm{Exp}(\lambda e_1 \wedge e_2) \;=\; e_2^* \wedge e_1^* \;+ \lambda \;=\; \lambda\,\mathrm{Exp}(\lambda^{-1} e_2^* \wedge e_1^*)\;;$$

thus we realize that $\mathcal{F}(\mathrm{Exp}(u)) = \lambda\,\mathrm{Exp}(w)$ if

$$w \;=\; \lambda^{-1} e_2^* \wedge e_1^* \;+ e_4^* \wedge e_3^* \;+ \cdots + e_r^* \wedge e_{r-1}^*\,.$$

Let us calculate d_u and d_w; here are their values on e_1^* and e_2^*, respectively e_1 and e_2 :

$$\begin{aligned} \mathrm{d}_u(e_1^*) &= \lambda e_2\,, & \mathrm{d}_w(e_1) &= \lambda^{-1} e_2^*\,,\\ \mathrm{d}_u(e_2^*) &= -\lambda e_1\,, & \mathrm{d}_w(e_2) &= -\lambda^{-1} e_1^*\,; \end{aligned}$$

their values on e_{2i-1}^* and e_{2i}^*, respectively e_{2i-1} and e_{2i}, are given by analogous equalities in which λ is replaced with 1. All this shows that $\det(\mathrm{d}_u) = \lambda^2$ and that $\mathrm{d}_w \circ \mathrm{d}_u = -\mathrm{id}$. □

In the next proposition the dimension r of M may be even or odd.

(5.9.6) **Proposition.** *Let u be an element of $\bigwedge^2(M)$, and w an element of $\bigwedge^2(M^*)$. If $\mathrm{Exp}(w)$ is treated as a linear form on $\bigwedge(M)$, its value on $\mathrm{Exp}(u)$ is a scalar κ such that*

$$\kappa^2 \;=\; \det(\mathrm{id}_M - \mathrm{d}_u \circ \mathrm{d}_w) \;=\; \det(\mathrm{id}_{M^*} - \mathrm{d}_w \circ \mathrm{d}_u)\,.$$

When $\kappa \neq 0$, there exists $v \in \bigwedge^2(M)$ such that $\mathrm{Exp}(w) \rfloor \mathrm{Exp}(u) = \kappa\,\mathrm{Exp}(v)$, and then

$$\mathrm{d}_v \;=\; (\mathrm{id}_M - \mathrm{d}_u \circ \mathrm{d}_w)^{-1} \circ \mathrm{d}_u \;=\; \mathrm{d}_u \circ (\mathrm{id}_{M^*} - \mathrm{d}_w \circ \mathrm{d}_u)^{-1}\,.$$

Of course there is an analogous statement involving $\mathrm{Exp}(w) \lfloor \mathrm{Exp}(u)$.

Proof. If (5.9.6) is true for a space M of dimension r, it is true for all its subspaces, and consequently for all spaces of dimension $\leq r$; therefore it suffices to prove (5.9.6) when r is even. It also suffices to prove it when the field K is infinite, because every finite field K has an infinite field extension (for instance the field $K(t)$ of rational functions). When V is a vector space of finite dimension over an *infinite* field K, the algebra of polynomial functions $V \to K$ contains no divisors of zero. Consequently if F and G are polynomial functions on V, if G does not vanish everywhere, and if $F(\xi) = 0$ for all $\xi \in V$ such that $G(\xi) \neq 0$, then $FG = 0$, whence $F = 0$ since $G \neq 0$. Here V is the vector space $\bigwedge^2(M) \times \bigwedge^2(M^*)$; the first statement in (5.9.6) means the vanishing of some polynomial function $F : V \to K$, and the second statement means the vanishing of 2^{r-1} rational functions (because $\dim(\bigwedge_0(M)) = 2^{r-1}$); but the vanishing of a rational function is equivalent to the vanishing of its numerator. Besides, the second statement implies the first one, since the value of the linear form $\mathrm{Exp}(w)$ on $\mathrm{Exp}(u)$ is the component in $K = \bigwedge^0(M)$ of their interior product. We shall prove the vanishing of these 2^{r-1} rational functions at every point (u, w) that does not annihilate the polynomial function G defined in this way:

$$G(u, w) = \det(\mathrm{d}_u)\,\det(\mathrm{id}_M - \mathrm{d}_u \circ \mathrm{d}_w)\ ;$$

since r is assumed to be even, $\bigwedge^2(M)$ contains an element u with support equal to M, whence $G(u, 0) \neq 0$; since G does not vanish everywhere, it suffices to consider points (u, w) such that $G(u, w) \neq 0$.

After these preliminaries we begin the calculation:

$$\mathrm{Exp}(w) \rfloor \mathrm{Exp}(u) \;=\; \mathcal{F}_* \circ \mathcal{F}(\mathrm{Exp}(w) \rfloor \mathrm{Exp}(u)) \;=\; \mathcal{F}_*(\mathrm{Exp}(w) \wedge \mathcal{F}(\mathrm{Exp}(u)))\ ,$$

and we apply (5.9.5) twice. First there exist $\kappa' \in K$ and $w' \in \bigwedge^2(M^*)$ such that

$$\mathcal{F}(\mathrm{Exp}(u)) = \kappa' \mathrm{Exp}(w')\,, \quad \kappa'^2 = \det(\mathrm{d}_u) \quad \text{and} \quad \mathrm{d}_{w'} = -\mathrm{d}_u^{-1}\ ;$$

whence $\mathrm{Exp}(w) \wedge \mathcal{F}(\mathrm{Exp}(u) = \kappa'\,\mathrm{Exp}(w + w')$. Secondly there exist $\lambda \in K$ and $v \in \bigwedge^2(M)$ such that

$$\mathcal{F}_*(\mathrm{Exp}(w{+}w')) = \lambda\,\mathrm{Exp}(v)\,, \quad \lambda^2 = \det(\mathrm{d}_w - \mathrm{d}_u^{-1}) \quad \text{and} \quad \mathrm{d}_v = (-\mathrm{d}_w + \mathrm{d}_u^{-1})^{-1}\ ;$$

now we observe that

$$(-\mathrm{d}_w + \mathrm{d}_u^{-1})^{-1} \;=\; (\mathrm{id}_M - \mathrm{d}_u \circ \mathrm{d}_w)^{-1} \circ \mathrm{d}_u \;=\; \mathrm{d}_u \circ (\mathrm{id}_{M^*} - \mathrm{d}_w \circ \mathrm{d}_u)^{-1}\ ;$$

all this gives the announced value of d_v, and also the announced value of κ^2 if we set $\kappa = \lambda\kappa'$; indeed, since r is even, $\mathrm{d}_u \circ (\mathrm{d}_w - \mathrm{d}_u^{-1})$ has the same determinant as $\mathrm{id}_M - \mathrm{d}_u \circ \mathrm{d}_w$. □

Two digressions

By using exponentials of bivectors (elements of $\bigwedge^2(M)$ or $\bigwedge^2(M^*)$), we avoid using pfaffians of skew symmetric matrices; the existence of pfaffians can be deduced from the equality $\kappa^2 = \det(\mathrm{d}_u)$ in (5.9.5), but from (5.9.6) we can deduce a still stronger result.

Let $(u_{i,j})$ and $(w_{i,j})$, with $1 \leq i < j \leq r$, be two families of $r(r-1)/2$ indeterminates, and t one more indeterminate; we consider the ring $\mathbb{Z}[t, (u_{i,j}), (w_{i,j})]$ of polynomials in these $1 + r(r-1)$ indeterminates with integer coefficients. Let U be the skew symmetric matrix of order r in which the entries above the diagonal are the indeterminates $u_{i,j}$, and W the analogous skew symmetric matrix containing all $w_{i,j}$. At last let I be the unit matrix of order r. We are interested in the determinant of $t^2 I - UW$.

(5.9.7) **Proposition.** *The determinant of the above matrix $t^2I - UW$ has a square root in the ring of polynomials $\mathbb{Z}[t, (u_{i,j}), (w_{i,j})]$; this square root $\Delta(t, (u_{i,j}), (w_{i,j}))$ is uniquely determined if we require it to be equal to t^r when U and W are replaced with* 0. *It is a homogeneous polynomial of degree r, and it contains only powers of t with an exponent of the same parity as r :*

$$\Delta(t, (u_{i,j}), (w_{i,j})) = \sum_{0 \leq k \leq r/2} t^{r-2k} \Delta_k((u_{i,j}), (w_{i,j})) \; ;$$

each polynomial $\Delta_k((u_{i,j}), (w_{i,j}))$ has degree k separately in $(u_{i,j})$ and in $(w_{i,j})$.

Besides, when t is replaced with 0 *and $(w_{i,j})$ with $(u_{i,j})$, then $\Delta(0, (u_{i,j}), (u_{i,j})) =$* $\det(U)$. *When r is odd, this means that* $\det(U) = 0$. *When r is even, this means that* $\det(U) = \Delta_{r/2}((u_{i,j}), (u_{i,j}))$.

Proof. Let B be the field of fractions of $A = \mathbb{Z}[t, (u_{i,j}), (w_{i,j})]$, and let M be the vector space B^r over B, provided with its usual basis. In $\bigwedge_B^2(M)$ and $\bigwedge_B^2(M^*)$ we can find elements u and w such that the matrices of d_u and d_w are respectively $t^{-1}U$ and $t^{-1}W$, whence

$$\det(t^2 I - UW) = t^{2r} \det(\mathrm{id}_M - \mathrm{d}_u \circ \mathrm{d}_w).$$

From (5.9.6) we deduce that $\det(t^2 I - UW)$ is the square of $t^r\kappa$ if κ is the value of the linear form $\mathrm{Exp}(w)$ on $\mathrm{Exp}(u)$. When an element $\det(t^2I - UW)$ of A has a square root $t^r\kappa$ in B, its square root is also in A; this can be proved by decompositions into products of irreducible elements, like the analogous theorem stating that every square root in $\mathbb{Q}$ of an element of $\mathbb{Z}$ belongs to $\mathbb{Z}$. Thus $\Delta(t, (u_{i,j}), (w_{i,j}))$ is this polynomial $t^r\kappa$. The above description of this polynomial is a matter of routine argument; only the last statement in (5.9.7) deserves an explanation, and only when r is even. It is clear that $\Delta_{r/2}((u_{i,j}), (u_{i,j}))$ is a square root of $\det(-U^2) = \det(U)^2$; consequently it is equal either to $\det(U)$ or to $-\det(U)$. To find out which is true, it suffices to make an experiment, and to replace U with a particular invertible skew symmetric matrix with integer coefficients; this experiment is a straightforward calculation. $\square$

By replacing the indeterminates $(w_{i,j})$ with suitable integers, we can get a skew symmetric matrix W_0 such that $\det(W_0) = 1$; the polynomial $\Delta_{r/2}(U, W_0)$ in the $r(r-1)/2$ indeterminates $(u_{i,j})$ is the square root of $\det(U)$ which takes the value 1 when U is replaced with W_0; this square root (determined up to a factor ± 1 depending on W_0) is called the *pfaffian*.

The proposition (5.9.5) invites another digression in a quite different direction. Here $K = \mathbb{R}$, and i is a square root of -1 in $\mathbb{C}$. It is worth knowing that with most calculations in the sections **5.9** and **5.10** are associated parallel calculations in which the transformations $\mathcal{F}$ and $\mathcal{F}_*$ are replaced with Fourier transformation, the interior multiplications with generalized differential operators, and the exponentials of bivectors with functions $\exp(iu)$ and $\exp(iw)$ determined by quadratic forms $u : M \to \mathbb{R}$ and $w : M^* \to \mathbb{R}$. The mappings $\mathrm{d}_u : M \to M^*$ and $\mathrm{d}_w : M^* \to M$ are defined as d_q in **2.3**. The Fourier transformation $\mathcal{F}$ maps every regular enough function $f : M \to \mathbb{C}$ to $\mathcal{F}(f) : M^* \to \mathbb{C}$ defined by

$$\mathcal{F}(f)(y) \;=\; (2\pi)^{-r/2} \int_M \mathrm{e}^{iy(x)}\, f(x)\, dx.$$

The following proposition is well known in functional analysis.

(5.9.8) **Proposition.** *If u is a nondegenerate quadratic form on M, there exist a nonzero $\kappa \in \mathbb{C}$ and a nondegenerate quadratic form w on M^* such that*

$$\mathcal{F}(\exp(iu)) = \kappa^{-1}\exp(iw)\,, \quad \kappa^2 = (-i)^r \det(\mathrm{d}_u) \quad \textit{and} \quad \mathrm{d}_w = -\mathrm{d}_u^{-1}\,.$$

Besides the presence of $i = \sqrt{-1}$ (which allows $\mathcal{F}(\exp(iu))$ to exist for all quadratic forms u) there are two essential discrepancies between (5.9.5) and (5.9.8). First, although both results lead to the calculation of the square root of a determinant (a skew symmetric determinant in (5.9.5), a symmetric one in (5.9.8)), in the former case this square root is given by a pfaffian, whereas in the latter case it is given by a formula involving the signature s of the quadratic form u:

$$\kappa \;=\; \exp(-is\pi/4)\, \sqrt{|\det(\mathrm{d}_u)|}\;;$$

it is worth observing that κ depends on the image of s in $\mathbb{Z}/8\mathbb{Z}$, like the Brauer–Wall class of $C\ell(M, u)$. Secondly there is an "inversion rule" that suggests writing κ^{-1} (instead of κ) in the final result $\kappa^{-1}\exp(iw)$. This "inversion rule" is corroborated by (5.ex.39), and (with another point of view) by [Sato, Miwa, Jimbo IV 1979]. The comment written just after (5.ex.41) summarizes the results that may be reached by means of Fourier transformation, and (5.9.8) in particular, in the cliffordian treatment of Weyl algebras.

5.10 The Lipschitz monoid Lip(M) when K is a field

The description of the Lie algebra associated with the Lipschitz monoid Lip(M) was achieved in (5.5.6); it was successful in many cases, in particular when M was a projective module. On the contrary, up to now a precise description of Lip(M) itself has been obtained only when K is a field; in this case it proves to be equal to the monoid $\ell ip(M)$ defined just before (5.9.3). Here and later in **7.2** the following lemma plays an important role.

(5.10.1) **Lemma.** *When M is a finite dimensional vector space over the field K, and x a nonzero element of $\bigwedge(M)$, there exists a bilinear form $\beta : M \times M \to K$ such that x is invertible in the algebra $\bigwedge(M;\beta)$.*

Proof. If the component of x in $K = \bigwedge^0(M)$ does not vanish, we can choose $\beta = 0$. Otherwise let k be the smallest degree such that the component of x in $\bigwedge^k(M)$ does not vanish; there exists a basis $(a_1, a_2, \dots, a_r)$ of M such that x is a sum $x' + x''$ in which $x' = a_1 \wedge a_2 \wedge \cdots \wedge a_k$, whereas x'' involves other exterior products of vectors of this basis. We choose β such that $\beta(a_i, a_i) = 1$ for $i = 1, 2, \dots, k$, and $\beta(a_i, a_j) = 0$ for all other couples (i, j); thus $x'^2 = \pm 1$ in $\bigwedge(M;\beta)$, whereas $x''x'$ can be written as $\sum_{j>k} a_j y_j$ for suitable elements $y_j \in \bigwedge(M)$. All a_j with $j > k$ belong to $\mathrm{Z}^g(\bigwedge(M;\beta))$, and their squares vanish; consequently $(x''x')^n$ vanishes as soon as $n > r - k$. This proves the invertibility of $xx' = \pm 1 + x''x'$ and the invertibility of x itself. □

(5.10.2) **Theorem.** *When M is a vector space over a field K, then* Lip(M) *is the multiplicative monoid $\ell ip(M)$ generated in $\bigwedge(M)$ by the elements of K and M, and the exponentials of elements of $\bigwedge^2(M)$.*

Proof. We already know that Lip(M) contains $\ell ip(M)$, and we must prove that every nonzero element x of Lip(M) belongs to $\ell ip(M)$. It suffices to prove this when M has finite dimension, because x belongs to the subalgebra generated by a finite dimensional subspace of M. There exists a bilinear form β on M such that x is invertible in $\bigwedge(M;\beta)$, and we can extend β to a bilinear form β' on the larger space $P = M^* \oplus M$ in such a way that the quadratic form $b \longmapsto \beta'(b, b)$ is nondegenerate; for instance we can set

$$\beta'((h_1, a_1), (h_2, a_2)) = h_1(a_2) + \beta(a_1, a_2) \qquad \text{for all } h_1, h_2 \in M^* \text{ and all } a_1, a_2 \in M\,;$$

thus we get a hyperbolic quadratic form $b \longmapsto \beta'(b, b)$ on P. If it belongs to Dieudonné's exceptional case (defined in **5.7**), we can still enlarge P so as to avoid this exceptional case. Since x belongs to Lip(M), it belongs to Lip(P) too, and if this implies that it belongs to $\ell ip(P)$, there is no difficulty in reaching the awaited conclusion: $x \in \ell ip(M)$. In other words, we can assume that the quadratic form $a \longmapsto \beta(a, a)$ is nondegenerate on M, and that the theorem (5.7.2) or (5.7.3) holds true for it.

If the lipschitzian element x is invertible in $\bigwedge(M;\beta)$, it belongs to the Lipschitz group $\mathrm{GLip}(M;\beta)$ because of the invariance property (5.4.1). The corresponding orthogonal transformation G_x is a product of reflections. Since the nondegeneracy of the quadratic form $a \longmapsto \beta(a,a)$ implies that the kernel of the morphism $\mathrm{GLip}(M;\beta) \to \mathrm{GO}(M;\beta)$ is reduced to $K^\times$, we know that x is a product in $\bigwedge(M;\beta)$ of elements of M, therefore a product of elements of $\ell ip(M)$.

Now comes the decisive moment of the proof: because of (5.9.4) in the previous section, $\ell ip(M)$ is also a monoid in the algebra $\bigwedge(M;\beta)$. Indeed let us prove that $\pi(\mathrm{Exp}(\beta_{\prime\prime}) \rfloor (y \otimes z))$ belongs to $\ell ip(M)$ whenever y and z belong to it. If we identify $\bigwedge(M) \hat{\otimes} \bigwedge(M)$ with $\bigwedge(M \oplus M)$, we can claim that $y \otimes z$ belongs to $\ell ip(M \oplus M)$. Then from (5.9.4) we deduce that $\mathrm{Exp}(\beta_{\prime\prime}) \rfloor (y \otimes z)$ also belongs to it. At last, π is the algebra morphism associated by the functor $\bigwedge$ with the following mapping:

$$M \oplus M \longrightarrow M \ , \quad (a,b) \longmapsto a+b \ ;$$

consequently π maps every element of $\ell ip(M \oplus M)$ to an element of $\ell ip(M)$. Since the above x is a product of elements of $\ell ip(M)$ in the algebra $\bigwedge(M;\beta)$, it also belongs to $\ell ip(M)$. □

(5.10.3) **Corollary.** *If q is a nonzero quadratic form on the vector space M, every element in the kernel of the surjective morphism $\mathrm{GLip}(M,q) \to \mathrm{GO}(M,q)$ is a product of elements of M. Consequently when $\mathrm{GO}(M,q)$ is generated by reflections (as it always is, except in the very few cases mentioned in (5.7.3) or (5.8.11)), then the group $\mathrm{GLip}(M,q)$ is generated by the elements $a \in M$ such that $q(a) \neq 0$.*

Proof. From (5.6.8) or (5.8.7) we know that $\mathrm{Z}^g(C\ell(M,q))$ is the subalgebra generated by $M_0 = \mathrm{Ker}(\mathrm{b}_q)$. Consequently the kernel of $\mathrm{GLip}(M,q) \to \mathrm{GO}(M,q)$ is equal to $\mathrm{GLip}(M_0,q_0)$ if q_0 is the restriction of q to M_0.

We consider two cases.

First case: $\mathrm{Ker}(\mathrm{b}_q) = \mathrm{Ker}(q)$; this assumption is always true when K does not have characteristic 2. In this case $C\ell(M_0,q_0)$ is the exterior algebra $\bigwedge(M_0)$. An invertible element of $\ell ip(M_0)$ is the product of an element of $K^\times$ and some other factors like $1+ab$ with a and b in M_0. From the factorization lemma (5.7.7) we deduce that each factor $1+ab$ is a product of four elements of M, since M contains elements c such that $q(c)$ is invertible.

Second case: $\mathrm{Ker}(\mathrm{b}_q) \neq \mathrm{Ker}(q)$; thus every element of M_0 is invertible if it is not in $\mathrm{Ker}(q)$. If x belongs to the subalgebra $\bigwedge(\mathrm{Ker}(q))$, the previous argument still proves that x is a product of vectors. If x does not belong to $\bigwedge(\mathrm{Ker}(q))$, we consider the smallest integer p such that $x \in C\ell^{\leq p}(M,q)$, and we prove that there is an invertible $a_1 \in M_0$ such that $a_1x \in C\ell^{\leq p-1}(M,q)$; if a_1x is not in $\bigwedge(\mathrm{Ker}(q))$, then $a_2a_1x \in C\ell^{\leq p-2}(M,q)$ for some invertible $a_2 \in M_0$; and so forth... until we get a product $a_j \cdots a_2a_1x$ that falls into $\bigwedge(\mathrm{Ker}(q))$, and that allows us to conclude. To prove the existence of an invertible $a_1x \in M_0$ such that $a_1x \in C\ell^{\leq p-1}(M,q)$, we replace $C\ell(M,q)$ with $\bigwedge(M;\beta)$, where β is any admissible scalar product, and from (5.4.1) and (5.10.2) we deduce that

$$x \;=\; \lambda\, d_1 \wedge \cdots \wedge d_n \wedge (1 + e_1 \wedge e_2) \wedge \cdots \wedge (1 + e_{2k-1} \wedge e_{2k})$$

for some scalar λ (only indispensable if $n = 0$) and some linearly independent vectors $d_1,\dots,d_n, e_1,\dots,e_{2k}$. It is clear that $p = n+2k$. Since x is not in $\bigwedge(\mathrm{Ker}(q))$, in the subspace spanned by $(d_1,\dots,d_n,e_1,\dots,e_{2k})$ there is an invertible vector a_1, and from (4.8.9) we deduce that $a_1 x \in C\ell^{\leq p-1}(M_0,q_0)$.

It remains to prove that the group $\mathrm{GLip}(M,q)$ is generated by the invertible elements of M, provided that $\mathrm{GO}(M,q)$ is generated by reflections. When x belongs to $\mathrm{GLip}(M,q)$, then G_x is a product of reflections and consequently $G_x = G_y$ for some product y of invertible elements of M; thus xy^{-1} belongs to the kernel of $\mathrm{GLip}(M,q) \to \mathrm{GO}(M,q)$ and is itself a product of invertible elements of M. □

All important results of this chapter have been reached. The end of this section is devoted to a rather specialized topic: we replace a Clifford algebra $C\ell(M,q)$ with an isomorphic algebra $\bigwedge(M;\beta)$, and we look for precise formulas describing some products in the monoid $\mathrm{Lip}(M;\beta)$. If τ_β is the reversion in $\bigwedge(M;\beta)$, an element x of $\mathrm{Lip}(M)$ belongs to $\mathrm{GLip}(M;\beta)$ (the Lipschitz group isomorphic to $\mathrm{GLip}(M,q)$) if and only if $x\,\tau_\beta(x) \neq 0$.

We shall only consider a space M of finite dimension, since infinite dimensions do not raise serious difficulties here; indeed every element of $\bigwedge(M;\beta)$ belongs to the subalgebra generated by a finite dimensional subspace of M. We begin with a technical lemma.

(5.10.4) **Lemma.** *Let M and N be finite dimensional vector spaces over the field K, $f \in \mathrm{Hom}(M,N)$ and $g \in \mathrm{Hom}(N,M)$. Endomorphisms of $M \oplus N$ are described by square matrices of order 2 in the usual way. First*

$$\det \begin{pmatrix} \mathrm{id}_M & g \\ f & \mathrm{id}_N \end{pmatrix} \;=\; \det(\mathrm{id}_M - gf) \;=\; \det(\mathrm{id}_N - fg)\ .$$

Secondly, when $\det(\mathrm{id}_M - gf)$ is invertible,

$$\begin{pmatrix} \mathrm{id}_M & g \\ f & \mathrm{id}_N \end{pmatrix}^{-1} \;=\; \begin{pmatrix} (\mathrm{id}_M - gf)^{-1} & -g(\mathrm{id}_N - fg)^{-1} \\ -f(\mathrm{id}_M - gf)^{-1} & (\mathrm{id}_N - fg)^{-1} \end{pmatrix}\ ;$$

and moreover

$$f(\mathrm{id}_M - gf)^{-1} = (\mathrm{id}_N - fg)^{-1} f\ ,$$
$$g(\mathrm{id}_N - fg)^{-1} = (\mathrm{id}_M - gf)^{-1} g\ .$$

Proof. The last two equalities are trivial. To prove the first three, we can assume that K is infinite; this assumption implies that the algebra of polynomial functions on $\mathrm{Hom}(M,N) \times \mathrm{Hom}(N,M)$ contains no divisors of zero. We begin with the second equality; if it is true for some couple of spaces (M, N), it remains true when M and N are replaced with subspaces; therefore we can assume that M and N have the same dimension. In this case f may be bijective, and when it is bijective, then $\mathrm{id}_N - fg = f(\mathrm{id}_M - gf)f^{-1}$, and consequently $\mathrm{id}_N - fg$ and $\mathrm{id}_M - gf$ have the same determinant. Since the product

$$\det(f)\ \big(\det(\mathrm{id}_M - gf) - \det(\mathrm{id}_N - fg)\big)$$

vanishes everywhere, the second factor must vanish everywhere. Then we observe that

$$\det\begin{pmatrix}\mathrm{id}_M & g\\ f & \mathrm{id}_N\end{pmatrix} = \det\begin{pmatrix}\mathrm{id}_M & -g\\ -f & \mathrm{id}_N\end{pmatrix}\ ;$$

indeed these matrices represent the following endomorphisms of $M \oplus N$:

$$(a,b) \longmapsto (a + g(b),\ f(a) + b) \quad \text{and} \quad (a,b) \longmapsto (a - g(b),\ -f(a) + b)\ ,$$

and they are conjugate by means of the transformation $(a,b) \longmapsto (a,-b)$. Now the first and third equalities in (5.10.4) are consequences of this one:

$$\begin{pmatrix}\mathrm{id}_M & g\\ f & \mathrm{id}_N\end{pmatrix}\begin{pmatrix}\mathrm{id}_M & -g\\ -f & \mathrm{id}_N\end{pmatrix} = \begin{pmatrix}\mathrm{id}_M - gf & 0\\ 0 & \mathrm{id}_N - fg\end{pmatrix}\ ;$$

indeed all the previous results prove that the determinants in the first equality of (5.10.4) have the same square, and consequently are equal because they are polynomial functions that take the same value 1 when $(f,g) = (0,0)$. At last in the above equality it is easy to find the inverse of the matrix in the right-hand member; the third equality in (5.10.4) follows immediately. □

(5.10.5) **Proposition.** *For every element u of $\bigwedge^2(M)$,*

$$\mathrm{Exp}(u)\ \tau_\beta(\mathrm{Exp}(u)) = \det(\mathrm{id}_M + \mathrm{d}_u\mathrm{d}_\beta) = \det(\mathrm{id}_M + \mathrm{d}_u\mathrm{d}_\beta^{to}).$$

When $\mathrm{Exp}(u)$ *belongs to* $\mathrm{GLip}(M;\beta)$, *the corresponding orthogonal transformation is*

$$G_{\mathrm{Exp}(u)} = (\mathrm{id}_M + \mathrm{d}_u\mathrm{d}_\beta)^{-1}\ (\mathrm{id}_M + \mathrm{d}_u\mathrm{d}_\beta^{to}).$$

Proof. Let us set $x = \mathrm{Exp}(u)$, whence

$$\tau(x) = \mathrm{Exp}(-u) \quad \text{and} \quad \tau_\beta(x) = \mathrm{Exp}([\beta]) \rfloor \mathrm{Exp}(-u)$$

(see (4.7.14)), and consequently

$$x\ \tau_\beta(x) = \pi\ \big(\mathrm{Exp}(\beta_{/\!/} + 1 \otimes [\beta])\ \rfloor\ \mathrm{Exp}(u \otimes 1 - 1 \otimes u)\big).$$

Remember that $x\,\tau_\beta(x)$ belongs to K and that π is the algebra morphism associated by the functor $\bigwedge$ with some morphism $M \oplus M \to M$; thus $x\,\tau_\beta(x)$ is merely the value of the linear form $\mathrm{Exp}(\beta_{/\!/} + 1 \otimes [\beta])$ on $\mathrm{Exp}(u \otimes 1 - 1 \otimes u)$. According to (5.9.6), its square is equal to the determinant of $\mathrm{id}_{M\oplus M} - \mathrm{d}_{u'}\mathrm{d}_{w'}$ if we set $w' = \beta_{//} + 1\otimes[\beta]$ and $u' = u\otimes 1 - 1\otimes u$. From the definitions (4.3.2) and (4.4.2) we derive:

$$\mathrm{d}_{w'}(a,b) = w' \lfloor (a,b) = (\mathrm{d}_\beta^{to}(b)\ ,\ \mathrm{d}_\beta(a)\ + (\mathrm{d}_\beta + \mathrm{d}_\beta^{to})(b))\ ,$$
$$\mathrm{d}_{u'}(g,h) = (g,h) \rfloor u' = (\mathrm{d}_u(g)\ ,\ -\mathrm{d}_u(h))\ ,$$

whence

$$(x\,\tau_\beta(x))^2 = \det(\Phi) \quad \text{with} \quad \Phi = \begin{pmatrix} \mathrm{id}_M & -\mathrm{d}_u\mathrm{d}_\beta^{to} \\ \mathrm{d}_u\mathrm{d}_\beta & \mathrm{id}_M + \mathrm{d}_u(\mathrm{d}_\beta + \mathrm{d}_\beta^{to}) \end{pmatrix}.$$

The calculation of $\det(\Phi)$ is very easy because

$$\Phi(a,-a) = ((\mathrm{id}_M + \mathrm{d}_u\mathrm{d}_\beta^{to})(a)\ ,\ -(\mathrm{id}_M + \mathrm{d}_u\mathrm{d}_\beta^{to})(a))\ ,$$
$$\Phi(a,0) = ((\mathrm{id}_M + \mathrm{d}_u\mathrm{d}_\beta)(a)\ ,\ 0)\ +\ (-\mathrm{d}_u\mathrm{d}_\beta(a)\ ,\ \mathrm{d}_u\mathrm{d}_\beta(a))\ ,$$

whence $\det(\Phi) = \det(\mathrm{id}_M + \mathrm{d}_u\mathrm{d}_\beta)\det(\mathrm{id}_M + \mathrm{d}_u\mathrm{d}_\beta^{to})$.

Now we must explain why the two determinants in the right-hand member are equal. The determinant of $\mathrm{id}_M + \mathrm{d}_u\mathrm{d}_\beta$ is equal to that of the endomorphism $(\mathrm{id}_M + \mathrm{d}_u\mathrm{d}_\beta)^*$ derived from it by the functor $\mathrm{Hom}(\dots,K)$; if the definition of d_β^* takes the twisting rule (4.2.1) into account, we must write $\mathrm{d}_\beta^*(a)(b) = -\mathrm{d}_\beta(b)(a)$, whence $\mathrm{d}_\beta^* = \mathrm{d}_\beta^{to}$; and similarly $\mathrm{d}_u^* = \mathrm{d}_u^{to} = \mathrm{d}_u$ because u induces an alternate bilinear form on M^*. From this argument and the beginning of (5.10.4) we deduce

$$\begin{aligned}\det(\mathrm{id}_M + \mathrm{d}_u\mathrm{d}_\beta) &= \det(\mathrm{id}_M^* + \mathrm{d}_\beta^*\mathrm{d}_u^*)\\ &= \det(\mathrm{id}_{M^*} + \mathrm{d}_\beta^{to}\mathrm{d}_u) \ = \ \det(\mathrm{id}_M + \mathrm{d}_u\mathrm{d}_\beta^{to}).\end{aligned}$$

Thus $x\tau_\beta(x)$ and $\det(\mathrm{id}_M + \mathrm{d}_u\mathrm{d}_\beta)$ are polynomial functions on $\bigwedge^2(M)$ that have the same square, and that take the same value 1 when $u = 0$. When K is infinite, this implies that they are equal; and when K is finite, we reach the same conclusion by means of an infinite field extension.

When $x\tau_\beta(x)$ is invertible, x belongs to $\mathrm{GLip}(M;\beta)$, and there is an orthogonal transformation G_x such that $G_x(a)x = xa$ for all $a \in M$. The following calculation of G_x is merely a generalization of Lipschitz's own argument. We set $b = G_x(a)$ and we calculate bx and xa by means of (4.8.9) and (4.5.4):

$$bx = b\wedge x + \mathrm{d}_\beta(b)\rfloor x \ = b\wedge x + (\mathrm{d}_\beta(b)\rfloor u)\wedge x \ = (b\ + \mathrm{d}_u\mathrm{d}_\beta(b))\wedge x\ ,$$
$$xa = x\wedge a + \mathrm{d}_\beta^{to}(a)\rfloor x = a\wedge x + (\mathrm{d}_\beta^{to}(a)\rfloor u)\wedge x = (a\ + \mathrm{d}_u\mathrm{d}_\beta^{to}(a))\wedge x\ ;$$

since x is $\wedge$-invertible, the equality $bx = xa$ implies

$$(\mathrm{id}_M + \mathrm{d}_u\mathrm{d}_\beta)(b) \ = \ (\mathrm{id}_M + \mathrm{d}_u\mathrm{d}_\beta^{to})(a)\ ;$$

since $b = G_x(a)$, we get the announced value of G_x. □

(5.10.6) **Proposition.** *When u and v are elements of $\bigwedge^2(M)$, and when κ is the component in $K = \bigwedge^0(M)$ of the product of* Exp(u) *and* Exp(v) *in* $\bigwedge(M;\beta)$, *then*

$$\kappa^2 \;=\; \det(\mathrm{id}_M - \mathrm{d}_v \mathrm{d}_\beta \mathrm{d}_u \mathrm{d}_\beta^{to}) \;=\; \det(\mathrm{id}_M - \mathrm{d}_u \mathrm{d}_\beta^{to} \mathrm{d}_v \mathrm{d}_\beta)\ .$$

When κ is invertible, there exists $w \in \bigwedge^2(M)$ such that $\mathrm{Exp}(u)\ \mathrm{Exp}(v) = \kappa\ \mathrm{Exp}(w)$ *and*

$$\mathrm{d}_w \;=\; (\mathrm{id}_M + \mathrm{d}_v\mathrm{d}_\beta)(\mathrm{id}_M - \mathrm{d}_u\mathrm{d}_\beta^{to}\mathrm{d}_v\mathrm{d}_\beta)^{-1}\mathrm{d}_u + (\mathrm{id}_M + \mathrm{d}_u\mathrm{d}_\beta^{to})(\mathrm{id}_M - \mathrm{d}_v\mathrm{d}_\beta\mathrm{d}_u\mathrm{d}_\beta^{to})^{-1}\mathrm{d}_v.$$

Proof. First $\mathrm{Exp}(u)\mathrm{Exp}(v) = \pi\big(\mathrm{Exp}(\beta_{/\!/}) \rfloor \mathrm{Exp}(u \otimes 1 + 1 \otimes v)\big)$. Since $\pi = \bigwedge(f)$ for some morphism $f : M \oplus M \to M$, we realize that κ is the value of the linear form $\mathrm{Exp}(\beta_{/\!/})$ on $\mathrm{Exp}(u \otimes 1 + 1 \otimes v))$, whence $\kappa^2 = \det(\Psi)$ with

$$\Psi \;=\; \mathrm{id}_{M\oplus M} \;-\; \begin{pmatrix} \mathrm{d}_u & 0 \\ 0 & \mathrm{d}_v \end{pmatrix}\begin{pmatrix} 0 & \mathrm{d}_\beta^{to} \\ \mathrm{d}_\beta & 0 \end{pmatrix} \;=\; \begin{pmatrix} \mathrm{id}_M & -\mathrm{d}_u\mathrm{d}_\beta^{to} \\ -\mathrm{d}_v\mathrm{d}_\beta & \mathrm{id}_M \end{pmatrix}.$$

From (5.10.4) we deduce at once the first two equalities in (5.10.6). When κ is invertible, the interior product of $\mathrm{Exp}(\beta_{/\!/})$ and $\mathrm{Exp}(u \otimes 1 + 1 \otimes v)$ is equal to $\kappa\,\mathrm{Exp}(w'')$, where w'' is the element of $\bigwedge^2(M \oplus M)$ such that $\mathrm{d}_{w''}$ is equal to

$$\Psi^{-1}\begin{pmatrix} \mathrm{d}_u & 0 \\ 0 & \mathrm{d}_v \end{pmatrix} = \begin{pmatrix} (\mathrm{id}_M - \mathrm{d}_u\mathrm{d}_\beta^{to}\mathrm{d}_v\mathrm{d}_\beta)^{-1}\mathrm{d}_u & \mathrm{d}_u\mathrm{d}_\beta^{to}(\mathrm{id}_M - \mathrm{d}_v\mathrm{d}_\beta\mathrm{d}_u\mathrm{d}_\beta^{to})^{-1}\mathrm{d}_v \\ \mathrm{d}_v\mathrm{d}_\beta(\mathrm{id}_M - \mathrm{d}_u\mathrm{d}_\beta^{to}\mathrm{d}_v\mathrm{d}_\beta)^{-1}\mathrm{d}_u & (\mathrm{id}_M - \mathrm{d}_v\mathrm{d}_\beta\mathrm{d}_u\mathrm{d}_\beta^{to})^{-1}\mathrm{d}_v \end{pmatrix}.$$

We know that $\pi = \bigwedge(f)$ with $f(a,b) = a + b$, whence $\pi(\mathrm{Exp}(w'')) = \mathrm{Exp}(\pi(w''))$ (see (4.5.5)); thus the element w mentioned in (5.10.6) is $\pi(w'')$. The transposed mapping $f^* : M^* \to M^* \oplus M^*$ is the diagonal mapping $h \longmapsto (h,h)$; therefore from (4.4.6) we deduce (for all $h \in M^*$)

$$\mathrm{d}_w(h) \;=\; h \rfloor \pi(w'') \;=\; \pi((h,h) \rfloor w'') \;=\; f \circ \mathrm{d}_{w''}(h,h)\ ;$$

this shows that d_w is the sum of the four entries in the matrix representing $\mathrm{d}_{w''}$ above. □

Remark. The expression of d_w in (5.10.6) may look complicated; nevertheless it plays an important and effective role in [Sato, Miwa, Jimbo II 1979] through its expansion as a series. Suppose that K is $\mathbb{R}$ or $\mathbb{C}$ and that u and v belong to a small enough neighbourhood of 0 in $\bigwedge^2(M)$; then the expansion $(1-\xi)^{-1} = \sum_k \xi^k$ can be used when ξ is $\mathrm{d}_u\mathrm{d}_\beta^{to}\mathrm{d}_v\mathrm{d}_\beta$ or $\mathrm{d}_v\mathrm{d}_\beta\mathrm{d}_u\mathrm{d}_\beta^{to}$. Here is the expansion of d_w up to the order 4 in (u,v) :

$$\begin{aligned}\mathrm{d}_w = \mathrm{d}_u + \mathrm{d}_v + \mathrm{d}_u\mathrm{d}_\beta^{to}\mathrm{d}_v + \mathrm{d}_v\mathrm{d}_\beta\mathrm{d}_u + \mathrm{d}_u\mathrm{d}_\beta^{to}\mathrm{d}_v\mathrm{d}_\beta\mathrm{d}_u + \mathrm{d}_v\mathrm{d}_\beta\mathrm{d}_u\mathrm{d}_\beta^{to}\mathrm{d}_v \\ + \mathrm{d}_u\mathrm{d}_\beta^{to}\mathrm{d}_v\mathrm{d}_\beta\mathrm{d}_u\mathrm{d}_\beta^{to}\mathrm{d}_v + \mathrm{d}_v\mathrm{d}_\beta\mathrm{d}_u\mathrm{d}_\beta^{to}\mathrm{d}_v\mathrm{d}_\beta\mathrm{d}_u + \cdots\ ;\end{aligned}$$

the expansion of d_w is the sum of all products beginning with d_u or d_v and obeying this very simple rule: when a factor d_u (resp. d_v) is not the last one, it is followed (on the right side) by $\mathrm{d}_\beta^{to}\mathrm{d}_v$ (resp. $\mathrm{d}_\beta\mathrm{d}_u$).

Let us apply (5.10.5) and (5.10.6) to the most classical case of a quadratic space (M,q) over a field K that does not have characteristic 2, when it is equipped with the canonical scalar product $\beta = \mathrm{b}_q/2$. Then $\tau_\beta = \tau$, $\bigwedge^2(M)$ is a Lie algebra in $\bigwedge(M, \mathrm{b}_q/2)$ (see (5.5.4)), and there is an isomorphism $u \longmapsto F_u$ from $\bigwedge^2(M)$ onto the Lie algebra of all infinitesimal automorphisms of (M,q) (see (5.5.3)); from the definition of F_u and from (4.4.12) we deduce (for all $a \in M$):

$$F_u(a) \;=\; ua - au \;=\; -\mathrm{d}_q(a) \rfloor u \;=\; -\mathrm{d}_u \mathrm{d}_q(a) \;=\; -2\mathrm{d}_u \mathrm{d}_\beta(a) \;=\; 2\mathrm{d}_u \mathrm{d}_\beta^{to}(a)\ ;$$

since q is nondegenerate, u is determined by F_u as well as by d_u. If we prefer F_u to d_u, we transform (5.10.5) and (5.10.6) into the following two corollaries. The first one (5.10.7) belongs to the earliest history of Clifford algebras, because it was already known by Lipschitz; nevertheless the lipschitzian elements $\mathrm{Exp}(u)$ that here are described as exponentials of bivectors, were defined in Lipschitz's works by means of pfaffians of skew symmetric matrices; the version presented in [Porteous 2000] under the name "pfaffian chart" or "Lipschitz chart" and the version presented in [Weil 1979] are closer to the historical truth. The subsequent corollary (5.10.8) can be derived either from (5.10.6) or from (5.10.7).

(5.10.7) **Corollary.** *We assume that the field K does not have characteristic 2 and that the bilinear form β is symmetric and nondegenerate. For every element u of $\bigwedge^2(M)$,*

$$\mathrm{Exp}(u)\ \tau(\mathrm{Exp}(u)) \;=\; \det\left(\mathrm{id}_M - \frac{1}{2}F_u\right) \;=\; \det\left(\mathrm{id}_M + \frac{1}{2}F_u\right).$$

When $\mathrm{Exp}(u)$ belongs to $\mathrm{GLip}((M;\beta)$, the corresponding orthogonal transformation is

$$G_{\mathrm{Exp}(u)} \;=\; \left(\mathrm{id}_M - \frac{1}{2}F_u\right)^{-1}\left(\mathrm{id}_M + \frac{1}{2}F_u\right) \;=\; \left(\mathrm{id}_M + \frac{1}{2}F_u\right)\left(\mathrm{id}_M - \frac{1}{2}F_u\right)^{-1}.$$

(5.10.8) **Corollary.** *With the same assumptions as in (5.10.7), let u and v be elements of $\bigwedge^2(M)$, and κ the component in $K = \bigwedge^0(M)$ of the product of $\mathrm{Exp}(u)$ and $\mathrm{Exp}(v)$ in $\bigwedge(M;\beta)$. Then*

$$\kappa^2 \;=\; \det\left(\mathrm{id}_M + \frac{1}{4}F_uF_v\right) \;=\; \det\left(\mathrm{id}_M + \frac{1}{4}F_vF_u\right).$$

When κ is invertible, there exists $w \in \bigwedge^2(M)$ such that $\mathrm{Exp}(u)\mathrm{Exp}(v) = \kappa\,\mathrm{Exp}(w)$ and

$$\mathrm{id}_M + \frac{1}{2}F_w = \left(\mathrm{id}_M + \frac{1}{2}F_u\right)\left(\mathrm{id}_M + \frac{1}{4}F_vF_u\right)^{-1}\left(\mathrm{id}_M + \frac{1}{2}F_v\right), \quad \textit{or equivalently}$$

$$\mathrm{id}_M - \frac{1}{2}F_w = \left(\mathrm{id}_M - \frac{1}{2}F_v\right)\left(\mathrm{id}_M + \frac{1}{4}F_uF_v\right)^{-1}\left(\mathrm{id}_M - \frac{1}{2}F_u\right).$$

Exercises

(5.ex.1) Let (M, q) be a quadratic module such that $\mathrm{Ker}(q) = 0$ and $\mathrm{Ker}(\mathrm{b}_q) \neq 0$. Prove that $g(a) = a$ for all $g \in \mathrm{Aut}(M, q)$ and all $a \in \mathrm{Ker}(\mathrm{b}_q)$.

(5.ex.2) Let V be a totally isotropic submodule of (M, q); assume that M and V are finitely generated, and that the ranks of M/V and V at every prime ideal are respectively $\leq m$ and $\leq n$. Prove that

$$C\ell(M, q; V)^{\leq m} = C\ell(M, q) \quad \text{and} \quad C\ell(M, q; V)^{\leq -n-1} = 0 \, .$$

(5.ex.3) Let (M, q) be a quadratic module, with M a finitely generated projective module, and V a totally isotropic direct summand of M. Let n be the rank of V at some prime ideal $\mathfrak{p}$ of K. Prove that $C\ell(M, q; V)^{\leq k}$ is a direct summand of $C\ell(M, q)$ which has the same rank as $C\ell^{\leq n+k}(M, q)$ at $\mathfrak{p}$.

Hint. Assume that K is a local ring, and that $(a_1, \dots, a_m, b_1, \dots, b_n)$ is a basis of M in which the last n elements generate V; with every subset F of $\{1, 2, \dots, m\}$ associate the product a_F of all a_i with $i \in F$, and define similarly b_G for every subset G of $\{1, 2, \dots, n\}$; then consider the linear automorphism of $C\ell(M, q)$ mapping every $a_F b_G$ to $a_F b_{G'}$ where G' is the subset complementary to G.

(5.ex.4) Let V be any submodule of M, and V^{an} the annihilator of V in M^*, that is the submodule of all linear forms vanishing on V. The algebra $\bigwedge(M^*)$ is filtered by the submodules $\bigwedge(M^*; V^{an})^{\leq k}$ whereas $\bigwedge^*(M)$ is filtered by the submodules $\bigwedge^*(M; V)^{\leq k}$. Prove that the canonical morphism $\bigwedge(M^*) \to \bigwedge^*(M)$ is a morphism of filtered algebras; it is even an isomorphism of filtered algebras when M is a finitely generated projective module, and V a direct summand.

(5.ex.5) Prove that the canonical morphism $\bigwedge(M^*) \to \bigwedge^*(M)$ maps $\mathrm{Lip}(M^*)$ into $\mathrm{Lip}^*(M)$; it induces an isomorphism of monoids when M is finitely generated and projective.

(5.ex.6) This exercise investigates the group $\mathrm{GLip}(M)$ of all invertible lipschitzian elements in $\bigwedge(M)$ when M is a projective module.

(a) Prove that an element of $\bigwedge(M)$ is invertible if and only if its component in $\bigwedge^0(M) = K$ is invertible.

(b) Let x be an element of $\mathrm{GLip}(M)$ that has components 1 and 0 respectively in $\bigwedge^0(M)$ and $\bigwedge^2(M)$; prove that $x = 1$ by means of the trick presented in the proof of (5.4.3).

(c) Prove that every element of $\mathrm{GLip}(M)$ can be written $\lambda \mathrm{Exp}(u)$ for some invertible $\lambda \in K$ and some $u \in \bigwedge^2(M)$.

(5.ex.7) The notation is the same as in (4.ex.8) where the interior products $x \rfloor y$ and $x \lfloor y$ of two elements of $\bigwedge(M; \beta)$ have been defined. Prove that these interior products are lipschitzian whenever x and y are lipschitzian. (*Hint:* (4.8.10), (5.3.13),...).

(5.ex.8)* Let M be a finitely generated module, the rank of which is everywhere ≤ 4. If x is any element of $\bigwedge(M)$, its component in $\bigwedge^k(M)$ (for $k = 0, 1, 2, 3, 4$) is denoted by x_k. For every $u \in \bigwedge^2(M)$ let $u^{[2]}$ be the component of $\mathrm{Exp}(u)$ in $\bigwedge^4(M)$; thus $u \wedge u = 2u^{[2]}$. Besides, let $\pi' : \bigwedge(M) \to \bigwedge(M) \hat{\otimes} \bigwedge(M)$ be the comultiplication of $\bigwedge(M)$, and $\pi'_{i,j}$ the component of π' in $\bigwedge^i(M) \otimes \bigwedge^j(M)$ for each $(i,j) \in \mathbb{N}^2$.

(a) Consider an even element $x \in \bigwedge_0(M)$, and set $\omega = x_0x_4 - {x_2}^{[2]}$. Prove that x belongs to $\mathrm{Lip}(M)$ if and only if both 2ω and $\pi'_{1,3}(\omega)$ vanish.
Hint. Consider $y = \pi'(x) \wedge (1 \otimes \tau(x))$ as in the proof of (5.4.3), and prove that its components in $\bigwedge^0(M) \otimes \bigwedge^4(M)$ and $\bigwedge^1(M) \otimes \bigwedge^3(M)$ are $2 \otimes \omega$ and $\pi'_{1,3}(\omega)$.

(b) Now consider an odd element $x \in \bigwedge_1(M)$, and $\omega = x_1 \wedge x_3$. Prove that x belongs to $\mathrm{Lip}(M)$ if and only if 2ω and $\pi'_{1,3}(\omega)$ both vanish.

(c) Let ω be an element of $\bigwedge^4(M)$. Prove that the vanishing of $\pi'_{1,3}(\omega)$ implies $\omega = 0$ when M is a projective module.

(5.ex.9) Let (M, q) be a quadratic module, with M a finitely generated module of rank ≤ 4 at every prime ideal, and such that the mapping $a \longmapsto 2a$ is bijective from M onto M. Let x be a locally homogeneous element of $C\ell(M, q)$. This exercise intends to prove that $x\tau(x)$ belongs to $C\ell^0(M,q) \oplus C\ell^4(M,q)$, and that x is lipschitzian if and only if $x\tau(x)$ belongs to $C\ell^0(M,q) = K$.

(a) Prove that $x \wedge \tau(x)$ and $x\tau(x)$ belong to $C\ell^0(M,q) \oplus C\ell^4(M,q)$.
Hint. See (4.ex.11)(a).

(b) Prove that $x \wedge \tau(x)$ and $x\tau(x)$ have the same component in $\bigwedge^4(M)$.
Hint. For $k = 0, 1, 2, 3, 4$, let x_k be the component of x in $C\ell^k(M,q)$; after localization, you can suppose x homogeneous and write either $x = x_0 + x_2 + x_4$ or $x = x_1 + x_3$; deduce from (4.8.10) that the component of $x\tau(x)$ in $\bigwedge^4(M)$ is $2x_0x_4 + x_2 \wedge x_2$ or $-2x_1 \wedge x_3$; you need to know that ${x_4}^2$ or ${x_3}^2$ belongs to K: see (4.ex.12).

(c) Prove that x is lipschitzian if and only if $x\tau(x)$ belongs to K.
Hint. Because of (a) and (b) above, this is equivalent to $x \wedge \tau(x) \in K$; and according to (5.4.1), we can replace q with 0; use the automorphism of $M \oplus M$ defined by $(a,b) \longmapsto (a+b,\ a-b)$, which maps Δ and Δ' respectively to $M \oplus 0$ and $0 \oplus M$; its extension as an automorphism of $\bigwedge(M) \hat{\otimes} \bigwedge(M)$ maps $x \otimes \tau(x)$ to $\pi'(x) \wedge \pi''(x)$ with $\pi'' = (\mathrm{id}_\wedge \otimes \sigma) \circ \pi' \circ \tau$; according to (5.3.10), x is lipschitzian if and only if the component of $\pi'(x) \wedge \pi''(x)$ in $\bigwedge^4(M) \otimes 1$ vanishes.

(5.ex.10)* Let p be a prime integer ≥ 2, r an exponent ≥ 2, and K the local ring $\mathbb{Z}/p^r\mathbb{Z}$. Let M be a free module of rank 4 over K. This exercise intends to give a precise description of the lipschitzian elements of $\bigwedge(M)$; it needs the results of (5.ex.8) (or (5.ex.9) if $p \neq 2$).

(a) Let x be a nonzero odd element of $\bigwedge(M)$, x_1 and x_3 its components in M and $\bigwedge^3(M)$, i and j the greatest integers such that $p^{r-i}x_1 = 0$ and $p^{r-j}x_3 = 0$. Prove that x is lipschitzian if and only if one of these two conditions is fulfilled: either $i+j \geq r$, or $i+j < r$ and there is a basis (a,b,c,d) of M such that $x_1 = p^i a$ and $x_3 = p^j(a+p^k d)\wedge b \wedge c$ with $i+j+k \geq r$.

(b) Let x be a nonzero even element of $\bigwedge(M)$, and x_0, x_2, x_4 its components in K, $\bigwedge^2(M)$, $\bigwedge^4(M)$. Prove the existence of a basis (a,b,c,d) of M and the existence of two exponents i and j both $\leq r$ such that $x_2 = p^i a\wedge b + p^j c\wedge d$. Let h and k be the greatest integers such that $p^{r-h}x_0 = 0$ and $p^{r-k}x_4 = 0$. Prove that x is lipschitzian if and only if one of these two conditions is fulfilled: either $i+j$ and $h+k$ are both $\geq r$, or $i+j = h+k < r$ and $x_0x_4 = p^{i+j}a\wedge b\wedge c\wedge d$.

(5.ex.11) Let M be a finitely generated projective module such that the mapping $a \longmapsto 2a$ is bijective from M onto M, and q a quadratic form on M. Prove that $xy\tau(x)$ belongs to $C\ell^k(M,q)$ for all $y \in C\ell^k(M,q)$ and all $x \in \mathrm{Lip}(M,q)$.
Hint. (5.3.2) states that $xy\tau(x)$ is in $C\ell^{\leq k}(M,q)$, and it remains to prove that it is in $C\ell^{\geq k}(M,q)$; when (M,q) is a quadratic space of constant rank r, use the bijection $C\ell^r(M,q)\otimes C\ell^{\geq k}(M,q) \to C\ell^{\leq r-k}(M,q)$ resulting from (4.8.15); in the general case, extend q to a nondegenerate quadratic form on $M^*\oplus M$.

Orthogonal transformations (and their infinitesimal transformations)

(5.ex.12) Let (M,q) be a tamely degenerate quadratic module as in **5.6**; it is the direct sum of $\mathrm{Ker}(\mathrm{b}_q)$ and a quadratic space (M',q'). Let f be an endomorphism of M, and f' the endomorphism of M' mapping every $a \in M'$ to the component of $f(a)$ in M'. Prove that f is an infinitesimal automorphism of (M,q) if and only if $f(a) \in \mathrm{Ker}(\mathrm{b}_q)$ for all $a \in \mathrm{Ker}(\mathrm{b}_q)$, and f' is an infinitesimal automorphism of (M',q').

Prove that an infinitesimal automorphism f of (M,q) is an infinitesimal orthogonal transformation if and only if $f(a) = 0$ for all $a \in \mathrm{Ker}(\mathrm{b}_q)$.
Comment. This shows that the infinitesimal version of (5.8.3) is true even without the hypothesis $\mathrm{Ker}(q) = \mathrm{Ker}(\mathrm{b}_q)$.

(5.ex.13) Let (M,q) be a quadratic space, M' a direct summand of M, and q' the restriction of q to M'. The subalgebra of $C\ell(M,q)$ generated by M' is identified with $C\ell(M',q')$. With each $y \in C\ell_0^{\leq 2}(M,q)$ is associated the infinitesimal orthogonal transformation F_y defined by $F_y(a) = [y,a]$. Prove that

$$\mathrm{Im}(F_y) \subset M' \iff y \in C\ell_0^{\leq 2}(M',q') \iff \mathrm{Ker}(F_y) \supset M'^{\perp} .$$

(5.ex.14) Let K be the quotient of the polynomial ring $(\mathbb{Z}/8\mathbb{Z})[T]$ by the ideal generated by $2T(T-1)$ and let t be the image of T in K. Thus every element of K can be written $\xi + \zeta t + t(t-1)\psi(t)$ with $\xi,\ \zeta \in \mathbb{Z}/8\mathbb{Z}$ and ψ a polynomial with coefficients in $\mathbb{Z}/2\mathbb{Z}$. Let (M,q) be a free quadratic module of rank 1 over K generated by an element b such that $q(b)=1$.

(a) Prove that the only idempotents in K are 0 and 1, but that the group $\mu_2(K)$ of square roots of 1 is the group of order 16 generated by -1, 3, $1-2t$ and $1+4t$. Consequently the group $\mathrm{Aut}(M,q)$ contains 16 elements.

(b) The sequence $(1-t,\ tb)$ satisfies the conditions that allow it to define an element of $\mathrm{G''Lip}(M,q)$ and an orthogonal transformation g; indeed $(1-t)^2+(tb)^2=1$, and $(1-t)tb \in \mathrm{Z}_1^r(\mathrm{C}\ell(M,q))$. Therefore

$$g(b) \;=\; (1-t)b(1-t)-(tb)b(tb) \;=\; (1-2t)b.$$

Prove that g does not belong to the image of $\mathrm{G'Lip}(M,q) \to \mathrm{GO}(M,q)$, and that there is no $e \in \mathrm{Ip}(K)$ such that $\det(g)=1-2e$.

(c) Let $\mu_2'(K)$ be the subgroup of all $s \in \mu_2(K)$ such that the fraction $s/1$ is equal to $1/1$ or $-1/1$ in every localization $K_{\mathfrak{p}}$. Prove that $\mu_2'(K)$ is the group of order 4 generated by -1 and $1-2t$. Consequently $\mathrm{GO}(M,q)$ is the group of order 4 generated by $-\mathrm{id}$ and g.

Hint. Let $\mathfrak{m}$ (resp. $\mathfrak{m}'$) be the kernel of the ring morphism $K \to \mathbb{Z}/2\mathbb{Z}$ that maps t to 0 (resp. 1); the fractions $3/1$ and $3(1+4t)/1$ are not equal to $\pm 1/1$ in $K_{\mathfrak{m}}$; the fractions $3/1$ and $(1+4t)/1$ are not equal to $\pm 1/1$ in $K_{\mathfrak{m}'}$.

(5.ex.15)* The quadratic module under consideration in (5.ex.14) is "almost nondegenerate" according to the definition given in (2.ex.14). Now let (M,q) be any almost nondegenerate quadratic module of constant odd rank over some ring K.

(a) When K is a local ring, it is known that (M,q) is the orthogonal sum of a quadratic space and a free quadratic module generated by an element e such that $q(e)$ is invertible. Prove that every $g \in \mathrm{GO}(M,q)$ is a product of reflections, and that $\mathrm{GO}(M,q)$ is the subgroup of all $g \in \mathrm{Aut}(M,q)$ such that $\det(g)=\pm 1$.

(b) When K is an arbitrary ring, prove that the morphism $\mathrm{G''Lip}(M,q) \to \mathrm{GO}(M,q)$ is bijective, and that $\mathrm{GO}(M,q)$ is the subgroup of all $g \in \mathrm{Aut}(M,q)$ such that an equality $\det(g)/1 = \pm 1/1$ holds in every localization of K.

(c) Now the mapping $\lambda \longmapsto 2\lambda$ is assumed to be injective from K into itself. Prove that $\mathrm{Z}^r(\mathrm{C}\ell(M,q))$ is reduced to K, that the mapping $\mathrm{G'Lip}(M,q) \to \mathrm{GO}(M,q)$ is bijective, and that $\mathrm{GO}(M,q)$ is the union of the subsets $\mathrm{GO}_e(M,q)$ with $e \in \mathrm{Ip}(K)$.

(5.ex.16) Here is an alternative proof of the surjectiveness of the morphism $\mathrm{GLip}(M,q) \to \mathrm{GO}(M,q)$ when M is a vector space of finite dimension over a field K; it ignores Lemmas (5.7.4) and (5.7.5), but uses (5.6.4) and the results of (5.ex.13) and (5.ex.12); it may also use (5.7.7) if decompositions into products of reflections are aimed at. Let g be an automorphism of (M,q) such that $\mathrm{Ker}(g-\mathrm{id}) \supset \mathrm{Ker}(\mathrm{b}_q)$.

(a) When $\mathrm{Im}(g-\mathrm{id})$ is not totally isotropic, prove the existence of a product g' of reflections such that $\mathrm{Im}(g'g-\mathrm{id})$ is a totally isotropic subspace of $\mathrm{Im}(g-\mathrm{id})$.
Hint. If $g(a)-a$ is not isotropic, $\mathrm{Ker}(G_{g(a)-a}g-\mathrm{id})$ is strictly larger than $\mathrm{Ker}(g-\mathrm{id})$.

(b) Prove that $\mathrm{Im}(g-\mathrm{id})$ is totally isotropic if and only if $g-\mathrm{id}$ is an infinitesimal orthogonal transformation. Thus the initial problem is reduced to this one: if f is an infinitesimal orthogonal transformation such that $\mathrm{Im}(f)$ is totally isotropic, then $\mathrm{id}+f=G_x$ for some $x \in \mathrm{GLip}(M,q)$.

(c) When $\mathrm{Ker}(\mathrm{b}_q)=0$, it follows from (5.ex.13) that $f=F_y$ for some $y=\sum_i b_i c_i$ with all b_i and c_i in $\mathrm{Im}(f)$. Prove that $\mathrm{id}+f=G_x$ if $x=\prod_i(1+b_ic_i)$.

(d) When $\mathrm{Ker}(\mathrm{b}_q)\neq 0$, prove that $f=F_y$ for some $y \in C\ell_0^{\leq 2}(M,q)$ like this:

$$y = \sum_i b_i c_i + \sum_j d_j e_j$$

with all b_i, $c_i \in \mathrm{Im}(f)$ and all $d_j \in \mathrm{Im}(f)\cap\mathrm{Ker}(\mathrm{b}_q)$; then $\mathrm{id}+f=G_x$ if x is the product of all $1+b_ic_i$ (on the left side) and all $1+d_je_j$ (on the right side).

Lipschitz monoids and orthogonal groups of quadratic spaces

(5.ex.17) Let (M,q) be a quadratic space of constant rank r, f an infinitesimal automorphism of (M,q), and g an automorphism of (M,q). We are interested in the characteristic polynomials of f and g, that are the determinants of $\lambda\,\mathrm{id}-f$ and $\lambda\,\mathrm{id}-g$ (with λ an indeterminate). We forget (5.6.3) and we use localizations and matrix calculus.

(a) Prove that $\det(\lambda\,\mathrm{id}-f)=\det(\lambda\,\mathrm{id}+f)$.

(b) Prove that $\det(g)$ is a square root of 1, and that

$$\det(\lambda\,\mathrm{id}-g) = \det(g)\ (-\lambda)^r\ \det(\lambda^{-1}\mathrm{id}-g)\ .$$

(c) Assume that r is odd (and consequently 2 is invertible in K). Prove that M contains nonzero elements a such that $g(a)=\det(g)\ a$.
Hint. An endomorphism of M cannot be injective when its determinant vanishes.

(d) Assume that r is even, that 2 is not a divisor of zero in K, and that $\det(g)=-1$. Prove that M contains nonzero elements a such that $g(a)=a$, and nonzero elements b such that $g(b)=-b$.

(5.ex.18) Let (M,q) be a quadratic space, M' a direct summand as in (5.4.5), and g' an automorphism of (M',q'). Prove that g' belongs to the image of $\mathrm{G'Lip}(M',q')\to\mathrm{GO}(M',q')$ if and only if g' can be extended to an automorphism g of (M,q) such that $g(a)=a$ for all $a\in M'^{\perp}$.

(5.ex.19) Let K be a ring containing two elements x and y such that $x^2+y^2=1$, and in which 2 is invertible. Let (M,q) be the module K^2 provided with the quadratic form $(\lambda,\mu)\longmapsto\lambda^2+\mu^2$, and g the orthogonal transformation defined in the canonical basis (e_1,e_2) of K^2 in this way:

$$g(e_1)=xe_1+ye_2 \quad \text{and} \quad g(e_2)=-ye_1+xe_2\ .$$

(a) Prove that $\mathrm{Z}^r(C\ell(g))$ is the submodule generated by

$$z_1=(1+x)-ye_1e_2 \quad \text{and} \quad z_2=y-(1-x)e_1e_2\ .$$

Hint. $z_1\tau(z_1)+z_2\tau(z_2)=4\in K^\times$.
Prove that the mapping $z\longmapsto z+\tau(z)$ induces an isomorphism from $\mathrm{Z}^r(C\ell(g))$ onto the ideal of K generated by $1+x$ and y provided that y is not a divisor of zero in K.

(b) Here K is the ring $F[X,Y]/(X^2+Y^2-1)$ derived from a field F of characteristic $\neq 2$ as it is explained in (1.ex.24). When is $\mathrm{Z}^r(C\ell(g))$ a free module?

(c) Here K is the ring of all continuous functions $f:\mathbb{R}\to\mathbb{R}$ such that $f(t+1)=f(t)$ for all $t\in\mathbb{R}$ (as in (1.ex.23)), and x and y are the functions defined by $x(t)=\cos(2\pi t)$ and $y(t)=\sin(2\pi t)$. Thus g represents a group morphism from the additive group $\mathbb{R}$ into $\mathrm{GO}(\mathbb{R}^2,q_0)$ if q_0 is the usual quadratic form on $\mathbb{R}^2$. Prove that $\mathrm{Z}^r(C\ell(g))$ is not a free module.

(d) Here K is the ring of all continuous functions $\mathbb{R}\to\mathbb{R}$, and x and y are defined as in (c). Prove that $\mathrm{Z}^r(C\ell(g))$ is the free module generated by $z=\cos(\pi t)-e_1e_2\sin(\pi t)$. The mapping $t\longmapsto z(t)$ is a group morphism $\mathbb{R}\to\mathrm{GLip}(\mathbb{R}^2,q_0)$, but $z(t+1)=-z(t)$ for all $t\in\mathbb{R}$.

(5.ex.20) If g is an automorphism of a quadratic space (M,q), it follows from (5.8.1) that every element of $\mathrm{Z}^g(C\ell(g))$ is lipschitzian. This exercise presents a short proof of this fact with these additional hypotheses: the mapping $a\longmapsto 2a$ is bijective from M onto M, and the mappings $a\longmapsto na$ are injective for all integers $n\geq 3$. Thus the algebra $C\ell(M,q)\,\hat{\otimes}\,C\ell(M,q)^{to}$ is graded by submodules $(C\ell(M,q)\,\hat{\otimes}\,C\ell(M,q)^{to};\Delta,\Delta')^k$ with $k\in\mathbb{Z}$.

(a) Let $\mathcal{B}(M,K)$ be the module of all bilinear forms $M\times M\to K$. Consider these two bijective arrows: $M\otimes M\to M^*\otimes M^*\to\mathcal{B}(M,K)$; the former arrow is $\mathrm{d}_q\otimes\mathrm{d}_q$, the latter is a natural mapping which here is bijective since M is projective and finitely generated. Let Γ be the reciprocal image of b_q in $M\otimes M$, and let $b_1,c_1,b_2,c_2,\dots,b_m,c_m$ be elements of M such that $\Gamma=\sum_{i=1}^m b_i\otimes c_i$. Prove that

$$\forall a\in M,\qquad a=\sum_{i=1}^m \mathrm{b}_q(a,b_i)c_i=\sum_{i=1}^m \mathrm{b}_q(a,c_i)b_i.$$

(b) Let $\Gamma_{//}$ be the natural image of Γ in $\mathrm{C}\ell(M,q) \hat{\otimes} \mathrm{C}\ell(M,q)^{to}$. Prove the following equalities:

$$\forall a \in M, \qquad (a \otimes 1^{to})\ \Gamma_{//} - \Gamma_{//}\ (a \otimes 1^{to}) = 1 \otimes a^{to}\ ,$$
$$\forall a \in M, \qquad (1 \otimes a^{to})\ \Gamma_{//} - \Gamma_{//}\ (1 \otimes a^{to}) = a \otimes 1^{to}.$$

Deduce from these equalities that an element ζ of $\mathrm{C}\ell(M,q) \hat{\otimes} \mathrm{C}\ell(M,q)^{to}$ has degree k for the grading determined by Δ and Δ' if and only if $\zeta\,\Gamma_{//} - \Gamma_{//}\,\zeta = k\zeta$.

(c) Let x be a homogeneous element of $\mathrm{Z}^g(\mathrm{C}\ell(g))$. Prove this equality for all b and $c \in M$:

$$(x \otimes \tau(x)^{to})\ (b \otimes c^{to}) \;=\; (g(b) \otimes g(c)^{to})\ (x \otimes \tau(x)^{to}).$$

Besides, let $b_1, c_1, \dots, b_m, c_m$ be as above, and prove that

$$\Gamma_{//} \;=\; \sum_{i=1}^{m} b_i \otimes c_i^{to} \;=\sum_{i=1}^{m} g(b_i) \otimes g(c_i)^{to}.$$

Conclude that x belongs to $\mathrm{Lip}(M,q)$.

(5.ex.21) Consider a quadratic space over K, and the group morphism $x \longmapsto x\tau(x)$ from $\mathrm{GLip}(M,q)$ into $K^\times$ defined in (5.6.3).

(a) When the morphism $\mathrm{GLip}(M,q) \to \mathrm{G'Lip}(M,q)$ is surjective, prove that the morphism $x \longmapsto x\tau(x)$ induces a morphism from $\mathrm{G'Lip}(M,q)$ into the quotient of $K^\times$ by the subgroup $K^{\times 2}$ of squares.

Comment. Because of the isomorphism $\mathrm{G'Lip}(M,q) \to \mathrm{GO}(M,q)$, we also get a group morphism $\mathrm{GO}(M,q) \to K^\times/K^{\times 2}$; it is called the *spinorial norm.*

(b) Let K be a ring such that the group $K^\times/K^{\times 2}$ is infinite (for instance $K = \mathbb{Q}$), let (M,q) be a quadratic space over K such that $\mathrm{GLip}(M,q) \to \mathrm{GO}(M,q)$ is surjective, and the image of the spinorial norm $\mathrm{GO}(M,q) \to K^\times/K^{\times 2}$ is infinite; for instance (M,q) may be any $\mathbb{Q}$-quadratic space of dimension ≥ 2. Prove that $\mathrm{GO}(M,q)$ contains infinitely many *normal* subgroups.

(5.ex.22) Let (M,q) be a quadratic space of rank everywhere ≤ 4, and x a locally homogeneous element of $\mathrm{C}\ell(M,q)$. Prove that x belongs to $\mathrm{GLip}(M,q)$ if and only if $x\tau(x)$ belongs to $K^\times$.

Hint. Since q is nondegenerate, $\mathrm{GLip}(M,q)$ is the group of all locally homogeneous and invertible x such that $xax^{-1} \in M$ for all $a \in M$; and since the rank of M is ≤ 4, M is the submodule of all $a \in \mathrm{C}\ell_1(M,q)$ such that $\tau(a) = a$ (see (3.ex.19)).

Lipschitz groups and orthogonal groups of real quadratic spaces

In the next five exercises (M,q) *is a quadratic space over* $\mathbb{R}$. *The first two exercises determine the number of connected components of* $\mathrm{GLip}(M,q)$ *and* $\mathrm{GO}(M,q)$ *for the topologies induced by the usual topologies of the real vector spaces* $C\ell(M,q)$ *and* $\mathrm{End}(M)$. *The following three exercises describe the even subgroup* $G_0\mathrm{Lip}(M,q)$ *when* (M,q) *is a real quadratic space of dimension* 3 *or* 4. *The most important results are summed up just after* (5.ex.27). *Only very elementary knowledge in topology is required.*

(5.ex.23) Let (M,q) be a quadratic space over the field $\mathbb{R}$ of real numbers. As in **2.7**, m (resp. n) is the maximal dimension of a positive definite (resp. negative definite) subspace, and we suppose $m+n > 0$. Here we are interested in the number of connected components of the group $\mathrm{GLip}(M,q)$, that is the group generated by the invertible elements of M (see (5.10.3)). The mappings $x \longmapsto \partial x$ and $x \longmapsto x\tau(x)$ are group morphisms from $\mathrm{GLip}(M,q)$ into $\mathbb{Z}/2\mathbb{Z}$ and $\mathbb{R}^\times$ (see (5.3.6)), from which we can deduce a lower bound (2 or 4) for the number of connected components, and we shall prove that this lower bound is the exact number when m or n is ≥ 2. Let M_+ (resp. M_-) be the subset of all $a \in M$ such that $q(a) > 0$ (resp. $q(a) < 0$).

(a) Prove that M_+ (resp. M_-) is pathwise connected if and only if $m \geq 2$ (resp. $n \geq 2$). How many connected components has it when $m = 1$ (resp. $n = 1$)?
(b) Verify that -1 is in the neutral connected component of $\mathrm{GLip}(M,q)$ when m or n is ≥ 2.
(c) Explain that every product ab with $a \in M_+$ and $b \in M_-$ is equal to a product $b'a'$ with $a' \in M_+$ and $b' \in M_-$ (and conversely).
(d) Prove that $\mathrm{GLip}(M,q)$ has always four pathwise connected components except in these cases:
   - it has eight connected components when $m = n = 1$;
   - it has two connected components when q is positive or negative definite, and $\dim(M) \geq 2$.

(5.ex.24) The notation is that of (5.ex.23). Besides the group $\mathrm{GLip}(M,q)$ we also consider the *spinorial group* $\mathrm{Spin}^\pm(M,q)$, that is the subgroup of all $x \in \mathrm{GLip}(M,q)$ such that $x\tau(x) = \pm 1$; this subgroup has as many connected components as $\mathrm{GLip}(M,q)$ since it is isomorphic to its quotient by the subgroup of real positive numbers. From (5.3.7) we can derive the exact sequence

$$1 \longrightarrow \{1,-1\} \longrightarrow \mathrm{Spin}^\pm(M,q) \longrightarrow \mathrm{GO}(M,q) \longrightarrow 1.$$

Here we are interested in the number of connected components of $\mathrm{GO}(M,q)$. Explain briefly what happens when (m,n) is $(1,0)$ or $(0,1)$ or $(1,1)$. In the other cases the neutral connected component of $\mathrm{Spin}^\pm(M,q)$ contains $\{1,-1\}$; prove that the groups $\mathrm{Spin}^\pm(M,q)$ and $\mathrm{GO}(M,q)$ have the same number of connected components; consequently $\mathrm{Spin}^\pm(M,q)$ is a "two-sheet covering group" over $\mathrm{GO}(M,q)$. *Hint.* The main difficulty is to prove that $\mathrm{GO}(M,q)$ has at least four connected components when $mn \neq 0$; decompose M into an orthogonal sum $M_1 \perp M_2$ with q positive definite on M_1 and negative definite on M_2; if g is an automorphism of

(M,q), let g_j (for $j = 1,2$) be the endomorphism of M_j defined in this way: $g_j(a)$ is the orthogonal projection of $g(a)$ onto M_j; prove that g_j is always bijective, and that the sign of its determinant is arbitrary. It is even true that $|\det(g_j)| \geq 1$.

(5.ex.25) Let (M,q) be a quadratic space of dimension 3 over $\mathbb{R}$.

(a) When q is (positive or negative) definite, prove the existence of an isomorphism Φ from $C\ell_0(M,q)$ onto the division ring of quaternions $\mathbb{H} = \mathbb{R} \oplus \mathbb{R}i \oplus \mathbb{R}j \oplus \mathbb{R}ij$ (where $i^2 = j^2 = -1$ and $ji = -ij$). By this isomorphism, the restriction τ_0 of τ to $C\ell_0(M,q)$ corresponds to the quaternionic conjugation that maps every $y = \kappa + \lambda i + \mu j + \nu ij$ to $\bar{y} = \kappa - \lambda i - \mu j - \nu ij$ (whence $y\bar{y} = \kappa^2 + \lambda^2 + \mu^2 + \nu^2$).
Hint. Standard involutions are unique (see (1.13.8)).

(b) When q is not definite, prove the existence of an isomorphism Φ from $C\ell_0(M,q)$ onto the matrix algebra $\mathcal{M}(2,\mathbb{R})$. By this isomorphism, τ_0 corresponds to the involution $y \longmapsto y^\dagger = \mathrm{tr}(y)I - y$ where $\mathrm{tr}(y)$ is the trace of y, and I the unit matrix (whence $yy^\dagger = \det(y)I$).

(c) In both cases describe the images by Φ of the groups $\mathrm{GLip}_0(M,q)$ and $\mathrm{Spin}_0^\pm(M,q)$; remember (5.4.3).

(d) Let ω be a nonzero element of $C\ell^3(M,q)$, and $f : M \to C\ell^2(M,q)$ the bijection defined by $f(a) = a\omega$ (see (4.8.15)). For all $x \in \mathrm{GLip}_0(M,q)$ prove that the orthogonal transformation G_x corresponds through f to the inner automorphism of $C\ell_0(M,q)$ determined by x: $G_x(a) = f^{-1}(x\,f(a)\,x^{-1})$.

(5.ex.26) Let (M,q) be a quadratic space of dimension 4 over $\mathbb{R}$ that is either (positive or negative) definite, or hyperbolic; thus the center Z of $C\ell_0(M,q)$ is isomorphic to $\mathbb{R}^2$ and contains an idempotent ε such that $Z = \mathbb{R}\varepsilon \oplus \mathbb{R}(1-\varepsilon)$. Consequently $C\ell_0(M,q)$ is the direct sum of the ideals C' and C'' respectively generated by ε and $1-\varepsilon$, and both invariant by τ since $\tau(\varepsilon) = \varepsilon$.

(a) When q is definite, prove the existence of isomorphisms $\Phi' : C' \to \mathbb{H}$ and $\Phi'' : C'' \to \mathbb{H}$; by these isomorphisms the restrictions of τ to C' and C'' correspond to the quaternionic conjugation in $\mathbb{H}$.

(b) When q is hyperbolic, prove the existence of isomorphisms Φ' and Φ'' from C' and C'' onto $\mathcal{M}(2,\mathbb{R})$. By these isomorphisms the restrictions of τ to C' and C'' correspond to the involution $y \longmapsto y^\dagger$ defined in (5.ex.25)(b).

(c) Let x be an element of $C\ell_0(M,q)$, with components $x' = \varepsilon x$ and $x'' = (1-\varepsilon)x$ in C' and C''. Deduce from (5.ex.22) which property x' and x'' must satisfy for x to belong to $\mathrm{GLip}_0(M,q)$. Let us set $\Phi(x) = (\Phi'(x'),\ \Phi''(x''))$; thus Φ is an isomorphism from $C\ell_0(M,q)$ onto $\mathbb{H}^2$ or $\mathcal{M}(2,\mathbb{R})^2$. Describe the images by Φ of the groups $\mathrm{GLip}_0(M,q)$ and $\mathrm{Spin}_0^\pm(M,q)$.

(d) Let e be an element of M such that $q(e) \neq 0$. Explain why the twisted inner automorphism Θ_e extending the reflection G_e permutes the ideals C' and C'' of $C\ell_0(M,q)$. For every $a \in M$ we set $f(a) = \varepsilon a e^{-1}$ and $g(a) = (1-\varepsilon)e^{-1}a$ (as in (3.ex.20)(e)). Prove that f (resp. g) is a bijection from M onto C' (resp. C''), and that $g \circ f^{-1}$ and $f \circ g^{-1}$ are the bijections $C' \longleftrightarrow C''$ induced by

Θ_e. Let x be an element of $\mathrm{GLip}_0(M,q)$ with components x' and x'' in C' and C''; thus the components of $\Theta_e(x)$ are $y' = ex''e^{-1}$ and $y'' = ex'e^{-1}$. Prove the following equalities:

$$G_x(a) \;=\; f^{-1}\big(x'\,f(a)\,y'^{-1}\big) \;=\; g^{-1}\big(y''\,g(a)\,x''^{-1}\big)\,.$$

(5.ex.27) Let (M,q) be a quadratic space of dimension 4 over $\mathbb{R}$ that is neither (positive or negative) definite, nor hyperbolic; in other words, its signature is ± 2. In the center Z of $C\ell_0(M,q)$ there is an element ω such that $\omega^2 = -1$, and consequently Z is isomorphic to the field $\mathbb{C}$ of complex numbers. We extend the standard involution φ of Z to automorphisms of $C\ell_0(M,q)$ and $\mathcal{M}(2,Z)$ still denoted by φ and defined in this way: first we choose $e \in M$ such that $q(e)$ has the same sign as the signature of q, and we set $\varphi(x) = exe^{-1}$ for all $x \in C\ell_0(M,q)$; secondly φ operates on an element of $\mathcal{M}(2,Z)$ just by operating on the four entries of this matrix.

(a) Prove the existence of a Z-linear isomorphism Φ from $C\ell_0(M,q)$ onto $\mathcal{M}(2,Z)$ such that $\Phi(\varphi(x)) = \varphi(\Phi(x))$ for all $x \in C\ell_0(M,q)$; by this isomorphism, the restriction of τ to $C\ell_0(M,q)$ (which is Z-linear) corresponds to the involution $y \longmapsto y^\dagger$ defined in (5.ex.25)(b).
Hint. Let (e_0, e_1, e_2, e) be an orthogonal basis of M such that $q(e) = q(e_2) = q(e_1) = -q(e_0)$ and $e_0e_1e_2e = q(e)^2\omega$; you may define Φ in such a way that

$$\Phi(e_0e^{-1}) = \begin{pmatrix} 0 & \omega \\ -\omega & 0 \end{pmatrix}, \qquad \Phi(e_1e^{-1}) = \begin{pmatrix} 0 & \omega \\ \omega & 0 \end{pmatrix},$$
$$\Phi(e_2e^{-1}) = \begin{pmatrix} \omega & 0 \\ 0 & -\omega \end{pmatrix}.$$

(b) Describe the images by Φ of the groups $\mathrm{GLip}_0(M,q)$ and $\mathrm{Spin}_0^{\pm}(M,q)$.
Hint. (5.ex.22).

(c) Let us set $f(a) = ae$ for all $a \in M$. Prove that f is a bijection from M onto the subspace M' of all $y \in C\ell_0(M,q)$ that are invariant by the involution $\tau \circ \varphi = \varphi \circ \tau$. Moreover the mapping $y \longmapsto \det(\Phi(y))$ is a *real* quadratic form of signature 2 on M'. For every $x \in \mathrm{GLip}_0(M,q)$ prove that $G_x(a) = f^{-1}\big(x\,f(a)\,\varphi(x)^{-1}\big)$.

Selected results from the five previous exercises

Some important results can be reformulated in another way. When $\dim(M) \geq 3$, $\mathrm{Spin}^{\pm}(M,q)$ and $\mathrm{GO}(M,q)$ both have two or four connected components according as q is definite or not. The neutral connected component of $\mathrm{Spin}^{\pm}(M,q)$ is the subgroup $\mathrm{Spin}_0(M,q)$ of all $x \in \mathrm{GLip}_0(M,q)$ such that $x\tau(x) = 1$; it contains the kernel $\{1,-1\}$ of the canonical surjective morphism $\mathrm{Spin}^{\pm}(M,q) \to \mathrm{GO}(M,q)$. Let $\mathrm{SL}(2,\mathbb{R})$ (resp. $\mathrm{SL}(2,\mathbb{C})$) be the group of real (resp. complex) square matrices of order 2 with determinant 1, and $\mathrm{SL}(1,\mathbb{H})$ the group of all $y \in \mathbb{H}$ such that $y\bar{y} = 1$. When $\dim(M) = 3$, $\mathrm{Spin}_0(M,q)$ is isomorphic either to $\mathrm{SL}(1,\mathbb{H})$ or to $\mathrm{SL}(2,\mathbb{R})$

according as q is definite or not. When $\dim(M) = 4$, this group is isomorphic to $\mathrm{SL}(1,\mathbb{H}) \times \mathrm{SL}(1,\mathbb{H})$, or to $\mathrm{SL}(2,\mathbb{R}) \times \mathrm{SL}(2,\mathbb{R})$, or to $\mathrm{SL}(2,\mathbb{C})$ according as q is definite, or hyperbolic, or otherwise.

Instead of $\mathrm{SL}(1,\mathbb{H}) = \mathrm{GL}(1,\mathbb{H}) \cap \mathrm{SL}(4,\mathbb{R})$, many people prefer the isomorphic group $\mathrm{SU}(2) = \mathrm{SL}(2,\mathbb{C}) \cap \mathrm{GO}(4,\mathbb{R})$; the latter is the image of the former by the algebra morphism

$$\mathbb{H} \longrightarrow \mathcal{M}(2,\mathbb{C}) \ , \qquad \kappa + \lambda i + \mu j + \nu ij \ \longmapsto \begin{pmatrix} \kappa + \lambda i & -\mu - \nu i \\ \mu - \nu i & \kappa - \lambda i \end{pmatrix} .$$

About the Sections 5.9. and 5.10.

(5.ex.28) Here M is a finite dimensional vector space over a field K. This exercise proposes another proof of (5.9.4) that ignores the transformations $\mathcal{F}$ and $\mathcal{F}_*$. It must be proved that $z \rfloor x$ is in $\ell ip(M)$ when z and x are respectively in $\ell ip(M^*)$ and $\ell ip(M)$. In all the calculations x is written in this way:

$$x \ = \ a_1 \wedge a_2 \wedge \cdots \wedge a_k \wedge \mathrm{Exp}(u) \ \text{ with } \ k \geq 0, \ \ a_1, \dots, a_k \in M \ \text{ and } \ u \in \bigwedge^2(M) \ .$$

As for z, explain why it suffices to consider two cases, either $z = h$ with $h \in M^*$, or $z = 1 + h_1 \wedge h_2$ with h_1 and $h_2 \in M^*$.

When $z = h$, explain why it suffices to consider the following two cases:

- either $k \geq 0$ and $h(a_j) = 0$ for $j = 1, 2, \dots, k$;
- or $k \geq 1$, $h(a_1) = 1$ and $h(a_j) = 0$ for $j = 2, 3, \dots, k$.

When $z = 1 + h_1 \wedge h_2$, explain why it suffices to consider the following three cases:

- either $k \geq 0$ and $h_i(a_j) = 0$ for $i = 1, 2$ and $j = 1, 2, \dots, k$;
- or $k \geq 1$, $h(a_1) = 1$ and $h_i(a_j) = 0$ if $(i, j) \neq (1, 1)$;
- or $k \geq 2$, $h_i(a_i) = 1$ for $i = 1, 2$ and $h_i(a_j) = 0$ if $i \neq j$.

Then achieve the five required verifications.

(5.ex.29) Here M is a finite dimensional vector space over a field K. This exercise proposes another proof of (5.9.6) that ignores the transformations $\mathcal{F}$ and $\mathcal{F}_*$.

(a) Reduce the problem to the case of an infinite field K.

(b) Let $\mathcal{W}$ be the subset of all $w \in \bigwedge^2(M^*)$ such that (5.9.6) is true for (u, w) with all $u \in \bigwedge^2(M)$. Prove that $w + w'$ belongs to $\mathcal{W}$ whenever w and w' belong to it.
Hint. Since K is infinite, it suffices to consider all $u \in \bigwedge^2(M)$ such that $\mathrm{id}_M - \mathrm{d}_u \circ \mathrm{d}_w$ is bijective; thus the exact calculation of $\mathrm{Exp}(w) \rfloor \mathrm{Exp}(u)$ is possible.

(c) Verify that (5.9.6) is true whenever w is decomposable (in other words, $w = h \wedge h'$ for some h and $h' \in M^*$), and conclude.

(5.ex.30) This exercise is devoted to the calculation of the polynomials Δ_k mentioned in (5.9.7). Here B is still the field of fractions of the polynomial ring $A = \mathbb{Z}[t, (u_{i,j}), (w_{i,j})]$; the vector spaces $M = B^r$ and M^* are provided with their natural dual bases $(b_1, b_2, \dots, b_r)$ and $(b_1^*, \dots, b_r^*)$, but the definition of u and w give them a meaning somewhat different from that in the proof of (5.9.7), since here we set

$$u = \sum_{i<j} u_{i,j}\ b_i \wedge b_j \quad \text{and} \quad w = \sum_{i<j} w_{i,j}\ b_j^* \wedge b_i^*.$$

(a) Verify that the matrices of d_u and d_w are respectively $-U$ and $-W$.
(b) For every integer k such that $0 \le 2k \le r$, let $u^{[k]}$ and $w^{[k]}$ be the components of $\mathrm{Exp}(u)$ and $\mathrm{Exp}(w)$ respectively in $\bigwedge^{2k}(M)$ and $\bigwedge^{2k}(M^*)$. Prove that

$$\Delta_k((u_{i,j}), (w_{i,j})) = w^{[k]}(u^{[k]}) .$$

(c) Prove for instance that

$$\Delta_1((u_{i,j}), (w_{i,j})) = \sum_{i<j} u_{i,j} w_{i,j} ,$$

$$\Delta_2((u_{i,j}), (w_{i,j}))$$
$$= \sum_{i<j<k<l} (u_{i,j}u_{k,l} - u_{i,k}u_{j,l} + u_{i,l}u_{j,k})\ (w_{i,j}w_{k,l} - w_{i,k}w_{j,l} + w_{i,l}w_{j,k}) ,$$

and so forth....

Remark. Equivalent results can be derived from (5.10.8); consider the quadratic form q on $M = B^r$ such that $(b_1, b_2, \dots, b_r)$ is an orthogonal basis and $q(b_i) = 1$ for $i = 1, 2, \dots, r$; then the matrix of $F_u/2$ is also $-U$; let $(v_{i,j})$ (with $1 \le i < j \le r$) be another family of indeterminates, from which an element v of $\bigwedge^2(M)$ is derived in the same way; it follows from (5.10.8) and (4.8.16) that

$$\Delta_k((u_{i,j}), (v_{i,j})) = (-1)^k\ \mathrm{Scal}(u^{[k]} v^{[k]}).$$

(5.ex.31)* Let K be an infinite field, and M a finite dimensional vector space over K. Here $\bigwedge(M)$ is provided with the Zariski topology, for which the closed subsets are the closed algebraic submanifolds (defined by means of polynomial equations). From the definition (5.3.1) it is clear that $\mathrm{Lip}(M)$ is a closed algebraic submanifold of $\bigwedge(M)$. Here it is proved that its even and odd components $\mathrm{Lip}_0(M)$ and $\mathrm{Lip}_1(M)$ are irreducible; in other words, neither is the union of two closed submanifolds in a nontrivial way.

(a) Prove that $\mathrm{Lip}_0(M)$ is an irreducible algebraic manifold.
Hint. It is clear that $\mathrm{GLip}(M)$ is irreducible, since it is the group of all $\lambda\,\mathrm{Exp}(u)$ with $\lambda \in K^\times$ and $u \in \bigwedge^2(M)$ (see either (5.10.2) or (5.ex.6)); its

closure in $\bigwedge_0(M)$ is a multiplicative monoid which contains all products $a\wedge b$ because of the equality $\lambda + a \wedge b = \lambda\ \mathrm{Exp}(\lambda^{-1}a \wedge b)$ (valid for all $\lambda \in K^\times$ and all a, $b \in M$); therefore its closure is $\mathrm{Lip}_0(M)$.

(b) Prove that $\mathrm{Lip}_1(M)$ is an algebraic manifold isomorphic to $\mathrm{Lip}_0(M)$.
Hint. Use the multiplication by a in some algebra $\bigwedge(M;\beta)$ such that $\beta(a,a) \neq 0$.

(5.ex.32)* As a corollary of (5.ex.31) and (5.4.1), prove the following statement: for every finite dimensional vector space M over an infinite field K, and for every nonzero quadratic form q on M, the Lipschitz monoid $\mathrm{Lip}(M,q)$ is the topological closure of the Lipschitz group $\mathrm{GLip}(M,q)$. But the closure of $\mathrm{GLip}(M)$ (with $q=0$) is the even component $\mathrm{Lip}_0(M)$.

(5.ex.33) Let M be a finite dimensional vector space over a field K of characteristic $\neq 2$. Let $\mathcal{U}$ be the subset of all $f \in \mathrm{End}(M)$ without eigenvalue equal to 1 or -1, and $\mathcal{V}$ the subset of all $g \in \mathrm{End}(M)$ without eigenvalue equal to 0 or -1. For every $f \in \mathcal{U}$ (resp. $g \in \mathcal{V}$) we set:

$$g = \left(\mathrm{id} - \frac{1}{2}f\right)^{-1}\left(\mathrm{id} + \frac{1}{2}f\right) \quad (\text{resp. } f \;=\; 2\ (g-\mathrm{id})\ (g+\mathrm{id})^{-1}).$$

(a) Verify that two reciprocal bijections $\mathcal{U} \longleftrightarrow \mathcal{V}$ are defined in this way.

(b) Let q be a nondegenerate quadratic form on M; suppose that $f \in \mathcal{U}$ and $g \in \mathcal{V}$ are associated with each other by the above relations, and prove that g is an automorphism of (M,q) if and only if f is an infinitesimal automorphism of (M,q).
Comment. This result, attributed to Cayley, plays an important role in Lipschitz's works about orthogonal transformations; the factors 1/2 and 2 in the above formulas are not usual, but they ensure that the differential of the mapping $f \longmapsto g$ at the point 0 is the identity mapping of $\mathrm{End}(M)$.

(c)* Let ψ be a symplectic form on M (a nondegenerate alternate bilinear form); prove that g is an automorphism of (M,ψ) if and only if f is an infinitesimal automorphism of (M,ψ).

(5.ex.34) Let g be an automorphism of a quadratic space (M,q) over a field of characteristic $\neq 2$. As explained in (5.ex.21), the "spinorial norm" maps g to the element of $K^\times/K^{\times 2}$ defined in this way: it is $x\tau(x)$ modulo $K^{\times 2}$ if $g = G_x$.

(a) Suppose that -1 is not an eigenvalue of g, and deduce from (5.10.7) and (5.ex.33) that the spinorial norm of g is equal to the determinant of $2(g+\mathrm{id})$ modulo $K^{\times 2}$.

(b) When -1 is an eigenvalue of g, prove that M is the orthogonal sum $M' \perp M''$ of two subspaces invariant by g and satisfying these properties: the restriction of $g+\mathrm{id}$ to M' is bijective, whereas its restriction to M'' is nilpotent. Let g' be the restriction of g to M', and q'' the restriction of q to M''; the determinant of q'' is independent modulo $K^{\times 2}$ of the basis of M'' in which it is calculated (this is a classical consequence of (2.ex.3)); prove that the spinorial norm of g is $\det(2(g' + \mathrm{id}_{M'}))\det(q'')$ modulo $K^{\times 2}$.

Comment. In [Zassenhaus 1962] the same results about spinorial norms are reached without Clifford algebras.

(5.ex.35) Let M be a finite dimensional vector space over a field K of characteristic $\neq 2$, and q an anisotropic quadratic form on M (such that q never vanishes on any nonzero element of M); consequently q is nondegenerate. Prove that $\mathrm{GLip}(M,q)$ is the subset of *all* nonzero elements of $\mathrm{Lip}(M,q)$.
Hint. Consider a nonzero lipschitzian element x in the algebra $\bigwedge(M;\mathrm{b}_q/2)$:

$$x = a_1 \wedge a_2 \wedge \cdots \wedge a_k \wedge \mathrm{Exp}(u) \quad \text{with } k \geq 0, \quad a_1, \ldots, a_k \in M \text{ and } u \in \bigwedge^2(M)\,;$$

you can suppose that $(a_1, a_2, \ldots, a_k)$ is an orthogonal basis in the subspace N it spans; since $M = N \oplus N^\perp$, you can suppose $u \in \bigwedge^2(N^\perp)$; thus x is the Clifford product $a_1 a_2 \cdots a_k \mathrm{Exp}(u)$; to prove that $\mathrm{Exp}(u)$ is invertible, it suffices to prove that $\mathrm{id} - F_u/2$ is bijective (see (5.10.7)).

(5.ex.36)* Let (M,q) be a quadratic space over a field K of characteristic 0, let L be the ring $K[[t]]$ of formal series, and $L' = K((t))$ its field of fractions. On $L' \otimes M$ we consider the nondegenerate L'-quadratic form $t \otimes q$ defined by $\lambda \otimes a \longmapsto t\lambda^2 q(a)$ (for all $\lambda \in L'$ and $a \in M$); its restriction to $L \otimes M$ (treated as a L-quadratic form) is also denoted by $t \otimes q$. Besides the exterior algebra $\bigwedge_L(L \otimes M) = L \otimes \bigwedge(M)$, we also consider the Clifford algebra $\bigwedge_L(L \otimes M;\ t \otimes \mathrm{b}_q/2)$, and we treat it as an L-subalgebra of $\bigwedge_{L'}(L' \otimes M;\ t \otimes \mathrm{b}_q/2)$. Let u be any element of $L \otimes \bigwedge^2(M)$.

(a) For every $n \in \mathbb{N}$, let $F^{\geq n}$ be the L-submodule of $\bigwedge_L(L \otimes M)$ containing all $t^j \otimes \bigwedge^{\geq k}(M)$ such that $2j + k \geq n$. Prove that these submodules $F^{\geq n}$ determine a decreasing filtration both for the exterior multiplication and the Clifford multiplication of $\bigwedge_L(L \otimes M;\ t \otimes \mathrm{b}_q/2)$. Consequently, besides the exterior exponential $\mathrm{Exp}(u)$, there is also a Clifford exponential $\exp(u)$.
Remember that F_u is the endomorphism of $L' \otimes M$ defined by $F_u(a) = [u, a]$ for all $a \in L' \otimes M$ (see (5.5.3)). It is the image of an element of $t \otimes \mathrm{End}(M)$ by the canonical isomorphism $L' \otimes \mathrm{End}(M) \to \mathrm{End}_{L'}(L' \otimes M)$. Consequently, for every formal series $P(t) \in K[[t]]$, it is possible to define $P(F_u/2)$. In (b) just beneath we shall meet these three formal series: $\sinh(t)$ and $\cosh(t)$ (the odd and even parts of $\exp(t)$) and their quotient $\tanh(t)$.

(b) Prove that $\exp(u)$ is a lipschitzian element of $\bigwedge_L(L \otimes M)$, that there exist $\kappa \in L$ and $v \in L \otimes \bigwedge^2(M)$ such that $\exp(u) = \kappa\,\mathrm{Exp}(v)$, and that moreover

$$\kappa^2 = \det\left(\cosh\left(\frac{1}{2}F_u\right)\right) \quad \text{and} \quad \frac{1}{2}F_v = \tanh\left(\frac{1}{2}F_u\right).$$

Hint. When $x = \exp(u)$, classical results of Lie theory say that the inner automorphism $y \longmapsto xyx^{-1}$ is the exponential of the derivation $y \longmapsto [u,y]$; since this inner automorphism leaves M invariant, x is lipschitzian; therefore you can write $x = \kappa \operatorname{Exp}(v)$ and deduce from (5.10.7) that

$$1 = \kappa^2 \det\left(\mathrm{id} - \frac{1}{2}F_v\right) \quad \text{and} \quad \exp(F_u) = \left(\mathrm{id} - \frac{1}{2}F_v\right)^{-1}\left(\mathrm{id} + \frac{1}{2}F_v\right).$$

Comments. When K is the field $\mathbb{R}$ or $\mathbb{C}$, routine arguments with power series show that the above equalities remain true when t is replaced with 1, and when exp, cosh, tanh are understood as analytical functions on $C\ell(M,q)$ or $\mathrm{End}(M)$, provided that $F_u/2$ has no eigenvalue λ such that $\cosh(\lambda) = 0$. Besides, the eigenvalues of $\cosh(F_u/2)$ other than 1 have even multiplicities because of (5.ex.17)(a).

(5.ex.37) Let β be any bilinear form on a module M, and q the quadratic form such that $q(a) = \beta(a,a)$ for all $a \in M$. Let $M^\dagger$ be a submodule of M^* satisfying these two conditions: $M^\dagger$ contains $\mathrm{Im}(\mathrm{d}_\beta)$ and $\mathrm{Im}(\mathrm{d}_\beta^{to})$, and $\mathrm{Hom}(M, M^\dagger)$ contains bijective elements. Of course $M^\dagger = M^*$ when (M,q) is a quadratic space. We treat d_β and d_β^{to} as elements of $\mathrm{Hom}(M, M^\dagger)$, and for every $u \in \bigwedge^2(M)$ the notation d_u means the element of $\mathrm{Hom}(M^\dagger, M)$ defined by $\mathrm{d}_u(h) = h \rfloor u$. Let $\mathrm{Hom}_\wedge(M^\dagger, M)$ be the submodule of all $\delta \in \mathrm{Hom}(M^\dagger, M)$ such that $h(\delta(h)) = 0$ for all $h \in M^\dagger$; it is clear that d_u belongs to $\mathrm{Hom}_\wedge(M^\dagger, M)$ for all $u \in \bigwedge^2(M)$. The notations id and $\mathrm{id}^\dagger$ mean the identity mappings of M and $M^\dagger$.

Let $\mathcal{U}$ be the subset of all $\delta \in \mathrm{Hom}(M^\dagger, M)$ such that $\mathrm{id} + \delta\mathrm{d}_\beta$ and $\mathrm{id} + \delta\mathrm{d}_\beta^{to}$ are bijective, and $\mathcal{V}$ the subset of all bijective $g \in \mathrm{End}(M)$ such that $\mathrm{d}_\beta g - \mathrm{d}_\beta^{to}$ is bijective from M onto $M^\dagger$. The proposition (5.10.5) suggests studying the mappings $\Phi : \mathcal{U} \to \mathrm{End}(M)$ and $\Psi : \mathcal{V} \to \mathrm{Hom}(M^\dagger, M)$ defined in this way:

$$\Phi(\delta) = (\mathrm{id} + \delta\mathrm{d}_\beta)^{-1}(\mathrm{id} + \delta\mathrm{d}_\beta^{to}) \quad \text{and} \quad \Psi(g) = (\mathrm{id} - g)\,(\mathrm{d}_\beta g - \mathrm{d}_\beta^{to})^{-1}.$$

(a) Verify that for all $(\delta, g) \in \mathrm{Hom}(M^\dagger, M) \times \mathrm{End}(M)$:

$$(\mathrm{id} + \delta\mathrm{d}_\beta)\, g = \mathrm{id} + \delta\mathrm{d}_\beta^{to} \iff \delta\,(\mathrm{d}_\beta g - \mathrm{d}_\beta^{to}) = \mathrm{id} - g\,;$$

consequently, if (δ, g) belongs to $\mathcal{U} \times \mathcal{V}$, the equalities $g = \Phi(\delta)$ and $\delta = \Psi(g)$ are equivalent.
Nonetheless the equality $g = \Phi(\delta)$ never holds for any $(\delta, g) \in \mathcal{U} \times \mathcal{V}$ when $\mathrm{Ker}(\mathrm{d}_q) \neq 0$. Indeed it is easy to verify that $\mathrm{Ker}(\mathrm{id} - \Phi(\delta))$ always contains $\mathrm{Ker}(\mathrm{d}_q)$; unfortunately when $\mathrm{Ker}(\mathrm{id} - g)$ contains $\mathrm{Ker}(\mathrm{d}_q)$, then $\mathrm{d}_\beta g - \mathrm{d}_\beta^{to}$ annihilates $\mathrm{Ker}(\mathrm{d}_q)$, and cannot be bijective.

(b) Suppose that $g = \Phi(\delta)$ and prove that

$$\forall a \in M\,, \quad q(a) - q(g(a)) = h(\delta(h)) \quad \text{if } h = (\mathrm{d}_\beta g - \mathrm{d}_\beta^{to})(a)\,;$$

conclude that Φ maps $\mathcal{U} \cap \mathrm{Hom}_\wedge(M^\dagger, M)$ into $\mathrm{Aut}(M,q)$.

Conversely suppose that $\delta = \Psi(g)$ and prove that

$$\forall h \in M^\dagger\ , \quad h(\delta(h)) = q(a) - q(g(a)) \quad \text{if } a = (\mathrm{d}_\beta g - \mathrm{d}_\beta^{to})^{-1}(h)\ ;$$

thus Ψ maps $\mathcal{V} \cap \mathrm{Aut}(M,q)$ into $\mathrm{Hom}_\wedge(M^\dagger, M)$.

(c) Here M is a finitely generated projective module and $M^\dagger = M^*$; thus the mapping $u \longmapsto \mathrm{d}_u$ is bijective from $\bigwedge^2(M)$ onto $\mathrm{Hom}_\wedge(M^\dagger, M)$. Let δ be an element of $\mathrm{Hom}_\wedge(M^\dagger, M)$, and consider

$$\mathrm{id} + \delta \mathrm{d}_\beta \text{ and } \mathrm{id} + \delta \mathrm{d}_\beta^{to} \text{ in } \mathrm{End}(M),\ \ \mathrm{id}^\dagger + \mathrm{d}_\beta \delta \text{ and } \mathrm{id}^\dagger + \mathrm{d}_\beta^{to}\delta \text{ in } \mathrm{End}(M^\dagger).$$

Prove that all these four endomorphisms are bijective when one of them is bijective.

(d) Verify that these two equalities are consequences of the equalities at the beginning of (a):

$$(\mathrm{id}^\dagger + \mathrm{d}_\beta\delta)\ (\mathrm{d}_\beta g - \mathrm{d}_\beta^{to}) = \mathrm{d}_\beta - \mathrm{d}_\beta^{to} \quad = \mathrm{d}_q\ ,$$
$$(\mathrm{id}^\dagger + \mathrm{d}_\beta^{to}\delta)\ (\mathrm{d}_\beta g - \mathrm{d}_\beta^{to}) = (\mathrm{d}_\beta - \mathrm{d}_\beta^{to})\ g\ .$$

Conclude that, when (M,q) is a quadratic space, Φ and Ψ are reciprocal bijections between $\mathcal{U} \cap \mathrm{Hom}_\wedge(M^\dagger, M)$ and $\mathcal{V} \cap \mathrm{Aut}(M,q)$.

(e) Suppose that $\delta = \mathrm{d}_u$ for some $u \in \bigwedge^2(M)$ and that δ belongs to $\mathcal{U}$; therefore $g = \Phi(\delta)$ belongs to $\mathrm{Aut}(M,q)$. Set $x = \mathrm{Exp}(u)$ and prove that, for every $a \in M$, $g(a)$ (resp. $g^{-1}(a)$) is the only $b \in M$ such that $bx = xa$ (resp. $ax = xb$). This implies that x belongs to $\mathrm{Z}^g(C\ell(g))$. When (M,q) is a quadratic space, prove that x belongs to $\mathrm{GLip}(M;\beta)$.

Weyl algebras (for interested readers)

(5.ex.38) Consider the Weyl algebra $\mathrm{W}(M,\psi)$ defined in (4.ex.18), and prove the theorem analogous to (5.5.3) involving the Lie subalgebra $\mathrm{W}_0^{\leq 2}(M,\psi)$ and the Lie algebra of infinitesimal automorphisms f of (M,ψ); the latter contains all $f \in \mathrm{End}(M)$ such that $\psi(f(a),b) + \psi(a,f(b)) = 0$ for all a and $b \in M$ (or equivalently, the bilinear form $(a,b) \longmapsto \psi(f(a),b)$ is symmetric). When M is finitely generated and projective, and ψ nondegenerate, you can use an admissible scalar product β and the associated deformation $\mathrm{S}(M;\beta)$ of the symmetric algebra (see (4.ex.20) or (4.ex.21)).

When $\beta = \psi/2$, the subspace $\mathrm{S}^2(M)$ is a Lie subalgebra of $\mathrm{S}(M;\psi/2)$, and $[\mathrm{S}^2(M),\ \mathrm{S}^k(M)] \subset \mathrm{S}^k(M)$ for every degree k (compare with (5.5.4)).

(5.ex.39)* Here M is a vector space of finite dimension over a field K of characteristic 0. Yet the case of a field K of characteristic ≥ 3 is treated in [Helmstetter

1982]. Let $L = K[[t]]$ be the ring of formal series derived from K. Besides the symmetric algebra $\mathrm{S}_L(L \otimes M) = L \otimes \mathrm{S}(M)$ we also consider the algebra $\bar{\mathrm{S}}_L(L \otimes M)$ that is the direct product of all subspaces $L \otimes \mathrm{S}^k(M)$; and similarly $\bar{\mathrm{S}}_L(L \otimes M^*)$. If f is any element of $\bar{\mathrm{S}}_L(L \otimes M^*)$, and x any element of $\bar{\mathrm{S}}_L(L \otimes M)$, it is not always possible to define the interior product $f \rfloor x$; nevertheless, if u and w belong respectively to $L \otimes \mathrm{S}^2(M)$ and to $tL \otimes \mathrm{S}^2(M^*)$, and if $\mathrm{Exp}(u)$ and $\mathrm{Exp}(w)$ are their exponential in $\bar{\mathrm{S}}_L(L \otimes M)$ and $\bar{\mathrm{S}}_L(L \otimes M^*)$, it is possible to define the interior product of $\mathrm{Exp}(w)$ and $\mathrm{Exp}(u)$ as a formally convergent infinite sum, because each symmetric power w^k belongs to $t^k L \otimes \mathrm{S}^{2k}(M^*)$. Prove the existence of $\kappa \in L^\times$ and $v \in L \otimes \mathrm{S}^2(M)$ such that this interior product is equal to $\kappa^{-1}\mathrm{Exp}(v)$, and prove that

$$\mathrm{d}_v = (\mathrm{id} - \mathrm{d}_u \circ \mathrm{d}_w)^{-1} \circ \mathrm{d}_u \quad \text{and} \quad \kappa^2 = \det(\mathrm{id} - \mathrm{d}_u \circ \mathrm{d}_w) .$$

Hint. This is the formal counterpart of (5.9.6) for symmetric algebras, with the intervention of the "inversion rule" already met in (5.9.8); to prove it, try an adaptation of (5.ex.29); there is no problem to adapt (5.ex.29)(b); then it suffices to consider the particular case $w = t\lambda \otimes h^2$ with $\lambda \in L$ and $h \in M^*$; let us set

$$a = (1 \otimes h) \rfloor u \in L \otimes M \quad \text{and} \quad \mu = (\tfrac{1}{2} \otimes h^2) \rfloor u \in L ;$$

first prove by induction on m that

$$\left(\frac{1}{m!} \otimes h^m\right) \rfloor \mathrm{Exp}(u) = \mathrm{Exp}(u) \vee \sum_{0 \le 2j \le m} \frac{\mu^j}{j!} \frac{a^{m-2j}}{(m-2j)!} ;$$

when $w = t\lambda \otimes h^2$, this leads to

$$\mathrm{Exp}(w) \rfloor \mathrm{Exp}(u) = \mathrm{Exp}(u) \vee \sum_{k \ge 0} P_k(t) \frac{(t\lambda a^2)^k}{k!}$$

with

$$P_k(t) = \sum_{j \ge 0} \frac{(2k+2j)!\ k!}{(k+j)!\ (2k)!} \frac{(t\lambda\mu)^j}{j!} = \frac{1}{(1-4t\lambda\mu)^k \sqrt{1-4t\lambda\mu}}.$$

(5.ex.40)* Let M be a vector space of finite even dimension over a field K of characteristic 0, ψ a nondegenerate alternate bilinear form on M, and $\beta = \psi/2$ the canonical scalar product. As in (5.ex.36) we use $L = K[[t]]$ and $L' = K((t))$, and we consider the algebra $\bar{\mathrm{S}}_L(L \otimes M;\ t \otimes \beta)$ defined in (4.ex.22), that must be understood as a "formal enlargement" of $\mathrm{W}_L(L \otimes M,\ t \otimes \psi)$. Every $u \in L \otimes \mathrm{S}^2(M)$ has an exponential $\mathrm{Exp}(u)$ in $\bar{\mathrm{S}}_L(L \otimes M)$, that must be distinguished from $\exp(u)$, its exponential in $\bar{\mathrm{S}}_L(L \otimes M;\ t \otimes \beta)$ (to which a treatment analogous to the one in (5.ex.36) can later be applied).

(a) Prove that an element of $\bar{\mathrm{S}}_L(L\otimes M;\ t\otimes\beta)$ is invertible if and only if it has a nonzero component in $1\otimes K$ (that is the subspace $t^j\otimes \mathrm{S}^k(M)$ corresponding to $(j,k)=(0,0)$). Consequently $\mathrm{Exp}(u)$ has an inverse $\mathrm{Exp}(u)^{-1}$ for the Weyl multiplication.

(b) Prove that the inner automorphism of $\bar{\mathrm{S}}_L(L\otimes M;\ t\otimes\beta)$ determined by $x=\mathrm{Exp}(u)$ leaves $L\otimes M$ invariant, and determines by restriction an automorphism G_x of $(L\otimes M,\ 1\otimes\psi)$; moreover, if we derive from u the infinitesimal automorphism F_u defined as usually by $F_u(a)=[u,a]$, we can still write

$$G_x=\left(\mathrm{id}-\frac{1}{2}F_u\right)^{-1}\left(\mathrm{id}+\frac{1}{2}F_u\right).$$

(c) Prove that the mapping $u\longmapsto G_x$ (with $x=\mathrm{Exp}(u)$) is a bijection from $L\otimes \mathrm{S}^2(M)$ onto the group $\mathrm{Aut}_{\mathrm{id}}(L\otimes M,\ 1\otimes\psi)$ of all automorphisms g that have a component in $1\otimes \mathrm{End}(M)$ equal to id_M (in other words, all automorphisms g that shrink to id_M when t is replaced with 0).
Hint. See (5.ex.33)(c) and (5.ex.38).

(d) Let $\mathrm{GLip}^\vee(L\otimes M,\ t\otimes\beta)$ be the subset of all products $\lambda\mathrm{Exp}(u)$ with $\lambda\in L^\times$ and $u\in L\otimes\mathrm{S}^2(M)$. Prove that $\mathrm{GLip}^\vee(L\otimes M,\ t\otimes\beta)$ is a group for the Weyl multiplication of $\bar{\mathrm{S}}(L\otimes M,\ t\otimes\beta)$, that the mapping $x\longmapsto G_x$ is a surjective morphism from this group onto the group $\mathrm{Aut}_{\mathrm{id}}((L\otimes M,\ 1\otimes\psi)$, and that its kernel is $L^\times$.

(e) With the help of (5.ex.39) prove that, for every $u\in L\otimes S^2(M)$,

$$\mathrm{Exp}(u)\mathrm{Exp}(-u)=\det\left(\mathrm{id}-\frac{1}{2}F_u\right)^{-1}=\det\left(\mathrm{id}+\frac{1}{2}F_u\right)^{-1}.$$

(f) From (b) and (e), derive a corollary analogous to (5.10.8) for the product of $\mathrm{Exp}(u)$ and $\mathrm{Exp}(v)$ in $\bar{\mathrm{S}}_L(L\otimes M;\ t\otimes\beta)$.

(5.ex.41) As in (4.ex.23), M is a finite dimensional vector space over $\mathbb{R}$, β is an element of $M\otimes M$, and ψ is derived from β by skew symmetrization. All this determines the Weyl algebra $\mathrm{W}_{\mathbb{C}}(\mathbb{C}\otimes M^*,\ i\otimes\psi)$ (with $i=\sqrt{-1}$), and the isomorphic algebra $\mathrm{S}_{\mathbb{C}}(\mathbb{C}\otimes M^*;\ i\otimes\beta)$. Let $\mathcal{L}$ be the direct sum of $i\otimes\mathrm{S}^2(M^*)$ and $\mathbb{R}$ identified with $1\otimes\mathrm{S}^0(M^*)$. Prove that $\mathcal{L}$ is a Lie subalgebra of $\mathrm{S}_{\mathbb{C}}(\mathbb{C}\otimes M^*;\ i\otimes\beta)$, and that with each $w\in\mathcal{L}$ is associated an infinitesimal automorphism F_w of (M^*,ψ) in this way:

$$\forall h\in M^*,\qquad 1\otimes F_w(h)\ =\ [\ w,\ 1\otimes h\ ].$$

When ψ determines a nondegenerate bilinear form on M^* (and consequently on M too, because of the isomorphism $\mathrm{d}_\psi: M^*\to M$), prove that we get an isomorphism from the quotient Lie algebra $\mathcal{L}/\mathbb{R}$ onto the Lie algebra of all infinitesimal automorphisms of (M^*,ψ). Besides, when $\beta=\psi/2$, then $i\otimes\mathrm{S}^2(M^*)$ is a Lie subalgebra supplementary to $\mathbb{R}$ in $\mathcal{L}$.

Comment. In (4.ex.23) a $\star$-multiplication has been defined at least for the functions $M \to \mathbb{C}$ that have a Fourier transform with compact support in M^*; this $\star$ multiplication does not work so easily when more general functions or distributions are involved; anyhow the ordinary multiplication of distributions is already a difficult theory. This explains why it is so difficult to find a Lie group $\mathrm{GLip}^\vee(M^*;\beta)$ involving this $\star$-multiplication and lying over the Lie algebra $\mathcal{L}$ defined in (5.ex.41), in the same way as the Lie group $\mathrm{GLip}(M,q)$ lies over the Lie algebra $C\ell_0^{\leq 2}(M,q)$. Here is a short report about the construction of $\mathrm{GLip}^\vee(M^*;\beta)$. Although it is devoted to a real Lie group, it cannot succeed if only real things are involved; it needs purely imaginary things in a $\mathbb{C}$-extension. Another noticeable feature (imposed by (5.9.8)) is the intervention of the group $(\mathbb{R}^\times)^{1/4}$ of all $\lambda \in \mathbb{C}$ such that $\lambda^4 \in \mathbb{R}^\times$; it is the direct product of the group $\mathbb{R}^{\times 2}$ of real positive numbers and the group $\mu_8(\mathbb{C})$ of eighth roots of 1. The group $\mu_8(\mathbb{C})$ is involved in various domains where an 8-periodicity appears (for instance in **6.8**), and it is probable that all these domains are somewhat related to one another.

First we must define the symplectic Lipschitz monoid $\mathrm{Lip}^\vee(M^*)$. By analogy with (5.9.3) we may guess that Fourier transformation must map $\mathrm{Lip}^\vee(M^*)$ onto $\mathrm{Lip}^\vee(M)$. Because of the "inversion rule" (see (5.9.8) and (5.ex.39)), it must contain an exceptional element ∞ that is the counterpart of the zero element of $\mathrm{Lip}(M)$. The other elements of $\mathrm{Lip}^\vee(M^*)$ are all the distributions f on M that can be written in this way, for some vector subspace N of M, for some Lebesgue measure dx_N on N, for some quadratic form $w : M \to \mathbb{R}$ and for some $\mu \in (\mathbb{R}^\times)^{1/4}$:

$$\int_M \varphi(x)\ f(x)\ dx \ = \ \mu \int_N \varphi(x)\ \exp(iw(x))\ dx_N .$$

With this f is associated $\tau(f)$ that is obtained by replacing μ with its conjugate $\bar{\mu}$, and w by $-w$.

It is worth explaining how much this set $\mathrm{Lip}^\vee(M^*)$ looks like $\mathrm{Lip}(M^*)$. According to (5.10.2) every nonzero element of $\mathrm{Lip}(M^*)$ can be written

$$\begin{aligned} &\lambda h_1 \wedge h_2 \wedge \cdots \wedge h_r \wedge \mathrm{Exp}(w') \\ &\qquad \text{with } \lambda \in K^\times,\ r \geq 0,\ h_1,\ h_2, \dots, h_r \in M^* \text{ and } w' \in \bigwedge^2(M^*) ; \end{aligned}$$

let H be the subspace of M^* spanned by $h_1, \dots, h_r$, and H^{an} its annihilator in M; on one side w' determines an alternate bilinear form on M, on the other side the above exterior product only depends on the image of w' in $\bigwedge^2(M^*/H)$; therefore this element of $\mathrm{Lip}(M^*)$ only depends on the restriction to H^{an} of the bilinear form determined by w'. Exactly like the above element f of $\mathrm{Lip}^\vee(M^*)$ which only depends on the restriction to N of the quadratic form w.

Now $\mathrm{Lip}^\vee(M^*)$ must become a monoid for some $\star$-multiplication that must be the ordinary multiplication (or $\vee$-multiplication) when $\beta = 0$. Of course the $\star$-product of two elements of $\mathrm{Lip}^\vee(M^*)$ is ∞ whenever a factor is ∞. When f and g are elements of $\mathrm{Lip}^\vee(M^*)$ other than ∞, their $\star$-product is defined by the equality written in (4.ex.23), when it is possible to give it a meaning; when it is possible, $f \star g$ still belongs to $\mathrm{Lip}^\vee(M^*)$; when it is impossible, by definition $f\star g = \infty$. Thus $\mathrm{Lip}^\vee(M^*)$ becomes a $\star$-monoid $\mathrm{Lip}^\vee(M^*;\beta)$. In this monoid there is an anti-automorphism τ_β (such that $\tau_\beta(f \star g) = \tau_\beta(g) \star \tau_\beta(f)$) which coincides with τ if $\beta = \psi/2$. Let $\mathrm{GLip}^\vee(M;\beta)$ be the subset of all $f \in \mathrm{Lip}^\vee(M^*)$ such that $f\star\tau_\beta(f) \neq \infty$; this implies that $f\star\tau_\beta(f)$ is a constant positive function on M; thus $\mathrm{GLip}^\vee(M^*;\beta)$ is a $\star$-group, and we get a group morphism $\mathrm{GLip}^\vee(M^*;\beta) \to \mathbb{R}^{\times 2}$ if we map every f to the constant value of $f \star \tau_\beta(f)$. When $\beta = 0$, the group $\mathrm{GLip}^\vee(M^*)$ is the subset of all $f \in \mathrm{Lip}^\vee(M^*)$ that are ordinary functions on M, in other words such that $N = M$ (if f and N are related as above); this group has eight connected components (because of the factor $\mu \in (\mathbb{R}^\times)^{1/4}$).

Then it appears that $f \star h \star \tau_\beta(f)$ is a linear form on M for all $h \in M^*$ and all $f \in \mathrm{GLip}^\vee(M^*;\beta)$. This leads to a group morphism $\mathrm{GLip}^\vee(M^*;\beta) \to \mathrm{Aut}(M^*;\psi)$. When ψ determines a nondegenerate bilinear form on M^*, the kernel of this morphism is $(\mathbb{R}^\times)^{1/4}$. In all cases its image is the subgroup of all elements of $\mathrm{Aut}(M^*,\psi)$ leaving invariant all elements of $\mathrm{Ker}(\mathrm{d}_\psi)$.

Now it is important to know the number of connected components of $\mathrm{GLip}^\vee(M^*;\beta)$ when $\psi \neq 0$. The set $\mathrm{Lip}^\vee(M^*) \setminus \{\infty\}$ has already two connected components; the component which the above f belongs to depends on the sign of the real number $\mu^4(-1)^{\dim(N)}$. Besides, there is a path between 1 and -1 inside $\mathrm{GLip}^\vee(M^*;\beta)$. Consequently the number of connected components is 2 or 4, and it requires more work to prove that it is exactly 4. Let $\mathrm{GLip}^\vee_{con}(M;\beta)$ be the neutral connected component; the other connected components are obtained by multiplying this one by $(1+i)/\sqrt{2}$ or i or $(-1+i)/\sqrt{2}$; consequently all connected components have the same image in $\mathrm{Aut}(M^*,\psi)$.

When ψ is nondegenerate, we get an exact sequence

$$1 \longrightarrow \mathbb{R}^\times \longrightarrow \mathrm{GLip}^\vee_{con}(M;\beta) \longrightarrow \mathrm{Aut}(M^*,\psi) \longrightarrow 1.$$

Thus the group of all $f \in \mathrm{GLip}^\vee_{con}(M;\beta)$ such that $f\tau_\beta(f) = 1$, is a two-sheet covering group over the "symplectic group" $\mathrm{Aut}(M^*,\psi)$, in the same way as the spinorial group (see (5.ex.24)) is a two-sheet covering group over the orthogonal group. It is worth recalling that the complex symplectic group $\mathrm{Aut}(\mathbb{C}\otimes M^*,\ 1\otimes\psi)$ is simply connected and does not admit nontrivial coverings.

Here it is impossible to define a "symplectic Clifford group" $\mathrm{GC}\ell^\vee(M,\psi)$ that should be the group of all even $\star$-invertible f such that $f \star h \star f^{-1} \in M^*$ for all $h \in M^*$, because the $\star$-product of two distributions (like the ordinary product) in general does not exist. The preliminary definition of a Lipschitz monoid $\mathrm{Lip}^\vee(M;\beta)$ is an indispensable step before the construction of the wanted group. This fact has been a strong encouragement to begin Chapter **5** with Lipschitz monoids rather than with traditional Clifford groups.

Chapter 6

Further Algebraic Developments

The main purpose of this chapter is to give more information about the graded Azumaya algebras which have been presented in Chapter **3**. Since Clifford algebras of quadratic spaces are graded Azumaya algebras (see (3.7.5)), it is necessary to know an honourable part of the classical theory about these algebras. This theory requires preliminary developments through other subjects: graded modules over noncommutative graded algebras (in **6.2**), graded semi-simple modules (in **6.3**), graded Morita theory (in **6.4**), graded separable algebras (in **6.5**) and graded central simple algebras (in **6.6**). A great part of the information expounded here comes from works by H. Bass and Ch. Small, but several modifications have been achieved, sometimes suggested by N. Jacobson's books on algebra.

6.1 Modules over a noncommutative algebra

As in the previous chapters, K is a commutative ring with unit 1. Let A be a noncommutative algebra over K, that is an object in the category $\mathcal{A}lg(K)$. Let us make precise the sense of the word "noncommutative": A may be commutative or not commutative. Since A is assumed to contain a unit element 1_A, there is a canonical morphism $K \to A$; it is injective if and only if A is a faithful K-module; in this case 1_A is identified with 1 and K becomes a subalgebra of A.

Let us remember that a K-module M is a left module (resp. right module) over A if there is a K-bilinear multiplication $A \times M \to M$ (resp. $M \times A \to M$) such that $a(bx) = (ab)x$ (resp. $(xa)b = x(ab)$) for all a and b in A, and all x in M. When it is not otherwise specified, every module is a left module; indeed every right module over A shall be treated as a left module over the opposite algebra

A^o (already defined in **3.1**) in which the equality $a^o b^o = (ba)^o$ holds. In a right module M over A, the notation $a^o x$ means xa for all $a \in A$ and all $x \in M$.

When M is a left module over A and a right module over B, we say that M is a *bimodule over* A *and* B if $(ax)b = a(xb)$ for all $a \in A$, all $b \in B$ and all $x \in M$. Because of the universal property of the algebra $A \otimes B^o$ (see (1.3.3)), it is equivalent to say that M is a left module over $A \otimes B^o$; the notation $(a \otimes b^o)x$ means axb. When $A = B$, a module over $A \otimes A^o$ is merely called a bimodule over A.

When M and N are modules over A, then $\mathrm{Hom}_A(M, N)$ is the subset of all $f \in \mathrm{Hom}(M, N)$ intertwining the operations in M and N of all $a \in A$. When M is a module over $A \otimes B^o$ and N a module over $A \otimes C^o$, then $\mathrm{Hom}_A(M, N)$ is a module over $B \otimes C^o$; indeed, if f, b, c belong respectively to $\mathrm{Hom}_A(M, N)$, B and C, then bfc (synonymous with $(b \otimes c^o)f$) is the A-linear mapping $M \to N$ defined in this way: $(bfc)(x) = (f(xb))c$. An element of the center $\mathrm{Z}(A)$ may operate on the right side as well as on the left side, in such a way that the notations ax and xa are synonymous for each $a \in \mathrm{Z}(A)$; therefore the previous considerations are relevant if we replace B or C with $\mathrm{Z}(A)$, and thus we can turn $\mathrm{Hom}_A(M, N)$ into a module over $\mathrm{Z}(A)$ in two different ways which anyhow give the same structure of module over $\mathrm{Z}(A)$.

The functor $\mathrm{Hom}_A(\dots, \dots)$ is left exact relative to both variables; this leads to the definition of projective modules P over A, for which the functor $\mathrm{Hom}_A(P, \dots)$ is exact. The statements in **1.7** about projectiveness are still true.

When M is a right module over A and N a left module over A, the tensor product $M \otimes_A N$ is the quotient of $M \otimes N$ (that is $M \otimes_K N$) by the submodule generated by all elements $xa \otimes y - x \otimes ay$, so that the equality $xa \otimes y = x \otimes ay$ holds in $M \otimes_A N$ for all $a \in A$, all $x \in M$ and all $y \in N$. When M is a module over $B \otimes A^o$ and N a module over $A \otimes C^o$, then $M \otimes_A N$ is a module over $B \otimes C^o$; indeed $b(x \otimes y)c$ (synonymous with $(b \otimes c^o)(x \otimes y)$) is by definition $(bx) \otimes (yc)$. This allows us (in two different but equivalent ways) to give $M \otimes_A N$ a structure of module over the center $\mathrm{Z}(A)$.

The functor $\cdots \otimes_A \cdots$ is still right exact. There is still an associativity property for such tensor products; if M, N, P are respectively modules over A^o, $A \otimes B^o$ and B, there are canonical isomorphisms

$$(M \otimes_A N) \otimes_B P \longleftrightarrow M \otimes_A (N \otimes_B P).$$

The commutativity property affords canonical isomorphisms between $M \otimes_A N$ and $N \otimes_{A^o} M$.

A module G over A is called a *generator* of modules over A (or shortly a *generator*) if for every module M over A there exists a set J and a surjective A-linear mapping from $\bigoplus_{j \in J} G$ (a direct sum of modules all isomorphic to G) onto M. Obviously A (considered as a module over itself) is a generator, since every subset J of M generating it as a module over A allows us to construct a surjective A-linear mapping from $\bigoplus_{j \in J} A$ onto M.

The following lemma shows that G *is a generator if and only if the functor* $\mathrm{Hom}_A(G,\dots)$ *is faithful* in the following sense: for every pair (M,N) of modules over A, the following mapping is injective:

$$\begin{aligned}\mathrm{Hom}_A(M,N) &\longrightarrow \mathrm{Hom}_K(\mathrm{Hom}_A(G,M),\ \mathrm{Hom}_A(G,N)),\\ f &\longmapsto \mathrm{Hom}_A(G,f) = (u \longmapsto f\circ u).\end{aligned}$$

(6.1.1) **Lemma.** *Let G and M be modules over A. The following four assertions are equivalent:*

(a) *there exists a set J and a surjective A-linear mapping from $\bigoplus_{j\in J} G$ onto M;*
(b) *there exists a K-module Q and a surjective A-linear mapping from $G\otimes Q$ onto M;*
(c) *the following canonical mapping is surjective:*

$$G\otimes_K \mathrm{Hom}_A(G,M) \longrightarrow M\ , \quad g\otimes u \longmapsto u(g)\ ;$$

(d) *the vanishing of any A-linear mapping $f : M \to N$ (with any target N) is equivalent to the vanishing of the K-linear mapping*

$$\mathrm{Hom}_A(G,f) : \mathrm{Hom}_A(G,M) \to \mathrm{Hom}_A(G,N).$$

Proof. The assertion (a) means the existence of a surjective mapping $G\otimes Q' \to M$ involving a free K-module Q' with a basis indexed by J; thus the equivalence (a)⇔(b) follows from the following fact: for every K-module Q there exists a surjective morphism $Q' \to Q$ with Q' a free K-module.

The equivalence (b)⇔(c) follows from the following fact: for every $v \in \mathrm{Hom}_A(G\otimes Q,\ M)$ and every $q \in Q$, the mapping $g \longmapsto v(g\otimes q)$ belongs to $\mathrm{Hom}_A(G,M)$, and consequently the images of all mappings like v are contained in the image of the canonical mapping obtained with $Q = \mathrm{Hom}_A(G,M)$.

Let us prove (b)⇒(d). Let $v : G\otimes Q \to M$ be a surjective A-linear mapping, and $f : M \to N$ an A-linear mapping such that $\mathrm{Hom}_A(G,f)$ vanishes. This implies the vanishing of all morphisms $G \to M \to N$ involving f; this implies the vanishing of $f\circ v : G\otimes Q \to M \to N$, because for every $q \in Q$, the mapping $g \longmapsto f(v(g\otimes q))$ vanishes. Since v is surjective, the vanishing of $f\circ v$ implies $f = 0$.

Let us prove (d)⇒(c). We set $Q = \mathrm{Hom}_A(G,M)$ and consider the image N of the canonical mapping $G\otimes Q \to M$. If f is the quotient mapping $M \to M/N$, it is clear that $\mathrm{Hom}_A(G,f)$ is the null mapping from $\mathrm{Hom}_A(G,M)$ into $\mathrm{Hom}_A(G,\ M/N)$; now (d) implies $f = 0$ and $M = N$. □

The following lemma is especially useful when $M = A$.

(6.1.2) **Lemma.** *If the four assertions in* (6.1.1) *are true, and if M is a generator (for instance if $M = A$), then G too is a generator.*

Proof. Let M' be any A-module; from two surjective A-linear mappings $G \otimes Q \to M$ and $M \otimes Q' \to M'$ we derive the following surjective A-linear mapping:

$$G \otimes (Q \otimes Q') \longleftrightarrow (G \otimes Q) \otimes Q' \longrightarrow M \otimes Q' \longrightarrow M' . \qquad \square$$

(6.1.3) **Remark.** When M is a finitely generated A-module, the set J in the assertion (a) of (6.1.1) can be required to be finite, and the module Q in (b) can be required to be finitely generated. Indeed in case of a surjective mapping $\bigoplus_{j \in J} G \to M$, each element $x \in M$ is the image of some $(g_j)_{j \in J}$ in which the nonzero components g_j only involve a finite subset of J.

(6.1.4) **Example.** Let P be a finitely generated projective K-module, $P^* = \mathrm{Hom}(P, K)$ and $B = \mathrm{End}(P)$. Thus P is a module over B in a natural way, P^* is a module over B^o, and the canonical mapping $P \otimes P^* \to B$ is an isomorphism of bimodules over B, since all localisations show that it is bijective. Here P is a generator of modules over B, whereas P^* is a generator of modules over B^o.

For a module over A, the property of being a generator may be understood as being somewhat stronger than the property of being faithful. Indeed every generator is obviously a faithful module over A, and in some cases, both properties are equivalent, as stated in the following proposition.

(6.1.5) **Proposition.** *When A is a commutative K-algebra, a finitely generated projective module over A is a generator if and only if it is faithful.*

Proof. We already know that every generator is faithful. Conversely let P be a finitely generated and faithful projective module over A. By localisations at the prime ideals of A we can prove the surjectiveness of the canonical mapping $P \otimes_A \mathrm{Hom}_A(P, A) \to A$. It suffices to remember that $P \otimes_A \mathrm{Hom}_A(P, A)$ is a quotient of $P \otimes \mathrm{Hom}_A(P, A)$, and to apply (6.1.2). $\square$

When M is a projective module over A, any surjective A-linear mapping $\bigoplus_{j \in J} G \to M$ makes M become a direct summand of $\bigoplus_{j \in J} G$. When the generator G itself is projective, then the projective modules over A are all the modules that are isomorphic to a direct summand of $\bigoplus_{j \in J} G$ for some set J. Of course A itself is a projective generator.

As preparation to the Morita theory later expounded in **6.4**, we add the following results.

(6.1.6) **Proposition.** *Let P be a module over A, and let us set $Q = \mathrm{Hom}_A(P, A)$ and $B = \mathrm{End}_A(P)$; thus P is a left module over $A \otimes B$, and Q a right module over it. Let us consider these two canonical mappings:*

$$P \otimes Q \longrightarrow A , \qquad z \otimes h \longmapsto h(z) ,$$
$$Q \otimes P \longrightarrow B , \qquad h \otimes z \longmapsto (x \longmapsto h(x)\ z).$$

The first mapping is $(A \otimes A^o)$-linear, and factorizes through $P \otimes_{B^o} Q$, whereas the second mapping is $(B^o \otimes B)$-linear and factorizes through $Q \otimes_A P$. Besides, P is a generator of A-modules if and only if the first mapping is surjective, whereas P is a a finitely generated projective module over A if and only if the second one is surjective.

Proof. Let us recall that Q is a right module over $A \otimes B$ in this way: $(h(a \otimes b))(x) = h(b(x))a$ for all $x \in P$. It is clear that the images of $(ha) \otimes z$ and $h \otimes (az)$ in B are equal, whence the factorization of the second canonical mapping through $Q \otimes_A P$. The factorization of the first one is obtained in the same way, provided that we remember that zb^o means $b(z)$, and $b^o h$ means $h \circ b$; thus $(zb^o) \otimes h$ and $z \otimes (b^o h)$ have the same image in A. The first mapping is $(A \otimes A^o)$-linear because the image of $(az) \otimes (ha')$ in A is $ah(z)a'$. Consequently the image of this mapping is an ideal of A. The second mapping is $(B^o \otimes B)$-linear because the image of $(h \circ b) \otimes b'(z)$ in B is the endomorphism $x \longmapsto b'(h(b(x))\, z)$. Consequently the image of this mapping is an ideal of B.

From (6.1.2) and (6.1.1) we know that P is a generator if and only if the first mapping is surjective. Now P is a finitely generated projective module over A if and only if it is isomorphic to a direct summand of A^n for some integer n. This means that id_P can be factorized as $P \to A^n \to P$. The mapping $P \to A^n$ is defined by n elements $h_1, \dots, h_n$ of Q, whereas the mapping $A^n \to P$ is defined by n elements $z_1, \dots, z_n$ of P. The composition of these two mappings is id_P if and only if the image of $\sum_{i=0}^n z_i \otimes h_i$ in B is id_P. This means that the second canonical mapping is surjective, since its image is an ideal of B. □

6.2 Graded modules over a graded algebra

Here all gradings are parity gradings over the group $\mathbb{Z}/2\mathbb{Z}$. As in the previous chapters, parities are indicated by lower indices 0 or 1. In **3.2** it has been explained that every graded algebra A gives rise to a family $(A,\ A^o,\ A^{to},\ A^t)$ of four graded algebras.

Let M be a graded left module over A; this means that M is an A-module, that A is graded as a K-module $(M = M_0 \oplus M_1)$ and that $\partial(ax) = \partial a + \partial x$ for all homogeneous $a \in A$ and $x \in M$. When it is not otherwise specified, all modules are left modules; graded right modules over A are treated as graded left modules either over A^o as in **6.1**, or over A^{to} when the twisting rule (4.2.1) must be respected; by definition $a^{to}x = (-1)^{\partial a \partial x} xa$.

Every graded module M over A gives rise to a family of eight graded modules:

$$(M,\ M^c,\ M^s,\ M^{cs},\ M^t,\ M^{ct},\ M^{st},\ M^{cst})\ ;$$

as modules over K, they are all isomorphic to M by means of canonical isomorphisms $x \longmapsto x^c$, $x \longmapsto x^s$, and so forth...; the first four objects are modules over A, and the last four are modules over A^t. Let us begin with the *conjugate module* M^c, which exists even when M is not graded; it comes from the grade automorphism $a \longmapsto (-1)^{\partial a}a$ in the algebra A; by definition $ax^c = (-1)^{\partial a}(ax)^c$. The module M^s only differs from M by a *shifted grading*: $(M^s)_0 = (M_1)^s$ and $(M^s)_1 = (M_0)^s$; in other words, $\partial x^s = 1 - \partial x$; this notwithstanding, $ax^s = (ax)^s$. The *twisted module* M^t involves the twisted algebra A^t in this way: $a^t x^t = (-1)^{\partial a \partial x}(ax)^t$. The other four modules result from various combinations of these three operations; for instance M^{st} means $(M^s)^t$, and thus $a^t x^{st} = (-1)^{\partial a(1-\partial x)}(ax)^{st}$. Of course there is a group of order 8 acting on the above family of eight modules, but this group is not commutative; it is generated by two elements s and t of order 2 such that st has order 4; the center of this group is generated by the element $c = stst = tsts$ of order 2. Consequently $M^{ts} = M^{cst}$.

When M is a graded module over A, the mapping $x \longmapsto (-1)^{\partial x}x^c$ is an isomorphism from M onto M^c. Nonetheless when M is a nongraded module over A, it may happen that M and M^c are not isomorphic modules: see (6.ex.1).

(6.2.1) **Remark.** When A is a Clifford algebra $C\ell(M,q)$, it is stated in (3.2.2) that the mapping $a \longmapsto a^t$ (with $a \in M$) extends to an isomorphism from $C\ell(M,-q)$ onto $C\ell(M,q)^t$. Here we had rather identify $C\ell(M,-q)$ with $C\ell(M,q)^t$ by means of this canonical isomorphism; this identification has two advantages. First the mapping $x \longmapsto x^t$ induces an algebra isomorphism $C\ell_0(M,q) \to C\ell_0(M,q)^t$ and demonstrates that $C\ell_0(M,q)$ and $C\ell_0(M,-q)$ are isomorphic algebras. Secondly to every graded module S over $C\ell(M,q)$ there corresponds a graded module S^t over $C\ell(M,-q)$, and conversely. In general (and in particular in the following example suggested by quantum mechanics) there is no such bijective correspondence (up to isomorphy) between modules over $C\ell(M,q)$ and modules over $C\ell(M,-q)$ when gradings are not imposed.

(6.2.2) **Example.** Let (M,q) be a quadratic space of dimension 4 and signature -2 over the field $\mathbb{R}$ of real numbers; when an "orthonormal" basis is chosen, the quadratic form q is often written $t^2 - x^2 - y^2 - z^2$ if t is the "time coordinate", and x, y, z the "space coordinates". It is possible to prove that $C\ell(M,q)$ (as a nongraded algebra) is isomorphic to the matrix algebra $\mathcal{M}(2,\mathbb{H})$ with coefficients in the division ring $\mathbb{H}$ of real quaternions (see (3.ex.29)); consequently all irreducible modules S over $C\ell(M,q)$ are isomorphic to each other (see (6.6.3)), they have dimension 8 over $\mathbb{R}$, and the centralizer of $C\ell(M,q)$ in $\mathrm{End}(S)$ is isomorphic to $\mathbb{H}$. In this centralizer we can choose an element i such that $i^2 = -\mathrm{id}_S$. If we set $\mathbb{C} = \mathbb{R} \oplus \mathbb{R}i$, then S becomes a module over $\mathbb{C} \otimes C\ell(M,q)$. It is often believed that this structure of module over this $\mathbb{C}$-algebra is necessary for S to become a genuine "spinor space", because the "Dirac equation" absolutely requires the presence of an endomorphism of S like this i. Nonetheless this opinion may be seriously questioned.

Indeed, if ω is the product of the elements of an "orthonormal" basis of M, it is easy to verify that $\omega^2 = -1$ and $\omega a = -a\omega$ for all $a \in M$ (in accordance with (4.8.15) and (3.5.13)). Let σ be the endomorphism of S defined by $\sigma(s) =$

$-i(\omega s) = -\omega(is)$. It is easy to verify that $\sigma^2 = \mathrm{id}_S$ and $\sigma(as) = -a\sigma(s)$ for all $a \in M$; consequently S becomes a graded module over $\mathrm{C}\ell(M,q)$ if we set $S_0 = \mathrm{Ker}(\sigma - id)$ and $S_1 = \mathrm{Ker}(\sigma + id)$. The elements of S_0 or S_1 are called "Weyl spinors" in the specialized literature. Conversely if S is a graded $\mathrm{C}\ell(M,q)$-module, we set $\sigma(s) = (-1)^{\partial s}s$, and $i(s) = \sigma(\omega s) = \omega\sigma(s)$; this implies (after some easy calculations) that $i^2 = -\mathrm{id}_S$ and that i belongs to the centralizer of $\mathrm{C}\ell(M,q)$ in $\mathrm{End}(S)$. Thus we can claim that a complex spinor space is the same thing as a real graded spinor space. Yet it may be preferable to explain the intrusion of imaginary numbers by means of an algebraic structure defined on $\mathbb{R}$.

It is usually admitted as evident that the opposite quadratic form $-t^2 + x^2 + y^2 + z^2$ leads to an equivalent physical theory. This is often explained by means of an isomorphism $\mathbb{C} \otimes \mathrm{C}\ell(M,q) \to \mathbb{C} \otimes \mathrm{C}\ell(M,-q)$ mapping every $1 \otimes a$ (with $a \in M$) to $i \otimes a$. Nevertheless if S is treated as a real graded module, there is a better explanation that uses the twisted module S^t which is a module over $\mathrm{C}\ell(M,q)^t = \mathrm{C}\ell(M,-q)$, and that avoids the imaginary vectors $i \otimes a$. Here there is no natural correspondence between modules over $\mathrm{C}\ell(M,q)$ and modules over $\mathrm{C}\ell(M,-q)$ because $\mathrm{C}\ell(M,-q)$ is isomorphic to the matrix algebra $\mathcal{M}(4,\mathbb{R})$, and its irreducible modules have dimension 4 over $\mathbb{R}$. As a module over $\mathrm{C}\ell(M,q)^t$, S^t is not irreducible, because it contains plenty of submodules of dimension 4 over $\mathbb{R}$; but S^t is irreducible as a graded module.

This example also shows how natural it is to associate with every graded Clifford module S two modules S^t and S^s. Indeed S^t appears when we replace q with $-q$, and S^s appears in this example when we replace i with $-i$, or ω with $-\omega$. As explained above, the operations denoted by the exponents s and t generate a group of order 8.

Remember that the grading of A is said to be regular if the multiplication mapping $A_1 \otimes A_1 \to A_0$ is surjective (see Definitions (3.5.2)). This property is equivalent to the existence of a sequence of odd elements $(a_1, b_1, a_2, b_2, \dots, a_m, b_m)$ of arbitrary even length $2m \geq 2$ such that $\sum_{i=1}^m a_i b_i = 1_A$. Such a sequence is called a *complete system of odd elements.*

(6.2.3) **Extension lemma.** *Let M and N be two graded modules over a regularly graded algebra A. Every A_0-linear mapping $f_0 : M_0 \to N_0$ extends in a unique way to a graded A-linear mapping $f : M \to N$. The extension f is injective (resp. surjective) if and only if f_0 is injective (resp. surjective).*

Proof. Let $(a_1, b_1, \dots, a_m, b_m)$ be a complete system of odd elements. If the extension f exists, for every $x \in M_1$ we can write

$$f(x) = \sum_{i=1}^m a_i b_i \ f(x) = \sum_{i=1}^m a_i \ f_0(b_i x) \ ;$$

this proves the unicity of the extension f. Now let us set $f(x) = \sum_i a_i f_0(b_i x)$ for all odd $x \in M_1$; thus we get a K-linear mapping $f : M \to N$, and we must prove that it is A-linear. In other words we must prove that $cf(y) = f(cy)$ for all homogeneous $c \in A$ and all homogeneous $y \in M$. This is evident when y is even; the verification is easy when c and y are odd; the only difficulty appears when c is even and y is odd. In this case the complete system of odd elements $(a_1, b_1, \dots, a_m, b_m)$ must be used twice:

$$\begin{aligned} f(cy) &= \sum_i a_i f_0(b_i cy) = \sum_i \sum_j a_i \; f_0(b_i c a_j b_j x) \\ &= \sum_i \sum_j a_i b_i c a_j \; f_0(b_j x) = \sum_j c a_j f_0(b_j x) = cf(y). \end{aligned}$$

It is clear that f_0 is injective (resp. surjective) whenever f is injective (resp. surjective). Conversely if f_0 is injective, every equality $f(x) = 0$ with an odd x implies $f_0(b_i x) = b_i f(x) = 0$ for $i = 1, 2, \dots, m$, whence $b_i x = 0$ and finally $x = \sum_i a_i b_i x = 0$. If f_0 is surjective, for every odd y in N and for $i = 1, 2, \dots, m$, there exists $x_i \in M_0$ such that $b_i y = f_0(x_i)$, whence $y = f(\sum_i a_i x_i)$. □

Whatever the module M_0 over A_0 may be, there always exists a graded module M over A that has this M_0 as its even component; indeed the tensor product $A \otimes_{A_0} M_0$ inherits from A a structure of graded module over A, and its even component $A_0 \otimes_{A_0} M_0$ is canonically isomorphic to M_0 as a module over A_0. When the grading of A is regular, the above extension lemma says that two graded modules over A are isomorphic if and only if their even components are isomorphic as modules over A_0; consequently there is, up to isomorphy, a bijective correspondence between the modules over A_0 and the graded modules over A. This agrees with the bijective correspondence between the graded modules over A and A^t (see above), because the even subalgebras A_0 and A_0^t are isomorphic. Here is another consequence of (6.2.3).

(6.2.4) **Corollary.** *When M is a graded module over a regularly graded algebra A, the mapping*

$$A \otimes_{A_0} M_0 \longrightarrow M \;, \quad a \otimes x \longmapsto ax \;,$$

is an isomorphism of graded modules over A.

(6.2.5) **Remark.** Many graded algebras A satisfy this much stronger regularity property: A is a faithful K-module and A_1 contains a submodule X invertible inside A according to the definition given just after (5.1.12). When such a submodule X exists, the evident K-linear mapping $X \otimes A_0 \to A_1$ is an isomorphism of K-modules; consequently, when A is a finitely generated module, the grading of A is balanced. If M is a graded module over A, the evident mapping $X \otimes M_0 \to M_1$ is also an isomorphism of K-modules, and consequently, when M is finitely generated, its grading too is balanced.

(6.2.6) **Examples.**

(a) When P is a graded finitely generated and faithful projective K-module, the grading of $A = \mathrm{End}(P)$ has already been considered in (3.5.8). It is regular if and only if P_0 and P_1 are both faithful. It is balanced if and only if the grading of P is balanced. When there exists a K-linear isomorphism $f : P_0 \to P_1$, we can derive from f an odd invertible element $g \in A_1$: $g(z_0+z_1) = f^{-1}(z_1)+f(z_0)$; obviously g generates a free K-submodule X of A_1 which is invertible inside A, and the grading of A is regular in the sense of (6.2.5).

(b) Let us now consider a Clifford algebra $C\ell(M,q)$. When the ideal of K generated by the image of q is K itself, the grading of $C\ell(M,q)$ is regular, because there are elements $x_1,\dots,x_m$ of M, and elements $\lambda_1,\dots,\lambda_m$ of K such that $(x_1, \lambda_1 x_1, \dots, x_m, \lambda_m x_m)$ is a complete system of odd elements of $C\ell(M,q)$. If (M,q) is a quadratic space, and if the subset $\mathrm{GO}_1(M,q)$ of the orthogonal group is not empty (see **5.6**), then $C\ell_1(M,q)$ contains submodules invertible inside $C\ell(M,q)$, and the grading satisfies the stronger regularity property presented in (6.2.5).

(c) Of course the grading of $A = A_0 \oplus A_1$ is not regular when $A_1A_1 = 0$; this means that A_0 is any trivially graded algebra, and A_1 any bimodule over A_0. In this case, no graded A-linear mapping $\bigoplus_{j\in J} A \to A^s$ can be surjective, because its image is contained in $(A_1)^s = (A^s)_0$.

(d) If the algebra $A = \bigoplus_{n\in\mathbb{N}} A^n$ is graded over $\mathbb{N}$, as an algebra graded over $\mathbb{Z}/2\mathbb{Z}$ it is never regularly graded; indeed A_1A_1 is contained in $\bigoplus_{n\geq 2} A^n$.

It must be emphasized that there are three categories of graded modules over A. Whereas the objects are always the graded left modules over A, there are three choices of morphisms. When no precision is given, the morphisms are only the graded A-linear mappings. But we may also accept all the A-linear mappings as morphisms; in this case, the module $\mathrm{Hom}_A(M,N)$ of all A-linear morphisms from M to N is a graded K-module; its even component $\mathrm{Hom}_{A,0}(M,N)$ is the module of all graded A-linear mappings $M \to N$, which formerly were the only accepted morphisms; besides, there is an evident bijection from the odd component $\mathrm{Hom}_{A,1}(M,N)$ onto $\mathrm{Hom}_{A,0}(M,N^s)$. Nonetheless when the twisting rule (4.2.1) must be respected, we must used the graded K-module $\mathrm{Hom}^g_A(M,N)$ of *A-g-linear mappings* defined in this way: $\mathrm{Hom}^g_{A,0}(M,N)$ is the same thing as $\mathrm{Hom}_{A,0}(M,N)$, but $\mathrm{Hom}^g_{A,1}(M,N)$ is the set of all K-linear mappings $f : M \to N$ such that $f(M_i) \subset N_{1-i}$ (for $i = 0,1$) and $f(ax) = (-1)^{\partial a} a f(x)$ (for all homogeneous $a \in A$ and $x \in M$); there is an evident bijection from $\mathrm{Hom}^g_{A,1}(M,N)$ onto $\mathrm{Hom}_{A,0}(M,N^{cs})$.

The proof of the following lemma is straightforward.

(6.2.7) **Lemma.** *Let M and N be graded modules over A. With every $f \in \mathrm{Hom}_K(M,N)$ we associate $f^t \in \mathrm{Hom}(M^t,N^t)$ defined by $f^t(x^t) = f(x)^t$. This mapping $f \longmapsto f^t$ induces a bijection from $\mathrm{Hom}^g_A(M,N)$ onto $\mathrm{Hom}_{A^t}(M^t,N^t)$.*

Let us suppose that M is a graded bimodule over A and B. When the twisting rule (4.2.1) is not relevant for the objects under consideration, we treat M as a graded module over $A \otimes B^o$ and we get graded algebra morphisms $A \to \mathrm{End}_{B^o}(M)$ and $B \to \mathrm{End}_A(M)^o$. But when the twisting rule must be respected, we treat M as a graded module over $A \hat{\otimes} B^{to}$ (twisted tensor-product) and we get graded algebra morphisms $A \to \mathrm{End}^g_{B^{to}}(M)$ and $B \to \mathrm{End}^g_A(M)^{to}$.

With every graded module P over A are now associated these three covariant functors with variables in the category of graded modules over A : $\mathrm{Hom}_{A,0}(P,\dots)$, $\mathrm{Hom}^g_A(P,\dots)$ and $\mathrm{Hom}_A(P,\dots)$; besides, there is also the functor $\mathrm{Hom}_A(P,\dots)$ with variables in the category of nongraded modules over A. These four functors are all left exact and lead to four concepts of projectiveness; fortunately these four concepts coincide, and thus we can say that a module is projective over A without more precision.

(6.2.8) **Proposition.** *Let P be a graded module over a graded algebra A. The following four assertions are equivalent:*

(a) *the functor* $\mathrm{Hom}_{A,0}(P,\dots)$ *is exact;*
(b) *the functor* $\mathrm{Hom}^g_A(P,\dots)$ *is exact;*
(c) *the functor* $\mathrm{Hom}_A(P,\dots)$ *with a graded variable module is exact;*
(d) *the functor* $\mathrm{Hom}_A(P,\dots)$ *with a nongraded variable module is exact.*

When A is regularly graded, these four assertions are also equivalent to this one:

(e) *the functor* $\mathrm{Hom}_{A_0}(P_0,\dots)$ *is exact.*

Proof in two steps. First step. We prove that (a)⇒(d). In all cases P is projective if for every surjective mapping $v : M \to N$ and every mapping $g : P \to N$ there exists $f : P \to M$ such that $g = vf$. But the precise meaning of the projectiveness of P depends on the category which M, N, v, g and f must belong to. Here M and N are nongraded modules over A, and the surjective v is merely A-linear. From M we derive a graded module $M \oplus M$ over K, with even component $M \oplus 0$ and odd component $0 \oplus M$, which becomes a graded A-module if we set $a(x,y) = (ax, ay)$ whenever a is an even element of A, but $a(x,y) = (ay, ax)$ whenever a is odd. We get an A-linear mapping $s : M \oplus M \to M$ by setting $s(x,y) = x+y$. In the same way we derive from N a graded module $N \oplus N$ over A, and a mapping $t : N \oplus N \to N$. From v we derive a graded surjective A-linear mapping $v' : M \oplus M \to N \oplus N$ in this way: $v'(x,y) = (v(x), v(y))$. Obviously $vs = tv'$. From $g : P \to N$ we derive a graded A-linear mapping $g' : P \to N \oplus N$ in this way: $g'(z)$ is equal to $(g(z), 0)$ or $(0, g(z))$ according to the parity of z in P. Obviously $g = tg'$. If the assertion (a) is true, there exists $f' : P \to M \oplus M$ such that $g' = v'f'$, and it suffices to set $f = sf'$ to get the desired conclusion:

$$vf \;=\; vsf' \;=\; tv'f' \;=\; tg' \;=\; g.$$

Second step. Now everything is more or less evident. Every one of the assertions (b), (c), (d) implies (a). Indeed when graded mappings v and g are given as above, every one of these assertions ensures the existence of some f such that $g = vf$, but this f may be *not* graded; let f_0 and f_1 be the homogeneous components of f; since v and g are graded, $vf_0 = g$ and $vf_1 = 0$, and thus the case is settled. Conversely from the first step we know that (a)⇒(d), and since the implication (d)⇒(c) is trivial, we have also got (a)⇒(c). When P is projective in the sense of (a), then P^t is also projective in the sense of (a) (after replacing A with A^t); consequently P^t is projective in the sense of (c), and because of Lemma (6.2.7) this implies that A is projective in the sense of (b). Now the first four assertions have become equivalent, and the equivalence (a)⇔(e) is a direct consequence of the extension lemma (6.2.3). □

The same reasons which have just led us to contemplate three graded concepts of "projective module" besides the nongraded one, now lead us to contemplate three graded concepts of "generator" besides the nongraded one. Indeed there are three graded versions of Lemmas (6.1.1) and (6.1.2), which involve respectively the functors $\mathrm{Hom}_{A,0}(G,\dots)$, $\mathrm{Hom}^g_A(G,\dots)$ and $\mathrm{Hom}_A(G,\dots)$; in each graded version of (6.1.1), Q is merely a trivially graded K-module. Each concept of "graded generator" means that the corresponding functor is faithful. From the example (c) in (6.2.6) we already know that A is not always a generator with respect to $\mathrm{Hom}_{A,0}(A,\dots)$; this explains the presence of $G \oplus G^s$ in the next proposition.

(6.2.9) **Proposition.** *Let G be a graded module over a graded algebra A. The following four assertions are equivalent:*

(a) *the functor* $\mathrm{Hom}_{A,0}(G \oplus G^s,\dots)$ *is faithful;*
(b) *the functor* $\mathrm{Hom}^g_A(G,\dots)$ *is faithful;*
(c) *the functor* $\mathrm{Hom}_A(G,\dots)$ *with a graded variable module is faithful;*
(d) *the functor* $\mathrm{Hom}_A(G,\dots)$ *with a nongraded variable module is faithful.*

When A is regularly graded, these four assertions are also equivalent to these:

(e) *the functor* $\mathrm{Hom}_{A,0}(G,\dots)$ *is faithful;*
(f) *the functor* $\mathrm{Hom}_{A_0}(G_0,\dots)$ *is faithful.*

When these assertions are true for G, we still say that G is a generator of modules over A without more precision.

Proof. Obviously A is a generator with respect to both functors $\mathrm{Hom}_A(\dots,\dots)$ (with graded or nongraded variables); this proves (c)⇔(d), because, as stated in (6.1.2) or (6.1.6), (c) and (d) are both equivalent to the surjectiveness of $G \otimes \mathrm{Hom}_A(G, A) \to A$. Now there are evident bijections

$$\mathrm{Hom}_A(G,M) \longleftrightarrow \mathrm{Hom}_{A,0}(G \oplus G^s,\ M) \quad \text{and}$$
$$\mathrm{Hom}^g_A(G,M) \longleftrightarrow \mathrm{Hom}_{A,0}(G \oplus G^{cs},\ M)\ ;$$

since G^s and G^{cs} are isomorphic modules (by $g^s \longmapsto (-1)^{\partial g} g^{cs}$), the assertions (a), (b), (c) are equivalent.

It is clear that (e)$\Rightarrow$(a) without any additional hypothesis. Conversely, if A contains a complete system of odd elements $(a_1, b_1, \dots, a_m, b_m)$, there is a *graded* surjective A-linear mapping from $\bigoplus_{i=0}^m G$ onto $G \oplus G^s$, which maps every $(x_0, x_1, \dots, x_m)$ to $(x_0\,, \sum_{i=1}^m (a_i x_i)^s)$; it is surjective because every $(x, y^s) \in G \oplus G^s$ is the image of $(x,\, b_1 y,\, \dots,\, b_m y)$. Because of (6.1.2), G is already a generator with respect to $\mathrm{Hom}_{A,0}(\dots,\dots)$ when $G \oplus G^s$ is a generator; in other words (a)$\Rightarrow$(e). At last the equivalence of (e) and (f) is a consequence of the extension lemma (6.2.3). $\square$

It is clear that $A \oplus A^s$ is always a projective generator with respect to the functor $\mathrm{Hom}_{A,0}(\dots,\dots)$; consequently a graded module over A is projective if and only if it is isomorphic to a *graded* direct summand of some $\bigoplus_{j\in J}(A \oplus A^s)$. When A is regularly graded, direct sums $\bigoplus_{j\in J} A$ are already sufficient.

Besides, there are graded versions of (6.1.6) in which the words "projective" and "generator" refer no longer to the functor $\mathrm{Hom}_A(\dots,\dots)$, but to

$$\mathrm{Hom}_A^g(\dots,\dots), \quad \text{or even to} \quad \mathrm{Hom}_{A,0}(\dots,\dots)$$

if A is regularly graded.

6.3 Graded semi-simple modules

Semi-simplicity is well treated in the literature, and the graded versions of the usual theorems do not raise serious difficulties; therefore this section only contains what here is strictly necessary.

Let M be a graded module over a graded algebra A. A graded submodule of M is a submodule N such that $N = (N \cap M_0) \oplus (N \cap M_1)$. We say that M is a *graded irreducible module* (or a *graded simple module*) over A if it contains exactly two graded submodules, namely 0 and M (whence $M \neq 0$). Every module that is irreducible and graded, is a graded irreducible module; but conversely a graded irreducible module is not necessarily irreducible, since it may contain nontrivial submodules that are not graded; at the end of (6.2.2) there is a graded irreducible module S^t that is not irreducible.

The module M is said to be *semi-simple* if it is generated by all the graded irreducible submodules it contains. Consequently every sum of semi-simple submodules is semi-simple. The module 0 is semi-simple. The next proposition (when N is replaced by 0) proves that a graded semi-simple module is the direct sum of a number of the graded irreducible submodules it contains.

(6.3.1) **Proposition.** *If N is a graded submodule of a semi-simple module M, there is a subset J of the set $\mathcal{S}$ of all graded irreducible submodules of M such that M is the direct sum of N and all elements of J.*

Proof. We suppose that $N \neq M$ since $J = \emptyset$ if $N = M$. If M' is any graded submodule of M, and P any graded irreducible submodule of M, then $M' \cap P$ is a graded submodule of P, therefore either 0 or P; consequently if the sum $M' + P$ is not direct, then $P \subset M'$. Let E be the set of all subsets $J \subset \mathcal{S}$ such that the sum of N and all elements of J is direct. Since every totally ordered subset of E has an upper bound in E, from Zorn's Lemma we deduce the existence of a maximal element J in E. Let M' be the direct sum of N and all elements of J. If P is a graded irreducible submodule of M, the sum $M' + P$ cannot be direct since J is maximal; therefore M' contains all graded irreducible submodules of M, whence $M' = M$. □

The next proposition states a characteristic property of graded semi-simple modules which is often used as an alternative definition of graded semi-simplicity. Since this property is obviously inherited by all graded submodules, it implies (after the proof of (6.3.2)) that every graded submodule of a graded semi-simple module is still graded semi-simple.

(6.3.2) **Proposition.** *A graded module M is semi-simple if and only if every graded submodule N admits a graded supplementary submodule.*

Proof. From (6.3.1) we deduce that this condition is necessary. Conversely let us suppose that every graded submodule of M admits a graded supplementary submodule; we can still suppose $M \neq 0$. Every graded submodule M' of M inherits this property; indeed if $N \subset M'$ and $M = N \oplus P$, then $M' = N \oplus (M' \cap P)$. Consequently, if $M' \neq 0$ and if M' is not graded irreducible, it is a direct sum of two graded nonzero submodules. Let us first prove that M contains graded irreducible submodules. Indeed if x is a nonzero element of M, by Zorn's Lemma there is a maximal element N in the set of all graded submodules not containing x, and if P is a graded submodule supplementary to N in M, it is clear that $P \neq 0$. If P were not graded irreducible, it should be a direct sum $P_1 \oplus P_2$ of nonzero graded submodules, and x could not belong both to $N \oplus P_1$ and $N \oplus P_2$ since x is not in N; this would contradict the maximality of N; therefore P is graded irreducible. Now let M' be the sum of all graded irreducible submodules of M, and M'' a graded supplementary submodule; since M'' inherits the property of M, it should contain a graded irreducible submodule if it were not reduced to 0; but this would contradict the definition of M'; consequently $M'' = 0$ and $M' = M$. □

Here is the graded version of Jacobson's density theorem for situations requiring the respect of the twisting rule (4.2.1).

(6.3.3) **Theorem.** *Let M be a graded semi-simple module over a graded algebra C, and $B = \mathrm{End}_C^g(M)$. Thus $\mathrm{End}_B^g(M)$ contains the image C_M of the natural algebra morphism $C \to \mathrm{End}(M)$. The subalgebra C_M is "dense" in $\mathrm{End}_B^g(M)$ in the following sense: if $(x_1, x_2, \dots, x_m)$ is any finite sequence of elements of M_0, and $(x_{m+1}, \dots, x_{m+n})$ any finite sequence of elements of M_1, for every $f \in \mathrm{End}_B^g(M)$ there exists $c \in C$ such that $f(x_i) = cx_i$ for $i = 1, 2, \dots, m+n$.*

Proof. Let ξ be this even element of $N = M^m \oplus (M^s)^n$:

$$\xi \;=\; (x_1, x_2, \dots, x_m, x^s_{m+1}, \dots, x^s_{m+n}) \;,$$

and let g be the element of $\mathrm{End}(N)$ defined in this way, for all $y_1, \dots, y_{m+n} \in M$:

$$g(y_1, \dots, y_m, y^s_{m+1}, \dots, y^s_{m+n}) \;=\; (f(y_1), \dots, f(y_m), f(y_{m+1})^s, \dots, f(y_{m+n})^s).$$

We must prove that $g(\xi)$ belongs to the submodule $C\xi$ generated by ξ over C. Since M is graded semi-simple, N is also graded semi-simple, and there is a graded submodule over C supplementary to $C\xi$. Consequently in $\mathrm{End}_{C,0}(N)$ there is a projector p from N onto $C\xi$. If we manage to prove that g commutes with p (in fact, with every element of $\mathrm{End}_{C,0}(N)$), we can reach the conclusion within one line:

$$g(\xi) \;=\; g(p(\xi)) \;=\; p(g(\xi)) \;\in C\xi.$$

Thus it only remains to prove that $gp = pg$. For all pairs (i, j) of indices between 1 and $m + n$ there are endomorphisms $p_{i,j} \in \mathrm{End}_C(M)$ that allow us to write, for all $y_1, \dots, y_{m+n} \in M$:

$$p(y_1, \dots, y^s_{m+n}) \;=\; (z_1, \dots, z^s_{m+n}) \quad \text{with } z_j = \sum_{i=1}^{m+n} p_{i,j}(y_i).$$

Let us denote briefly by σ the grade automorphism of M (that is $y \longmapsto (-1)^{\partial y} y$) since here we need no other grade automorphism, and let $\partial(i, j)$ be the parity of $p_{i,j}$; it is even if i and j are both $\le m$, or both $> m$, it is odd if i and j are on both sides of m. If $p_{i,j}$ is even, it belongs to B_0. But if it is odd, it belongs to $\mathrm{End}_{C,1}(M)$ which is not in general the same thing as $B_1 = \mathrm{End}^g_{C,1}(M)$; yet an easy verification shows that $p_{i,j}\sigma$ is in B_1 in this case. Consequently $p_{i,j}\sigma^{\partial(i,j)} \in B$ for all (i, j). Moreover we can assume that f is homogeneous, whence $\sigma f = (-1)^{\partial f} f\sigma$. The equality $pg = gp$ is equivalent to $p_{i,j} f = f p_{i,j}$ for all (i, j); as a matter of fact,

$$\begin{aligned} p_{i,j} f &= (p_{i,j}\sigma^{\partial(i,j)})\, \sigma^{\partial(i,j)} f = (-1)^{\partial f \partial(i,j)} (p_{i,j}\sigma^{\partial(i,j)})\, f\, \sigma^{\partial(i,j)} \\ &= f\, (p_{i,j}\sigma^{\partial(i,j)})\, \sigma^{\partial(i,j)} = f p_{i,j} \;. \end{aligned}$$

□

We say that A is a *graded semi-simple algebra* if every graded module over A is semi-simple. In (6.ex.6) there is a precise description of graded semi-simple algebras.

(6.3.4) **Proposition.** *A graded algebra A is a graded semi-simple algebra if and only if every graded A-module is projective.*

Indeed every graded submodule M' of every graded A-module M admits a supplementary submodule if and only if every exact sequence $0 \to M' \to M \to M'' \to 0$ splits in $\mathcal{M}od_0(A)$, or equivalently, if and only if every graded A-module M'' is projective.

□

6.4 Graded Morita theory

Besides the classical Morita theory involving merely modules over noncommutative algebras, there are other kinds of Morita theories involving special subcategories. The here expounded theory is devoted to graded algebras (with a grading over $\mathbb{Z}/2\mathbb{Z}$) and graded modules over them, in situations that require the respect of the twisting rule (4.2.1). For every graded K-algebra A we denote by $\mathcal{M}od_0(A)$ the subcategory of $\mathcal{M}od(A)$ containing the graded modules over A and the graded A-linear mappings. Because of the twisting rule, we shall also use the A-g-linear mappings defined just before (6.2.7), which determine a category $\mathcal{M}od^g(A)$ with the same objects but with more morphisms than $\mathcal{M}od_0(A)$.

The graded right modules over A are treated as objects of the category $\mathcal{M}od_0(A^{to})$, and the notation $a^{to}x$ (with $a \in A$ and x in a right module) is synonymous with $(-1)^{\partial a \partial x}xa$. Moreover a graded bimodule over the algebras A and B is treated as an object of $\mathcal{M}od_0(A \hat{\otimes} B^{to})$; indeed the associativity equality $a(xb) = (ax)b$ is equivalent to the commutation equality $a(b^{to}x) = (-1)^{\partial a \partial b} b^{to}(ax)$. As an object of $\mathcal{M}od_0(A \hat{\otimes} A^{to})$, A^{to} is the same thing as A : $(a_1 \otimes a_2^{to})\ a^{to} = (-1)^{\partial a_2 \partial a}(a_1 a a_2)^{to}$.

(6.4.1) **Definition.** We say that $(A, B; P, Q)$ is a *graded Morita context* over K if the following five conditions are fulfilled:

(a) A and B are graded K-algebras;

(b) P is an object of $\mathcal{M}od_0(A \hat{\otimes} B^{to})$, and Q an object of $\mathcal{M}od_0(B \hat{\otimes} A^{to})$;

(c) there are two *pairing mappings* $P \otimes_B Q \to A$ and $Q \otimes_A P \to B$ which are merely treated as multiplications $p \otimes q \longmapsto pq$ and $q \otimes p \longmapsto qp$; thus the following associativity equalities hold (for all $a \in A$, $b \in B$, $p \in P$, $q \in Q$):

$$(pb)q = p(bq) \quad \text{and} \quad (qa)p = q(ap).$$

(d) the pairing mappings are graded and respectively $(A \hat{\otimes} A^{to})$-linear and $(B \hat{\otimes} B^{to})$-linear, so that the following associativity equalities also hold:

$$a(pq)a' = (ap)(qa') \quad \text{and} \quad b(qp)b' = (bq)(pb').$$

(e) at last the pairing mappings themselves respect the following associativity equalities:

$$(pq)p' = p(qp') \quad \text{and} \quad (qp)q' = q(pq').$$

In a first theorem we shall learn what happens when one pairing mapping is surjective; in a second theorem we shall learn which new facts appear when both pairing mappings are surjective; in a third theorem we shall learn how to construct a graded Morita context when one algebra (A or B) and one module (P or Q) are given. A fourth and last theorem will make precise the relations between graded Morita contexts and some equivalences of categories. In these theorems, when it is stated that some morphism is an isomorphism, the description of this morphism is followed by the mention of the category which it belongs to.

(6.4.2) **Theorem.** *Let $(A, B; P, Q)$ be a graded Morita context in which the pairing mapping $P \otimes_B Q \to A$ is surjective; then*

(a) *this pairing mapping is bijective;*

(b) *these two algebra-morphisms are isomorphisms:*

$$\begin{aligned} A^{to} &\longrightarrow \mathrm{End}^g_B(Q), & a^{to} &\longmapsto (q \longmapsto (-1)^{\partial a \partial q} qa) & &\text{in } \mathcal{M}od_0(A \hat{\otimes} A^{to}); \\ A &\longrightarrow \mathrm{End}^g_{B^{to}}(P), & a &\longmapsto (p \longmapsto ap), & &\text{in } \mathcal{M}od_0(A \hat{\otimes} A^{to}); \end{aligned}$$

(c) *these two morphisms are isomorphisms:*

$$\begin{aligned} P &\longrightarrow \mathrm{Hom}^g_B(Q, B), & p &\longmapsto (q \longmapsto (-1)^{\partial p \partial q} qp)), & &\text{in } \mathcal{M}od_0(A \hat{\otimes} B^{to}); \\ Q &\longrightarrow \mathrm{Hom}^g_{B^{to}}(P, B), & q &\longmapsto (p \longmapsto qp)), & &\text{in } \mathcal{M}od_0(B \hat{\otimes} A^{to}); \end{aligned}$$

(d) *P and Q are generators respectively in $\mathcal{M}od(A)$ and $\mathcal{M}od(A^{to})$; and they are finitely generated projective modules respectively in $\mathcal{M}od(B^{to})$ and $\mathcal{M}od(B)$;*

(e) *for every N and N' respectively in $\mathcal{M}od_0(B)$ and $\mathcal{M}od_0(B^{to})$ there are isomorphisms*

$$P \otimes_B N \longrightarrow \mathrm{Hom}^g_B(Q, N), \qquad p \otimes y \longmapsto (q \longmapsto (-1)^{\partial q(\partial p + \partial y)}(qp)y), \quad \text{in } \mathcal{M}od_0(A);$$

$$N' \otimes_B Q \longrightarrow \mathrm{Hom}^g_{B^{to}}(P, N'), \quad y' \otimes q \longmapsto (p \longmapsto y'(qp)), \quad \text{in } \mathcal{M}od_0(A^{to}).$$

Proof. The reader is assumed to be able to verify by himself that the six morphisms described in (6.4.2) are well defined and belong to the announced categories and should also observe that the twisting rule has been carefully respected. The surjectiveness of $P \otimes_B Q \to A$ implies the existence of a sequence $(p_1, p_2, \dots, p_m)$ of homogeneous elements of P, and a sequence $(q_1, q_2, \dots, q_m)$ of homogeneous elements of Q such that $\sum_{i=1}^m p_i q_i = 1_A$; we can assume $\partial p_i = \partial q_i$ for $i = 1, 2, \dots, m$. To prove (a) (the injectiveness of this pairing mapping), we suppose that $\sum_{j=1}^n p'_j q'_j = 0$ for some homogeneous $p'_1, \dots, p'_n$ in P and some homogeneous $q'_1, \dots, q'_n$ in Q; then the various associativity hypotheses allow us to write

$$\sum_j p'_j \otimes q'_j = \sum_j \sum_i p'_j \otimes (q'_j p_i) q_i = \sum_i (\sum_j p'_j q'_j) p_i \otimes q_i = 0 .$$

Now let us consider the first algebra morphism in (b); if the image in $\mathrm{End}^g_B(Q)$ of some $a \in A$ vanishes, then $q_i a = 0$ for $i = 1, 2, \dots, m$, and consequently

$a = \sum_i p_i q_i a = 0$. And if f is any homogeneous element of $\mathrm{End}_B^g(Q)$, then for all $q' \in Q$,

$$f(q') = \sum_i f(q'p_iq_i) = \sum_i (-1)^{\partial f(\partial q' + \partial p_i)} q'p_i\ f(q_i) = (-1)^{\partial q' \partial a} q'a$$

if we set $a = \sum_i (-1)^{\partial f \partial p_i} p_i f(q_i)$, whence $\partial a = \partial f$. For the first morphism in (c) the proof is exactly the same (the source A^{to} is replaced with P, and the target Q with B).

For the second morphism in (b) or in (c), the proof requires merely symmetrically presented calculations. For instance let us consider the second morphism in (c); if the image in $\mathrm{Hom}_{B^{to}}^g(P, B)$ of some $q' \in Q$ vanishes, then $q' = \sum_i q'p_iq_i = 0$ since all $q'p_i$ vanish. And for every $g \in \mathrm{Hom}_{B^{to}}^g(P, B)$ we have:

$$g(p') = \sum_i g(p_iq_ip') = \sum_i g(p_i)\ q_ip' = q'p' \quad \text{with } q' = \sum_i g(p_i)q_i\ .$$

Since the pairing mapping $P \otimes_B Q \to A$ is surjective, there is a surjective mapping $P \otimes Q \to A$ which proves that P and Q are generators as modules over A or A^{to} (see (6.1.2)). From (6.1.6) we deduce that they are finitely generated and projective as modules over B or B^{to}; indeed all the following morphisms are surjective (and even bijective, except the second one):

$$P \otimes \mathrm{Hom}_{B^{to}}^g(P, B) \longleftrightarrow P \otimes Q \longrightarrow P \otimes_A Q \longrightarrow A \longrightarrow \mathrm{End}_{B^{to}}^g(P)\ ;$$

truly these surjective mappings involve the functor $\mathrm{Hom}_{B^{to}}^g(P, \dots)$ whereas (6.1.6) would here require $\mathrm{Hom}_{B^{to}}(P, \dots)$, but because of (6.2.8) we do not have to worry about this detail.

At last we observe that all the following arrows are bijective:

$$P \otimes_B N \longrightarrow \mathrm{Hom}_B^g(Q, B) \otimes_B N \longrightarrow \mathrm{Hom}_B^g(Q,\ B \otimes_B N) \longrightarrow \mathrm{Hom}_B^g(Q, N)\ ;$$

only the bijectiveness of the second arrow needs a justification; since Q is projective in $\mathcal{M}od_0(B)$, it is isomorphic to a graded direct summand in some $\bigoplus_{j \in J}(B \oplus B^s)$, and this allows us to prove the bijectiveness of all morphisms like this one for every graded bimodule N'' over B :

$$\mathrm{Hom}_B^g(Q, N'') \otimes_B N \longrightarrow \mathrm{Hom}_B^g(Q,\ N'' \otimes_B N).$$

For the second morphism in (e), the proof is similar. □

Of course one could state a theorem parallel to (6.4.2) when the other pairing mapping is surjective. When both pairing mappings are surjective (and consequently bijective), as an immediate corollary we get the bijectiveness of all these eight morphisms:

$$\begin{array}{rcccl} \mathrm{End}^g_{B^{to}}(P) & \longleftarrow & A & \longrightarrow & (\mathrm{End}^g_B(Q))^{to} \ , \\ (\mathrm{End}^g_A(P))^{to} & \longleftarrow & B & \longrightarrow & \mathrm{End}^g_{A^{to}}(Q) \ , \\ \mathrm{Hom}^g_{A^{to}}(Q,A) & \longleftarrow & P & \longrightarrow & \mathrm{Hom}^g_B(Q,B) \ , \\ \mathrm{Hom}^g_A(P,A) & \longleftarrow & Q & \longrightarrow & \mathrm{Hom}^g_{B^{to}}(P,B). \end{array}$$

Moreover P (resp. Q) is a finitely generated projective generator both as a module over A (resp. A^{to}) and as a module over B^{to} (resp. B).

Besides, it is time to explain that the first purpose of Morita theory is the research into equivalences between two categories; here we are interested in graded K-linear equivalences between $\mathcal{M}od^g(A)$ and $\mathcal{M}od^g(B)$. An *equivalence of categories* between these categories is given by a functor $\mathcal{F}$ from the former to the latter, and a functor $\mathcal{G}$ from the latter to the former, such that $\mathcal{G}\circ\mathcal{F}$ and $\mathcal{F}\circ\mathcal{G}$ are isomorphic to the identity functors; in other words, there are functor morphisms that systematically give isomorphisms

$$M \longleftrightarrow \mathcal{G}(\mathcal{F}(M)) \text{ in } \mathcal{M}od_0(A) \quad \text{and} \quad N \longleftrightarrow \mathcal{F}(\mathcal{G}(N)) \text{ in } \mathcal{M}od_0(B) \ .$$

Here $\mathcal{F}$ and $\mathcal{G}$ will be *graded K-linear functors* in the following sense: for each couple of objects (M_1, M_2), $\mathcal{F}$ determines a graded K-linear mapping from $\mathrm{Hom}^g_A(M_1, M_2)$ into $\mathrm{Hom}^g_B(\mathcal{F}(M_1),\ \mathcal{F}(M_2))$, and the same for $\mathcal{G}$. When $\mathcal{F}$ and $\mathcal{G}$ satisfy these conditions, we say that we have a *graded K-linear equivalence of categories.*

It is worth noticing that such a graded functor $\mathcal{F}$ induces a functor $\mathcal{F}_0$ from $\mathcal{M}od_0(A)$ to $\mathcal{M}od_0(B)$; and the same for $\mathcal{G}$; consequently a graded equivalence between $\mathcal{M}od^g(A)$ and $\mathcal{M}od^g(B)$ affords an equivalence between $\mathcal{M}od_0(A)$ and $\mathcal{M}od_0(B)$. Conversely a functor $\mathcal{F}_0$ from $\mathcal{M}od_0(A)$ to $\mathcal{M}od_0(B)$, together with an isomorphism between the functors $M \longmapsto \mathcal{F}_0(M^{cs})$ and $M \longmapsto (\mathcal{F}_0(M))^{cs}$, extends to a graded functor $\mathcal{F}$ from $\mathcal{M}od^g(A)$ to $\mathcal{M}od^g(B)$; indeed $\mathrm{Hom}^g_A(M_1, M_2)$ is canonically isomorphic to $\mathrm{Hom}_{A,0}(M_1,\ M_2 \oplus M_2^{cs})$.

A graded Morita context $(A, B; P, Q)$ allows us to define four functors:

$$\begin{array}{llcll} \mathcal{F} & : \mathcal{M}od^g(A) & \longrightarrow & \mathcal{M}od^g(B) \ , & M \longmapsto Q \otimes_A M \ , \\ \mathcal{G} & : \mathcal{M}od^g(B) & \longrightarrow & \mathcal{M}od^g(A) \ , & N \longmapsto P \otimes_B N \ , \\ \mathcal{F}' & : \mathcal{M}od^g(A^{to}) & \longrightarrow & \mathcal{M}od^g(B^{to}) \ , & M' \longmapsto M' \otimes_A P \ , \\ \mathcal{G}' & : \mathcal{M}od^g(B^{to}) & \longrightarrow & \mathcal{M}od^g(A^{to}) \ , & N' \longmapsto N' \otimes_B Q. \end{array}$$

Of course the twisting rule (4.2.1) must be respected in the definition of $\mathcal{F}$ and $\mathcal{G}$; for instance for a homogeneous $f \in \mathrm{Hom}^g_A(M_1, M_2)$ we must write $\mathcal{F}(f)(q\otimes x) = (-1)^{\partial f \partial q} q \otimes f(x)$.

When both pairing mappings are bijective, $\mathcal{F}$ and $\mathcal{G}$ determine a graded K-linear equivalence between $\mathcal{M}od^g(A)$ and $\mathcal{M}od^g(B)$, and $\mathcal{F}'$ and $\mathcal{G}'$ determine a graded K-linear equivalence between $\mathcal{M}od^g(A^{to})$ and $\mathcal{M}od^g(B^{to})$. Indeed for left modules (for instance) we have the following canonical isomorphisms:

$$P \otimes_B (Q \otimes_A M) \longleftrightarrow (P \otimes_B Q) \otimes_A M \longrightarrow A \otimes_A M \longleftrightarrow M.$$

This equivalence is the first result mentioned in the next theorem. Besides, it is worth mentioning that the assertion (e) in (6.4.2) affords an alternative definition for each of the above four functors.

(6.4.3) **Theorem.** *When both pairing mappings are surjective, then*

(a) *the above defined functors $\mathcal{F}$ and $\mathcal{G}$ determine a graded K-linear equivalence between the categories $\mathcal{M}od^g(A)$ and $\mathcal{M}od^g(B)$, whereas $\mathcal{F}'$ and $\mathcal{G}'$ determine a graded K-linear equivalence between $\mathcal{M}od^g(A^{to})$ and $\mathcal{M}od^g(B^{to})$;*

(b) *the graded centers of A and B are isomorphic; more precisely, all these four canonical algebra morphisms are isomorphisms:*

$$\mathrm{Z}^g(A) \longrightarrow \mathrm{End}^g_{A\otimes B^{to}}(P) \longleftarrow \mathrm{Z}^g(B) \longrightarrow \mathrm{End}^g_{B\otimes A^{to}}(Q) \longleftarrow \mathrm{Z}^g(A)\ ;$$

(c) *when $\mathfrak{a}$ (resp. $\mathfrak{a}'$) runs through the sets of graded left (resp. right) ideals of A, and $\mathfrak{b}$ (resp. $\mathfrak{b}'$) through the sets of graded left (resp. right) ideals of B, the four mappings*

$$\mathfrak{a} \longmapsto Q\mathfrak{a} \quad (\textit{resp. } \mathfrak{a}' \longmapsto \mathfrak{a}'P), \qquad \mathfrak{b} \longmapsto P\mathfrak{b} \quad (\textit{resp. } \mathfrak{b}' \longmapsto \mathfrak{b}'Q),$$

determine bijections onto the following four targets: the set of graded B-submodules of Q (resp. the set of graded B^{to}-submodules of P), the set of graded A-submodules of P (resp. the set of graded A^{to}-submodules of Q). By restriction, each of these four mappings gives a bijection from the subset of graded two-sided ideals of A or B onto the subset of graded subbimodules of P or Q. Consequently A and B have isomorphic lattices of graded two-sided ideals, which are also isomorphic to the lattices of graded subbimodules of P and Q.

Proof. The evident assertion (a) has already been explained. To prove (b), it is sufficient to prove the bijectiveness of $\mathrm{Z}^g(B) \to \mathrm{End}^g_{A\otimes B^{to}}(P)$ (for instance). Since $\mathrm{Z}^g(B)$ is isomorphic to $\mathrm{Z}^g(B^{to})$ (by $b \longmapsto b^{to}$), we consider the isomorphism $B^{to} \to \mathrm{End}^g_A(P)$ and observe that by restriction it gives an injective morphism from $\mathrm{Z}^g(B^{to})$ into $\mathrm{End}^g_A(P)$. Let us suppose that some $f \in \mathrm{End}^g_A(P)$ is the image of $b^{to} \in B^{to}$, in other words, $f(p) = (-1)^{\partial b \partial p} pb$ for all p; then f belongs to the subalgebra $\mathrm{End}^g_{A\otimes B^{to}}(P)$ if and only if for every $b' \in B$ the mapping $p \longmapsto f(pb') - f(p)b'$ vanishes; now this mapping also belongs to $\mathrm{End}^g_A(P)$ and is the image of the element of B^{to} that can be deduced from the following calculation:

$$f(pb') - f(p)b' = (-1)^{\partial b \partial p}\ p\ ((-1)^{\partial b \partial b'} b'b - bb')\ ;$$

now it is clear that f belongs to $\mathrm{End}^g_{A\otimes B^{to}}(P)$ if and only if b belongs to $\mathrm{Z}^g(B)$.

To prove (c), it is sufficient to consider the mapping $\mathfrak{a} \longmapsto Q\mathfrak{a}$ (for instance); obviously $Q\mathfrak{a}$ is a B-submodule of Q for every K-submodule $\mathfrak{a}$ of A, and is graded if $\mathfrak{a}$ is graded. Conversely if Q' is a K-submodule of Q, let $(Q' : Q)_A$ be the subset of all $a \in A$ such that $Qa \subset Q'$; obviously $(Q' : Q)_A$ is a left ideal of A, and is graded if Q' is graded. It is also obvious that $(Q\mathfrak{a} : Q)_A \supset \mathfrak{a}$ and $Q(Q : Q')_A \subset Q'$; we shall prove that these inclusions are equalities when $\mathfrak{a}$ is a left ideal and Q' a B-submodule. Let us suppose that $Qa' \subset Q\mathfrak{a}$ for some $a' \in A$, and let us prove that $a' \in \mathfrak{a}$ if $\mathfrak{a}$ is a left ideal; since the pairing mapping $P \otimes_B Q \to A$ is surjective, we can write $1_A = \sum_i p_i q_i$ as in the proof of (6.4.2), and thus

$$a' = \sum_i p_i(q_i a') \in \sum_i p_i Q\mathfrak{a} \subset \mathfrak{a} .$$

Since the other pairing mapping $Q \otimes_A P \to B$ is also surjective, we can also write $1_B = \sum_j q_j p_j$ for some $q_j \in Q$ and some $p_j \in P$. When Q' is a B-submodule, for every $q' \in Q'$ all $p_j q'$ belong to $(Q' : Q)_A$ since qp_jq' belongs to Q' for all $q \in Q$; consequently

$$q' = \sum_j q_j(p_j q') \in Q\,(Q' : Q)_A .$$

All this settles the discussion about the bijection $\mathfrak{a} \longmapsto Q\mathfrak{a}$. It is an isomorphism of lattices because every inclusion $\mathfrak{a}_1 \subset \mathfrak{a}_2$ implies $Q\mathfrak{a}_1 \subset Q\mathfrak{a}_2$. Obviously $Q\mathfrak{a}$ is a subbimodule when $\mathfrak{a}$ is a two-sided ideal; and conversely $\mathfrak{a}$ is a two-sided ideal when $Q\mathfrak{a}$ is a subbimodule, because $\mathfrak{a} = (Q\mathfrak{a} : Q)_A$. □

Remark. The assertion (c) in (6.4.3) could have been treated as a particular case of (a). Indeed when the functor $\mathcal{F}$ is a graded equivalence of categories, it must be exact and faithful, and for every object M in $\mathcal{M}od^g(A)$ it determines a bijection between the graded A-submodules of M and the graded B-submodules of $Q \otimes_A M$. When $M = A$, this is precisely a bijection between the graded left ideals $\mathfrak{a}$ of A and the graded B-submodules of $Q \otimes_A A$. If we identify this last object with Q itself, $Q \otimes_A \mathfrak{a}$ is identified with the submodule $Q\mathfrak{a}$; since Q is projective and consequently flat over A^{to}, there is no objection to this identification. Nonetheless the bijections like $\mathfrak{a} \longmapsto Q\mathfrak{a}$ deserve a very precise treatment, revealing the converse bijection $Q' \longmapsto (Q' : Q)_A$ (described in the above proof), and precisely showing the correspondence between the graded two-sided ideals of A and B. If $\mathfrak{a}$ and $\mathfrak{b}$ are graded two-sided ideals of respectively A and B, they correspond to each other if and only if $Q\mathfrak{a} = \mathfrak{b}Q$, or equivalently $\mathfrak{a}P = P\mathfrak{b}$.

Now it will be explained how to construct a Morita context $(A, B; P, Q)$ when only one algebra and one module are given; the next theorem starts with the couple (A, P), but there are three parallel theorems starting with (B, Q) or (B, P) or (A, Q), which might be derived from (6.4.4) and this evident assertion: if $(A, B; P, Q)$ is a Morita context, the same is true for $(B, A; Q, P)$, $(B^{to}, A^{to}; P, Q)$, and $(A^{to}, B^{to}; Q, P)$.

(6.4.4) **Theorem.** *Let A be a graded algebra, P a graded module over A, and let us set*

$$B = (\mathrm{End}_A^g(P))^{to} \quad \textit{and } Q = \mathrm{Hom}_A^g(P, A) \ ;$$

thus P is a graded module over $A \hat{\otimes} B^{to}$, and Q a graded module over $B \hat{\otimes} A^{to}$. If p is a homogeneous element of P, and q a homogeneous element of Q, we define two products $pq \in A$ and $qp \in B$ in this way:

$$pq = (-1)^{\partial p \partial q} q(p) \textit{ and } (qp)^{to} \textit{ maps every } p' \in P \textit{ to } (-1)^{\partial p \partial p'} q(p') \ p \ ;$$

thus we get K-bilinear multiplications $P \times Q \to A$ and $Q \times P \to B$. It is stated that:

(a) *these multiplications induce two pairing mappings $P \otimes_B Q \to A$ and $Q \otimes_A P \to B$ which make $(A, B; P, Q)$ become a graded Morita context;*
(b) *the pairing mapping $P \otimes_B Q \to A$ is surjective if and only if P is a generator of modules over A, whereas the pairing mapping $Q \otimes_A P \to B$ is surjective if and only if P is a finitely generated projective module over A.*

Proof. This is almost the same thing as (6.1.6), since this proposition remains valid when gradings and twistings are involved as is explained at the end of **6.2**. Nonetheless there is a little discrepancy in the definition of B, since $B = \mathrm{End}_A(P)$ in (6.1.6) whereas here $B = \mathrm{End}_A^g(P)^{to}$, but the subsequent modifications raise no difficulty. The associativity properties $(pq)p' = p(qp')$ and $(qp)q' = q(pq')$ are not yet mentioned in (6.1.6), but they only require a straightforward verification. □

Of course when P is a finitely generated projective generator over A, we can add to (6.4.4) all the conclusions of (6.4.2) and (6.4.3). Here is an evident example.

(6.4.5) **Corollary.** *When P is a graded, finitely generated and faithful projective K-module, then $\mathrm{Z}^g(\mathrm{End}(P)) = K$ and the mapping $\mathfrak{a} \longmapsto \mathfrak{a}\,\mathrm{End}(P)$ is a bijection from the set of ideals of K onto the set of graded (two-sided) ideals of* End(P).

Proof. From K and P we derive a graded Morita context $(\mathrm{End}(P), K; P, P^*)$, and P satisfies the conditions that ensure the pairing mappings to be bijective (see (6.1.5)). The conclusions follow from (6.4.3). □

It remains to prove that graded Morita theory yields all the graded K-linear equivalences of categories of graded modules.

(6.4.6) **Theorem.** *Let A and B be graded algebras, and $\mathcal{F}$ and $\mathcal{G}$ two graded and K-linear functors defining an equivalence between the categories $\mathcal{M}od^g(A)$ and $\mathcal{M}od^g(B)$. If we set $Q = \mathcal{F}(A)$ and $P = \mathcal{G}(B)$, then $(A, B; P, Q)$ with suitable pairing mappings is a graded Morita context; moreover the pairing mappings are surjective and the functors $\mathcal{F}$ and $\mathcal{G}$ are respectively isomorphic to $Q \otimes_A \cdots$ and $P \otimes_B \cdots$.*

Proof. Since $\mathcal{G}\circ\mathcal{F}$ and $\mathcal{F}\circ\mathcal{G}$ are isomorphic to the identity functors, $\mathcal{F}$ determines an isomorphism from each $\mathrm{Hom}_A^g(M_1, M_2)$ onto the corresponding $\mathrm{Hom}_B^g(\mathcal{F}(M_1),\ \mathcal{F}(M_2))$, and the same for $\mathcal{G}$. Consequently $\mathcal{F}$ and $\mathcal{G}$ map injective (resp. surjective) morphisms to injective (resp. surjective) ones, they map projective objects to projective ones, and generators to generators. They also map finitely generated objects to finitely generated ones, because finitely generated objects are characterized by this property: they are not the union of a strictly ascending sequence of graded submodules. All this proves that P (resp. Q) is a finitely generated projective generator in $\mathcal{M}od(A)$ (resp. $\mathcal{M}od(B)$). Besides, if M is a graded module over $A\hat{\otimes} C^{to}$ (for some graded algebra C), then $\mathcal{F}(M)$ is a graded module over $B\hat{\otimes} C^{to}$ because of these algebra morphisms:

$$C^{to} \longrightarrow \mathrm{End}_A^g(M) \longrightarrow \mathrm{End}_B^g(\mathcal{F}(M)) \ ;$$

this shows that Q (resp. P) is a module over $B\hat{\otimes} A^{to}$ (resp. $A\hat{\otimes} B^{to}$).

For every object N of $\mathcal{M}od_0(B)$ there is an isomorphism $N\to \mathrm{Hom}_B^g(B,N)$ defined by $y \longmapsto \ (b\longmapsto (-1)^{\partial b\partial y}\ by)$. Such an isomorphism is involved at the beginning of this chain of isomorphisms in the category $\mathcal{M}od_0(B\hat{\otimes} A^{to})$:

$$Q\to \mathrm{Hom}_B^g(B,Q)\to \mathrm{Hom}_A^g(\mathcal{G}(B),\mathcal{G}(Q))\to \mathrm{Hom}_A^g(P,\mathcal{G}(\mathcal{F}(A)))\to \mathrm{Hom}_A^g(P,A).$$

Let us also consider this chain of algebra isomorphisms:

$$B^{to} \longrightarrow \mathrm{End}_B^g(B) \longrightarrow \mathrm{End}_A^g(\mathcal{G}(B)) \longrightarrow \mathrm{End}_A^g(P) \ ;$$

because of Theorem (6.4.4), the two previous chains of isomorphisms prove that $(A,B;P,Q)$ is a graded Morita context. Moreover P satisfies the conditions that ensure the surjectivity of the pairing mappings. At last for every object M in $\mathcal{M}od_0(A)$ we have the following isomorphisms in $\mathcal{M}od_0(B)$:

$$\mathcal{F}(M) \longrightarrow \mathrm{Hom}_B^g(B,\mathcal{F}(M)) \longrightarrow \mathrm{Hom}_A^g(\mathcal{G}(B),\ \mathcal{G}\circ\mathcal{F}(M)) \longrightarrow \mathrm{Hom}_A^g(P,M) \ ;$$

from the statement (e) in (6.4.2) we deduce the existence of an isomorphism $Q\otimes_A M\to \mathrm{Hom}_A^g(P,M)$ and we conclude that $\mathcal{F}$ is isomorphic to the functor $M\longmapsto Q\otimes_A M$. For $\mathcal{G}$ the proof is similar. □

6.5 Graded separable algebras

We begin at once with graded separability, since nongraded separability is automatically obtained when nongraded algebras are treated as trivially graded ones. If A is a graded algebra over K, it is said to be a *graded separable algebra* over K if it is a projective module over $A\hat{\otimes} A^{to}$ (tensor product over K when no other ring is specified).

With every graded bimodule M over A we associate the *graded centralizer of A in M*, that is the K-submodule $\mathrm{Z}^g(A,M)$ generated by the homogeneous $x\in M$ such that $(a\otimes 1_A^{to} - 1_A\otimes a^{to})\,x = 0$, or equivalently $ax = (-1)^{\partial a\partial x}xa$ for all homogeneous $a\in A$. Thus $\mathrm{Z}^g(A)$ (defined in (3.5.2)) is the same thing as

$Z^g(A,A)$. There are canonical bijections

$$\text{(6.5.1)} \qquad \mathrm{Hom}^g_{A\,\hat{\otimes}\,A^{to}}(A,M) \longleftrightarrow Z^g(A,M) \quad \text{in} \quad \mathcal{M}od(K);$$

the arrow from the left side to the right side is defined by $f \longmapsto f(1_A)$; the converse arrow is defined by $x \longmapsto (a \longmapsto xa)$. It is easy to verify that the above definition of $Z^g(A,M)$ leads to a graded functor $Z^g(A,\dots)$ from the category $\mathcal{M}od^g(A\,\hat{\otimes}\,A^{to})$ toward the category $\mathcal{M}od^g(K)$. Thus (6.5.1) means that this functor is isomorphic to $\mathrm{Hom}^g_{A\,\hat{\otimes}\,A^{to}}(A,\dots)$.

Previously the multiplication in A was represented by a linear mapping $\pi : A\otimes A \to A$; but now π shall be treated as a mapping $A\,\hat{\otimes}\,A^{to} \to A$. Now $ab = \pi(a\otimes b^{to})$, and thus π is $(A\,\hat{\otimes}\,A^{to})$-linear because, for all homogeneous a, b, x, y in A,

$$\pi((a\otimes b^{to})(x\otimes y^{to})) \;=\; \pm\,\pi(ax\otimes(yb)^{to}) \;=\; \pm\,axyb \;=\; (a\otimes b^{to})\;\pi(x\otimes y^{to}).$$

(6.5.2) **Theorem.** *The following assertions are equivalent:*

(a) *A is a graded separable algebra over K;*

(b) *the functor $Z^g(A,\dots)$ is exact;*

(c) *the functor $Z^g(A,\dots)$ transforms the surjective mapping $\pi : A\,\hat{\otimes}\,A^{to} \to A$ into a surjective mapping $Z^g(A,\ A\,\hat{\otimes}\,A^{to}) \to Z^g(A)$.*

When A is finitely generated as a K-algebra, they are also equivalent to these two assertions:

(d) *for every prime ideal $\mathfrak{p}$ of K, $A_{\mathfrak{p}}$ is graded separable over $K_{\mathfrak{p}}$;*

(e) *for every maximal ideal $\mathfrak{m}$ of K, $A_{\mathfrak{m}}$ is graded separable over $K_{\mathfrak{m}}$.*

Proof. The equivalence (a)⇔(b) and the implication (b)⇒(c) are immediate consequences of (6.5.1) and the definition of separability. Let us prove (c)⇒(a). If w is an even element of $A\,\hat{\otimes}\,A^{to}$ such that $\pi(w) = 1_A$, we set $\psi(a) = (a\otimes 1_A^{to})w$ for all $a \in A$, whence $\pi\circ\psi = \mathrm{id}_A$, which implies that A is isomorphic to a direct summand of $A\,\hat{\otimes}\,A^{to}$ in the category $\mathcal{M}od_0(A)$, since ψ is A-linear. Nonetheless if $Z^g(A,\pi)$ is surjective, we can require w to be in $Z^g(A,\ A\,\hat{\otimes}\,A^{to})$, so that $(a\otimes 1_A^{to})w = (1_A\otimes a^{to})w$. Thus ψ is $(A\,\hat{\otimes}\,A^{to})$-linear, A is isomorphic to a direct summand of $A\,\hat{\otimes}\,A^{to}$ in $\mathcal{M}od_0(A\,\hat{\otimes}\,A^{to})$, and A is projective in this category.

Let us assume that A is generated as a K-algebra by the homogeneous elements $a_1, a_2, \dots, a_r$; then $Z^g(A,M)$ is the intersection of the kernels of these r mappings:

$$M \longrightarrow M\;, \qquad x \longmapsto a_i x\;-(-1)^{\partial a_i \partial x} x a_i\;;$$

indeed for every homogeneous $x \in M$ the subset of all $a \in A$ such that $ax = (-1)^{\partial a \partial x} xa$ is a graded subalgebra of A. Since there is no problem in localizing a finite intersection of kernels, it is sure that

$$(Z^g(A,M))_{\mathfrak{p}} \;=\; Z^g(A_{\mathfrak{p}}, M_{\mathfrak{p}}) \quad \text{for every prime ideal of } K.$$

Since the separability of A depends on the surjectivity of some K-linear mapping, it is a local property, at least when A is a finitely generated algebra. □

(6.5.3) **Corollary.** *If $f : A \to B$ is a surjective morphism of graded K-algebras, and if A is a graded separable algebra, then B too is a graded separable algebra, and $\mathrm{Z}^g(B) = f(\mathrm{Z}^g(A))$.*

Proof. For every graded bimodule M over B, $\mathrm{Z}^g(B, M)$ is the same thing as $\mathrm{Z}^g(A, M)$. Therefore the graded separability of B follows from that of A. Moreover f induces a surjective morphism from $\mathrm{Z}^g(A, A) = \mathrm{Z}^g(A)$ onto $\mathrm{Z}^g(A, B) = \mathrm{Z}^g(B)$. □

(6.5.4) **Corollary.** *Let $K \to K'$ be an extension of K (as in* **1.9**). *If A is graded separable over K, then $K' \otimes A$ is graded separable over K', and the natural morphism $K' \otimes \mathrm{Z}^g(A) \to \mathrm{Z}^g(K' \otimes A)$ is an isomorphism. Conversely, if the extension $K \to K'$ is faithfully flat, and if A is finitely generated as a K-algebra, the graded separability of $K' \otimes A$ over K' implies the graded separability of A.*

Proof. If P is a graded projective module over some graded K-algebra B, then for every graded module M over B the natural morphism

$$K' \otimes \mathrm{Hom}^g_B(P, M) \longrightarrow \mathrm{Hom}^g_{K' \otimes B}(K' \otimes P,\ K' \otimes M)$$

is bijective; as in the proof of (1.9.7), it suffices to verify its bijectiveness when P is B or B^s, and then to remember that $\mathrm{Hom}^g_B(B, M)$ and $\mathrm{Hom}^g_B(B^s, M)$ are respectively isomorphic to M and M^{cs}. When B and P are replaced with $A \hat{\otimes} A^{to}$ and A (assumed to be separable), then we get this result: the natural mapping $K' \otimes \mathrm{Z}^g(A, M) \to \mathrm{Z}^g(K' \otimes A,\ K' \otimes M)$ is an isomorphism. The conclusion follows from the assertion (c) in (6.5.2) and the right exactness of the functor $K' \otimes \ldots$. The converse assertion is proved like (e)⇒(c) in (6.5.2): indeed a faithfully flat extension behaves nicely with respect to finite intersections of kernels and surjective mappings. □

(6.5.5) **Corollary.** *If A_1 and A_2 are graded separable algebras over K, then $A_1 \hat{\otimes} A_2$ is also graded separable, and the natural morphism from $\mathrm{Z}^g(A_1) \hat{\otimes} \mathrm{Z}^g(A_2)$ into $\mathrm{Z}^g(A_1 \hat{\otimes} A_2)$ is an isomorphism.*

Proof. For $i = 1, 2$, let P_i be a graded projective module over some graded K-algebra B_i, and M_i a graded module over B_i. The natural morphism

$$\mathrm{Hom}^g_{B_1}(P_1, M_1) \otimes \mathrm{Hom}^g_{B_2}(P_2, M_2) \longrightarrow \mathrm{Hom}^g_{B_1 \hat{\otimes} B_2}(P_1 \otimes P_2,\ M_1 \otimes M_2)$$

is an isomorphism; this is proved as in the proof of (1.9.7). Now replace B_i with $A_i \hat{\otimes} A_i^{to}$, and P_i with A_i for $i = 1, 2$. Observe that $B_1 \hat{\otimes} B_2$ is canonically isomorphic to $(A_1 \hat{\otimes} A_2) \hat{\otimes} (A_1 \hat{\otimes} A_2)^{to}$. All this proves the bijectiveness of the natural morphism

$$\mathrm{Z}^g(A_1, M_1) \otimes \mathrm{Z}^g(A_2, M_2) \longrightarrow \mathrm{Z}^g(A_1 \hat{\otimes} A_2,\ M_1 \otimes M_2)\ ;$$

the proof ends with the assertion (c) in (6.4.2) and the right exactness of tensor products. □

Examples

As a first example we consider the graded quadratic extensions defined in **3.4**.

(6.5.6) **Proposition.** *When A is a graded K-algebra that is finitely generated, projective of constant rank 2 as a K-module, the following assertions are equivalent:*

(a) *A is a graded quadratic extension of K;*
(b) *A is a graded separable algebra over K.*

Proof. For such an algebra, separability is a local property; since by localization A_0 and A_1 give free modules, it is sufficient to prove (6.5.6) when A_0 and A_1 are free modules; the rank of A_1 is either 0 or 1. Let $(1, z)$ be a basis of A, and let us set $z^2 = \beta z - \gamma$. We already know that A is a quadratic extension if and only if $\beta^2 - 4\gamma$ is invertible in K. Let us first suppose that $A_1 = 0$; thus we can replace $A \hat{\otimes} A^{to}$ with $A \otimes A^o$. Easy calculations show that $\mathrm{Z}(A,\ A \otimes A^o)$ is the free submodule of $A \otimes A^o$ generated by the two elements

$$z \otimes 1^o + 1 \otimes z^o - \beta \otimes 1^o \quad \text{and} \quad z \otimes z^o - \gamma \otimes 1^o \ ;$$

their images by π are $2z - \beta$ and $\beta z - 2\gamma$. Thus A is separable if and only if these two elements generate $\mathrm{Z}(A) = A$, and Lemma (1.13.5) shows that they generate A if and only if they constitute a basis of A. This property is equivalent to the invertibility of some determinant, which is precisely $\beta^2 - 4\gamma$.

When $A_1 \neq 0$, we can choose z in A_1 and set $\beta = 0$. Analogous calculations, taking into account the twisting caused by the odd element z, show that $\mathrm{Z}^g(A,\ A \hat{\otimes} A^{to})$ is now the free submodule of $A \hat{\otimes} A^{to}$ generated by

$$z \otimes 1^{to} - 1 \otimes z^{to} \quad \text{and} \quad z \otimes z^{to} - \gamma \otimes 1^{to} \ ;$$

their images by π are 0 and 2γ. Thus A is separable if and only if 2γ generates $\mathrm{Z}^g(A) = K$. This means the invertibility of 2 and γ, as desired. □

(6.5.7) **Remark.** Let K be a field, F an element of $K[t]$, and (F) the ideal generated by this polynomial F. It has been proved that the quotient $K[t]/(F)$ is separable if and only if F has no multiple root in any field extension of K, in other words, if and only if F and its derivative are coprime. When F is an irreducible polynomial, $K[t]/(F)$ is also a field, and the here presented concept of separability coincides with the usual one in Galois theory. Besides, when K is a field, it has been proved that a finite dimensional K-algebra is graded separable if and only if it is isomorphic to a direct product of simple algebras, the centers of which are separable field extensions of K; as in (6.5.6) this condition ignores the grading of A.

The algebras $\mathrm{End}(P)$ already present in **3.5** are our second example.

(6.5.8) **Proposition.** *If P is a graded finitely generated projective K-module, $\mathrm{End}(P)$ is a graded separable algebra over K.*

Proof. Let us set $A = \mathrm{End}(P)$, and let e be the idempotent of K such that $(1-e)P = 0$ whereas P is a faithful module over Ke. From (3.5.8) we know that the natural morphism $Ke \to \mathrm{Z}^g(A)$ is an isomorphism; this is also a consequence of (6.4.3)(b), since $(Ke, A^{to}; P, P^*)$ is a graded Morita context (see (6.4.4)). We can prove the separability of A by localization, and it suffices to localize at prime ideals not containing e. Thus we reduce the problem to the case of free modules P_0 and P_1 that are not both reduced to 0. Let $(b_1, b_2, \dots, b_r)$ be a basis of P made of homogeneous elements, and let $(h_1, h_2, \dots, h_r)$ be the dual basis of P^*. We get a bijection $f : P \otimes P^* \to A$ if we map every $b \otimes h$ to the endomorphism $x \longmapsto h(x)b$. Let us set

$$w = \sum_{i,j} f(x_i \otimes h_j) \otimes f(x_j \otimes h_i)^{to} \in (A \hat{\otimes} A^{to})_0 .$$

An easy calculation shows that $\pi(w) = \mathrm{id}_P$. Moreover w belongs to $\mathrm{Z}^g(A,\ A \hat{\otimes} A^{to})$ because the products $f(x_m \otimes h_n)\, w$ and $w\, f(x_m \otimes h_n)$ are both equal to $\sum_j f(x_m \otimes h_j) \otimes f(x_j \otimes h_n)$ for all $m,\ n \in \{1, 2, \dots, r\}$. This proves the surjectiveness of $\mathrm{Z}^g(A,\ A \hat{\otimes} A^{to}) \to \mathrm{Z}(A)$. □

Graded derivations of graded algebras in graded bimodules

Now we explore the relations between graded separability and graded derivations. Take notice that the "graded derivations" are not "graded morphisms" according to the definition at the beginning of **4.2**, because graded derivations may have even and odd components.

(6.5.9) **Definitions.** Let M be a graded bimodule over A. A homogeneous $d \in \mathrm{Hom}(A, M)$ is called a *graded K-derivation of A in M* (of degree 0 or 1) if

$$d(ab) = d(a)b + (-1)^{\partial a \partial d} a d(b) \quad \text{for all homogeneous } a \text{ and } b \text{ in } A.$$

The K-modules of even and odd graded derivations of A in M are respectively denoted by $\mathrm{Der}_0^g(A, M)$ and $\mathrm{Der}_1^g(A, M)$, and their direct sum by $\mathrm{Der}^g(A, M)$. With every homogeneous $x \in M$ is associated a *graded inner derivation* D_x of A in M defined by

$$D_x \ : \ a \longmapsto (-1)^{\partial x \partial a}(a \otimes 1_A^{to} - 1_A \otimes a^{to})\ x = (-1)^{\partial x \partial a} ax - xa\ .$$

The quotient of $\mathrm{Der}^g(A, M)$ by the submodule $\mathrm{In}^g(A, M)$ generated by the inner derivations is denoted by $(\mathrm{Der/In})^g(A, M)$. Moreover let $\mathrm{J}(A)$ be the kernel of the multiplication $\pi : A \hat{\otimes} A^{to} \to A$. The mapping δ defined by $a \longmapsto a \otimes 1_A^{to} - 1_A \otimes a^{to}$ is an even derivation of A in $\mathrm{J}(A)$, called the *canonical derivation* of A.

Of course the validity of these definitions depends on two verifications: the inner derivations and the canonical derivation are actually graded derivations; but these verifications are done by straightforward calculations. In a context that does not require the respect of the twisting rule (4.2.1), we would have defined a module

$\mathrm{Der}(A, M)$ with an even component equal to $\mathrm{Der}_0^g(A, M)$, but an odd component in general different from $\mathrm{Der}_1^g(A, M)$.

The definition of $\mathrm{J}(A)$ gives the exact sequence

$$(6.5.10) \qquad 0 \longrightarrow \mathrm{J}(A) \longrightarrow A \hat{\otimes} A^{to} \longrightarrow A \longrightarrow 0 \quad \text{in } \mathcal{M}od_0(A \hat{\otimes} A^{to});$$

the importance of this exact sequence lies in the fact that *the graded separability of A is equivalent to its splitting*. Indeed if A is separable, its projectiveness makes (6.5.10) split; and if it splits, A is projective as a direct summand of $A \hat{\otimes} A^{to}$.

Another immediate consequence of (6.5.9) is the following exact sequence in $\mathcal{M}od^g(K)$, which is valid for every graded bimodule M over A :

$$(6.5.11) \quad 0 \longrightarrow \mathrm{Z}^g(A, M) \longrightarrow M \longrightarrow \mathrm{Der}^g(A, M) \longrightarrow (\mathrm{Der}/\mathrm{In})^g(A, M) \longrightarrow 0 \ ;$$

the third arrow maps every $x \in M$ to the associated inner derivation D_x.

(6.5.12) **Theorem.** *The following assertions are equivalent:*

(a) *A is a graded separable algebra;*

(b) *for every graded bimodule M over A, the first three arrows of* (6.5.11) *give a* splitting *exact sequence*

$$0 \longrightarrow \mathrm{Z}^g(A, M) \longrightarrow M \longrightarrow \mathrm{Der}^g(A, M) \longrightarrow 0 \ ;$$

(c) *all derivations of A in any graded bimodule M are inner derivations;*

(d) *the canonical derivation $\delta : A \to \mathrm{J}(A)$ is an inner derivation.*

When A is actually a graded separable algebra, every bimodule M contains $\mathrm{Z}^g(A, M)$ as a direct summand; in particular, $\mathrm{Z}^g(A)$ is a direct summand of A.

The proof of (6.5.12) requires two preliminary lemmas.

(6.5.13) **Lemma.** *As a left ideal of $A \hat{\otimes} A^{to}$, $\mathrm{J}(A)$ is generated by the image of δ.*

Indeed $\mathrm{J}(A)$ is even generated by $\mathrm{Im}(\delta)$ as an A-module, because, for every element $\sum_i a_i \otimes b_i^{to}$ of $\mathrm{J}(A)$, the equality $\sum_i a_i b_i = 0$ implies

$$\sum_i a_i \otimes b_i^{to} = -\sum_i (a_i \otimes 1_A^{to}) \ (b_i \otimes 1_A^{to} - 1_A \otimes b_i^{to}). \qquad \square$$

(6.5.14) **Lemma.** *For every bimodule M over A, the mapping*

$$\mathrm{Hom}(\delta, M) \ : \ \mathrm{Hom}(\mathrm{J}(A), M) \longrightarrow \mathrm{Hom}(A, M) \ , \quad f \longmapsto f \circ \delta \ ,$$

induces by restriction an isomorphism

$$\textit{from} \quad \mathrm{Hom}^g_{A \hat{\otimes} A^{to}}(\mathrm{J}(A), M) \quad \textit{onto} \quad \mathrm{Der}^g(A, M).$$

The inner derivations correspond to the mappings $\mathrm{J}(A) \to M$ that can be extended to an $(A \hat{\otimes} A^{to})$-g-linear mapping from $A \hat{\otimes} A^{to}$ into M.

Proof. If $\delta' : A \to J'$ is a graded derivation of A in some bimodule J', and if $f' : J' \to M$ is an $(A \hat{\otimes} A^{to})$-g-linear mapping, it is easy to verify that $f' \circ \delta'$ is a graded derivation. In particular this is true when $J' = \mathrm{J}(A)$ and $\delta' = \delta$. The vanishing of $f \circ \delta$ implies the vanishing of f when f is $(A \hat{\otimes} A^{to})$-g-linear, because $\mathrm{J}(A)$ is generated by $\mathrm{Im}(\delta)$ as a left ideal of $A \hat{\otimes} A^{to}$; thus the mapping $f \longmapsto f \circ \delta$ (with an $(A \hat{\otimes} A^{to})$-g-linear f) is injective. It is also easy to prove the assertion involving inner derivations. Indeed the inner derivation D_x is equal to $f \circ \delta$ if f is the $(A \hat{\otimes} A^{to})$-g-linear mapping $A \hat{\otimes} A^{to} \to M$ such that $f(1_A \otimes 1_A^{to}) = x$. Conversely, for every $(A \hat{\otimes} A^{to})$-g-linear mapping $f : A \hat{\otimes} A^{to} \to M$, it soon appears that $f \circ \delta$ is the inner derivation determined by $f(1_A \otimes 1_A^{to})$.

Now let us prove that the above mapping $f \longmapsto f \circ \delta$ is surjective onto $\mathrm{Der}^g(A, M)$. Let $d : A \to M$ be any graded derivation; with it we associate the K-linear mapping $f : A \hat{\otimes} A^{to} \to M$ such that $f(a \otimes b^{to}) = -(-1)^{\partial a \partial d} a d(b)$ for all a and b in A. First we verify that $d = f \circ \delta$; this is done by an easy calculation, involving the equality $d(1_A) = 0$, an immediate consequence of (6.5.9). Secondly we verify that the restriction of f to $\mathrm{J}(A)$ is $(A \hat{\otimes} A^{to})$-g-linear; for this purpose we take $u = a \otimes b^{to}$ in $A \hat{\otimes} A^{to}$, and $v = \sum_i a_i \otimes b_i^{to}$ in $\mathrm{J}(A)$, and we verify that $f(uv) = (-1)^{\partial f \partial u} u f(v)$. Indeed on one side,

$$f(uv) = -\sum_i (-1)^{\partial b \partial v + (\partial a + \partial a_i)\partial d} a a_i \, d(b_i) b - \sum_i (-1)^{\partial b \partial v + (\partial a + \partial v)\partial d} a a_i b_i \, d(b) \ ;$$

the equality $\sum_i a_i b_i = 0$ implies the vanishing of the second sum; on the other side,

$$u f(v) = -\sum_i (-1)^{\partial b \partial v + (\partial b + \partial a_i)\partial d} \, a a_i \, d(b_i) b \ ;$$

since $\partial f = \partial d$ and $\partial u = \partial a + \partial b$, the verification has been successful. □

Proof of (6.5.12). The only exactness of the sequence mentioned in (b) already implies the vanishing of $(\mathrm{Der}/\mathrm{In})^g(A, M)$ by comparison with (6.5.11); consequently (b)⇒(c). Obviously (c)⇒(d). Now we prove (d)⇒(a)⇒(b) by means of the exact sequence (6.5.10), the splitting of which is equivalent to (a). When δ is an inner derivation, according to (6.5.14) there is an $(A \hat{\otimes} A^{to})$-g-linear mapping $f : A \hat{\otimes} A^{to} \to A$ such that $\delta = f \circ \delta$; since $\mathrm{J}(A)$ is generated by the image of δ as a left ideal of $A \hat{\otimes} A^{to}$, the restriction of f to $\mathrm{J}(A)$ is the identity mapping; since the identity mapping of $J(A)$ factorizes through $A \hat{\otimes} A^{to}$, $\mathrm{J}(A)$ is a direct summand of $A \hat{\otimes} A^{to}$ in the category $\mathcal{M}od_0(A \hat{\otimes} A^{to})$, and (6.5.10) splits. In other words, (d)⇒(a). Now if (6.5.10) splits, for every bimodule M it gives another splitting exact sequence:

$$0 \longrightarrow \mathrm{Hom}^g_{A \hat{\otimes} A^{to}}(A, M) \longrightarrow \mathrm{Hom}^g_{A \hat{\otimes} A^{to}}(A \hat{\otimes} A^{to}, \ M)$$
$$\longrightarrow \mathrm{Hom}^g_{A \hat{\otimes} A^{to}}(\mathrm{J}(A), M) \longrightarrow 0 \ ;$$

in this sequence, the nontrivial objects are respectively isomorphic to $\mathrm{Z}^g(A, M)$ (see (6.5.1)), to M (classical isomorphism) and to $\mathrm{Der}^g(A, M)$ (see (6.5.14)). Thus we have obtained a splitting exact sequence equivalent to the one mentioned in (b); in other words, (a)⇒(b). At last, the final statement in (6.5.12) is an immediate consequence of (b). □

Remark. The readers that are acquainted with homological algebra, would easily deduce from the exact sequence (6.5.11), and from the isomorphisms revealed by (6.5.1) and (6.5.14), that $(\mathrm{Der}/\mathrm{In})^g(A,M)$ is isomorphic to a cohomology module that might be denoted by $(\mathrm{Ext}^g)^1_{A\,\hat{\otimes}\,A^{to}}(A,M)$. For them it is consequently evident that the assertions (a) and (c) in (6.5.12) are equivalent. It is worth recalling that the more classical K-modules $\mathrm{Ext}^k_{A\otimes A^o}(A,M)$ (which do not take the gradings into account) make up the Hochschild cohomology of the algebra A with coefficients in M.

(6.5.15) **Corollary.** *The direct product of two graded separable algebras is still a graded separable algebra.*

Proof. Instead of a direct product, we can consider a graded algebra A such that $\mathrm{Z}_0(A)$ contains a nontrivial idempotent e'; we set $e'' = 1_A - e'$, and we prove that A is graded separable if (and only if) the ideals $e'A$ and $e''A$ are both graded separable. Let d be a (graded) derivation of A in a graded bimodule M. Since M is the direct sum of the graded bimodules $e'Me'$, $e'Me''$, $e''Me'$ and $e''Me''$, this derivation is an inner derivation if (and only if) its four components $e'de'$, $e'de''$, $e''de'$ and $e''de''$ are inner derivations. Since $e'de'$ vanishes on $e''A$ (indeed $e'd(e''ae'')e' = 0$ for all $a \in A$), it is a derivation of $e'A$ in $e'Me'$ extended by 0 on $e''A$. Similarly $e''de''$ is a derivation of $e''A$ in $e''Me''$ extended by 0 on $e'A$. Just below it is proved that $e'de''$ is the inner derivation associated with $d(e')e''$; similarly $e''de'$ too is an inner derivation. Consequently d is an inner derivation of A if and only if $e'de'$ is an inner derivation of $e'A$, and $e''de''$ an inner derivation of $e''A$; the conclusion follows.

It remains to verify that $e'd(a)e'' = (-1)^{\partial a \partial d} ad(e')e'' - d(e')e''a$ for all $a \in A$. The following facts allow us to verify it: first e' and e'' commute with all elements of A (but not necessarily with the elements of M); secondly the equality $e'e'' = 0$ implies $d(e')e'' = -e'd(e'')$ (and by the way $d(e'') = -d(e')$ since $d(1_A) = 0$); finally the equality $e'ae'' = 0$ implies $e'd(a)e'' = -d(e')ae'' - (-1)^{\partial a \partial d} e'ad(e'')$.

There is an alternative proof based on (6.5.2)(c), in which the conclusion follows from $\mathrm{Z}^g(A,\ A\,\hat{\otimes}\,A^{to}) = \mathrm{Z}^g(e'A,\ e'A\,\hat{\otimes}\,(e'A)^{to}) \oplus \mathrm{Z}^g(e''A,\ e''A\,\hat{\otimes}\,(e''A)^{to})$. □

(6.5.16) **Corollary.** *Let A be a graded algebra over a local ring K with maximal ideal $\mathfrak{m}$. If A is a free K-module of finite rank, these two assertions are equivalent:*

(a) *A is graded separable over the local ring K;*
(b) *$A/\mathfrak{m}A$ is graded separable over the field $K/\mathfrak{m}$.*

Proof. The notations K', A',... mean $K/\mathfrak{m}$, $A/\mathfrak{m}A$,.... Since the canonical morphism $K \to K'$ is surjective, for A' to be separable over K or K' it is the same thing; thus (6.5.3) shows that (a)⇒(b). Conversely if A' is separable, its canonical derivation $\delta' : A' \to \mathrm{J}(A')$ is an inner derivation associated with some even element $x' \in \mathrm{J}(A')$, which comes from some even element $x \in \mathrm{J}(A)$, because (6.5.13) implies the surjectiveness of $\mathrm{J}(A) \to \mathrm{J}(A')$. Consequently

$$\forall a \in A, \quad \delta(a) \equiv \delta(a)x \quad \text{modulo } \mathfrak{m}\mathrm{J}(A);$$

whence $\mathrm{J}(A) = \mathrm{J}(A)x + \mathfrak{m}\mathrm{J}(A)$, still because of (6.5.13). The projectiveness of A implies the splitting in $\mathcal{M}od(K)$ of the exact sequence (6.5.10); thus $\mathrm{J}(A)$ is a direct summand of $A \hat{\otimes} A^{to}$, and is also a finitely generated projective K-module. Now Nakayama's Lemma implies $\mathrm{J}(A) = \mathrm{J}(A)x$. The graded $(A \hat{\otimes} A^{to})$-linear mapping $\mathrm{J}(A) \to A \hat{\otimes} A^{to} \to \mathrm{J}(A)$ defined by $u \longmapsto u \longmapsto ux$ is surjective, and consequently bijective (see (1.13.5)); this proves that the identity mapping of $\mathrm{J}(A)$ factorizes through $A \hat{\otimes} A^{to}$ in $\mathcal{M}od_0(A \hat{\otimes} A^{to})$. All this results in the splitting of (6.5.10) in $\mathcal{M}od_0(A \hat{\otimes} A^{to})$, and leads to the conclusion. □

(6.5.17) **Comment.** If the algebra A is a finitely generated projective K-module, from (6.5.2) and (6.5.16) we deduce that A is separable over K if and only if the extension $(K/\mathfrak{m}) \otimes A$ is separable over the *field* $K/\mathfrak{m}$ for every maximal ideal $\mathfrak{m}$ of K.

This section ends with the so-called transitivity of separability.

(6.5.18) **Proposition.** *Let $K \to L$ be an extension of K by a trivially graded commutative ring L that is separable over K. Every graded separable algebra A over L is also graded separable over K.*

Proof. Since L is separable over K, the exact sequence $0 \to \mathrm{J}(L) \to L \otimes L \to L \to 0$ splits in $\mathcal{M}od(L \otimes L)$. Thus we get a splitting exact sequence

$$\begin{aligned} 0 \longrightarrow (A \hat{\otimes} A^{to}) \otimes_{L\otimes L} \mathrm{J}(L) &\longrightarrow (A \hat{\otimes} A^{to}) \otimes_{L\otimes L} L \otimes L \\ &\longrightarrow (A \hat{\otimes} A^{to}) \otimes_{L\otimes L} L \longrightarrow 0 \end{aligned}$$

in $\mathcal{M}od(A \hat{\otimes} A^{to})$. The third object in this sequence is canonically isomorphic to $A \hat{\otimes} A^{to}$. In the fourth object the following equality holds (for all $a, b \in A$ and all $\lambda, \mu \in L$):

$$\lambda a \otimes b^{to} \otimes \mu \;=\; a \otimes b^{to} \otimes \lambda\mu \;=\; a \otimes b^{to} \otimes \mu\lambda \;=\; a \otimes (\lambda b)^{to} \otimes \mu \,;$$

from this fact it is easy to deduce that this fourth object is isomorphic to $A \hat{\otimes}_L A^{to}$. Consequently the canonical mapping $A \hat{\otimes} A^{to} \to A \hat{\otimes}_L A^{to}$ makes $A \hat{\otimes}_L A^{to}$ become a direct summand of $A \hat{\otimes} A^{to}$ in $\mathcal{M}od(A \hat{\otimes} A^{to})$. Because of the separability of A over L, the multiplication mapping $A \hat{\otimes}_L A^{to} \to A$ makes A become a direct summand of $A \hat{\otimes}_L A^{to}$ in this category. Therefore A is a direct summand of $A \hat{\otimes} A^{to}$ in this category, whence the separability of A over K. □

6.6 Graded central simple algebras over a field

In all this section K is a field, and all spaces and algebras have finite dimensions over K. A graded algebra A of finite dimension over K is said to be a *graded simple algebra* if it contains exactly two graded (two-sided) ideals, namely 0 and A. It is a *graded central simple algebra* if moreover $\mathrm{Z}^g(A) = K$. When A is trivially graded, there are similar definitions without the word "graded".

When K does not have characteristic 2 and A is graded simple, then $Z_1^g(A) = 0$ and the requirement $\mathrm{Z}^g(A) = K$ is equivalent to $Z_0(A) = K$. Indeed the equality $xy = -yx$ is valid for each couple (x, y) of elements of $Z_1^g(A)$, whence $x^2 = -x^2$ and $x^2 = 0$; consequently the left ideal generated by x, which is even a two-sided ideal, cannot be A, therefore must be 0, whence $x = 0$.

(6.6.1) **Proposition.** *If $A \hat{\otimes} B$ is a graded central simple algebra over K, then A and B are both graded central simple over K.*

Proof. Obviously A and B have finite nonzero dimensions. If $\mathrm{Z}^g(A)$ (for instance) contained an element x that would not be in K, then $x \otimes 1$ should be an element of $\mathrm{Z}^g(A \hat{\otimes} B)$ outside K, contrary to the hypothesis in (6.6.1). And if A contained a nontrivial graded ideal $\mathfrak{a}$ (other than 0 and A), then $\mathfrak{a} \otimes B$ should be a nontrivial graded ideal of $A \hat{\otimes} B$, also contrary to the hypothesis. □

Now several examples are presented. When A is a graded algebra of finite dimension over K, we say that A is a *graded division algebra* if every homogeneous element is invertible except 0; this implies that A is graded simple. If moreover $\mathrm{Z}^g(A) = K$, we say that A is a *graded central division algebra*. When the grading of A is trivial, there are similar definitions without the word "graded".

For instance when K does not have characteristic 2, every graded quadratic extension A of K is a graded central division algebra if $A_1 \neq 0$; indeed $\mathrm{Z}^g(A) = A_0 = K$ and A_1 is generated by an invertible element. In particular this is true for $A = (K^2)^g$, that is $K \times K$ with its only nontrivial grading for which $(1, 1)$ is even and $(1, -1)$ odd; it is a graded division algebra although it contains plenty of nonhomogeneous elements $(\lambda, 0)$ and $(0, \lambda)$ that are not invertible.

Now let r be a positive integer, m and n two nonnegative integers such that $m+n = r$. The matrix algebra $\mathcal{M}(r, K)$ is naturally isomorphic to $\mathrm{End}(K^r)$. If we provide K^r with the grading for which the first m elements of its canonical basis are even, whereas the last n ones are odd, and if $\mathcal{M}(r, K)$ is provided with the corresponding grading, then it is denoted by $\mathcal{M}(m, n; K)$. Here are the dimensions of the even and odd components of $\mathcal{M}(m, n; K)$:

$$\dim(\mathcal{M}_0(m, n; K)) = m^2 + n^2 \ , \quad \dim(\mathcal{M}_1(m, n; K)) = 2mn \ ;$$

the dimensions of the even and odd components of $\mathcal{M}(m, n; K)$ determine (m, n) up to the order (in other words, only (n, m) gives them the same dimensions as

$(m,n))$ because

$$(m \pm n)^2 = \dim(\mathcal{M}_0(m,n;K)) \pm \dim(\mathcal{M}_1(m,n;K)) \ ;$$

the grading is trivial if and only if m or n vanishes; in all other cases the grading is regular (see (3.5.8)); the grading is balanced if and only if $m = n$. All these algebras $\mathcal{M}(m,n;K)$ are graded central simple: see (6.4.5). Moreover from (3.5.4) we deduce that there exists an isomorphism

$$\mathcal{M}(m,n;K) \,\hat{\otimes}\, \mathcal{M}(m',n';K) \longrightarrow \mathcal{M}(mm'+nn',\ mn'+nm';\ K).$$

When A is any K-algebra, we can also define $\mathcal{M}(r,A)$ which is canonically isomorphic to $A \otimes \mathcal{M}(r,K)$ and to $\mathrm{End}_{A^o}(A^r)$. Let us precisely describe the latter isomorphism; we treat A^r as a free *right* module of rank r over A; every element $(a_{i,j})$ of $\mathcal{M}(r,A)$ is mapped to the endomorphism $(x_1,\dots,x_r) \longmapsto (y_1,\dots,y_r)$ such that $y_i = \sum_{j=1}^r a_{i,j}x_j$ for $i = 1,2,\dots,r$. If A is graded, there is an obvious grading on $\mathcal{M}(r,A)$ for which $\mathcal{M}_0(r,A) = \mathcal{M}(r,A_0)$. With this grading, $\mathcal{M}(r,A)$ is isomorphic both to $\mathrm{End}_{A^o}(A^r)$ and $\mathrm{End}^g_{A^{to}}(A^r)$, since $a^o x$ and $(-1)^{\partial a \partial x} a^{to} x$ (with $a \in A$ and $x \in A^r$) are by definition equal to xa. Very soon (after (6.6.4)) it shall become clear that a twisted tensor product of graded central simple algebras is still graded central simple; therefore when A is graded central simple, this property is inherited by $\mathcal{M}(r,A)$.

When A is trivially graded, the definition of $\mathcal{M}(m,n;A)$ is quite evident; after (6.6.4) it will prove to be graded central simple if A is central simple. When A is not trivially graded, it is possible to define an algebra $\mathcal{M}(m,n;A)$ naturally isomorphic to $A \otimes \mathcal{M}(m,n;K)$ and to $\mathrm{End}_{A^o}(A^r)$, and also a twisted algebra $\mathcal{M}^g(m,n;A)$ naturally isomorphic to $A \,\hat{\otimes}\, \mathcal{M}(m,n;K)$ and to $\mathrm{End}^g_{A^{to}}(A^r)$. But when A_1 contains invertible elements, $\mathcal{M}(m,n;A)$ and $\mathcal{M}^g(m,n;A)$ are both isomorphic to $\mathcal{M}(m+n,\ A)$ (see (6.ex.3)), and this explains why we will never use them here.

(6.6.2) **Theorem.** *When A is a graded central simple algebra over the field K, one (and only one) of the following two assertions is true:*

either there exists a (trivially graded) central division algebra B and two integers m and n such that A is isomorphic to $\mathcal{M}(m,n;B)$;
or there exists a graded central division algebra B (such that $B_1 \neq 0$) and an integer r such that A is isomorphic to $\mathcal{M}(r,B)$.

In the former case, B is determined by A up to isomorphy, and the couple (m,n) is unique up to the order; when $mn = 0$, then $A_1 = 0$; but when $mn \neq 0$, then $A_1 \neq 0$ and $\mathrm{Z}(A_0)$ is isomorphic to K^2. In the latter case, B is determined by A up to isomorphy, and r is unique; moreover the gradings of A and B are balanced and $\mathrm{Z}(A_0)$ is a field.

Proof. If A is actually isomorphic to some $\mathcal{M}(m,n;B)$ or $\mathcal{M}(r,B)$ as stated in (6.6.2), there is a graded Morita context $(A,B;P,Q)$ in which P is $B^m \oplus B^n$ or B^r treated as a right module over B (see (6.4.4)); obviously P satisfies the conditions in $\mathcal{M}od^g(B^{to})$ that ensure the pairing mappings to be bijective; their bijectiveness requires $\mathrm{Z}^g(B) = \mathrm{Z}^g(A) = K$ (see (6.4.3)(b)). Moreover the categories $\mathcal{M}od^g(A)$ and $\mathcal{M}od^g(B)$ are equivalent. Let us describe the objects of $\mathcal{M}od^g(B)$. When $B_1 = 0$ and B is a division ring, the homogeneous components of any object of $\mathcal{M}od^g(B)$ are free modules over B; consequently it is a direct sum of graded irreducible submodules isomorphic either to B (trivially graded) or to B^s (in which all elements are odd). When $B_1 \neq 0$ and all nonzero elements of B_1 are invertible, then the grading of B is regular, and every graded module over B is determined by its even component (see (6.2.4)) which is a free module over the division ring B_0; therefore the whole module is a direct sum of graded irreducible submodules isomorphic to B. Because of the above-mentioned equivalence of categories, every object in $\mathcal{M}od^g(A)$ is a direct sum of irreducible objects, all isomorphic to P or to P^s when $B_1 = 0$, all isomorphic to P when $B_1 \neq 0$. This proves the unicity of B up to isomorphy, since B^{to} is isomorphic to $\mathrm{End}^g_A(P)$ for every graded irreducible module P over A. Then the couple (m,n) is determined up to order by the dimensions of A_0 and A_1 when $B_1 = 0$, and r is determined by the dimension of A when $B_1 \neq 0$. Moreover when $B_1 \neq 0$, then B_1 contains invertible elements, the grading of B is balanced, and the grading of A too.

When A is isomorphic to $\mathcal{M}(r,B)$ with $B_1 \neq 0$, then A_0 is isomorphic to $\mathcal{M}(r,B_0)$ and the equality $\mathrm{Z}(A_0) = \mathrm{Z}(B_0)$ follows from a suitable Morita context $(A_0, B_0; B_0^r, \dots)$; this proves that $\mathrm{Z}(A_0)$ is a field. When A is isomorphic to $\mathcal{M}(m,n;B)$ with $B_1 = 0$ and $mn \neq 0$, then A_0 is isomorphic to $\mathcal{M}(m,B) \times \mathcal{M}(n,B)$, the center of which is isomorphic to the center of $B \times B$, that is $K \times K$. When $mn = 0$, then $A_1 = 0$ and $\mathrm{Z}(A) = \mathrm{Z}^g(A) = K$. All this shows that it is easy to foresee whether B must be trivially graded or not.

Now we prove the existence of B and (m,n) or r. If every homogeneous nonzero element of A is invertible, there is nothing to prove. If some nonzero homogeneous elements are not invertible, A contains nontrivial graded left ideals; let P be a minimal graded left ideal; since A has finite dimension, such an ideal P does exist; it is generated by one homogeneous element. We set $B = (\mathrm{End}^g_A(P))^{to}$ and we prove that conversely $A = \mathrm{End}^g_{B^{to}}(P)$ by means of Morita theory. First $(A, B; P, Q)$ is a graded Morita context for a suitable Q. Secondly we prove that P is projective over A because it is a direct summand of A in $\mathcal{M}od_0(A)$. Indeed the ideal PA is graded and must be equal to A (since A is graded simple); since P is irreducible as a graded module over A, each left ideal Pa (with a homogeneous in A) is either 0 or isomorphic to P; consequently A is semi-simple in $\mathcal{M}od_0(A)$ and from (6.3.1) or (6.3.2) we deduce the existence of a graded left ideal supplementary to P. Thirdly, since P is projective over A, the pairing mapping $P \otimes Q \to A$ is not zero, and since its image is a graded ideal of A, it must be A; thus P is a generator of modules over A, and from (6.4.4) and (6.4.2) we deduce $A = \mathrm{End}^g_{B^{to}}(P)$.

Since P is a graded minimal left ideal of A, every nonzero homogeneous element b of B is invertible; indeed $\mathrm{Ker}(b^{to})$ (resp. $\mathrm{Im}(b^{to})$) is a graded left ideal contained in P, and consequently equal to 0 (resp. P).

When $B_1 = 0$, we choose a basis of P over the division ring B, made of homogeneous elements, let us say m even ones and n odd ones; this leads to an isomorphism $\mathcal{M}(m,n;B) \to \mathrm{End}_{B^o}(P)$. When $B_1 \neq 0$, P_0 is a free module over the division ring B_0 and every basis of P_0 over B_0 is a basis of P over B, because of (6.2.4); thus we get an isomorphism $\mathcal{M}(r,B) \to \mathrm{End}^g_{B^{to}}(P)$. In both cases A is isomorphic to the announced matrix algebra. □

The beginning of the proof of (6.6.2) deserves to be emphasized in a corollary.

(6.6.3) **Corollary.** *Let B be a graded central division algebra over K. When $B_1 \neq 0$, every graded module over $\mathcal{M}(r,B)$ is a direct sum of graded irreducible submodules, which are all isomorphic to B^r. When $B_1 = 0$, every graded module over $\mathcal{M}(m,n;B)$ is also a direct sum of graded irreducible submodules; there is a graded irreducible module P such that $P_0 = B^m$ and $P_1 = B^n$, and all other ones are isomorphic either to P or to P^s (the module P with shifted grading).*

It is also worth remembering that $\mathrm{End}^g_A(P)$ is isomorphic to B^{to} when P is a graded irreducible module over a graded algebra A isomorphic to $\mathcal{M}(r,B)$ or $\mathcal{M}(m,n;B)$ as in (6.6.3). See (6.ex.4) for the calculation of $\mathrm{End}^g_A(M)$ when M is any finitely generated module over A; it is isomorphic to $\mathcal{M}(j,B^{to})$ when $B_1 \neq 0$ and $M = P^j$; it is isomorphic to $\mathcal{M}(j,k;B^o)$ when $B_1 = 0$ and $M = P^j \oplus (P^s)^k$.

In (3.5.10) it has been observed that every graded Azumaya algebra over a field is a graded central simple algebra; now we shall prove the converse statement.

(6.6.4) **Theorem.** *A finite dimensional algebra over K is a graded central simple algebra if and only if it is a graded Azumaya algebra.*

Proof. Since we already know that a graded Azumaya algebra is graded central simple, it suffices to prove the surjectiveness of the mapping $A \hat{\otimes} A^{to} \to \mathrm{End}(A)$ when A is a graded central simple algebra. This is a consequence of the density theorem (6.3.3). Indeed, since A is a graded simple algebra, it is a graded irreducible module over $C = A \hat{\otimes} A^{to}$; moreover $\mathrm{End}^g_C(A) = K$ since $\mathrm{Z}^g(A) = K$ (see (6.5.1)). Consequently, if $(x_1, \dots, x_{m+n})$ is a basis of A over K in which the first m elements are even and the last n ones are odd, for every $f \in \mathrm{End}(A)$ there exists $c \in C$ such that $cx_i = f(x_i)$ for $i = 1, 2, \dots, m+n$. □

Now we can apply to graded central simple algebras all the results of **3.5** obtained for graded Azumaya algebras, except (3.5.14) (not yet proved in all cases) and (3.5.15) (for which no general proof is provided in this book). The next theorem implies that (3.5.15) is true when K is a field.

(6.6.5) **Theorem.** *If A is a graded central simple algebra over K, there exists a (finite-dimensional) field extension $K \to L$ such that one of the following assertions is true:*

- *either there exists a couple* (m,n) *such that* $L\otimes A$ *is isomorphic to* $\mathcal{M}(m,n;L)$;
- *or there exists* r *such that* $L\otimes A$ *is isomorphic to* $\mathcal{M}(r,\ (L^2)^g)$.

In the latter case K *cannot have characteristic* 2.

Proof. Let B be a graded K-algebra such that B_0 contains an element z outside K. The morphism from $K[Z]$ (algebra of polynomials) into B_0 defined by $Z\longmapsto z$ shows that the subalgebra $K[z]$ generated by z is isomorphic to the quotient of $K[Z]$ by the ideal (F) generated by some polynomial F of degree >1; it is a field if and only if F is irreducible on K. Let us assume that B_0 is a division ring. If $K\rightarrow K'$ is a field extension with K' isomorphic to the field $K[z]$, then F is no longer irreducible on K' since K' contains a root of F (the image of z by the isomorphism $K[z]\rightarrow K'$), and $K'[Z]/(F)$ is no longer a field; consequently the subalgebra generated by $1\otimes z$ in $K'\otimes B$ is not a field, and $K'\otimes B_0$ is not a division ring.

After this preliminary remark we come back to the algebra B obtained in (6.6.2); when $B_0\neq K$, there is a field extension $K\rightarrow K'$ such that $K'\otimes B_0$ is not a division ring; nevertheless, because of (6.6.4), $K'\otimes B$ is a graded Azumaya algebra over K', consequently graded central simple, and isomorphic either to some $\mathcal{M}(m_1,n_1;B')$ with $B'_1=0$ (and even $n_1=0$ if $B_1=0$), or to some $\mathcal{M}(r_1,B')$ with $B'_1\neq 0$; in both cases B' is a graded Azumaya algebra over K' and all its nonzero homogeneous elements are invertible; moreover $\dim_{K'}(B'_0)$ is strictly smaller than $\dim_K(B_0)$. As for $K'\otimes A$, in all cases it is isomorphic to some algebra $\mathcal{M}(m',n';B')$ or $\mathcal{M}(r',B')$. If $B'_0\neq K'$, we make a new suitable field extension $K'\rightarrow K''$ that will show that $K''\otimes A$ is isomorphic to some algebra $\mathcal{M}(m'',n'';B'')$ or $\mathcal{M}(r'',B'')$ such that $\dim_{K''}(B''_0)$ is still strictly smaller than $\dim_{K'}(B'_0)$. This process must end with a field extension $K\rightarrow L$ such that $L\otimes A$ is isomorphic to some $\mathcal{M}(p,q;C)$ or $\mathcal{M}(s,C)$ with $C_0=L$.

When $C_1=0$, the proof is ended. Otherwise C_1 contains invertible elements, and has the same dimension as C_0; consequently C has dimension 2 over L and is commutative. This cannot happen if K has characteristic 2, because the equalities $\mathrm{Z}^g(C)=\mathrm{Z}(C)=C$ contradict the fact that C is a graded Azumaya algebra over L. Consequently C is a graded quadratic extension of L, generated (as an algebra) by some invertible $z\in C_1$ such that $z^2\in L$. If z^2 has a square root in L, C is isomorphic to $(L^2)^g$ and the proof is ended. Otherwise we use a last field extension $L\rightarrow L'$ where L' is a quadratic extension of L, isomorphic to C without grading; at last $L'\otimes C$ is isomorphic to $(L'^2)^g$, and the proof is definitively ended. □

(6.6.6) **Remark.** Remember that the great theorem (3.5.14) has been already proved in **3.5** for every graded Azumaya algebra A for which (3.5.15) is true. Because of (6.6.5) this theorem is now proved for every graded central simple algebra over a field.

The general proof of (3.5.14) needs many results about graded separability, in particular the next one.

(6.6.7) **Proposition.** *Let A be a graded algebra of finite dimension over K, such that $Z^g(A) = K$. It is graded central simple over K if and only if it is graded separable over K. Besides, if it is actually graded central simple, the even subalgebra A_0 and the whole algebra A without grading are both separable over K.*

Proof. When A is graded central simple, then its graded separability is a consequence of (6.6.5), with the help of the following results of the previous section: (6.5.8), (6.5.6), (6.5.5), (6.5.4). Since in (6.6.5) the algebras $L \otimes A_0$ and $L \otimes A$ without grading are isomorphic either to some $\mathcal{M}(r, L)$ or to some $\mathcal{M}_0(m, n; L)$ or to some $\mathcal{M}(r, L)^2$, the separability of A_0 and A without grading is also a consequence of these results of **6.5**, and yet (6.5.15) when a direct product appears.

Conversely let us assume that A is graded separable. We first prove that A is a graded semi-simple algebra; in other words, every graded A-module M is projective (see (6.3.4)). If M and N are graded modules over A, then $\mathrm{Hom}_K(M, N)$ is a graded module over $C = A \hat{\otimes} A^{to}$, because for all $f \in \mathrm{Hom}_K(M, N)$ and all $a, b \in A$ we can define afb in this way: $(afb)(x) = a\,f(bx)$. It is easy to verify that $\mathrm{Hom}_A^g(M, N)$ is equal to $Z^g(A,\ \mathrm{Hom}_K(M, N))$. Now the functor $\mathrm{Hom}_K(M, \dots)$ is exact because K is a field; and then the functor $Z^g(A,\ \mathrm{Hom}_K(M, \dots))$ is exact because A is graded separable; consequently the functor $\mathrm{Hom}_A^g(M, \dots)$ is exact and M is projective over A.

When A is graded separable over K, then A^{to} and $C = A \hat{\otimes} A^{to}$ too are graded separable over K; therefore C too is a graded semi-simple algebra. Let us treat A as a module over C. If A contained a nontrivial graded (two-sided) ideal $\mathfrak{a}$, then A should be the direct sum of $\mathfrak{a}$ and another nontrivial graded ideal $\mathfrak{b}$, and the projections of 1 in $\mathfrak{a}$ and $\mathfrak{b}$ should be two linearly independent elements of $Z_0(A)$; but this is impossible when $Z_0(A) = K$. □

6.7 More information about graded Azumaya algebras

Let A be a graded algebra over K, that is projective and finitely generated as a K-module. In (3.5.6) it is stated that A is a graded Azumaya algebra if and only if $A/\mathfrak{m}A$ is a graded Azumaya over the field $K/\mathfrak{m}$ for every maximal ideal $\mathfrak{m}$ of K. Because of (6.6.4), this condition precisely means that $A/\mathfrak{m}A$ is a graded central simple algebra over $K/\mathfrak{m}$. This fact and the results of **6.5** about graded separability allow us to complete the proof of the great theorem (3.5.14) devoted especially to $Z(A_0, A)$, $Z(A)$ and $Z(A_0)$.

End of the proof of (3.5.14). In (6.6.6) it has been observed that (3.5.14) is now proved for all graded central simple algebras over fields. Let us come back to our previous ring K, and let A be a graded Azumaya algebra over K such that A_0 and A_1 have constant ranks. The statements in (3.5.14) about the ranks of A_0 and A_1 are already obtained by means of one maximal ideal $\mathfrak{m}$ and the corresponding extension $(K/\mathfrak{m}) \otimes A = A/\mathfrak{m}A$. When A has odd type, the invertibility of 2 in K follows from its invertibility in $K/\mathfrak{m}$ for every maximal ideal $\mathfrak{m}$.

From (6.6.7) we know that all the extensions $A/\mathfrak{m}A$ are separable with and without their gradings, and the even subalgebras $A_0/\mathfrak{m}A_0$ too; because of (6.5.17) we can assert that A is separable with and without its grading, and A_0 too. From the last assertion in (6.5.12) we deduce that $\mathrm{Z}^g(A)$, $\mathrm{Z}(A)$ *and* $\mathrm{Z}(A_0, A)$ *are direct summands of* A, *whereas* $\mathrm{Z}_0(A)$ *and* $\mathrm{Z}(A_0)$ *are direct summands of* A_0. Consequently they are projective submodules, the ranks of which appear in the extensions $A/\mathfrak{m}A$. All this agrees with what is announced in (3.5.14); in particular $\mathrm{Z}^g(A) = \mathrm{Z}_0(A) = K$ whereas $\mathrm{Z}(A_0, A)$ has constant rank 2 if $A_1 \neq 0$. Since $\mathrm{Z}(A_0)$ and $\mathrm{Z}(A)$ are direct summands of $\mathrm{Z}(A_0, A)$, their ranks show that they are respectively equal to $\mathrm{Z}(A_0, A)$ and K when A has even type and $A_1 \neq 0$, and that they are equal to K and $\mathrm{Z}(A_0, A)$ when A has odd type.

To prove that $\mathrm{Z}(A_0, A)$ is a graded quadratic extension of K, we can either put forward (2.3.4) if we argue as in **3.4**, or put forward (6.5.6) and again (6.5.17).

It remains to prove (3.5.13); it is a triviality when A has odd type, because in this case $\varphi(z) = (-1)^{\partial z} z$, and consequently (3.5.13) is the same thing as $yz = zy$, which agrees with the equality $\mathrm{Z}(A_0, A) = \mathrm{Z}(A)$. When A has even type and $A_1 \neq 0$, then $\mathrm{Z}(A_0, A) = \mathrm{Z}(A_0)$, and (3.5.13) means $yz = zy$ if y is even, and $yz = \varphi(z)y$ if y is odd. A proof is only necessary when y is odd; let us treat A_1 as a bimodule over the commutative algebra $\mathrm{Z}(A_0)$; the action of z on A_1 is the multiplication by z on one side, and the multiplication by $\varphi(z)$ on the other side; thus we must prove that $\mathrm{Z}(\mathrm{Z}(A_0), A_1)$ is equal to A_1; now it is a direct summand of A_1 (see (6.5.12)) because $\mathrm{Z}(A_0)$ is separable (see (6.5.6)); consequently this assertion about $\mathrm{Z}(\mathrm{Z}(A_0), A_1)$ is true in A because it is true in all algebras $A/\mathfrak{m}A$. □

Our first study of graded Azumaya algebras was based on Definition (3.5.1), the most convenient at that moment; but there are other possible definitions which might be chosen in the following theorem.

(6.7.1) **Theorem.** *Let A be a graded algebra over K. The following assertions are equivalent, and mean that A is a graded Azumaya algebra over K :*

(a) *A is a finitely generated and faithful projective K-module and the canonical morphism $A \hat{\otimes} A^{to} \to \mathrm{End}(A)$ is bijective;*

(b) *A is a finitely generated projective K-module, and for each maximal ideal $\mathfrak{m}$ of K, $A/\mathfrak{m}A$ is a graded central simple algebra over $K/\mathfrak{m}$;*

(c) *A is a finitely generated K-module, it is a graded separable algebra over K, and $\mathrm{Z}^g(A) = K$;*

(d) *$\mathrm{Z}^g(A) = K$ and A is a generator of modules over $A \hat{\otimes} A^{to}$;*

(e) *these two functors determine an equivalence between the categories $\mathcal{M}od^g(K)$ and $\mathcal{M}od^g(A \hat{\otimes} A^{to})$:*

$$\mathcal{M}od^g(K) \longrightarrow \mathcal{M}od^g(A \hat{\otimes} A^{to}) , \qquad M \longmapsto A \otimes_K M ,$$
$$\mathcal{M}od^g(A \hat{\otimes} A^{to}) \longrightarrow \mathcal{M}od^g(K) , \qquad N \longmapsto \mathrm{Z}^g(A, N).$$

Proof. As explained above, the equivalence (a)$\Leftrightarrow$(b) is a consequence of (3.5.6) and (6.6.4). When the assertion (b) is true, every $A/\mathfrak{m}A$ is graded separable over $K/\mathfrak{m}$ (see (6.6.7)); consequently A is graded separable over K (see (6.5.17)) and $\mathrm{Z}^g(A)$ is a direct summand of A; since $\mathrm{Z}^g(A)$ has constant rank 1, it is equal to K, as stated in (c).

To prove (c)$\Rightarrow$(d), more work is necessary. Let M be a graded maximal (two-sided) ideal of A, and $\mathfrak{m} = M \cap K$. Since $\mathrm{Z}^g(A/M) = K/\mathfrak{m}$ (see (6.5.3)), and since every homogeneous nonzero element in $\mathrm{Z}^g(A/M)$ is invertible (because it generates a graded ideal that must be equal to A/M), $K/\mathfrak{m}$ is a field, and $\mathfrak{m}$ is a maximal ideal of K. Since $A/\mathfrak{m}A$ is graded separable over $K/\mathfrak{m}$ (see again (6.5.3)), it is a graded central simple algebra over $K/\mathfrak{m}$. Consequently the kernel of the graded algebra morphism $A/\mathfrak{m}A \rightarrow A/M$ must be 0, and this proves that $M = \mathfrak{m}A$. From (6.5.5) we deduce that $A \hat{\otimes} A^{to}$ too is a graded separable algebra, the graded center of which is K. Therefore every graded maximal ideal of $A \hat{\otimes} A^{to}$ too is generated by a maximal ideal of K. Now let us consider this mapping:

$$A \otimes \mathrm{Z}^g(A,\ A \hat{\otimes} A^{to}) \longrightarrow A \hat{\otimes} A^{to}\ , \qquad a \otimes z \longmapsto az = (-1)^{\partial a \partial z} za\ ;$$

its image is a graded ideal of $A \hat{\otimes} A^{to}$. If this mapping is surjective, A is a generator in $\mathcal{M}od^g(A \hat{\otimes} A^{to})$ and (d) is proved. If it were not surjective, its image should be contained in some graded maximal ideal generated by some maximal ideal $\mathfrak{m}$ of K; this would imply $\mathrm{Z}^g(A,\ A \hat{\otimes} A^{to}) \subset \mathfrak{m}A \hat{\otimes} A^{to}$, contrary to the surjectiveness of the mapping $\mathrm{Z}^g(A,\ A \hat{\otimes} A^{to}) \rightarrow \mathrm{Z}^g(A) = K$ (see (6.5.2)).

When $\mathrm{Z}^g(A) = K$ as in the assertion (d), then $\mathrm{End}^g_{A \hat{\otimes} A^{to}}(A) = K$. With the help of (6.4.4) we get a graded Morita context $(A \hat{\otimes} A^{to},\ K;\ A, Q)$ in which $Q = \mathrm{Z}^g(A,\ A \hat{\otimes} A^{to})$ because of (6.5.1). When A is a generator in $\mathcal{M}od^g(A \hat{\otimes} A^{to})$, the pairing mapping $A \hat{\otimes}_K Q \rightarrow A \hat{\otimes} A^{to}$ is bijective. From (6.4.2)(d) we deduce that A is finitely generated and projective in $\mathcal{M}od^g(K)$, and from (6.4.2)(b) we deduce that $A \hat{\otimes} A^{to}$ is isomorphic to $\mathrm{End}_K(A)$. Therefore (d)$\Rightarrow$(a).

When (a) is true, by using (6.4.4) in another way we get a graded Morita context $(A \hat{\otimes} A^{to},\ K;\ A, A^*)$, in which $A \hat{\otimes} A^{to}$ has taken the place of $\mathrm{End}_K(A)$, and A^* means $\mathrm{Hom}_K(A, K)$. Since A is a finitely generated projective generator in $\mathcal{M}od^g(K)$, the pairing mappings are bijective. By means of (6.4.2) and (6.4.3) we can prove again that (a) implies (c) and (d); but now we want to prove (a)$\Rightarrow$(e), and we deduce from (6.4.3) an equivalence of categories between $\mathcal{M}od^g(A \hat{\otimes} A^{to})$ and $\mathcal{M}od^g(K)$. This equivalence requires two functors, one involving $A \otimes_K M$, and the other one involving $\mathrm{Z}^g(A, N)$ because of the following isomorphisms coming from (6.4.2)(e) and (6.5.1):

$$A^* \otimes_{A \hat{\otimes} A^{to}} N \longrightarrow \mathrm{Hom}_{A \hat{\otimes} A^{to}}(A, N) \longrightarrow \mathrm{Z}^g(A, N)\ .$$

Conversely when the assertion (e) is true, we can put forward (6.4.6) because the equivalence under consideration is determined by graded K-linear functors; thus we know that there is a graded Morita context $(A \hat{\otimes} A^{to},\ K;\ A, Q)$ with bijective pairing mappings; this implies that A is a faithful finitely generated projective K-module, and that $A \hat{\otimes} A^{to}$ is isomorphic to $\mathrm{End}_K(A)$ (see (6.4.2)(b)); thus (e)⇒(a) and the proof is complete. □

Here are other consequences of the previous three sections.

(6.7.2) **Proposition.** *If $A \hat{\otimes} B$ is a graded Azumaya algebra, and if A is a finitely generated projective module, then A is a graded Azumaya algebra.*

Indeed from (6.6.1) we deduce that $A/\mathfrak{m}A$ is a graded central simple algebra over $K/\mathfrak{m}$ for each maximal ideal $\mathfrak{m}$ of K. □

Remark. In (6.7.2) the assumption that A is a finitely generated projective module is not necessary. Indeed when M is a finitely generated faithful projective module, there exists N such that $M \otimes N$ is a *free* module of finite nonzero rank; applying this to $M = A \otimes B$, we get a free module $A \otimes (B \otimes N)$ of finite nonzero rank, which allows us to prove that A is a finitely generated faithful projective module: see (1.ex.14).

(6.7.3) **Proposition.** *A graded Azumaya algebra A is separable with and without its grading, and its even subalgebra A_0 too. Consequently if N is a bimodule over A, $\mathrm{Z}(A,N)$ and $\mathrm{Z}(A_0,N)$ are direct summands of N, and also $\mathrm{Z}^g(A,N)$ if N is a graded bimodule over A.*

Indeed this is a consequence of (6.6.7), (6.5.17) and (6.5.12). □

(6.7.4) **Proposition.** *If A is a graded Azumaya algebra over K, the mapping $\mathfrak{a} \longmapsto \mathfrak{a}A$ is a bijection from the set of ideals $\mathfrak{a}$ of K onto the set of graded (two-sided) ideals of A.*

This is a consequence of (6.4.3)(c), because $(A \hat{\otimes} A^{to},\ K;\ A, A^*)$ is a graded Morita context with bijective pairing mappings (as observed in the proof of (6.7.1)), and the graded submodules of A in the category $\mathcal{M}od^g(A \hat{\otimes} A^{to})$ are its graded ideals. □

(6.7.5) **Proposition.** *If M is a graded K-module, and A a graded Azumaya algebra, the mapping $x \longmapsto 1 \otimes x$ is a bijection from M onto $\mathrm{Z}^g(A,\ A \otimes_K M)$.*

Proof. From the assertion (e) in (6.7.1) we deduce that $\mathrm{Z}^g(A,\ A \otimes_K M)$ must be isomorphic to M; a precise examination of the arguments yields this sequence of isomorphisms:

$$M \longrightarrow K \otimes_K M \longrightarrow (A^* \otimes_{A \hat{\otimes} A^{to}} A) \otimes_K M \longrightarrow A^* \otimes_{A \hat{\otimes} A^{to}} (A \otimes_K M)$$
$$\longrightarrow \mathrm{Hom}^g_{A \hat{\otimes} A^{to}}(A,\ A \otimes_K M) \longrightarrow \mathrm{Z}^g(A,\ A \otimes_K M).$$

Because of (1.13.2) there exists a linear form $f \in A^*$ such that $f(1) = 1$. The image in $A^* \otimes_{A \hat{\otimes} A^{to}} A \otimes_K M$ of any element x of M is $f \otimes 1 \otimes x$; now the image in $\mathrm{Z}^g(A,\ A \otimes_K M)$ of any $g \otimes a \otimes x$ (with $g \in A^*$ and $a \in A$) is $g(1)a \otimes x$; thus the image of $f \otimes 1 \otimes x$ is $1 \otimes x$. □

(6.7.6) **Proposition.** *If N is a graded bimodule over A, the mapping $x \otimes z \longmapsto xz$ is a bijection from $A \otimes_K \mathrm{Z}^g(A,N)$ onto N.*

Proof. Because of the equivalence of categories announced in (6.7.1)(e), N is isomorphic to $A \otimes M$ with $M = \mathrm{Z}^g(A,N)$. Consequently it suffices to verify the bijectiveness of the mapping

$$A \otimes_K \mathrm{Z}^g(A,\ A \otimes_K M) \longrightarrow A \otimes_K M\ .$$

its bijectiveness is an immediate consequence of (6.7.5). □

(6.7.7) **Corollary.** *Let C be a graded algebra that contains a graded Azumaya algebra A as a graded subalgebra; thus C is a bimodule over A, and we can consider the graded subalgebra $B = \mathrm{Z}^g(A,C)$. The natural algebra morphism $A \hat{\otimes} B \to C$ is an isomorphism. If moreover C is a graded Azumaya algebra, then B too is a graded Azumaya algebra, and conversely $A = \mathrm{Z}^g(B,C)$.*

Proof. It is clear that $\mathrm{Z}^g(A,C)$ is a graded subalgebra and that we get an algebra morphism $A \hat{\otimes} B \to C$. Its bijectiveness follows from (6.7.6). Since K is a direct summand of A (see (1.13.2)), B is a direct summand of C, therefore a finitely generated projective K-module if C too is a graded Azumaya algebra; thus (6.7.2) implies that B is a graded Azumaya algebra. The evident inclusion $A \subset \mathrm{Z}^g(B,C)$ is an equality because C is also isomorphic to $B \hat{\otimes} \mathrm{Z}^g(B,C)$. □

Graded automorphisms of a graded Azumaya algebra

In (5.8.1) there is an isomorphism $\mathrm{G'Lip}(M,q) \to \mathrm{Aut}(M,q)$ when (M,q) is a quadratic space, and $C\ell(M,q)$ a graded Azumaya algebra. The fact that all automorphisms of (M,q) extend to generalized twisted inner automorphisms of $C\ell(M,q)$ is also a consequence of the next theorem. See (5.1.5) for the definition of $\mathrm{Z}^g(\theta)$; here $\mathrm{Z}^r(\theta) = \mathrm{Z}^g(\theta)$ because $\mathrm{Z}^r(A) = \mathrm{Z}^g(A) = K$ (see (5.1.8)).

(6.7.8) **Theorem.** *If θ is a graded automorphism of a graded Azumaya algebra A, then θ is a generalized twisted inner automorphism and the graded submodule $\mathrm{Z}^g(\theta)$ is invertible inside A.*

Proof. The automorphism θ allows us to give to A another structure of bimodule over A; now an element $a \otimes b^{to}$ of $A \hat{\otimes} A^{to}$ operates in A in this way: $x \longmapsto (-1)^{\partial b \partial x}\theta(a)xb$. The graded centralizer for this new structure of bimodule is precisely the submodule $\mathrm{Z}^g(\theta)$ defined in (5.1.5). We know that $\mathrm{Z}^g(\theta)$ is a direct summand of A (see (6.5.12)), and that the natural mapping $A \otimes \mathrm{Z}^g(\theta) \to A$ is bijective (see (6.7.6)); therefore $\mathrm{Z}^g(\theta)$ is a projective module of constant rank 1. The same assertions are true for $\mathrm{Z}^g(\theta^{-1})$. The bijectiveness of the natural mapping $\mathrm{Z}^g(\theta) \otimes \mathrm{Z}^g(\theta^{-1}) \to K$ (that is $x \otimes x' \longmapsto xx'$) follows from the bijectiveness of $A \otimes \mathrm{Z}^g(\theta) \otimes \mathrm{Z}^g(\theta^{-1}) \to A$. □

As an immediate consequence of (6.7.8) we get the exact sequence of multiplicative groups

$$(6.7.9) \qquad 1 \longrightarrow K^\times \longrightarrow \mathrm{Aut}^g_{in}(A) \longrightarrow \mathrm{Aut}^g(A) \longrightarrow \mathrm{Pic}(K) \ ;$$

$\mathrm{Aut}^g_{in}(A)$ is the subgroup of twisted inner automorphisms Θ_x determined by locally homogeneous invertible elements x (see **5.1**); remember that $\Theta_x = \mathrm{id}_A$ if and only if x belongs to the multiplicative group of $\mathrm{Z}^r(A)$, that is $K^\times$ in the present case. The last morphism in (6.7.9) maps every θ to the isomorphy class of $\mathrm{Z}(\theta)$ in the Picard group of K; it is the neutral class if and only if $\mathrm{Z}^r(\theta)$ is a free submodule generated by an invertible locally homogeneous element. When K is a local ring, $\mathrm{Pic}(K)$ is a trivial group and thus we find again the graded version of the Skolem–Noether theorem, stating that every graded automorphism is a twisted inner automorphism.

To find the image of the last morphism in (6.7.9) a preliminary lemma is necessary.

(6.7.10) **Lemma.** *Every graded algebra morphism $\theta : A \to B$ between graded Azumaya algebras is injective, and it is bijective if and only if A and B have the same rank at every maximal ideal.*

Proof. Since graded Azumaya algebras are faithful modules, the restriction of θ to K is injective; since $\mathrm{Ker}(\theta)$ is a graded ideal of A, from (6.7.4) we deduce that $\mathrm{Ker}(\theta) = 0$ and θ is injective. Consequently $\theta(A)$ is also a graded Azumaya algebra, and from (6.7.7) we deduce that the multiplication morphism $\theta(A) \hat{\otimes} \mathrm{Z}^g(\theta(A), B) \to B$ is bijective. Consequently θ is bijective if and only if A and B have everywhere the same rank; indeed $\mathrm{Z}^g(\theta(A), B) = K$ if they have the same rank. □

(6.7.11) **Proposition.** *When P is an invertible module over K, and A a graded Azumaya algebra, the following assertions are equivalent:*

(a) *there exists $\theta \in \mathrm{Aut}^g(A)$ such that P and $\mathrm{Z}^g(\theta)$ are isomorphic in $\mathcal{M}od_0(K)$;*
(b) *A and $A \otimes P$ are isomorphic in $\mathcal{M}od_0(A^{to})$;*
(c) *A and $A \otimes P$ are isomorphic in $\mathcal{M}od_0(A)$.*

Proof. The isomorphism of right A-modules $A \otimes \mathrm{Z}^r(\theta) \to A$ has already been observed in the proof of (6.7.8). Conversely if $f : A \otimes P \to A$ is an isomorphism of graded right modules over A, by means of f we carry onto A the structure of left A-module of $A \otimes P$. In other words we get an algebra morphism $A \to \mathrm{End}_{A^{to}}(A)$ if we map every $y \in A$ to $x \longmapsto f(yf^{-1}(x))$. Yet we already get an isomorphism $A \to \mathrm{End}_{A^{to}}(A)$ by mapping y to $x \longmapsto yx$. Consequently there is an algebra morphism $\theta : A \to A$ such that $f(yf^{-1}(x)) = \theta(y)x$ for all x and y in A. Because of (6.7.10), θ is an automorphism of A. From the equalities $f^{-1}(\theta(y)x) = yf^{-1}(x)$ and $f^{-1}(xy) = f^{-1}(x)y$, it is easy to deduce that $\mathrm{Z}^g(\theta)$ is the image of $\mathrm{Z}^g(A,\ A \otimes P)$

by f. Since $\mathrm{Z}^g(A,\ A\otimes P)$ is isomorphic to P (see (6.7.5)), we have proved that $\mathrm{Z}^g(\theta)$ is isomorphic to P, and that conversely (b)⇒(a).

Now the equivalence (a)⇔(c) follows from these two facts: the canonical mapping $A \to A^{to}$, that is $a \longmapsto a^{to}$, is an isomorphism in $\mathcal{M}od_0(A \hat{\otimes} A^{to})$, and every graded automorphism of A^{to} is equal to θ^{to} (that is $a^{to} \longmapsto \theta(a)^{to}$) for some $\theta \in \mathrm{Aut}^g(A)$. □

More information about Brauer–Wall groups

With every graded Azumaya algebra A over K is associated a class in the group $\mathrm{Br}^g(K)$; in **3.5** this group has been defined as a quotient of a monoid by a submonoid, and this definition corresponds to the property (a) in the next theorem; yet other properties may allow us to recognize whether two graded Azumaya algebras have the same class in $\mathrm{Br}^g(K)$.

(6.7.12) **Proposition.** *Let A and B be graded Azumaya algebras over K. The following assertions are equivalent (and mean that A and B have the same class in $\mathrm{Br}^g(K)$):*

(a) *there exist graded finitely generated and faithful projective K-modules P and Q such that $A \hat{\otimes} \mathrm{End}(P)$ is isomorphic to $B \hat{\otimes} \mathrm{End}(Q)$;*
(b) *there is a graded K-linear equivalence of categories between $\mathcal{M}od^g(A)$ and $\mathcal{M}od^g(B)$;*
(c) *in $\mathcal{M}od^g(B^{to})$ there exists a graded finitely generated projective generator P such that A is isomorphic to $\mathrm{End}_{B^{to}}(P)$;*
(d) *$A \hat{\otimes} B^{to}$ is isomorphic to $\mathrm{End}(P)$ for some graded finitely generated projective module P.*

Proof. To prove (a)⇒(b), we construct a graded Morita context with bijective pairing mappings in which the algebras are A and $A \hat{\otimes} \mathrm{End}(P)$, whence a graded K-linear equivalence of categories between $\mathcal{M}od^g(A)$ and $\mathcal{M}od^g(A \hat{\otimes} \mathrm{End}(P))$. Similarly there is a graded K-linear equivalence of categories between $\mathcal{M}od^g(B)$ and $\mathcal{M}od^g(B \hat{\otimes} \mathrm{End}(Q))$, and thus (b) is a consequence of (a). Because of (6.4.4) such a graded Morita context exists if there is a finitely generated projective generator P' in $\mathcal{M}od^g(A)$ such that $(A \hat{\otimes} \mathrm{End}(P))^{to}$, or equivalently $A^{to} \hat{\otimes} \mathrm{End}(P)^{to}$, is isomorphic to $\mathrm{End}^g_A(P')$. We can choose $P' = A \otimes P^*$ with $P^* = \mathrm{Hom}(P, K)$, because A^{to} is isomorphic to $\mathrm{End}_A(A)$, and $\mathrm{End}(P)^{to}$ is isomorphic to $\mathrm{End}(P^*)$; thus there is an algebra morphism $A^{to} \hat{\otimes} \mathrm{End}(P)^{to} \to \mathrm{End}^g_A(A \otimes P^*)$, and its bijectiveness follows from the projectiveness of P by the same argument as in the proof of (1.9.7).

The implication (b)⇒(c) is a consequence of (6.4.6) which affords a graded Morita context $(A, B; P, Q)$ with surjective pairing mappings, and (6.4.2) which affords an isomorphism $A \to \mathrm{End}_{B^{to}}(P)$.

Now we prove (c)⇒(d). When (c) is true, we get the following chain of isomorphisms, the end of which shall be explained at once:

$$A \hat{\otimes} B^{to} \longrightarrow B^{to} \hat{\otimes} A \longrightarrow B^{to} \hat{\otimes} \mathrm{End}^g_{B^{to}}(P) \longrightarrow \mathrm{End}^g_{B \hat{\otimes} B^{to}}(B \otimes P) \longrightarrow \mathrm{End}_K(P).$$

Here $B \otimes P$ is a module over $B \hat{\otimes} B^{to}$ in the following way:

$$(b' \otimes (b'')^{to})\ (b''' \otimes x) \;=\; (-1)^{\partial b''(\partial b''' + \partial x)}\ b'b''' \otimes xb''\ ;$$

and the third algebra morphism above is defined in this way:

$$b^{to} \otimes f \;\longmapsto\; (b''' \otimes x \;\longmapsto\; (-1)^{\partial b'''(\partial f + \partial b)}\ b'''b \otimes f(x))\ ;$$

it is an isomorphism because P is projective over B^{to} (for reasons similar to those in the proof of (1.9.7)). The fourth algebra isomorphism comes from the equivalence between the categories $\mathcal{M}od^g(B \hat{\otimes} B^{to})$ and $\mathcal{M}od^g(K)$. It remains to notice that P is faithful in $\mathcal{M}od^g(K)$ because it is a generator in $\mathcal{M}od^g(B^{to})$, that it is projective in $\mathcal{M}od^g(K)$ because it is projective in $\mathcal{M}od^g(B^{to})$ and B is projective over K, and that it is finitely generated over K because $B \otimes P$ is finitely generated in the equivalent category $\mathcal{M}od^g(B \hat{\otimes} B^{to})$.

At last we prove (d)⇒(a). When (d) is true, we get the following isomorphisms:

$$B \hat{\otimes} \mathrm{End}(P) \longrightarrow B \hat{\otimes} A \hat{\otimes} B^{to} \longrightarrow A \hat{\otimes} B \hat{\otimes} B^{to} \longrightarrow A \hat{\otimes} \mathrm{End}(B). \qquad \square$$

6.8 Involutions on graded central simple algebras

The following developments are motivated by the fact that Clifford algebras are not just graded algebras, they are also provided with a reversion τ that often plays an important role. The definition of involutions (which other authors rather call "anti-involutions") is given in (1.13.7); when τ is said to be an involution of a graded algebra A, it must be understood that τ too is graded: $\tau(A_i) = A_i$ for $i = 0, 1$.

(6.8.1) **Proposition.** *Let A and A' be graded algebras provided with involutions τ and τ'. There is a unique involution τ'' on $A \hat{\otimes} A'$ such that (for all $x \in A$ and all $x' \in A'$)*

$$\tau''(x \otimes 1) = \tau(x) \otimes 1 \quad \textit{and} \quad \tau''(1 \otimes x') = 1 \otimes \tau'(x')\ ;$$

this involution τ'' is denoted by $\tau \tilde{\otimes} \tau'$ and is defined in this way:

$$(\tau \tilde{\otimes} \tau')(x \otimes x') \;=\; (-1)^{\partial x \partial x'}\ \tau(x) \otimes \tau'(x').$$

This statement can be proved like the particular case presented in (3.2.8). Observe that $\tau \tilde{\otimes} \tau'$ is not the same thing as $\tau \hat{\otimes} \tau'$ defined in **4.2**; the definition of $\tau \,\tilde{\otimes}\, \tau'$ cannot be explained by the twisting rule (4.2.1); as a matter of fact, this rule is already violated by the equality $\tau(xy) = \tau(y)\tau(x)$ where the reversion of the letters x and y raises no twisting sign.

The involutions of graded central simple algebras over a field K are classified by a cyclic group of order 8 when K does not have characteristic 2. But when K has characteristic 2, this group shrinks and becomes a group of order 2. In the following definition and lemmas we forget the algebra structure of A, and treat it momentarily as a space of finite dimension over K.

(6.8.2) **Definition.** Let τ be an involutive endomorphism of a finite-dimensional space A. The *dimensional trace* of τ is

$$\begin{aligned}\mathrm{dtr}(\tau) &= \dim(\mathrm{Ker}(\tau - \mathrm{id})) - \dim(\mathrm{Im}(\tau - \mathrm{id})) \\ &= 2\dim(\mathrm{Ker}(\tau - \mathrm{id})) - \dim(A) = \dim(A) - 2\dim(\mathrm{Im}(\tau - \mathrm{id})).\end{aligned}$$

The image of $\mathrm{dtr}(\tau)$ by the canonical morphism $\mathbb{Z} \to K$ is actually the trace of τ. Indeed when K does not have characteristic 2, then A is the direct sum of $\mathrm{Ker}(\tau - \mathrm{id})$ and $\mathrm{Ker}(\tau + \mathrm{id}) = \mathrm{Im}(\tau - \mathrm{id})$; and when K has characteristic 2, all eigenvalues of τ are equal to 1 since $(\tau - \mathrm{id})^2 = 0$. Moreover when K does not have characteristic 2, then $\mathrm{dtr}(-\tau) = -\mathrm{dtr}(\tau)$ and $\mathrm{dtr}(\tau)$ may be any element of $\mathbb{Z}$. But when K has characteristic 2, then $\mathrm{dtr}(-\tau) = \mathrm{dtr}(\tau)$ and $\mathrm{dtr}(\tau)$ is always nonnegative.

(6.8.3) **Lemma.** *Let A and A' be spaces of finite dimensions over K. If τ and τ' are involutive endomorphisms of respectively A and A', then $\mathrm{dtr}(\tau \otimes \tau')$ is the product of $\mathrm{dtr}(\tau)$ and $\mathrm{dtr}(\tau')$.*

Proof. When K does not have characteristic 2, this can be verified by easy calculations of dimensions, because A is the direct sum of $\mathrm{Ker}(\tau - \mathrm{id})$ and $\mathrm{Ker}(\tau + \mathrm{id})$, and the same for (A', τ'). When K has characteristic 2, and m and n are the dimensions of $\mathrm{Ker}(\tau - \mathrm{id})$ and $\mathrm{Ker}(\tau' - \mathrm{id})$, there is a basis $(b_1, b_2, \dots)$ of A such that $\tau(b_j) = b_j$ if $j \le m$, and $\tau(b_j) = b_j + b_{j-m}$ if $j > m$, and similarly a basis $(b'_1, b'_2, \dots)$ of M' such that $\tau'(b'_k) = b'_k$ if $k \le n$, and $\tau'(b'_k) = b'_k + b'_{k-n}$ if $k > n$. Thus we realize that $\mathrm{Ker}(\tau \otimes \tau' - \mathrm{id})$ is the subspace spanned by all $b_j \otimes b'_k$ with $j \le m$ and $k \le n$, and all $b_j \otimes b'_{k-n} - b_{j-m} \otimes b'_k$ with $j > m$ and $k > n$. Therefore the dimension of $\mathrm{Ker}(\tau \otimes \tau' - \mathrm{id})$ is again $mn + (\dim(A) - m)(\dim(A') - n)$. □

Now we introduce gradings, and tackle the difficulty raised by the unusual twisting in $\tau \,\tilde{\otimes}\, \tau'$. This difficulty requires embedding of $\mathbb{Z}$ into the field $\mathbb{C}$ of complex numbers; here i means $\sqrt{-1}$.

(6.8.4) **Definition.** Let τ be a graded involutive endomorphism of a graded space A of finite dimension over K; let τ_0 and τ_1 be its restrictions to A_0 and A_1. When K does not have characteristic 2, the *complex divided trace* of τ is the complex number

$$\mathrm{cp.dv.tr}(\tau) \;=\; \frac{\mathrm{dtr}(\tau_0) \;+\; i\;\mathrm{dtr}(\tau_1)}{\sqrt{\dim(A)}}.$$

When K has characteristic 2, its *divided trace* and *twisted divided trace* are the real numbers

$$\mathrm{dv.tr}(\tau) = \frac{\mathrm{dtr}(\tau)}{\sqrt{\dim(A)}} \quad \text{and} \quad \mathrm{tw.dv.tr}(\tau) = \frac{\mathrm{dtr}(\tau_0) - \mathrm{dtr}(\tau_1)}{\sqrt{\dim(A)}}.$$

When K does not have characteristic 2, the grading of A is determined by the involutive endomorphism σ defined by $\sigma(x) = (-1)^{\partial x}x$, and since τ commutes with σ, we actually obtain a pair $(\tau,\ \sigma\tau)$ of graded involutive endomorphisms; obviously the complex divided traces of τ and $\sigma\tau$ are conjugate complex numbers. Such pairs do not appear when K has characteristic 2 since $\sigma = \mathrm{id}_A$.

In the following two examples we can observe that the division by the square root of $\dim(A)$ gives complex numbers of module 1 (precisely, eighth roots of 1 in $\mathbb{C}$) and this will prove to be true for all the complex or twisted divided traces that shall be calculated later.

(6.8.5) **Example.** Let S be a graded space of finite dimension over K, and $\mathcal{B}''$ the space of all bilinear forms $\beta : S \times S \to K$. It is graded in this way: $\mathcal{B}''_i$ (for $i = 0, 1$) in the subspace of all β that vanish on $S_j \times S_k$ if $i \neq j + k$ (in $\mathbb{Z}/2\mathbb{Z}$). Let ρ'' be the graded involutive endomorphism of $\mathcal{B}''$ that maps every β to the symmetrically opposite bilinear form β^o defined by $\beta^o(s,t) = \beta(t,s)$ for all s and $t \in S$. *When K does not have characteristic 2, the complex divided trace of ρ'' is always equal to 1, and when K has characteristic 2, then* $\mathrm{dv.tr}(\rho'') = \mathrm{tw.dv.tr}(\rho'') = 1$.

Indeed let m and n be the dimensions of S_0 and S_1, and for every $(j,k) \in (\mathbb{Z}/2\mathbb{Z})^2$ let $\mathcal{B}''_{j,k}$ be the subspace of all β vanishing on $S_{j'} \times S_{k'}$ if $(j',k') \neq (j,k)$. Since the odd component ρ''_1 induces bijections between $\mathcal{B}''_{0,1}$ and $\mathcal{B}''_{1,0}$, we realize that $\mathrm{dtr}(\rho''_1) = 0$. The even component ρ''_0 leaves $\mathcal{B}''_{0,0}$ and $\mathcal{B}''_{1,1}$ invariant, whence

$$2\ \dim(\mathrm{Ker}(\rho''_0 - \mathrm{id})) \ = \ m(m+1) + n(n+1)\ .$$

Since $\dim(\mathcal{B}''_0) = m^2 + n^2$, we realize that $\mathrm{dtr}(\rho''_0) = m + n$. And since $\dim(\mathcal{B}'') = (m+n)^2$, all the announced conclusions follow. □

(6.8.6) **Example.** Now K does not have characteristic 2, and A is a graded quadratic extension of A with a nontrivial grading. The only automorphisms of A are id_A and the standard involution φ, and both are (graded) involutions of A; obviously

$$\mathrm{cp.dv.tr}(\mathrm{id}) = \frac{1+i}{\sqrt{2}} \quad \text{and} \quad \mathrm{cp.dv.tr}(\varphi) = \frac{1-i}{\sqrt{2}}.$$

The very justification of the definition (6.8.4) lies in the next theorem, which is the first step toward the classification of involutions of graded central simple algebras. This classification can be achieved immediately after (6.8.7) by means of (6.6.5) and (6.7.8) (as is explained in (6.ex.17)), or by means of a graded module S over A as in (6.8.14) and (6.8.15) below.

(6.8.7) **Theorem.** *Let the pair (A,τ) be as above in (6.8.4), and let (A',τ') be a similar pair. When K does not have characteristic 2, the complex divided trace of $\tau\,\tilde{\otimes}\,\tau'$ is the product in $\mathbb{C}$ of the complex divided traces of τ and τ'. When K has characteristic 2, the same property is true for their divided traces and twisted divided traces.*

Proof. Let t_0, t_1, t'_0, t'_1 be the dimensional traces of τ_0, τ_1, τ'_0, τ'_1, and let t''_0 and t''_1 be the dimensional traces of the homogeneous components of $\tau\,\tilde{\otimes}\,\tau'$. We calculate t''_0 and t''_1 with the help of (6.8.3). When K does not have characteristic 2, there is actually a twisting in $\tau\,\tilde{\otimes}\,\tau'$:

$$t''_0 = t_0t'_0 - t_1t'_1\,, \quad t''_1 = t_0t'_1 + t_1t'_0\,, \quad \text{whence} \quad t''_0 + it''_1 = (t_0+it_1)(t'_0+it'_1)\,.$$

When K has characteristic 2, the twisting in $\tau\,\tilde{\otimes}\,\tau'$ disappears:

$$t''_0 = t_0t'_0 + t_1t'_1\,, \quad t''_1 = t_0t'_1 + t_1t'_0\,, \quad \text{whence} \quad t''_0 \pm t''_1 = (t_0 \pm t_1)(t'_0 \pm t'_1)\,.$$

In both cases we have discovered a multiplicative property, and it is respected by the divisions by the square roots of the dimensions, since $\dim(A\otimes A')$ is the product of $\dim(A)$ and $\dim(A')$. □

(6.8.8) **Example.** Let (M,q) be a quadratic space of dimension r over the field K, and τ the reversion in $C\ell(M,q)$. When K does not have characteristic 2, then

$$\text{cp.dv.tr}(\tau) = \kappa^r \quad \text{and} \quad \text{cp.dv.tr}(\sigma\tau) = \kappa^{-r} \quad \text{if } \kappa = (1+i)/\sqrt{2}.$$

Indeed, if we identify the Clifford algebra of the orthogonal sum of (M,q) and some other quadratic space (M',q') with $C\ell(M,q)\,\hat{\otimes}\,C\ell(M',q')$, then its reversion is identified with $\tau\,\tilde{\otimes}\,\tau'$ (in accordance with (3.2.8)); therefore the multiplicative property stated in (6.8.7) allows us to deduce the general formulas from the particular case $r=1$. In this case, τ is the identity mapping, $\sigma\tau$ the standard involution, and the result can be read in (6.8.6). □

When K has characteristic 2, then r is even,

$$\text{dv.tr}(\tau) = 1 \quad \text{and} \quad \text{tw.dv.tr}(\tau) = (-1)^{r/2}.$$

Indeed it suffices to verify these general formulas in the case $r=2$, when τ_0 is the standard involution on $C\ell_0(M,q)$ (whence $\text{dtr}(\tau_0)=0$), whereas τ_1 is the identity mapping of M (whence $\text{dtr}(\tau_1)=2$). □

Scalar products on graded modules

Let A be a graded central simple algebra provided with a (graded) involution τ, and S a graded module over A that has finite nonzero dimension over K. Since A is graded simple, S is a faithful module over A, and A is isomorphic to its image in $\text{End}(S)$ (see (6.7.10)). If we set

$$A'' = \text{End}(S) \quad \text{and} \quad A' = \text{End}^g_A(S),$$

from (6.7.7) we deduce that the natural algebra morphism $A \hat{\otimes} A' \to A''$ is bijective, that A' too is a graded central simple algebra, and that $A \cong \mathrm{End}^g_{A'}(S)$. Now let us define $\mathcal{B}''$ and ρ'' as in the example (6.8.5). It is clear that $\mathcal{B}''$ is a right module over A'' in this way:

$$\forall f \in A'',\ \forall \beta \in \mathcal{B}'',\ \forall s,\ t \in S, \qquad (\beta f)(s,t) = \beta(f(s),\ t).$$

If β_n is a nondegenerate homogeneous element of $\mathcal{B}''$, for every $f \in A''$ there exists a unique $\tau''(f) \in A''$ such that, for all $s,\ t \in S$,

$$\beta_n(f(s),\ t) \ = \ (-1)^{\partial f \partial \beta_n}\ \beta_n(s,\ \tau''(f)(t))\ ; \tag{6.8.9}$$

it is easy to prove that τ'' is a graded anti-automorphism of A'', and even an involution of A'' if β_n is symmetric or skew symmetric.

Now we turn $\mathcal{B}''$ into a graded bimodule over A. For all $a,\ b \in A$, for all s, $t \in S$ and for all $\beta \in \mathcal{B}''$,

$$(a\beta b)(s,t) \ = \ \beta(bs,\ \tau(a)t). \tag{6.8.10}$$

As a matter of fact, we can turn $\mathcal{B}''$ into a bimodule over A in different ways, but the discussion about the other actions of $A \hat{\otimes} A^{to}$ in $\mathcal{B}''$ can wait till the next chapter. Let us set $\mathcal{B}' = \mathrm{Z}^g(A, \mathcal{B}'')$; by definition the elements of $\mathcal{B}'$ are the *scalar products on S associated with τ* for the action defined in (6.8.10). Since the natural mapping $A \otimes \mathrm{Z}^g(A, \mathcal{B}'') \to \mathcal{B}''$ is bijective (see (6.7.6)), and since $\mathcal{B}''$ and A'' have the same dimension, we realize that $\mathcal{B}'$ and A' too have the same dimension. Because of (6.8.10), the homogeneous elements of $\mathcal{B}'$ are the homogeneous elements of $\mathcal{B}''$ such that, for all $a \in A$ and all $s,\ t \in S$,

$$\beta(as,\ t) \ = \ (-1)^{\partial a \partial \beta}\ \beta(s,\ \tau(a)t). \tag{6.8.11}$$

Obviously ρ'' leaves $\mathcal{B}'$ invariant; let ρ' be the restriction of ρ'' to $\mathcal{B}'$. When the characteristic of K is $\neq 2$, our first main purpose is $\mathrm{cp.dv.tr}(\rho')$ which allows us to calculate the dimensions of the kernels of $\rho'_0 \pm \mathrm{id}$ and $\rho'_1 \pm \mathrm{id}$, in other words, the dimensions of the spaces of homogeneous scalar products that are symmetric or skew symmetric. When K has characteristic 2, only $\mathrm{tw.dv.tr}(\rho')$ is meaningful since later $\mathrm{dv.tr}(\rho')$ proves to be always equal to 1. From (6.8.11) we also deduce that βf belongs to $\mathcal{B}'$ whenever β belongs to $\mathcal{B}'$ and f to A'; in other words, $\mathcal{B}'$ *is a right module over* A'; the structure of bimodule defined in (6.8.10) has been chosen precisely for this property to be true.

(6.8.12) **Lemma.** *At least one of the components $\mathcal{B}'_0$ or $\mathcal{B}'_1$ contains a nondegenerate bilinear form β_n that is symmetric or skew symmetric.*

Proof. From (6.6.3) we know that S is a direct sum of graded submodules all isomorphic to some graded irreducible module P or (in some cases) to P^s. There is a natural bijection between the scalar products on P (associated with τ) and the scalar products on P^s and by this bijection symmetric (resp. skew symmetric) scalar products correspond to symmetric (resp. skew symmetric) ones. Therefore it suffices to prove (6.8.12) when $S = P$. If β is a nonzero homogeneous scalar product on P (associated with τ), from (6.8.11) we deduce that the kernel of $d_\beta : P \to P^*$ is a graded submodule of P; since P is graded irreducible, it is reduced to 0, and β is nondegenerate. Consequently any nonzero eigenvector of ρ' in $\mathcal{B}'_0$ or $\mathcal{B}'_1$ can become the wanted β_n. □

Up to the end, β_n is a nondegenerate homogeneous scalar product on S associated with τ that is symmetric or skew symmetric. It induces an involution τ'' on A'' as is explained in (6.8.9); from (6.8.9) and (6.8.11) it follows that τ'' leaves A' invariant; in other words, β_n induces an involution τ' in A'. The calculation of cp.dv.tr(τ') (or tw.dv.tr(τ')) is another main purpose. If we compare (6.8.9) and (6.8.11), we realize that τ'' corresponds to $\tau \tilde{\otimes} \tau'$ by the natural isomorphism $A \hat{\otimes} A' \to A''$; this allows us to apply (6.8.7) and to reach the next theorem. For every integer $n \geq 2$, we denote by $\mu_n(\mathbb{C})$ the group of nth roots of 1 in $\mathbb{C}$; for instance $\mu_4(\mathbb{C}) = \{1,\ i, -1, -i\}$.

(6.8.13) **Theorem.** *When K does not have characteristic 2, then*

$$\text{cp.dv.tr}(\tau)\ \text{cp.dv.tr}(\rho') \;=\; 1.$$

Moreover there exists $k \in \mu_4(\mathbb{C})$ such that

$$\text{cp.dv.tr}(\rho') \;=\; k\ \text{cp.dv.tr}(\tau')\ ,$$

and the value of k is given by these equalities in which $\pm$ means $+$ or $-$ according as β_n is symmetric or skew symmetric: $k = \pm 1$ if β_n is even, and $k = \pm i$ if β_n is odd.

Proof. From (6.8.7) we deduce

$$\text{cp.dv.tr}(\tau)\ \text{cp.dv.tr}(\tau') \;=\; \text{cp.dv.tr}(\tau'').$$

Let us momentarily admit that $\text{cp.dv.tr}(\rho') = k\ \text{cp.dv.tr}(\tau')$ for some $k \in \mathbb{C}^\times$ that only depends on β_n; if we replace A with K, then A' is replaced with A'', and ρ' and τ' with ρ'' and τ''; consequently $\text{cp.dv.tr}(\rho'') = k\ \text{cp.dv.tr}(\tau'')$ with the same k since k depends only on β_n. It follows that

$$\text{cp.dv.tr}(\tau)\ \text{cp.dv.tr}(\rho') \;=\; \text{cp.dv.tr}(\rho'')\ ,$$

and thus we have proved the first statement in (6.8.13) since $\text{cp.dv.tr}(\rho'') = 1$ (see (6.8.5)). Now let us prove that cp.dv.tr(ρ') and cp.dv.tr(τ') are related as announced. Since β_n is nondegenerate, the mapping $f \longmapsto \beta_n f$ is injective from A' into $\mathcal{B}'$; it is even bijective since A' and $\mathcal{B}'$ have the same dimension; it respects or changes the parities according as β_n is even or odd. Because of (6.8.9), $\beta_n f$ is symmetric or skew symmetric if and only if f is an eigenvector of τ'. All this allows us to derive cp.dv.tr(ρ') from cp.dv.tr(τ'). Let us treat in detail the case of an odd and skew symmetric β_n. In this case the mapping $f \longmapsto \beta_n f$ induces bijections from $\mathrm{Ker}(\tau'_0 - \mathrm{id})$ onto $\mathrm{Ker}(\rho'_1 + \mathrm{id})$, from $\mathrm{Ker}(\tau'_0 + \mathrm{id})$ onto $\mathrm{Ker}(\rho'_1 - \mathrm{id})$, from $\mathrm{Ker}(\tau'_1 - \mathrm{id})$ onto $\mathrm{Ker}(\rho'_0 - \mathrm{id})$ (mind the twisting exponent $\partial f \partial \beta_n$ in (6.8.9)), and from $\mathrm{Ker}(\tau'_1 + \mathrm{id})$ onto $\mathrm{Ker}(\rho'_0 + \mathrm{id})$; it follows that cp.dv.tr(ρ') = $-i$ cp.dv.tr(τ'). □

The next step in the classification of involutions of graded central simple algebras is to prove that cp.dv.tr(τ) $\in \mu_8(\mathbb{C})$; this can be deduced from (6.8.13).

(6.8.14) **Theorem.** *When K does not have characteristic 2, then* cp.dv.tr(τ) $\in \mu_8(\mathbb{C})$ *for every involution τ of a graded central simple algebra A. Moreover* (cp.dv.tr(τ))^4 *is equal to* 1 *or* -1 *according as A has even or odd type, and* (cp.dv.tr(τ))^2 *belongs to* $\{1,\ i\}$ *or to* $\{-1, -i\}$ *according as τ operates trivially or nontrivially on* $\mathrm{Z}(A_0, A)$.

Proof. When A is isomorphic to some $\mathcal{M}(r, B)$ or $\mathcal{M}(m, n; B)$ as in (6.6.2), and S is the graded irreducible module B^r or $B^m \oplus B^n$ over A, then the graded Morita context $(A, B, S, \dots)$ shows that $\mathrm{End}^g_A(S) \cong B^{to}$. When A has a trivial class in $\mathrm{Br}^g(K)$, in other words when $B \cong K$, then $A' = \mathrm{End}^g_A(S) = K$, whence cp.dv.tr($\tau'$) = 1, and consequently cp.dv.tr(ρ') and cp.dv.tr(τ) both belong to $\mu_4(\mathbb{C})$. When A does not have a trivial class, anyhow $A \hat{\otimes} A^{to}$ has a trivial class since it is isomorphic to $\mathrm{End}(A)$; with τ is associated an involution τ^{to} of A^{to} defined by $\tau^{to}(a^{to}) = \tau(a)^{to}$, which has the same complex divided trace as τ. Therefore cp.dv.tr($\tau \tilde{\otimes} \tau^{to}$) = (cp.dv.tr($\tau$))^2 belongs to $\mu_4(\mathbb{C})$, and cp.dv.tr(τ) to $\mu_8(\mathbb{C})$. When A has even type, then $\dim(A)$ is a square and cp.dv.tr(τ) belongs to $\mathbb{Q} \oplus \mathbb{Q}i$, whence cp.dv.tr($\tau$) $\in \mu_4(\mathbb{C})$. But when A has odd type, then $2\dim(A)$ is a square and $\sqrt{2}$ cp.dv.tr(τ) belongs to $\mathbb{Q} \oplus \mathbb{Q}i$, with the result that cp.dv.tr(τ) is a primitive eighth root of 1.

It is clear that τ leaves $\mathrm{Z}(A_0, A)$ invariant. When A is trivially graded, then $\mathrm{Z}(A_0, A) = K$ and cp.dv.tr(τ) is real, therefore equal to 1 or -1; thus τ acts trivially on $\mathrm{Z}(A_0, A)$ and (cp.dv.tr(τ))^2 = 1. When $A_1 \neq 0$, then $\mathrm{Z}(A_0, A)$ is a quadratic extension $K \oplus Kd$ with $d^2 \in K^\times$, and the restriction of τ to $\mathrm{Z}(A_0, A)$ may be the identity mapping or the standard involution according as it maps d to itself or to $-d$. The type of A is the parity of d, and moreover $ad = da$ for all $a \in A$ if d is odd, whereas $ad = d\sigma(a)$ if d is even (see (3.5.13)). With τ is associated another involution $\sigma\tau$, another space $\mathcal{B}'_\sigma$ of scalar products and another mapping $\rho'_\sigma : \beta \longmapsto \beta^o$. Since the complex divided traces of τ and $\sigma\tau$ are conjugate, the same is true for the complex divided traces of ρ' and ρ'_σ. The mapping $\beta \longmapsto \beta d$ (defined by $(\beta d)(s,t) = \beta(ds,t)$) is a bijection from $\mathcal{B}'$ onto $\mathcal{B}'_\sigma$ because an easy calculation shows that

$$(\beta d)(as,\ t) \ = \ (-1)^{\partial a \partial(\beta d)}\ (\beta d)(s,\ \sigma\tau(a)t)\ ;$$

it is funny to observe that the emergence of σ in the right-hand member is due to quite different reasons according as d is even or odd; when d is even, then

$da = \sigma(a)d$, but when d is odd, then $(-1)^{\partial a \partial \beta} a = (-1)^{\partial a \partial(\beta d)} \sigma(a)$. This bijection $\beta \longmapsto \beta d$ respects or changes the symmetry property of β according to the next equality, in which $\pm$ means $+$ or $-$ according as β is symmetric or skew symmetric:

$$(\beta d)(s,t) \;=\; \pm(-1)^{\partial\beta\partial d}\,(\beta\tau(d))(t,s).$$

All this allows us to prove that $\text{cp.dv.tr}(\rho'_\sigma) = k'\,\text{cp.dv.tr}(\rho')$ for some $k' \in \mu_4(\mathbb{C})$, and the calculation of k' gives precisely the value of $(\text{cp.dv.tr}(\tau))^2$ because

$$(\text{cp.dv.tr}(\tau))^2 = (\text{cp.dv.tr}(\rho'))^{-2} = (\text{cp.dv.tr}(\rho'))^{-1}\text{cp.dv.tr}(\rho'_\sigma) = k'.$$

Let us treat in detail the case of an odd d when $\tau(d) = -d$. In this case the mapping $\beta \longmapsto \beta d$ induces bijections from $\text{Ker}(\rho'_0 - \text{id})$ onto $\text{Ker}(\rho'_{\sigma,1} + \text{id})$, from $\text{Ker}(\rho'_0 + \text{id})$ onto $\text{Ker}(\rho'_{\sigma,1} - \text{id})$, from $\text{Ker}(\rho'_1 - \text{id})$ onto $\text{Ker}(\rho'_{\sigma,0} - \text{id})$, and from $\text{Ker}(\rho'_1 + \text{id})$ onto $\text{Ker}(\rho'_{\sigma,0} + \text{id})$. Thus $\text{cp.dv.tr}(\rho'_\sigma) = -i\ \text{cp.dv.tr}(\rho')$, and we conclude that $(\text{cp.dv.tr}(\tau))^2 = -i$ in accordance with (6.8.14). □

Thus the involutions of graded central simple algebras are classified by the group $\mu_8(\mathbb{C})$ through the complex divided traces, and the product in the classifying group $\mu_8(\mathbb{C})$ corresponds to the special tensor product $\tilde{\otimes}$ defined in (6.8.1).

When K has characteristic 2, the results are much poorer and can be collected in a single theorem. Now the involutions of graded central simple algebras are classified by the group $\mu_2(\mathbb{C})$ through the twisted divided traces.

(6.8.15) **Theorem.** *When K has characteristic 2, then* $\text{dv.tr}(\tau)$, $\text{dv.tr}(\rho')$ *and* $\text{dv.tr}(\tau')$ *are all equal to 1, whereas* $\text{tw.dv.tr}(\tau)$, $\text{tw.dv.tr}(\rho')$ *and* $\text{tw.dv.tr}(\tau')$ *belong to* $\mu_2(\mathbb{C})$. *Moreover*

$$\begin{aligned}
&\text{tw.dv.tr}(\tau)\ \text{tw.dv.tr}(\rho') = 1\ , \\
&\qquad \text{tw.dv.tr}(\rho') = \text{tw.dv.tr}(\tau') && \textit{if } \beta_n \textit{ is even,}\\
&\qquad \text{tw.dv.tr}(\rho') = -\text{tw.dv.tr}(\tau') && \textit{if } \beta_n \textit{ is odd,}\\
&\qquad \text{tw.dv.tr}(\tau) = 1 && \textit{if } \tau \textit{ operates trivially on } \text{Z}(A_0),\\
&\qquad \text{tw.dv.tr}(\tau) = -1 && \textit{if } \tau \textit{ operates nontrivially on } \text{Z}(A_0).
\end{aligned}$$

Proof. By means of the bijection $f \longmapsto \beta_n f$ from A' onto $\mathcal{B}'$ it is easy to prove that $\text{dv.tr}(\rho') = \text{dv.tr}(\tau')$, whereas $\text{tw.dv.tr}(\rho')$ is equal to $\text{tw.dv.tr}(\tau')$ or to $-\text{tw.dv.tr}(\tau')$ according to the parity of β_n. When we know this, from (6.8.7) and (6.8.5) we deduce that

$$\text{dv.tr}(\tau)\ \text{dv.tr}(\rho') = \text{dv.tr}(\rho'') = 1\ ,$$
$$\text{tw.dv.tr}(\tau)\ \text{tw.dv.tr}(\rho') = \text{tw.dv.tr}(\rho'') = 1.$$

Then we prove that $\text{dv.tr}(\tau) = 1$ and $\text{tw.dv.tr}(\tau) \in \mu_2(\mathbb{C})$ by the argument presented in the beginning of the proof of (6.8.14): when A has a trivial class in $\text{Br}^g(K)$, and when S is a graded irreducible module P, then $A' = K$ whence

$\mathrm{dv.tr}(\tau') = \mathrm{tw.dv.tr}(\tau') = 1$, and the general case follows from the fact that $A \hat{\otimes} A^{to}$ has a trivial class; in this way we prove that $\mathrm{dv.tr}(\tau) \in \mu_2(\mathbb{C})$ and $\mathrm{tw.dv.tr}(\tau) \in \mu_4(\mathbb{C})$, but we also know that $\mathrm{dv.tr}(\tau) \geq 0$ and $\mathrm{tw.dv.tr}(\tau) \in \mathbb{R}$.

When A is trivially graded, then $\mathrm{tw.dv.tr}(\tau) = \mathrm{dv.tr}(\tau) = 1$, and τ operates trivially on $\mathrm{Z}(A_0) = K$. Consequently it remains to prove this assertion: when $A_1 \neq 0$, the value of $\mathrm{tw.dv.tr}(\tau)$ depends on the restriction of τ to the quadratic extension $\mathrm{Z}(A_0, A) = \mathrm{Z}(A_0)$; unfortunately the argument presented in the proof of (6.8.14) for the analogous assertion is no longer suitable. Since we already know that $\mathrm{tw.dv.tr}(\tau) \in \mu_2(\mathbb{C})$ and $\mathrm{dv.tr}(\tau) = 1$, we must prove that $\mathrm{dtr}(\tau_0) \neq 0$ (resp. $\mathrm{dtr}(\tau_0) = 0$) if the restriction of τ to $\mathrm{Z}(A_0)$ is the identity mapping (resp. the standard involution). By means of a field extension of K (as in (6.6.5)) we reduce the problem to the case of an algebra $A = \mathcal{M}(m, n; K)$. In this case A_0 is the direct sum of two ideals respectively isomorphic to $\mathcal{M}(m, K)$ and $\mathcal{M}(n, K)$. If τ operates nontrivially on $\mathrm{Z}(A_0)$, it permutes these two ideals, whence $\mathrm{dtr}(\tau_0) = 0$ (and by the way, $m = n$). If τ operates trivially on $\mathrm{Z}(A_0)$, it leaves invariant these two ideals of A_0, and $\mathrm{dtr}(\tau_0) \neq 0$ because we know that the divided traces of the restrictions of τ_0 to these ideals are equal to 1. □

Extensions of scalar products

When A is a graded central simple algebra, there exists a linear form $h : A \to K$ (unique up to an invertible factor in K) such that $h(A_1) = 0$ and the bilinear form $(a, b) \longmapsto h(ab)$ is symmetric and nondegenerate. Of course the existence of h follows from (3.6.6) and (3.6.7); but when K is a field, we do not need these difficult propositions. Indeed by means of a suitable field extension $K \to L$ (see (6.6.5)) we can prove that $A_1 + [A, A]$ (where $[A, A]$ is the subspace generated by all Lie brackets $[a, b] = ab - ba$) is a hyperplane of A; this hyperplane determines a linear form h which is unique modulo $K^\times$; the announced properties of h follow from similar properties of $L \otimes h$ which are proved by means of well-known theorems about traces of matrices. In particular the equality $h \circ w = h$ holds for every automorphism w of A (see (3.6.4)), and for every anti-automorphism w too, since the transposition of matrices does not modify the traces. When $\dim(A)$ is invertible in K, then A is the direct sum of K and $A_1 + [A, A]$, and we can normalize h so that $h(1) = 1$; after this normalization $h(a)$ (with $a \in A$) is called the *scalar component* of a and denoted by $\mathrm{Scal}(a)$; the title "Extensions of scalar products" is especially justified when $h = \mathrm{Scal}$.

Let $\beta : S \times S \to K$ be a homogeneous element of $\mathcal{B}''$. The *extension of β with values in A* is the bilinear mapping $E_A : S \times S \to A$ defined in this way:

$$(6.8.16) \qquad \forall s,\ t \in S,\ \forall c \in A, \qquad h(c\, E_A(s, t)) = (-1)^{\partial c \partial \beta}\ \beta(cs,\ t).$$

It is easy to prove that E_A is the only bilinear mapping $S \times S \to A$ that has the same parity as β (in other words, $\partial E_A(s, t) = \partial s + \partial t + \partial \beta$), that is A-g-linear with respect to the left side variable (in other words, $E_A(as,t) = (-1)^{\partial a \partial \beta} a\, E_A(s,t)$), and such that $h \circ E_A = \beta$ (whence the name "extension of β" when $h = \mathrm{Scal}$).

(6.8.17) **Proposition.** *If β is a homogeneous scalar product associated with τ, then*

$$\forall a,\ b \in A,\ \forall s,\ t \in S, \qquad E_A(as,\ bt) \;=\; (-1)^{\partial a \partial \beta} a\ E_A(s,t)\ \tau(b)\ ,$$

and the linear mapping $S \otimes S \to A$ defined by $s \otimes t \longmapsto E_A(s,t)$ is surjective if $\beta \neq 0$. If this scalar product β is symmetric or skew symmetric, then

$$\forall s,\ t \in S, \qquad E_A(t,s)) \;=\; \pm\ \tau(E_A(s,t))\ ,$$

where $\pm$ means $+$ or $-$ according as β is symmetric or skew symmetric.

Indeed the equalities written in (6.8.17) mean that, for every $c \in A$,

$$h(c\ E_A(as,bt)) = (-1)^{\partial a \partial \beta} h(ca\ E_A(s,t)\ \tau(b))\ ,$$
$$h(c\ E_A(s,t)) = h(c\ \tau(E_A(t,s)))\ ;$$

their validity follows from straightforward verifications using (6.8.11), $h([A,A]) = 0$ and $h \circ \tau = h$. Only the surjectiveness of the mapping $S \otimes S \to A$ needs an explanation: its image is a graded ideal of A, consequently 0 or A. □

We can turn $S \otimes S$ into a bimodule over A if we set

(6.8.18) $$\forall a,\ b \in A,\ \forall s,\ t \in S, \qquad a(s \otimes t)b \;=\; as \otimes \tau(b)t.$$

The first statement in (6.8.17) can be interpreted in this way: *the mapping $s \otimes t \longmapsto E_A(s,t)$ is $(A \hat{\otimes} A^{to})$-g-linear from $S \otimes S$ onto A.* The precise meaning of this statement requires that the actions of $A \hat{\otimes} A^{to}$ in $S \otimes S$ and A are defined in accordance with the twisting rule (4.2.1); for instance $(a \otimes b^{to})(s \otimes t)$ is synonymous with $(-1)^{\partial b(\partial s + \partial t)} a(s \otimes t)b$, which is now defined by (6.8.18).

Let us suppose that β is the above nondegenerate scalar product β_n associated with τ, which has been assumed to be symmetric or skew symmetric, and let us remember that β_n induces an involution τ' in A'. Since S is a module over A' too, we can treat β_n as a scalar product associated with τ', and thus we get an extension $E_{A'}$ with values in A'. Many authors forget the parity grading of A and consider a nongraded module M, which they treat as a right module over the opposite algebra $\mathrm{End}_A(M)^o$; then they define scalar products $M \times M \to \mathrm{End}_A(M)^o$ with special properties that actually mean that they are in some way extensions of scalar products $M \times M \to K$ associated with τ or $\sigma\tau$.

We can even treat β_n as a scalar product associated with the involution $\tau \tilde{\otimes} \tau'$ of $A \hat{\otimes} A'$; thus we get an extension $E : S \times S \to A \hat{\otimes} A'$ in which the properties of E_A and $E_{A'}$ appear together. It determines a surjective linear mapping $S \otimes S \to A \hat{\otimes} A'$, in which the source and the target have the same dimension as $A'' = \mathrm{End}(S)$; consequently it is bijective and we have proved what follows.

(6.8.19) **Proposition.** *If E is the extension of β_n with values in $A \hat{\otimes} A'$, the linear mapping $S \otimes S \to A \hat{\otimes} A'$ induced by E is an isomorphism in the category $\mathcal{M}od^g(A \hat{\otimes} A' \hat{\otimes} A^{to} \hat{\otimes} A'^{to})$.*

Often people are interested in the action on S of some multiplicative group G made of homogeneous elements of A. This group may be for instance the Clifford–Lipschitz group in the Clifford algebra of a quadratic space; then the elements of S are called spinors, and the elements of $S \otimes S$ bispinors. The action of G on $S \otimes S$ is actually easier to analyse than its action on S since its action on $S \otimes S$ looks like its action on $A \hat{\otimes} A'$, all the more because G acts trivially on A'.

In our study of scalar products on S it is clear that the parity grading of S plays an important role; it has enabled us to present a *unified theory* in which the essential results are independent of the type of A (even or odd) and of the nature of $\mathrm{Z}(A_0, A)^{ng}$ (that is $\mathrm{Z}(A_0, A)$ without grading, either K if $A_1 = 0$, or K^2, or a quadratic field extension of K). Since A is a graded algebra, the natural modules over A are the graded modules; even the usual spinor spaces in quantum mechanics are graded modules (see (6.2.2)), although their parity grading is often overlooked. In (6.ex.21) and (6.ex.22) you can discover how this unified theory of scalar products can be applied to the usual spinor spaces of quantum mechanics; no calculations with "Dirac matrices" are necessary.

If really the study of scalar products on a nongraded module M is ever necessary, it is always possible to use the trick that has been successful in the first step of the proof of (6.2.8): we use the graded module $S = M \oplus M$ with homogeneous components $M \oplus 0$ and $0 \oplus M$, in which $a(x, y)$ means (ax, ay) for all $a \in A_0$, but (ay, ax) for all $a \in A_1$, and we derive the scalar products on M from the scalar products on S. See (6.ex.20) for more details.

The classification of the couples (A, τ)

Here A is a graded central simple algebra over a field K of characteristic $\neq 2$, and τ is a (graded) involution of A. The classification of such couples (A, τ) was first imagined in the noteworthy article [Wall 1968]. Here we get the same results by different ideas: with every couple (A, τ) we associate the class $([A], \mathrm{cp.dv.tr}(\tau))$ in the group $\mathrm{Br}^g(A) \times \mu_8(\mathbb{C})$. This raises only one question: which is the subgroup of $\mathrm{Br}^g(A) \times \mu_8(\mathbb{C})$ made of all these classes? There is a canonical morphism $\mathrm{Br}^g(K) \to \mathbb{Z}/2\mathbb{Z}$ mapping $[A]$ to the type of A, there is also a canonical morphism $\mu_8(\mathbb{C}) \to Z/2\mathbb{Z}$ with kernel $\mu_4(\mathbb{C})$, and (6.8.14) says that $[A]$ and $\mathrm{cp.dv.tr}(\tau)$ have the same image in $\mathbb{Z}/2\mathbb{Z}$. *We actually get the subgroup of all elements of* $\mathrm{Br}^g(K) \times \mu_8(\mathbb{C})$ *in which both components have the same image in* $\mathbb{Z}/2\mathbb{Z}$. To prove this, it suffices to find a couple (A, τ) such that $[A]$ is trivial whereas $\mathrm{cp.dv.tr}(\tau)$ is a primitive fourth root of 1. Actually the complex divided trace of the standard involution of $\mathcal{M}(1, 1; K)$ is equal to $-i$.

Complex divided traces allow us to define a canonical isomorphism $\mathrm{Br}^g(\mathbb{R}) \to \mu_8(\mathbb{C})$. Indeed every graded central simple algebra over $\mathbb{R}$ admits a *positive involu-*

tion τ_+ satisfying this property: $\mathrm{Scal}(a\,\tau_+(a)) \geq 0$ for all $a \in A$, if Scal is defined as above before (6.8.16). This is proved in (6.ex.16), and the following statements too: if A admits several positive involutions, they all have the same complex divided trace, which only depends on the class of A in $\mathrm{Br}^g(\mathbb{R})$; moreover $\tau \,\tilde{\otimes}\, \tau'$ is positive if both τ and τ' are positive. All this gives a group morphism $\mathrm{Br}^g(\mathbb{R}) \to \mu_8(\mathbb{C})$ defined by $[A] \longmapsto \mathrm{cp.dv.tr}(\tau_+)$. It is bijective because $\mathrm{Br}^g(\mathbb{R})$ is a cyclic group of order 8, and $[A]$ and $\mathrm{cp.dv.tr}(\tau_+)$ have the same image in $\mathbb{Z}/2\mathbb{Z}$.

Consequently the couples (A,τ) over $\mathbb{R}$ are classified by the couples $(t_+, t) \in \mu_8(\mathbb{C}) \times \mu_8(\mathbb{C})$ such that $t_+^4 = t^4$; of course $t = \mathrm{cp.dv.tr}(\tau)$ and $t_+ = \mathrm{cp.dv.tr}(\tau_+)$ if τ_+ is any positive involution on A. Let us calculate (t_+, t) when A is the Clifford algebra of a quadratic space (M,q), and τ the reversion in $C\ell(M,q)$; let us assume that the maximal positive definite (resp. negative definite) subspaces have dimension m (resp. n) as in (2.8.1). If we set $\kappa = (1+i)/\sqrt{2}$, then

$$(t_+,\ t) \;=\; (\kappa^{m-n},\ \kappa^{m+n}) \,. \tag{6.8.20}$$

Indeed, because of the multiplicative property stated in (6.8.7), it suffices to verify (6.8.20) when $m+n = 1$. In such a case, τ is the identity mapping. If $(m,n) = (1,0)$, then $C\ell(M,q)$ is isomorphic to $(\mathbb{R}^2)^g$ which admits the identity mapping as a positive involution, whence $(t_+, t) = (\kappa, \kappa)$ (see (6.8.6)). If $(m,n) = (0,1)$, then $C\ell(M,q)$ is isomorphic to $\mathbb{C}^g$ which admits the standard involution (or complex conjugation) as a positive involution, whence $(t_+, t) = (\kappa^{-1}, \kappa)$. □

The property described by (6.8.20) is often called the 32-periodicity of real Clifford algebras; of course this periodicity does not refer to a cyclic group of order 32, but to a subgroup of $\mu_8(\mathbb{C}) \times \mu_8(\mathbb{C})$. This subgroup is sometimes compared to the set of the 32 white squares in a chess-board of dimension 8×8, and this explains the name "cliffordian chess-board" given to various tables that display some detail of the information concentrated in the short formula (6.8.20).

When K has characteristic 2, the couples (A,τ) are classified by the whole group $\mathrm{Br}^g(K) \times \mu_2(K)$. It is probably impossible to refine this classification by means of a larger group, although refinements (without classifying groups) have been proposed (see [Knus 1991] or [Knus et al. 1998]).

Exercises

(6.ex.1) Let K be a field that does not have characteristic 2, and let $A = K \times K$ be provided with its unique nontrivial parity grading; this means that A_1 is spanned by $(1,-1)$. As explained in **6.2**, with each module M over A is associated a conjugate module M^c. Prove that the following assertions are equivalent:

(a) M is a free module over A.
(b) $(1,0)M$ and $(0,1)M$ are isomorphic vector spaces over K.
(c) M and M^c are isomorphic modules over A.
(d) There is a parity grading on M for which M is a graded module over A.

(6.ex.2) Let M and N be graded modules over A. Assume that M is a direct sum of m graded submodules $M_1,\dots,M_m$, whereas N is a direct sum of n graded submodules $N_1,\dots,N_n$. Prove that there is a natural bijection between $\mathrm{Hom}^g_A(M,N)$ (resp. $\mathrm{Hom}_{A,0}(M,N)$) and the set of matrices $(f_{j,i})$ with n lines and m columns, in which every entry $f_{j,i}$ belongs to $\mathrm{Hom}^g_A(M_i,N_j)$ (resp. $\mathrm{Hom}_{A,0}(M_i,N_j)$).

(6.ex.3) Let B be a graded K-algebra, σ the automorphism of B defined by $b \longmapsto (-1)^{\partial b}b$, m and n nonnegative integers, $\mathcal{M}(m,n;B)$ the usual graded matrix algebra isomorphic to $B \otimes \mathcal{M}(m,n;K)$, and $\mathcal{M}^g(m,n;B)$ the twisted tensor product $B \hat{\otimes} \mathcal{M}(m,n;K)$.

(a) Identify $\mathcal{M}^g(m,n;B)$ and $\mathcal{M}(m,n;B)$ as graded modules, and prove that the former is provided with this twisted multiplication:

$$\begin{pmatrix} a & c \\ b & d \end{pmatrix} \begin{pmatrix} a' & c' \\ b' & d' \end{pmatrix} = \begin{pmatrix} aa' + c\sigma(b') & ac' + c\sigma(d') \\ b\sigma(a') + db' & b\sigma(c') + dd' \end{pmatrix} ;$$

here a and a' (resp. d and d') are square matrices of order m (resp. n), whereas b and b' (resp. c and c') are rectangular matrices of dimensions $n \times m$ (resp. $m \times n$), and σ operates on a matrix by operating on all its entries.

(b) Prove that $\mathcal{M}^g(m,n;B)$ is isomorphic to $\mathcal{M}(m,n;B)$ through this mapping:

$$\begin{pmatrix} a & c \\ b & d \end{pmatrix} \longmapsto \begin{pmatrix} a & c \\ \sigma(b) & \sigma(d) \end{pmatrix}.$$

(c) Prove that $\mathcal{M}(m,n;B)$ is isomorphic to $\mathcal{M}(m+n,B)$ through the following graded mapping when there is an odd invertible element w in B_1:

$$\begin{pmatrix} a & c \\ b & d \end{pmatrix} \longmapsto \begin{pmatrix} a & cw^{-1} \\ wb & wdw^{-1} \end{pmatrix}.$$

Therefore it is not sensible to use $\mathcal{M}^g(m,n;B)$ or $\mathcal{M}(m,n;B)$ when B_1 contains invertible elements.

(6.ex.4) Let P be a graded module over A, and σ' the automorphism of P defined by $x \longmapsto (-1)^{\partial x}x$. We set $B = \mathrm{End}^g_A(P)$.

(a) Prove that we get bijections $B \to \mathrm{Hom}^g_A(P,P^s)$ and $B \to \mathrm{Hom}^g_A(P^s,P)$ if we map every $f \in B$ to the morphisms $x \longmapsto (f \circ \sigma'(x))^s$ and $x^s \longmapsto f \circ \sigma'(x)$.

(b) Let (m,n) be a couple of nonnegative integers, and $M = P^m \oplus (P^s)^n$. Prove that $\mathrm{End}^g_A(M)$ is isomorphic to the algebra $\mathcal{M}^g(m,n;B)$ defined in (6.ex.3).

Hint. Map every matrix $(f_{j,i}) \in \mathcal{M}^g(m,n;B)$ to the endomorphism

$$(x_1,\dots,x^s_{m+n}) \longmapsto (y_1,\dots,y^s_{m+n})$$

of M determined by

$$y_j = \sum_{i=1}^{m} f_{j,i}(x_i) + \sum_{i=m+1}^{m+n} f_{j,i}\sigma'(x_i) \quad \text{if } j \leq m,$$

$$y_j = \sum_{i=1}^{m} f_{j,i}\sigma'(x_i) + \sum_{i=m+1}^{m+n} f_{j,i}(x_i) \quad \text{if } j > m.$$

Comment. If B_1 contains an invertible element, P^s is isomorphic to P, M to P^{m+n}, and $\mathrm{End}^g(M)$ to $\mathcal{M}(m+n,\ B)$, in accordance with (6.ex.3).

(6.ex.5) This exercise contains the graded version of Schur's lemma. Let M and N be two graded modules over A.

(a) Prove that every nonzero homogeneous element of $\mathrm{Hom}^g_A(M,N)$ is injective (resp. surjective) if M (resp. N) is graded irreducible.

(b) Assume that M is graded irreducible. Prove that $\mathrm{End}^g_A(M)$ is a graded division ring (that is a ring in which every nonzero homogeneous element is invertible, as in **6.6**); its grading is nontrivial or trivial according as M is or is not isomorphic to M^s.

(c) Now assume that M and N are both graded irreducible. Determine in which cases $\mathrm{Hom}_{A,0}(M,N)$ or $\mathrm{Hom}^g_{A,1}(M,N)$ or both are not reduced to 0.

(6.ex.6)* Here every graded semi-simple algebra is proved to be isomorphic to a finite direct product of algebras that are each one isomorphic to some $\mathcal{M}(m,n;B)$ or $\mathcal{M}(r,B)$ with B a graded division algebra (with a trivial grading in the former case, a nontrivial one in the latter case). Obviously $\mathrm{Z}^g_1(B) = 0$ when the field $\mathrm{Z}_0(B)$ does not have characteristic 2, and consequently $\mathcal{M}(m,n;B)$ or $\mathcal{M}(r,B)$ is a graded central simple algebra over $\mathrm{Z}_0(B)$ (which is an extension of the basic ring K). But when the field $\mathrm{Z}_0(B)$ has characteristic 2, it may happen that $\mathrm{Z}^g_1(B) \neq 0$.

(a) Prove that a finite direct product of graded semi-simple algebras is still semi-simple. Prove that the algebra $\mathcal{M}(m,n;B)$ or $\mathcal{M}(r,B)$ is semi-simple when B is a graded division algebra (in which every nonzero homogeneous element is invertible).

(b) Assume that M is a graded semi-simple and finitely generated module over a graded algebra A. Decompose M into a finite direct sum of graded irreducible submodules $P_{i,j}$ in such a way that the first index i gives the isomorphy classes of $P_{i,j}$ and $P_{i,j}^s$; in other words, $P_{h,k}$ is isomorphic to $P_{i,j}$ or $P_{i,j}^s$ if and only if $h = i$. Consider $C = \mathrm{End}^g_A(M)$ and deduce from (6.ex.4) and (6.ex.5) that C is isomorphic to a finite direct product of algebras C_i which are each one isomorphic to some $\mathcal{M}(m,n;B)$ or $\mathcal{M}(r,B)$ with B a graded division algebra over K. Consequently C is a graded semi-simple algebra.

Let A_M be the image of $A \to \mathrm{End}_K(M)$. Deduce from the density theorem (6.3.3) that conversely $A_M = \mathrm{End}^g_C(M)$. Consequently, if M is a faithful A-module, A too is a graded semi-simple algebra.

(c) Assume that A is a graded semi-simple module over itself (for the natural action of A on itself by multiplications on the left side). Thus $A^{to} = \mathrm{End}_A^g(A)$. Prove that A and A^{to} are graded semi-simple algebras, and that A is isomorphic to a finite direct product of graded algebras A_i that are each one isomorphic to some $\mathcal{M}(m,n;B)$ or $\mathcal{M}(r,B)$ with B a graded division algebra.

Graded Azumaya algebras. Graded central simple algebras

(6.ex.7) Let A be a graded algebra of finite dimension over a field K, such that A_0 is a division ring, and $\mathrm{Z}_0(A) = K$. When K does not have characteristic 2, prove that A is a graded central simple algebra over K if and only if every nonzero element of A_1 is invertible. When K has characteristic 2, you must add the hypothesis $\mathrm{Z}_1^g(A) = 0$.

(6.ex.8) Let A be a graded Azumaya algebra over a *local* ring K. Prove that its grading is balanced if and only if A_1 contains invertible elements.
Hint. First suppose that K is a field, and remember (6.6.2).

(6.ex.9) Let N be a graded bimodule over a graded Azumaya algebra A; for each $y \in N$, we wish to construct its inverse image by the isomorphism $A \otimes \mathrm{Z}^g(A,N) \rightarrow N$.

(a) For each linear form $f \in A^*$ prove the existence of $w = \sum_j b_j \otimes c_j^{to} \in A \hat{\otimes} A^{to}$ such that

$$\forall x \in A, \quad f(x) = wx = \sum_j (-1)^{\partial c_j \partial x}\, b_j x c_j \ .$$

Prove that wy belongs to $\mathrm{Z}^g(A,N)$ for all $y \in N$.

(b) There exist $a_1, a_2, \dots$ in A and $f_1, f_2, \dots$ in A^* such that $x = \sum_i a_i f_i(x)$ for all $x \in A$; and for each i we can write $f_i(x) = w_i x$ for some $w_i \in A \hat{\otimes} A^{to}$; prove that $\sum_i a_i \otimes w_i y$ is the inverse image of y in $A \otimes \mathrm{Z}^g(A,N)$.

(6.ex.10) Let A be a graded Azumaya algebra such that $\mathrm{Z}(A_0, A)$ is a quadratic extension, and let D be its discriminant module.

Remember that $da = (-1)^{\partial a(1+\partial d)} ad$ when a and d are homogeneous elements of respectively A and D. With every graded bimodule M over A we can associate the graded bimodule $D \otimes M$ and the graded bimodule $M^{c\cdot}$ conjugate on the left side:

$$a(d \otimes x)b = (-1)^{\partial a \partial d}\, d \otimes axb \ ,$$
$$a\ x^{c\cdot} b = (-1)^{\partial a}\ (axb)^{c\cdot}.$$

Prove that the mapping $d \otimes x \longmapsto (dx)^{c\cdot}$ is an isomorphism of graded bimodules from $D \otimes M$ onto $M^{c\cdot}$. Consequently it determines an isomorphism of graded K-modules $D \otimes \mathrm{Z}^g(A,M) \rightarrow \mathrm{Z}^g(A, M^{c\cdot})$.

In the same way there is an isomorphism from $M \otimes D$ onto the bimodule $M^{\cdot c}$ conjugate on the right side. Anyhow $M^{c\cdot}$ and $M^{\cdot c}$ are isomorphic through the mapping $x^{c\cdot} \longmapsto (-1)^{\partial x} x^{\cdot c}$.

(6.ex.11) The dual module A^* of a graded Azumaya algebra A is a bimodule over A in this way:

$$\forall h \in A^*,\ \forall a,\ b,\ c \in A, \quad (ahb)(c) \ = \ (-1)^{\partial a}\ h(bca)\ ;$$

the twisting exponent ∂a is equal to the awaited exponent $\partial a(\partial h + \partial b + \partial c)$ if $\partial h = \partial b + \partial c + \partial a$. As explained in (6.ex.10) there is also a bimodule $(A^*)^{c\cdot}$:

$$\forall h \in A^*,\ \forall a,\ b,\ c \in A, \quad (ah^{c\cdot}b)(c) \ = \ h(bca).$$

We are interested in $H = \mathrm{Z}^g(A, A^*)$ and in the submodule H' of A^* such that $(H')^{c\cdot} = \mathrm{Z}^g(A, (A^*)^{c\cdot})$. From (6.7.3) and (6.7.6) we deduce that H and H' are direct summands of A^* of constant rank 1.

(a) Verify that a homogeneous $h \in A^*$ belongs to H (resp. H') if and only if

$$\forall a,\ c \in A, \quad h(ac) = (-1)^{\partial a \partial c} h(ca) \quad (\text{resp. } h(ac) = (-1)^{\partial a \partial h} h(ca)).$$

(b) Assume that H' contains an even element h_0 such that the symmetric bilinear form $(a, c) \longmapsto h_0(ac)$ is nondegenerate, and prove that H' is the free submodule generated by h_0.

(c) Assume that $\mathrm{Z}(A_0, A)$ is a quadratic extension, and let D be its discriminant module. Deduce from (6.ex.10) that H' is isomorphic to $D \otimes H$ (whence $H \cong D$ if H' is free and even as in (b)).

Comments. From (3.6.6) and (3.6.7) we know that H' contains an element h_0 satisfying the hypotheses of (b); the validity of these propositions is ensured by (3.5.15) in general, and by (3.7.4) if A is the Clifford algebra of a quadratic space (M, q); when 2 is invertible in K we can also refer to (4.8.16). From the constructions presented in (4.ex.6) with $A = C\ell(M, q)$, we can deduce that the graded modules H and D are isomorphic to $\bigwedge^{\max}(M)$.

(6.ex.12) We know that $\mathrm{Br}^g(\mathbb{R})$ is a cyclic group of order 8 generated by the class of $(\mathbb{R}^2)^g$ (see (3.ex.22)). From (6.6.2) we deduce that, up to isomorphism, there are eight graded central division algebras B of finite dimension over $\mathbb{R}$. Give the list of these eight algebras B according to the cyclic order of $\mathrm{Br}^g(\mathbb{R})$.

Suggestion. If the $G_{m,n}$ are defined as in **2.7**, the answer may be

$$\mathbb{R},\ C\ell(G_{1,0}),\ C\ell(G_{2,0}),\ C\ell(G_{3,0}),\ \mathbb{H},\ C\ell(G_{0,3}),\ C\ell(G_{0,2}),\ C\ell(G_{0,1}),\ \text{again } \mathbb{R}, \dots;$$

up to isomorphy, it is the same thing as

$$\mathbb{R},\ (\mathbb{R}^2)^g,\ \mathrm{End}_{\mathbb{R}}(\mathbb{C}),\ \mathbb{C}^g \otimes \mathbb{H},\ \mathbb{H},\ (\mathbb{R}^2)^g \otimes \mathbb{H},\ \mathbb{H}^g,\ \mathbb{C}^g,\ \text{again } \mathbb{R}, \dots;$$

an exponent ^{g} means a nontrivial grading; the even (resp. odd) elements of $\mathrm{End}_{\mathbb{R}}(\mathbb{C})$ are the $\mathbb{C}$-linear (resp. $\mathbb{C}$-semilinear) endomorphisms.

(6.ex.13)* Here is an example of a division ring B that has dimension 9 over its center which is the field $\mathbb{Q}$ of rational numbers; it comes from [Blanchard 1972]. Let $\mathbb{Q}[a]$ be the field generated over $\mathbb{Q}$ by a root a of the polynomial t^3-2, and $\mathbb{Q}[b]$ the field generated by a root b of $P(t)=t^3+t^2-2t-1$. It is easy to verify that the polynomial $P(t^2-2)$ is divisible by $P(t)$.

(a) Prove that $P(t)$ has three roots b, b' and b'' in $\mathbb{Q}[b]$, such that $b'=b^2-2$, $b''=b'^2-2$ and $b=b''^2-2$. Prove that (b,b',b'') is a basis of $\mathbb{Q}[b]$ over $\mathbb{Q}$, and that $\mathbb{Q}[b]$ admits an automorphism φ that maps b, b', b'' respectively to b', b'', b.

Remark. If r is a root of t^7-1 other than 1, it is easy to prove that $r+r^6$, r^2+r^5 and r^3+r^4 are the roots of $P(t)$.

(b) The norm of an element $x=\lambda b+\mu b'+\nu b''$ of $\mathbb{Q}[b]$ is by definition

$$\begin{aligned}\mathcal{N}(x)&=x\,\varphi(x)\,\varphi^2(x)\\&=\lambda^3+\mu^3+\nu^3-\lambda\mu\nu+3(\lambda^2\mu+\mu^2\nu+\nu^2\lambda)-4(\lambda^2\nu+\mu^2\lambda+\nu^2\mu).\end{aligned}$$

Verify that $\mathcal{N}(x)$ is an odd integer whenever the components λ, μ, ν of x are three integers not all even.

(c) On the vector space $B=\mathbb{Q}[a]\otimes\mathbb{Q}[b]$ we define a new $\mathbb{Q}$-bilinear multiplication such that

$$(a^i\otimes x)\;(a^j\otimes y)\;=\;a^{i+j}\otimes\varphi^j(x)y$$

for all exponents i and j in $\mathbb{Z}$, and for all elements x and y of $\mathbb{Q}[b]$. Prove that B is an associative algebra, and even a division ring. Besides, every commutative subalgebra of B other than $\mathbb{Q}$ is a field of dimension 3 over $\mathbb{Q}$; consequently $\mathrm{Z}(B)=\mathbb{Q}$.

Hint. B is a vector space over $\mathbb{Q}[b]$ when this field acts on B by multiplications on the *right* side; if $1\otimes x+a\otimes y+a^2\otimes z$ is a nonzero element of B, prove the invertibility of the multiplication by this element on the *left* side; since this multiplication is $\mathbb{Q}[b]$-linear, it suffices to verify the invertibility of its determinant

$$\Delta\;=\;\det\begin{pmatrix}x&2\varphi(z)&2\varphi^2(y)\\y&\varphi(x)&2\varphi^2(z)\\z&\varphi(y)&\varphi^2(x)\end{pmatrix}\;\in\mathbb{Q}[b]\quad(\text{actually }\Delta\in\mathbb{Q})\,;$$

after multiplications by suitable factors (among them perhaps a^{-1} or a^{-2}) reduce the problem to this case: the components of x, y, z in the basis (b,b',b'') are integers, and the components of x are not all even; then Δ is an odd integer.

(d) Let us set $M = \mathbb{Q}[a] \otimes \mathbb{Q}[a] \otimes \mathbb{Q}[b]$. Prove the existence of an algebra isomorphism from $B \otimes B \otimes B$ onto $\mathrm{End}_{\mathbb{Q}}(M)$ mapping every $(a^i \otimes x) \otimes (a^j \otimes y) \otimes (a^k \otimes z)$ to

$$a^m \otimes a^n \otimes w \longmapsto a^{m+i-k} \otimes a^{n+j+2k} \otimes \varphi^{-k}\big(\varphi^m(x)\varphi^n(y)\, zw\big) \;;$$

conclude that the Brauer class of B has order 3 in $\mathrm{Br}(\mathbb{Q})$.

(6.ex.14)* Let A be a graded Azumaya algebra over K, and L a graded subalgebra of A; when L acts on A by multiplications on the left side (resp. on both sides), A is an L-module (resp. a bimodule over L).

(a) Assume that L is graded separable over K, and prove that A is projective over L.
Hint. From (6.5.10) deduce a splitting exact sequence

$$0 \longrightarrow \mathrm{J}(L) \otimes_L A \longrightarrow L \otimes_K A \longrightarrow A \longrightarrow 0 \quad \text{in } \mathcal{M}od_0(L).$$

(b) Prove that $A \hat{\otimes} \mathrm{Z}^g(L, A)$ is isomorphic to $\mathrm{End}_L^g(A)$.
Hint. $\mathrm{End}_K(A)$ is a bimodule over $A \hat{\otimes} A$ in this way:

$$((a \otimes b)\, f\, (a' \otimes b'))(x) \;=\; \pm\, a\, f(a'xb)\, b' \quad \text{for all } f \in \mathrm{End}_K(A) \;;$$

thus $\mathrm{End}_L^g(A) = \mathrm{Z}^g(L \otimes 1,\ \mathrm{End}(A))$; and there is an isomorphism

$$\mathrm{Z}^g(L, A) \longrightarrow \mathrm{Z}^g(L \hat{\otimes} A,\ \mathrm{End}(A)) = \mathrm{Z}^g(1 \otimes A,\ \mathrm{End}_L^g(A)).$$

(c) Prove the existence of graded subalgebras L such that $\mathrm{Z}^g(L, A) = L$.

(d) Let L be a subalgebra of A_0 such that $\mathrm{Z}(L, A) = L$; thus L is a commutative extension of K. Suppose moreover that A is a projective L-module. Prove that the class of A in $\mathrm{Br}^g(K)$ belongs to the kernel of the group morphism $\mathrm{Br}^g(K) \to \mathrm{Br}^g(L)$.
Let $\mathfrak{q}$ be a prime ideal of L, and $\mathfrak{p} = K \cap \mathfrak{q}$ its image in $\mathrm{Spec}(K)$. Prove that $\mathrm{rk}(\mathfrak{p}, A) = \mathrm{rk}(\mathfrak{q}, A)^2$.

(e) To the hypotheses of (d) we add this one: L is a finitely generated projective K-module. Prove that $\mathrm{rk}(\mathfrak{p}, L) = \mathrm{rk}(\mathfrak{q}, A)$.
Hint. Reduce the problem to the case of constant ranks, and remember (1.ex.22).

Involutions of graded central simple algebras

(6.ex.15) Let τ be an involution of a graded Azumaya algebra A over a ring K that is not a field. Let $\mathcal{V}(2)$ be the closed subset of $\mathrm{Spec}(K)$ made of all prime ideals $\mathfrak{p}$ containing 2. On the open subset $\mathrm{Spec}(K) \setminus \mathcal{V}(2)$ we define a function cp.dv.tr with values in $\mu_8(\mathbb{C})$; if $\hat{\tau}_{\mathfrak{p}}$ is the involution of $(K_{\mathfrak{p}}/\mathfrak{p}K_{\mathfrak{p}}) \otimes A$ induced by τ, then cp.dv.tr$(\mathfrak{p},\tau)$ is cp.dv.tr$(\hat{\tau}_{\mathfrak{p}})$. On Spec$(K)$ itself we define a function $(\text{cp.dv.tr})^2$ with values in $\mu_4(\mathbb{C})$; on Spec$(K) \setminus \mathcal{V}(2)$ it is actually the square of the previous function; and on $\mathcal{V}(2)$ it is by definition tw.dv.tr$(\hat{\tau}_{\mathfrak{p}})$. Prove that both functions are locally constant on their domains of definition.

(6.ex.16) Here K is the field $\mathbb{R}$ of real numbers. Every graded central simple algebra A over $\mathbb{R}$ is provided with a graded projection Scal : $A \to K$ such that the bilinear form $(a,b) \longmapsto \text{Scal}(ab)$ is symmetric and nondegenerate. An involution τ of A is said to be positive if the bilinear form $(a,b) \longmapsto \text{Scal}(a\tau(b))$ is positive definite; this bilinear form too is symmetric because every involution leaves Scal invariant.

(a) Prove that the transposition of matrices in $\mathcal{M}(m,n;\mathbb{R})$ is a positive involution, and that its complex divided trace is 1.

(b) Let τ and τ' be involutions of A and A', and $\tau'' = \tau \tilde{\otimes} \tau'$ as in (6.8.1). Prove that τ'' is positive whenever τ and τ' are positive.
Hint. Let Scal$'$ and Scal$''$ be the natural projections $A' \to \mathbb{R}$ and $A \hat{\otimes} A' \to \mathbb{R}$; the bilinear form $(x,y) \longmapsto \text{Scal}''(x\tau''(y))$ is the tensor product of the bilinear forms $(a,b) \longmapsto \text{Scal}(a\tau(b))$ and $(a',b') \longmapsto \text{Scal}'(a'\tau'(b'))$.

(c) Prove that every graded central simple algebra A over $\mathbb{R}$ admits a positive involution.
Remark. The positive involution is unique only when A is a graded central division algebra; in (6.ex.12) you find the eight isomorphy classes of graded central division algebras over $\mathbb{R}$.

(d) Let τ be a positive involution of A, and τ_0 and τ_1 its restrictions to A_0 and A_1. Prove that the traces of τ_0 and τ_1 are equal to the signatures of the restrictions to A_0 and A_1 of the bilinear form $(a,b) \longmapsto \text{Scal}(ab)$. Consequently cp.dv.tr$(\tau)$ is determined by A.

(e) Deduce from (a) that the complex divided trace of all positive involutions of A only depends on the class of A in $\text{Br}^g(\mathbb{R})$. This leads to an isomorphism $\text{Br}^g(\mathbb{R}) \to \mu_8(\mathbb{C})$.

(6.ex.17) There is an alternative proof of (6.8.14) that is based on the generalized Skolem–Noether Theorem (6.7.8) and not on (6.8.13). Because of (6.6.5) you can assume that τ is an involution on an algebra $\mathcal{M}(m,n;K)$ or $(\mathcal{M}(r,K)^2)^g$. The transposition of matrices gives a fundamental involution $\hat{\tau}$ on this algebra, and because of (6.7.8) there is an invertible homogeneous x such that $\tau(a) = (-1)^{\partial x \partial a} x\,\hat{\tau}(a)\,x^{-1}$ for every homogeneous a. Moreover $\hat{\tau}(x) = \pm x$ because $\tau^2 = \text{id}$. According to the type of A, according to the parity of x, and according to the sign $\pm$ in the previous equality, cp.dv.tr(τ) is one of the elements of $\mu_8(\mathbb{C})$.

When K has characteristic 2, similar methods should allow you to prove the part of (6.8.15) that deals with dv.tr(τ) and tw.dv.tr(τ).

(6.ex.18) Suppose that A is isomorphic to $\mathcal{M}(m,n;B)$ or to $\mathcal{M}(r,B)$ as in (6.6.2), and prove this theorem (attributed to Albert in the nongraded case): if A admits

involutions, then the graded central division algebra B and all algebras with the same class in $\mathrm{Br}^g(K)$ admit involutions.
Hint. B^{to} is isomorphic to $A' = \mathrm{End}^g_A(S)$ if S is a graded irreducible module over A.

(6.ex.19) Let K be a field that does not have characteristic 2, A a graded central simple algebra over K provided with an involution τ, and M a module over A^{ng} (that is A without grading). The grading of A is assumed to be nontrivial, (whence a pair of involutions $(\tau,\ \sigma\tau)$ on A), and M is assumed to have finite nonzero dimension over K. We consider $C'' = \mathrm{End}(M)$, $C' = \mathrm{End}_{A_0}(M)$, the space $\mathcal{P}''$ of all bilinear forms $\varpi : M \times M \to K$ and the subspace $\mathcal{P}'$ of all $\varpi \in \mathcal{P}''$ such that $\varpi(ax,\ y) = \varpi(x,\ \tau(a)y)$ for all $a \in A_0$ and all $x,\ y \in M$. If d is a nonzero element of the discriminant module D of $\mathrm{Z}(A_0, A)$, its image d_M in C'' belongs to C'.

(a) Prove that M is a semi-simple module over A^{ng}. If $\mathrm{Z}(A)$ is a field (either K or a quadratic field extension), then all irreducible modules over A^{ng} are isomorphic. But if $\mathrm{Z}(A) \cong (K^2)^g$, and if N is an irreducible module over A^{ng}, then every irreducible module is isomorphic to N or to the conjugate module N^c; in this case M is a faithful A^{ng}-module if and only if $\mathrm{Hom}_A(M, M^c) \neq 0$; and when it is not faithful, there is some $d \in D$ such that $d_M = \mathrm{id}_M$.

(b) Let $(b_1, b'_1, b_2, b'_2, \dots)$ be a family of elements of A_1 such that $\sum_i b_i b'_i = 1$. With every $g \in C'$ (resp. $\varpi \in \mathcal{P}'$) we associate $g^\dagger \in C''$ (resp. $\varpi^\dagger \in \mathcal{P}''$) defined in this way:

$$g^\dagger(x) = \sum_i b_i g(b'_i x) \quad (\text{resp. } \varpi^\dagger(x,y) = \sum_i \varpi(b'_i x,\ \tau(b_i)y)).$$

Prove that $g^{\dagger\dagger} = g$ and $(gh)^\dagger = g^\dagger h^\dagger$ for all $g,\ h \in C'$, that $\varpi^{\dagger\dagger} = \varpi$ for all $\varpi \in \mathcal{P}'$, and that, for all $b \in A_1$ and all $x,\ y \in M$,

$$g^\dagger(bx) = bg(x) \quad \text{and} \quad \varpi^\dagger(bx,\ y) = \varpi(x,\ \tau(b)y)\ ;$$

therefore $g^\dagger \in C'$ and $\varpi^\dagger \in \mathcal{P}'$. Thus C' becomes a graded algebra if C'_0 (resp. C'_1) is the subset of all $g \in C'$ such that $g^\dagger = g$ (resp. $g^\dagger = -g$), and $C'_0 = \mathrm{End}_A(M)$ whereas $C'_1 \cong \mathrm{Hom}_A(M, M^c)$. Similarly we define $\mathcal{P}'_0$ and $\mathcal{P}'_1$; thus $\mathcal{P}'_0$ (resp. $\mathcal{P}'_1$) is the space of scalar products on M associated with τ (resp. with $\sigma\tau$).
Let ρ''_M be the involutive endomorphism of $\mathcal{P}''$ that maps every ϖ to ϖ^o defined by $\varpi^o(x,y) = \varpi(y,x)$; obviously ρ''_M leaves $\mathcal{P}'_0$ and $\mathcal{P}'_1$ invariant; the calculation of the complex divided trace of the restriction ρ'_M of ρ''_M to $\mathcal{P}'$ is the main purpose of (6.ex.20).

(c) Prove that C' is a graded central simple algebra, that $A_0 \cong \mathrm{End}_{C'}(M)$ and that $\dim(A)\dim(C') = 2\dim(M)^2$. Moreover $\mathcal{P}'$ is a right module over C' that has the same dimension as C' over K. When M is a faithful A^{ng}-module, the quadratic extension $\mathrm{Z}(C'_0, C')$ is the intersection $K\mathrm{id}_M \oplus K d_M$ of C' with the image of A in C''; moreover $A \cong \mathrm{End}_{C'_0}(M)$. When M is not a faithful

A^{ng}-module, then $C'_1 = 0$, and $\mathcal{P}'_0$ (resp. $\mathcal{P}'_1$) is reduced to 0 if $\tau(d) = -d$ (resp. if $\tau(d) = d$).

Hint. When A has even type, deduce from (6.7.6) and (6.7.7) that $A \otimes \mathcal{P}'_0 \cong \mathcal{P}'' \cong A \otimes \mathcal{P}'_1$, that $A \otimes C'_0 \cong C''$ and that C'_0 is central simple; when A has odd type, then $A_0 \otimes \mathcal{P}' \cong \mathcal{P}''$, $A_0 \otimes C' \cong C''$ and $(C')^{ng}$ is central simple. Take notice that $\partial d_M = 1 - \partial d$.

(d) Prove that $\mathcal{P}'$ contains a homogeneous nondegenerate element ϖ_n that is symmetric or skew symmetric. It induces an involution τ'_M on C' in this way:

$$\forall g \in C', \ \forall x, \ y \in M, \quad \varpi_n(g(x), \ y) = (-1)^{\partial g \partial \varpi_n} \varpi_n(x, \ \tau'_M(g)(y)).$$

Prove that $\text{cp.dv.tr}(\rho'_M) = k \ \text{cp.dv.tr}(\tau'_M)$ for some $k \in \mu_4(\mathbb{C})$.

(6.ex.20) This is the continuation of (6.ex.19). To calculate $\text{cp.dv.tr}(\rho'_M)$ we consider the graded module $S = M \oplus M$ in which $S_0 = M \oplus 0$, $S_1 = 0 \oplus M$ and $a(x, y) = (ax, ay)$ for all $a \in A_0$ whereas $b(x, y) = (by, bx)$ for all $b \in A_1$. With S we also use all the notation explained in **6.8**. Now $A'' = \text{End}(S)$ can be identified with $\mathcal{M}(2, C'')$ where $C'' = \text{End}(M)$, and similarly $\mathcal{B}''$ can be identified with $\mathcal{M}(2, \mathcal{P}'')$. We distinguish two families of parity gradings: the gradings mentioned in **6.8** are called main gradings; by definition the main gradings of C' and $\mathcal{P}'$ are trivial, and the gradings of C' and $\mathcal{P}'$ defined in (6.ex.19) are called auxiliary gradings. Besides its main grading, S receives an auxiliary grading for which the homogeneous components are the diagonals of $M \oplus M$. Each grading of S determines a grading of A'' (resp. $\mathcal{B}''$), but on A'' (resp. $\mathcal{B}''$) we get two *compatible gradings* which make it split into a direct sum of four subspaces $A''_{i,j}$ (resp. $\mathcal{B}''_{i,j}$) with $(i, j) \in (\mathbb{Z}/2\mathbb{Z})^2$. Every $g \in C''$ has an image $g_{i,j}$ in each $A''_{i,j}$:

$$g_{0,0} = \begin{pmatrix} g & 0 \\ 0 & g \end{pmatrix}, \quad g_{0,1} = \begin{pmatrix} g & 0 \\ 0 & -g \end{pmatrix}, \quad g_{1,0} = \begin{pmatrix} 0 & g \\ g & 0 \end{pmatrix}, \quad g_{1,1} = \begin{pmatrix} 0 & -g \\ g & 0 \end{pmatrix}.$$

Similarly every $\varpi \in \mathcal{P}''$ has an image $\varpi_{i,j}$ in each $\mathcal{B}''_{i,j}$. The image of id_M in $A''_{1,1}$ is denoted by θ, in other words, $\theta(x, y) = (-y, x)$. It is easy to verify that $\theta \in A'$. Since $\theta^2 = -\text{id}_S$, we get a graded Azumaya algebra $\Theta = K\text{id}_S \oplus K\theta$ contained in A'.

(a) Prove that we get an algebra isomorphism from $\Theta \hat{\otimes} C'$ onto A' in this way:

$$\text{id}_S \otimes g \ + \theta \otimes h \ \longmapsto \ \begin{pmatrix} g & -h^\dagger \\ h & g^\dagger \end{pmatrix};$$

the twisting of the tensor product refers to the auxiliary gradings since the main grading is trivial on C'. This isomorphism $\Theta \hat{\otimes} C' \to A'$ is graded over the double group $(\mathbb{Z}/2\mathbb{Z})^2$.

Hint. The extension lemma (6.2.3) reveals a bijection $C' \to A'_0$; and $A'_1 = \theta A'_0$.

(b) Similarly there is a bijection from $\Theta \otimes \mathcal{P}'$ onto $\mathcal{B}'$:

$$\mathrm{id}_S \otimes \varpi_1 \;+ \theta \otimes \varpi_2 \;\longmapsto\; \begin{pmatrix} \varpi_1 & -\varpi_2^\dagger \\ \varpi_2 & \varpi_1^\dagger \end{pmatrix}.$$

(c) From this bijection deduce that

$$\mathrm{cp.dv.tr}(\rho') \;=\; \frac{1-i}{\sqrt{2}}\;\mathrm{cp.dv.tr}(\rho'_M).$$

The spinor spaces used in quantum mechanics

(6.ex.21) As in (6.2.2) let (M,q) be a quadratic space over $\mathbb{R}$ of dimension 4 and signature -2, and S a graded irreducible module over $A = C\ell(M,q)$; this is an algebra provided with a reversion τ and a conjugation $\sigma\tau$ as in (6.8.8). All other notation like A', $\mathcal{B}'$,... has been explained in **6.8**.

(a) Prove that A has the same class in $\mathrm{Br}^g(\mathbb{R})$ as $\mathbb{H}^g$. Consequently $A' \cong (\mathbb{H}^g)^{to}$. The choice of an element $i \in A'_0$ such that $i^2 = -\mathrm{id}_S$ allows us to identify A'_0 with $\mathbb{C}$. Then A'_1 is the subspace of all elements of A' that anticommute with i. Let j be any element of A'_1 such that $j^2 = \mathrm{id}_S$.

(b) For every $(u,v,w) \in (\mathbb{Z}/2\mathbb{Z})^3$ let $\mathcal{B}'_{u,v,w}$ be the space of scalar products on S associated with $\sigma^v\tau$, with the parity u and the symmetry corresponding to w; its elements $\beta : S \times S \to \mathbb{R}$ are characterized by these properties (for all s and $t \in S$ and all $a \in A$):

$$\beta(s,t) = 0 \;\text{ if }\; \partial s + \partial t \neq u,\;\; \beta(as,t) = \beta(s,\; \sigma^{u+v}\tau(a)t)\;,$$
$$\text{and}\quad \beta(t,s) = (-1)^w \beta(s,t)\;.$$

Prove that $\mathcal{B}'_{u,v,w}$ has dimension 1 if $u = 1$, dimension 0 if $(u,w) = (0,0)$, dimension 2 if $(u,w) = (0,1)$.

Let β_n be a nonzero (therefore nondegenerate) element of $\mathcal{B}'_{1,1,0}$. It induces an involution τ' on A'; prove that $\tau'(i) = -i$ and $\tau'(j) = j$.

(c) Let $J : S \otimes S \to A \hat{\otimes} A'$ be the isomorphism mentioned in (6.8.19). To calculate J we use an orthogonal basis (e_0, e_1, e_2, e_3) of M such that $q(e_0) = 1$ and $q(e_k) = -1$ for $k = 1,2,3$, and we derive from it a basis (e_F) of A indexed by the set $\mathcal{P}$ of all subsets $F \subset \{0,1,2,3\}$. In A' we use the basis $(\varepsilon_0, \varepsilon_1, \varepsilon_2, \varepsilon_3) = (1, i, j, ij)$. Verify that

$$J(s\otimes t) \;=\; \sum_{F\in\mathcal{P}}\sum_{k=0}^{3} (e_F\,\sigma\tau(e_F))\;\beta_n(s,\; e_F(\varepsilon_k t))\; e_F \otimes \varepsilon_k\;.$$

Hint. $e_F\sigma\tau(e_F) = \pm 1$ for all $F \in \mathcal{P}$, and $\varepsilon_k\tau'(\varepsilon_k) = 1$ for $k = 0,1,2,3$.

Comment. It is worth explaining how J allows us to study the action of the Clifford–Lipschitz group on the space $S \otimes S$ of bispinors. With every

$x \in \mathrm{GLip}(M, q)$ is associated a scalar $x\sigma\tau(x) \in \mathbb{R}^\times$ and an orthogonal transformation g defined by $g(a) = xa\sigma(x)^{-1}$ for all $a \in M$; from (6.8.17) and (5.1.3) (which defines $C\ell(g) = \Theta_x$) it is easy to deduce that

$$J(xs \otimes xt) \ = \ (x\sigma\tau(x)) \ (\sigma \otimes \sigma')^{\partial x} \circ (C\ell(g) \otimes \mathrm{id}_{A'}) \circ J(s, t).$$

(d) The mapping $s \otimes t \longmapsto is \otimes it$ is an involutive endomorphism of $S \otimes S$; the eigenspace of the eigenvalue -1 is identified with the tensor product $S \otimes_{\mathbb{C}} S$ over $\mathbb{C} = A'_0$, whereas the other eigenspace is denoted by $S \,\bar{\otimes}\,_{\mathbb{C}} S$. Thus $s \otimes_{\mathbb{C}} t$ and $s \,\bar{\otimes}\, t$ are identified respectively with $(s \otimes t - is \otimes it)/2$ and $(s \otimes t + is \otimes it)/2$. Since physicists are interested only in $\mathbb{C}$-sesquilinear mappings defined on $S \times S$, with a symmetric real part and a skew symmetric imaginary part, here we are interested especially in symmetric tensors that belong to $S \,\bar{\otimes}\,_{\mathbb{C}} S$. Prove that an element of $S \otimes S$ belongs to $S \,\bar{\otimes}\,_{\mathbb{C}} S$ (resp. $S \otimes_{\mathbb{C}} S$) if and only if it is mapped by J to an element of $A \otimes A'_0$ (resp. $A \otimes A'_1$).

Let us set $A^k = C\ell^k(M, q)$ for $k = 0, 1, 2, 3, 4$ (according to definitions in **4.8**); this means that A^k is spanned by all e_F such that F is a subset of cardinal k. Now $A \otimes A'_0$ becomes a direct sum of 10 subspaces $A^k \otimes 1$ and $A^k \otimes i$. Prove that an element of $S \,\bar{\otimes}\,_{\mathbb{C}} S$ is a symmetric (resp. skew symmetric) tensor in $S \otimes S$ if and only if it is mapped by J to an element of

$$\begin{aligned} &((A_0 \oplus A_1 \oplus A_4) \otimes 1) \oplus ((A_2 \oplus A_3) \otimes i) \\ &\big(\text{resp. } ((A_0 \oplus A_1 \oplus A_4) \otimes i) \oplus ((A_2 \oplus A_3) \otimes 1)\ \big). \end{aligned}$$

(e) For physicists, all the physical information given by an isolated spinor $s \in S$ comes from the five elements $\xi_k \in A^k$ (with $k = 0, 1, 2, 3, 4$) such that the component of $J(s \otimes s)$ in $A \otimes A'_0$ is

$$\xi_0 \otimes 1 \ - \xi_1 \otimes 1 \ + \xi_2 \otimes i \ - \xi_3 \otimes i \ + \xi_4 \otimes 1 \ ;$$

if we set $\alpha_k = 1$ for $k = 0, 1, 4$ and $\alpha_k = i$ for $k = 2, 3$, then

$$\forall c \in A^k, \quad \mathrm{Scal}(c\xi_k) = \beta_n(cs, \ \alpha_k s) \ .$$

Consider the sequence (ξ'_k) derived from another spinor s', and prove that $\xi_k = \xi'_k$ for $k = 0, 1, 2, 3, 4$ if and only if there exists $\theta \in \mathbb{R}$ such that $s' = (\cos\theta \ + i \sin\theta) \ s$.

Hint. The five equalities $\xi_k = \xi'_k$ are equivalent to the equality $s \,\bar{\otimes}\, s = s' \,\bar{\otimes}\, s'$ in $S \,\bar{\otimes}\,_{\mathbb{C}} S$; now if s, t, s', t' are nonzero elements of S, the equality $s \,\bar{\otimes}\, t = s' \,\bar{\otimes}\, t'$ is equivalent to the existence of $\nu \in \mathbb{C}$ such that $s' = \nu s$ and $t = \bar{\nu} t'$.

(6.ex.22) In order to get more information about the sequence (ξ_k) of "observable quantities" derived from a spinor s in (6.ex.21)(e), we use a particular spinor space S. It is a space A_0^{sp} that is canonically isomorphic to A_0 through a bijection

$x \longmapsto x^{sp}$; the notation x^{sp} means "x treated as a spinor". Let (e_0, e_1, e_2, e_3) be an orthonormal basis of M as in (6.ex.21)(c); the discriminant module $C\ell^4(M,q)$ of $QZ(M,q)$ is spanned by $\omega = e_0e_1e_2e_3$. Since ${e_0}^2 = 1$, the notation ${e_0}^n$ is meaningful when $n \in \mathbb{Z}/2\mathbb{Z}$.

(a) A spinor x^{sp} is said to be even (resp. odd) if $\omega x = xe_2e_1$ (resp. $\omega x = -xe_2e_1$); moreover if a is any (homogeneous) element of A, we set $ax^{sp} = (ax\,{e_0}^{\partial a})^{sp}$. Prove that A_0^{sp} is now a graded irreducible module S over A, and that we get an element $i \in A_0'$ such that $i^2 = -1$ if we set $ix^{sp} = (xe_2e_1)^{sp}$.

(b) Prove that $x\tau(y) + y\tau(x)$ and $\tau(x)y + \tau(y)x$ are equal and belong to $QZ(M,q)$ for all $x, y \in A_0$. Consequently there are two bilinear forms β_n and β' on $S = A_0^{sp}$ such that

$$x\tau(y) + y\tau(x) \;=\; 2\,\beta_n(x^{sp}, y^{sp}) \;+ 2\omega\;\beta'(x^{sp}, y^{sp})\;.$$

The factor 2 in this definition ensures that $\beta_n(1^{sp}, 1^{sp}) = 1$. Prove that β_n and β' belong respectively to $\mathcal{B}'_{1,1,0}$ and $\mathcal{B}'_{1,0,0}$. Moreover $\beta'(x^{sp}, y^{sp}) = -\beta_n(\omega x^{sp},\; y^{sp})$.

Hint. Since the algebra $C\ell(M,q)$ is generated by M, it suffices to prove that

$$\forall a \in M, \quad 2(\beta_n + \omega\beta')(ax^{sp},\; y^{sp}) \;=\; 2(\beta_n - \omega\beta')(x^{sp},\; ay^{sp})\;;$$

remember (4.8.16) and $a\omega = -\omega a$ for all $a \in M$.

(c) The above β_n enables us to associate a sequence (ξ_k) of five elements $\xi_k \in A_k$ with every spinor $s = x^{sp} \in A_0^{sp}$ as is explained in (6.ex.21)(e). Prove the following equalities (discovered by Hestenes in a quite different way):

$$\begin{aligned}
\xi_0 + \xi_4 = x\,\tau(x) \;&=\; \tau(x)\,x\;,\\
\xi_1 &= x\,e_0\,\tau(x)\;,\\
\xi_2 &= x\,e_1e_2\,\tau(x)\;,\\
\xi_3 &= x\,e_0e_1e_2\,\tau(x).
\end{aligned}$$

Prove that $\xi_1 \neq 0$ if $x \neq 0$. (*Hint:* prove that $b_q(e_0, \xi_1) > 0$.)

(d) Prove the following relations:

$$\begin{aligned}
&{\xi_1}^2 = -{\xi_3}^2 = (\xi_0 + \xi_4)(\xi_0 - \xi_4)\;,\\
&\xi_1\xi_3 = \xi_3\xi_1 = (\xi_0 - \xi_4)\xi_2\;,\\
&\xi_1\xi_2 = (\xi_0 - \xi_4)\xi_3 \qquad \text{or equivalently} \quad \xi_2\xi_1 = (\xi_0 + \xi_4)\xi_3\;,\\
&\xi_3\xi_2 = -(\xi_0 - \xi_4)\xi_1 \qquad \text{or equivalently} \quad \xi_2\xi_3 = -(\xi_0 + \xi_4)\xi_1\;,\\
&{\xi_2}^2 = -(\xi_0 + \xi_4)^2\;.
\end{aligned}$$

Comments. The four relations in the first two lines are called the Fierz identities. When $(\xi_1, \xi_4) \neq (0,0)$, all other relations are consequences of the Fierz identities. In the usual version of the Fierz identities, ξ_3 is replaced with $\omega\xi_3 \in M$, and

ξ_4 with $-\omega\xi_4 \in \mathbb{R}$. The third equality implies that ξ_1 and $\omega\xi_3$ are orthogonal vectors in M, and the first one shows that $q(\xi_1) = -q(\omega\xi_3)$. Observe that $\xi_0 + \xi_4$ and $\xi_0 - \xi_4$ are conjugate elements in $\mathrm{QZ}(M, q)$ (isomorphic to $\mathbb{C}$), and that their product is a nonnegative real number; consequently ξ_1 belongs to the subset of all $a \in M$ such that $q(a) \geq 0$. When 0 is removed from this subset, there remain two connected components called "past" and "future". Since ξ_1 never vanishes when $x \neq 0$, it remains in the same connected component.

In the general setting of (6.ex.21), J depends on the choice of a nonzero $\beta_n \in \mathcal{B}'_{1,1,0}$; replacing β_n with $-\beta_n$ would change J into $-J$, and ξ_1 into $-\xi_1$; consequently the two connected components of $\mathcal{B}'_{1,1,0} \setminus \{0\}$ correspond respectively to "past" and "future"; in other words, the orientation of the line $\mathcal{B}'_{1,1,0}$ is equivalent to the orientation of time. It is always assumed that e_0 is oriented toward "future", and that β_n is chosen so that ξ_1 too is oriented toward "future"; this assumption is compatible with the choice of β_n in (6.ex.22) because $\xi_1 = e_0$ when $x = 1$. Since the orientations of e_0 and β_n determine each other, and since $\mathcal{B}'_{1,0,0}$ is generated by the scalar product $(x^{sp}, y^{sp}) \longmapsto \beta_n(\omega x^{sp}, y^{sp})$, the orientation of the line $\mathcal{B}'_{1,0,0}$ depends on $e_0\omega = e_1e_2e_3$; therefore it is equivalent to the orientation of space. Unlike the orientation of space which is a pure convention, the orientation of time has a physical meaning; this explains why S becomes a genuine spinor space only after the choice of a nonzero element β_n in $\mathcal{B}'_{1,1,0}$ as above.

Chapter 7

Hyperbolic Spaces

The main object in this chapter (from Section **7.2** onwards) is a quadratic space (M, q) that is decomposed into a direct sum of a totally isotropic submodule U such that $U = U^\perp$, and *any* supplementary submodule V. Thus (M, q) is hyperbolic (see (2.5.5)) even if V is not totally isotropic, and d_q induces canonical isomorphisms $U \to (M/U^\perp)^* \to V^*$ and $V \to M/U^\perp \to U^*$ (see (2.3.7)). As a quadratic module, V will also interest us, and the short notation (V, q) (instead of $(V, q_{|V})$) will be preferred; it is clear that the quadratic form q on M is determined by its restriction to V and by the above isomorphism $U \to V^*$.

Most of this chapter is devoted to a generalization of Chevalley's theory of hyperbolic spaces (see [Chevalley 1954], Chap. III). Nonetheless it must be recalled that Chevalley's results involve only hyperbolic spaces over fields, and decompositions $M = U \oplus V$ in which both U and V are totally isotropic. Therefore the large part of the section **7.5** that is devoted to an arbitrary quadratic module (V, q) (with V a finitely generated projective K-module and $q(V)$ in general $\neq 0$) is completely foreign to Chevalley's ideas.

The main result is the existence of a canonical bijection between all totally isotropic direct summands T of (M, q) such that $T^\perp = T$ and all direct summands of $C\ell(V, q)$ of constant rank 1 contained in $\mathrm{Lip}(V, q)$. Therefore the concept of "simple spinor" in [Cartan 1938], which turned into "pure spinor" in [Chevalley 1954], becomes here superfluous, since the simple or pure spinors will prove to be the lipschitzian elements in $C\ell(V, q)$. Here we use the ideas of both Lipschitz and Chevalley, and we are happy to demonstrate how advantageous it is to respect the whole cliffordian tradition without exclusion.

7.1 Some representations of Clifford algebras

Let (M, q) be a quadratic module over the ring K. We assume that M is the direct sum $U \oplus V$ of two submodules satisfying these properties: U is totally isotropic (in other words, $q(U) = 0$) and the restriction of q to V admits a scalar product (see

(4.8.6)). Here are immediate consequences of these hypotheses; first the canonical injections $U \to M$ and $V \to M$ extend to injective algebra morphisms $\bigwedge(U) \to C\ell(M,q)$ and $C\ell(V,q) \to C\ell(M,q)$ because of (4.8.5); the notation $C\ell(V,q)$ means the Clifford algebra associated with the restriction of q to V. The algebras $\bigwedge(U)$ and $C\ell(V,q)$ are identified with their images in $C\ell(M,q)$. Moreover q itself admits scalar products; we can find scalar products $\beta : M \times M \to K$ such that $\beta(M,U) = 0$ (or equivalently $\beta(b,a) = \mathrm{b}_q(b,a)$ whenever $b \in U$), and scalar products β such that $\beta(U,M) = 0$ (or equivalently $\beta(a,b) = \mathrm{b}_q(a,b)$ whenever $b \in U$). Whatever β may be, we always require $\beta(U,U) = 0$.

(7.1.1) **Theorem.** *The algebra $C\ell(M,q)$ is the direct sum of the subalgebra $C\ell(V,q)$ and the left ideal $C\ell(M,q)U$ generated by U. Thus $C\ell(V,q)$ is canonically isomorphic to the quotient $C\ell(M,q)/C\ell(M,q)U$. If we transport into $C\ell(V,q)$ the action of $C\ell(M,q)$ in this quotient, then $C\ell(M,q)$ acts in $C\ell(V,q)$ in this way: every $z \in C\ell(V,q)$ operates in $C\ell(V,q)$ by the Clifford multiplication $v \longmapsto zv$, and every $y \in \bigwedge(U)$ operates by the interior multiplication $v \longmapsto \bigwedge(\mathrm{d}_q)(y) \rfloor v$; here $\bigwedge(\mathrm{d}_q)$ means the algebra morphism $\bigwedge(U) \to \bigwedge(M^*) \to \bigwedge^*(M)$ extending the restriction of d_q to U.*

Proof. Let β be an admissible scalar product for q such that $\beta(M,U) = 0$; we can replace $C\ell(M,q)$ with the isomorphic algebra $\bigwedge(M;\beta)$. Since $xb = x \wedge b$ for every $x \in \bigwedge(M;\beta)$ and every $b \in U$, the left ideal $\bigwedge(M;\beta)U$ is also the ideal generated by U in the exterior algebra $\bigwedge(M)$; this algebra is isomorphic to $\bigwedge(V) \hat{\otimes} \bigwedge(U)$, and is obviously the direct sum of $\bigwedge(V)$ and the ideal generated by U. As a submodule of $\bigwedge(M)$, the subalgebra of $\bigwedge(M;\beta)$ generated by V is the same thing as $\bigwedge(V)$; consequently $C\ell(M,q)$ is the direct sum of $C\ell(V,q)$ and $C\ell(M,q)U$.

The operations of every $x \in C\ell(M,q)$ in $C\ell(V,q)$ maps every $v \in C\ell(V,q)$ to the unique $v' \in C\ell(V,q)$ such that $v' - xv$ belongs to $C\ell(M,q)U$. If x is in $C\ell(V,q)$, it is clear that $v' = xv$. When b is in U, from (4.4.12) we derive $bv = \mathrm{d}_q(b) \rfloor v + \sigma(v)b$; this shows that the operation of b maps v to $v' = \mathrm{d}_q(b) \rfloor v$; for a general element $y \in \bigwedge(U)$, the conclusion follows from (4.4.3). □

The next lemma and its corollary corroborate the stability of the lipschitzian property by a large number of natural operations (for instance interior multiplications as in (5.3.13)).

(7.1.2) **Lemma.** *If x is a lipschitzian element of $C\ell(M,q)$, its projection in $C\ell(V,q)$ with respect to the left ideal $C\ell(M,q)U$ is also lipschitzian.*

Proof. When $\beta(M,U) = 0$ as in the proof of (7.1.1), the left ideal generated by U in $\bigwedge(M,\beta)$ is the ideal generated by U in $\bigwedge(M)$; therefore the projection from $\bigwedge(M;\beta)$ onto $\bigwedge(V;\beta)$ with respect to this ideal coincides with the algebra morphisms $\bigwedge(M) \to \bigwedge(V)$ associated by the functor $\bigwedge$ to the projection $M \to V$. Since x is also lipschitzian in $\bigwedge(M)$ (by the invariance property (5.4.1)), the conclusion follows from the stability of the lipschitzian property stated in (5.3.14). □

(7.1.3) **Corollary.** *If x and z are lipschitzian elements respectively in* $C\ell(M,q)$ *and* $C\ell(V,q)$*, the representation of* $C\ell(M,q)$ *in* $C\ell(V,q)$ *defined in* (7.1.1) *maps* (x,z) *to a lipschitzian element of* $C\ell(V,q)$.

Proof. The product xz is still lipschitzian (see (5.3.2)), and also its projection onto $C\ell(V,q)$ with respect to $C\ell(M,q)U$ (see (7.1.2)). This projection is exactly the image of (x,z) in $C\ell(V,q)$. □

To the previous hypotheses we add this one: U is a finitely generated projective module; then the representation of $C\ell(M,q)$ in its quotient by the left ideal $C\ell(M,q)U$ is equivalent to its representation in the left ideal generated by $\bigwedge^{\max}(U)$; this is the submodule of all $\omega \in \bigwedge(U)$ such that $b \wedge \omega = 0$ for all $b \in U$ (see(3.2.6)); it is a direct summand of $\bigwedge(U)$ of constant rank 1.

(7.1.4) **Proposition.** *The multiplication mapping $z \otimes \omega \longmapsto z\omega$ induces a bijection from* $C\ell(V,q) \otimes \bigwedge^{\max}(U)$ *onto the left ideal of* $C\ell(M,q)$ *generated by* $\bigwedge^{\max}(U)$*. Thus there is a representation of* $C\ell(M,q)$ *in* $C\ell(V,q)$ *for which the operation of every $x \in C\ell(M,q)$ maps every v to v' such that $v'\omega = xv\omega$ for all $\omega \in \bigwedge^{\max}(U)$; this representation is the same as the one defined in* (7.1.1).

Proof. Since $C\ell(M,q)$ is the direct sum of $C\ell(V,q)$ and $C\ell(M,q)U$, and since every multiplication by an element of U annihilates $\bigwedge^{\max}(U)$, the left ideal $C\ell(M,q)\bigwedge^{\max}(U)$ is generated by the products $z\omega$ with $z \in C\ell(V,q)$. Let us use the same scalar product β as in the proof of (7.1.1), so that $z\omega = z \wedge \omega$. From the canonical isomorphism $\bigwedge(V) \hat{\otimes} \bigwedge(U) \to \bigwedge(M)$ we deduce that $C\ell(V,q) \otimes \bigwedge^{\max}(U)$ is mapped bijectively onto the left ideal generated by $\bigwedge^{\max}(U)$.

The action of the algebra $C\ell(M,q)$ in its left ideal $C\ell(M,q)\bigwedge^{\max}(U)$ gives an action of $C\ell(M,q)$ in $C\ell(V,q)$ because this left ideal is isomorphic to $C\ell(V,q) \otimes \bigwedge^{\max}(U)$, and the module $\bigwedge^{\max}(U)$ is invertible. This new action of $C\ell(M,q)$ in $C\ell(V,q)$ coincides with the previous action defined in (7.1.1). Indeed for every $x \in C\ell(M,q)$, the previous operation of x in $C\ell(V,q)$ maps every v to the element v' such that $v' - xv$ belongs to $C\ell(M,q)U$, and this implies $(v' - xv)\omega = 0$. □

Hurried readers may skip the end of this section which is devoted to precise calculations involving the representation of $C\ell(M,q)$ in $C\ell(V,q)$ defined in (7.1.1). We choose an admissible scalar product β for q, and we replace $C\ell(M,q)$ with $\bigwedge(M;\beta)$; we require $\beta(U,U) = 0$, so that two elements of $\bigwedge(U)$ have the same product in $\bigwedge(M;\beta)$ as in $\bigwedge(M)$. An element of M is now written $b + c$, with b and c its components in U and V. The exterior multiplication mapping π : $\bigwedge(M) \hat{\otimes} \bigwedge(M) \to \bigwedge(M)$ is the algebra morphism associated by the functor $\bigwedge$ with the mapping $M \oplus M \to M$ defined by $(b + c,\ b' + c') \longmapsto b + b' + c + c'$.

We also use the algebra morphism $\varpi : \bigwedge(M) \to \bigwedge(M)$ associated by the functor $\bigwedge$ with the parallel projection from M onto V with respect to U, that is $b + c \longmapsto c$. It already appeared in the proof of (7.1.2).

At last from β we derive an element $\beta^\dagger$ of $\bigwedge^{*2}(M)$:

$$\beta^\dagger((b+c)\wedge(b'+c')) \ = \ \beta(c',b)-\beta(c,b') \ ;$$

obviously $\beta^\dagger$ only depends on the restriction of β to $V\times U$; if we choose β such that $\beta(M,U)=0$ (as in the proof of (7.1.1)), then $\beta^\dagger=0$. But $\beta^\dagger\neq 0$ when (for instance) 2 is invertible in K and $\beta=\mathrm{b}_q/2$ (the canonical scalar product).

(7.1.5) **Proposition.** *For every $x\in\bigwedge(M;\beta)$, the projection of x onto $\bigwedge(V;\beta)$ with respect to the left ideal $\bigwedge(M;\beta)U$ is equal to* $\varpi(\mathrm{Exp}(\beta^\dagger)\rfloor x)$.

Proof. We identify $\bigwedge(M)$ with $\bigwedge(U)\hat{\otimes}\bigwedge(V)$, and we write $u\otimes 1$ and $1\otimes v$ for elements that are respectively in $\bigwedge(U)$ and $\bigwedge(V)$. Thus the left ideal $C\ell(M,q)U$ is generated by the products $(1\otimes v)(u\otimes 1)$ with $u\in\bigwedge^{>0}(U)$, and ϖ is the parallel projection from $\bigwedge(U)\hat{\otimes}\bigwedge(V)$ onto $1\otimes\bigwedge(V)$ with respect to $\bigwedge^{>0}(U)\otimes\bigwedge(V)$. We must prove that, for all u and v respectively in $\bigwedge^{>0}(U)$ and $\bigwedge(V)$;

$$\varpi(\ \mathrm{Exp}(\beta^\dagger)\ \rfloor\ (1\otimes v))=1\otimes v,$$
$$\varpi(\ \mathrm{Exp}(\beta^\dagger)\ \rfloor\ (1\otimes v)(u\otimes 1))=0\ .$$

The first equality comes from the fact that the interior multiplication by $\mathrm{Exp}(\beta^\dagger)$ leaves $1\otimes v$ invariant: this follows from (4.5.9) since $\beta^\dagger$ is a linear form on $\bigwedge(U)\hat{\otimes}\bigwedge(V)$ vanishing on all $\bigwedge^i(U)\otimes\bigwedge^j(V)$ except $U\otimes V$. Now we calculate $(1\otimes v)(u\otimes 1)$ in $\bigwedge(M;\beta)$ according to (4.7.1):

$$\begin{aligned}(1\otimes v)(u\otimes 1)&=\pi(\ \mathrm{Exp}(\beta_{\prime\prime})\ \rfloor\ (1\otimes v)\otimes(u\otimes 1))\\ &=(-1)^{\partial u\partial v}\ \pi(\ \mathrm{Exp}(\beta_{\prime\prime}^{to})\ \rfloor\ (u\otimes 1)\otimes(1\otimes v)).\end{aligned}$$

We shall use (4.4.6) with the following morphism:

$$w\ :\ U\oplus V\oplus U\oplus V\longrightarrow U\oplus V\ ,\quad (b,c,b',c')\longmapsto(b,c')\ ;$$

on one side, π and $\bigwedge(w)$ have the same restriction to $\bigwedge(U)\otimes 1\otimes 1\otimes\bigwedge(V)$ since they map $(u\otimes 1)\otimes(1\otimes v)$ to $u\otimes v$; on the other side, it is easy to verify that $\beta_{\prime\prime}^{to}$ and $\bigwedge^*(w)(-\beta^\dagger)$ have the same restriction to $\bigwedge(U)\otimes 1\otimes 1\otimes\bigwedge(V)$; thus from (4.4.6) we derive

$$\begin{aligned}(1\otimes v)(u\otimes 1)&=(-1)^{\partial u\partial v}\ \bigwedge(w)\ (\mathrm{Exp}(\bigwedge^*(w)(-\beta^\dagger))\ \rfloor\ (u\otimes 1)\otimes(1\otimes v))\\ &=(-1)^{\partial u\partial v}\ \mathrm{Exp}(-\beta^\dagger)\ \rfloor\ (u\otimes v)\ .\end{aligned}$$

The calculation ends in this way:

$$\begin{aligned}\varpi(\ \mathrm{Exp}(\beta^\dagger)\ \rfloor\ (1\otimes v)(u\otimes 1))&=\pm\ \varpi(\ \mathrm{Exp}(\beta^\dagger)\ \rfloor\ (\mathrm{Exp}(-\beta^\dagger)\ \rfloor\ (u\otimes v)))\\ &=\pm\ \varpi(u\otimes v)\\ &=0\quad\text{because } u\in\textstyle\bigwedge^{>0}(U).\end{aligned}$$

□

(7.1.6) **Corollary.** *For every $x \in C\ell(M,q)$ let R_x be the endomorphism of $C\ell(M,q)$ defined in this way: it maps the left ideal $C\ell(M,q)U$ to 0, and its restriction to $C\ell(V,q)$ is the operation of x in $C\ell(V,q)$ corresponding to the representation defined in (7.1.1). It maps every $z \in C\ell(M,q)$ to*

$$R_x(z) \;=\; \varpi\pi\big(\,\mathrm{Exp}(\beta_{/\!/} + \pi^*(\beta^\dagger)) \;\rfloor\; (x \otimes z)\big).$$

Proof. Since $R_x(z)$ is the projection of xz onto $\bigwedge(V;\beta)$ with respect to $\bigwedge(M;\beta)U$, we begin the calculation by means of (7.1.5) and pursue it with the help of (4.4.6):

$$\begin{aligned} R_x(z) &= \varpi(\,\mathrm{Exp}(\beta^\dagger) \;\rfloor\; xz) \\ &= \varpi(\,\mathrm{Exp}(\beta^\dagger) \;\rfloor\; \pi(\mathrm{Exp}(\beta_{/\!/}) \;\rfloor\; (x \otimes z))) \\ &= \varpi\pi(\,\mathrm{Exp}(\pi^*(\beta^\dagger)) \;\rfloor (\mathrm{Exp}(\beta_{/\!/}) \;\rfloor\; (x \otimes z))) \\ &= \varpi\pi(\,\mathrm{Exp}(\beta_{/\!/} + \pi^*(\beta^\dagger)) \;\rfloor\; (x \otimes z)). \end{aligned}$$ □

It is worth giving the precise expression of the element of $\bigwedge^{*2}(M \oplus M)$ that appears in (7.1.6):

$$\begin{aligned} &(\beta_{/\!/} + \pi^*(\beta^\dagger))\;((b_1 + c_1,\; b_2 + c_2) \wedge (b_1' + c_1',\; b_2' + c_2')) \\ &\quad = \beta(b_1 + c_1,\; b_2' + c_2') - \beta(b_1' + c_1',\; b_2 + c_2) \\ &\qquad + \beta(c_1' + c_2',\; b_1 + b_2) - \beta(c_1 + c_2,\; b_1' + b_2') \\ &\quad = \mathrm{b}_q(b_1, c_2') - \mathrm{b}_q(b_1', c_2) \;+\; \beta(c_1, c_2') - \beta(c_1', c_2) \\ &\qquad - \beta(c_1, b_1') + \beta(c_1', b_1) \;-\; \beta(c_2, b_2') + \beta(c_2', b_2)\;; \end{aligned}$$

thus $\beta_{/\!/} + \pi^*(\beta^\dagger)$ is decomposed into a sum of four terms; the first term (which gives $\mathrm{b}_q(b_1, c_2') - \mathrm{b}_q(b_1', c_2)$) does not depend on the choice of β; when later (M,q) is a hyperbolic space and $U \cong V^*$, this term represents the duality between U and V. The second term (which gives $\beta(c_1, c_2') - \beta(c_1', c_2)$) only depends on the multiplication inside the subalgebra $\bigwedge(V,\beta)$. The third and fourth terms are $\beta^\dagger \otimes 1$ and $1 \otimes \beta^\dagger$, and they vanish when β is chosen in such a way that $\beta(M,U) = 0$. But even when $\beta^\dagger \neq 0$, the last term $1 \otimes \beta^\dagger$ can be dropped when we calculate $R_x(z)$ with z already in $\bigwedge(V,\beta)$; indeed the interior multiplication by $\mathrm{Exp}(1 \otimes \beta^\dagger)$ leaves $x \otimes z$ invariant for all $z \in \bigwedge(V,\beta)$ (again (4.5.9)). Only the first three terms are involved in the restriction of R_x to $C\ell(V,q)$.

7.2 The Cartan–Chevalley mapping

In this section and in all the following ones, we assume that (M,q) is a quadratic space, that M is a direct sum $U \oplus V$ of two submodules, that U is totally isotropic, and that $U^\perp = U$; according to (2.5.5), (M,q) is a hyperbolic space isomorphic to $\mathbf{H}(U)$, and also to $\mathbf{H}(V)$. Although the definition of $\mathbf{H}(V)$ ignores the restriction of q to V, this must not suggest that this restriction should vanish. On the contrary, since the purpose of this chapter is double: first the study of a hyperbolic space like (M,q), and secondly the study of any quadratic module (V,q) (with V a finitely generated projective module) by means of an embedding into a hyperbolic space $(U \oplus V;\ q)$ with $U \cong V^*$.

According to (7.1.1) there is a representation of the algebra $C\ell(M,q)$ in the module $C\ell(V,q)$; the operation in $C\ell(V,q)$ of any $x \in C\ell(M,q)$ is denoted by R_x. This R_x is not exactly the same as the one defined in (7.1.6), because the latter has been extended by 0 on the supplementary left ideal $C\ell(M,q)U$; but this discrepancy is tiny enough to be overlooked.

(7.2.1) **Theorem.** *When (M,q) is a hyperbolic space as above, the mapping $x \longmapsto R_x$ is a graded algebra isomorphism from $C\ell(M,q)$ onto $\mathrm{End}(C\ell(V,q))$.*

Proof. When V is totally isomorphic, (7.2.1) is the same thing as (3.7.2), because the isomorphism $C\ell(M,q) \to \mathrm{End}(\bigwedge(V))$ constructed in the proof of (3.7.2) is exactly the present mapping $x \longmapsto R_x$. When V is not totally isotropic, (2.5.4) will still state the existence of a totally isotropic submodule V' supplementary to U. As modules over $C\ell(M,q)$, both $C\ell(V,q)$ and $\bigwedge(V')$ are isomorphic to the quotient of $C\ell(M,q)$ by the left ideal $C\ell(M,q)U$. Since we get an isomorphism $C\ell(M,q) \to \mathrm{End}(\bigwedge(V'))$ for the module $\bigwedge(V')$ over $C\ell(M,q)$, we also get an isomorphism $C\ell(M,q) \to \mathrm{End}(C\ell(V,q))$ for the isomorphic module $C\ell(V,q)$. □

Now we must prove that $C\ell(M,q)$ is the direct sum of the left ideal $C\ell(M,q)U$ and some other left ideal; this follows immediately from the next lemma when V is totally isotropic; but it is true even when V is not totally isotropic, since M is the direct sum of U and some other totally isotropic submodule (see (2.5.4)). When V is totally isotropic, $\bigwedge^{\max}(V)$ is defined according to (3.2.6).

(7.2.2) **Lemma.** *When the hyperbolic space (M,q) is the direct sum of two totally isotropic submodules U and V, then $C\ell(M,q)$ is the direct sum of the left ideals generated by U and $\bigwedge^{\max}(V)$. The left ideal $C\ell(M,q)\bigwedge^{\max}(V)$ has the same rank as $\bigwedge(U)$ at any prime ideal of K.*

Proof. By inverting the roles of U and V, from (7.1.4) we deduce that the left ideal generated by $\bigwedge^{\max}(V)$ is the image of the multiplication mapping $\bigwedge(U) \otimes \bigwedge^{\max}(V) \to C\ell(M,q)$, and that it has everywhere the same rank as $\bigwedge(U)$. From (7.1.1) we know that $\bigwedge(V)$ is supplementary to the left ideal $C\ell(M,q)U$; to prove that the left ideal $\bigwedge(U)\bigwedge^{\max}(V)$ also is supplementary to $C\ell(M,q)U$, it suffices to find a bijection $f : \bigwedge(U)\bigwedge^{\max}(V) \to \bigwedge(V)$ such that $f(x) - x$ belongs to $C\ell(M,q)U$ for all $x \in \bigwedge(U)\bigwedge^{\max}(V)$. Since $y\omega' - R_y(\omega')$ belongs to $C\ell(M,q)U$ (by definition of R_y) for all $y \in \bigwedge(U)$ and $\omega' \in \bigwedge^{\max}(V)$, it suffices to prove the bijectiveness of the mapping

$$f' : \bigwedge(U) \otimes \bigwedge^{\max}(V) \longrightarrow \bigwedge(V)\,, \quad y \otimes \omega' \longmapsto R_y(\omega').$$

Because of (1.13.5) it suffices to prove that f' is surjective; besides, we can suppose that K is a local ring, and that $\bigwedge^{\max}(V)$ is generated by one element ω'. For every $z \in \bigwedge(V)$, there exists $x \in C\ell(M,q)$ such that $R_x(\omega') = z$, since the mapping $x \longmapsto R_x$ is bijective onto $\mathrm{End}(\bigwedge(V))$. This x is the sum of some $y \in \bigwedge(U)$ and some element of $C\ell(M,q)V$ (see (7.1.1)), whence $R_y(\omega') = R_x(\omega') = z$. This proves the surjectiveness of f'. □

Now we define the two objects that are the source and the target of the *Cartan–Chevalley mapping* χ involved in the next theorem. First $\mathcal{T}(M,q)$ is the set of all totally isotropic direct summands T of M such that $T = T^\perp$. For every direct summand T of M, the sum of the ranks of T and $T^\perp$ is the rank of M (see (2.3.7)); consequently *a totally isotropic direct summand T satisfies the equality $T = T^\perp$ if and only if its rank at every prime ideal of K (or equivalently at every maximal ideal) is equal to the rank of U or V*; the elements of $\mathcal{T}(M,q)$ are called *totally isotropic direct summands of maximal rank.* When K is a field, they are also called *maximal totally isotropic subspaces.*

Secondly $\mathrm{BLip}(V,q)$ is the set of all *lipschitzian direct summands of constant rank 1*, in other words, all graded direct summands of $C\ell(V,q)$ of constant rank 1 that are contained in $\mathrm{Lip}(V,q)$. The letter B in the notation $\mathrm{BLip}(V,q)$ recalls that it is the base of the bundle in which the total space is the subset of all $(z,Z) \in C\ell(V,q) \times \mathrm{BLip}(V,q)$ such that $z \in Z$. Of course $\mathcal{T}(M,q)$ may also be considered as the base of the bundle in which the total space is the subset of all $(a,T) \in M \times \mathcal{T}(M,q)$ such that $a \in T$. It must be observed that *the group* $\mathrm{G'Lip}(V,q)$ *is a subset of* $\mathrm{BLip}(V,q)$, in other words, every $Z \in \mathrm{G'Lip}(V,q)$ is a graded direct summand of $C\ell(V,q)$. Indeed K is a direct summand of $C\ell(V,q)$ (see (1.13.2)), and since Z is invertible inside $C\ell(V,q)$, the multiplication mapping $Z \otimes C\ell(V,q) \to C\ell(V,q)$ is a bijection that maps the direct summand $Z \otimes K$ onto the direct summand Z.

(7.2.3) **Theorem.** *For every $T \in \mathcal{T}(M,q)$, let $\chi(T)$ be the submodule of all $z \in C\ell(V,q)$ such that $R_a(z) = 0$ for all $a \in T$. For instance $\chi(U) = K$. This mapping χ is a bijection from $\mathcal{T}(M,q)$ onto* $\mathrm{BLip}(V,q)$. *Conversely every $Z \in \mathrm{BLip}(V,q)$ is the image of the submodule $\chi^{-1}(Z)$ of all $a \in M$ such that $R_a(z) = 0$ for all $z \in Z$.*

Beginning of the proof. Everything in (7.2.3) can be settled at once, except the fact that the image of χ is precisely $\mathrm{BLip}(V,q)$; this fact shall be proved later with the help of the next four propositions. For every $T \in \mathcal{T}(M,q)$, the graded left ideal $C\ell(M,q)T$ admits a supplementary graded left ideal J'; indeed this is asserted in (7.2.2) for $C\ell(M,q)U$, but is also valid for $C\ell(M,q)T$. Let ε and ε' be the projections of 1 in the supplementary graded left ideals $C\ell(M,q)T$ and J'; thus R_ε and $R_{\varepsilon'}$ are even idempotent elements of $\mathrm{End}(C\ell(V,q))$, the images of which are supplementary graded submodules of $C\ell(V,q)$, and R_ε and $R_{\varepsilon'}$ are the projections onto these submodules. We can write $\mathrm{Im}(R_{\varepsilon'}) = \mathrm{Ker}(R_\varepsilon) = \chi(T)$ because the left ideal generated by ε is precisely the left ideal generated by T.

The rank of the left ideal J' is equal to the rank of $\bigwedge(T)$ (see (7.2.2)), that is also the rank of $C\ell(V,q)$. The images of the elements of J' in $\mathrm{End}(C\ell(M,q))$ are all linear mappings $\chi(T) \to C\ell(V,q))$ extended by 0 on the supplementary submodule $\mathrm{Im}(R_\varepsilon)$; this allows us to calculate the rank of $\chi(T)$ and to prove that it is equal to 1. It is clear that $\chi(U)$ contains K, and is equal to K because K too is a direct summand of $C\ell(V,q)$.

Conversely let Z be a graded direct summand of $C\ell(V,q)$ of rank 1, and N the submodule of all $a \in M$ such that $R_a(Z) = 0$. This N is totally isotropic because $q(a)z = R_a(R_a(z)) = 0$ for all $a \in N$ and all $z \in Z$, and Z is a faithful module. If Z belongs to the image of χ, then N contains a submodule T that belongs to $\mathcal{T}(M,q)$, whence $N = T$ because $T = T^\perp$. This implies the injectiveness of χ. □

(7.2.4) **Historical comment.** Chevalley's definition of the Cartan–Chevalley mapping was somewhat different (see (7.ex.4)(a)), but he showed that his definition was equivalent to (7.2.3); he called "spinors" all the elements of $C\ell(V,q)$, and "pure spinors" those belonging to some line $\chi(T)$ with $T \in \mathcal{T}(M,q)$. He always assumed V to be totally isotropic like U. He could not describe the image of the Cartan–Chevalley mapping with reference to a Lipschitz monoid (as we are soon going to do) since Lipschitz's ideas were foreign to his own comprehension. But he discovered two properties characterizing the image of this mapping: first a property equivalent to Theorem (5.10.2), which is only valid when K is a field and V is totally isotropic, and secondly the Cartan–Chevalley criterion suggested by Elie Cartan's pioneering work. Later, when the existence of Lipschitz monoids began to be contemplated, it was conjectured that the image of the Cartan–Chevalley mapping should be closely related to $\mathrm{Lip}(V,q)$, and that the Cartan–Chevalley criterion should characterize the lipschitzian elements of $C\ell(V,q)$ even when V was not totally isotropic and when K was not a field. Under this conjecture the definition (5.3.1) of the Lipschitz monoid was derived from the Cartan–Chevalley criterion under the hypothesis that 2 was invertible in K, as is explained in (7.ex.7), and the relevance of the definition (5.3.1) was confirmed by the developments here expounded in Chapter **5**. Still later the Cartan–Chevalley criterion proved to be equivalent to the definition (5.3.1) even when 2 was not invertible, as is explained farther in Section **7.4**.

Lipschitz monoids appeared at several places and under various definitions, but always without reference to Lipschitz; indeed Lipschitz's work had meanwhile fallen into oblivion, and probably Chevalley much contributed to this oblivion. The renewal of Lipschitz's ideas was first due to the solving of new problems which needed them; only later did historical researches (for instance those of Lounesto) show that the solving of these problems should be understood as a continuation of Lipschitz's work.

To complete the proof of (7.2.3) we need the group $\mathrm{G'Lip}(M,q)$. By means of the representation $x \longmapsto R_x$ it acts on the set of all graded direct summands of $C\ell(V,q)$ of constant rank 1, and because of (7.1.3) it leaves invariant the subset $\mathrm{BLip}(V,q)$. Moreover $\mathrm{G'Lip}(M,q)$ acts in $\mathcal{T}(M,q)$ through the isomorphism $\mathrm{G'Lip}(M,q) \to \mathrm{GO}(M,q)$ (see (5.8.1)).

(7.2.5) **Proposition.** *The Cartan Chevalley mapping is equivariant when the group* $\mathrm{G'Lip}(M,q)$ *acts on one side on* $\mathcal{T}(M,q)$, *on the other side on the set of all direct summands of* $\mathrm{C}\ell(V,q)$ *of constant rank 1.*

Proof. Let X be an element of $\mathrm{G'Lip}(M,q)$, G_X its image in $\mathrm{GO}(M,q)$, T an element of $\mathcal{T}(M,q)$, and Z a direct summand of $\mathrm{C}\ell(V,q)$ of constant rank 1. By definition of G_X every $a \in T$ has an image $a' \in G_X(T)$ such that $a'x = (-1)^{\partial x} xa$ for all $x \in X$. The following six assertions are equivalent:

$$Z = \chi(T) \;;$$
$$\forall z \in Z,\quad \forall a \in T,\qquad R_a(z) = 0 \;;$$
$$\forall x \in X,\quad \forall z \in Z,\quad \forall a \in T,\qquad R_x(R_a(z)) = R_{xa}(z) = 0 \;;$$
$$\forall x \in X,\quad \forall z \in Z,\quad \forall a' \in G_X(T),\quad R_{a'}(R_x(z)) = R_{a'x}(z) = 0 \;;$$
$$\forall x \in X,\quad \forall z \in Z,\qquad R_x(z) \in \chi(G_X(T)) \;;$$
$$R_X(Z) = \chi(G_X(T)) \,.$$

□

(7.2.6) **Proposition.** *When K is a* local *ring, the group* $\mathrm{G'Lip}(M,q)$ *acts transitively in* $\mathcal{T}(M,q)$.

Proof. Let T be an element of $\mathcal{T}(M,q)$. It suffices to prove the existence of some $g \in \mathrm{GO}(M,q)$ such that $g(U) = T$. Because of (2.5.4) we know that (M,q) contains totally isotropic submodules U' and T' respectively supplementary to U and T. Since K is a local ring, U and T are free modules of the same rank; therefore there exists an isomorphism $f : U \to T$. The functor $\mathrm{Hom}(\dots, K)$ associates with f a morphism f^* between the dual spaces, and since b_q makes U' and T' become canonically isomorphic to the dual spaces U^* and T^*, we can consider f^* as an isomorphism $T' \to U'$. Let g be the endomorphism of M defined in this way for all $(b,b') \in U \times U'$: $g(b+b') = f(b) + f^{*-1}(b')$. It is easy to prove that g is an automorphism of (M,q), and consequently an element of $\mathrm{GO}(M,q)$ mapping U onto T. □

(7.2.7) **Proposition.** *When K is a* local *ring, the group* $\mathrm{G'Lip}(M,q)$ *acts transitively in* $\mathrm{BLip}(V,q)$.

Proof. Let Z be an element of $\mathrm{BLip}(V,q)$ generated by some element z. We must prove the existence of some $x \in \mathrm{GLip}(M,q)$ such that $R_x(1) = z$. For this purpose we use a scalar product β on M that is admissible for q, we replace $\mathrm{C}\ell(M,q)$ with $\bigwedge(M;\beta)$, and we first prove that there exists a bilinear form $\beta' : V \times V \to K$ such that z is invertible in $\bigwedge(V,\beta')$, or in other words, such that $z \in \mathrm{GLip}(V;\beta')$. Since anyhow z belongs to $\mathrm{Lip}(V;\beta')$, the product of z and $\tau_{\beta'}(z)$ in $\bigwedge(V,\beta')$ belongs to K, and we must choose β' in such a way that this product is invertible in K. But its invertibility in K is equivalent to the invertibility of its image in $K/\mathfrak{m}$, if $\mathfrak{m}$ is the maximal ideal of K. Besides, since V is a free module, every bilinear form $V/\mathfrak{m}V \times V/\mathfrak{m}V \to K/\mathfrak{m}$ comes from a bilinear form $V \times V \to K$. Thus it suffices to prove the existence of β' when K is a field. When K is a field, it has been proved in (5.10.1) that z is actually invertible in some algebra $\bigwedge(V;\beta')$.

Since we get a bijection $U \to V^*$ if we map every $a \in U$ to the restriction of $\mathrm{d}_q(a)$ to V, there exists a mapping $f : V \to U$ such that

$$\forall c,\ c' \in V\ , \quad \beta'(c,c') \ = \ \beta(c,c') + \mathrm{b}_q(f(c),\, c')\ ;$$

according to (4.8.9), this means that the left multiplication by c in $\bigwedge(V;\beta')$ is equal to $R_{c+f(c)}$. Let V' be the submodule of all $c + f(c)$ with $c \in V$. Since $q(c + f(c)) = \beta'(c,c)$ for all $c \in V$, the bijection $c \longmapsto c + f(c)$ from V onto V' extends to an algebra isomorphism $F : \bigwedge(V;\beta') \to \bigwedge(V';\beta)$. For every $v \in \bigwedge(V)$, the left multiplication by v in $\bigwedge(V;\beta')$ is equal to $R_{F(v)}$. Moreover F maps z to some element of $\mathrm{GLip}(V';\beta)$, which is a subgroup of $\mathrm{GLip}(M;\beta)$. Since the left multiplication by z in $\bigwedge(V;\beta')$ maps 1 to z, we conclude that $R_{F(z)}(1) = z$. □

(7.2.8) **Proposition.** *When T belongs to $\mathcal{T}(M,q)$, a submodule of T that generates the same left ideal in $\mathrm{C}\ell(M,q)$ as T, is equal to T.*

Proof. It suffices to prove (7.2.8) when K is a local ring, and even when K is a field, because of Nakayama's Lemma (see (1.12.2)). Besides, it suffices to prove it when $T = U$. Let us suppose that K is a field and that U_1 is a subspace of U strictly smaller than U. Since b_q determines a duality between U and V, there exists a nonzero $c \in V$ such that $\mathrm{b}_q(b,c) = 0$ for all $b \in U_1$. This means that the subspace of all $z \in \mathrm{C}\ell(V,q)$ such that $R_b(z) = 0$ for all $b \in U_1$ is strictly larger than $\chi(U) = K$, since it contains c. Consequently U_1 cannot generate the same left ideal as U. □

End of the proof of (7.2.3). We must still prove that the image of χ is $\mathrm{BLip}(V,q)$. When K is a local ring, this is an immediate consequence of Propositions (7.2.5), (7.2.6), (7.2.7). When K is not a local ring, we already know that all localizations of $\chi(T)$ are contained in the corresponding Lipschitz monoid, and consequently $\chi(T)$ is contained in $\mathrm{Lip}(V,q)$ (see (5.3.5)); it follows that χ is an injective mapping from $\mathcal{T}(M,q)$ into $\mathrm{BLip}(V,q)$, and only its surjectiveness remains in question.

Let Z be an element of $\mathrm{BLip}(V,q)$, and T the submodule of all $a \in \mathrm{C}\ell(M,q)$ such that $R_a(Z) = 0$. As explained above, T is a totally isotropic submodule; we already know that, for every prime ideal $\mathfrak{p}$ of K, the localization $T_\mathfrak{p}$ belongs to $\mathcal{T}(M_\mathfrak{p}, q_\mathfrak{p})$. If we manage to prove that T is finitely generated, we can conclude that T is a projective module (see (1.12.9)), that T is a direct summand of M (see (1.13.1)), that its rank is that of U and V, and finally that T belongs to $\mathcal{T}(M,q)$, and $\chi(T) = Z$.

Let us prove that T is finitely generated. Since Z is a direct summand of $\mathrm{C}\ell(V,q)$, the left ideal J of all $x \in \mathrm{C}\ell(M,q)$ such that $R_x(Z) = 0$ is a direct summand of $\mathrm{C}\ell(M,q)$, and consequently is finitely generated. By localization it is easy to proved that the multiplication mapping $C\ell(M,q)\otimes T \to J$ is surjective. Since J is finitely generated, this mapping is already surjective when we replace T with some finitely generated submodule T_1. Every localization of T_1 generates the same left ideal as the corresponding localization of T. Because of (7.2.8), T and T_1 have the same localizations, and are equal. □

The other Cartan–Chevalley mapping $\chi^\perp : \mathcal{T}(M,q) \to \mathrm{BLip}(V^\perp, q)$

The equality $M = U \oplus V$ implies $M = U \oplus V^\perp$ (see (2.3.7)); consequently we can do with $V^\perp$ all we have done with V. There is an algebra isomorphism $x \longmapsto R_x^\perp$ from $C\ell(M,q)$ onto $\mathrm{End}(C\ell(V^\perp,q))$ and there is a Cartan–Chevalley bijection $\chi^\perp$ from $\mathcal{T}(M,q)$ onto $\mathrm{BLip}(V^\perp,q)$. First we shall compare R_x and $R_x^\perp$.

Since V and $V^\perp$ are both supplementary to U, there exists a unique isomorphism $p : V \to V^\perp$ such that $p(c)-c$ belongs to U for all $c \in V$. Since $q(p(c)-c)$ and $\mathrm{b}_q(c,p(c))$ both vanish, we realize that $q(p(c)) = -q(c)$ for all $c \in V$. Consequently p extends to an algebra isomorphism $C\ell(V,-q) \to C\ell(V^\perp,q)$. Since $C\ell(V,-q)$ is isomorphic to $C\ell(V,q)^{to}$ (see (3.2.2)), the mapping $c^{to} \longmapsto p(c)$ extends to an isomorphism $C\ell(V,q)^{to} \to C\ell(V^\perp,q)$.

(7.2.9) **Proposition.** *There is a unique bijection $\psi : Cl(V,q) \to C\ell(V^\perp,q)$ such that $\psi(z) - z$ belongs to the left ideal $C\ell(M,q)U$ for all $z \in C\ell(V,q)$ (whence $\psi(c) = p(c)$ if $c \in V$). It is an isomorphism of $C\ell(M,q)$-modules:*

$$\forall x \in C\ell(M,q),\ \forall z \in C\ell(V,q), \quad \psi(R_x(z)) = R_x^\perp(\psi(z)).$$

Besides, the mapping $z^{to} \longmapsto \psi(z)$ is an algebra isomorphism from $C\ell(V,q)^{to}$ onto $C\ell(V^\perp,q)$.

This last property of ψ shows that it commutes with the reversion τ, and that consequently $\psi(z) - z$ belongs to the right ideal $U\,C\ell(M,q)$ too.

Proof. The beginning of (7.2.9) follows from the fact that $C\ell(V,q)$ and $C\ell(V^\perp,q)$ are both supplementary to $C\ell(M,q)U$, and consequently isomorphic to $C\ell(M,q)/C\ell(M,q)U$ as modules over $C\ell(M,q)$. To prove the last assertion in (7.2.9), it suffices to prove that $p(c)\psi(z) = (-1)^{\partial z}\psi(zc)$ for all $c \in V$ and $z \in C\ell(V,q)$. This follows from $\mathrm{d}_q(p(c)) \rfloor z = 0$ and $\mathrm{d}_q(c) \rfloor z = cz - (-1)^{\partial z} zc$ (see (4.4.12)):

$$\begin{aligned} p(c)\,\psi(z) = R_{p(c)}^\perp(\psi(z)) &= \psi(R_{p(c)}(z)) = \psi(cz + \mathrm{d}_q(p(c)-c) \rfloor z) \\ &= \psi(cz - \mathrm{d}_q(c) \rfloor z) = (-1)^{\partial z}\psi(zc). \end{aligned}$$ □

As an immediate corollary of (7.2.9) we can state

(7.2.10) $$\forall T \in \mathcal{T}(M,q), \quad \chi^\perp(T) = \psi(\chi(T))\ .$$

A first application of (7.2.10) occurs in the description of the submodules $T \in \mathcal{T}(M,q)$ such that $T = (T \cap U) \oplus (T \cap V)$; they are in bijection with the totally isotropic direct summands N of V.

(7.2.11) **Proposition.** *If N is a totally isotropic direct summand of V, the submodule $(N^\perp \cap U) \oplus N$ belongs to $\mathcal{T}(M,q)$, and for every $T \in \mathcal{T}(M,q)$ these assertions are equivalent:*

(a) $T = (T \cap U) \oplus (T \cap V)$ *and* $T \cap V = N$.
(b) $T = (N^\perp \cap U) \oplus N$.
(c) $\chi(T) = \bigwedge^{\max}(N)$.
(d) $\chi^\perp(T) = \bigwedge^{\max}(p(N))$.
(e) $T = (N^\perp \cap U) \oplus p(N)$.
(f) $T = (T \cap U) \oplus (T \cap V^\perp)$ *and* $T \cap V^\perp = p(N)$.

Proof. Since V is the direct sum of N and some other submodule, V^* is the direct sum of the annihilator of N (the submodule of all linear forms vanishing on N) and some other submodule that is isomorphic to N^*. Since d_q induces an isomorphism between U and V^*, U is the direct sum of $N^\perp \cap U$ (isomorphic to $(V/N)^*$) and some submodule isomorphic to N^*. Since $(N^\perp \cap U) \oplus N$ is a totally isotropic direct summand that has the same rank as V, it belongs to $\mathcal{T}(M,q)$. Now it is clear that (b)⇒(a). Conversely (a) implies the inclusion $T \subset (N^\perp \cap U) \oplus N$ which must be an equality because it involves two direct summands that have the same rank. Now (b) implies the inclusion $\chi(T) \supset \bigwedge^{\max}(N)$ which must be an equality for the same reason, and conversely (c)⇒(b) because of the bijectiveness of χ. The equivalence of (c) and (d) follows from (7.2.10) because $\psi(\bigwedge^{\max}(N)) = \bigwedge^{\max}(p(N))$. Since V and $V^\perp$ can play the same role, and since the equality $N^\perp \cap U = p(N)^\perp \cap U$ is clear, the end of the proof is evident. □

When V is totally isotropic, then $V = V^\perp$ and moreover $\chi(V) = \bigwedge^{\max}(V)$.

It has been noticed that $\mathrm{G}'\mathrm{Lip}(V,q)$ is a subset of $\mathrm{BLip}(V,q)$; with the help of $V^\perp$ we can determine which subset of $\mathcal{T}(M,q)$ is mapped onto $\mathrm{G}'\mathrm{Lip}(V,q)$.

(7.2.12) **Proposition.** *For every $T \in \mathcal{T}(M,q)$, the following assertions are equivalent:*

(a) $M = T \oplus V$.
(b) $M = T \oplus V^\perp$.
(c) $\chi(T) \in \mathrm{G}'\mathrm{Lip}(V,q)$.

Proof. The equivalence (a)⇔(b) follows from (2.3.7). Let us set $Z = \chi(T)$ and suppose that $Z \in \mathrm{G}'\mathrm{Lip}(V,q)$. From the equality $Z = R_Z(K)$ we derive $T = G_Z(U)$ (see (7.2.5)). And from (5.4.5) we deduce that $G_Z(d) = d$ for all $d \in V^\perp$. Consequently G_Z maps U and $V^\perp$ respectively to T and $V^\perp$, and the equality $M = T \oplus V^\perp$ follows from $M = U \oplus V^\perp$. Thus we have proved (c)⇒(b). Conversely let us suppose that U and T are both supplementary to $V^\perp$. By means of b_q they become both canonically isomorphic to $(V^\perp)^*$, whence a bijection $f : U \to T$. If we set $g(b+d) = f(b) + d$ for all $(b,d) \in U \times V^\perp$, we get an automorphism g of (M,q), and there exists $Z \in \mathrm{G'Lip}(M,q)$ such that $g = G_Z$. Since $T = g(U)$, from (7.2.5) we derive $\chi(T) = R_Z(K)$. Now from (5.4.5) we deduce that Z belongs to $\mathrm{G'Lip}(V,q)$. Consequently $\chi(T) = Z \in \mathrm{G'Lip}(V,q)$. □

7.3 The bijection $C\ell(V) \otimes \bigwedge^{\max}(U) \otimes C\ell(V) \to C\ell(M)$

The hypotheses are the same as in the previous section; the submodule $\bigwedge^{\max}(U)$ (already mentioned in (7.1.4)) here plays an essential role. We also use the reversions in the algebras $C\ell(M,q)$ and $C\ell(V,q)$ (see (3.1.4)); both reversions are denoted by τ since the latter is the restriction of the former; it is clear that $\bigwedge^{\max}(U)$ is invariant by τ. Following Chevalley, we are interested in the mapping Ω defined in this way:

$$\Omega \;:\; C\ell(V,q) \otimes \bigwedge^{\max}(U) \otimes C\ell(V,q) \longrightarrow C\ell(M,q)\;, \quad z \otimes \omega \otimes z' \longmapsto z\;\omega\;\tau(z').$$

Let us prove at once this equality for all x and $x' \in C\ell(M,q)$:

$$R_x(z)\;\omega\;\tau(R_{x'}(z')) \;=\; x\;\big(z\;\omega\;\tau(z')\big)\;\tau(x')\;; \tag{7.3.1}$$

indeed $R_x(z)$ is the sum of xz and some element of the left ideal $C\ell(M,q)U$ that annihilates $\bigwedge^{\max}(U)$ by multiplication on the left side, whereas $\tau(R_{x'}(z'))$ is the sum of $\tau(z')\tau(x')$ and some element in the right ideal $UC\ell(M,q)$ that annihilates $\bigwedge^{\max}(U)$ by multiplication on the right side. □

(7.3.2) **Theorem.** *The mapping Ω is bijective.*

Proof. Because of (7.3.1) the image of Ω is a graded ideal of $C\ell(M,q)$. Since $C\ell(M,q)$ is a graded Azumaya algebra, its graded ideals are determined by the ideals of K in a bijective way (see (6.7.4)). The image of Ω contains $\bigwedge^{\max}(U)$ that is a graded direct summand of constant rank 1. Therefore the corresponding ideal of K is K itself, and we have proved the surjectiveness of Ω. Besides, the source and the target of Ω are finitely generated projective modules that have the same rank at every prime ideal of K. The bijectiveness of Ω now follows from (1.13.5). □

(7.3.3) **Theorem.** *Let T be an element of $\mathcal{T}(M,q)$, and $Z = \chi(T)$ its image in* $\mathrm{BLip}(V,q)$. *The mapping Ω induces a bijection*

$$Z \otimes \bigwedge^{\max}(U) \otimes Z \longrightarrow \bigwedge^{\max}(T).$$

Proof. This is trivial when $T = U$ and $Z = K$. To prove it for every T, we can suppose that K is a local ring; then there exists $X \in \mathrm{G'Lip}(M,q)$ such that

$T = G_X(U)$ (see (7.2.6)), whence $Z = R_X(K)$ (see (7.2.5)). Thus Ω maps $Z \otimes \bigwedge^{\max}(U) \otimes Z$ on the submodule generated by all products $R_x(1)\ \omega\ \tau(R_{x'}(1))$ with x and x' in X. Because of (7.3.1) these products are equal to $x\omega\tau(x')$. Since the submodule $\tau(X)$ is the inverse of the submodule X in the algebra $C\ell(M,q)$, we realize that the submodule generated by all $x\omega\tau(x')$ is the image of $\bigwedge^{\max}(U)$ by the automorphism $C\ell(G_X)$ of $C\ell(M,q)$ derived from G_X; it is equal to $\bigwedge^{\max}(T)$ since $G_X(U) = T$. □

The following proposition has been stated here merely because its proof uses Propositions (7.2.5) and (7.2.6) in the same way as the proof of (7.3.3).

(7.3.4) **Proposition.** *Let T and T' be elements of $\mathcal{T}(M,q)$, $Z = \chi(T)$ and $Z' = \chi(T')$ their images in* BLip(V,q). *The intersection $T \cap T'$ is the submodule of all $a \in M$ such that*

$$\forall z \in Z\ ,\ \forall z' \in Z'\ ,\ \forall \omega \in \bigwedge^{\max}(U)\ ,\quad \mathrm{d}_q(a) \rfloor (z\ \omega\ \tau(z')) = 0.$$

Proof. It suffices to prove (7.3.4) when K is a local ring; by using the transitive action of G′Lip(M,q) on $\mathcal{T}(M,q)$, we can reduce the problem to the case $T' = U$. In other words, if r is the rank of U, and z and ω are generators of Z and $\bigwedge^r(U)$ respectively, it suffices to prove that $\mathrm{d}_q(a) \rfloor (z\omega) = 0$ if and only if a belongs to $T \cap U$. This preliminary argument needs the equality $\mathrm{d}_q(g(a)) \rfloor C\ell(g)(\xi) = C\ell(g)(\mathrm{d}_q(a) \rfloor \xi)$ (an easy consequence of (4.4.6)) which holds for all $g \in \mathrm{GO}(M,q)$, $a \in M$ and $\xi \in C\ell(M,q)$.

In the equality $\mathrm{d}_q(a) \rfloor (z\omega) = (\mathrm{d}_q(a) \rfloor z)\omega + \sigma(z)(\mathrm{d}_q(a) \rfloor \omega)$ (see (4.4.4)) the factors ω and $\mathrm{d}_q(a) \rfloor \omega$ belong respectively to $\bigwedge^r(U)$ and $\bigwedge^{r-1}(U)$. Since the multiplication mapping $C\ell(V,q) \otimes \bigwedge(U) \to C\ell(M,q)$ is bijective, this allows us to prove that $\mathrm{d}_q(a) \rfloor (z\omega)$ vanishes if and only if $\mathrm{d}_q(a) \rfloor z$ and $\mathrm{d}_q(a) \rfloor \omega$ both vanish. The vanishing of $\mathrm{d}_q(a) \rfloor \omega$ means that a belongs to $U^\perp = U$; indeed by means of a basis of U it is easy to prove the injectiveness of the mapping $U^* \to \bigwedge(U)$ defined by $h \longmapsto h \rfloor \omega$. Now for every $a \in U$ we can write $\mathrm{d}_q(a) \rfloor z = R_a(z)$, and the vanishing of $R_a(z)$ means that a belongs to $T \cap U$. □

With the help of (4.8.12) we get this immediate corollary.

(7.3.5) **Corollary.** *The hypotheses are the same as in (7.3.4). If N is a direct summand of M, then $z\ \omega\ \tau(z')$ belongs to the subalgebra generated by N in $C\ell(M,q)$ for all z, z', ω respectively in Z, Z' and $\bigwedge^{\max}(U)$ if and only if $N^\perp$ is contained in $T \cap T'$.*

The components of Ω when 2 is invertible

In **4.8** it is explained that, when 2 is invertible, there is a canonical linear bijection $\bigwedge(M) \to C\ell(M,q)$ which provides the module $C\ell(M,q)$ (but not the algebra $C\ell(M,q)$) with a grading by submodules $C\ell^k(M,q)$. We suppose that M

has a nonzero constant rank $2r$; therefore $C\ell^{2r}(M,q)$ is the discriminant module of $QZ(M,q)$ (see (4.8.15)). Since (M,q) is hyperbolic, this quadratic extension $QZ(M,q)$ is trivial, in accordance with the next lemma.

(7.3.6) **Lemma.** *$C\ell^{2r}(M,q)$ is a free module generated by an element ζ such that*

$$\zeta^2 = 1 \; ; \quad \forall z \in C\ell(V,q), \quad R_\zeta(z) = \sigma(z) \; ;$$
$$\textit{and} \quad \forall \omega \in \bigwedge^r(U) \; , \quad \omega = \zeta\omega = \sigma(\omega)\zeta.$$

Proof. Because of Theorem (7.2.1) there exists an element $\zeta \in C\ell(M,q)$ such that $R_\zeta = \sigma$. The submodule generated by σ in $\mathrm{End}(C\ell(V,q))$ is a direct summand because the canonical images of K in $\mathrm{End}(C\ell_0(V,q))$ and $\mathrm{End}(C\ell_1(V,q))$ are direct summands (see (1.13.2)); consequently ζ generates a direct summand (of constant rank 1) in $C\ell(M,q)$. The equality $\zeta^2 = 1$ follows from $\sigma^2 = \mathrm{id}$. Since σ anticommutes with all odd elements, and since the isomorphism $x \longmapsto R_x$ is graded, ζ is an even element that anticommutes with all odd elements; because of (4.8.13), this implies $c \wedge \zeta = 0$ for all $c \in M$, and consequently ζ belongs to $C\ell^{2r}(M,q)$, and generates it. At last R_ω is the interior multiplication by $\bigwedge(\mathrm{d}_q)(\omega)$, which belongs to $\bigwedge^{*r}(V)$ and must be given the degree $-r$. Therefore R_ω maps $C\ell^r(V,q)$ onto $C\ell^0(V,q) = K$, and annihilates $C\ell^{<r}(V,q)$. Consequently $R_\omega = R_\zeta R_\omega = (-1)^r R_\omega R_\zeta$ and the last announced equalities follow. □

If V is also totally isotropic, for every $\omega' \in \bigwedge^r(V)$ the operator $R_{\omega'}$ maps $\bigwedge^0(V)$ onto $\bigwedge^r(V)$, and annihilates $\bigwedge^{>0}(M)$. A similar argument involving R_ζ and $R_{\omega'}$ shows that $\omega' = \omega'\zeta = \zeta\sigma(\omega')$.

It is worth looking at the components Ω_k of Ω in the submodules $C\ell^k(M,q)$; they satisfy the following properties (for $k = 0, 1, \ldots, 2r$, for all z and z' in $C\ell(V,q)$, and for all $\omega \in \bigwedge^r(U)$):

(7.3.7) $$\Omega_k(z \otimes \omega \otimes z') = 0 \quad \text{if} \quad \partial z + \partial z' \neq r - k \text{ modulo } 2 \; ;$$

(7.3.8) $$\Omega_k(z' \otimes \omega \otimes z) = (-1)^{r(r-1)/2}(-1)^{k(k-1)/2}\, \Omega_k(z \otimes \omega \otimes z') \; ;$$

(7.3.9) $$\Omega_k(z \otimes \omega \otimes z') = \zeta\, \Omega_{2r-k}(\sigma(z) \otimes \omega \otimes z')$$

if ζ is defined as in (7.3.6);

(7.3.10) $$\Omega_0(R_x(z) \otimes \omega \otimes z') = \Omega_0(z \otimes \omega \otimes R_{\tau(x)}(z')),$$
$$\Omega_{2r}(R_x(z) \otimes \omega \otimes z') = \Omega_{2r}(z \otimes \omega \otimes R_{\sigma\tau(x)}(z'))$$

for all $x \in C\ell(M,q)$;

(7.3.11) $$\Omega_k(R_x(z) \otimes \omega \otimes R_x(z')) = (-1)^{k\partial x}(x\tau(x))\; C\ell(G_X) \circ \Omega_k(z \otimes \omega \otimes z')$$

for all $X \in \mathrm{G'Lip}(M,q)$ and all $x \in X$ (whence $G_X \in \mathrm{GO}(M,q)$ and $x\tau(x) \in K$).

Proof. First (7.3.7) follows from the fact that the parity of $\Omega(z \otimes \omega \otimes z')$ is $\partial z + \partial\omega + \partial z'$. Since the reversion τ of $C\ell(M,q)$ is also described by (3.1.5), we deduce (7.3.8) from the equalities

$$\tau(\Omega_k(z \otimes \omega \otimes z')) = (-1)^{k(k-1)/2}\Omega_k(z \otimes \omega \otimes z') \quad \text{and} \quad \tau(\omega) = (-1)^{r(r-1)/2}\omega\ .$$

Since the multiplication by ζ maps $C\ell^{2r-k}(M,q)$ bijectively onto $C\ell^k(M,q)$ (see (4.8.15)), the equality (7.3.9) follows from the equalities $\zeta z = \sigma(z)\zeta$ and $\zeta\omega = \omega$. The equality in (7.3.10) involving the scalar component Ω_0 is a consequence of (7.3.1) and (4.8.16):

$$\Omega_0(R_x(z) \otimes \omega \otimes z') = \mathrm{Scal}(xz\omega\tau(z')) = \mathrm{Scal}(z\omega\tau(z')x) = \Omega_0(z \otimes \omega \otimes R_{\tau(x)}(z'))\ .$$

With (7.3.9) we deduce the equality involving Ω_{2r} from the preceding one. At last (7.3.11) is another consequence of (7.3.1); remember that $C\ell(G_X) = \Theta_X$ is defined in this way: $\Theta_X(y)x = (-1)^{\partial x \partial y}xy$ for all $x \in X$ and all $y \in C\ell(M,q)$. □

The equalities (7.3.10) involve the space $\mathcal{B}_V$ of all bilinear forms on $C\ell(V,q)$, because every $\omega \in \bigwedge^r(U)$ determines two elements φ_3 and φ_4 of $\mathcal{B}_V$:

$$\Omega_0(z \otimes \omega \otimes z') = \varphi_3(z,z') \quad \text{and} \quad \Omega_{2r}(z \otimes \omega \otimes z') = \varphi_4(z,z')\zeta\ .$$

Although we do not find in (7.3.10) the twisting sign mentioned in the analogous equality (6.8.11), it is probable that (7.3.10) means that φ_3 and φ_4 belong to the graded centralizer of $C\ell(M,q)$ in $\mathcal{B}_V$ for a suitable structure of graded bimodule over $C\ell(M,q)$. To make it clear, let us consider any graded algebra $A = A_0 \oplus A_1$ provided with an involution τ, any graded module S over A, and the graded module $\mathcal{B}$ of all bilinear forms $\varphi : S \otimes S \to K$. How can $\mathcal{B}$ become a graded bimodule over A by means of a formula

$$(x\varphi x')(z,z') \;=\; (-1)^{\xi}\ \varphi(x'z,\ \tau(x)z')$$

in which the twisting exponent ξ is a function of ∂x, $\partial x'$, ∂z, $\partial z'$? It is not necessary to take $\partial\varphi$ into account since both members of the above equality vanish if the equality $\partial\varphi = \partial x + \partial x' + \partial z + \partial z'$ does not hold. In (7.ex.6) it is proved that there are exactly eight twisting exponents ξ that give $\mathcal{B}$ a structure of bimodule whatever A and S may be; they depend on an element $(\lambda,\mu,\nu) \in (\mathbb{Z}/2\mathbb{Z})^3$ in the following way:

$$\xi \;=\; \lambda(\partial x\partial x' + \partial x\partial z + \partial x'\partial z') + \mu\partial x + \nu\partial x'\ .$$

It is easy to verify that $\mathrm{Z}^g(A,\mathcal{B})$ is not modified when we add $\partial x + \partial x'$ to ξ. Therefore, according to the value of (λ,μ,ν), we get four graded centralizers of A in $\mathcal{B}$. When (λ,μ,ν) is equal to $(0,0,0)$ (resp. $(0,1,0)$, resp. $(1,1,0)$, resp. $(1,0,0)$), then the homogeneous elements of $\mathrm{Z}^g(A,\mathcal{B})$ are the homogeneous bilinear forms φ_1 (resp. φ_2, resp. φ_3, resp. φ_4) such that

$$\varphi_1(xz,\ z') = (-1)^{\partial x\partial\varphi}\varphi_1(z,\ \tau(x)z')\ , \qquad \varphi_3(xz,\ z') = \varphi_3(z,\ \tau(x)z')\ ,$$

$$\varphi_2(xz,\ z') = (-1)^{\partial x\partial\varphi}\varphi_2(x,\ \sigma\tau(x)z')\ , \quad \varphi_4(xz,\ z') = \varphi_4(x,\ \sigma\tau(x)z')\ ,$$

for all $x \in A$ and all $z, z' \in S$. Now if φ_1 is any (homogeneous) bilinear form on S, it is easy to verify that the previous four equalities are equivalent if we set

$$\varphi_2(z,z') = (-1)^{\partial z}\varphi_1(z,z') \,, \qquad \varphi_3(z,z') = (-1)^{\partial z(1+\partial z')}\varphi_1(z,z')$$
$$\text{and} \quad \varphi_4(z,z') = (-1)^{\partial z \partial z'}\varphi_1(z,z') \quad \text{for all (homogeneous) } z,\ z' \in S.$$

Consequently all four graded centralizers of A in $\mathcal{B}$ are known when one of them is known. It is also worth noticing that $\varphi_1 = \varphi_3$ and $\varphi_2 = \varphi_4$ when φ_1 is even, and that $\varphi_1 = \varphi_4$ and $\varphi_2 = \varphi_3$ when φ_1 is odd.

Besides, all the eight twisting exponents ξ are amenable to the twisting rule (4.2.1) by means of suitable subterfuges. In particular the action of $\tau(x)$ on the left side must be interpreted as an action of x^o (the image of x in the opposite algebra A^o) on the right side; in other words, $\tau(x)z'$ becomes $z'x^o$. Instead of variables z and z' in S we may also consider variables z^c and z'^c in the conjugate module S^c (defined in **6.2**); for instance let us suppose that ψ is a bilinear mapping $S \times S^c \to K$; then (4.2.1) is compatible with

$$(x\psi x')(z, z'^c) \;=\; (-1)^{\partial x(\partial\psi+\partial x'+\partial z+\partial z')}\psi(x'z,\ z'^c x^o) \;;$$

since $\partial\psi$ must be replaced with $\partial x + \partial x' + \partial z + \partial z'$, and since $z'^c x^o$ means $(-1)^{\partial x}(\tau(x)z')^c$, this equality is equivalent to the equality defining $(x\varphi x')(z,z')$ with the twisting exponent $\xi = 0$. In this way we get the four exponents ξ with $\lambda = 0$. We get the four exponents ξ with $\lambda = 1$ if we permute the positions of z and z'. For instance if ψ is a bilinear mapping $S^c \times S \to K$, then (4.2.1) is also compatible with

$$(x\psi x')(z'^c, z) \;=\; (-1)^{\partial x(\partial\psi+\partial z')+\partial x'\partial z'}\psi(z'^c x^o,\ x'z) \;;$$

this equality is equivalent to the equality defining $(x\varphi x')(z,z')$ with the twisting exponent $\xi = \partial x\partial x' + \partial x\partial z + \partial x'\partial z'$.

Let us come back to $\mathcal{B}_V$ and to the equalities (7.3.10). Since $\mathcal{B}_V$ has the same rank as $C\ell(M,q)$ (see (7.2.1)), and since the multiplication mapping $C\ell(M,q) \otimes Z^g(C\ell(M,q),\mathcal{B}_V) \to \mathcal{B}_V$ is always bijective (see (6.7.6)), we realize that the graded centralizer of $C\ell(M,q)$ in $\mathcal{B}_V$ is a direct summand of constant rank 1. Now it is easy to prove that the above defined mapping $\omega \longmapsto \varphi_3$ or $\omega \longmapsto \varphi_4$ is an isomorphism $\bigwedge^r(U) \to Z^g(C\ell(M,q),\mathcal{B}_V)$ for a suitable structure of graded bimodule on $\mathcal{B}_V$. The bilinear forms φ_3 are the elements of $Z^g(C\ell(M,q),\mathcal{B}_V)$ when we set

$$(x\varphi x')(z,z') = (-1)^{\partial x(1+\partial x'+\partial z)+\partial x'\partial z'}\ \varphi(R_{x'}(z),\ R_{\tau(x)}(z')).$$

Nevertheless it is also possible to ignore this discussion about the twisting exponents ξ, and strictly to respect the terminology presented in **6.8**. If we maintain that the equality (6.8.11) characterizes the scalar products associated with an involution τ, then we observe that Ω_0 determines bilinear forms φ_3 which have the same parity as r, and we realize that they are the scalar products associated with the involution $\sigma^r\tau$. These scalar products admit extensions $C\ell(V)\times C\ell(V) \to$

$C\ell(M)$ which are defined as in (6.8.16), and the comparison between (6.8.17) and (7.3.1) shows that these extensions are the bilinear mappings $(z, z') \longmapsto \sigma^r \circ \Omega(z \otimes \omega \otimes z')$. If we overlook the factor σ^r which comes from the conventions previously chosen in **6.8**, we can say that Ω is the extension of its scalar component Ω_0 according to the following equality which is an easy consequence of (7.3.1):

$$\text{(7.3.12)} \qquad \mathrm{Scal}(x\ \Omega(z \otimes \omega \otimes z')) = \mathrm{Scal}(xz\omega\tau(z')) \\ = \Omega_0(R_x(z) \otimes \omega \otimes z')\ ;$$

since the symmetric bilinear form $(x, y) \longmapsto \mathrm{Scal}(xy)$ is nondegenerate on $C\ell(M, q)$ (see (4.8.16)), this equality shows how Ω_0 determines Ω.

In (7.ex.4)(b) and (7.ex.5) (resp. (7.ex.8)(a)) the scalar products φ_3 (resp. φ_4) are constructed without the hypothesis that 2 is invertible.

At last (7.3.11) describes the action of the Lipschitz group $\mathrm{GLip}(M, q)$ in the module $C\ell(V, q) \otimes C\ell(V, q)$. This description is a classical problem when K is the field $\mathbb{R}$ or $\mathbb{C}$ of real or complex numbers. When $K = \mathbb{C}$, a *spinorial group* $\mathrm{Spin}(M, q)$ is extracted from $\mathrm{GLip}(M, q)$, it is the subgroup of all $x \in \mathrm{GLip}(M, q)$ such that $x\tau(x) = 1$. The morphism $\mathrm{Spin}(M, q) \to \mathrm{GO}(M, q)$ is still surjective, but its kernel is reduced to $\{1, -1\}$. When $K = \mathbb{R}$, the spinorial group $\mathrm{Spin}(M, q)$ defined in the same way does not give a surjective morphism $\mathrm{Spin}(M, q) \to \mathrm{GO}(M, q)$, unless q is positive definite, contrary to the present assumption that q is hyperbolic. But we get a surjective morphism $\mathrm{Spin}^{\pm}(M, q) \to \mathrm{GO}(M, q)$ with kernel $\{1, -1\}$ if we use the subgroup $\mathrm{Spin}^{\pm}(M, q)$ of all $x \in \mathrm{GLip}(M, q)$ such that $x\tau(x) = \pm 1$. Groups $\mathrm{Spin}^{\pm}(M, q)$ with an arbitrary q have been studied in (5.ex.24). The requirement $x\tau(x) = \pm 1$ simplifies the equality (7.3.11) and all other equalities involving the scalar $x\tau(x)$.

Spinor spaces and scalar products of spinors

When (M, q) is any quadratic space over a field K, usually a *spinor space* is a minimal faithful module over $C\ell(M, q)$; this definition does not take the parity grading of $C\ell(M, q)$ into account. When the center of $C\ell(M, q)$ is a field (either K or a quadratic extension of K), $C\ell(M, q)$ is a simple algebra, and all irreducible modules are faithful, and isomorphic to one another; in this case a spinor space is merely an irreducible module. But when the center of $C\ell(M, q)$ is isomorphic to K^2, then $C\ell(M, q)$ (without its grading) is the direct sum of two simple ideals both isomorphic to $C\ell_0(M, q)$; therefore there are two isomorphy classes of irreducible modules over $C\ell(M, q)$, and every minimal faithful module is the direct sum of two nonisomorphic irreducible modules called *half spinor spaces*. Nevertheless this terminology may be questioned because it ignores the parity gradings; in (6.2.2) it is explained that the spinor spaces required by the Dirac equation in physics, which are at the origin of this terminology, are graded; and on the side of purely algebraic studies, the most recent progress (which often stems from H. Bass's works) show that the "good objects" are the graded modules, whether irreducible or not.

All graded irreducible modules over $C\ell(M,q)$ are faithful, and they are isomorphic to one another when the center of $C\ell_0(M,q)$ is a field (see (6.6.2) and (6.6.3)); when $Z(C\ell_0(M,q))$ is isomorphic to K^2, every graded irreducible module over $C\ell(M,q)$ is isomorphic either to some particular module S, or to the module S^s with shifted grading; but S^s is isomorphic to S in a weaker sense, since the odd bijections $z \longmapsto z^s$ and $z \longmapsto (-1)^{\partial z} z^s$ are respectively $C\ell(M,q)$-linear and $C\ell(M,q)$-g-linear. If we come back to the case of a hyperbolic space (M,q) as above, the representation $x \longmapsto R_x$ of $C\ell(M,q)$ in $C\ell(V,q)$ (defined in (7.2.1)) is graded, faithful and irreducible; in this case we get the easiest example of a spinor space, especially when V is totally isotropic.

Let S be a module over $C\ell(M,q)$, and $\mathcal{B}$ the space of all bilinear mappings $\varphi : S\times S \to K$; it is a bimodule over $C\ell(M,q)$ for different possible actions of $C\ell(M,q)$. As suggested by the developments around the Dirac equation in physics, particular elements of $\mathcal{B}$, called *scalar products of spinors*, must be emphasized; they must be nondegenerate, either symmetric or skew symmetric, and they must be associated in some way to the involution τ (the reversion) or $\sigma\tau$ (the conjugation). When parity gradings are not taken into account, the resulting theory splits into the study of many particular cases, especially depending on the parity of the dimension of M, and on the nature of the quadratic extension $QZ(M,q)$.

When S is a *graded* module, we get a unified theory as explained in **6.8**. The above discussion about the scalar products φ_1, φ_2, φ_3 and φ_4 shows that we do not have to worry about the twisting signs, which can always be adapted to any sensible system of conventions (not necessarily the conventions preferred in **6.8**). Besides, with the unified graded theory the study of nongraded modules becomes much easier: this is explained in (6.ex.19) and (6.ex.20).

7.4 The Cartan–Chevalley criterion

Let us come back to (7.3.3). We suppose that V has a nonzero constant rank r, so that $\bigwedge^{\max}(U) = \bigwedge^r(U)$. The case of a nonconstant rank can be reduced to this case by means of suitable idempotents of K. For every z that belongs to some $Z = \chi(T)$, the product $z\,\omega\,\tau(z)$ belongs to $\bigwedge^r(T)$, which is contained in $C\ell^{\leq r}(M,q)$. We are going to prove that the lipschitzian elements z are actually characterized by the fact that $z\,\omega\,\tau(z)$ belongs to $C\ell^{\leq r}(M,q)$; this is the Cartan–Chevalley criterion. As explained in (7.2.4), the Cartan–Chevalley criterion has played a capital role in the discovery of the definition (5.3.1) of the Lipschitz monoid. Nevertheless hurried readers are advised to skip the long proof of the next theorem.

(7.4.1) **Theorem.** *A locally homogeneous $z \in C\ell(V,q)$ belongs to $\mathrm{Lip}(V,q)$ if and only if*

$$\forall \omega \in \bigwedge^r(U)\,, \quad z\,\omega\,\tau(z) \in C\ell^{\leq r}(M,q)\,.$$

Proof. When z is lipschitzian, from (5.3.3) we immediately deduce that $z\omega\tau(z)$ belongs to $C\ell^{\leq r}(M,q)$. The main difficulty is to prove the converse statement; indeed it is much harder to derive useful consequences from the hypothesis $z\omega\tau(z) \in C\ell^{\leq r}(M,q)$. Chevalley's proof depends on the assumption that V is totally isotropic, and cannot help us when $q(V) \neq 0$. Therefore we must go the opposite way, and reduce the general case to the case of a nondegenerate restriction to V; this is the aim of the first step.

First step. Let us suppose that V is the direct sum of two submodules V' and V'' of constant ranks r' and r''. We know that $\mathrm{Lip}(V',q)$ is the intersection of $\mathrm{Lip}(V,q)$ and $C\ell(V',q)$ (see (5.4.4)). On the other side, V^* can be identified with the direct sum of V'^* and V''^*, and the isomorphism $U \to V^*$ induced by d_q makes U split into the direct sum of two submodules U' and U'' isomorphic to the dual spaces of V' and V'' respectively; thus U'' is orthogonal to V', and U' to V''. There is a canonical isomorphism from $\bigwedge^{r'}(U') \otimes \bigwedge^{r''}(U'')$ onto $\bigwedge^{r}(U)$. If we set $M' = U' \oplus V'$ and $M'' = U'' \oplus V''$, we can use the bijective multiplication mapping coming from (4.8.5):

$$C\ell(M',q) \otimes C\ell(M'',q) \longrightarrow C\ell(M,q) \ , \quad x' \otimes x'' \longmapsto x'x'' \ ;$$

by means of localizations and (4.8.11), it is easy to verify that this bijection induces an injection

$$\big(C\ell^{\leq k}(M',q) \ / \ C\ell^{<k}(M',q)\big) \ \otimes \ \bigwedge^{r''}(U'') \ \longrightarrow \ C\ell^{\leq k+r''}(M,q) \ / \ C\ell^{<k+r''}(M,q)$$

for every degree $k = 0, 1, \ldots, 2r'$. Now let z' be a homogeneous element of $C\ell(V',q')$, and let ω' and ω'' be elements of respectively $\bigwedge^{r'}(U')$ and $\bigwedge^{r''}(U'')$, so that $\omega'\omega''$ (the same thing as $\omega' \wedge \omega''$) belongs to $\bigwedge^{r}(U)$. Since V' and U'' are orthogonal, we can write

$$z'(\omega'\omega'')\tau(z') \ = \ (-1)^{r''\partial z'} \ (z'\omega'\tau(z')) \ \omega'' \ ;$$

all this proves that $z'\omega'\tau(z')$ belongs to $C\ell^{\leq r'}(M',q)$ for all $\omega' \in \bigwedge^{r'}(U')$ if and only if $z'\omega\tau(z')$ belongs to $C\ell^{\leq r}(M,q)$ for all $\omega \in \bigwedge^{r}(U)$. Consequently, if Theorem (7.4.1) is true for (V,M,q), it is also true for (V',M',q).

Now (V,q) can always be embedded as a direct summand into a quadratic space (for instance (M,q)). Consequently it suffices to prove (7.4.1) when the restriction of q to V is nondegenerate.

Second step. When the restriction of q to V is nondegenerate, M is the direct sum of V and $V^\perp$ (see (2.3.8)). Just before (7.2.9) it is explained that the equality $M = U \oplus V$ implies $M = U \oplus V^\perp$, that there is a bijection $p : V \to V^\perp$ such that $c - p(c)$ belongs to U for every $c \in V$, whence $q(p(c)) = -q(c)$, and that p extends to a bijection $\psi : C\ell(V,q) \to C\ell(V^\perp,q)$ such that $z^{to} \longmapsto \psi(z)$ is an algebra isomorphism from $C\ell(V,q)^{to}$ onto $C\ell(V^\perp,q)$.

The mapping $(c, c') \longmapsto c + p(c')$ is an isomorphism from $(V, q) \perp (V, -q)$ onto (M, q), and it maps the diagonal Δ'_V of $V \oplus V$ (that is the submodule of all $(c, -c)$) bijectively onto U. Thus we catch sight of the objects that play essential roles in the definition of lipschitzian elements of (V, q) according to (5.3.1): a locally homogeneous element z of $C\ell(V, q)$ is called lipschitzian if $z \otimes \tau(z)^{to}$ belongs to $C\ell((V, q) \perp (V, -q);\ \Delta'_V)^{\leq 0}$; in this definition $C\ell((V, q) \perp (V, -q))$ is identified with $C\ell(V, q) \hat{\otimes} C\ell(V, q)^{to}$. Because of the above isomorphism $C\ell(V, q)^{to} \rightarrow C\ell(V^\perp, q)$ we can state that z is lipschitzian if and only if $z\,\psi\tau(z)$ belongs to the subalgebra $C\ell(M, q; U)^{\leq 0}$.

This modified version of the definition (5.3.1) lets V and $V^\perp$ play symmetric roles; therefore we must manage to let V and $V^\perp$ also play symmetric roles in the Cartan–Chevalley criterion. This can be achieved in an easy way: indeed in (7.2.9) it is stated that $\psi(v) - v$ belongs to $C\ell(M, q)U$, and consequently to $UC\ell(M, q)$ too, for all $v \in C\ell(V, q)$, whence $\omega v = \omega\psi(v)$; thus we can replace $z\omega\,\tau(z)$ with $z\omega\,\psi\tau(z)$ in the Cartan–Chevalley criterion.

Third step. Since M is the orthogonal sum of V and $V^\perp$, there is an isomorphism

$$C\ell(V, q) \hat{\otimes} C\ell(V^\perp, q) \longrightarrow C\ell(M, q)\ , \qquad v \otimes w \longmapsto vw.$$

From the bijectiveness of Ω (see (7.3.2)) and the equality $\omega v = \omega\psi(v)$ (see above)), we deduce the bijectiveness of the mapping

$$C\ell(V, q) \otimes \bigwedge^r(U) \otimes C\ell(V^\perp, q) \longrightarrow C\ell(M, q)\ , \qquad v \otimes \omega \otimes w \longmapsto v\omega w.$$

These two bijections imply the existence of another bijection

$$J\ :\ C\ell(M, q) \otimes \bigwedge^r(U) \longrightarrow C\ell(M, q) \quad \text{such that} \quad J(vw \otimes \omega) \ = \ (-1)^{r\partial w}\ v\omega w$$

whenever v, w and ω belong respectively to $C\ell(V, q)$, $C\ell(V^\perp, q)$ and $\bigwedge^r(U)$. If we manage to prove that, for every $k \in \mathbb{Z}$, J induces a bijection

$$C\ell(M, q; U)^{\leq k} \otimes \bigwedge^r(U) \longrightarrow C\ell^{\leq k+r}(M, q)\ ,$$

the proof is complete, because $z\,\psi\tau(z)$ then belongs to $C\ell(M, q; U)^{\leq 0}$ if and only if $z\omega\,\psi\tau(z)$ belongs to $C\ell^{\leq r}(M, q)$.

We know that the submodules $C\ell^{\leq k+r}(M, q)$ are direct summands of $C\ell(M, q)$ (see (4.8.7)). From the assumption that U is a totally isotropic direct summand of constant rank r, we can deduce that $C\ell(M, q; U)^{\leq k}$ is also a direct summand with the same rank as $C\ell^{\leq k+r}(M, q)$: see (5.ex.3). Consequently it suffices to prove that J maps $C\ell(M, q; U)^{\leq k} \otimes \bigwedge^r(U)$ into $C\ell^{\leq k+r}(M, q)$.

Fourth step. Let γ be the symmetric bilinear form on M that coincides with b_q on $V^\perp \times V^\perp$, but vanishes on $V \times M$ and $M \times V$. From γ we derive an element $\gamma_{\prime\prime}$ of $\bigwedge^{*2}(M \times M)$ in the same way as we have derived $\beta_{\prime\prime}$ from β in **4.7**. Besides π_q

is the multiplication mapping $C\ell(M,q)\otimes C\ell(M,q)\to C\ell(M,q)$ as in **4.7**. Let us prove in this fourth step that, for all $x\in C\ell(M;q)$ and all $\omega\in\bigwedge^r(U)$,

$$J(x\otimes\omega) \;=\; \pi_q\big(\,\mathrm{Exp}(-\gamma_{\prime\prime})\;\rfloor\;(x\otimes\omega)\big)\ .$$

Indeed the right-hand member of this equality is the product of x and ω in the algebra $C\ell(M,q;-\gamma)$. We can replace $C\ell(M,q)$ with the canonically isomorphic algebra $C\ell(V,q)\,\hat{\otimes}\,C\ell(V^\perp,q)$, and since V and $V^\perp$ are orthogonal with respect to q and γ, we can replace $C\ell(M,q;-\gamma)$ with the twisted tensor product of $C\ell(V,q;-\gamma)$ and $C\ell(V^\perp,q;-\gamma)$. But these two subalgebras are respectively equal to $C\ell(V,q)$ and $C\ell(V^\perp,q;-\mathrm{b}_q)$ because of the definition of γ. Besides, the mapping $w\longmapsto w^{to}$ is an isomorphism from $C\ell(V^\perp,q;-\mathrm{b}_q)$ onto $C\ell(V^\perp,q)^{to}$ (see (4.7.8)). Consequently, when v, w and x' belong respectively to $C\ell(V,q)$, $C\ell(V^\perp,q)$ and $C\ell(M,q)$, the product of vw and x' in $C\ell(M,q;-\gamma)$ is $(-1)^{\partial x'\partial w}vx'w$; this agrees with the equality $J(vw\otimes\omega)=(-1)^{r\partial w}v\omega w$.

Fifth step. To complete the proof, we must explain why $J(x\otimes\omega)$ belongs to $C\ell^{\leq k+r}(M)$ whenever x belongs to $C\ell(M,q;U)^{\leq k}$. First we prove that

$$x\otimes\omega^{to}\quad\text{belongs to}\quad(C\ell(M,q)\,\hat{\otimes}\,C\ell(M,q)^{to};\ \Delta'_M)^{\leq k+r}\ ;$$

as in **5.3**, $C\ell(M,q)\,\hat{\otimes}\,C\ell(M,q)^{to}$ is treated as the Clifford algebra of the hyperbolic space $(M,q)\perp(M,-q)$, which contains Δ'_M as a totally isotropic direct summand. Indeed we can suppose that U is a free module with basis $(b_1,b_2,\dots,b_r)$ since by localization we can reduce the problem to this case. Then $\omega=\lambda b_1b_2\cdots b_r$ for some $\lambda\in K$. Since a permutation on the basis $(b_1,\dots,b_r)$ leaves the product $b_1b_2\cdots b_r$ invariant up to a sign $\pm$, it suffices to prove that $x\otimes\omega^{to}$ has a degree $\leq k+r$ for the filtration determined by Δ'_M whenever

$$x\;=\;x'b_sb_{s-1}\cdots b_2b_1\quad\text{with}\quad x'\in C\ell^{\leq k+s}(M,q)\ ;$$

this assertion is proved by the equality

$$x\otimes\omega^{to}\;=\;(x'\otimes\lambda^{to})\,(b_s\otimes b_s^{to})\cdots(b_2\otimes b_2^{to})\,(b_1\otimes b_1^{to})\,(1\otimes(b_{s+1}\cdots b_r)^{to})\ ,$$

because every factor $b_j\otimes b_j^{to}$ has degree ≤ 0 for the above filtration:

$$b_j\otimes b_j^{to}\;=\;(b_j\otimes 1^{to}-1\otimes b_j^{to})\,(1\otimes b_j^{to})\ .$$

Then the interior multiplication by $\mathrm{Exp}(-\gamma_{\prime\prime})$ leaves invariant the filtering degrees determined by Δ'_M; indeed, if we prove that $\gamma_{\prime\prime}$ has a filtering degree ≤ 0, this is a consequence of (5.2.9) and (5.2.8). By definition, for all a_1, a'_1, a_2, a'_2 in M,

$$\gamma_{\prime\prime}((a_1,a'_1)\wedge(a_2,a'_2))\;=\;\gamma(a_1,a'_2)-\gamma(a_2,a'_1)\ ;$$

$\gamma_{\prime\prime}$ has a filtering degree ≤ 0 (for the filtration determined by Δ'_M) if and only if the right-hand member of the above equality vanishes whenever $a'_1=-a_1$ and

$a'_2 = -a_2$; but this means that γ is symmetric, as it actually is. It should also be mentioned that the interior multiplication by $\mathrm{Exp}(-\gamma_{\prime\prime})$ has the same effect on $x \otimes \omega$ and $x \otimes \omega^{to}$, because $C\ell(M,q)^{to}$ is a deformation of $C\ell(M,q)$ (see (4.7.8)), and all deformations of $C\ell(M,q)$ are equivalent to it as comodules over $\bigwedge(M)$ (see (4.7.5)).

Let u and v be any elements of $C\ell(M,q)$; it remains to prove that $\pi_q(u \otimes v)$ falls into $C\ell^{\leq k+r}(M,q)$ whenever $u \otimes v^{to}$ has a degree $\leq k+r$ for the filtration determined by Δ'_M. It suffices to repeat the argument in the proof of (5.3.3). Indeed $u \otimes v^{to}$ is a sum of two kinds of terms; first there are terms that have degree $\leq k+r$ for the natural filtration of $C\ell(M,q) \hat{\otimes} C\ell(M,q)^{to}$ that ignores Δ'_M; obviously π_q maps them into $C\ell^{\leq k+r}(M,q)$. Secondly there are terms like $(u' \otimes v'^{to})(a \otimes 1^{to} - 1 \otimes a^{to})$ which are all mapped to 0 by π_q :

$$(u' \otimes v'^{to})(a \otimes 1^{to} - 1 \otimes a^{to}) = (-1)^{\partial v'} \ (u'a \otimes v'^{to} - u' \otimes (av')^{to}) \ ,$$
$$\pi_q(u'a \otimes v' - u' \otimes av') = 0 \ .$$

This ends the proof of (7.4.1). □

Improvements when 2 is invertible

When 2 is invertible in K (or when the mapping $a \longmapsto 2a$ is bijective from M onto M), the natural filtration of $C\ell(M,q)$ comes from a grading by submodules $C\ell^k(M,q)$ (see **4.8**). When T is a totally isotropic direct summand of M of constant rank r, it is clear that $\bigwedge^r(T) \subset C\ell^r(M,q)$. Thus (7.3.3) and (7.4.1) lead to a more precise result.

(7.4.2) **Corollary.** *A locally homogeneous $z \in C\ell(V,q)$ belongs to* $\mathrm{Lip}(V,q)$ *if and only if*

$$\forall \omega \in \bigwedge^r(U) \ , \quad z \ \omega \ \tau(z) \ \in \ C\ell^r(M,q) \ .$$

Proof. It remains to explain why $z\omega\tau(z)$ belongs to $C\ell^r(M,q)$ whenever z belongs to $\mathrm{Lip}(V,q)$, even if it does not belong to an element of $\mathrm{BLip}(V,q)$. From (5.3.3) it follows that it belongs to $C\ell^{\leq r}(M,q)$, and we must still prove that it belongs to $C\ell^{\geq r}(M,q)$. According to (4.8.15), there is a bijective multiplication mapping from $C\ell^{2r}(M,q) \otimes C\ell^k(M,q)$ onto $C\ell^{2r-k}(M,q)$ for $k = 0,1,2,\ldots,2r$, and consequently it suffices to prove that $\zeta(z\omega\tau(z))$ belongs to $C\ell^{\leq r}(M,q)$ if ζ is defined as in (7.3.6). Since z is locally homogeneous, it suffices to prove this for each of his homogeneous components. Since $\zeta z = \sigma(z)\zeta$ and $\zeta\omega = \omega$ (see (7.3.6)), we observe that $\zeta(z\omega\tau(z)) = \sigma(z)\omega\tau(z)$ and the conclusion follows. □

The equalities (7.3.7), (7.3.8), (7.3.9) give more information about the Cartan–Chevalley criterion. Indeed, if z is a locally homogeneous element of $C\ell(V,q)$, from (7.3.7) we deduce that the component of $z\omega\tau(z)$ in $C\ell^k(M,q)$ vanishes whenever $r-k$ is odd. From (7.3.8) and the equality

$$\frac{1}{2}r(r-1) - \frac{1}{2}k(k-1) = \frac{1}{2}(r-k)(r-k-1) + (r-k)k$$

we deduce that it vanishes whenever $r-k$ is even but not divisible by 4. And from (7.3.9) we deduce that it vanishes if and only if the component in $C\ell^{2r-k}(M,q)$ vanishes. Consequently *z is lipschitzian if and only if the component of $z\omega\tau(z)$ in $C\ell^{r+4j}(M,q)$ vanishes for all $\omega \in \bigwedge^r(U)$ and for all $j > 0$ (or equivalently, for all $j < 0$).*

When the rank of V is everywhere ≤ 3, then $C\ell^{r+4j}(M,q)$ is reduced to 0 whenever $j \neq 0$; thus we prove again that in this case every locally homogeneous z is lipschitzian (see (5.4.3)).

Since the submodules $C\ell^k(M,q)$ are orthogonal for the symmetric bilinear form $(x,y) \longmapsto \mathrm{Scal}(xy)$, from (7.3.12) we can still deduce that a locally homogeneous z is lipschitzian if and only if $\Omega_0(R_x(z) \otimes \omega \otimes z) = 0$ for all $\omega \in \bigwedge^r(U)$ and for all $x \in C\ell^{r+4j}(M,q)$ with $j > 0$ (or equivalently with $j < 0$).

7.5 Applications to Lipschitz monoids

It is clear that every quadratic module (V,q), with V a finitely generated projective module, can be embedded in a hyperbolic space (M,q) in such a way that M is the direct sum of V and a totally isotropic submodule U isomorphic to V^*. Then the Cartan–Chevalley mapping determines a bijection between $\mathrm{BLip}(V,q)$ and $\mathcal{T}(M,q)$, in such a way that the group $\mathrm{G'Lip}(V,q)$ corresponds to the elements of $\mathcal{T}(M,q)$ supplementary to V (see (7.2.12); thus every fact about $\mathcal{T}(M,q)$ may give information about $\mathrm{BLip}(V,q)$. Although $\mathrm{Lip}(V,q)$ is not always the union of all elements of $\mathrm{BLip}(V,q)$, it is probable that an element of $\mathrm{Lip}(V,q)$ is that more interesting if it belongs to an element of $\mathrm{BLip}(V,q)$; in a similar way an isotropic element of M is that more interesting if it belongs to an element of $\mathcal{T}(M,q)$. Unfortunately up to now it is still difficult to get information about $\mathcal{T}(M,q)$ when K is not a field; therefore only the first three statements in this section are valid with an arbitrary ring K. First we prove the bijectiveness of the natural mapping $\bigwedge^2(V) \to \mathrm{G'Lip}(V)$ by means of Chevalley's method; another proof is supplied in (5.ex.6).

(7.5.1) **Proposition.** *Here V is assumed to be totally isotropic. Let $\mathcal{T}_V(M,q)$ be the subset of all $T \in \mathcal{T}(M,q)$ such that $M = T \oplus V$, and $\mathrm{Hom}_\wedge(U,V)$ the submodule of all $\delta \in \mathrm{Hom}(U,V)$ such that $\mathrm{b}_q(b,\delta(b)) = 0$ for all $b \in U$. There is a commutative diagram*

$$\begin{array}{ccc} \bigwedge^2(V) & \longrightarrow & \mathrm{G'Lip}(V) \\ \downarrow & & \downarrow \\ \mathrm{Hom}_\wedge(U,V) & \longrightarrow & \mathcal{T}_V(M,q) \end{array}$$

in which the four arrows are bijective, and are defined in this way:

$$\begin{array}{rclrcl}
\bigwedge^2(V) & \longrightarrow & \mathrm{G'Lip}(V)\,, & v & \longmapsto & K\ \mathrm{Exp}(v)\,,\\
\bigwedge^2(V) & \longrightarrow & \mathrm{Hom}_\wedge(U,V)\,, & v & \longmapsto & (b \longmapsto [v,b] - -\mathrm{d}_q(b)\rfloor v)\,,\\
\mathrm{G'Lip}(V) & \longrightarrow & \mathcal{T}_V(M,q)\,, & Z & \longmapsto & G_Z(U)\,,\\
\mathrm{Hom}_\wedge(U,V) & \longrightarrow & \mathcal{T}_V(M,q)\,, & \delta & \longmapsto & \{\, b+\delta(b) \mid b \in U\,\}\,.
\end{array}$$

Proof. The main statement in this proposition is the bijectiveness of $\bigwedge^2(V) \to \mathrm{G'Lip}(V)$; all the remainder serves to master this mapping. We already know that $\mathrm{Exp}(v)$ belongs to $\mathrm{GLip}(V)$ for every $v \in \bigwedge^2(V)$. The mapping $b \longmapsto -\mathrm{d}_q(b)\rfloor v$ belongs to $\mathrm{Hom}_\wedge(U,V)$ because

$$\mathrm{b}_q(b,\ \mathrm{d}_q(b)\rfloor v) \;=\; \mathrm{d}_q(b)\rfloor\ (\mathrm{d}_q(b)\rfloor v) \;=\; (\mathrm{d}_q(b)\wedge \mathrm{d}_q(b))\ \rfloor v \;=\; 0\,.$$

By means of localizations it is easy to prove the bijectiveness of $\bigwedge^2(V) \to \mathrm{Hom}_\wedge(U,V)$; indeed with every basis $(c_1, c_2, \dots, c_r)$ of V is associated a dual basis $(b_1, b_2, \dots, b_r)$ of U such that $\mathrm{b}_q(b_i, b_j) = 1$ if $i = j$, 0 if $i \neq j$; then every element of $\bigwedge^2(V)$ is described by a skew symmetric (or rather alternate) matrix of order r, and it is mapped to the element of $\mathrm{Hom}_\wedge(U,V)$ that is described by the same matrix. The existence of a bijective mapping $\mathcal{T}_V(M,q) \to \mathrm{G'Lip}(V)$ is stated in (7.2.12); it maps T to $\chi(T)$. Conversely the equality $\chi(T) = Z$, with $Z \in \mathrm{G'Lip}(V)$, implies $\chi(T) = R_Z(K) = R_Z(\chi(U)) = \chi(G_Z(U))$ (see (7.2.5)), whence $T = G_Z(U)$. The submodules T of M supplementary to V are in bijection with $\mathrm{Hom}(U,V)$ in this way: with each $\delta \in \mathrm{Hom}(U,V)$ is associated a graph in $U \times V$ (the subset of all $(b, \delta(b))$); if we treat this graph as a subset of $M = U \oplus V$, it is the same thing as the submodule T of all $b + \delta(b)$ with $b \in U$. Since $q(b+\delta(b)) = \mathrm{b}_q(b, \delta(b))$, this submodule T is totally isotropic if and only if δ belongs to $\mathrm{Hom}_\wedge(U,V)$; whence the bijection from $\mathrm{Hom}_\wedge(U,V)$ onto $\mathcal{T}_V(M,q)$.

It remains to prove the commutativity of the diagram; since the bijectiveness of three arrows is sure, its commutativity implies the bijectiveness of the fourth arrow $\bigwedge^2(V) \to \mathrm{G'Lip}(V)$. Let us set $z = \mathrm{Exp}(v)$ with $v \in \bigwedge^2(V)$, whence $z^{-1} = \mathrm{Exp}(-v)$. From (4.4.12) and (4.5.4) we derive (for all $b \in U$)

$$bz - zb \;=\; (\mathrm{d}_q(b)\rfloor v)\ z \quad \text{whence} \quad G_z(b) \;=\; zbz^{-1} \;=\; b\ -(\mathrm{d}_q(b)\rfloor v)\ ;$$

this implies that the two images of v in $\mathcal{T}_V(M,q)$ are equal. □

Now we are especially interested in elements of $\mathrm{BLip}(V,q)$ that are not in the group $\mathrm{G'Lip}(V,q)$. In (7.2.11) we discovered the elements $\bigwedge^{\max}(N)$ with N a totally isotropic direct summand of V. How can we recognize whether an element Z of $\mathrm{BLip}(V,q)$ is equal to the product of some $X \in \mathrm{G'Lip}(V,q)$ and some $\bigwedge^{\max}(N)$ derived from a totally isotropic direct summand of V? Let us begin with three observations; first

$$X \bigwedge^{\max}(N) \;=\; \bigwedge^{\max}(G_X(N))\ X$$

if G_X is the orthogonal transformation of (V, q) derived from X. Secondly if Z is actually equal to some product $X \bigwedge^{\max}(N)$ as above, it is easy to guess which are the submodules N and $G_X(N)$; this is explained in (7.5.2) below. Thirdly, when N is known, and also $T = \chi^{-1}(Z)$, then the existence of $X \in \mathrm{G'Lip}(V, q)$ such that $Z = X \bigwedge^{\max}(N)$ is equivalent to the existence of $g \in \mathrm{GO}(M, q)$ such that

$$T \;=\; g((N^\perp \cap U) \oplus N) \quad \text{and} \quad g(w) = w \quad \text{for all} \quad w \in V^\perp;$$

indeed $X \bigwedge^{\max}(N)$ is the same thing as $R_X(\bigwedge^{\max}(N))$; because of (7.2.5) the equality $Z = R_X(\bigwedge^{\max}(N))$ is equivalent to $T = G_X(T_N)$ if T and T_N are the elements of $\mathcal{T}(M, q)$ associated with Z and $\bigwedge^{\max}(N)$; because of (5.4.5), G_X (that is now the element of $\mathrm{GO}(M, q)$ derived from X) is any orthogonal transformation of (M, q) that leaves invariant all elements of $V^\perp$; and in (7.2.11) we read $T_N = (N^\perp \cap U) \oplus N$. In (7.5.3) there is a criterion allowing us to recognize whether such an element g of $\mathrm{GO}(M, q)$ exists. Therefore the next two propositions constitute a solution to our problem.

The bijection $p : V \to V^\perp$ involved in (7.5.2) is defined before (7.2.9).

(7.5.2) **Proposition.** *Let Z be any element of* $\mathrm{BLip}(V, q)$*, T the associated element of $\mathcal{T}(M, q)$, and let* $\mathrm{RKer}(Z)$ *(resp.* $\mathrm{LKer}(Z)$*) be the submodule of all $c \in V$ such that $zc = 0$ (resp. $cz = 0$) for all $z \in Z$; it is called the* right kernel *(resp. the* left kernel*) of Z. It is a totally isotropic submodule of V and moreover*

$$\mathrm{RKer}(Z) = p^{-1}(T \cap V^\perp) \quad (\textit{resp. } \mathrm{LKer}(Z) = T \cap V).$$

If $Z = X \bigwedge^{\max}(N)$ for some $X \in \mathrm{G'Lip}(V, q)$ and some totally isotropic direct summand N of V, then $N = \mathrm{RKer}(Z)$ and $G_X(N) = \mathrm{LKer}(Z)$.

Proof. The equality $Zc = 0$ implies $Zq(c) = (Zc)c = 0$, whence $q(c) = 0$. Consequently $\mathrm{RKer}(Z)$ is totally isotropic, and the same for $\mathrm{LKer}(Z)$. Since $cz = R_c(z)$, the equality $\mathrm{LKer}(Z) = T \cap V$ is evident. The equality $zc = 0$ is equivalent to $p(c)\psi(z) = 0$ if ψ is the twisted anti-isomorphism $C\ell(V, q) \to C\ell(V^\perp, q)$ involved in (7.2.9). Since $\psi(Z) = \chi^\perp(T)$, the vanishing of $p(c)\psi(Z)$ means that $p(c)$ belongs to $T \cap V^\perp$, whence $\mathrm{RKer}(Z) = p^{-1}(T \cap V^\perp)$. If $Z = X \bigwedge^{\max}(N)$ with X and N as in (7.5.2), the equality $Zc = 0$ is equivalent to $\bigwedge^{\max}(N)c = 0$. If P is a submodule supplementary to N in V, the multiplication mapping $\bigwedge(N) \otimes C\ell(P, q) \to C\ell(V, q)$ is bijective (see (4.8.5)), and allows us to prove that $\bigwedge^{\max}(N)c = 0$ if and only if $c \in N$. Since $X \bigwedge^{\max}(N) = \bigwedge^{\max}(G_X(N))X$, similarly $cZ = 0$ if and only if $c \in G_X(N)$. □

Before (7.5.3) is stated, three remarks might be helpful. First if $T \cap V^\perp$ is a direct summand of $V^\perp$, then it is a direct summand of M, and consequently a direct summand of T. Secondly if $T \cap V^\perp$ is a direct summand of T, then $T \cap V$ is not necessarily a direct summand of T, as it appears in the example (7.5.5) below. Thirdly d_q induces injective mappings

$$T/(T \cap V) \longrightarrow (V^\perp/(T \cap V^\perp))^* \quad \text{and} \quad V^\perp/(T \cap V^\perp) \longrightarrow (T/(T \cap V))^*$$

because $T = T^\perp$; if $T \cap V^\perp$ is a direct summand, and if the first mapping is bijective, then $V^\perp/(T\cap V^\perp)$ and $T/(T\cap V)$ are projective, consequently $T\cap V$ too is a direct summand; moreover b_q induces a duality between $T/(T\cap V)$ and $V^\perp/(T\cap V^\perp)$, and consequently both mappings are bijective; analogous conclusions would follow from the hypotheses that $T \cap V$ is a direct summand and that the second mapping is bijective.

(7.5.3) **Proposition.** *Let T be an element of $\mathcal{T}(M,q)$ and N a totally isotropic direct summand of V. There exists $g \in \mathrm{GO}(M,q)$ such that $T = g((N^\perp \cap U) \oplus N)$ and $g(w) = w$ for all $w \in V^\perp$ if and only if these three conditions are all fulfilled:*

- *$T \cap V^\perp = p(N)$;*
- *T contains a submodule that is supplementary both to $T \cap V$ and to $T \cap V^\perp$;*
- *b_q induces a duality between $T/(T\cap V)$ and $V^\perp/(T\cap V^\perp)$.*

Proof. Let us set $W_2 = p(N)$ and let W_3 be a submodule of $V^\perp$ supplementary to W_2. Here the lower index 2 (resp. 3) is given to direct summands of M that have the same rank as N (resp. V/N). Since b_q induces a duality between U and $V^\perp$, U is the direct sum of $U_2 = U\cap W_3^\perp$ (isomorphic to W_2^*) and $U_3 = U\cap W_2^\perp = N^\perp\cap U$. From (7.2.11)(equivalence of the assertions (b) and (e)) we know that $N \oplus U_3 = W_2 \oplus U_3$. If $T = g(N \oplus U_3)$ for some $g \in \mathrm{GO}(M,q)$ such that $V^\perp \subset \mathrm{Ker}(g - \mathrm{id})$, then $T = g(N) \oplus g(U_3) = g(W_2) \oplus g(U_3)$ with $g(N) \subset V$ (because $g(V) = V$) and $g(W_2) = W_2$; moreover $g(U)$ is supplementary in M both to V and to $V^\perp$. Therefore $g(N) = T\cap V$, $W_2 = T\cap V^\perp$ and $g(U_3)$ is supplementary in T both to $T\cap V$ and to $T\cap V^\perp$. Since b_q induces a duality between U_3 and W_3, it induces a duality also between $g(U_3)$ and $W_3 = g(W_3)$, or equivalently between $T/(T\cap V)$ and $V^\perp/(T\cap V^\perp)$.

Conversely let us assume that $T \cap V^\perp = W_2$, that U_3' is a submodule of T supplementary to W_2 and to $V_2 = T\cap V$, and that b_q induces a duality between U_3' and W_3. If we manage to prove the existence of a totally isotropic submodule U_2' such that $\mathrm{b}_q(U_2', U_3') = 0$ and $M = (U_2' \oplus U_3') \oplus V^\perp$, then we can claim the existence of an orthogonal transformation g such that $V^\perp \subset \mathrm{Ker}(g - \mathrm{id})$ and $g(U) = U_2' \oplus U_3'$, because b_q induces a duality both between $V^\perp$ and U, and between $V^\perp$ and the totally isotropic submodule $U_2' \oplus U_3'$. When the existence of U_2' is sure, the proof ends in this way: on one side $U_3 = U \cap W_2^\perp$; on the other side the inclusion $U_3' \subset (U_2' \oplus U_3') \cap W_2^\perp$ is an equality because it involves direct summands that have the same rank; consequently $g(U_3) = U_3'$, whence $T = W_2 \oplus U_3' = g(W_2 \oplus U_3) = g((N^\perp \cap U) \oplus N)$.

Since b_q induces a duality between U_3' and W_3, its restriction to $U_3' \oplus W_3$ is nondegenerate (even hyperbolic), and its restriction to $(U_3'\oplus W_3)^\perp$ too is nondegenerate. Since $(U_3'\oplus W_3)^\perp$ contains the totally isotropic direct summand $V_2 = T\cap V$, and since the rank of V_2 is half of the rank of $(U_3'\oplus W_3)^\perp$, this quadratic subspace too is hyperbolic (see (2.5.5)), and it is the direct sum of V_2 and some other totally isotropic direct summand U_2' (see (2.5.4)). Thus $M = (U_2'\oplus V_2) \perp (U_3'\oplus W_3)$.

Since $T = V_2 \oplus U_3' = W_2 \oplus U_3'$, we can claim that $M = U_2' \oplus W_2 \oplus U_3' \oplus W_3 = (U_2' \oplus U_3') \oplus V^\perp$. □

(7.5.4) **Example.** Here V and U are both totally isotropic submodules of (M, q), and they are free modules of rank 2; let (c_1, c_2) be a basis of V and (b_1, b_2) the dual basis of U, such that $\mathrm{b}_q(b_i, c_j) = 1$ if $i = j$, 0 if $i \neq j$. Let λ be a nonzero element of K, and $\mathfrak{n}$ the ideal of all $\nu \in K$ such that $\lambda\nu = 0$. With λ we associate the totally isotropic submodule T with basis (a_1, a_2) such that

$$a_1 = c_1 + \lambda b_2 \ , \quad a_2 = c_2 - \lambda b_1 \ ;$$

it is an element of $\mathcal{T}(M, q)$ because $M = T \oplus U$. Obviously $T \cap V = \mathfrak{n}c_1 \oplus \mathfrak{n}c_2$; consequently $T \cap V$ is a direct summand of M if and only if $\mathfrak{n}$ is a direct summand of K. Now $\mathfrak{n}$ is a direct summand of K if and only if there is an idempotent e such that λ belongs to Ke and is not a divisor of zero in Ke. If e exists, then $T \cap V = (1-e)T = (1-e)V$ and $\bigwedge^{\max}(T \cap V)$ is the submodule generated by $e + (1-e)c_1 \wedge c_2$. Then b_q induces a duality between $T/(T \cap V)$ and $V/(T \cap V)$ if and only if it induces a duality between eT and eV; this occurs if and only if λ is invertible in Ke.

On the other hand it is easy to calculate that, whatever λ may be, $\chi(T)$ is the submodule of $\bigwedge(V)$ generated by $\lambda + c_1 \wedge c_2$. Consequently, if λ belongs to the ideal Ke generated by some $e \in \mathrm{Ip}(K)$, and if there exists $\mu \in Ke$ such that $\lambda\mu = e$, then $\lambda + c_1 \wedge c_2$ must be the product of $e + (1-e)c_1 \wedge c_2$ and some $x \in \mathrm{GLip}(V)$. Because of (7.5.1), x is the exponential of an element of $\bigwedge^2(V)$ multiplied by an invertible element of K. All this is corroborated by this equality:

$$\lambda + c_1 \wedge c_2 \ = \ (\lambda + (1-e))\ \mathrm{Exp}(\mu c_1 \wedge c_2) \ \wedge \ (e + (1-e)c_1 \wedge c_2) \ .$$

(7.5.5) **Example.** Let F be a field, and $K = F[\lambda, \mu, \lambda', \mu']$ the ring of polynomials in four indeterminates λ, μ, λ', μ'. Whenever it seems opportune, we will accept a ring extension $K \to K'$ with K' a ring of fractions of K. Here V is still a free module with basis (c_1, c_2) over K, (b_1, b_2) is the dual basis in U, but the restriction of q to V does not vanish:

$$\forall \xi,\ \zeta \in K, \quad q(\xi c_1 + \zeta c_2) = (\lambda\xi + \mu\zeta)(\lambda'\xi + \mu'\zeta) \ .$$

Therefore we need the bijection $p : V \to V^\perp$ and we calculate

$$\begin{aligned} p(c_1) &= c_1 - 2\lambda\lambda' b_1 - (\lambda\mu' + \mu\lambda')b_2 \ , \\ p(c_2) &= c_2 - (\lambda\mu' + \mu\lambda')b_1 - 2\mu\mu' b_2. \end{aligned}$$

Let T be the totally isotropic submodule with basis (a_1, a_2) such that

$$\begin{aligned} a_1 = c_1 - \lambda\lambda' b_1 - \lambda\mu' b_2 \ &= \ p(c_1) + \lambda\lambda' b_1 + \mu\lambda' b_2 \ , \\ a_2 = c_2 - \mu\lambda' b_1 - \mu\mu' b_2 \ &= \ p(c_2) + \lambda\mu' b_1 + \mu\mu' b_2. \end{aligned}$$

Since $M = T \oplus U$, we have an element of $\mathcal{T}(M,q)$. It is clear that $T \cap V$ and $T \cap V^\perp$ are the submodules respectively generated by

$$\mu a_1 - \lambda a_2 - \mu c_1 - \lambda c_2 \quad \text{and} \quad \mu' a_1 - \lambda' a_2 - p(\mu' c_1 - \lambda' c_2).$$

A localization at a maximal ideal containing λ and μ shows that $T \cap V$ is not a direct summand; neither is $T \cap V^\perp$. If $\xi a_1 + \zeta a_2$ is any element of T, in general by a suitable extension $K \to K'$ we can make the submodule $K(\xi a_1 + \zeta a_2)$ become supplementary both to $T \cap V$ and $T \cap V^\perp$ in T; an easy calculation of determinants shows that this happens if and only if we make $\lambda\xi + \mu\zeta$ and $\lambda'\xi + \mu'\zeta$ become invertible; of course this imposes a little constraint on the choice of $\xi a_1 + \zeta a_2$, since neither $\lambda\xi + \mu\zeta$ nor $\lambda'\xi + \mu'\zeta$ may be the null polynomial. It is worth observing that if we make just $\lambda'\xi + \mu'\zeta$ become invertible, then $T \cap V^\perp$ becomes a direct summand, but not $T \cap V$. Does b_q induce a duality between $T/(T \cap V)$ and $V^\perp/(T \cap V^\perp)$? When $\lambda\xi + \mu\zeta$ and $\lambda'\xi + \mu'\zeta$ are invertible, it is equivalent to ask whether b_q induces a duality between the submodules generated by $\xi a_1 + \zeta a_2$ and $p(\xi c_1 + \zeta c_2)$, and the answer is positive because

$$\mathrm{b}_q(\xi a_1 + \zeta a_2,\ p(\xi c_1 + \zeta c_2)) = -(\lambda\xi + \mu\zeta)(\lambda'\xi + \mu'\zeta) = -q(\xi c_1 + \zeta c_2).$$

On the other side it is easy to calculate that $\chi(T)$ is the submodule generated by $\lambda\mu' - c_1c_2 = -\mu\lambda' + c_2c_1$. Consequently, when $\lambda\xi + \mu\zeta$ and $\lambda'\xi + \mu'\zeta$ are invertible, then $\lambda\mu' - c_1c_2$ is equal to $\mu'c_1 - \lambda'c_2$ (generator of $p^{-1}(T \cap V^\perp)$) multiplied on the left side by some $x \in \mathrm{GLip}(K' \otimes (V,q))$; it is also equal to $\mu c_1 - \lambda c_2$ multiplied on the right side by a factor x' colinear with x. Since x and x' are odd, they belong to $K' \otimes V$, and it is sensible to conjecture that they may be colinear with $\xi c_1 + \zeta c_2$ which is obviously invertible. All this is corroborated by these calculations:

$$(\xi c_1 + \zeta c_2)(\mu' c_1 - \lambda' c_2) = (\lambda'\xi + \mu'\zeta)(\lambda\mu' - c_1c_2)\ ,$$
$$(\mu c_1 - \lambda c_2)(\xi c_1 + \zeta c_2) = -(\lambda\xi + \mu\zeta)(\lambda\mu' - c_1c_2).$$

When K is a field, the criterion presented in (7.5.3) gives a positive answer whenever the first condition $T \cap V^\perp = p(N)$ is fulfilled.

(7.5.6) **Lemma.** *Let us assume that K is a field. When T is an element of $\mathcal{T}(M,q)$, then T contains a submodule supplementary both to $T \cap V$ and to $T \cap V^\perp$, and b_q induces a duality between $T/(T \cap V)$ and $V^\perp/(T \cap V^\perp)$.*

Proof. Let r be the common dimension of T, U and V, and s the dimension of $T \cap V$. The dimension of $T + V^\perp = (T \cap V)^\perp$ is $2r - s$, and consequently $T \cap V^\perp$ has the same dimension s as $T \cap V$. By means of a basis of T that contains a basis of $T \cap V \cap V^\perp$, a basis of $T \cap V$ and a basis of $T \cap V^\perp$, it is easy to prove that T contains a subspace supplementary both to $T \cap V$ and $T \cap V^\perp$. At last the mapping $T/(T \cap V) \to (V^\perp/(T \cap V^\perp))^*$ is bijective because it is injective, and the source and the target have the same dimension $r - s$. □

The next theorem (the main result of this section) is an immediate consequence of (7.5.6), (7.5.3) and (7.5.2). When an element $z \in \mathrm{Lip}(V,q)$ generates an element Z of $\mathrm{BLip}(V,q)$, the notations $\mathrm{RKer}(z)$ and $\mathrm{LKer}(z)$ have the same meaning as $\mathrm{RKer}(Z)$ and $\mathrm{LKer}(Z)$ defined in (4.7.8).

(7.5.7) **Theorem.** *Let us assume that K is a field, and that z is a nonzero element of $\mathrm{Lip}(V,q)$. If y is any nonzero element of $\bigwedge^{\max}(\mathrm{RKer}(z))$, there exists $x \in \mathrm{GLip}(V,q)$ such that $z = xy$. Moreover $\mathrm{LKer}(z) = G_x(\mathrm{RKer}(z))$, and there is a nonzero $y' \in \bigwedge^{\max}(\mathrm{LKer}(z))$ such that $z = y'x$.*

When the group $\mathrm{GLip}(V,q)$ operates in the set $\mathrm{Lip}(V,q)$ by multiplications on the left side (resp. on the right side), the orbits other than the trivial orbit $\{0\}$ are in bijection with the totally isotropic subspaces of V, and the orbit $\mathrm{GLip}(V,q)$ corresponds to the totally isotropic subspace reduced to 0; if z is a nonzero element of $\mathrm{Lip}(V,q)$, and N a totally isotropic subspace of V, then z is in the same orbit as the nonzero elements of $\bigwedge^{\max}(N)$ if and only if $N = \mathrm{RKer}(z)$ (resp. $N = \mathrm{LKer}(z)$).

Every element of $\bigwedge^{\max}(N)$ (when N is totally isotropic) is a product of isotropic elements of V. Of course it must be understood that a product of 0 factor is equal to 1. Whence the following corollary.

(7.5.8) **Corollary.** *When K is a field, $\mathrm{Lip}(V,q)$ is generated as a monoid by the group $\mathrm{GLip}(V,q)$ and all isotropic vectors of V. In particular,*

- *if the restriction of q to V is anisotropic, then every nonzero element of $\mathrm{Lip}(V,q)$ is invertible;*
- *if $\mathrm{GLip}(V,q)$ is generated by the invertible elements of V (as it almost always is when $q(V) \neq 0$), then $\mathrm{Lip}(V,q)$ is the monoid generated by all elements of V;*
- *if $q(V) = 0$, then $\mathrm{Lip}(V)$ is the monoid generated by all elements of K and V, and by all exponentials of elements of $\bigwedge^2(V)$.*

Proposition (7.5.1) is involved in the last statement of (7.5.8), which is exactly the same thing as Theorem (5.10.2).

When $q(V) \neq 0$, then $\mathrm{GLip}(V,q)$ is generated by the invertible elements of V except in the cases pointed out in (5.7.3); let us recall them with the notation of this chapter. The quadratic module (V,q) over the field K belongs to an exceptional case when these three conditions are fulfilled: first $K \cong \mathbb{Z}/2\mathbb{Z}$, secondly $\dim(V) \geq 3$ whereas $V/(V \cap V^\perp)$ has dimension 2 or 4, thirdly $q(V \cap V^\perp) = 0$ and q induces on $V/(V \cap V^\perp)$ a hyperbolic quadratic form. Let $\Gamma(V,q)$ be the subgroup of $\mathrm{GLip}(V,q)$ generated by the invertible elements of V; from (5.7.8), (5.7.9) and (5.10.3) it is possible to deduce a subgroup of $\mathrm{GLip}(V,q)$ supplementary to $\Gamma(V,q)$. In Dieudonné's exceptional case, in other words when $\dim(V/(V \cap V^\perp)) = 4$, we can find c_1 and c_2 such that $Kc_1 \oplus Kc_2 \oplus (V \cap V^\perp)$ is a maximal totally isotropic subspace; then $\{1,\ 1 + c_1c_2\}$ is a subgroup of order 2 supplementary to $\Gamma(V,q)$. When $\dim(V/(V \cap V^\perp)) = 2$, then V contains an isotropic element c' outside

$V \cap V^\perp$, and the mapping $c \longmapsto 1 + cc'$ is an injective morphism from the additive group $V \cap V^\perp$ onto a subgroup supplementary to $\Gamma(V, q)$.

Although the main purpose of this section is the study of $\mathrm{Lip}(V, q)$, we may still state a geometrical property involving elements of $\mathcal{T}(M, q)$.

(7.5.9) **Proposition.** *When K is a field, for every pair (T, T') of elements of $\mathcal{T}(M, q)$ these three assertions are equivalent:*

(a) $T + V = T' + V$;
(b) $T \cap V^\perp = T' \cap V^\perp$;
(c) *there exists $x \in \mathrm{GLip}(V, q)$ such that $T' = G_x(T)$.*

Proof. The equivalence (a)⇔(b) comes from the equality $(T + V)^\perp = T \cap V^\perp$. The implication (c)⇒(b) is trivial, because every G_x with $x \in \mathrm{GLip}(V, q)$ leaves invariant every element of $V^\perp$ (see (5.4.5)). Conversely let us assume that there is a totally isotropic subspace N of V such $T \cap V^\perp$ and $T' \cap V^\perp$ are both equal to $p(N)$. Then (7.5.3) and (7.5.6) show that T and T' are in the same orbit as $(N^\perp \cap U) \oplus N$ under the action of the group $\mathrm{GLip}(V, q)$. □

7.6 Applications to totally isotropic direct summands of maximal rank

Whereas the previous section was devoted to the quadratic module (V, q), here we are again interested in the hyperbolic space (M, q) and in the set $\mathcal{T}(M, q)$; we assume M to be a faithful module; thus the center of $C\ell_0(M, q)$ is a quadratic extension denoted by $\mathrm{QZ}(M, q)$ in **3.7**.

With each $T \in \mathcal{T}(M, q)$ is associated a graded direct summand $\chi(T)$ in $C\ell(V, q)$; since $\chi(T)$ has constant rank 1, every localization of $\chi(T)$ is either even or odd; thus at every prime ideal $\mathfrak{p}$ of K, T has a parity which is an element of $\mathbb{Z}/2\mathbb{Z}$ and which is denoted by $\mathrm{par}(\mathfrak{p}; U, T)$. For instance $\mathrm{par}(\mathfrak{p}; U, U)$ is always even because $\chi(U) = K$. Although the definition of $\mathrm{par}(\mathfrak{p}; U, T)$ involves the decomposition $M = U \oplus V$, the proposition (7.6.3) shows that it does not depend on V, that $\mathrm{par}(\mathfrak{p}; U, T) = \mathrm{par}(\mathfrak{p}; T, U)$, and that, for all T and $T' \in \mathcal{T}(M, q)$,

$$\text{(7.6.1)} \qquad \mathrm{par}(\mathfrak{p}; T, T') = \mathrm{par}(\mathfrak{p}; U, T') - \mathrm{par}(\mathfrak{p}; U, T).$$

Therefore the best name for $\mathrm{par}(\mathfrak{p}; U, T)$ should be *the difference of parity between U and T at the prime ideal $\mathfrak{p}$*.

Obviously every ring extension $f : K \to L$ gives a hyperbolic space $L \otimes (M, q)$ which is the direct sum of $L \otimes U$ and $L \otimes V$, and a mapping $\mathcal{T}(M, q) \to \mathcal{T}(L \otimes (M, q))$. Since $\chi(T)$ is a direct summand, it is clear that $\chi(L \otimes T) = L \otimes \chi(T)$, whence, for every prime ideal $\mathfrak{q}$ of L,

$$\mathrm{par}(\mathfrak{q}; (L \otimes U), (L \otimes T)) = \mathrm{par}(f^{-1}(\mathfrak{q}); U, T).$$

In particular $\mathrm{par}(\mathfrak{p}; U, T)$ can be calculated by means of the extension $K \to K_{\mathfrak{p}}$ and even by means of the extension to the residue field $F_{\mathfrak{p}} = K_{\mathfrak{p}}/\mathfrak{p}K_{\mathfrak{p}}$:

$$\mathrm{par}(\mathfrak{p}; U, T) = \mathrm{par}(U_{\mathfrak{p}}, T_{\mathfrak{p}}) = \mathrm{par}((F_{\mathfrak{p}} \otimes U), (F_{\mathfrak{p}} \otimes T)) ;$$

the extension $K \to F_{\mathfrak{p}}$ allows us to apply (7.6.5), (7.6.7) and (7.5.9) which are only valid when K is a field.

If $\mathfrak{p} \subset \mathfrak{p}'$, there is a ring morphism $K_{\mathfrak{p}'} \to K_{\mathfrak{p}}$ and consequently $\mathrm{par}(\mathfrak{p}; U, T) = \mathrm{par}(\mathfrak{p}'; U, T)$. This justifies the short notation $\mathrm{par}(U, T)$ when K is a local ring.

When U and T remain constant, the function $\mathfrak{p} \longmapsto \mathrm{par}(\mathfrak{p}; U, T)$ is locally constant on $\mathrm{Spec}(K)$ (see (1.12.7)); consequently there is a unique idempotent $e \in \mathrm{Ip}(K)$ such that $\mathrm{par}(\mathfrak{p}; U, T)$ is 0 or 1 according as $\mathfrak{p}$ contains e or not. Remember that every set of idempotents is provided with the boolean addition $(e, e') \longmapsto e \tilde{+} e' = e + e' - 2ee'$.

(7.6.2) **Lemma.** *The quadratic extension $Z = \mathrm{QZ}(M, q)$ contains an idempotent ε such that $Z = K\varepsilon \oplus K(1-\varepsilon)$ and the mapping $e \longmapsto e \tilde{+} \varepsilon$ is a bijection from $\mathrm{Ip}(K)$ onto the subset of all $\varepsilon' \in \mathrm{Ip}(Z)$ such that $Z = K\varepsilon' \oplus K(1-\varepsilon')$. The converse mapping is $\varepsilon' \longmapsto \varepsilon \tilde{+} \varepsilon'$.*

Proof. The existence of ε follows from the fact that Z is a trivial quadratic extension. The idempotents $\varepsilon' \in \mathrm{Ip}(Z)$ such that $Z = K\varepsilon' \oplus K(1-\varepsilon')$ are in bijection with the isomorphism $f' : K^2 \to Z$, and if f is the isomorphism $K^2 \to Z$ associated with ε, they are in bijection with the automorphisms of Z : with ε' is associated $f' \circ f^{-1}$. From (3.4.15) we know that the automorphisms of Z are in bijection with the idempotents $e \in \mathrm{Ip}(K)$; the automorphism associated with e induces the identity mapping on $(1-e)Z$, the standard involution on eZ, and maps ε to $\varepsilon' = (1-e)\varepsilon + e(1-\varepsilon')$. This equality is equivalent to $\varepsilon' = e \tilde{+} \varepsilon$, and the conclusion follows. □

(7.6.3) **Proposition.** *For every $T \in \mathcal{T}(M, q)$ there is a unique $\varepsilon(T) \in \mathrm{Ip}(Z)$ (with $Z = \mathrm{QZ}(M, q)$) satisfying these two conditions:*

$$Z = K\varepsilon(T) \oplus K(1-\varepsilon(T)) \ ,$$
$$\bigwedge^{\max}(T) \subset \varepsilon(T)\ \mathrm{C}\ell(M, q) \quad (\textit{or equivalently } (1-\varepsilon(T)) \bigwedge^{\max}(T) = 0).$$

In particular $\varepsilon(U)$ is the element of $\mathrm{C}\ell(M, q)$ such that $R_{\varepsilon(U)}$ is the parallel projection $\mathrm{C}\ell(V, q) \to \mathrm{C}\ell_0(V, q)$ with respect to $\mathrm{C}\ell_1(V, q)$. Moreover $\varepsilon(U) \tilde{+} \varepsilon(T)$ is the unique $e \in \mathrm{Ip}(K)$ such that $\mathrm{par}(\mathfrak{p}; U, T)$ is even or odd according as $\mathfrak{p}$ contains e or not.

Proof. Since the algebra morphism $x \longmapsto R_x$ is bijective from $\mathrm{C}\ell(M, q)$ onto $\mathrm{End}(\mathrm{C}\ell(V, q))$, there is a unique $\varepsilon \in \mathrm{C}\ell(M, q)$ such that R_ε is the projection onto the even component $\mathrm{C}\ell_0(V, q)$. The center of $\mathrm{End}_0(\mathrm{C}\ell(V, q))$ is $KR_\varepsilon \oplus KR_{1-\varepsilon}$ and consequently the center of $\mathrm{C}\ell_0(M, q)$, that is Z, is equal to $K\varepsilon \oplus K(1-\varepsilon)$. For all $\omega \in \bigwedge^{\max}(U)$ we deduce from (7.3.1) that $\varepsilon\omega = R_\varepsilon(1)\,\omega = \omega$; this equality is equivalent to $\omega \in \varepsilon\mathrm{C}\ell(M,q)$ and to $(1-\varepsilon)\omega = 0$. If ε_2 is an idempotent of Z such that similarly $Z = K\varepsilon_2 \oplus K(1-\varepsilon_2)$ and $\bigwedge^{\max}(U) \subset \varepsilon_2\mathrm{C}\ell(M,q)$, then $\varepsilon_2 = (1-e_2)\varepsilon + e_2(1-\varepsilon)$ for some $e_2 \in \mathrm{Ip}(K)$, and for all $\omega \in \bigwedge^{\max}(U)$,

$$\omega = \varepsilon_2\omega = (1-e_2)\varepsilon\omega + e_2(1-\varepsilon)\omega = (1-e_2)\omega\ ;$$

since $\bigwedge^{\max}(U)$ is a faithful module, this implies $e_2 = 0$ and $\varepsilon_2 = \varepsilon$. Since ε is characterized by the properties $Z = K\varepsilon \oplus K(1-\varepsilon)$ and $\bigwedge^{\max}(U) \subset \varepsilon\mathrm{C}\ell(M,q)$, it does not depend on V and can be denoted by $\varepsilon(U)$.

Since U is any element of $\mathcal{T}(M,q)$, with every $T \in \mathcal{T}(M,q)$ is associated in the same way an idempotent $\varepsilon(T)$ in Z. We must find a relation between $\varepsilon(T)$ and the idempotent e of K such that $(1-e)\chi(T)$ is even whereas $e\chi(T)$ is odd.

Since the standard involution of Z maps ε to $1-\varepsilon$, from (3.5.13) we derive that

$$\forall x \in \mathrm{C}\ell_0(M,q), \quad \varepsilon x = x\varepsilon\,, \quad \text{and} \quad \forall x \in \mathrm{C}\ell_1(M,q), \quad (1-\varepsilon)x = x\varepsilon\,.$$

From (7.3.2) we know that $\bigwedge^{\max}(T)$ is generated by the products $z\omega\tau(z)$ with $z \in \chi(T)$ and $\omega \in \bigwedge^{\max}(U)$. If we set $\varepsilon' = (1-e)\varepsilon + e(1-\varepsilon)$ (with $\varepsilon = \varepsilon(U)$ as above), then for all $z \in \chi(T)$ and all $\omega \in \bigwedge^{\max}(U)$ we can write

$$\varepsilon(1-e)z\omega = (1-e)z\varepsilon\omega = (1-e)z\omega \quad \text{and} \quad (1-\varepsilon)ez\omega = ez\varepsilon\omega = ez\omega\ ,$$

whence $\varepsilon' z\omega\tau(z) = z\omega\tau(z)$ and $\varepsilon' = \varepsilon(T)$. □

The equality (7.6.1) follows from (7.6.3) and

$$\varepsilon(T)\tilde{+}\varepsilon(T') \;=\; (\varepsilon(U)\tilde{+}\varepsilon(T'))\;\tilde{+}\;(\varepsilon(U)\tilde{+}\varepsilon(T)).$$

When 2 is invertible in K, the element ζ mentioned in (7.3.6) is equal to $2\varepsilon(U)-1$; when moreover M has constant rank $2r$, then the inclusion $\bigwedge^{\max}(T) \subset \varepsilon(T)\mathrm{C}\ell(M,q)$ (with $T \in \mathcal{T}(M,q)$) is equivalent to $\bigwedge^r(T) \subset \varepsilon(T)\mathrm{C}\ell^r(M,q)$.

In the remainder of this section we suppose that K is a field and we derive consequences from (7.3.4). Every element in a Clifford algebra (or exterior algebra) over a field K has a *support* that is defined in the next lemma; the definition of the support of a bivector just before (5.9.2) is compatible with this more general definition.

(7.6.4) **Lemma.** *Let $(N,\tilde{q})$ be a quadratic module over a field K, and let x be a nonzero element of $\mathrm{C}\ell(N,\tilde{q})$. There is a smallest subspace of N among all the subspaces P such that x belongs to the subalgebra generated by P; it is called the support of x. If x is lipschitzian and if m is the integer such that x belongs to $\mathrm{C}\ell^{\leq m}(N,\tilde{q})$ but not to $\mathrm{C}\ell^{<m}(N,\tilde{q})$, then m is the dimension of the support of x.*

Proof. If P and P' are subspaces of N, the intersection of the subalgebras generated by P and P' is the subalgebra generated by $P \cap P'$; this can be proved by means of a basis of N that contains a basis of $P \cap P'$, a basis of P and a basis of P'; remember that by definition every subalgebra contains K. Among the subspaces P such that x belongs to the subalgebra generated by P, there is a subspace P_0 of finite minimal dimension; let us prove that $P_0 \subset P$ whenever the subalgebra generated by P contains x. Indeed x belongs to the subalgebra generated by $P_0 \cap P$, whence $\dim(P_0 \cap P) \geq \dim(P_0)$ and $P_0 \subset P$. If β is a bilinear form on N, the subalgebras generated by any subspace P in $C\ell(N,\tilde{q})$ and in its deformation $C\ell(N,\tilde{q};\beta)$ are equal as subspaces of $C\ell(N,\tilde{q})$; consequently the support of x is invariant by deformation. There is a deformation $C\ell(N,\tilde{q};\beta)$ that is isomorphic to $\bigwedge(N)$, and if x is lipschitzian in $C\ell(N,\tilde{q})$, its image x' in $\bigwedge(N)$ is lipschitzian in $\bigwedge(N)$ (see (5.4.1)). Theorem (5.10.2) (corroborated by (7.5.8) just above) states that x' is the exterior product of a decomposable element (a scalar or a product of elements of N) and the exponential of an element of $\bigwedge^2(N)$; therefore

$$x' \;=\; \lambda d_1 \wedge d_2 \wedge \wedge d_n \wedge (1 + e_1 \wedge e_2) \wedge \cdots \wedge (1 + e_{2k-1} \wedge e_{2k})$$

with $(d_1, d_2, \ldots, d_n, e_1, e_2, \ldots, e_{2k})$ a linearly independent family of vectors, and λ a scalar that is only indispensable when $n = 0$. The support of x' (which is also the support of x) is the subspace spanned by this family of vectors; this can be proved in a rigorous way by means of (4.8.12). If we set $m = n + 2k$, it is clear that x' belongs to $\bigwedge^{\leq m}(N)$ but not to $\bigwedge^{<m}(N)$, whence the conclusion. □

(7.6.5) **Proposition.** *Let $M = U \oplus V$ be a hyperbolic space of dimension $2r$ over a field K, and let T and T' be maximal totally isotropic subspaces like U. If ω, z and z' are generators of the lines $\bigwedge^r(U)$, $\chi(T)$ and $\chi(T')$, then $z\omega_T(z')$ is a lipschitzian element with support $T + T'$ in M.*

Proof. If P is any subspace of M, from (7.3.5) we know that $z\omega_T(z')$ belongs to $C\ell(P^\perp, q)$ if and only if $P \subset T \cap T'$. Consequently the support of $z\omega_T(z')$ is $(T \cap T')^\perp$. Since $T = T^\perp$ and $T' = T'^\perp$, by means of (2.3.6) and (2.3.7) we realize that $(T \cap T')^\perp = T + T'$. □

(7.6.6) **Corollary.** *With the same notation as in* (7.6.5) *the following assertions are equivalent:*

(a) *T and T' have the same parity;*
(b) $\dim(T + T')$ *has the same parity as r ;*
(c) $\dim(T \cap T')$ *has the same parity as r.*

Proof. The parities of T and T' are equal if and only if the parity of $z\omega_T(z')$ is the same as the parity of ω, that is the parity of r. From (7.6.4) we deduce that the parity of $z\omega_T(z)$ is the parity of the dimension of its support; as stated in (7.6.5), its support is $T + T'$; whence (a)⇔(b). Since the sum of the dimensions of $T + T'$ and $T \cap T'$ is $2r$, it is clear that (b)⇔(c). □

When the field K does not have characteristic 2, it is sensible to associate with every nonzero $x \in \mathrm{Lip}(M, q)$ two subspaces of M, first its support $\mathrm{Sup}(x)$ defined in (7.6.4), and a smaller subspace $\mathrm{Ker}^\wedge(x)$ defined by means of the canonical isomorphism $C\ell(M,q) \to \bigwedge(M; \mathrm{b}_q/2)$; if x' is the canonical image of x in $\bigwedge(M)$, then $\mathrm{Ker}^\wedge(x) = \mathrm{LKer}(x') = \mathrm{RKer}(x')$; in other words, $\mathrm{Ker}^\wedge(x)$ is the subspace of all $a \in M$ such that $a \wedge x' = 0$. If x' is written as above in the proof of (7.6.4), then $\mathrm{Ker}^\wedge(x)$ is the subspace spanned by $(d_1, d_2, \dots, d_n)$; its dimension n has the same parity as the dimension m of the support, all nonzero components of x' have a degree between m and n, and the nonzero components of highest and lowest degrees are decomposable elements in $\bigwedge^m(\mathrm{Sup}(x))$ and $\bigwedge^n(\mathrm{Ker}^\wedge(x))$.

(7.6.7) **Proposition.** *With the same notation as in* (7.6.5) *we can write*

$$\mathrm{Ker}^\wedge(z\omega\tau(z')) = T \cap T'$$

if the field K does not have characteristic 2.

Proof. This is a consequence of (7.6.5) and (7.3.9), because the multiplication by ζ (an element of $C\ell^{2r}(M,q)$) maps every $C\ell^k(M,q)$ bijectively onto $C\ell^{2r-k}(M,q)$ (see (4.8.15)). Indeed let us set $x = z\omega\tau(z')$ and $m = \dim(T+T')$; thus x belongs to $C\ell^{\le m}(M,q)$, and its component in $C\ell^m(M,q)$ is the product of the m elements of an orthogonal basis $(a_1, a_2, \dots, a_m)$ of $T + T'$; they have the same product in $C\ell(M,q)$ and in the deformation $C\ell(M,q; -\mathrm{b}_q/2)$ isomorphic to $\bigwedge(M)$. From (7.3.9) we deduce $x = \pm\zeta x$. Consequently, if we set $n = 2r - m$, we know that x belongs to $C\ell^{\ge n}(M,q)$, and that the dimension of $\mathrm{Ker}^\wedge(x)$ is n, like that of $T \cap T'$. Up to an invertible scalar, the component of x in $C\ell^n(M,q)$ is $\zeta a_1 a_2 \cdots a_m$; to prove that it is an element of $\bigwedge^n(T\cap T')$, it suffices to verify that $a' \wedge (\zeta a_1 a_2 \cdots a_m) = 0$ for all $a' \in T \cap T'$. From (4.8.13) we deduce

$$2a' \wedge (\zeta a_1 a_2 \cdots a_m) \;=\; a'\zeta a_1 a_2 \cdots a_m \;+\; (-1)^m\, \zeta a_1 a_2 \cdots a_m a'\;;$$

now a' anticommutes with each a_i $(i = 1, 2, \dots, m)$ because a' is orthogonal to $T + T'$, and moreover a' anticommutes with ζ (see (3.5.13)); therefore $a' \wedge (\zeta a_1 a_2 \cdots a_m) = 0$ as predicted. □

Exercises

(7.ex.1) Exceptionally in this exercise we accept that the canonical mappings $K \to C\ell(M,q)$ and $\rho : M \to C\ell(M,q)$ may be not injective; consequently we use again the notations 1_q and ρ. We suppose that M is a direct sum of a totally isotropic submodule U and any other submodule V. Let q' be the restriction of q to V, and ρ' the canonical mapping $V \to C\ell(V,q')$. By means of the "universal property" of $C\ell(M,q)$ (see **3.1**), prove the existence of an algebra morphism $f : C\ell(M,q) \to \mathrm{End}(C\ell(V,q'))$ such that $f(\rho(c))$ is the multiplication $z \longmapsto \rho'(c)z$ for every $c \in V$, whereas, for every $b \in U$, $f(\rho(b))$ is the interior multiplication $z \longmapsto \mathrm{d}'_q(b) \rfloor z$ by the restriction $\mathrm{d}'_q(b)$ of $\mathrm{d}_q(b)$ to V.

(7.ex.2)* It may be interesting to find out which features ensure the equality $R_{xy}(z) = R_x R_y(z)$ (with x, y, $z \in C\ell(M,q)$) when both members are calculated by means of the formula discovered in (7.1.6). Straightforward calculations (like those in the proof of (4.7.2)) show that

$$\begin{aligned} R_{xy}(z) &= \varpi\pi \circ (\pi \otimes \mathrm{id}_\wedge)\ \big(\mathrm{Exp}(f)\ \rfloor\ (x \otimes y \otimes z)\big) \\ &\qquad \text{with } f \ = \ (\pi \otimes \mathrm{id}_\wedge)^*(\beta_{\prime\prime} + \pi^*(\beta^\dagger))\ +\ \beta_{\prime\prime} \otimes 1\ , \\ R_x R_y(z) &= \varpi\pi \circ (\mathrm{id}_\wedge \otimes \varpi\pi)\ \big(\mathrm{Exp}(g)\ \rfloor\ (x \otimes y \otimes z)\big) \\ &\qquad \text{with } g \ = \ (\mathrm{id}_\wedge \otimes \varpi\pi)^*(\beta_{\prime\prime} + \pi^*(\beta^\dagger))\ +\ 1 \otimes (\beta_{\prime\prime} + \pi^*(\beta^\dagger))\ ; \end{aligned}$$

the presence of $(\pi\otimes\mathrm{id}_\wedge)^*$ (resp. $(\mathrm{id}_\wedge\otimes\varpi\pi)^*$) in the definition of f (resp. g) comes from an application of (4.4.6). Obviously $\varpi\pi \circ (\pi \otimes \mathrm{id}_\wedge)$ and $\varpi\pi \circ (\mathrm{id}_\wedge \otimes \varpi\pi)$ are the algebra morphisms associated by the functor $\bigwedge$ with the same mapping $M \oplus M \oplus M \to M$. But the verification of the equality $f = g$ is a much tougher exercise.

(7.ex.3) Let A be a graded Azumaya algebra that is isomorphic to $\mathrm{End}(P)$, with P some graded finitely generated projective module of constant nonzero rank r. Let S (resp. S') be a graded left ideal (resp. a graded right ideal) of constant rank r. It is also assumed that A contains a graded left ideal (resp. a graded right ideal) that is supplementary to S (resp. S').

(a) Prove that $S\cap S'$ is a graded direct summand R of A, and that R has constant rank 1.
Hint. You can suppose that $A = \mathrm{End}(P)$; prove that P contains a graded direct summand H of rank $r-1$ such that S is the subset of all $f \in A$ such that $\mathrm{Ker}(f) \supset H$, and that P contains a graded direct summand L' of rank 1 such that S' is the subset of all $f \in A$ such that $\mathrm{Im}(f) \subset L'$.

(b) Let $\mathcal{B}'$ be the graded module of all bilinear mappings $\psi : S' \times S \to K$. It is a graded module over $A \hat{\otimes} A^{to}$ according to this equality (in which x and x' run through A, z runs through S and z' through S'):

$$(x\psi x')(z', z) \ = \ (-1)^{\partial x(\partial\psi+\partial z')+\partial x'\partial z'}\ \psi(z'x,\ x'z).$$

For every $(z', z) \in S' \times S$ it is clear that $z'z$ belongs to $R = S \cap S'$; therefore every $\omega^* \in R^*$ determines an element Ψ_{ω^*} of $\mathcal{B}'$: $\Psi_{\omega^*}(z', z) = \omega^*(z'z)$. Prove that $\mathrm{Z}^g(A, \mathcal{B}')$ is the submodule of all Ψ_{ω^*} .
Hint. The submodule of all Ψ_{ω^*} is a direct summand of $\mathcal{B}'$ because the multiplication mapping $S' \otimes S \to R$ is surjective, and R is projective.

(c) Now we assume that A is provided with an involution τ (see (1.13.7)), and that $S' = \tau(S)$. We consider $R = S \cap \tau(S)$ and the module $\mathcal{B}$ of all bilinear mappings $\varphi : S \times S \to K$, which is a bimodule over A according to the equality

$$(x\varphi x')(z', z) \ = \ (-1)^{\partial x(\partial\varphi+\partial z')+\partial x'\partial z'}\ \varphi(\tau(x)z',\ x'z)\ ,$$

in which the twisting exponent is equivalent to $\partial x(1 + \partial x' + \partial z) + \partial x'\partial z'$. Explain how to get the elements of $\mathrm{Z}^g(A, \mathcal{B})$.

Comment. This construction of $Z^g(A, \mathcal{B})$, which probably stems from Chevalley, has been borrowed by many other mathematicians; when K is a field, it has been generalized to algebras which (unlike the previous algebra A) have a nontrivial Brauer–Wall class, by means of a graded minimal left ideal S.

(7.ex.4) This exercise is an application of the previous one to the Cartan–Chevalley mapping. As in **7.3**, (M, q) is a hyperbolic quadratic space, U belongs to $\mathcal{T}(M, q)$, and V is any supplementary submodule. If T is any element of $\mathcal{T}(M, q)$, we can apply the results of (7.ex.3) to the left ideal S_U generated by $\bigwedge^{\max}(U)$ in $C\ell(M, q)$, and to the right ideal S'_T generated by $\bigwedge^{\max}(T)$ in $C\ell(M, q)$; in particular $S_U \cap S'_T$ is a graded direct summand of $C\ell(M, q)$ of constant rank 1.

(a) In (7.1.4) there is a bijection $C\ell(V, q) \otimes \bigwedge^{\max}(U) \to S_U$. Prove that $\chi(T)$ is the submodule of $C\ell(V, q)$ such that $\chi(T) \otimes \bigwedge^{\max}(U)$ is mapped bijectively onto $S_U \cap S'_T$.
Comment. This property of $\chi(T)$ underlies Chevalley's original definition of $\chi(T)$.

(b) Obviously $\tau(S_U)$ is the right ideal S'_U generated by $\bigwedge^{\max}(U)$, and $\bigwedge^{\max}(U) = S_U \cap S'_U$. As suggested in (7.ex.3)(c), every element $\omega^* \in (\bigwedge^{\max}(U))^*$ determines a bilinear form Φ_{ω^*} on S_U; but here we are interested in the module $\mathcal{B}_V$ of all bilinear forms on $C\ell(V, q)$; consequently with every $w \in (\bigwedge^{\max}(U))^* \otimes \bigwedge^{\max}(U) \otimes \bigwedge^{\max}(U)$ we associate a bilinear form F_w on $C\ell(V, q)$, which is defined in this way when $w = \omega^* \otimes \omega_1 \otimes \omega_2$:

$$F_w(z', z) = \omega^*(\omega_1 z' z \omega_2) .$$

Prove that $Z^g(C\ell(M, q), \mathcal{B}_V)$ is the submodule of all F_w.
Remark. $(\bigwedge^{\max}(U))^* \otimes \bigwedge^{\max}(U) \otimes \bigwedge^{\max}(U)$ is canonically isomorphic to $\bigwedge^{\max}(U)$ and to $(\bigwedge^{\max}(V))^*$.

(7.ex.5) Here is a construction of $Z^g(C\ell(M, q), \mathcal{B}_V)$ quite different from the one in (7.ex.4)(b). We assume that V has constant rank r. Since the graded algebra $\mathrm{Gr}(C\ell(V, q))$ derived from $C\ell(V, q)$ is canonically isomorphic to $\bigwedge(V)$ (see (4.8.7)), there is a canonical surjective mapping $p : C\ell(V, q) \to \bigwedge^r(V)$ with kernel $C\ell^{<r}(V, q)$. Every $w \in \bigwedge^{*r}(V)$ determines a bilinear form F_w on $C\ell(V, q)$ in this way:

$$F_w(z', z) = w \circ p(\tau(z')\, z).$$

Prove that the mapping $w \longmapsto F_w$ is a bijection from $\bigwedge^{*r}(V)$ onto the submodule of all $\varphi \in \mathcal{B}_V$ such that

$$\forall x \in C\ell(M, q), \quad \forall z',\ z \in C\ell(V, q), \quad \varphi(R_x(z'),\ z) = \varphi(z',\ R_{\tau(x)}(z)).$$

Hint. It suffices to prove that $\tau(R_a(z'))\, z - \tau(z') R_a(z)$ belongs to $C\ell^{<r}(V, q)$ for all $a \in M$, and all z' and $z \in C\ell(V, q)$; this is trivial if $a = c \in V$, and follows from (4.4.4) and other properties when $a = b \in U$.

(7.ex.6) Let A be a graded algebra provided with an involution τ. As usual we set $\sigma(x) = (-1)^{\partial x}x$ for all $x \in A$. Moreover let S be a graded module over A, and $\mathcal{B}$ the graded module of all bilinear forms $\varphi : S \times S \to K$. We try to make $\mathcal{B}$ become a graded bimodule over A by means of a formula of this kind:

$$(x\varphi x')(z,z') \;=\; (-1)^{\xi}\,\varphi(x'z,\;\tau(x)z')\;;$$

a function ξ of ∂x, $\partial x'$, ∂z, $\partial z'$ (with values in $\mathbb{Z}/2\mathbb{Z}$) is said to be an admissible twisting exponent if it makes $\mathcal{B}$ become a bimodule over A for all triplets (A, τ, S) over any ring K.

(a) Let $\xi(\partial x, \partial x', \partial z, \partial z')$ be an admissible twisting exponent. Prove that $\xi(0, 0, \partial z, \partial z') = 0$, that $\xi(1, 0, \partial z, \partial z')$ is a function $\zeta(\partial z)$ of the only variable ∂z, and that $\xi(0, 1, \partial z, \partial z')$ is a function $\zeta'(\partial z')$ of the only variable $\partial z'$; moreover

$$\begin{aligned}\xi(\partial x, \partial x', \partial z, \partial z') &= \partial x\ \zeta(\partial z) + \partial x'\ \zeta'(\partial x + \partial z')\\ &= \partial x\ \zeta(\partial x' + \partial z)\ +\ \partial x'\ \zeta'(\partial z').\end{aligned}$$

Conversely two functions $\zeta(\partial z)$ and $\zeta'(\partial z')$ determine an admissible twisting exponent ξ if and only if $\zeta(0) + \zeta(1) = \zeta'(0) + \zeta'(1)$.

(b) Consequently all admissible twisting exponents ξ are in bijection with the elements (λ, μ, ν) of $(\mathbb{Z}/2\mathbb{Z})^3$ according to this formula:

$$\xi = \lambda(\partial x\partial x' + \partial x\partial z + \partial x'\partial z') + \mu\partial x + \nu\partial x'.$$

The notation $\mathcal{B}(\lambda, \mu, \nu)$ means the bimodule $\mathcal{B}$ over A obtained with the corresponding twisting exponent ξ. Replacing μ with $\mu + 1$ is equivalent to replacing τ with $\sigma\tau$.

(c) Let σ_1 (resp. σ_2, resp. σ_3) be the automorphism of the K-module $\mathcal{B}$ that maps every element φ to

$$(z,z') \longmapsto (-1)^{\partial z\partial z'}\varphi(z,z') \quad (\text{resp. } (-1)^{\partial z}\varphi(z,z'),\ \text{resp. } (-1)^{\partial z'}\varphi(z,z')).$$

Prove that σ_1 (resp. σ_2, resp. σ_3) is an isomorphism from $\mathcal{B}(\lambda, \mu, \nu)$ onto $\mathcal{B}(\lambda + 1,\ \mu, \nu)$ (resp. $\mathcal{B}(\lambda,\ \mu + 1,\ \nu)$, resp. $\mathcal{B}(\lambda, \mu,\ \nu + 1)$).

(d) For every $(i, j) \in (\mathbb{Z}/2\mathbb{Z})^2$, let $\mathcal{Z}_i(j)$ be the submodule of all $\varphi \in \mathcal{B}_i$ such that

$$\forall x \in A, \quad \forall z,\ z' \in S, \quad \varphi(xz,\ z') \;=\; \varphi(z,\ \sigma^j\tau(x)z')\ .$$

Verify that σ_1 (resp. σ_2 or σ_3) induces a bijection from $\mathcal{Z}_i(j)$) onto $\mathcal{Z}_i(i{+}j{+}1)$ (resp. $\mathcal{Z}_i(j + 1)$).

(e) Prove that the homogeneous elements φ of the centralizer $Z^g(A,\ \mathcal{B}(\lambda, \mu, \nu))$ are characterized by this condition:

$$\forall x \in A, \quad \forall z,\ z' \in, \quad \varphi(xz,\ z') = (-1)^{(\lambda+\mu+\nu)\partial x}(-1)^{(\lambda+1)\partial x\partial\varphi}\varphi(z,\ \tau(x)z')\ ;$$

consequently this centralizer is $\ \mathcal{Z}_0(\lambda + \mu + \nu) \oplus \mathcal{Z}_1(\mu + \nu + 1)$.

(7.ex.7) Here is a direct proof of (7.4.2) (without the help of (7.4.1)) when 2 is invertible in K. As reported in (7.2.4), the definition (5.3.1) has been first derived from the Cartan–Chevalley criterion under the assumption that 2 is invertible, and with the argument presented just below. As in (7.4.2) we assume that V has constant rank r. We set $\beta = \mathrm{b}_q/2$ (the canonical scalar product on M) and we identify $C\ell(M,q)$ with $\bigwedge(M;\beta)$. We also write q_2 for the quadratic form on $V\oplus V$ defined by $q_2(c,c') = \beta(c,c')$. Obviously $(V\oplus V,\ q_2)$ is the direct sum of the totally isotropic subspaces $V\oplus 0$ and $0\oplus V$, whence a grading of the algebra $C\ell(V\oplus V,\ q_2)$ according to (5.2.7).

(a) Prove the existence of an algebra morphism F from $C\ell(V\oplus V,\ q_2)$ into $\mathrm{End}(\bigwedge(M))$ mapping every $(c,c')\in V\oplus V$ to the endomorphism $x\longmapsto c\wedge x + \mathrm{d}_\beta(c')\rfloor x$. Prove the existence of an algebra morphism D from $C\ell(V,q)$ into $C\ell(V\oplus V,\ q_2)$ mapping every $c\in V$ to (c,c). Then deduce from (4.8.9) that for every $z\in C\ell(V,q)$ the left multiplication $x\longmapsto zx$ in $C\ell(M,q)=\bigwedge(M;\beta)$ is the endomorphism of $\bigwedge(M)$ equal to $F\circ D(z)$.
Similarly there is an algebra morphism D' from $C\ell(V,q)^{to}$ into $C\ell(V\oplus V,\ q_2)$ that maps every $c^{to}\in V^{to}$ to $(c,-c)$. And the twisted right multiplication $x\longmapsto (-1)^{\partial x\partial z}xz$ is the endomorphism $F\circ D'(z^{to})$. Besides, if z and z' are homogeneous elements of $C\ell(V,q)$, then $D(z)$ and $D'(z'^{to})$ commute or anticommute in $C\ell(V\oplus V,\ q_2)$ according to the sign $(-1)^{\partial z\partial z'}$.

(b) Prove that for every degree $k\in\mathbb{Z}$ the mapping $\zeta\otimes\omega\longmapsto F(\zeta)(\omega)$ induces a bijection

$$C\ell(V\oplus V,\ q_2\ ;\ V\oplus 0,\ 0\oplus V)^k\otimes\bigwedge^r(U)\ \longrightarrow\ C\ell^{r+k}(M,q).$$

Hint. Since $(V\oplus V;\ q_2)$ is the orthogonal sum of its diagonals Δ_V and Δ'_V, the bijectiveness of $C\ell(V\oplus V,\ q_2)\otimes\bigwedge^r(U)\to C\ell(M,q)$ follows from (7.3.2).

(c) Observe that the mapping $(c,c')\longmapsto(c+c',\ c-c')$ is an isomorphism from $(V,q)\perp(V,-q)$ onto $(V\oplus V,\ q_2)$ that maps the diagonals Δ_V and Δ'_V respectively onto $V\oplus 0$ and $0\oplus V$. Conclude that, for every homogeneous $z\in C\ell(V,q)$, all products $z\omega\tau(z)$ (with $\omega\in\bigwedge^r(U)$) are in $C\ell^r(M,q)$ if and only if $z\otimes\tau(z)^{to}$ belongs to $(C\ell(V,q)\hat{\otimes}C\ell(V,q)^{to};\Delta,\Delta')^0$.
Hint. $z\omega\tau(z) = \pm F(D(z)D'(\tau(z)^{to}))(\omega)$.

(7.ex.8) Let (M,q) be a hyperbolic space of constant rank $2r$, let U be an element of $\mathcal{T}(M,q)$, and V a submodule supplementary to U. Since the graded algebra $\mathrm{Gr}(C\ell(M,q))$ is isomorphic to $\bigwedge(M)$ (see (4.8.7)), there is a canonical surjective mapping $C\ell(M,q)\to\bigwedge^{2r}(M)$ with kernel $C\ell^{<2r}(M,q)$; therefore, even when 2 is not invertible in K, from the mapping Ω defined in **7.3** we can derive a mapping

$$\Omega_{2r}\ :\ C\ell(V,q)\otimes\bigwedge^r(U)\otimes C\ell(V,q)\longrightarrow\bigwedge^{2r}(M).$$

Take notice that $\bigwedge^{2r}(M)$ is canonically isomorphic to K because there are canonical isomorphisms $\bigwedge^r(U)\otimes\bigwedge^r(V)\to\bigwedge^{2r}(M)$ and $\bigwedge^r(U)\otimes\bigwedge^r(V)\to K$ (resulting from $M=U\oplus V$ and $U\cong V^*$).

(a) Prove that, for all $x \in C\ell(M,q)$ and all z and z' in $C\ell(V,q)$,

$$\Omega_{2r}(z \otimes \omega \otimes z') = (-1)^{r(r+1)/2}\,\Omega_{2r}(z' \otimes \omega \otimes z)\ ;$$
$$\Omega_{2r}(R_x(z) \otimes \omega \otimes z') = \Omega_{2r}(z \otimes \omega \otimes R_{\sigma\tau(x)}(z'))\ .$$

Consequently Ω_{2r} gives the graded centralizer $Z^g(C\ell(M,q),\ \mathcal{B}_V)$ when $\mathcal{B}_V$ is a bimodule over $C\ell(M,q)$ according to a formula that you shall specify.

(b) Prove that an element x' of $C\ell(M,q)$ belongs to $C\ell^{\leq r}(M,q)$ if and only if xx' belongs to $C\ell^{<2r}(M,q)$ for all $x \in C\ell^{<r}(M,q)$. Prove that a locally homogeneous $z \in C\ell(V,q)$ belongs to $\mathrm{Lip}(V,q)$ if and only if $\Omega_{2r}(R_x(z)\otimes\omega\otimes z)$ vanishes for all $x \in C\ell^{<r}(M,q)$ and all $\omega \in \bigwedge^r(U)$.

(c) Prove the existence of a bilinear mapping ψ from $C\ell(V,q) \times C\ell(V,q)$ into $\bigwedge^r(V)$ such that, for all $z,\ z' \in C\ell(V,q)$ and all $\omega \in \bigwedge^r(U)$,

$$\psi(z,z') \wedge \omega \ = \ (-1)^{r\partial z'}\ \Omega_{2r}(z \otimes \omega \otimes z').$$

Prove that ψ is nondegenerate, either symmetric or skew symmetric.
Hint. The nondegeneracy of ψ depends on the bijectiveness of $\mathrm{d}_\psi : C\ell(V,q) \to \mathrm{Hom}(C\ell(V,q),\ \bigwedge^r(V))$; since $C\ell(V,q)^* \otimes \bigwedge^r(U)$ is faithfully flat, you can replace d_ψ with a mapping from $C\ell(V,q) \otimes C\ell(V,q)^* \otimes \bigwedge^r(U)$ into some module that proves to be isomorphic to $\mathcal{B}_V \otimes \bigwedge^{2r}(M)$; because of (7.2.1) there are isomorphisms

$$C\ell(V,q) \otimes C\ell(V,q)^* \longleftrightarrow \mathrm{End}(C\ell(V,q)) \longleftrightarrow C\ell(M,q)\ ;$$

finally you have to examine this mapping:

$$C\ell(M,q) \otimes \bigwedge^r(U) \longrightarrow \mathcal{B}_V \otimes \bigwedge^{2r}(M)\ ,$$
$$x \otimes \omega \longmapsto \big((z,z') \longmapsto \pm\Omega_{2r}(R_x(z) \otimes \omega \otimes z')\big)\ ;$$

its bijectiveness follows from (6.7.6) and (a) just above.

(d) For every $T \in \mathcal{T}(M,q)$, let $\tilde{\chi}(T)$ be the submodule of $C\ell(V,q)$ generated by all $R_a(z)$ with $a \in T$ and $z \in C\ell(V,q)$. Prove that $\tilde{\chi}(T)$ is a graded direct summand of $C\ell(V,q)$ of constant rank $2^r - 1$. Prove that $\chi(T)$ and $\tilde{\chi}(T)$ are orthogonal to each other for the bilinear mapping ψ defined in (c). Prove that $\chi(T) \subset \tilde{\chi}(T)$.

(7.ex.9) Let (M,q) be a hyperbolic space, U an element of $\mathcal{T}(M,q)$, and V and V' two submodules supplementary to U; thus there exists a unique bijection $p : V \to V'$ such that $p(c) - c \in U$ for all $c \in V$. Since $C\ell(V,q)$ and $C\ell(V',q)$ are both supplementary to the left ideal $C\ell(M,q)U$, this mapping p extends to an isomorphism $\psi : C\ell(V,q) \to C\ell(V',q)$ of modules over $C\ell(M,q)$; by definition, $\psi(z) - z$ belongs to $C\ell(M,q)U$ for every $z \in C\ell(V,q)$. We intend to calculate ψ.

(a) We set $\gamma(c,c') = \mathrm{b}_q(p(c) - c,\ c')$ and $q'(c) = q(c) + \gamma(c,c)$ for all c and c' in V. Verify that p is an isomorphism from (V, q') onto (V', q). Consequently p extends to an algebra isomorphism $C\ell(p) : C\ell(V, q') \to C\ell(V', q)$.

(b) As explained in **4.7**, there is an isomorphism Φ_γ from $C\ell(V, q')$ onto the deformation $C\ell(V, q; \gamma)$. Prove that $\psi = C\ell(p) \circ \Phi_\gamma^{-1}$.

Hint. Prove that $p(c)\psi(z) = \psi(c \star z)$ (with $c \star z = cz + \mathrm{d}_\gamma(c) \rfloor z$) for all $c \in V$ and $z \in C\ell'V, q)$.

(c) When $V' = V^\perp$, deduce from (4.7.8) the description of ψ given in (7.2.9).

(7.ex.10) Let (M, q) be a hyperbolic space over a field K, and T, U, V three elements of $\mathcal{T}(M, q)$. The purpose of this exercise is to prove that M is the orthogonal sum $M_1 \perp M_2 \perp \cdots \perp M_n$ of hyperbolic subspaces M_i such that each one satisfies one of these two properties:

either M_i has dimension 4, whereas $M_i \cap T$, $M_i \cap U$, $M_i \cap V$ all have dimension 2, and are pairwise supplementary in M_i;

or M_i has dimension 2, whereas $M_i \cap T$, $M_i \cap U$, $M_i \cap V$ all have dimension 1; therefore these three totally isotropic lines cannot be pairwise distinct.

(a) Prove this statement when $M = U \oplus V$.

Hint. Let U' (resp. V') be the projection of T on U (resp. V) with respect to the decomposition $M = U \oplus V$; or equivalently $U' = U \cap (T \cap V)^\perp$ and $V' = V \cap (T \cap U)^\perp$; let U_∞ be a subspace of U supplementary to U'; observe that T determines an alternate bilinear form f on U' according to this definition: for all b and b' in U', $f(b, b') = \mathrm{b}_q(c, b')$ if c is any element of V such that $b + c$ belongs to T; the kernel of f is $U_0 = T \cap U$, and consequently U' is the direct sum of U_0 and some planes $U_1, U_2, \dots, U_m$ that are pairwise orthogonal with respect to f; to the decomposition of U as the direct sum of U_0, U_∞ and the m planes $U_1, \dots, U_m$, there corresponds the dual decomposition of V as the direct sum of V_∞, V_0 and m planes $V_1, \dots, V_m$; this means that V_∞ (resp. V_0) is orthogonal to all U_j except U_0 (resp. U_∞), whereas V_i (for $i = 1, 2, \dots, m$) is orthogonal to all U_j except U_i; at the end, the subspaces M_i of dimension 4 are the m subspaces $U_i \oplus V_i$.

(b)* Prove this statement without the above additional assumption.

Hint. Let T_0, T_1 and T_2 be subspaces of M such that

$$U \cap V = T_0 \oplus (T \cap U \cap V)\,, \qquad T = T_2 \oplus (T \cap (U + V))\,,$$
$$T \cap (U + V) = T_1 \oplus ((T \cap U) + (T \cap V))\,,$$

and let U_0, U_1, U_2, V_0, V_1, V_2 be chosen in a similar way; therefore

$$T = (T \cap U \cap V) \oplus U_0 \oplus V_0 \oplus T_1 \oplus T_2 \quad \text{and so forth } \dots;$$

observe that T_1, U_1, V_1 have the same dimension, and can be chosen in such a way that $U_1 \oplus V_1 = V_1 \oplus T_1 = T_1 \oplus U_1$; besides, $U_1 \oplus V_1$ is a hyperbolic subspace of (M, q); and $T_0 \oplus T_2$ and the two similar subspaces are hyperbolic too; moreover it is possible to choose T_2, U_2, V_2 inside $(U_1 \oplus V_1)^\perp$; then you are brought back to the previous case, because

$$T+U+V \;=\; (T\cap U\cap V) \perp (T_0 \oplus T_2) \perp (U_0 \oplus U_2) \perp (V_0 \oplus V_2) \perp (U_1 \oplus V_1).$$

(7.ex.11) Let (M,q) be a hyperbolic space over a ring K, and $K \to L$ an extension of this ring. By means of counterexamples, show that the natural mapping $\mathcal{T}(M,q) \to \mathcal{T}(L \otimes (M,q))$ may be not injective, and that it may be not surjective. Prove that it is surjective if K is a local ring, and L its residue field; *hint:* (7.ex.10)(a).

(7.ex.12) Let (M,q) be a hyperbolic space of constant rank $2r$ over a ring K, and V a direct summand of constant rank r in M; does there exist any $U \in \mathcal{T}(M,q)$ such that $M = U \oplus V$? By means of a counterexample (for instance with $K = \mathbb{Z}$ and $r = 1$) show that the answer may be negative. Deduce from (2.7.8) that the answer is always positive when K is a field. Then deduce from (7.ex.11) that the answer is still positive when K is a local ring.

(7.ex.13) Let V be a vector space of dimension r over a field, and ω^* a nonzero element of $\bigwedge^r(V^*)$; we set $\mathcal{F}(z) = \omega^* \lfloor z$ as in **5.9**, whence a transformation $\mathcal{F} : \bigwedge(V) \to \bigwedge(V^*)$. With every nonzero $z \in \mathrm{Lip}(V)$ we associate its support $\mathrm{Sup}(z)$ and the subspace $\mathrm{Ker}^\wedge(z)$ of all $b \in V$ such that $b \wedge z = 0$. Besides, with every subspace N of V we associate its annihilator N^{an} in V^*. Prove that

$$\mathrm{Sup}(\mathcal{F}(z)) = (\mathrm{Ker}^\wedge(z))^{an} \quad \text{and} \quad \mathrm{Ker}^\wedge(\mathcal{F}(z)) = (\mathrm{Sup}(z))^{an}.$$

Therefore the quotients $P = \mathrm{Sup}(z)/\mathrm{Ker}^\wedge(z)$ and $P' = \mathrm{Sup}(\mathcal{F}(z))/\mathrm{Ker}^\wedge(\mathcal{F}(z))$ can be treated as dual spaces. Explain that z (resp. $\mathcal{F}(z)$) determines a nondegenerate alternate bilinear form f (resp. g) on P' (resp. P); deduce from (5.9.5) that f and $-g$ induce reciprocal mappings $P' \longleftrightarrow P$.

(7.ex.14)* Let (M,q) be a hyperbolic space over a field K, and let U and V be two supplementary totally isotropic subspaces of (M,q). Thus b_q determines a duality between U and V. Let $\mathcal{A}(U)$ (resp. $\mathcal{A}(V)$) be the set of all alternate bilinear forms $U' \times U' \to K$ (resp. $V' \times V' \to K$) with U' (resp. V') any subspace of U (resp. of V). Consider the diagram

$$\begin{array}{ccccc}
\mathrm{BLip}(V) & & \dashrightarrow & & \mathrm{BLip}(U) \\
\Big| & \nwarrow & & \nearrow & \Big| \\
\Big| & & \mathcal{T}(M,q) & & \Big| \\
\downarrow & \swarrow & & \searrow & \downarrow \\
\mathcal{A}(U) & & \dashrightarrow & & \mathcal{A}(V)
\end{array}$$

in which the arrows are defined as follows:

- the arrow $\mathcal{T}(M,q) \to \mathrm{BLip}(V)$ is the Cartan–Chevalley mapping $T \longmapsto \chi(T)$, and the arrow $\mathcal{T}(M,q) \to \mathrm{BLip}(U)$ is defined by permuting the roles of U and V;

- the arrow $\mathrm{BLip}(V) \to \mathrm{BLip}(U)$ comes from the isomorphism $\bigwedge(V) \to \bigwedge(U^*)$ induced by d_q and from the transformation $\mathcal{F}_* : \bigwedge(U^*) \to \bigwedge(U)$ defined as in **5.9**; according to (5.3.14) d_q induces a bijection $\mathrm{Lip}(V) \to \mathrm{Lip}(U^*)$, and according to (5.9.3) $\mathcal{F}_*$ induces a bijection $\mathrm{Lip}(U^*) \to \mathrm{Lip}(U)$,
- the arrow $\mathrm{BLip}(V) \to \mathcal{A}(U)$ maps the line spanned by any nonzero lipschitzian element

$$c_1 \wedge c_2 \wedge \cdots \wedge c_k \wedge \mathrm{Exp}(v) \quad \text{with } k \geq 0, \quad c_1, c_2, \ldots, c_k \in V \quad \text{and} \quad v \in \bigwedge^2(V),$$

  to the alternate linear form f induced by v on $U' = U \cap (Kc_1 \oplus \cdots \oplus Kc_k)^\perp$; in other words, an element b of U belongs to U' if and only if $\mathrm{b}_q(b, c_j) = 0$ for $j = 1, 2, \ldots, k$, and

$$\forall b,\ b' \in U', \quad f(b, b') = (\mathrm{d}_q(b) \wedge \mathrm{d}_q(b'))(v) \ ;$$

  of course $U' = U$ when $k = 0$; the arrow $\mathrm{BLip}(U) \to \mathcal{A}(V)$ is defined in the same way by permuting the roles of U and V;
- the arrow $\mathcal{T}(M, q) \to \mathcal{A}(U)$ maps any maximal totally isotropic subspace T to the alternate bilinear form f defined on the projection U' of T on U in this way: $f(b, b') = \mathrm{b}_q(c, b')$ if c is any element of V such that $b + c$ belongs to T; prove that an alternate bilinear form on U' is actually defined in this way; of course $\mathcal{T}(M, q) \to \mathcal{A}(V)$ is defined in the same way by permuting the roles of U and V;
- at last the arrow $\mathcal{A}(U) \to \mathcal{A}(V)$ is the mapping $f \longmapsto g$ defined in this way: every alternate bilinear form f on a subspace U' of U determines a nondegenerate alternate bilinear form f' on the quotient of U' by the kernel U_0 of d_f; whence an isomorphism $U'/U_0 \to (U'/U_0)^*$; then b_q induces a nondegenerate duality between U'/U_0 and the quotient V'/V_0 where $V' = V \cap U_0^\perp$ and $V_0 = V \cap U'^\perp$; with f' is associated a nondegenerate alternate bilinear form g' on V'/V_0 such that f' and $-g'$ induce reciprocal isomorphisms $U'/U_0 \longleftrightarrow V'/V_0$; and finally g' determines an alternate bilinear form g on V'.

Prove that the eight arrows in the above diagram are all bijective, and that the four triangles in this diagram are all commutative.
Hint. (7.ex.10)(a) and (7.ex.13).

(7.ex.15) Let (P, q) be a quadratic space of dimension r over a field K; let s be the maximal dimension of totally isotropic subspaces contained in (P, q).

(a) Prove that $0 \leq 2s \leq r$. The equality $s = 0$ means that (P, q) is anisotropic, whereas the equality $2s = r$ means that it is hyperbolic.

(b) Prove that every totally isotropic subspace of (P, q) is contained in a totally isotropic subspace of dimension s.

(c) Let T and U be two totally isotropic subspaces of dimension s in (P, q); prove that T and U are contained in a hyperbolic subspace of dimension $2s$ in (P, q).

(7.ex.16) Let (M,q) be a hyperbolic space of rank $2r$ over a local ring K, and U an element of $\mathcal{T}(M,q)$. For each parity $i = 0,1$, let $\mathcal{T}_i(M,q)$ be the subset of all $T \in \mathcal{T}(M,q)$ such that $\mathrm{par}(U,T) = i$, and let $\bigwedge^r_{|i}(M)$ be the subspace of $\bigwedge^r(M)$ spanned by all elements of all $\bigwedge^r(T)$ with $T \in \mathcal{T}_i(M,q)$. Prove that $\bigwedge^r(M)$ is the direct sum of $\bigwedge^r_{|0}(M)$ and $\bigwedge^r_{|1}(M)$.

Let g be an orthogonal transformation of (M,q). Prove that $\bigwedge(g)$ maps each $\bigwedge^r_{|i}(M)$ onto itself (resp. onto $\bigwedge^r_{|1-i}(M)$) if g belongs to the subgroup $\mathrm{SO}(M,q)$ (resp. if g does not belong to $\mathrm{SO}(M,q)$).

(7.ex.17) Consider a quadratic space (M',q') of odd dimension $2r-1$ (with $r \geq 2$) over a field K of characteristic $\neq 2$, and suppose that (M',q') contains totally isotropic subspaces of dimension $r-1$. Let $\mathcal{T}'(M',q')$ be the set of all totally isotropic subspaces of dimension $r-1$.

(a) Prove that (M',q') can be embedded into a hyperbolic space (M,q) of dimension $2r$.

(b) Prove that every element of $\mathcal{T}'(M',q')$ is contained in exactly two elements of $\mathcal{T}(M,q)$.

(c) Define $\mathcal{T}_0(M,q)$ and $\mathcal{T}_1(M,q)$ as in (7.ex.16). Prove that, for $i = 0,1$, the mapping $T \longmapsto T \cap M'$ is a bijection from $\mathcal{T}_i(M,q)$ onto $\mathcal{T}'(M',q')$.

Remark. When M' has odd dimension over a field K of characteristic 2, the above results are still valid when q' is "almost nondegenerate" (see (2.ex.14)).

(7.ex.18)* Consider a quadratic space of dimension 4 over a field K, and suppose that q is not anisotropic. Let a_0 be a nonzero element of M such that $q(a_0) = 0$. The "projective space" of dimension 3 derived from M is the set P of all vector lines (or vector subspaces of dimension 1) contained in M; if b is a nonzero element of M, $p(b)$ is the line spanned by b, considered as an element of P. The projective lines (resp. projective planes) in P are the images by p of the vector subspaces of dimension 2 (resp. 3) in M. The image of $q^{-1}(0)$ in P is denoted by $\mathcal{Q}$ and is called a *projective quadric* ; let $x_0 = p(a_0)$ be a point of $\mathcal{Q}$.

(a) A projective line L containing x_0 is said to be tangent to $\mathcal{Q}$ at the point x_0 if the restriction of q to the plane $p^{-1}(L)$ is degenerate. Prove that the lines tangent to $\mathcal{Q}$ at x_0 are the projective lines passing through x_0 and contained in some plane, called the plane tangent to $\mathcal{Q}$ at x_0.

(b) Prove that the intersection of $\mathcal{Q}$ with any projective line L passing through x_0 is a set of two points (x_0 and another one), except when L is tangent to $\mathcal{Q}$ at this point. What happens when L is tangent to $\mathcal{Q}$? Remember that L may be contained in $\mathcal{Q}$ when (M,q) is hyperbolic!

(c) Now (M,q) is assumed to be a hyperbolic quadratic space, and $\mathcal{L}(\mathcal{Q})$ is the set of all projective lines contained in $\mathcal{Q}$. Prove that $\mathcal{L}(\mathcal{Q})$ is the union of two subsets $\mathcal{L}_i(\mathcal{Q})$ with $i = 0, 1$, in such a way that these assertions are true for every pair (L, L') of elements of $\mathcal{L}(\mathcal{Q})$:

if L and L' belong to the same subset $\mathcal{L}_i(\mathcal{Q})$ (with $i = 0, 1$), either $L \cap L'$ is empty, or $L = L'$,

if L and L' do not belong to the same $\mathcal{L}_i(\mathcal{Q})$, then $L \cap L'$ contains exactly one point.

Hint. See (7.6.6).

(d) With the same assumptions as in (c), prove that there are two elements of $\mathcal{L}(\mathcal{Q})$ passing through x_0, and that the plane tangent to $\mathcal{Q}$ at x_0 is the projective plane generated by these two tangent lines.

Weyl algebras (for interested readers)

(7.ex.19) Let ψ be an alternate bilinear form on a finitely generated projective module M that is the direct sum of two submodules U and V such that $\psi(U, U) = 0$. Thus $\mathrm{S}(U)$ and $\mathrm{W}(V, \psi)$ can be identified with subalgebras of $\mathrm{W}(M, \psi)$.

(a) Prove that $\mathrm{W}(M,\psi)$ is the direct sum of $\mathrm{W}(V,\psi)$ and the left ideal $\mathrm{W}(M,\psi)U$. This leads to an algebra morphism $\mathrm{W}(M, \psi) \to \mathrm{End}(\mathrm{W}(V, \psi))$. Describe the operations in $\mathrm{W}(V, \psi)$ of the elements of $\mathrm{S}(U)$ and $\mathrm{W}(V, \psi)$.

Hint. Use a scalar product β such that $\psi(a, a') = \beta(a, a') - \beta(a', a)$ for all a, $a' \in M$, and $\beta(M, U) = 0$.

(b) Suppose that there is a positive integer n such that $nV = 0$; prove that b^n belongs to the kernel of $\mathrm{W}(M, \psi) \to \mathrm{End}(\mathrm{W}(V, \psi))$ for all $b \in U$. Let r be a positive integer such that the rank of V is always $\leq r$; prove that the kernel of this morphism contains $\mathrm{S}^k(U)$ for all $k > (n - 1)r$.

(7.ex.20) The hypotheses are the same as in (7.ex.19). Let $\beta : M \times M \to K$ be a scalar product such that $\psi(a, a') = \beta(a, a') - \beta(a', a)$ for all a, $a' \in M$, and $\beta(U, U) = 0$; this β gives an element $\beta_{/\!/}$ of $\mathrm{S}^{*2}(M \oplus M)$, and its restriction to $V \times U$ gives an element $\beta^{\dagger}$ of $\mathrm{S}^{*2}(M)$. State and prove the assertions analogous to (7.1.5) and (7.1.6). To define exponentials of elements of $\mathrm{S}^{*2}(M)$ or $\mathrm{S}^{*2}(M \oplus M)$, you may either assume that K contains a subring isomorphic to the field $\mathbb{Q}$, or define these exponentials as in (4.ex.2)(b).

(7.ex.21) Let ψ be a nondegenerate alternate bilinear form on a finitely generated projective module M. A *lagrangian submodule* of M is a direct summand U such that $\psi(U, U) = 0$ and $\mathrm{rk}(\mathfrak{p}; M) = 2\mathrm{rk}(\mathfrak{p}; U)$ at every prime ideal $\mathfrak{p}$ (or equivalently $U^{\perp} = U$).

(a) Prove that every lagrangian submodule U admits a supplementary lagrangian submodule V.

Hint. Follow the proof of (2.5.4).

(b) Prove that lagrangian submodules always exist when K is a local ring.

(7.ex.22) Here we assume that the canonical morphism $\mathbb{Z} \to K$ extends to a morphism $\mathbb{Q} \to K$. Let M be a finitely generated projective module of even rank over K, and ψ a nondegenerate alternate form on M. We also assume that M is the direct sum of a lagrangian submodule U and a supplementary submodule V; thus we get an algebra morphism $x \longmapsto R_x$ from $\mathrm{W}(M,\psi)$ into $\mathrm{End}(\mathrm{W}(V,\psi))$.

(a) Prove that the image of $\mathrm{W}(M,\psi) \to \mathrm{End}(\mathrm{W}(V,\psi))$ is dense in the following sense: for every integer $k > 0$ and for every $f \in \mathrm{End}(\mathrm{W}(V,\psi))$, there exists $x \in \mathrm{W}(M,\psi)$ such that $R_x(z) = f(z)$ for all $z \in W^{\leq k}(V,\psi)$.
Hint. There is a lagrangian submodule V' supplementary to U (see (7.ex.21) (a)); if this statement is true for V', it is also true for V; therefore you can assume V to be lagrangian; now follow the proof of (3.7.2).

(b) Prove that $\mathrm{W}(V,\psi)$ is an irreducible module over $\mathrm{W}(M,\psi)$ when K is a field of characteristic 0.

(c) Prove that the morphism $\mathrm{W}(M,\psi) \to \mathrm{End}(\mathrm{W}(V,\psi))$ is not surjective.
Hint. It suffices to prove that for every $x \in \mathrm{W}(M,\psi)$ there exists an integer j such that $R_x(W^{\leq k}(V,\psi)) \subset W^{\leq k+j}(V,\psi)$ for all $k \in \mathbb{N}$. If you are more courageous, you can even prove this more difficult assertion: if x is any nonzero element of $\mathrm{W}(M,\psi)$, there is no integer k such that $\mathrm{Im}(R_x) \subset W^{\leq k}(V,\psi)$.

(d) Prove that the morphism $\mathrm{W}(M,\psi) \to \mathrm{End}(\mathrm{W}(V,\psi))$ is injective.

(7.ex.23)* Let U be a vector space of finite dimension r over $\mathbb{R}$ and $V = U^*$ the dual space. We set $M = U \oplus V$ and $M^* = V \oplus U$. We provide M^* with the usual symplectic form ψ defined by $\psi(c+b,\ c'+b') = c(b') - c'(b)$. As announced in (4.ex.23), we are interested in the Weyl algebra $\mathrm{W}_{\mathbb{C}}(\mathbb{C} \otimes M^*,\ i \otimes \psi)$ (where $i = \sqrt{-1}$). Let $\mathrm{A}(U)$ be a space of functions or distributions $U \to \mathbb{C}$ that is stable by multiplication by polynomial functions and by derivations; thus every $g \in A(U)$ has a derivative $\partial_b(g)$ in the direction b for every $b \in U$. Prove the existence of a $\mathbb{C}$-algebra morphism $\mathrm{W}_{\mathbb{C}}(\mathbb{C} \otimes M^*,\ i \otimes \psi) \to \mathrm{End}_{\mathbb{C}}(\mathrm{A}(U))$ such that for every $v \in \mathrm{S}(V)$ the operation of $1 \otimes v$ in $\mathrm{A}(U)$ is the multiplication by v (considered as a polynomial function on U), whereas for every $b \in U$ the operation of $1 \otimes b$ is $i\partial_b$. The elements in the image of this morphism are called "differential operators with polynomial coefficients".

Comment. As announced in (4.ex.23), we are interested in "enlargements" of the Weyl algebra under consideration; here they should afford "generalized differential operators". We choose an admissible scalar product β such that $\beta(U,U) = 0$ (for instance $\beta = \psi/2$); thus $\mathrm{W}_{\mathbb{C}}(\mathbb{C}\otimes M^*,\ i\otimes\psi)$ can be identified with $\mathrm{S}_{\mathbb{C}}(\mathbb{C}\otimes M^*;\ i\otimes \beta)$, and the elements of the wanted enlargement can be treated as functions or distributions on M. Let f be an element of this enlargement; to derive from it an operator R_f we use Fourier transformation; for a suitable $g \in \mathrm{A}(U)$, we try to define $R_f(g)$ by means of the following equality which is suggested by (7.1.6) and by the subsequent decomposition of $\beta_{/\!/} + \pi^*(\beta^\dagger)$ into a sum of four terms, among which only the first three are here relevant; the letters t, t_1 and t_2 are variables running through V, whereas s runs through U, and $\varphi(t)$ is any "test function" on V :

$$(2\pi)^r \int_V \varphi(t)\ \mathcal{F}(R_f(g))(t)\ dt$$
$$= \int_{M^* \oplus V} \varphi(t_1 + t_2) \exp\big(- i\psi(s, t_2) - i\beta(t_1, t_2) + i\beta(t_1, s)\big)$$
$$\times \mathcal{F}(f)(t_1, s)\, \mathcal{F}(g)(t_2)\, dt_1\, ds\, dt_2.$$

When the Fourier transform $\mathcal{F}(f)$ or $\mathcal{F}(g)$ has a compact support, the existence of $R_f(g)$ is beyond doubt; but its existence is already ensured by much weaker hypotheses which are not discussed here. Besides, we can accept a symplectic form ψ that induces a duality between a lagrangian subspace U and any supplementary subspace V, even not lagrangian, since only the condition $\beta(U, U) = 0$ is required. The following exercise (analogous to (7.ex.14)) shows that a Cartan–Chevalley mapping also exists in this context.

(7.ex.24)* Let U, V, M and ψ be as in (7.ex.23); let $\mathrm{Lag}(M^*; \psi)$ be the set of lagrangian subspaces of M^* (among which there are U and V); here $\mathrm{A}(U)$ is the space of all distributions on U admitting a Fourier transform. With every $c + b \in M^*$ is associated a differential operator R_{c+b} on $\mathrm{A}(U)$: $R_{c+b}(g) = cg + i\partial_b(g)$. It is known that if T is any lagrangian subspace of M^*, the subset $\chi(T)$ of all $g \in \mathrm{A}(U)$ such that $R_a(g) = 0$ for all $a \in T$ is a vector space of dimension 1 over $\mathbb{C}$. For instance $\chi(U)$ is the subset of all constant functions on U, whereas $\chi(V)$ is the line spanned by the Dirac distribution at the null element of U. By definition $\mathrm{BLip}^\vee(V)$ is the image of this mapping χ.

Describe the elements of $\mathrm{BLip}^\vee(V)$, and explain which are the arrows in the following diagram, that make it become commutative; besides $\mathrm{Lag}(M^*, \psi)$, $\mathrm{BLip}^\vee(V)$ and $\mathrm{BLip}^\vee(U)$ which are already defined, it also involves the set $\mathcal{Q}(U)$ of all quadratic forms $U' \to \mathbb{R}$ defined on any vector subspace U' of U, and similarly $\mathcal{Q}(V)$:

$$\begin{array}{ccccc}
\mathrm{BLip}^\vee(V) & & \dashrightarrow & & \mathrm{BLip}^\vee(U) \\
| & \nwarrow & & \nearrow & | \\
| & & \mathrm{Lag}(M^*, \psi) & & | \\
\downarrow & \swarrow & & \searrow & \downarrow \\
\mathcal{Q}(U) & & \dashrightarrow & & \mathcal{Q}(V).
\end{array}$$

Comments. An element of $\mathrm{BLip}^\vee(V)$ is determined by a subspace U' of U and a quadratic form $q : U' \to \mathbb{R}$, and is made of all the distributions g on U that can be written in this way, for some Lebesgue measure $ds_{U'}$ on U' and some constant $\mu \in \mathbb{C}$:

$$\int_U \varphi(s)\ g(s)\ ds \ = \ \mu \int_{U'} \varphi(s)\ \exp(iq(s))\ ds_{U'}\ ;$$

thus the bijection $\mathrm{BLip}^{\vee}(V) \to \mathcal{Q}(U)$ is obvious. The steady presence of the factor i beside ψ accounts for its presence beside q, and implies that all these distributions g have a Fourier transform. The bijection $\mathrm{BLip}^{\vee}(V) \to \mathrm{BLip}^{\vee}(U)$ (with $U = V^*$) is merely Fourier transformation. It is worth explaining how $\mathrm{BLip}^{\vee}(V)$ is related to the Lipschitz monoid $\mathrm{Lip}^{\vee}(V)$ defined in the comment following (5.ex.41); except for the element ∞ that must be inserted into $\mathrm{Lip}^{\vee}(V)$ because of the "inversion rule", all the other elements of $\mathrm{Lip}^{\vee}(V)$ are distributions like the above g, with the additional requirement that μ must belong to the group $(\mathbb{R}^{\times})^{1/4}$ (in other words, $\mu^4 \in \mathbb{R}^{\times}$); thus the elements of $\mathrm{BLip}^{\vee}(V)$ are no longer subsets of $\mathrm{Lip}^{\vee}(V)$, but they are in bijection with the orbits of the group $(\mathbb{R}^{\times})^{1/4}$ in $\mathrm{Lip}^{\vee}(V) \setminus \{\infty\}$. In an analogous way the elements of $\mathrm{BLip}(V)$ are in bijection with the orbits of the group $\mathbb{R}^{\times}$ in $\mathrm{Lip}(V) \setminus \{0\}$.

Chapter 8

Complements about Witt Rings and Other Topics

This chapter contains some complements to previous chapters, which the authors have preferred to postpone until the last chapter. The quadratic Witt ring defined in **2.7** opens this chapter. In **8.1** we get precise results about this Witt ring when K is a local ring in which 2 is invertible. But in **8.2** where 2 is not invertible in the local ring K, we do not reach so effective and general results.

When K is an integral domain and L its field of fractions, it often occurs that the canonical morphism $\mathrm{WQ}(K) \to \mathrm{WQ}(L)$ is injective; this property, and still stronger ones, are valid for instance when K is a Prüfer ring. The section **8.3** devoted to Prüfer rings gives the opportunity for a digression about $\mathrm{W}(\mathbb{Q})$, the Witt ring of the field of rational numbers, although it is not possible here to get very deep into so sophisticated a subject.

The elementary properties of quadratic forms over fields of characteristic $\neq 2$ are very well known; but with a field of characteristic 2 special properties appear, especially because of the existence of "additive quadratic forms"; this is explained in **8.4**. And in **8.5** we study their Clifford algebras.

The section **8.6** is devoted to the group $\mathcal{H}(K)$ of classes of Clifford algebras, and involves both Witt rings and Brauer–Wall groups; this group $\mathcal{H}(K)$ is the image of the canonical morphism $\mathrm{WQ}(K) \to \mathrm{Br}^g(K)$, and gives the opportunity to revisit some results obtained in **3.8**.

8.1 Witt rings over local rings when 2 is invertible

The semiring $\mathrm{WIQ}(K)$ of isomorphy classes of quadratic spaces, the quadratic Witt–Grothendieck ring $\mathrm{WGQ}(K)$ and the quadratic Witt ring $\mathrm{WQ}(K)$ have been defined in **2.7**; there are canonical morphisms $\mathrm{WIQ}(K) \to \mathrm{WGQ}(K) \to \mathrm{WQ}(K)$.

When 2 is invertible in K, they are respectively isomorphic to $\mathrm{WIB}(K)$, $\mathrm{WGB}(K)$ and $\mathrm{WB}(K)$, whence the short notations $\mathrm{WI}(K)$, $\mathrm{WG}(K)$ and $\mathrm{W}(K)$; moreover the quadratic space K with quadratic form $\lambda \longmapsto \lambda^2/2$ affords a unit element in each of them.

Here we suppose that K is a local ring with maximal ideal $\mathfrak{m}$, and that 2 is not in $\mathfrak{m}$, and we search a precise description of the rings $\mathrm{WG}(K)$ and $\mathrm{W}(K)$. As in **2.6**, when $a_1, a_2, \dots, a_n$ are elements of $K^\times$, the notation $\langle a_1, a_2, \dots, a_n\rangle$ means the free module K^n provided with the quadratic form

$$(\lambda_1, \lambda_2, \dots, \lambda_n) \longmapsto \sum_{i=1}^{n} \frac{1}{2}\, a_i \lambda_i^2 \ .$$

We begin with the so-called cancellation theorem.

(8.1.1) **Theorem.** *Let N, N', P, P' be quadratic spaces over the local ring K. When $N \perp P$ and $N' \perp P'$ are isomorphic, and also P and P', then N and N' too are isomorphic.*

Proof. Since P admits orthogonal bases (see (2.6.2)), it suffices to prove that (8.1.1) is true when P and P' have rank 1. We can also assume that $N \perp P$ and $N' \perp P'$ are the same quadratic space M. Thus we reduce the problem to the proof of this statement: if the quadratic space (M, q) contains two elements a and a' such that $q(a)$ and $q(a')$ are the same element of $K^\times$, there exists an automorphism of (M, q) that maps a to $\pm a'$, and consequently $N = (Ka)^\perp$ onto $N' = (Ka')^\perp$. The existence of such an automorphism is already stated in (5.7.4). □

(8.1.2) **Corollary.** *The canonical mapping* $\mathrm{WI}(K) \to \mathrm{WG}(K)$ *is injective.*

Proof. Two quadratic spaces N and N' have the same class in $\mathrm{WG}(K)$ if and only if there exists a quadratic space P such such that $N \perp P$ and $N' \perp P$ are isomorphic; this implies that N and N' have already the same class in $\mathrm{WI}(K)$. □

For all $a \in K^\times$ we denote by $(\!(a)\!)$ and $[a]$ the classes of the quadratic space $\langle a\rangle$ in $\mathrm{WG}(K)$ and $\mathrm{W}(K)$. Besides, let $\mathbb{Z}^{(K^\times)}$ be the additive group freely generated by $K^\times$ (see **1.3**); it is a free group with basis $(e_a)_{a\in K^\times}$; since the tensor product of $\langle a\rangle$ and $\langle b\rangle$ is isomorphic to $\langle ab\rangle$, we provide $\mathbb{Z}^{(K^\times)}$ with the $\mathbb{Z}$-bilinear multiplication such that $e_a e_b = e_{ab}$; with this multiplication, $\mathbb{Z}^{(K^\times)}$ becomes the group ring of $K^\times$ over $\mathbb{Z}$, usually denoted by $\mathbb{Z}[K^\times]$. Since every K-quadratic space admits orthogonal bases, there are surjective ring morphisms $\mathbb{Z}[K^\times] \to \mathrm{WG}(K) \to \mathrm{W}(K)$ mapping every e_a successively to $(\!(a)\!)$ and to $[a]$. The main purpose of this section is a precise description of the kernels of these ring morphisms, that shall allow us to construct $\mathrm{WG}(K)$ and $\mathrm{W}(K)$ as quotients of $\mathbb{Z}[K^\times]$. Although these kernels are ideals, we look for subsets that generate them as additive subgroups. For the kernel of the second morphism, the next lemma is sufficient; it involves the determinant of b_q which is well defined modulo the subgroup $K^{\times 2}$ of all squares (see (2.ex.3)).

(8.1.3) **Lemma.** *If K is any ring (not necessarily a local ring), these two assertions are equivalent for every quadratic space (M, q) of rank 2 :*

(a) *(M, q) is a hyperbolic spacc.*
(b) *M contains a totally isotropic direct summand of rank 1.*

When K is a local ring, they are equivalent to this one:

(c) *there exists a basis (e_1, e_2) of M such that $q(e_1)=q(e_2)=0$ and $\mathrm{b}_q(e_1,e_2)=1$.*

When 2 is invertible in the local ring K, they are still equivalent to these two assertions:

(d) *there exists an orthogonal basis (e'_1, e'_2) of M such that $q(e'_1) = -q(e'_2) = 1/2$.*
(e) *the determinant of b_q is equal to -1 modulo $K^{\times 2}$.*

Proof. The equivalence (a)⇔(b) is a consequence of (2.5.5); indeed, if N is a totally isotropic direct summand of constant rank 1, the evident inclusion $N \subset N^\perp$ is an equality because $N^\perp$ too is a direct summand of rank 1. When K is a local ring, the implications $(a) \Rightarrow (c) \Rightarrow (b)$ are trivial. When 2 is invertible in K and (c) is true, we get a basis (e'_1, e'_2) with the desired properties if we set $e'_1 = e_1 + \frac{1}{2}e_2$ and $e'_2 = e_1 - \frac{1}{2}e_2$. The determinant of q in such a basis (e'_1, e'_2) is equal to -1. At last we prove that (e)⇒(b). If (b_1, b_2) is an orthogonal basis of M, the assertion (e) implies $q(b_1)q(b_2) = -\lambda^2$ for some $\lambda \in K^\times$; consequently $\lambda b_1 + q(b_1)b_2$ generates a totally isotropic submodule, that is supplementary to Kb_2. □

Every hyperbolic space $\mathbf{H}[N]$ over K is an orthogonal sum of hyperbolic spaces of rank 2 because N is assumed to be finitely generated and projective, consequently free; because of (8.1.3), $\mathbf{H}[N]$ is an orthogonal sum of hyperbolic spaces isomorphic to $\langle 1, -1\rangle$. This leads to this immediate corollary.

(8.1.4) **Corollary.** *The kernel of $\mathrm{WG}(K) \to \mathrm{W}(K)$ is generated by $(\!(1)\!) + (\!(-1)\!)$ as an additive subgroup; it contains $(\!(a)\!) + (\!(-a)\!)$ for all $a \in K^\times$. The kernel of $\mathbb{Z}[K^\times] \to \mathrm{W}(K)$ is generated by $e_1 + e_{-1}$ and the kernel of $\mathbb{Z}[K^\times] \to \mathrm{WG}(K)$.*

Thus we realize that the serious problem is the description of the kernel of $\mathbb{Z}[K^\times] \to \mathrm{WG}(K)$.

(8.1.5) **Proposition.** *The kernel of $\mathbb{Z}[K^\times] \to \mathrm{WG}(K)$ is the additive subgroup generated by the elements $e_{ab^2} - e_a$ (for all a, $b \in K^\times$) and the elements $e_a + e_b - e_c - e_d$ corresponding to all (a, b, c, d) such that $\langle a, b\rangle$ and $\langle c, d\rangle$ are isomorphic quadratic spaces.*

Indeed this is an immediate corollary of the following lemma.

(8.1.6) **Lemma.** *Let (M, q) be a quadratic space of rank r over the above local ring K, and let $\mathcal{B} = (e_1, e_2, \dots, e_r)$ and $\mathcal{B}' = (e'_1, e'_2, \dots, e'_r)$ be two orthogonal bases of (M, q). There exists a finite sequence $(\mathcal{B}_0, \mathcal{B}_1, \dots, \mathcal{B}_n)$ of orthogonal bases such that $\mathcal{B}_0 = \mathcal{B}$, $\mathcal{B}_n = \mathcal{B}'$ and each $\mathcal{B}_{j-1}$ (for $j = 1, 2, \dots, n$) can be carried onto $\mathcal{B}_j$ by means of an operation that only modifies one or two elements.*

Proof. We prove this by induction on r. There is nothing to prove when r is ≤ 2; therefore we assume $r \geq 3$. Because of the induction hypothesis it suffices to prove that we can transform $\mathcal{B}$ into an orthogonal basis beginning with e'_1 by means of operations modifying one or two elements. Let us write $e'_1 = \sum_{i=1}^{r} \lambda_i e_i$ with all λ_i in K.

First we suppose that $q(e'_1) - q(\lambda_j e_j)$ is invertible for at least one $j \in \{1, 2, \ldots, r\}$. We can suppose $j = r$, because it is permitted to permute e_j and e_r if $j \neq r$. Let us set $e''_1 = \sum_{i=1}^{r-1} \lambda_i e_i$ so that $e'_1 = e''_1 + \lambda_r e_r$. Since $q(e''_1) = q(e'_1) - q(\lambda_r e_r)$ is invertible, there are orthogonal bases beginning with e''_1 in the submodule generated by $(e_1, e_2, \ldots, e_{r-1})$, and the induction hypothesis says that we can transform $\mathcal{B}$ into an orthogonal basis beginning with e''_1 by means of operations modifying one or two elements, and leaving e_r unchanged. Then it suffices to replace e''_1 and e_r respectively with e'_1 and the orthogonal vector $\lambda_r q(e_r) e''_1 - q(e''_1) e_r$.

Now we suppose that $q(e'_1) - q(\lambda_j e_j)$ belongs to the maximal ideal $\mathfrak{m}$ for $j = 1, 2, \ldots, r$. This implies that $q(\lambda_1 e_1) \equiv q(\lambda_2 e_2) \equiv q(e'_1)$ modulo $\mathfrak{m}$. Consequently $q(e'_1) - q(\lambda_1 e_1 + \lambda_2 e_2)$ cannot fall into $\mathfrak{m}$. If we replace e_1 and e_2 respectively with $\lambda_1 e_1 + \lambda_2 e_2$ and $-\lambda_2 q(e_2) e_1 + \lambda_1 q(e_1) e_2$, we are brought back to the previous case. $\square$

The elements generating the kernel of $\mathbb{Z}[K^\times] \to \mathrm{WG}(K)$ in (8.1.5) can be understood as relations imposed on the generators $(\!(a)\!)$ of $\mathrm{WG}(K)$. Nevertheless these relations are not yet very practical, because it is not always evident to reckon whether $\langle a, b\rangle$ and $\langle c, d\rangle$ are isomorphic. Consequently we must improve (8.1.6).

(8.1.7) **Lemma.** *In the lemma* (8.1.6) *we can demand every operation carrying an orthogonal basis onto the subsequent one to be a permutation of two elements, or to belong to one of the following three types which have respectively this effect when operating (for instance) on the first basis* $\mathcal{B}$:

(i) *to replace* (e_1, e_2) *with* $(\lambda e_1, e_2)$ *with any* $\lambda \in K^\times$;

(ii) *to replace* (e_1, e_2) *with* $(e_1 + e_2\ ,\ -q(e_2)e_1 + q(e_1)e_2)$ *when* $q(e_1) + q(e_2) \in K^\times$;

(iii) *to replace* (e_1, e_2) *with* $(e_1 + \mu q(e_1) e_2\ ,\ -\mu q(e_2) e_1 + e_2)$ *with* $\mu \in \mathfrak{m}$, *when* $q(e_1) + q(e_2) \in \mathfrak{m}$.

Moreover the operations of type (iii) *are indispensable only when the residue field* $K/\mathfrak{m}$ *is isomorphic to* $\mathbb{Z}/3\mathbb{Z}$ *and* $\mathfrak{m} \neq 0$.

Proof. It is clear that the improvement proposed in (8.1.7) involves a quadratic space of rank 2 with two orthogonal bases (e_1, e_2) and (e'_1, e'_2). Let us write $e'_1 = \lambda_1 e_1 + \lambda_2 e_2$. First we suppose that λ_1 and λ_2 are both invertible. Then we replace e_1 and e_2 respectively with $\lambda_1 e_1$ and $\lambda_2 e_2$; this requires two operations of type (i) (and also two permutations). An operation of type (ii) transforms $(\lambda_1 e_1,\ \lambda_2 e_2)$ into a basis beginning with e'_1 in which the second element must generate the same submodule as e_2'. The conclusion follows after another operation of type (i) (and two permutations).

At least one of the two factors λ_1 or λ_2 must be invertible; we can assume that λ_1 (for instance) is invertible, whereas $\lambda_2 \in \mathfrak{m}$. Here we moreover assume that $q(e_1)+q(e_2)$ is invertible. In this case we first perform an operation of type (ii) that carries (e_1, e_2) onto (e_1'', e_2'') such that $e_1'' = e_1 + e_2$ and $e_2'' = -q(e_2)e_1 + q(e_1)e_2$, whence $e_1' = \lambda_1''e_1'' + \lambda_2''e_2''$ with

$$\lambda_1'' = (q(e_1) + q(e_2))^{-1}(\lambda_1 q(e_1) + \lambda_2 q(e_2)) \ , \quad \lambda_2'' = (q(e_1) + q(e_2))^{-1}(-\lambda_1 + \lambda_2) \ ;$$

since λ_1'' and λ_2'' are both invertible, we are brought back to the previous case.

Now we assume that $q(e_1) + q(e_2)$ is not invertible, and that $K/\mathfrak{m}$ is not isomorphic to $\mathbb{Z}/3\mathbb{Z}$; therefore $K^\times$ contains an element κ such that $\kappa^2 - 1$ is invertible. Then we replace e_1 with κe_1 (by an operation of type (i)) and we observe that $q(\kappa e_1) + q(e_2)$ is invertible, because it is congruent to $(\kappa^2 - 1)q(e_1)$ modulo $\mathfrak{m}$. We are again brought back to the previous case.

It remains to consider the case in which $K/\mathfrak{m}$ is isomorphic to $\mathbb{Z}/3\mathbb{Z}$, λ_2 (for instance) is a nonzero but noninvertible element, and $q(e_1) + q(e_2) \in \mathfrak{m}$. In this case every $e \in M$ such that $q(e)$ is invertible, is congruent to $\pm e_1$ or $\pm e_2$ modulo $\mathfrak{m}M$, and no operation of type (ii) is possible. Now we replace e_1 with $\lambda_1 e_1$, and we set $\mu = q(\lambda_1 e_1)^{-1}\lambda_2$ so that the operation (iii) transforms $(\lambda_1 e_1,\ e_2)$ into a basis beginning with e_1'. Since the second vector generates the same submodule as e_2', an operation of type (i) leads to the conclusion. □

From (8.1.6) and (8.1.7) we can derive a precise description of the kernel of $\mathbb{Z}[K^\times] \to \mathrm{WG}(K)$. We state the resulting theorem in the most usual way: instead of showing elements generating this kernel (as an additive subgroup), we write the corresponding relations imposed in $\mathrm{WG}(K)$.

(8.1.8) **Theorem.** *As an additive group,* $\mathrm{WG}(K)$ *is the group generated by all symbols* $(\!(a)\!)$ *with* $a \in K^\times$*, constrained to the relations of these three types:*

(i) $(\!(ab^2)\!) = (\!(a)\!)$ *for all* $a,\ b \in K^\times$*;*
(ii) $(\!(a)\!) + (\!(b)\!) = (\!(a + b)\!) + (\!(ab(a + b)\,)\!)$ *whenever* $a + b \in K^\times$*;*
(iii) $(\!(a)\!) + (\!(b)\!) = (\!(\,(1+ab\mu^2)a)\!) + (\!(\,(1+ab\mu^2)b)\!)$ *whenever* $a+b$ *and* μ *are in* $\mathfrak{m}$*.*

Moreover the relations of type (iii) *are consequences of the previous ones if one of these two hypotheses is true: if* $K/\mathfrak{m}$ *is* not *isomorphic to* $\mathbb{Z}/3\mathbb{Z}$*, or if the mapping* $\nu \longmapsto \nu - \nu^2$ *is surjective from* $\mathfrak{m}$ *onto itself.*

Proof. First let us verify that all these relations are necessary; this is evident for the type (i). Let (M, q) be a quadratic space with orthogonal basis (e_1, e_2) such that $q(e_1) = a/2$ and $q(e_2) = b/2$. If $a+b$ is invertible, then $(e_1+e_2,\ -be_1+ae_2)$ is an orthogonal basis on which q takes the values $(a+b)/2$ and $ab(a+b)/2$, whence the relations (ii). If $\mu \in \mathfrak{m}$, then $(e_1 + \mu ae_2,\ -\mu be_1 + e_2)$ is an orthogonal basis on which q takes the values $(1 + ab\mu^2)a/2$ and $(1 + ab\mu^2)b/2$, whence the relations (iii), which actually are valid even when $a + b$ is not in $\mathfrak{m}$.

It remains to prove that the kernel of the ring morphism $\mathbb{Z}[K^\times] \to \mathrm{WG}(K)$ is generated by the elements corresponding to the relations (i), (ii) and (iii); this is an immediate consequence of Lemma (8.1.7), which also shows that the relations (iii) with an invertible $a+b$ are always consequences of the previous ones. Moreover, when $K/\mathfrak{m}$ is not isomorphic to $\mathbb{Z}/3\mathbb{Z}$, it also shows that all relations (iii) are consequences of the previous ones.

When the mapping $\nu \longmapsto \nu - \nu^2$ is surjective from $\mathfrak{m}$ onto itself, we first prove the surjectiveness of the mapping $\lambda \longmapsto \lambda^2$ from $1+\mathfrak{m}$ onto itself. Indeed, for every $\kappa \in 1+\mathfrak{m}$ there is a $\nu \in \mathfrak{m}$ such that $\nu - \nu^2 = (1-\kappa)/4$, whence $(1-2\nu)^2 = 1 - 4(\nu - \nu^2) = \kappa$. Now let us consider a relation of type (iii); since $1+ab\mu^2$ belongs to $1+\mathfrak{m}$, we can write $1+ab\mu^2 = \lambda^2$ for some $\lambda \in 1+\mathfrak{m}$, and with the relations (i) we already get $(\!(\,(1+ab\mu^2)a\,)\!) = (\!(a\lambda^2)\!) = (\!(a)\!)$ and $(\!(\,(1+ab\mu^2)b\,)\!) = (\!(b)\!)$. □

Remarks

(a) The kernel of $\mathbb{Z}[K^\times] \to \mathrm{WG}(K)$ is generated as an additive group by the elements corresponding to the relations mentioned in Theorem (8.1.8); yet as an *ideal* it may be generated by a smaller set of generators; it is an easy exercise to prove that we can replace a with 1 in the relations (i), (ii) (and (iii) when they are indispensable) if we wish to generate this kernel as an ideal.

(b) This remark and the following one are also valid when 2 is not invertible in the local ring K. When ν runs through $\mathfrak{m}$, it is easy to prove that the mapping $\nu \longmapsto \nu - \nu^2$ is injective; therefore its surjectiveness is equivalent to its bijectiveness. And when $\mathfrak{m}$ is finite, it is necessarily bijective; this applies to all local rings $\mathbb{Z}/p^k\mathbb{Z}$ where p is a prime integer, and $k>0$.

(c) The mapping $\nu \longmapsto \nu - \nu^2$ is actually surjective from $\mathfrak{m}$ onto itself when the local ring K is *henselian*. This means that the following condition is fulfilled for all polynomials $P(x) = x^n + \sum_{i=1}^n \alpha_i x^{n-i}$ in $K[x]$: if $Q(y)$ is the image of $P(x)$ in $(K/\mathfrak{m})[y]$ and if $Q(y)$ admits a simple root y_0 (such that $Q(y_0) = 0$ and $Q'(y_0) \neq 0$), then $P(x)$ admits a root x_0 in K with image y_0 in $K/\mathfrak{m}$. Indeed, when K is henselian, for every $\mu \in \mathfrak{m}$, the polynomial $x^2 - x + \mu$ must have a root ν with image 0 in $K/\mathfrak{m}$; in other words, $\nu \in \mathfrak{m}$ and $\nu - \nu^2 = \mu$. All the rings $\mathbb{Z}/p^k\mathbb{Z}$ (with p a prime integer) are henselian; the ring of p-adic integers (defined in (1.ex.29)) is also henselian. When F is a field, the ring $F[[t]]$ of formal series is also henselian. But the localization of the ring of polynomials $F[t]$ at the maximal ideal generated by t is not henselian; for instance if we set $a=1$, $b=-1+t$ and $\mu=t$, the polynomial $1+ab\mu^2 = 1-t^2+t^3$ has no square root in $F[t]_{(t)}$.

At last, let us compare the Witt rings of K and its residue field.

(8.1.9) **Proposition.** *The following assertions are equivalent:*

(a) *The mapping* $\nu \longmapsto \nu - \nu^2$ *is surjective from* $\mathfrak{m}$ *onto itself.*

(b) *The natural ring morphism* $\mathrm{WG}(K) \to \mathrm{WG}(K/\mathfrak{m})$ *is bijective.*

(c) *The natural ring morphism* $\mathrm{W}(K) \to \mathrm{W}(K/\mathfrak{m})$ *is bijective.*

Proof. The surjectiveness of the group morphism $K^\times \to (K/\mathfrak{m})^\times$ implies the surjectiveness of the ring morphisms involved in the assertions (b) and (c). When the assertion (a) is true, it suffices to prove that $\langle a\rangle$ and $\langle b\rangle$ are isomorphic whenever $a \equiv b$ modulo $\mathfrak{m}$, because this implies the injectiveness of the ring morphisms in (b) and (c). At the end of the proof of (8.1.8) it has been explained that every element of $1+\mathfrak{m}$ admits a square root in $1+\mathfrak{m}$; this allows us to write $ab^{-1} = \lambda^2$ for some $\lambda \in 1+\mathfrak{m}$, and the equality $a = \lambda^2 b$ leads to the conclusion.

Conversely let μ be an element of $\mathfrak{m}$. When the assertion (b) is true, the quadratic space $\langle 1-4\mu, -1\rangle$ must be hyperbolic; when the assertion (c) is true, it is still hyperbolic because (8.1.1) implies that the Witt class of a quadratic space vanishes if and only if it is hyperbolic. Consequently the equality $(1-4\mu)x^2 - y^2 = 0$ is true for some couple (x, y) of elements of K, not both in $\mathfrak{m}$. This implies that x and y are invertible, and that y/x or $-y/x$ belongs to $1+\mathfrak{m}$, and can be written $1-2\nu$ for a suitable $\nu \in \mathfrak{m}$. The equalities $(1-4\mu) = y^2/x^2 = (1-2\nu)^2$ imply that $\mu = \nu - \nu^2$. □

8.2 Continuation when 2 is not invertible

In this section we try to do the same as in the previous one when 2 is not invertible in the local ring K. After some successes, unfortunately the difficulties become so grave that we shall continue with the hypothesis that K is a field. Again we begin with the cancellation theorem.

(8.2.1) **Theorem.** *Let N, N', P, P' be quadratic spaces over the local ring K. When $N \perp P$ and $N' \perp P'$ are isomorphic, and also P and P', then N and N' too are isomorphic.*

Proof. Since P is an orthogonal sum of quadratic subspaces of rank 2 (see (2.6.2)), it suffices to prove (8.2.1) when P has rank 2. As in the proof of (8.1.1), we can reduce the problem to the proof of this assertion: if a quadratic space (M, q) contains four elements a, b, a', b' such that $q(a) = q(a') \in K^\times$, $q(b) = q(b')$ and $\mathrm{b}_q(a, b) = \mathrm{b}_q(a', b') \in K^\times$, there exists an automorphism of (M, q) that maps a to a' and b to b'. This has been already stated in (5.7.5). □

As in the previous section, this theorem implies the injectiveness of

$$\mathrm{WIQ}(K) \to \mathrm{WGQ}(K)$$

(see (8.1.2)). It also implies that the Witt class of a quadratic space is trivial if and only if it is hyperbolic.

The fact that every quadratic space over K is an orthogonal sum of quadratic subspaces of rank 2 leads to the following definitions. Let (M, q) be a quadratic space of rank $2n$ over K; a basis $(e_1, f_1; e_2, f_2; \ldots; e_n, f_n)$ is said to be *almost orthogonal* if every e_i and f_i is orthogonal to e_j and f_j whenever $i \neq j$. This implies that $\mathrm{b}_q(e_i, f_i)$ is invertible for $i = 1, 2, \ldots, n$. By replacing every f_i with

$f_i' = \mathrm{b}_q(e_i, f_i)^{-1} f_i$, we obtain an *almost orthonormal* basis $(e_1, f_1'; \dots; e_n, f_n')$ in which $\mathrm{b}_q(e_i, f_i') = 1$ for $i = 1, 2, \dots, n$. When the equality $2 = 0$ holds in K, such a basis is often called a *symplectic basis*, because in this case b_q is actually a symplectic form, for which this concept of symplectic basis already exists. With any sequence $(a_1, b_1, a_2, b_2, \dots, a_n, b_n)$ of $2n$ elements of K we associate a quadratic space denoted by $\langle a_1, b_1; a_2, b_2; \dots; a_n, b_n \rangle$ and provided with an almost orthonormal basis $(e_1, f_1; \dots; e_n, f_n)$ such that, for $i = 1, 2, \dots, n$, the quadratic form takes the values a_i and b_i respectively on e_i and f_i. The class of $\langle a, b \rangle$ in $\mathrm{WGQ}(K)$ (resp. $\mathrm{WQ}(K)$) is denoted by $(\!(a, b)\!)$ (resp. $[a, b]$). It is clear that $\mathrm{WGQ}(K)$ (resp. $\mathrm{WQ}(K)$) is generated by the classes $(\!(a, b)\!)$ (resp. $[a, b]$).

Consequently we consider the additive group $\mathbb{Z}^{(K \times K)}$ freely generated by the set $K \times K$, with basis $(\varepsilon_{a,b})_{(a,b) \in K \times K}$; there is a surjective group morphism from this free additive group onto $\mathrm{WGQ}(K)$ that maps every $\varepsilon_{a,b}$ to $(\!(a, b)\!)$. Here we shall not try to define a multiplication on this free additive group so that we get a ring morphism; the following lemma shows that in general such a multiplication would be very cumbersome.

(8.2.2) **Lemma.** *The tensor product of $\langle a, b \rangle$ and $\langle c, d \rangle$ is isomorphic to*

$$\langle\, 2ac,\ 2bd\ ;\ 2ad(1 - 4ab)(1 - 4cd),\ 2bc(1 - 4ab)^{-1}(1 - 4cd)^{-1} \rangle.$$

Proof. If (e_1, e_2) and (e_3, e_4) are the canonical bases of $\langle a, b \rangle$ and $\langle c, d \rangle$, we can write

$$q(e_1) = a\,, \quad q(e_2) = b\,, \quad \mathrm{b}_q(e_1, e_2) = 1\,,$$
$$q'(e_3) = c\,, \quad q'(e_4) = d\,, \quad \mathrm{b}_{q'}(e_3, e_4) = 1\,.$$

Let us set

$$e_1' = e_1 - 2ae_2\,, \quad e_2' = e_2 - 2be_1\,, \quad \lambda = 1 - 4ab\,,$$
$$e_3' = e_3 - 2ce_4\,, \quad e_4' = e_4 - 2de_3\,, \quad \mu = 1 - 4cd\,.$$

These elements e_j' (with $j = 1, 2, 3, 4$) have been chosen so that

$$\mathrm{b}_q(e_1, e_1') = \mathrm{b}_q(e_2, e_2') = \mathrm{b}_{q'}(e_3, e_3') = \mathrm{b}_{q'}(e_4, e_4') = 0\ ;$$

and it is easy to verify that

$$\mathrm{b}_q(e_1, e_2') = \mathrm{b}_q(e_1', e_2) = \mathrm{b}_q(e_1', e_2') = \lambda\ ,$$
$$\mathrm{b}_{q'}(e_3, e_4') = \mathrm{b}_{q'}(e_3', e_4) = \mathrm{b}_{q'}(e_3', e_4') = \mu.$$

All this allows us to construct an almost orthonormal basis of $\langle a, b \rangle \otimes \langle c, d \rangle$:

$$(\ e_1 \otimes e_3\ ,\quad e_2 \otimes e_4\ ;\quad e_1' \otimes e_4'\ ,\ (\lambda\mu)^{-1} e_2' \otimes e_3'\ ).$$

A last calculation shows that

$$q(e_1') = -a\lambda\ , \quad q(e_2') = -b\lambda\ , \quad q'(e_3') = -c\mu\ , \quad q'(e_4') = -d\mu$$

and allows us to calculate the values of $q \otimes q'$ on the above almost orthonormal basis, and to conclude that the tensor product is isomorphic to

$$\langle\ 2ac,\ 2bd\ ;\ 2ad\lambda\mu,\ 2bc(\lambda\mu)^{-1}\rangle. \qquad \square$$

Let $\mathrm{WIQ}_2(K)$ be the set of isomorphy classes of quadratic spaces of rank 2; it can be considered as a subset of $\mathrm{WGQ}(K)$ that generates it as an additive group.

(8.2.3) **Lemma.** *Let (M, q) be a quadratic space of rank 2. Every $\alpha \in K^\times$ determines an operation on the set of almost orthonormal bases of (M, q) by carrying every such a basis (e_1, e_2) onto $(\alpha e_1,\ \alpha^{-1}e_2)$. And every $\beta \in K$ determines an operation on the same set by carrying every (e_1, e_2) onto $((1-2\beta q(e_2))e_1+\beta e_2\ ,\ e_2)$. When (e_1, e_2) and (e_1', e_2') are two almost orthonormal bases of (M, q), there exists a finite sequence of operations that finally carries the former onto the latter, and which contains only operations of the two previous types, and trivial operations that merely permute the two elements of the basis.*

Proof. It is clear that the two operations defined in the first assertion of (8.2.3) carry every almost orthonormal basis of (M, q) onto a basis of the same kind. Only the second assertion requires serious explanations. Let us set $e_1' = \lambda e_1 + \mu e_2$. When $e_2' = e_2$, the equality $\mathrm{b}_q(e_1', e_2') = 1$ requires $\lambda = 1 - 2\mu q(e_2)$; thus (e_1, e_2) can be carried onto (e_1', e_2') by means of the operation of the second type in which $\beta = \mu$. It remains to prove that we can carry (e_1, e_2) onto a basis containing e_2' by means of the allowed operations; since we can permute e_1' and e_2', it suffices to explain how to carry (e_1, e_2) onto a basis containing $e_1' = \lambda e_1 + \mu e_2$. Since λ and μ cannot be both in $\mathfrak{m}$, and since we can permute e_1 and e_2, we can suppose that λ is invertible. In this case we perform the operation of the first type determined by some $\alpha \in K^\times$, followed by the operation of the second type determined by some $\beta \in K$, and thus we carry (e_1, e_2) onto

$$(\ (1 - 2\beta\alpha^{-2}q(e_2))\alpha e_1 + \beta\alpha^{-1}e_2\ ,\ \ \alpha^{-1}e_2\ )\ ,$$

and we try to determine (α, β) so that

$$(1 - 2\beta\alpha^{-2}q(e_2))\alpha = \lambda \quad \text{and} \quad \beta\alpha^{-1} = \mu\ .$$

These equations are equivalent to

$$\beta = \alpha\mu \quad \text{and} \quad \alpha = 2\mu q(e_2) + \lambda\ ;$$

obviously there is a (unique) solution (α, β). $\square$

Observe that in Lemma (8.2.3) the quadratic form q takes the value $q(e_1) + \beta(1 - \beta q(e_2))(1 - 4q(e_1)q(e_2))$ on $(1 - 2\beta q(e_2))e_1 + \beta e_2$; this leads immediately to this corollary.

(8.2.4) **Corollary.** *There is a natural bijection between the subset* $\mathrm{WIQ}_2(K)$ *and the quotient of the set* $K \times K$ *by the equivalence relation generated by the following equivalences (in which* a, b, β *run through* K, *and* α *through* $K^\times$*):*

$$(a,b) \sim (b,a) ,$$
$$(a,b) \sim (\alpha^2 a,\ \alpha^{-2} b) ,$$
$$(a,b) \sim (a + \beta(1-\beta b)(1-4ab)\, ,\ b) .$$

This knowledge about $\mathrm{WIQ}_2(K)$ is sufficient to extend the validity of (8.1.9).

(8.2.5) **Proposition.** *The following assertions are equivalent:*

(a) *The mapping* $\nu \longmapsto \nu - \nu^2$ *is surjective from* $\mathfrak{m}$ *onto itself.*
(b) *The natural ring morphism* $\mathrm{WGQ}(K) \to \mathrm{WGQ}(K/\mathfrak{m})$ *is bijective.*
(c) *The natural ring morphism* $\mathrm{WQ}(K) \to \mathrm{WQ}(K/\mathfrak{m})$ *is bijective.*

Proof. It is clear that the morphisms mentioned in (b) and (c) are always surjective. When (M,q) is a quadratic space over K, every almost orthonormal basis $(\bar e_1, \bar f_1; \dots; \bar e_n, \bar f_n)$ of $(K/\mathfrak{m}) \otimes (M,q)$ is the image of an almost orthonormal basis of (M,q); indeed $(\bar e_1, \bar f_1)$ is the image of some pair (e_1, f_1) in M, which can be normalized so that $\mathrm{b}_q(e_1,f_1) = 1$, and which generates an orthogonal summand of (M,q); then in $(Ke_1 \oplus Kf_1)^\perp$ we can find a pair (e_2,f_2) above $(\bar e_2, \bar f_2)$; we normalize it, and consider the submodule orthogonal to (e_1,f_1,e_2,f_2), and so forth.... After this preliminary observation, the injectiveness of the morphisms mentioned in (b) and (c) follows from this elementary assertion: if $a \equiv a'$ and $b \equiv b'$ modulo $\mathfrak{m}$, then $\langle a,b\rangle$ and $\langle a',b'\rangle$ are isomorphic quadratic spaces.

Let us prove that (a) implies this elementary assertion. Since $\langle a,b\rangle$ contains elements on which the quadratic form takes an invertible value (see (2.6.2)), we can assume that b is invertible, and we first prove that $\langle a,b\rangle$ and $\langle a',b\rangle$ are isomorphic. From (8.2.4) we know that $\langle a,b\rangle$ is isomorphic to $\langle a+\beta(1-\beta b)(1-4ab)\, ,\ b\rangle$ for every $\beta \in K$; we wish to find some β such that

$$a + \beta(1-\beta b)(1-4ab) = a' \ ;$$

since b and $1-4ab$ are invertible, this is equivalent to

$$\beta b - (\beta b)^2 = (a'-a)\ b(1-4ab)^{-1} \ ;$$

when the assertion (a) is true, such a β exists, because the right-hand member belongs to $\mathfrak{m}$.

When a is invertible (and therefore a' too), in the same way we prove that $\langle a',b\rangle$ and $\langle a',b'\rangle$ are isomorphic. When a and a' belong to $\mathfrak{m}$, our first result proves that $\langle a,b\rangle$ and $\langle a',b'\rangle$ are respectively isomorphic to $\langle 0,b\rangle$ and $\langle 0,b'\rangle$; then Lemma (8.1.3) (in which we only use the assertions (a), (b), (c)) proves that these quadratic spaces are hyperbolic and isomorphic to each other. Instead of (8.1.3), we might alternatively take the equivalence $(a,b) \sim (a+\beta(1-\beta b)(1-4ab),\ b)$

from (8.2.4), replace the variables (a, b, β) with $(b, 0, -b)$, and thus obtain the equivalence $(b, 0) \sim (0, 0)$, and similarly $(b', 0) \sim (0, 0)$.

Now we suppose that the assertion (b) or (c) is true. If μ is any element of $\mathfrak{m}$, the quadratic space $\langle -\mu, -1 \rangle$ must be hyperbolic; consequently $-\mu x^2 + xy - y^2 = 0$ for some x and y in K but not both in $\mathfrak{m}$. Thus $y(x - y) = \mu x^2$, and either y or $x - y$ belongs to $\mathfrak{m}$, whereas x is invertible. If y belongs to $\mathfrak{m}$, we realize that the equality $\nu - \nu^2 = \mu$ holds for $\nu = yx^{-1}$; if $x - y$ belongs to $\mathfrak{m}$, we realize that it holds for $\nu = (x - y)x^{-1}$. □

From now on we suppose that K is a field of characteristic 2. Our next purpose, that is the description of $\mathrm{WGQ}(K)$, inclines us to accept this simplification, because it is a tremendous problem when K is merely assumed to be a local ring such that $2 \in \mathfrak{m}$. When K is a field of characteristic 2, the equality $\mathrm{b}_q(x, x) = 0$ holds for every x in every quadratic space (M, q), and Lemma (8.2.2) shows that every tensor product is hyperbolic.

First we try to get a lemma similar to (8.1.6).

(8.2.6) **Lemma.** *Let (M, q) be a quadratic space of rank $2r$ over the field K of characteristic 2, and let $\mathcal{B} = (e_1, f_1; \dots; e_r, f_r)$ and $\mathcal{B}' = (e'_1, f'_1; \dots; e'_r, f'_r)$ be two almost orthogonal bases of (M, q). There exists a finite sequence of almost orthogonal bases, in which the first one is equal to $\mathcal{B}$ and the last one to $\mathcal{B}'$, and such that two consecutive bases only differ by modifications in one or two among the r pairs of elements constituting these bases.*

Proof. We proceed by induction on r. When r is 1 or 2, there is nothing to prove. Therefore we assume $r \geq 3$. Because of the induction hypothesis, it suffices to prove that we can transform the basis $\mathcal{B}$ into another one that is also almost orthogonal and begins with the pair (e'_1, f'_1). For $i = 1, 2, \dots, r$, let g_i and h_i be the components of e'_1 and f'_1 in the subspace $Ke_i \oplus Kf_i$.

First we suppose that $\mathrm{b}_q(e'_1, f'_1) \neq \mathrm{b}_q(g_j, h_j)$ for some $j \in \{1, 2, \dots, r\}$; we can assume that $j = r$. We set $e''_1 = \sum_{i=1}^{r-1} g_i$ and $f''_1 = \sum_{i=1}^{r-1} h_i$; the inequality $\mathrm{b}_q(e'_1, f'_1) \neq \mathrm{b}_q(g_r, h_r)$ implies that $\mathrm{b}_q(e''_1, f''_1)$ does not vanish, and consequently (e''_1, f''_1) can be the first pair in an almost orthogonal basis of the subspace spanned by $(e_1, f_1; \dots; e_{r-1}, f_{r-1})$. The induction hypothesis implies that we can transform the basis $\mathcal{B}$ into another one that is also almost orthogonal, begins with (e''_1, f''_1) and still ends with (e_r, f_r). Since $e'_1 = e''_1 + g_r$ and $f'_1 = f''_1 + h_r$, by only modifying the first pair (e''_1, f''_1) and the last pair (e_r, f_r), we get a new basis beginning with (e'_1, f'_1).

Now we suppose that $\mathrm{b}_q(e'_1, f'_1) = \mathrm{b}_q(g_i, h_i)$ for $i = 1, 2, \dots, r$; consequently r is odd. Then we set $g'_3 = \sum_{i=3}^{r} g_i$ and $h'_3 = \sum_{i=3}^{r} h_i$; this implies $e'_1 = g_1 + g_2 + g'_3$ and $f'_1 = h_1 + h_2 + h'_3$, and moreover

$$\mathrm{b}_q(e'_1, f'_1) = \mathrm{b}_q(g_1, h_1) = \mathrm{b}_q(g_2, h_2) = \mathrm{b}_q(g'_3, h'_3) \neq 0.$$

The induction hypothesis (applied to the subspace spanned by $(e_3, f_3, \dots, e_r, f_r)$) implies that we can transform the basis $\mathcal{B}$ into a new basis beginning with

$$(e_1, f_1; e_2, f_2; g_3', h_3').$$

Then we can replace the pairs (e_1, f_1) and (e_2, f_2) respectively with (g_1, h_1) and (g_2, h_2), and we perform these three modifications only involving the first three pairs:

$$\begin{array}{rcl} (\ g_1\ ,\ h_1 & ; & g_2\ ,\ h_2 \quad ; \quad g_3'\ ,\ h_3'\ )\ , \\ (\ g_1\ ,\ h_1 & ; & g_2 + g_3'\ ,\ h_2 \quad ; \quad g_3'\ ,\ h_2 + h_3')\ , \\ (g_1 + g_2 + g_3'\ ,\ h_1 & ; & g_2 + g_3'\ ,\ h_1 + h_2\ ; \quad g_3'\ ,\ h_2 + h_3')\ , \\ (g_1 + g_2 + g_3'\ ,\ h_1 + h_2 + h_3' & ; & g_2 + g_3'\ ,\ h_1 + h_2\ ; \quad g_1 + g_2\ ,\ h_2 + h_3')\ ; \end{array}$$

since the first pair is now (e_1', f_1'), the proof is finished. □

It was convenient to prove (8.2.6) with almost orthogonal bases; but since it is easy to normalize them, it is also valid for almost orthonormal (or symplectic) bases. Now we try to get a lemma similar to (8.1.7); in this lemma we find again the three types of operation already discovered in (8.2.3.), and a fourth type that has just appeared at the end of the proof of (8.2.6).

(8.2.7) **Lemma.** *When the bases $\mathcal{B}$ and $\mathcal{B}'$ in Lemma* (8.2.6) *are almost orthonormal, we can demand every operation carrying a basis onto the subsequent one to be either a permutation of two among the r pairs of elements, or an operation belonging to one of the following four types which have respectively this effect when operating (for instance) on $\mathcal{B}$:*

(i) *to replace (e_1, f_1) with (f_1, e_1) ;*
(ii) *to replace (e_1, f_1) with $(\alpha e_1,\ \alpha^{-1} f_1)$ with any $\alpha \in K^\times$;*
(iii) *to replace (e_1, f_1) with $(e_1 + \beta f_1,\ f_1)$ for some $\beta \in K$;*
(iv) *to replace $(e_1, f_1; e_2, f_2)$ with $(e_1 + e_2,\ f_1;\ e_2,\ f_1 + f_2)$ when $r \geq 2$.*

All these operations carry every almost orthonormal basis onto an almost orthonormal one.

Proof. Lemma (8.2.3) shows that (8.2.7) is true for quadratic spaces of rank 2, and Lemma (8.2.6) shows that it suffices to prove it for quadratic spaces of rank 4. Let $(e_1, f_1; e_2, f_2)$ and $(e_1', f_1'; e_2', f_2')$ be two almost orthonormal bases of a quadratic space of rank 4; it suffices to prove that we can transform the former into an almost orthonormal basis beginning with (e_1', f_1'), and then (8.2.3) ensures that we can also reach the second pair (e_2', f_2') without modifying the first one.

First let us suppose that $e_1 = e_1'$. Then $f_1' = \lambda e_1 + f_1 + \mu e_2 + \nu f_2$ for some λ, μ, ν in K. If $\mu = \nu = 0$, the conclusion follows immediately from (8.2.3); otherwise we deduce from (8.2.3) that we can transform (e_2, f_2) into a pair (e_2'', f_2'') such that

$e_2'' = \mu e_2 + \nu f_2$. Then we perform these four operations:

$$\begin{array}{rcl} (\ e_1\ ,\ f_1 & ; & e_2''\ ,\ f_2''\ )\ , \\ (\ f_1\ ,\ e_1 & ; & e_2''\ ,\ f_2''\ )\ , \\ (\ \lambda e_1 + f_1\ ,\ e_1 & ; & e_2''\ ,\ f_2''\ )\ , \\ (\lambda e_1 + f_1 + e_2'',\ e_1 & ; & e_2'',\ e_1 + f_2'')\ , \\ (\ e_1\ ,\ \lambda e_1 + f_1 + e_2'' & ; & e_2'',\ e_1 + f_2'')\ ; \end{array}$$

since $e_1' = e_1$ and $f_1' = \lambda e_1 + f_1 + e_2''$, this particular case is settled.

It remains to prove that we can transform $(e_1, f_1; e_2, f_2)$ into an almost orthonormal basis beginning with e_1' by means of the operations described in (8.2.7). Let g_1 and g_2 be the components of e_1' in $Ke_1 \oplus Kf_1$ and $Ke_2 \oplus Kf_2$. If $g_2 = 0$, from (8.2.3) we know that we can transform (e_1, f_1) into some (g_1, h_1) with a suitable h_1; if $g_1 = 0$, we first replace $(e_1, f_1; e_2, f_2)$ with $(e_2, f_2; e_1, f_1)$ and we are brought back to the previous case. If neither g_1 nor g_2 vanishes, from (8.2.3) we know that we can transform (e_1, f_1) into some (g_1, h_1), and (e_2, f_2) into some (g_2, h_2), and then an operation of type (iv) reveals $e_1' = g_1 + g_2$. □

From (8.2.7) we immediately deduce the kernel of $\mathbb{Z}^{(K \times K)} \to \mathrm{WGQ}(K)$; instead of giving a set of elements generating it as an additive subgroup, we rather give the relations in $\mathrm{WGQ}(K)$ corresponding to these generators.

(8.2.8) **Theorem.** *The kernel of the ring morphism $\mathbb{Z}^{(K \times K)} \to \mathrm{WGQ}(K)$ is generated as an additive subgroup by the elements that correspond to these four types of relations in* $\mathrm{WGQ}(K)$:

(i) $((a, b)) = ((b, a))$ *for all* $a,\ b \in K$;
(ii) $((a, b)) = ((\alpha^2 a,\ \alpha^{-2} b))$ *for all* $a,\ b \in K$ *and* $\alpha \in K^\times$;
(iii) $((a, b)) = ((a + \beta + \beta^2 b,\ b))$ *for all* $a,\ b,\ \beta \in K$;
(iv) $((a, b)) + ((c, d)) = ((a + c,\ b)) + ((c,\ b + d))$ *for all* $a,\ b,\ c,\ d \in K$.

Now we consider $\mathrm{WQ}(K)$; it is the quotient of $\mathrm{WGQ}(K)$ by the subgroup made of the classes of hyperbolic spaces. This subgroup is generated by the isomorphy class of hyperbolic spaces of dimension 2, that is the isomorphy class of $\langle 0, 0 \rangle$ (see (8.1.3)). Consequently the kernel of $\mathbb{Z}^{(K \times K)} \to \mathrm{WQ}(K)$ is generated by the elements of the four types described in (8.2.8), to which we must add the element $\varepsilon_{0,0}$ that gives the relation $[0, 0] = 0$. It is worth noticing that the awaited relation $[a, 0] = 0$ is a consequence of this relation together with the relation $[a, b] = [a + \beta + \beta^2 b,\ b]$ of type (iii) in which (a, b, β) is replaced with $(a, 0, a)$.

Nevertheless there is another description of $\mathrm{WQ}(K)$ that avoids $\mathbb{Z}^{(K \times K)}$. Indeed on one side it is clear that K is a vector space over its minimal subfield that we identify with $\mathbb{Z}/2\mathbb{Z}$; on the other side $\mathrm{WQ}(K)$ is also a vector space over $\mathbb{Z}/2\mathbb{Z}$, because the relations (iv) show that

$$[a, b] + [a, b] = [2a, b] + [a, 2b] = [0, b] + [a, 0] = 0\ ;$$

this fact might also be deduced from (2.5.8). Now it appears that the mapping $(a,b) \longmapsto [a,b]$ is a $(\mathbb{Z}/2\mathbb{Z})$-bilinear mapping from $K \times K$ into WQ(K); indeed

$$[a,c] + [b,c] = [a+b,\ c] + [b,\ 2c] = [a+b,\ c];$$

this symmetric bilinear mapping induces a linear mapping from the second symmetric power $\mathrm{S}^2_{\mathbb{Z}/2\mathbb{Z}}(K)$ into WQ(K) (see (1.4.3)); and this linear mapping is surjective because WQ(K) is generated as an additive group by the classes of quadratic spaces of dimension 2. Consequently it is possible to describe WQ(K) as a quotient of $\mathrm{S}^2_{\mathbb{Z}/2\mathbb{Z}}(K)$.

(8.2.9) **Theorem.** *There is a surjective $(\mathbb{Z}/2\mathbb{Z})$-linear mapping $\mathrm{S}^2_{\mathbb{Z}/2\mathbb{Z}}(K) \to \mathrm{WQ}(K)$ that maps every $a \vee b$ to $[a,b]$. Its kernel is the additive subgroup generated by all elements of these two types (with a, b running through K, and α through $K^\times$):*

$$a \vee b - \alpha^2 a \vee \alpha^{-2} b \quad \textit{and} \quad a \vee b - a \vee ab^2.$$

Proof. The existence of this $(\mathbb{Z}/2\mathbb{Z})$-linear mapping has been explained above; the universal property of $\mathbb{Z}^{(K\times K)}$ gives this commutative diagram:

$$\begin{array}{ccc} K \times K & \longrightarrow & \mathrm{S}^2_{\mathbb{Z}/2\mathbb{Z}}(K) \\ \downarrow & \nearrow & \downarrow \\ \mathbb{Z}^{(K\times K)} & \longrightarrow & WQ(K). \end{array}$$

The three arrows other than the arrows coming from $K \times K$ are surjective group morphisms. Consequently the kernel of $\mathrm{S}^2_{\mathbb{Z}/2\mathbb{Z}}(K) \to \mathrm{WQ}(K)$ is the image of the kernel of $\mathbb{Z}^{(K\times K)} \to \mathrm{WQ}(K)$. The image of $\varepsilon_{0,0}$ in $\mathrm{S}^2_{\mathbb{Z}/2\mathbb{Z}}(K)$ is 0, and also the images of all elements corresponding to the relations of types (i) or (iv):

$$a \vee b - b \vee a = 0 \quad \text{and} \quad a \vee b + c \vee d - (a+c) \vee b - c \vee (b+d) = 0.$$

The images of the elements corresponding to the relations of type (ii) are merely $a \vee b - \alpha^2 a \vee \alpha^{-2} b$. At last, for the type (iii) we get these images:

$$a \vee b - (a + \beta + \beta^2 b) \vee b = \beta \vee b - \beta^2 b \vee b\ ;$$

if we replace (b,β) with (a,b), we get the elements announced in (8.2.9). □

Example. When $K = \mathbb{Z}/2\mathbb{Z}$, we find again some results presented in **2.7**. Indeed the equalities $a \vee b = \alpha^2 a \vee \alpha^{-2} b$ and $a \vee b = a \vee ab^2$ hold for all a, b, and all invertible α; consequently WQ$(\mathbb{Z}/2\mathbb{Z})$ is isomorphic to $\mathrm{S}^2_{\mathbb{Z}/2\mathbb{Z}}(\mathbb{Z}/2\mathbb{Z})$ which is itself isomorphic to $\mathbb{Z}/2\mathbb{Z}$ as an additive group.

8.3 Witt rings of Prüfer rings

The ring K is said to be a *Prüfer ring* if it is an integral domain, and if every finitely generated torsionless K-module is projective. The first condition means that K contains no divisors of zero; it is well known that an integral domain K and all its rings of fractions $S^{-1}K$ (when the multiplicative subset S does not contain 0) can be identified with subrings of its field of fractions L. A K-module M is said to be torsionless if any equality $\lambda x = 0$ (with $\lambda \in K$ and $x \in M$) implies $\lambda = 0$ or $x = 0$; it is well known that every flat module over an integral domain (therefore every projective module too) is torsionless.

There are many other properties characterizing Prüfer rings; here is a sample of such properties, in which K is always assumed to be an integral domain:

- for every λ in L (the field of fractions), $K + K\lambda$ is an invertible K-submodule in L according to the definition in (1.ex.25);
- the nonzero finitely generated K-submodules of L constitute a group under multiplication;
- every ring K' such that $K \subset K' \subset L$ is flat over K;
- every ring K' such that $K \subset K' \subset L$ is a ring of fraction of K if it is a valuation ring (in other words, if it contains λ or λ^{-1} or both for every $\lambda \in L^\times$);
- the equality $\mathfrak{a} + (\mathfrak{b} \cap \mathfrak{c}) = (\mathfrak{a} \cap \mathfrak{b}) + (\mathfrak{a} \cap \mathfrak{c})$ holds for all ideals $\mathfrak{a}$, $\mathfrak{b}$, $\mathfrak{c}$ of K.

Besides, a module over a Prüfer ring is flat if and only if it is torsionless. The concept of Prüfer ring has been suggested by the study of Dedekind rings (which play an important role in some topics); indeed Dedekind rings are noetherian Prüfer rings. See [Bourbaki 1965, *Algèbre commutative*, Chap. 7, §3] or [Fontana, Huckaba, Papick 1997] for more information and for proof.

Of course every field is a Prüfer ring. Every principal ring (in other words, every integral domain in which every ideal is generated by one element) is also a Prüfer ring, because every finitely generated module over this ring is isomorphic to a direct sum of quotients of K by some ideals, and consequently it is free as soon as it is assumed to be torsionless. It follows that $\mathbb{Z}$ is a Prüfer ring. Other examples of Prüfer rings can be derived from the next proposition.

(8.3.1) **Proposition.** *If K is a Prüfer ring, every ring K' between K and its field of fractions L is a Prüfer ring.*

Proof. Let M' be a finitely generated torsionless module over K', and let M be the K-submodule of M' generated by any finite subset $\{x_1, x_2, \dots, x_m\}$ that generates M' as a K'-module. This K-module M is finitely generated, and torsionless too, since M' is torsionless over K'. Since K is a Prüfer ring, M is projective; consequently $K' \otimes_K M$ is a projective K'-module (see **1.9**). If we manage to prove the bijectiveness of the K'-linear mapping $K' \otimes_K M \to M'$ defined by $\lambda' \otimes x \longmapsto \lambda' x$, the proof is complete; the surjectiveness of this mapping is already evident. Every element x' of $K' \otimes M$ can be written $x' = \sum_i \lambda'_i \otimes x_i$ for some $\lambda'_1, \lambda'_2, \dots, \lambda'_m$ in K'; we must prove that $x' = 0$ when its image $\sum_i \lambda'_i x_i$ in M' vanishes. There are elements $\lambda_1, \lambda_2, \dots, \lambda_m$ and s in K such that $\lambda'_i = \lambda_i/s$ for $i = 1, 2, \dots, m$; the vanishing of $\sum_i \lambda'_i x_i$ in M' implies the vanishing of $\sum_i \lambda_i x_i$ in M :

$$\sum_i \lambda_i x_i = \sum_i \left(s\frac{\lambda_i}{s}\right) x_i = s \sum_i \frac{\lambda_i}{s} x_i = 0 \, ;$$

the vanishing of $\sum_i \lambda_i x_i$ in M implies the vanishing of $\sum_i \lambda'_i \otimes x_i$ in $L \otimes M$:

$$\sum_i \frac{\lambda_i}{s} \otimes x_i = \sum_i \frac{1}{s} \otimes \lambda_i x_i = \frac{1}{s} \otimes \sum_i \lambda_i x_i = 0.$$

Since M is projective over K, it is flat, and the mapping $K' \otimes M \to L \otimes M$ is injective like the morphism $K' \to L$. Thus the vanishing of $\sum_i \lambda'_i \otimes x_i$ in $L \otimes M$ implies the vanishing of x' in $K' \otimes M$. □

Here we are especially interested in this theorem.

(8.3.2) **Theorem.** *If K is a Prüfer ring, and L its field of fractions, the canonical morphisms*

$$\mathrm{WQ}(K) \longrightarrow \mathrm{WQ}(L) \quad \textit{and} \quad \mathrm{WB}(K) \longrightarrow \mathrm{WB}(L)$$

induced by the ring morphism $K \to L$ are injective.

Proof. Let (M, q) be a quadratic space over K such that the Witt class of $L \otimes (M, q)$ vanishes; this means that the orthogonal sum of this L-quadratic space and some hyperbolic L-quadratic space is hyperbolic; because of the cancellation theorem (8.1.1) or (8.2.1) this implies that already $L \otimes (M, q)$ is hyperbolic, and isomorphic to some $V^* \oplus V$, with V a finite dimensional vector space over L. Since M is a flat K-module, we can identify M with a K-submodule of $V^* \oplus V$. If P is the image of M by the projection $V^* \oplus V \to V$, we have an exact sequence

$$0 \longrightarrow M \cap V^* \longrightarrow M \longrightarrow P \longrightarrow 0 \, .$$

As a K-submodule of a L-vector space, P is torsionless; as an image of M, it is finitely generated; consequently P is projective, and the above exact sequence splits. Thus $M \cap V^*$ is a direct summand of M; it is totally isotropic because V^* is totally isotropic in $V^* \oplus V$.

As a projective module over an integral domain, $M \cap V^*$ has a constant rank, and its rank is the dimension of $L \otimes (M \cap V^*)$ over L (the localization of K at the prime ideal (0)). If we manage to prove that the rank of $M \cap V^*$ is half of the rank of M, we can claim that the totally isotropic direct summand $M \cap V^*$ is equal to its orthogonal submodule in M, and conclude that (M, q) is hyperbolic because of (2.5.5). From the above splitting exact sequence we deduce an exact sequence of L-vector spaces:

$$0 \longrightarrow L \otimes (M \cap V^*) \longrightarrow V^* \oplus V \longrightarrow L \otimes P \longrightarrow 0 \;;$$

since the extension $K \to L$ is flat, the injection $P \to V$ gives an injection $L \otimes P \to L \otimes V$. We can identify $L \otimes V$ (the module of fractions of V) with V, and thus we have proved the injectiveness of the L-linear mapping $L \otimes P \to V$ defined by $\lambda \otimes x \longmapsto \lambda x$. Consequently we can now identify $L \otimes P$ with the L-vector subspace P' of V generated by P. And similarly we can identify $L \otimes (M \cap V^*)$ with the L-vector subspace N' of V^* generated by $M \cap V^*$. Because of the above exact sequence, $V^* \oplus V$ is the direct sum of N' (subspace of V^*) and P' (subspace of V), whence $N' = V^*$ and $P' = V$. Now it is clear that the dimension of $L \otimes M$ is the double of the dimension of $L \otimes (M \cap V^*)$. It follows that the rank of M is the double of the rank of $M \cap V^*$, and that (M, q) is hyperbolic.

In this proof we have used the fact that V^* is totally isotropic in $V^* \oplus V$, without worrying about V. Thus this proof still works with a metabolic bilinear space $V^* \oplus V$, and allows us to prove the injectiveness of $\mathrm{WB}(K) \to \mathrm{WB}(L)$. □

The proof of (8.3.2) also implies this statement: *if the Witt class of a quadratic* (*resp. bilinear*) *space over a Prüfer ring vanishes, this space is hyperbolic* (*resp. metabolic*). Indeed its Witt class still vanishes after the ring extension to the field of fractions, and then the above proof shows that the quadratic or bilinear space under consideration is hyperbolic or metabolic.

Because of Theorem (8.3.2) we can identify $\mathrm{WQ}(K)$ with a subring of $\mathrm{WQ}(L)$, and $\mathrm{WB}(K)$ with a subring of $\mathrm{WB}(L)$. If K' is a ring such that $K \subset K' \subset L$, then K' too is a Prüfer ring (see (8.3.1)) and we get a sequence $\mathrm{WQ}(K) \to \mathrm{WQ}(K') \to \mathrm{WQ}(L)$ of two injective morphisms; thus we can identify $\mathrm{WQ}(K')$ with a subring of $\mathrm{WQ}(L)$, and $\mathrm{WQ}(K)$ with a subring of $\mathrm{WQ}(K')$. And the same for the bilinear Witt rings. This suggests paying attention to this question: is it possible to present $\mathrm{WQ}(K)$ as the intersection of some subrings $\mathrm{WQ}(K')$ derived from an interesting family of intermediate rings K'? Since K is the intersection in L of all its localizations $K_{\mathfrak{m}}$ at maximal ideals, it is sensible to choose these rings $K_{\mathfrak{m}}$ as intermediate rings. The following theorem has been proved in [Hoestmaelingen 1975] for quadratic Witt rings, but the proof should also work with bilinear Witt rings.

(8.3.3) **Theorem.** *When K is a Prüfer ring, then* $\mathrm{WQ}(K)$ *is the intersection in* $\mathrm{WQ}(L)$ (*the Witt ring of the field of fractions*) *of all subrings* $\mathrm{WQ}(K_{\mathfrak{m}})$ *associated with the localizations of K at all maximal ideals.*

Although the proof of this theorem deserves to be revisited, it requires so specialized knowledge that it is outside the scope of this book. The detailed study of an example seems to be here more profitable, especially if it is so interesting an example as the injective morphisms $\mathrm{WQ}(\mathbb{Z}) \to \mathrm{W}(\mathbb{Q})$ or $\mathrm{WB}(\mathbb{Z}) \to \mathrm{W}(\mathbb{Q})$. It will reveal a great sophistication because of the extreme complexity of $\mathrm{W}(\mathbb{Q})$.

The Witt ring $\mathrm{W}(\mathbb{Q})$ of the field of rational numbers

From (8.1.8) and (8.1.4) we know that $\mathrm{W}(\mathbb{Q})$ is the group generated by all symbols $[a]$ with $a \in \mathbb{Q}^\times$ constrained to the relations $[1] + [-1] = 0$,

$$[ab^2] = [a] \quad \text{and} \quad [a] + [b] = [a+b] + [ab(a+b)] \ \text{ if } a+b \neq 0 \ ;$$

consequently each class $[a]$ is equal to some $[b]$ with $b \in \mathbb{Z}$, and it is easy to prove that $\mathrm{W}(\mathbb{Q})$ is also the group generated by all symbols $[a]$ with $a \in \mathbb{Z} \setminus \{0\}$, constrained to the same relations. With every element p of the set

$$\mathcal{P} = \{ \ \infty, \ 2, \ 3, \ 5, \ 7, \ 11, \ 13, \ 17, \ 19, \ \dots \ \dots \}$$

we associate a field K_p and a group morphism $\delta_p : \mathrm{W}(\mathbb{Q}) \to \mathrm{W}(K_p)$. When $p = \infty$, then $K_\infty = \mathbb{R}$ and δ_∞ is the ring morphism $\mathrm{W}(\mathbb{Q}) \to \mathrm{W}(\mathbb{R})$ derived from the field extension $\mathbb{Q} \to \mathbb{R}$; among all group morphisms δ_p it is the only one that is a ring morphism; we can write $\mathrm{W}(\mathbb{R}) = \mathbb{Z}$ and say that δ_∞ maps the Witt class of every $\mathbb{Q}$-quadratic space (M, q) to the signature of the real quadratic space $\mathbb{R} \otimes (M, q)$. When p is a finite element of $\mathcal{P}$, then K_p is the finite field $\mathbb{Z}/p\mathbb{Z}$. When $p = 2$, both groups $\mathrm{WQ}(\mathbb{Z}/2\mathbb{Z})$ and $\mathrm{WB}(\mathbb{Z}/2\mathbb{Z})$ are isomorphic to $\mathbb{Z}/2\mathbb{Z}$, and here it suffices to write $\mathrm{W}(K_2)$ without more precision; but the "natural" construction of δ_p shows that it takes its values in $\mathrm{WB}(K_p)$. This "natural" construction requires quite long explanations, which are already available in the specialized literature, and here we propose to reach δ_p by a short way. For every $a \in \mathbb{Z}$, the notation a_p means the image of a in $\mathbb{Z}/p\mathbb{Z}$, and when a is not divisible by the odd prime integer p, the notation $[a_p]$ means the corresponding element of $\mathrm{W}(\mathbb{Z}/p\mathbb{Z})$. When we write $a = p^\alpha a'$, (resp. $b = p^\beta b'$,...), it is silently assumed that α (resp. β,...) is the largest exponent in $\mathbb{N}$ such that p^α (resp. p^β,...) divides a (resp. b,...).

(8.3.4) **Proposition.** *For every prime integer* $p \geq 3$, *there is a unique group morphism* $\delta_p : \mathrm{W}(\mathbb{Q}) \to \mathrm{W}(K_p)$ *satisfying this property for every nonzero integer* $a = p^\alpha a'$:

$$\delta_p([a]) = 0 \qquad \textit{if } \alpha \textit{ is even,}$$
$$\delta_p([a]) = [a'_p] \quad \textit{if } \alpha \textit{ is odd.}$$

Similarly there is a unique group morphism $\delta_2 : \mathrm{W}(\mathbb{Q}) \to \mathrm{W}(K_2)$ *that maps* $[a]$ *to* 0 *if and only if* $a = 2^\alpha a'$ *with* α *even and* a' *odd.*

Proof. We must prove that the definition of δ_p proposed in (8.3.4) is compatible with the relations between the generators $[a]$. There is no problem with the relations $[1] + [-1] = 0$ and $[ab^2] = [a]$. We have only to worry about the relations $[a] + [b] = [a+b] + [ab(a+b)]$, with $a + b \neq 0$. Let us set $a = p^\alpha a'$ and $b = p^\beta b'$; we must verify that the definition of δ_p gives four values $\delta_p([a])$, $\delta_p([b])$, $\delta_p([a+b])$ and $\delta_p([ab(a+b)])$ satisfying the desired relation in $\mathrm{W}(K_p)$. We distinguish three cases: either $\alpha \neq \beta$ (for instance $\alpha < \beta$), or $\alpha = \beta$ and $a' + b'$ is not divisible by p, or $\alpha = \beta$ and $a' + b'$ is divisible by p.

When $\alpha < \beta$, the definition of δ_p implies

$$\delta_p([a+b]) = \delta_p([a]) \quad \text{and} \quad \delta_p([ab(a+b)]) = \delta_p([b]) \ ,$$

whence the desired relation in $\mathrm{W}(K_p)$.

When $\alpha = \beta$, we set $a + b = p^\gamma c'$, and we first assume that $\alpha = \gamma$. When α is even, δ_p maps everything to 0, and when α is odd, we remember the following relation in $\mathrm{W}(K_p)$:

$$[a'_p] + [b'_p] = [a'_p + b'_p] + [a'_p b'_p (a'_p + b'_p)].$$

When $\alpha = \beta < \gamma$, then $-a'_p b'_p$ is a square in K_p; since the relation $[a'_p] + [-a'_p] = 0$ holds in $\mathrm{W}(K_p)$, we observe that δ_p maps $[a]$ and $[b]$ to opposite elements; thus it suffices to observe that δ_p also maps $[a+b]$ and $[ab(a+b)]$ to opposite elements.

For the group morphism δ_2 the proof is still easier because there are only two cases to distinguish; indeed when the odd prime integer p is replaced with 2, the equality $\alpha = \beta$ implies $\alpha < \gamma$. □

The definition of δ_p in (8.3.4) allows us to calculate the image by δ_p of the class of any $\mathbb{Q}$-quadratic space (M, q) as soon as we know an orthogonal basis $(e_1, e_2, \dots, e_r)$; this image is $\sum_{i=1}^r \delta_p([2q(e_i)])$.

The essential progress about quadratic forms over $\mathbb{Q}$ was achieved by Minkowski and other mathematicians (whose works are already involved in Theorem (2.8.7)), and later more properties were discovered by Milnor [1970] and other mathematicians: see [Milnor, Husemoller 1973], [Scharlau 1985]. Among the new results there is a description of the additive group $\mathrm{W}(\mathbb{Q})$.

(8.3.5) **Theorem.** *The group morphisms $\delta_p : \mathrm{W}(\mathbb{Q}) \to \mathrm{W}(K_p)$ (with p equal to ∞ or to a prime integer ≥ 2) determine a group isomorphism*

$$\Delta \ : \ \mathrm{W}(\mathbb{Q}) \longrightarrow \bigoplus_{p \in \mathcal{P}} \mathrm{W}(K_p) \ .$$

As explained above, $K_\infty = \mathbb{R}$ and $\delta_\infty : \mathrm{W}(\mathbb{Q}) \to \mathbb{Z}$ gives the signature. For a finite p, $K_p = \mathbb{Z}/p\mathbb{Z}$. When p is an odd prime, $\mathrm{W}(K_p)$ is a group of order 4, isomorphic to $\mathbb{Z}/4\mathbb{Z}$ if $p \equiv 3$ modulo 4, isomorphic to $(\mathbb{Z}/2\mathbb{Z})^2$ if $p \equiv 1$ modulo 4 (see (8.ex.11)). When $p = 2$, then $\mathrm{W}(K_2) = \mathrm{WB}(\mathbb{Z}/2\mathbb{Z})$ is a group of order 2. The proof of (8.3.5) is well expounded in the literature; the surjectiveness of Δ can be proved by means of the inductive argument presented in the proof of (8.3.8) below, and the main difficulty lies in the proof of its injectiveness, for which we refer to the above-mentioned books.

(8.3.6) **Example.** It is easy to verify that Δ takes the same value on $[10] - [2] + [15] + [3]$ and $[1] + [1]$; only the elements p of $\{\infty, 2, 3, 5\}$ are involved. Since Δ is injective, the $\mathbb{Q}$-quadratic space $\langle 10, -2, 15, 3 \rangle$ is isomorphic to $\langle 1, 1, 1, -1 \rangle$. In other words the $\mathbb{Q}$-quadratic space (M, q) with an orthogonal basis (b_1, b_2, b_3, b_4) such that $(2q(b_1), 2q(b_2), 2q(b_3), 2q(b_4)) = (1, 1, 1, -1)$ contains an orthogonal basis

(c_1, c_2, c_3, c_4) such that $(2q(c_1), 2q(c_2), 2q(c_3), 2q(c_4) = (10, -2, 15, 3)$. Let us find such a basis $(c_1, \dots, c_4)$. Because of the cancellation theorem (8.1.1) we can begin with any c_1 such that $2q(c_1) = 10$; then the research of an orthogonal c_3 such that $2q(c_3) = 15$ is still easy (easier than the research of c_2), and at last the research of (c_2, c_4) in the plane $(\mathbb{Q}c_1 \oplus \mathbb{Q}c_3)^\perp$ raises no difficulty:

$$c_1 = 3b_1 + b_2 \ , \qquad c_2 = -b_1 + 3b_2 + 2b_3 + 4b_4 \ ,$$
$$c_3 = -2b_1 + 6b_2 + 5b_4 \ , \qquad c_4 = -b_1 + 3b_2 + 3b_3 + 4b_4 \ . \qquad \square$$

Here we are especially interested in the canonical injections of $\mathrm{WB}(\mathbb{Z})$ and $\mathrm{WQ}(\mathbb{Z})$ into $\mathrm{W}(\mathbb{Q})$.

(8.3.7) **Proposition.** *If (M, φ) or (M, q) is a bilinear or quadratic module over $\mathbb{Z}$, for every $p \neq \infty$ the group morphism δ_p maps the Witt class of its $\mathbb{Q}$-extension to 0. In other words, the isomorphism Δ maps it to $(s, 0, 0, 0, \dots)$ if s is its signature.*

Proof. Since the Witt class of a bilinear space (M, φ) is not modified when a metabolic orthogonal summand is added, we can assume that φ is indefinite of odd type. Then from (2.8.11) we deduce that (M, φ) contains an orthogonal basis $(e_1, \dots, e_r)$ such that $\varphi(e_j, e_j) = \pm 1$ for $j = 1, 2, \dots, r$. The conclusion follows immediately because $\delta_p([\pm 1]) = 0$ if $p \neq \infty$. Since $\mathrm{W}(\mathbb{Q})$ stands for both $\mathrm{WQ}(\mathbb{Q})$ and $\mathrm{WB}(\mathbb{Q})$, the morphism $\mathrm{WQ}(\mathbb{Z}) \to \mathrm{W}(\mathbb{Q})$ can be factorized through $\mathrm{WB}(\mathbb{Z})$, and we get the same conclusion for quadratic spaces over $\mathbb{Z}$. $\square$

Now we consider the Witt rings of the localizations of $\mathbb{Z}$ at its maximal ideals; because of (8.3.2) we identify them with subrings of $\mathrm{W}(\mathbb{Q})$. If p is a finite element of $\mathcal{P}$, the localization $\mathbb{Z}_{(p)}$ is the subring of $\mathbb{Q}$ in which all prime elements of $\mathbb{Z}$ are invertible except p. When $p \geq 3$, it is clear that $\mathrm{W}(\mathbb{Z}_{(p)})$ *is the subgroup of* $\mathrm{W}(\mathbb{Q})$ *generated by all* $[a]$ *associated with an integer* a *that is not divisible by* p. When $p = 2$, we must distinguish $\mathrm{WB}(\mathbb{Z}_{(2)})$ and $\mathrm{WQ}(\mathbb{Z}_{(2)})$; as in **2.8**, we can say that $\mathrm{WQ}(\mathbb{Z}_{(2)})$ classifies the bilinear spaces (M, φ) of even type, in other words, such that $\varphi(x, x)$ is divisible by 2 in $\mathbb{Z}_{(2)}$ for all $x \in M$; indeed this implies the existence of a unique quadratic form q such that $\mathrm{b}_q = \varphi$. From (2.6.3) we know that every bilinear space (M, φ) of odd type admits an orthogonal basis; and if (M, φ) has even type, anyhow the orthogonal sum of (M, φ) and any discriminant module admits an orthogonal basis; consequently $\mathrm{WB}(\mathbb{Z}_{(2)})$ *is the subgroup of* $\mathrm{W}(\mathbb{Q})$ *generated by all* $[a]$ *associated with an odd integer* a. Still from (2.6.3) we know that every bilinear space (M, φ) of even type is an orthogonal sum of subspaces of rank 2; in each subspace there is a basis (e_1, e_2) such that $\varphi(e_1, e_1)$ and $\varphi(e_2, e_2)$ are even integers $2a$ and $2c$, whereas $\varphi(e_1, e_2)$ is an odd integer b (because $b^2 - 4ac$ must be invertible in $\mathbb{Z}_{(2)}$); if $ac = 0$, this subspace generated by (e_1, e_2) is hyperbolic, and if $ac \neq 0$, the equalities

$$\varphi(e_1,\ 2ae_2 - be_1) = 0 \quad \text{and} \quad \varphi(2ae_2 - be_1,\ 2ae_2 - be_1) = 2a(4ac - b^2)$$

show that after the extension $\mathbb{Z}_{(2)} \to \mathbb{Q}$ the Witt class of this subspace becomes $[2a] + [2a(4ac - b^2)]$. Consequently $\mathrm{WQ}(\mathbb{Z}_{(2)})$ *is the subgroup of* $\mathrm{W}(\mathbb{Q})$ *generated by all elements* $[2a]+[2a(4ac-b^2)]$ *associated with an odd integer* b *and with nonzero integers* a *and* c.

From the next proposition we shall soon deduce that $\mathrm{WB}(\mathbb{Z})$, as a subring of $\mathrm{W}(\mathbb{Q})$, is the intersection of all subrings $\mathrm{WB}(\mathbb{Z}_{(p)})$, in accordance with (8.3.3).

(8.3.8) **Proposition.** *For every prime integer* $p \geq 2$, $\mathrm{WB}(\mathbb{Z}_{(p)})$ *is the kernel of* $\delta_p : \mathrm{W}(\mathbb{Q}) \to \mathrm{W}(\mathbb{Z}/p\mathbb{Z})$.

Proof. Since δ_p vanishes on $[a]$ if the integer a is not divisible by p, it vanishes on every element of $\mathrm{WB}(\mathbb{Z}_{(p)})$. Conversely let w be an element of $\mathrm{W}(\mathbb{Q})$ such that $\delta_p(w) = 0$, and let ϖ' be the smallest prime integer ≥ 2 such that $\delta_{p'}(w) = 0$ for all $p' \geq \varpi'$. We will prove by induction on ϖ' that w belongs to $\mathrm{WB}(\mathbb{Z}_{(p)})$. If $\varpi' = 2$, then $\Delta(w)$ is the same thing as $\Delta(s\,[1])$ with $s = \delta_\infty(w)$; the injectiveness of Δ implies that $w = s\,[1]$, and that w belongs to $\mathrm{WB}(\mathbb{Z})$, therefore to $\mathrm{WB}(\mathbb{Z}_{(p)})$. Now let us suppose $\varpi' \geq 3$, and let ϖ be the next prime integer smaller than ϖ'; thus $\delta_\varpi(w) \neq 0$, and $\delta_\varpi(w)$ is a sum of terms $[a_\varpi]$ associated with integers a such that $0 < a < \varpi$. For each such integer a there is an integer b satisfying these three conditions: first $-\varpi < b < \varpi$; secondly $a + b \equiv 0$ modulo ϖ; and thirdly b is not divisible by p (the third condition is void if $p > \varpi$); indeed we can set $b = -a$ if a is not divisible by p, and $b = \varpi - a$ in the other case. Thus $[a_\varpi] + \delta_\varpi([b\varpi]) = 0$, and since we can do this for each term $[a_\varpi]$, we have proved the existence of an element $w' \in \mathrm{WB}(\mathbb{Z}_{(p)})$ such that $\delta_{p'}(w + w') = 0$ for all $p' \geq \varpi$. Because of the induction hypothesis we know that $w + w'$ belongs to $\mathrm{WB}(\mathbb{Z}_{(p)})$, and w too. □

If an element $w \in \mathrm{W}(\mathbb{Q})$ belongs to all subrings $\mathrm{WB}(\mathbb{Z}_{(p)})$, then $\Delta(w) = \Delta(s\,[1])$ with $s = \delta_\infty(w)$, and the injectiveness of Δ implies that w belongs to $\mathrm{WB}(\mathbb{Z})$. If w belongs to all subrings $\mathrm{WQ}(\mathbb{Z}_{(p)})$, it is not yet evident that $w \in \mathrm{WQ}(\mathbb{Z})$; to verify it, we need the group morphism $\theta : \mathrm{W}(\mathbb{Q}) \to \mathbb{Z}/8\mathbb{Z}$ which is defined in (8.ex.5). This morphism θ satisfies these two properties: first $\theta([a])$ only depends on the image of a in $\mathbb{Z}/8\mathbb{Z}$ when the integer a is not divisible by 4; secondly $\theta([1]) = 1$ modulo 8. These properties do not determine θ in a unique way; indeed, if δ_2' is the group morphism $\mathrm{W}(\mathbb{Q}) \to \mathbb{Z}/8\mathbb{Z}$ resulting from $\delta_2 : \mathrm{W}(\mathbb{Q}) \to \mathbb{Z}/2\mathbb{Z}$ and from the morphism $\mathbb{Z}/2\mathbb{Z} \to \mathbb{Z}/8\mathbb{Z}$ induced by the multiplication by 4 in $\mathbb{Z}$, then $\theta + \delta_2'$ too satisfies the properties required from θ; but only θ and $\theta + \delta_2'$ satisfy them (see (8.ex.5)). In (8.ex.6) it is proved that $\mathrm{WQ}(\mathbb{Z}_{(2)})$ is the intersection of the kernels of δ_2 and θ. Consequently, if w belongs to all subrings $\mathrm{WQ}(\mathbb{Z}_{(p)})$, we first write $w = s\,[1]$ as above, and since $\theta(w) = 0$ and $\theta([1]) = 1$ modulo 8, we conclude that $s \equiv 0$ modulo 8, and that $w \in \mathrm{WQ}(\mathbb{Z})$ (see (2.8.13)).

(8.3.9) **Example.** It is clear that δ_2 maps $[14]-[6]$ to 0; since $14 \equiv 6$ modulo 8, even θ maps it to 0. Therefore $[14] - [6]$ belongs to the image of $\mathrm{WQ}(\mathbb{Z}_{(2)}) \to \mathrm{W}(\mathbb{Q})$; let us verify it directly. The induction presented in the proof of (8.3.8) leads us to consider $[14]-[6]+[35]$ which is annihilated by δ_7, and then $[14]-[6]+[35]+[15]$ which is annihilated by δ_5 and even by δ_3, and which is consequently equal to $[1]+[1]$. Thus we discover that $[14]-[6] = [1]+[1]-[15]-[35]$ and that $[14]-[6]$ is the image in $\mathrm{W}(\mathbb{Q})$ of the $\mathbb{Z}_{(2)}$-bilinear space (M,φ) provided with an orthogonal basis (b_1,b_2,b_3,b_4) such that $(\varphi(b_1,b_1),\varphi(b_2,b_2),\varphi(b_3,b_3),\varphi(b_4,b_4)) = (1,1,-15,-35)$. This result can be improved because (M,φ) contains metabolic subspaces. Indeed if we set

$$c_1 = 5b_1 + 5b_2 + b_3 + b_4 \quad \text{and} \quad c_2 = b_2 \ ,$$

the plane P generated by c_1 and c_2 is metabolic because c_1 is isotropic and $\varphi(c_1,c_2)$ is invertible in $\mathbb{Z}_{(2)}$. This plane contains two isotropic lines respectively generated by c_1 and $c_1 - 10c_2$; since they are not supplementary in P, this plane is not hyperbolic. We can forget P and focus our attention on $P^\perp$ which is generated by

$$c_3 = 3b_1 + b_3 \quad \text{and} \quad c_4 = 7b_1 + b_4 \ .$$

On one side the Witt class of $\mathbb{Q} \otimes (M,\varphi)$ is $[14] - [6]$ because $\varphi(c_4,c_4) = 14$, $\varphi(2c_3 - 3c_4,\, 2c_3 - 3c_4) = -6 \times 25$, and c_4 is orthogonal to $2c_3 - 3c_4 = -15b_1 + 2b_3 - 3b_4$. On the other side the restriction of φ to $P^\perp$ has even type in accordance with the prediction:

$$\varphi(\lambda c_3 + \mu c_4,\ \lambda c_3 + \mu c_4) \ = \ 2\,(-3\lambda^2 + 21\lambda\mu + 7\mu^2). \qquad \square$$

(8.3.10) **Historical comment.** The description of $\mathrm{W}(\mathbb{Q})$ by means of the group isomorphism Δ in (8.3.5) was first expounded in [Scharlau 1972], but the description of the groups associated with the field $\mathbb{Q}$ has a somewhat longer story, some steps of which are briefly reported here. Let us begin with [Albert 1939] who proved that the subgroup $\mathrm{Br}_2(\mathbb{Q})$ of all elements of $\mathrm{Br}(\mathbb{Q})$ of order 1 or 2 is generated by Brauer classes of quaternion algebras. Consequently every element of $\mathrm{Br}_2(\mathbb{Q})$ is the class of a Clifford algebra $C\ell(M,q)$ such that $\mathrm{QZ}(M,q) = \mathbb{Q}^2$, as explained in [Micali, Villamayor 1970]. Efficient tools for the study of $\mathrm{W}(\mathbb{Q})$ were elaborated already in 1969 and led eventually to a calculation of $\mathrm{W}(\mathbb{Q})$ that was independent of that of Scharlau, and that appeared in [Micali, Villamayor 1971] §5.5, and in [Larotonda, Micali, Villamayor 1973]. These works were first concerned with the groups $\mathcal{H}(\mathbb{Q})$ and $\mathcal{H}_0(\mathbb{Q})$ that here are defined farther in **8.6**; but they also produced an elaborate formula for the group $\mathrm{W}(\mathbb{Q})$, which involved $\mathrm{Br}_2(\mathbb{Q})$ among other objects. In a notice added to [Larotonda, Micali, Villamayor 1973], it is explained that the comparison of this formula and that of Scharlau could be performed in different ways; either one accepts both formulas and then discovers again Albert's theorem about $\mathrm{Br}_2(\mathbb{Q})$, together with a set of quaternion algebras generating $\mathrm{Br}_2(\mathbb{Q})$; or one accepts Albert's theorem and proves the equivalence of both formulas describing $\mathrm{W}(\mathbb{Q})$. See [Kientega 1986] for the details of the calculations needed in this comparison. The above-mentioned notice in [Larotonda, Micali, Villamayor 1973] also explained the calculation of the Witt group of a "field of numbers" (a finite dimensional field extension of $\mathbb{Q}$). This calculation was achieved by [Knebusch, Scharlau 1971], and later independently by [Laborde 1974]. In [DeMeyer, Harrison, Miranda 1989] the multiplication in $\mathrm{W}(\mathbb{Q})$ is also taken into account, but this work requires more arithmetical knowledge (for instance Hilbert symbols, which anyhow are already involved in the proof of (2.8.7) expounded in [Serre]).

8.4 Quadratic forms in characteristic 2

Let K be a field of characteristic 2, Φ the *Frobenius morphism* $K \to K$ defined by $\lambda \longmapsto \lambda^2$, and $K^2 = \mathrm{Im}(\Phi)$; since K^2 is a subfield of K, we can treat K as a vector space over K^2. Later Φ will often be considered as an isomorphism $K \to K^2$, and this leads to the concept of Φ-linear mapping from a vector space over K into a vector space over K^2: a Φ-linear mapping f must satisfy the condition $f(\lambda x) = \Phi(\lambda)\, f(x)$ for all λ and x.

If V is a vector space over K, a quadratic form $q : V \to K$ is said to be *additive* if $\mathrm{b}_q = 0$, or equivalently $q(x+y) = q(x) + q(y)$ for all x and y in V. Beside this condition, an additive quadratic form must also satisfy the condition $q(\lambda x) = \Phi(\lambda) q(x)$, and both conditions together mean that q is a Φ-linear form on V. Therefore the image $q(V)$ is a K^2-subspace of K, denoted by $\mathrm{Im}(q)$. At last, let us remember that q is said to be *anisotropic* if the equality $q(x) = 0$ implies $x = 0$.

(8.4.1) **Proposition.** *Let (V, q) be a quadratic module over the field K of characteristic 2; let V_1 be a subspace supplementary to* $\mathrm{Ker}(\mathrm{b}_q)$ *in V, V_2 a subspace supplementary to* $\mathrm{Ker}(q)$ *in* $\mathrm{Ker}(\mathrm{b}_q)$*, and let q_1, q_2 and q_3 be the restrictions of q to V_1, V_2 and $V_3 =$* $\mathrm{Ker}(q)$*. Then (V, q) is the orthogonal sum of the quadratic submodules (V_1, q_1), (V_2, q_2), (V_3, q_3); moreover q_1 is weakly nondegenerate (consequently nondegenerate if V_1 has finite dimension), q_2 is an additive and anisotropic quadratic form, whereas q_3 is the null form. Besides, the isomorphy classes of (V_2, q_2) and $(V_1, q_1) \perp (V_2, q_2)$ only depend on (V, q), and not on the choices of V_1 and V_2.*

Proof. It is clear that the subspaces V_1, V_2 and V_3 are pairwise orthogonal and that q_3 is the null form. It is clear that the restriction of q to $\mathrm{Ker}(\mathrm{b}_q)$ is additive; and since the equality $q(x) = 0$ with x in $\mathrm{Ker}(\mathrm{b}_q)$ means that x belongs to $\mathrm{Ker}(q)$, q_2 is anisotropic. The restriction of $\mathrm{d}_q : V \longrightarrow V^*$ to V_1 is injective, and $\mathrm{d}_q(x)$ always vanishes on $\mathrm{Ker}(\mathrm{b}_q)$; consequently d_q induces an injection $V_1 \to V_1^*$; in other words, q_1 is weakly nondegenerate. At last, q induces quadratic forms on the quotients $V/\mathrm{Ker}(q)$ and $\mathrm{Ker}(\mathrm{b}_q)/\mathrm{Ker}(q)$; the former quotient (with the quadratic form induced by q) is isomorphic to $(V_1, q_1) \perp (V_2, q_2)$, and the latter quotient is isomorphic to (V_2, q_2). □

Let us pay more attention to additive quadratic forms, since they are quite different from the quadratic forms studied up to now. When q is an additive quadratic form on V, then (8.4.1) decomposes (V, q) into an orthogonal sum $(V_2, q_2) \perp (V_3, q_3)$, and (V_2, q_2), which is well defined up to isomorphy, is called *the anisotropic space derived from* (V, q). Since q is a Φ-linear mapping, it induces a bijection from $V/\mathrm{Ker}(q)$ onto $\mathrm{Im}(q)$, as stated in the next corollary.

(8.4.2) **Corollary.** *An additive quadratic form $q : V \to K$ induces a Φ-linear isomorphism from the anisotropic space (V_2, q_2) derived from (V, q) onto* $\mathrm{Im}(q)$*. A* *family of elements of* V_2 *is linearly independent over* K *if and only if its image by* q *is linearly independent over* K^2. *Moreover,*

$$\dim_K(V) \; = \; \dim_K(\mathrm{Ker}(q)) + \dim_{K^2}(\mathrm{Im}(q)) \; .$$

(8.4.3) **Corollary.** *The upper bound of the dimensions of the anisotropic spaces of additive quadratic forms is the dimension of* K *over* K^2.

Proof. The dimension of V_2 over K is the dimension of $\mathrm{Im}(q)$ over K^2, which cannot exceed $\dim_{K^2}(K)$. Now let $\mathcal{B}$ be a (perhaps infinite) basis of K over K^2, and $K^{(\mathcal{B})}$ the vector space over K with basis $(e_b)_{b\in\mathcal{B}}$; every element of $K^{(\mathcal{B})}$ is a finite sum $\sum_b \lambda_b e_b$ in which all scalars λ_b vanish except a finite number; and if we map every such a sum to $\sum_b \lambda_b^2 b$ in K, we get an additive quadratic form that is anisotropic, because all b in $\mathcal{B}$ are linearly independent over K^2. □

(8.4.4) **Examples.** A field F of characteristic 2 is said to be *perfect* if $F^2 = F$. This condition is always fulfilled when F is finite. When F is perfect, the field $F(t)$ of rational functions has dimension 2 over $F(t)^2$, because $(1,t)$ is a basis of $F(t)$ over $F(t)^2$. If we consider the field $F(t_1,t_2)$ with two indeterminates, it has dimension 4 over $F(t_1,t_2)^2$, because $(1,t_1,t_2,\ t_1t_2)$ is a basis of $F(t_1,t_2)$ over $F(t_1,t_2)^2$. More generally the field $F(t_1,t_2,\dots,t_n)$ has dimension 2^n over $F(t_1,t_2,\dots,t_n)^2$.

(8.4.5) **Theorem.** *Two quadratic modules* (V,q) *and* (V',q') *with additive quadratic forms* q *and* q' *are isomorphic if and only if*

$$\dim_K(\mathrm{Ker}(q)) = \dim_K(\mathrm{Ker}(q')) \quad \text{and} \quad \mathrm{Im}(q) = \mathrm{Im}(q') \; .$$

Proof. It is clear that these equalities are true when there is an isomorphism $(V,q) \to (V',q')$. Conversely let us suppose that they are true. Let V_2 be a subspace supplementary to $V_3 = \mathrm{Ker}(q)$ in V, so that (V,q) is the orthogonal sum of (V_2,q_2) and (V_3,q_3), and similarly $(V',q') = (V_2',q_2') \perp (V_3',q_3')$. Since V_3 and V_3' have the same dimension, there is a K-linear isomorphism $f_3 : V_3 \to V_3'$. Moreover q induces a Φ-linear bijection $\varphi : V_2 \to \mathrm{Im}(q)$, and q' a Φ-linear bijection $\varphi' : V_2' \to \mathrm{Im}(q')$. Since $\mathrm{Im}(q) = \mathrm{Im}(q')$, we get a K-linear bijection $f_2 : V_2 \to V_2'$ if we set $f_2 = \varphi'^{-1} \circ \varphi$. Obviously the couple (f_2,f_3) determines an isomorphism from (V,q) onto (V',q'). □

Now let us examine how additive quadratic forms behave in case of orthogonal sums or tensor products. As explained in **2.4**, a tensor product $q \otimes q'$ of quadratic forms is the same thing as $\mathrm{b}_q \otimes q'$ or $q \otimes \mathrm{b}_{q'}$; consequently $q \otimes q'$ is the null form whenever q or q' is an additive quadratic form.

(8.4.6) **Proposition.** *Let* (V,q) *and* (V',q') *be two quadratic modules in which* q *and* q' *are additive quadratic forms, and let* (V'',q'') *be their orthogonal sum. Then*

$$\begin{aligned} \dim_K(\mathrm{Ker}(q'')) &= \dim_K(\mathrm{Ker}(q)) + \dim_K(\mathrm{Ker}(q')) + \dim_{K^2}(\mathrm{Im}(q) \cap \mathrm{Im}(q')) \\ \textit{and} \quad \mathrm{Im}(q'') &= \mathrm{Im}(q) + \mathrm{Im}(q') \; . \end{aligned}$$

Proof. It is trivial that $\mathrm{Im}(q'')$ is the sum of $\mathrm{Im}(q)$ and $\mathrm{Im}(q')$ in K, in general not a direct sum. Let us set, according to the notation of (8.4.1),

$$(V,q) = (V_2,q_2) \perp (V_3,q_3) \quad \text{and} \quad (V',q') = (V_2',q_2') \perp (V_3',q_3') \;;$$

we can treat q'' as a quadratic form over $V_2 \oplus V_2' \oplus V_3 \oplus V_3'$; let us denote by q_2'' and q_3'' its restrictions to $V_2 \oplus V_2'$ and $V_3 \oplus V_3'$; of course q_2'' is not in general anisotropic; but since q_3'' is a null form, the kernel of q'' is the direct sum of $V_3 \oplus V_3'$ and the kernel of q_2''. Let us consider the exact sequence

$$0 \longrightarrow \mathrm{Im}(q) \cap \mathrm{Im}(q') \longrightarrow \mathrm{Im}(q) \oplus \mathrm{Im}(q') \longrightarrow \mathrm{Im}(q) + \mathrm{Im}(q') \longrightarrow 0 \;;$$

the second arrow maps x to $(x,-x)$, and the third arrow maps (x,y) to $x+y$. Since q (resp. q') induces a bijection $V_2 \to \mathrm{Im}(q)$ (resp. $V_2' \to \mathrm{Im}(q')$), this exact sequence yields another exact sequence of additive groups, in which the third arrow represents the additive quadratic form q_2'' :

$$0 \longrightarrow \mathrm{Im}(q) \cap \mathrm{Im}(q') \longrightarrow V_2 \oplus V_2' \longrightarrow \mathrm{Im}(q'') \longrightarrow 0 \;;$$

in this exact sequence the second arrow is Φ^{-1}-linear; it shows that the dimension of $\mathrm{Im}(q) \cap \mathrm{Im}(q')$ over K^2 is equal to the dimension of $\mathrm{Ker}(q_2'')$ over K. □

Now we consider again quadratic forms that may have nondegenerate components; when are two such quadratic forms isomorphic? This question is not trivial because in the orthogonal decomposition presented in (8.4.1), the isomorphy class of (V_1,q_1) in general depends on the choice of V_1. Nevertheless b_q induces a bilinear form on $V/\mathrm{Ker}(\mathrm{b}_q)$; consequently the isomorphy class of the bilinear module (V_1,b_{q_1}) does not depend on the choice of V_1.

(8.4.7) **Proposition.** *Let (V,q) and (V',q') be two quadratic modules over the field K of characteristic 2, and let us decompose them as explained in (8.4.1):*

$$\begin{aligned} (V,q) &= (V_1,q_1) \perp (V_2,q_2) \perp (V_3,q_3) \\ \textit{and} \quad (V',q') &= (V_1',q_1') \perp (V_2',q_2') \perp (V_3',q_3') \;; \end{aligned}$$

(V,q) and (V',q') are isomorphic if and only if these three conditions are fulfilled:

$$\begin{aligned} \dim_K(V_3) &= \dim_K(V_3') \;; \\ \mathrm{Im}(q_2) &= \mathrm{Im}(q_2') \;; \end{aligned}$$

there exists an isomorphism of bilinear modules $f_1 : (V_1,\mathrm{b}_{q_1}) \to (V_1',\mathrm{b}_{q_1'})$ such that

$$\forall x_1 \in V_1 \,, \quad q_1'(f_1(x_1)) - q_1(x_1) \;\in\; \mathrm{Im}(q_2) \,.$$

Proof. From (8.4.5) we already know that the first two conditions mean that $(V_2,q_2) \perp (V_3,q_3)$ and $(V_2',q_2') \perp (V_3',q_3')$ are isomorphic. When there is an isomorphism $f : (V,q) \to (V',q')$, then f maps every $x_1 \in V_1$ to a sum $x_1' + x_2' + x_3'$

in which $x'_j \in V'_j$ for $j = 1, 2, 3$; since $q_1(x_1) = q'_1(x'_1) + q'_2(x_2)$, it follows that f determines a bijection $f_1 : V_1 \to V'_1$ satisfying the condition written at the end of (8.4.7); and it is clear that f_1 must be an isomorphism of bilinear modules. Conversely let us suppose that all the conditions in (8.4.7) are fulfilled, and let $f_3 : V_3 \to V'_3$ and $f_2 : V_2 \to V'_2$ be the K-linear bijections obtained in the proof of (8.4.5). Let us consider the composition g of these two mappings:

$$g \; : \; V_1 \longrightarrow \mathrm{Im}(q_2) = \mathrm{Im}(q'_2) \longrightarrow V'_2 \; ;$$

the first arrow is defined by $x_1 \longmapsto q'_1(f_1(x_1)) - q_1(x_1)$; since f_1 is an isomorphism of bilinear modules, it is easy to verify that this first arrow is Φ-linear; the second arrow is the Φ^{-1}-linear bijection reciprocal to $V'_2 \to \mathrm{Im}(q'_2)$; thus g is a K-linear mapping $g : V_1 \to V'_2$. We get an isomorphism from (V, q) onto (V', q') if we map every $x_1 + x_2 + x_3 \in V$ to $(f_1(x_1),\ f_2(x_2) - g(x_1),\ f_3(x_3))$. □

8.5 Clifford algebras in characteristic 2

Let (V, q) be a finite dimensional quadratic module over a field K of characteristic 2. Let us recall the orthogonal decomposition of (8.4.1):

$$(V, q) \;=\; (V_1, q_1) \perp (V_2, q_2) \perp (V_3, q_3) \; ;$$

accordingly the Clifford algebra can be decomposed in this way:

$$C\ell(V, q) \;=\; C\ell(V_1, q_1) \otimes C\ell(V_2, q_2) \otimes \bigwedge(V_3).$$

The first factor $C\ell(V_1, q_1)$ is a graded Azumaya algebra; the last factor is a well-known exterior algebra, which here is commutative (in the ordinary sense) since the field K has characteristic 2. The next theorem is essentially concerned with the structure of $C\ell(V_2, q_2)$, the Clifford algebra of an additive anisotropic quadratic form. It is already clear that $C\ell(V_2, q_2)$ is also a commutative algebra; consequently, if the parity gradings are forgotten, $C\ell(V, q)$ is isomorphic to the Clifford algebra of the quadratic space $K' \otimes (V_1, q_1)$ over the commutative ring $K' = C\ell(V_2, q_2) \otimes \bigwedge(V_3)$.

Let us first consider a field K of any positive characteristic p, and $K \to L$ an algebraic field extension of K (such that every element of L generates a subfield of finite dimension over K); we say that L is a *purely inseparable extension* of K if for every $x \in L$ there exists an integer $n \geq 0$ such that x^{p^n} is in K, or equivalently, if the minimal polynomial of every element of L over K is equal to $X^{p^n} - a$ for some integer n and some $a \in K$.

A purely inseparable extension of a purely inseparable extension of K is still a purely inseparable extension of K, because the equalities $x^{p^m} = y$ and $y^{p^n} = a \in K$ imply $x^{p^{m+n}} = a$.

(8.5.1) **Example.** Let K be a field of characteristic 2, and $A = K[X]/(X^2 - \beta X + \gamma)$ the quotient of $K[X]$ by the ideal generated by the polynomial $X^2 - \beta X + \gamma$. The discriminant of this polynomial is β^2, and when $\beta \neq 0$, we get a quadratic extension of K (see **3.4**) that is separable over K (see (6.5.6)). When $\beta = 0$, then A is an inseparable extension, that contains nilpotent elements when γ belongs to the subfield K^2; but when γ is not in K^2, then A is a field, and it is a purely inseparable extension of K; indeed, if x is the image of X in A, for all λ and μ in K we can write $(\lambda + \mu x)^2 = \lambda^2 + \gamma\mu^2 \in K$.

The next theorem is especially interesting when the quadratic form q is additive and anisotropic; but it works as well with any additive quadratic form, whether anisotropic or not.

(8.5.2) **Theorem.** *Let K be a field of characteristic 2, V a vector space of finite nonzero dimension r over K, and q an additive quadratic form on V. There exists a purely inseparable extension L of K, of dimension 2^s over K for some $s \in \{0, 1, 2, \dots, r\}$, such that $C\ell_K(V, q)$ is isomorphic to $\bigwedge_L(L^{r-s})$ as a nongraded algebra.*

Proof. Let $(e_1, e_2, \dots, e_r)$ be a basis of V. After a suitable permutation of the vectors of this basis, we can manage to make these assertions become true for some s such that $0 \leq s \leq r$:

- $q(e_1)$ does not belong to K^2, and consequently $K[X]/(X^2 - q(e_1))$ is a purely inseparable extension K_1 of dimension 2 over K;
- $q(e_2)$ does not belong to ${K_1}^2$, and consequently $K_1[X]/(X^2 - q(e_2))$ is a purely inseparable extension K_2 of dimension 2 over K_1, therefore of dimension 4 over K;
- $q(e_3)$ does not belong to ${K_2}^2$, and consequently $K_2[X]/(X^2 - q(e_3))$ is a purely inseparable extension K_3 of dimension 2 over K_2 , therefore of dimension 8 over K;

and so forth ... up to

- $q(e_s)$ that does not belong to K_{s-1}^2, and consequently determines a purely inseparable extension K_s of dimension 2^s over K;

and then

$q(e_{s+1}), \dots, q(e_r)$ all belong to ${K_s}^2$.

For $j = 1, 2, \dots, s$, let V_j be the subspace spanned by $(e_1, e_2, \dots, e_j)$, and q_j the restriction of q to V_j; we also set $V_0 = 0$ and $K_0 = K$. By induction on j we prove that $C\ell(V_j, q_j)$ is isomorphic to K_j. This is clear if $j = 0$. Let us suppose that $C\ell(V_j, q_j)$ is isomorphic to K_j, and let f be the restriction of q to the subspace U generated by e_{j+1}; since $C\ell(V_{j+1}, q_{j+1})$ is isomorphic to $C\ell(V_j, q_j) \otimes C\ell(U, f)$, it is also isomorphic to $K_j \otimes_K C\ell_K(U, f)$ or equivalently $C\ell_{K_j}(K_j \otimes_K (U, f))$. The tensor algebra of $K_j \otimes_K U$ (considered as a K_j-module) is isomorphic to the ring of polynomials $K_j[X]$, and the Clifford algebra of $K_j \otimes_K (U,f)$ is isomorphic to the quotient of $K_j[X]$ by the ideal generated by $X^2 - q(e_{j+1})$; this is exactly K_{j+1}. Consequently $C\ell(V_{j+1}, q_{j+1})$ is isomorphic to K_{j+1}, and finally $C\ell(V_s, q_s)$ is isomorphic to K_s.

Now let V' be the subspace spanned by $(e_{s+1}, \dots, e_r)$, and q' the restriction of q to V'. By a similar argument we realize that $C\ell(V,q)$ is isomorphic to $K_s \otimes_K C\ell_K(V', q')$ or equivalently $C\ell_{K_s}(K_s \otimes_K (V', q'))$. We want to prove that this is isomorphic to $\bigwedge_{K_s}({K_s}^{r-s})$; it suffices to prove it when $s = 0$, since this proof remains valid when K and r are replaced respectively with K_s and $r - s$.

Therefore we suppose that $q(e_1)$, $q(e_2)$, ..., $q(e_r)$ all belong to K^2, and we prove that $C\ell(V,q)$ is isomorphic to $\bigwedge(K^r)$. Indeed let us write $q(e_j) = {\nu_j}^2$ with $\nu_j \in K$ for $j = 1, 2, \dots, r$; thus the equality $(e_j - \nu_j)^2 = 0$ holds in $C\ell(V,q)$ for $j = 1, 2, \dots, r$. These nilpotent elements $e_j - \nu_j$ generate $C\ell(V,q)$ as an algebra (with unit), and the universal property of $\bigwedge(K^r)$ implies the existence of a surjective algebra morphism $\bigwedge(K^r) \to C\ell(V,q)$. It is an isomorphism because both algebras have dimension 2^r over K. □

(8.5.3) **Example.** Let V be a vector space of dimension 4 over the field of rational functions $K = F(t_1, t_2)$ (see (8.4.4)), let (e_1, e_2, e_3, e_4) be a basis of V, and $q : V \to K$ the quadratic form defined by

$$q(\lambda_1 e_1 + \lambda_2 e_2 + \lambda_3 e_3 + \lambda_4 e_4) \;=\; t_1 {\lambda_1}^2 + t_2 {\lambda_2}^2 + {\lambda_3}^2 + t_1 t_2 {\lambda_4}^2 \;;$$

q is an additive and anisotropic quadratic form. The subalgebra generated by e_1 and e_2 in $C\ell(V,q)$ is a purely inseparable extension L of dimension 4 over K; it is isomorphic to $F(\sqrt{t_1}, \sqrt{t_2})$, and the images of $\sqrt{t_1}$ and $\sqrt{t_2}$ in $C\ell(V,q)$ are e_1 and e_2. The Frobenius morphism of L maps L bijectively onto $L^2 = K$, because F is assumed to be perfect (as in (8.4.4)). Obviously $q(e_3)$ and $q(e_4)$ belong to L^2, whence there exist two nilpotent elements $e_3 - 1$ and $e_4 - e_1 e_2$ in $C\ell(V,q)$. Therefore $C\ell(V,q)$ is isomorphic to $\bigwedge_L(L^2)$.

From the above theorem we can derive some corollaries stating that all idempotents in some Clifford algebras are trivial.

(8.5.4) **Corollary.** *The hypotheses being the same as in* (8.5.2), *every idempotent element of* $C\ell(V,q)$ *is equal to* 0 *or* 1.

Proof. Because of (8.5.2), it suffices to prove that in the exterior algebra of any vector space W over any field every idempotent element ε is equal to 0 or 1. Indeed the component of ε in $\bigwedge^0(W)$ must be an idempotent element of the basic field, consequently 0 or 1. When it is 0, and yet $\varepsilon \neq 0$, there is a positive integer k such that ε belongs to $\bigwedge^{\geq k}(W)$ but not to $\bigwedge^{>k}(W)$. Since ε^2 belongs to $\bigwedge^{\geq 2k}(W)$, and $\varepsilon^2 = \varepsilon$, there is a contradiction; therefore $\varepsilon = 0$. And when the scalar component of ε is 1, then $1 - \varepsilon$ is idempotent too, and the same argument proves that $1 - \varepsilon = 0$. □

(8.5.5) **Corollary.** *Let K be a local ring with maximal ideal $\mathfrak{m}$, such that $K/\mathfrak{m}$ is a field of characteristic 2. If M is a free K-module of finite rank, and q a quadratic form $M \rightarrow K$ that induces an additive quadratic form on $(K/\mathfrak{m}) \otimes M$, then the only idempotents of $\mathrm{C}\ell(M, q)$ are 0 and 1.*

Proof. More generally we prove this assertion: let C be an algebra over a local ring K, and let us suppose that C is a free K-module of finite rank; if the only idempotents of $(K/\mathfrak{m}) \otimes C$ are 0 and 1, then the same property is true for C. Indeed let $(e_1, e_2, \dots, e_n)$ be a basis of the free module C. If $x = \sum_j \lambda_j e_j$ is an idempotent of C, we know that the image of x in $(K/\mathfrak{m}) \otimes C$ is 0 or 1; if it is 1, we replace x with the idempotent $1 - x$, so that the image of x is again 0. Since the family (e_j) yields a basis of $(K/\mathfrak{m}) \otimes C$ over $K/\mathfrak{m}$, this implies that all coefficients λ_j belong to $\mathfrak{m}$. If J is the ideal generated by all these λ_j, the equality $x^2 = x$ implies $J^2 = J$, whence $\mathfrak{m}J = J$ (since $J \subset \mathfrak{m}$), and finally $J = 0$ because of Nakayama's lemma (1.12.1). It follows that $x = 0$. □

(8.5.6) **Corollary.** *Let K be a commutative ring such that* $\mathrm{Rad}(K)$ *contains* 2, *and such that* $\mathrm{Spec}(K)$ *is connected, and let M be a finitely generated projective K-module. If q is a quadratic form $M \rightarrow K$ such that* $\mathrm{b}_q(x, y)$ *belongs to* $\mathrm{Rad}(K)$ *for all x and y in M, then the only idempotents of* $\mathrm{C}\ell(M, q)$ *are* 0 *and* 1.

Proof. For every prime ideal $\mathfrak{p} \in \mathrm{Spec}(K)$, the residue field $F_\mathfrak{p}$ has characteristic 2, and q induces an additive quadratic form on $F_\mathfrak{p} \otimes (M, q)$. Consequently $F_\mathfrak{p} \otimes \mathrm{C}\ell(M, q)$ has no other idempotents than 0 or 1 (see (8.5.4)). Because of (8.5.5), the algebra $\mathrm{C}\ell(M, q)_\mathfrak{p}$ over the local ring $K_\mathfrak{p}$ contains no other idempotents than 0 or 1. Thus we are led to prove this assertion: if C is a K-algebra such that for every prime ideal $\mathfrak{p}$ there is no idempotent in $C_\mathfrak{p}$ other than 0 or 1, then the same is true for C itself, provided that $\mathrm{Spec}(K)$ is connected . Indeed let x be an idempotent of C, and U (resp. V) the subset of all $\mathfrak{p} \in \mathrm{Spec}(K)$ such that the image of x in $C_\mathfrak{p}$ is 1 (resp. 0). We know that $\mathrm{Spec}(K)$ is the disjoint union of U and V, and that these subsets are closed; indeed $U = \mathcal{V}(\mathfrak{a})$ (resp. $V = \mathcal{V}(\mathfrak{b})$) if $\mathfrak{a}$ (resp. $\mathfrak{b}$) is the ideal of all $\lambda \in K$ such that $\lambda x = 0$ (resp. $\lambda(1 - x) = 0$). Consequently V or U is empty, and x is equal to 0 or 1. □

Remark. In (8.5.6) we may replace $\mathrm{Rad}(K)$ with the Jacobson radical $\mathrm{J}(K)$ which is the intersection of all maximal ideals, and also the subset of all $\lambda \in K$ such that $1 + \lambda\mu$ is invertible for all $\mu \in K$. If we use $\mathrm{J}(K)$, we must replace $\mathrm{Spec}(K)$ with the subset $\mathrm{Max}(K)$ of all maximal ideals. On one side $\mathrm{Rad}(K)$ may be much smaller than $\mathrm{J}(K)$, on the other side $\mathrm{Spec}(K)$ may be connected even when $\mathrm{Max}(K)$ is not connected (as it happens for instance when K is the intersection of the localized rings $\mathbb{Z}_{(2)}$ and $\mathbb{Z}_{(3)}$ in $\mathbb{Q}$).

8.6 The group of classes of Clifford algebras

When (M,q) is a quadratic space, (3.7.8) says that the Brauer class of $C\ell(M,q)$ only depends on the Witt class of (M,q), and that there is a group morphism

$$\mathrm{WQ}(K) \longrightarrow \mathrm{Br}^g(K)\ , \quad [(M,q)] \longmapsto [C\ell(M,q)].$$

By definition the group of classes of Clifford algebras over K is the image $\mathcal{H}(K)$ of this morphism. Remember that $\mathrm{WQ}(K)$ is even a ring (sometimes without unit element); since very soon we speculate about an operation in $\mathcal{H}(K)$ that might correspond to the multiplication in $\mathrm{WQ}(K)$, here *we shall use additive notation for the group* $\mathrm{Br}^g(K)$, *and also for the group* $\mathrm{Q}^g(K)$ which is related to it by a surjective canonical morphism $\mathrm{Br}^g(K) \to \mathrm{Q}^g(K)$ defined by $[A] \longmapsto [\mathrm{QZ}(A)]$ (see (3.5.19)). It is worth observing that, when we proved the surjectiveness of this morphism just after (3.8.14), as a matter of fact we proved the surjectiveness of its restriction $\mathcal{H}(K) \to \mathrm{Q}^g(K)$.

From the definition of $\mathcal{H}(K)$ it follows that $\mathcal{H}$ is a functor from the category $\mathcal{C}om(\mathbb{Z})$ toward the category $\mathcal{M}od(\mathbb{Z})$. Every ring extension $K \to L$ determines a group morphism $\mathrm{Br}^g(K) \to \mathrm{Br}^g(L)$, and by restriction a group morphism $\mathcal{H}(K) \to \mathcal{H}(L)$, because $C\ell_L(L \otimes_K (M,q)) = L \otimes_K C\ell_K(M,q)$ for every quadratic space (M,q) over K. Besides, it is also clear that the group morphisms $\mathrm{WQ}(K) \to \mathcal{H}(K)$ and all the group morphisms involved in the exact sequences (3.5.19) and (3.4.13) determine morphisms between the functors WQ, $\mathcal{H}$, Br^g, Br, Q^g, Q, Ip', which are all functors from $\mathcal{C}om(\mathbb{Z})$ toward $\mathcal{M}od(\mathbb{Z})$.

The group $\mathcal{H}(K)$ allows us to formalize some results obtained in the previous chapters, especially in **3.8**. For instance the formulas presented below in (8.6.4) and (8.6.6) are essentially formalizations of (3.8.6) and (3.8.13). As for (3.8.15), now it means that $\mathcal{H}(K)$ is contained in the subgroup of all elements of $\mathrm{Br}^g(K)$ of order 1, 2, 4 or 8; anyhow this property also appears below as a corollary of (8.6.4); and yet a third proof is proposed in (8.ex.15).

Let us recall the essential facts that underlie the above definition of $\mathcal{H}(K)$. A graded Azumaya algebra S is said to be *trivial* if it is isomorphic to a graded algebra $\mathrm{End}(P)$ derived from some graded finitely generated and faithful projective module $P = P_0 \oplus P_1$. Two graded Azumaya algebras A and A' are said to be *equivalent* if there exist trivial algebras S and S' such that $A \hat{\otimes} S$ and $A' \hat{\otimes} S'$ are isomorphic. Since the quadratic extension $\mathrm{QZ}(S)$ is trivial, the tensor products $A \hat{\otimes} S$ and $A \otimes S$ are isomorphic (see (3.8.6)); consequently we get the same definition if we require $A \otimes S$ and $A' \otimes S'$ to be isomorphic. Since the tensor product (twisted or not) of two trivial algebras is still trivial, we have actually defined and equivalence relation for graded Azumaya algebras. The set of equivalence classes is a group because $A \hat{\otimes} A^{to}$ is always a trivial algebra. It is worth observing that *every algebra that is equivalent to a trivial algebra, is also trivial;* this is not at all evident, it is a consequence of (6.7.12): if A is equivalent to a trivial S, it is also equivalent to K, and (6.7.12)(c) or (6.7.12)(d) shows that it is isomorphic to some $\mathrm{End}_K(P)$.

Thus it suffices to remember that a Clifford algebra has a trivial class in $\mathcal{H}(K)$ if and only if it isomorphic to some trivial algebra S, and that the class of a twisted tensor product of Clifford algebras (that is the Clifford algebra of an orthogonal sum) is the sum of the classes of the factors. Besides, the class of $C\ell(M,-q)$ is the opposite of the class of $C\ell(M,q)$, because $C\ell(M,-q)$ is isomorphic to $C\ell(M,q)^{to}$.

A first conjecture

The group $\mathcal{H}(K)$ suggests two conjectures, which here are proved only when K is a local ring.

(8.6.1) **Conjecture.** *If the Clifford algebra of the quadratic space (M,q) has a trivial class, then the same is true for the Clifford algebra of $(M,q)\otimes(N,\varphi)$ whatever the bilinear space (N,φ) may be.*

If this conjecture is true for the ring K, the kernel of the group morphism $\mathrm{WQ}(K)\to\mathcal{H}(K)$ is an ideal, and therefore $\mathcal{H}(K)$ inherits a structure of ring (without unit element when 2 is not invertible in K).

Besides, the classes in $\mathrm{WB}(K)$ of all bilinear spaces (N,φ) such that $C\ell((M,q)\otimes(N,\varphi))$ has a trivial class for all quadratic spaces (M,q), constitute an ideal, and the quotient of $\mathrm{WB}(K)$ by this ideal is a ring $\mathcal{H}_B(K)$ (with unit element) that has the same relations with the ring $\mathcal{H}(K)$ as $\mathrm{WB}(K)$ with $\mathrm{WQ}(K)$: first the ring $\mathcal{H}_B(K)$ acts in the additive group $\mathcal{H}(K)$, secondly there is a ring morphism $h:\mathcal{H}(K)\to\mathcal{H}_B(K)$ that is compatible with the action of the latter in the former (in other words, $xy=h(x)y$ for all $x,\ y\in\mathcal{H}(K)$), and thirdly, when 2 is invertible in K, this ring morphism is an isomorphism that allows us to identify $\mathcal{H}_B(K)$ with $\mathcal{H}(K)$.

The next lemma prepares a partial proof of (8.6.1).

(8.6.2) **Lemma.** *Let D be a discriminant module. If the Clifford algebra of (M,q) has a trivial class, then the same is true for the Clifford algebra of $D\otimes(M,q)$.*

Proof. The Clifford algebra of $D\otimes(M,q)$ is isomorphic to $C\ell(M,q)_D$ (see (3.8.7)). Since $C\ell(M,q)$ is assumed to be isomorphic to some $\mathrm{End}(P)$, it suffices to prove that $\mathrm{End}(P)_D$ is isomorphic to some $\mathrm{End}(Q)$. This is true if we set $Q_0=P_0$ and $Q_1=D\otimes P_1$; indeed we get an isomorphism $\mathrm{End}(P)_D\to\mathrm{End}(Q)$ if we map every even $f\in\mathrm{End}_0(P)$ to the endomorphism g of Q defined in this way (for all $d\in D$, all $x\in P_0$ and all $y\in P_1$):

$$g(x)=f(x)\quad\text{and}\quad g(d\otimes y)=d\otimes f(y)\ ,$$

and if we map $d'\otimes f$ for every $d'\in D$ and every odd $f\in\mathrm{End}_1(P)$ to this endomorphism g of Q :

$$g(x)=d'\otimes f(x)\quad\text{and}\quad g(d\otimes y)=(dd')\,f(y)\ .$$

□

Proof of (8.6.1) *when* K *is a local ring.* If (N, φ) is the orthogonal sum of two bilinear spaces (N_1, φ_1) and (N_2, φ_2), then

$$C\ell((M,q) \otimes (N,\varphi)) \cong C\ell((M,q) \otimes (N_1,\varphi_1)) \hat{\otimes} C\ell((M,q) \otimes (N_2,\varphi_2)) ;$$

therefore if (8.6.1) is true for (N_1, φ_1) and (N_2, φ_2), it is also true for (N, φ). Yet we must also realize that (8.6.1) is true for (N_1, φ_1) if it is already established for (N, φ) and (N_2, φ_2). Now Lemma (8.6.2) states that (8.6.1) is true for all bilinear spaces of rank 1; therefore (8.6.1) is true for (N, φ) whenever it admits an orthogonal basis. And if (N, φ) does not, from (2.6.3) we know that anyhow the orthogonal sum of (N, φ) and any bilinear space D of rank 1 admits an orthogonal basis; since (8.6.1) is true for $(N, \varphi) \perp D$, it is also true for (N, φ). □

The three components of a graded Brauer class

Let $\mathrm{Ip}'(K)$ be the subset of all idempotents e of K such that $2e$ is invertible in Ke; it is easy to realize that $\mathrm{Ip}'(K)$ is an ideal of the boolean ring $\mathrm{Ip}(K)$. In the sequel several definitions and statements may become simpler when $\mathrm{Ip}'(K)$ is a principal ideal (generated by one element); in a boolean ring every finitely generated ideal is principal; unfortunately in (8.ex.23) there is an example of a ring K such that $\mathrm{Ip}'(K)$ *is not* a principal ideal. Let us remember the two exact sequences (3.5.19) and (3.4.13), here written again with additive notation:

$$0 \longrightarrow \mathrm{Br}(K) \longrightarrow \mathrm{Br}^g(K) \longrightarrow \mathrm{Q}^g(K) \longrightarrow 0 ,$$
$$0 \longrightarrow \mathrm{Q}(K) \longrightarrow \mathrm{Q}^g(K) \longrightarrow \mathrm{Ip}'(K) \longrightarrow 0 .$$

These exact sequences suggest the existence of a bijection from $\mathrm{Br}^g(K)$ onto $\mathrm{Ip}'(K) \times \mathrm{Q}(K) \times \mathrm{Br}(K)$ which should associate with every graded Brauer class $[A]$ its three components e, α, β respectively in $\mathrm{Ip}'(K)$, $\mathrm{Q}(K)$ and $\mathrm{Br}(K)$. Of course e is the idempotent such that $(1-e)A$ is a graded Azumaya algebra of even type over $K(1-e)$ (in other words its rank is a square), whereas eA has odd type over Ke (its rank is the double of a square). From A we have derived a quadratic extension $\mathrm{QZ}(A)$ (defined just before (3.5.18)), and e is also the idempotent such that $(1-e)\mathrm{QZ}(A)$ is a trivially graded quadratic extension of $K(1-e)$, whereas the odd component of $e\mathrm{QZ}(A)$ has constant rank 1 over Ke.

The choice of the third component β is not difficult, since a graded Azumaya algebra of even type is still an Azumaya algebra when its grading is ignored, whereas the even subalgebra of a graded Azumaya algebra of odd type is an Azumaya algebra; consequently β is the class in $\mathrm{Br}(K)$ of $(1-e)A^{ng} \oplus eA_0$; as in **3.8**, the notation A^{ng} means "A without grading". It is sure that the choice of the third component determines a bijection $\mathrm{Br}^g(K) \to \mathrm{Q}^g(K) \times \mathrm{Br}(K)$; indeed, when A has even type, the classes of $\mathrm{QZ}(A)$ and A^{ng} determine $[A]$ because of (3.8.14), and when A has odd type, the classes of $\mathrm{QZ}(A)$ and A_0 determine $[A]$ because of the multiplication isomorphism $\mathrm{QZ}(A) \otimes A_0 \to A$. Then we must choose the component α of $\mathrm{QZ}(A)$ in $\mathrm{Q}(K)$, and the most natural choice is the isomorphy class of $\mathrm{QZ}(A)^{ng}$; it is a good choice because $\mathrm{QZ}(A)$ is determined up to isomorphy by e and α.

When the three component mappings are defined on $\mathrm{Br}^g(K)$, a question quite naturally appears: if we know the three components of the classes of A and B, which are the three components of $[A]+[B]=[A\hat{\otimes} B]$? Besides, in **3.8** with every couple (A,D'), where A is a graded algebra and D' a discriminant module, we have associated an algebra $A_{D'}=A_0\oplus(D'\otimes A_1)$; if we know the components of the class of a graded Azumaya algebra A, which are the components of $[A_{D'}]$?

Before tackling these questions, some notation must be explained. First we can define $e\alpha$ and $e\beta$ for all $e\in\mathrm{Ip}(K)$, $\alpha\in\mathrm{Q}(K)$ and $\beta\in\mathrm{Br}(K)$. When α is the class of Z, then $e\alpha$ is the class of $(1-e)K^2\oplus eZ$; and when β is the class of A, then $e\beta$ is the class of $(1-e)K\oplus eA$. The following equalities are evident:

$$(ee')\alpha=e(e'\alpha)\ ,\quad e(\alpha+\alpha')=e\alpha+e\alpha'\ ,\quad (e\tilde{+}e')\alpha=e\alpha+e'\alpha\ ,$$
$$(ee')\beta=e(e'\beta)\ ,\quad e(\beta+\beta')=e\beta+e\beta'\ ,\quad (e\tilde{+}e')\beta=e\beta+e'\beta-2(ee'\beta)\ .$$

Thus $\mathrm{Q}(K)$ is a module over $\mathrm{Ip}(K)$. Although $\mathrm{Br}(K)$ is not always a module over $\mathrm{Ip}(K)$, the subgroup $\mathrm{Br}_2(K)$ of all β such that $2\beta=0$ is actually a module over it. And it is worth observing that the third component of an element of $\mathcal{H}(K)$ always belongs to the subgroup $\mathrm{Br}_2(K)$; indeed every Clifford algebra C is provided with a reversion τ that gives an isomorphism $C\to C^o$, and the well-known equality $[C]+[C^{to}]=0$ becomes $2[C]=0$ when C is trivially graded and isomorphic to C^o.

Secondly we get an $\mathrm{Ip}(K)$-linear mapping $\mathcal{J}:\mathrm{Ip}'(K)\to\mathrm{Q}(K)$ if we map every e (such that $2e$ is invertible in Ke) to the class of the nongraded quadratic extension $(1-e)K^2\oplus e(K\oplus Kj)$ where $K\oplus Kj$ is the K-algebra generated by an element j such that $j^2=-1$; thus $e(K\oplus Kj)$ is a quadratic extension of Ke for every $e\in\mathrm{Ip}'(K)$. It is easy to verify that $e\mathcal{J}(e')=\mathcal{J}(ee')$ and $\mathcal{J}(e'\tilde{+}e'')=\mathcal{J}(e')+\mathcal{J}(e'')$ for all $e\in\mathrm{Ip}(K)$ and all e', $e''\in\mathrm{Ip}'(K)$. When $\mathrm{Ip}'(K)$ is a principal ideal generated by one element e_0, it suffices to define $\mathcal{J}(e_0)$ because $\mathcal{J}(e)=e\mathcal{J}(e_0)$. The same observation is valid for the other $\mathrm{Ip}(K)$-linear mapping $\mathcal{D}:\mathrm{Ip}'(K)\to\mathrm{Q}(K)$ that shall be introduced later, just before (8.6.7).

Thirdly we need the $\mathbb{Z}$-bilinear mapping $\mathcal{Q}$ defined in **3.8**. When δ is the class in $\mathrm{Disc}(K)$ of a discriminant module D, and when α' is the class of Z' in $\mathrm{Q}(K)$, then $\mathcal{Q}(\delta,\alpha')$ is the Brauer class of $C\ell(D\otimes Z')^{ng}$; this notation is meaningful because D is a bilinear space and Z' a quadratic space; indeed the standard involution of Z' determines a norm $Z'\to K$ which is a nondegenerate quadratic form. To avoid the intrusion of a new notation, we will write $\mathcal{Q}(\alpha,\alpha')$ when we mean $\mathcal{Q}(\delta,\alpha')$ with δ depending on α in this way: it is the class of the discriminant module of a quadratic extension representing α. As explained in (3.8.16), in this way we get a symmetric $\mathbb{Z}$-bilinear mapping $\mathrm{Q}(K)\times\mathrm{Q}(K)\to\mathrm{Br}(K)$ which obviously takes its values in the subgroup $\mathrm{Br}_2(K)$. There is an alternative definition of $\mathcal{Q}(\alpha,\alpha')$ according to which it is the Brauer class of $C\ell(Z\perp Z')^{ng}$ if

Z and Z' are quadratic extensions representing α and α'; indeed $C\ell(Z \perp Z')$ is isomorphic to $C\ell(Z) \hat{\otimes} C\ell(Z')$, therefore isomorphic to $C\ell(Z) \otimes C\ell(D \otimes Z')$ (see (3.8.8)), and the class of $C\ell(Z)^{ng}$ is trivial (see (3.8.1)). Now $\mathrm{Disc}(K)$, $\mathrm{Q}(K)$ and $\mathrm{Br}_2(K)$ are modules over $\mathrm{Ip}(K)$, and obviously $\mathcal{Q}$ is bilinear over $\mathrm{Ip}(K)$: $\mathcal{Q}(e\delta, \alpha') = e\mathcal{Q}(\delta, \alpha') = \mathcal{Q}(\delta, e\alpha')$.

The mappings $\mathcal{J}$ and $\mathcal{Q}$ are related by this equality (for all $e \in \mathrm{Ip}'(K)$, and all $\alpha \in \mathrm{Q}(K)$):

$$(8.6.3) \qquad \mathcal{Q}(\mathcal{J}(e),\ \alpha) \ = \ e\ \mathcal{Q}(\alpha, \alpha).$$

Proof. We treat every algebra over K as the direct product of an algebra over $K(1-e)$ and an algebra over Ke. Since we compare the Brauer classes of two quaternion algebras that have trivial components over $K(1-e)$, it suffices to compare their components over Ke. Therefore we can assume that $e = 1$, and thus (8.6.3) becomes an immediate consequence of the statement (d) in (3.8.16). Indeed if Z is a quadratic extension with class α, and D its discriminant module, then $\mathcal{Q}(\alpha, \alpha)$ is the class of $C\ell(D \otimes Z)^{ng}$ which is also the class of $C\ell(J \otimes Z)^{ng}$ (see (3.8.16)) if J is the free discriminant module generated by j such that $j^2 = -1$; this is precisely the discriminant module of the quadratic extension $K \oplus Kj$ with class $\mathcal{J}(1)$. □

There is no impropriety in the fact that the $\mathrm{Ip}(K)$-linear mapping $\alpha \longmapsto \mathcal{Q}(\mathcal{J}(e), \alpha)$ is equal to the $\mathrm{Ip}(K)$-quadratic mapping $\alpha \longmapsto \mathcal{Q}(\alpha, \alpha)$, because every mapping that is linear over a boolean ring is also quadratic over it. Now we can tackle the addition of two Brauer–Wall classes.

(8.6.4) **Theorem.** *Let (e, α, β) and (e', α', β') give the three components of the Brauer classes of the graded Azumaya algebras A and A'. The three components of the Brauer class of $A \hat{\otimes} A'$ are given by the addition formula*

$$\begin{aligned}(e, \alpha, \beta) + (e', \alpha', \beta') \ = \ \big(\ & e \tilde{+} e'\ ,\ \ \alpha + \alpha' + \mathcal{J}(ee')\ , \\ & \beta + \beta'\ + \mathcal{Q}(\,\alpha + \mathcal{J}(e(1-e')),\ \alpha' + \mathcal{J}((1-e)e')\,)\ \big).\end{aligned}$$

Proof. The ring K is the direct sum of the ideals $K(1-e)(1-e')$, $Ke(1-e')$, $K(1-e)e'$ and Kee'; by means of the similar decomposition of $A \hat{\otimes} A'$, we can reduce the general case to four particular cases in which (e, e') is either $(0,0)$ or $(1,0)$ or $(0,1)$ or $(1,1)$. Consequently it suffices to prove these three addition formulas:

$$\begin{aligned}(0, \alpha, \beta) + (0, \alpha', \beta') &= (\,0\ ,\ \alpha + \alpha'\ ,\ \beta + \beta' + \mathcal{Q}(\alpha, \alpha')\,)\ , \\ (1, \alpha, \beta) + (0, \alpha', \beta') &= (\,1\ ,\ \alpha + \alpha'\ ,\ \beta + \beta' + \mathcal{Q}(\alpha + \mathcal{J}(1),\ \alpha')\,)\ , \\ (1, \alpha, \beta) + (1, \alpha', \beta') &= (\,0\ ,\ \alpha + \alpha' + \mathcal{J}(1)\ ,\ \beta + \beta' + \mathcal{Q}(\alpha, \alpha')\,)\ ;\end{aligned}$$

of course, the second and third equalities are only meaningful when 2 is invertible in K. The first equality is trivial when A is trivially graded, because in this case

$\alpha = 0$ and $A \hat{\otimes} A' = A \otimes A'$; consequently we can suppose that the grading of A is not trivial; and the same for A'.

First step. The first equality is an immediate consequence of (3.8.6) and (3.8.13). Indeed (3.8.6) implies that $A \hat{\otimes} A' = A_{D'} \otimes A'$ if D' is the discriminant module of $\mathrm{QZ}(A')$. And since $\mathrm{Br}(K)$ is here treated as an additive group, (3.8.13) implies that the class of $A_{D'}^{ng}$ is the sum of the classes of A^{ng} and $C\ell(D' \otimes Z)^{ng}$ if $Z = \mathrm{QZ}(A)$. The class of $C\ell(D' \otimes Z)^{ng}$ is $\mathcal{Q}(\alpha', \alpha) = \mathcal{Q}(\alpha, \alpha')$.

From this first result it follows that

$$-(0, \alpha, \beta) \;=\; (0,\ \alpha,\ -\beta + \mathcal{Q}(\alpha, \alpha))\ .$$

The *second step* is devoted to two very particular cases, which both require 2 to be invertible:

$$(1, \alpha, 0) + (0, 0, \beta) = (1, \alpha, \beta)\ ;$$
$$(1, \alpha, 0) + (1, \alpha', 0) = (0,\ \alpha + \alpha' + \mathcal{J}(1),\ \mathcal{Q}(\alpha, \alpha'))\ .$$

The first formula is an immediate consequence of the natural isomorphism $\mathrm{Z}(A) \otimes A_0 \to A$ which is valid for every graded Azumaya algebra A of odd type. The second formula is the most wearisome patch in this proof; here A and A' are graded quadratic extensions in which the odd components are the discriminant modules D and D'; thus A and A' are also graded Azumaya algebras, and we must find the three components of the class of $A \hat{\otimes} A'$. Because of (3.5.17), the first two components represent the class of $Z'' = A \star A'$ in $\mathrm{Q}^g(K)$; this trivially graded quadratic extension is not isomorphic to $A^{ng} \star A'^{ng}$ because of the twisting rule, according to which $(x \otimes x')^2 = -x^2 x'^2$ for odd elements $x \in D$ and $x' \in D'$; it is actually isomorphic to $A^{ng} \star A'^{ng} \star (K \oplus J)$ if J (the discriminant module of $K \oplus J$) is generated by an element j such that $j^2 = -1$. This shows that the second component of $A \hat{\otimes} A'$ is $\alpha + \alpha' + \mathcal{J}(1)$.

We can identify $A = K \oplus D$ with the Clifford algebra $C\ell(D_2)$ if D_2 is the quadratic space D with quadratic form $x \longmapsto x^2$; similarly $A' = C\ell(D'_2)$, and consequently $A \hat{\otimes} A' = C\ell(D_2 \perp D'_2)$. If we prove that the quadratic spaces $D_2 \perp D'_2$ and $D \otimes Z''$ are isomorphic, the conclusion follows, because the class of $C\ell(D \otimes Z'')^{ng}$ is $\mathcal{Q}(\alpha,\ \alpha + \alpha' + \mathcal{J}(1))$, which is the same thing as $\mathcal{Q}(\alpha, \alpha')$ because of (8.6.3). In the following calculations the lower index 2 transforms every discriminant module Δ into the quadratic module Δ_2 with quadratic form $\xi \longmapsto \xi^2$, and with associated bilinear form $(\xi, \xi') \longmapsto 2\xi\xi'$ (instead of $\xi\xi'$). Thus K_2 is the quadratic space K with quadratic form $\lambda \longmapsto \lambda^2$. It is clear that $\Delta \otimes \Delta'_2 = (\Delta \otimes \Delta')_2$ for all Δ and Δ'. Moreover if Δ is the discriminant module of a quadratic extension, as a quadratic space this quadratic extension is isomorphic to $K_2 \perp (J \otimes \Delta)_2$ because the standard involution operates as -1 on Δ. All this implies that Z'', as a quadratic space, is isomorphic to $K_2 \perp (D \otimes D')_2$. This shows that $D \otimes Z''$ is actually isomorphic to $D_2 \perp D'_2$.

Third step, in which we prove the above second and third equalities corresponding to the cases $(e,e')=(1,0)$ and $(e,e')=(1,1)$. First we notice that

$$-(1,\alpha,0) \;=\; (1,\ \alpha+\mathcal{J}(1),\ 0)\ ;$$

indeed the sum of $(1,\alpha,0)$ and $(1,\ \alpha+\mathcal{J}(1),\ 0)$ is equal to $(0,0,\ \mathcal{Q}(\alpha+\mathcal{J}(1),\ \alpha))$, which vanishes because of (8.6.3). Then we verify the vanishing of

$$(1,\ \alpha+\alpha',\ \beta+\beta'+\mathcal{Q}(\alpha+\mathcal{J}(1),\ \alpha')) - (1,\alpha,\beta) - (0,\alpha',\beta')\ ;$$

of course we replace the first term with the sum of $(1,\ \alpha+\alpha',\ 0)$ and $(0,0,\ \beta+\beta'+\mathcal{Q}(\alpha+\mathcal{J}(1),\ \alpha'))$, and we replace $-(1,\alpha,\beta)$ with the sum of $(1,\ \alpha+\mathcal{J}(1),\ 0)$ and $(0,0,-\beta)$; the sum of $(1,\ \alpha+\alpha',\ 0)$ and $(1,\ \alpha+\mathcal{J}(1),\ 0)$, which we calculate as explained in the second step, has component 0 in $\mathrm{Ip}'(K)$, and thus we get a sum of four terms with component 0 in $\mathrm{Ip}'(K)$, which we calculate as explained in the first step; this verification ends after another intervention of the equality $\mathcal{Q}(\alpha+\mathcal{J}(1),\ \alpha)=0$.

Finally the calculation of $(1,\alpha,\beta)+(1,\alpha',\beta')$ raises no difficulty, since it is equal to

$$(1,\alpha,0)+(1,\alpha',0)\ +(0,0,\beta)+(0,0,\beta'). \qquad \square$$

(8.6.5) **Corollary.** *If (e,α,β) gives the components of the class of A, the components of the class of A^{to} are given by*

$$-(e,\alpha,\beta) \;=\; (\ e\ ,\ \alpha+\mathcal{J}(e)\ ,\ -\beta+(1-e)\mathcal{Q}(\alpha,\alpha)\ )\ .$$

Proof. As a direct consequence of (8.6.4) we find

$$-(e,\alpha,\beta) \;=\; (\ e\ ,\ \alpha+\mathcal{J}(e)\ ,\ -\beta+\mathcal{Q}(\alpha,\ \alpha+\mathcal{J}(e))\ )\ ;$$

then (8.6.3) shows that

$$\mathcal{Q}(\alpha,\ \alpha+\mathcal{J}(e)) \;=\; \mathcal{Q}(\alpha,\alpha)+e\mathcal{Q}(\alpha,\alpha) \;=\; (1-e)\mathcal{Q}(\alpha,\alpha)\ . \qquad \square$$

After the class of $A\,\hat{\otimes}\,A'$ we calculate the class of $A_{D'}$.

(8.6.6) **Proposition.** *Let D' be a discriminant module with class δ' in* $\mathrm{Disc}(K)$, *and A a graded Azumaya algebra with class (e,α,β) in* $\mathrm{Br}^g(K)$. *The class of $A_{D'}=A_0\oplus(D'\otimes A_1)$ is*

$$(e,\alpha,\beta)_{\delta'} \;=\; (\ e\ ,\ \alpha+\delta'_Q(e)\ ,\ \beta+\ (1-e)\mathcal{Q}(\delta',\alpha)\ )\ ,$$

if $\delta'_Q(e)$ is the class of the quadratic extension $(1-e)K^2\oplus e(K\oplus D')$.

Here $K\oplus D'$ is the trivially graded algebra provided with the multiplication

$$(\lambda_1,d'_1)\,(\lambda_2,d'_2) \;=\; (\lambda_1\lambda_2+d'_1d'_2\ ,\ \lambda_1d'_2+\lambda_2d'_1).$$

Proof. Again it suffices to treat separately the cases $e = 0$ and $e = 1$ and to prove:

$$(0, \alpha, \beta)_{\delta'} = (0, \alpha, \beta + \mathcal{Q}(\delta', \alpha)) \quad \text{and} \quad (1, \alpha, \beta)_{\delta'} = (1, \alpha + \delta'_Q(1), \beta).$$

In the former case the same quadratic extension is derived from A and $A_{D'}$ whereas (3.8.13) says that the class of $A_{D'}^{ng}$ is the sum of the classes of A^{ng} and $C\ell(D' \otimes QZ(A))^{ng}$, which are respectively β and $\mathcal{Q}(\delta', \alpha)$. In the latter case A and $A_{D'}$ have the same even subalgebra, whereas $Z(A_{D'})$ is isomorphic to $Z(A) \star (K \oplus D')$, because its discriminant module is $D \otimes D'$ (if D is the discriminant module of $Z(A)$); thus the latter case immediately follows from the definitions. □

Let us derive some interesting consequences of (8.6.4). For every positive integer n,

$$2n\ (0, \alpha, \beta) = (0,\ 0,\ 2n\beta + n\mathcal{Q}(\alpha, \alpha))\ ;$$

consequently $2n(0, \alpha, \beta) = 0$, when n is even and $2n\beta = 0$. More generally

$$4n\ (e, \alpha, \beta) = (0,\ 0,\ 4n\beta + n\mathcal{Q}(\mathcal{J}(e),\ \mathcal{J}(e)))\ ;$$

consequently $4n(e, \alpha, \beta) = 0$ when n is even and $4n\beta = 0$. This gives another proof of (3.8.15), which states that the order of every element of $\mathcal{H}(K)$ is a divisor of 8.

It is also worth observing that the elements (e, α, β) such that $2n\beta = 0$ for some given positive integer n constitute a subgroup of $\mathrm{Br}^g(K)$.

The groups $E' \times Q \times B$ and the rings $E' \times Q \times B_2$

More generally let E be a boolean ring with unit element 1. It is well known that the equality $e^2 = e$ characterizing boolean rings implies $ee' = e'e$ and $e \tilde{+} e = 0$ for all elements e and e'. To avoid misunderstandings, we still use the notation $e \tilde{+} e'$ for sums in E, and we also write $1 - e$ instead of $1 \tilde{+} e$. Let E' be an ideal of E, Q a module over E, and B an additive group containing a subgroup B_2 that is a module over E. Besides, let $\mathcal{J} : E' \to Q$ be any E-linear mapping, and $\mathcal{Q} : Q \times Q \to B_2$ any symmetric E-bilinear mapping. It is easy to verify that *the addition defined in Theorem* (8.6.4) *determines a structure of group on the set* $E' \times Q \times B$. The property (8.6.3) is not indispensable to make this set become a group. Because of the equality $1 \tilde{+} 1 = 0$, each element of the additive group Q or B_2 has order 1 or 2, and each element of the subgroup $E' \times Q \times B_2$ has an order dividing 8.

Now let us suppose that $\mathcal{J}$ and $\mathcal{Q}$ satisfy the property (8.6.3), and let $\mathcal{D} : E' \to Q$ be any E-linear mapping. We define a multiplication on $E' \times Q \times B_2$ in the following way:

$$\text{(8.6.7)} \quad (e, \alpha, \beta)\ (e', \alpha', \beta') = \big(\ ee'\ ,\ \ e\alpha' + e'\alpha + \mathcal{D}(ee')\ ,\\ e\beta' + e'\beta\ + (1 - ee')\ \mathcal{Q}(\alpha + \mathcal{D}(e),\ \alpha' + \mathcal{D}(e'))\ \big).$$

It is easy to verify the associativity of this multiplication:

$$\begin{aligned}&(e,\alpha,\beta)(e',\alpha',\beta')(e'',\alpha'',\beta'')\\ &\quad=(ee'e'',\ e'e''\alpha+e''e\alpha'+ee'\alpha'',\ \ e'e''\beta+e''e\beta'+ee'\beta''\\ &\qquad+(1-e'e'')e\mathcal{Q}(\alpha',\alpha'')+(1-e''e)e'\mathcal{Q}(\alpha'',\alpha)+(1-ee')e''\mathcal{Q}(\alpha,\alpha')\,)\ ;\end{aligned}$$

but more patience is needed to verify that

$$(e,\alpha,\beta)\ \big((e',\alpha',\beta')+(e'',\alpha'',\beta'')\big)\ =\ (e,\alpha,\beta)(e',\alpha',\beta')+(e,\alpha,\beta)(e'',\alpha'',\beta'')\ ;$$

the verification of this distributivity law needs the relation

$$\mathcal{Q}(\alpha+\mathcal{D}(e),\ \mathcal{J}(e'e''))\ =\ \mathcal{Q}(e'\alpha+\mathcal{D}(ee'),\ e''\alpha+\mathcal{D}(ee''))\ ,$$

and it is the only place where (8.6.3) is absolutely indispensable.

(8.6.8) **Proposition.** *With the addition discovered in* (8.6.4) *and the multiplication defined in* (8.6.7)*, the set* $E'\times Q\times B_2$ *is a ring (perhaps without unit element). For each* $e\in E$*, the subset* $eE'\times eQ\times eB_2$ *is an ideal, and if* e *belongs to* E'*, it is the principal ideal generated by the idempotent* $(e,\mathcal{D}(e),0)$*. In particular if* $E=E'$*, then* $(1,\mathcal{D}(1),0)$ *is a unit element. Moreover the mapping* $e\longmapsto(e,\mathcal{D}(e),0)$ *is an isomorphism from* E' *onto the boolean ring of all idempotents of* $E'\times Q\times B_2$*.*

The proof of (8.6.8) is a matter of straightforward verifications. Obviously $eE'\times eQ\times eB_2$ is an ideal for all $e\in E$, and it is generated by the idempotent $(e,\mathcal{D}(e),0)$ when e belongs to E' because

$$(e,\mathcal{D}(e),0)\ (e',\alpha',\beta')\ =\ (ee',e\alpha',e\beta').$$

It is clear that $(e,\mathcal{D}(e),0)(e',\mathcal{D}(e'),0)=(ee',\mathcal{D}(ee'),0)$ but it takes more time to verify the equality

$$(e,\mathcal{D}(e),0)+(e',\mathcal{D}(e'),0)-2(ee',\mathcal{D}(ee'),0)\ =\ (e\tilde{+}e',\ \mathcal{D}(e)+\mathcal{D}(e'),\ 0)$$

which with the previous one means that $e\longmapsto(e,\mathcal{D}(e),0)$ is a morphism of boolean rings. At last, the equality $(e,\alpha,\beta)^2=(e,\mathcal{D}(e),\ (1-e)\mathcal{Q}(\alpha,\alpha))$ shows that every idempotent of $E'\times Q\times B_2$ is equal to $(e,\mathcal{D}(e),0)$ for some $e\in E'$, since $(1-e)\mathcal{Q}(\alpha,\alpha)=0$ when $\alpha=\mathcal{D}(e)$. □

Let us calculate the other powers $(e,\alpha,\beta)^n$ for $n>2$; they are periodical:

$$\begin{aligned}(e,\alpha,\beta)^n&=(e,\ e\alpha,\ e\beta)&&\text{if } n \text{ is odd }\geq 3,\\ (e,\alpha,\beta)^n&=(e,\ \mathcal{D}(e),\ 0)&&\text{if } n \text{ is even }\geq 4.\end{aligned}$$

A second conjecture

Our second conjecture is devoted to the structure of ring on $\mathcal{H}(K)$, the existence of which has followed from the first conjecture.

(8.6.9) **Conjecture.** *If* (e,α,β) *and* (e',α',β') *are the classes of the algebras* $C\ell(M,q)$ *and* $C\ell(M',q')$, *then the class of* $C\ell((M,q)\otimes(M',q'))$ *is given by the multiplication* (8.6.7) *in which* $\mathcal{D}$ *is the mapping* $\mathrm{Ip}'(K)\to \mathrm{Q}(K)$ *defined in this way:* $\mathcal{D}(e)$ *is the class of the quadratic extension* $(1-e)K^2\oplus e(K\oplus Kd)$, *where* $K\oplus Kd$ *is the algebra generated by an element* d *such that* $d^2=2$.

Proof of (8.6.9) *when* K *is a local ring.* When 2 is invertible in this local ring, (M',q') is an orthogonal sum of quadratic spaces of rank 1; since the multiplication (8.6.7) is distributive, it suffices to prove (8.6.9) when the rank of M' is 1. In this case $(M,q)\otimes(M',q')$ is the tensor product of (M,q) and the discriminant module $D'=(M',\mathrm{b}_{q'})$, and consequently its Clifford algebra is isomorphic to $C\ell(M,q)_{D'}$, the class of which is given by (8.6.6). We must remember that $\mathrm{b}_{q'}(a',a')=2q'(a')$ for all $a'\in M'$; therefore the quadratic extension $K\oplus D'$ mentioned in (8.6.6) is here isomorphic to $C\ell(M',q')^{ng}\star(K\oplus Kd)$ with $d^2=2$. This explains why we need the class $\mathcal{D}(1)$ of $K\oplus Kd$. Now it is easy to verify that $(e,\alpha,\beta)_{\delta'}$ (with δ' the image of $\alpha'+\mathcal{D}(1)$ in $\mathrm{Disc}(K)$) coincides with the result announced by (8.6.7) when $e'=1$ and $\beta'=0$; both (8.6.6) and (8.6.7) give the class $(e,\ \alpha+e\alpha'+\mathcal{D}(e),\ \beta+(1-e)\mathcal{Q}(\alpha,\ \alpha'+\mathcal{D}(1))\,)$.

When 2 is not invertible, then $e=e'=0$, and the distributivity of the multiplication (8.6.7) allows us only to consider the case of a quadratic module (M',q') of rank 2. Let (e_1,e_2) be a basis of M', and let us set $a=q'(e_1)$, $c=q'(e_2)$ and $b=\mathrm{b}_{q'}(e_1,e_2)$. We must prove that the class of the Clifford algebra of $(M,q)\otimes(M',q')$ is $(0,0,\ \mathcal{Q}(\alpha,\alpha'))$. Let D' be the discriminant module of $C\ell_0(M',q')$, and δ' its class in $\mathrm{Disc}(K)$; thus $\mathcal{Q}(\alpha,\alpha')$ is the same thing as $\mathcal{Q}(\delta',\alpha)$. Since the algebra $C\ell_0(M',q')$ is generated by the element $z=e_1e_2$ such that $z^2=bz-ac$, its discriminant module is generated by an element the square of which is b^2-4ac. Now $(M,q)\otimes(M',q')$ is the same thing as $(M,q)\otimes(M',\mathrm{b}_{q'})$, and from (2.6.3) we know that the orthogonal sum of $(M',\mathrm{b}_{q'})$ and any discriminant module D_0 admits an orthogonal basis (e'_1,e'_2,e'_3); in other words, it is the orthogonal sum of the discriminant modules D_k (with $k=1,2,3$)) generated by the elements of this orthogonal basis. Consequently the class of the Clifford algebra of $(M,q)\otimes(M',\mathrm{b}_{q'})$ is the sum of the classes of the three algebras $C\ell((M,q)\otimes D_k)$ with $k=1,2,3$, minus the class of $C\ell((M,q)\otimes D_0)$. Instead of subtracting this last class, we can add the class of $C\ell((M,q)\otimes J\otimes D_0)$, if J is the discriminant module J generated by an element j such that $j^2=-1$.

Let φ be the bilinear form on $(M',\mathrm{b}_{q'})\perp D_0$, and let us set $\lambda_k=\varphi(e'_k,e'_k)$ for $k=1,2,3$, and also $\lambda_0=\varphi(e_0,e_0)$ if e_0 is a generator of D_0. For $k=0,1,2,3$, let δ_k be the class of D_k in $\mathrm{Disc}(K)$, and δ_{-1} the class of J. Because of (8.6.6) we know that the class of $C\ell((M,q)\otimes D_k)$ is $(0,\alpha,\ \beta+\mathcal{Q}(\delta_k,\alpha))$; and the class of $C\ell((M,q)\otimes J\otimes D_0)$ is $(0,\alpha,\ \beta+\mathcal{Q}(\delta_{-1}+\delta_0,\ \alpha))$. Now we apply three times (8.6.4) to add four classes, the sum of which is the class of $\mathrm{C}\ell((M,q)\otimes(M',q'))$, and because of the $\mathbb{Z}$-bilinearity of $\mathcal{Q}$ the result is $(0,0,\ \mathcal{Q}(\delta_{-1}+\delta_0+\delta_1+\delta_2+\delta_3,\ \alpha))$. The sum of all δ_k with $k=-1,0,1,2,3$, is the class of $J\otimes D_0\otimes D_1\otimes D_2\otimes D_3$; this is a discriminant module generated by an element the square of which is $-\lambda_0\lambda_1\lambda_2\lambda_3$, and we must compare it with the discriminant module D' of $\mathrm{C}\ell_0(M',q')$. It is isomorphic to D' if the quotient $-\lambda_0\lambda_1\lambda_2\lambda_3\,/\,(b^2-4ac)$ is a square in $K^\times$. Since $\lambda_1\lambda_2\lambda_3$ is the determinant of φ in the basis (e'_1,e'_2,e'_3), and since $\lambda_0(4ac-b^2)$ is the determinant of φ in the basis (e_1,e_2,e_0), the quotient $\lambda_1\lambda_2\lambda_3\,/\,\lambda_0(4ac-b^2)$ is actually the square of the determinant of the elements (e'_1,e'_2,e'_3) relative to the basis (e_1,e_2,e_0), and thus we realize that $D'\cong J\otimes D_0\otimes D_1\otimes D_2\otimes D_3$. □

(8.6.10) **Historical comments.** The first attempts to study $\mathcal{H}(K)$ were directed to the subgroup $\mathcal{H}_0(K)$ that is the kernel of $\mathcal{H}(K)\to\mathrm{Ip}(K)$, and to the smaller subgroup $\mathcal{H}_{00}(K)$ that is the kernel of $\mathcal{H}(K)\to\mathrm{Q}^g(K)$; with the above notation, $\mathcal{H}_0(K)$ (resp. $\mathcal{H}_{00}(K)$) is the subgroup of all $(e,\alpha,\beta)\in\mathcal{H}(K)$ such that $e=0$ (resp. $e=0$ and $\alpha=0$). The injective group morphism $\mathcal{H}_{00}(K)\longmapsto\mathrm{Br}_2(K)$ (that is $(0,0,\beta)\longmapsto\beta$) appeared in [Micali, Villamayor 1968]; as explained in (8.ex.15), the existence of this morphism implies that the order of every element of $\mathcal{H}_0(K)$ (resp. $\mathcal{H}(K)$) is a divisor of 4 (resp. 8); when $K=\mathbb{Q}$, the morphism $\mathcal{H}_{00}(\mathbb{Q})\to\mathrm{Br}_2(\mathbb{Q})$ is even bijective (see (8.3.10)). The injective mapping $\mathcal{H}_0(K)\to\mathrm{Q}(K)\times\mathrm{Br}_2(K)$ appeared in [Micali, Villamayor 1970], together with an addition formula that is equivalent to (8.6.4) when $e=e'=0$, and that can be translated here by the formula $(\alpha,\beta)+(\alpha',\beta')=(\alpha+\alpha',\ \beta+\beta'+\mathcal{Q}(\alpha,\alpha'))$. The same topic was surveyed in [Revoy 1971] and [Micali, Revoy 1979]. The complete addition formula (8.6.4) appeared in [Helmstetter, Micali 1993], unfortunately with an incomplete multiplication formula (8.6.7) in which the mapping $\mathcal{D}$ was still missing; consequently the invertibility of 2 was still necessary, and the tensor product of (M,q) and (M',q') was defined in the Bourbaki fashion (which gives the quadratic space here denoted by $(M\otimes M',\ q\otimes q'/2)$). Now it remains to find a complete proof of Conjectures (8.6.1) and (8.6.9), and to consider this puzzling question: if (e,α,β) and (e',α',β') are the classes of the algebras A and A', and if β and β' belong to $\mathrm{Br}_2(K)$, is there a "natural operation" that would derive from A and A' a new algebra, the class of which is represented by the product in (8.6.7)?

Exercises

(8.ex.1) Let K be a local ring with maximal ideal $\mathfrak{m}$, in which 2 is invertible. Give a direct proof of the following fact: the relation $(\!(a)\!)+(\!(-a)\!)=(\!(1)\!)+(\!(-1)\!)$ (valid for every $a\in K^\times$) is a consequence of the relations of types (i), (ii) and (iii) listed in (8.1.8).
Hint. When $a^2-1\notin\mathfrak{m}$, you can set $c=a+1$ and $d=a-1$, and write

$$(\!(a)\!)+(\!(-a)\!)\ =\ (\!(ac^2)\!)+(\!(-ad^2)\!)\ =\ (\!(a(c^2-d^2)\,)\!)+(\!(-a^3c^2d^2(c^2-d^2)\,)\!)$$

and so forth ...; when $a^2-1\in\mathfrak{m}$ and 3 is invertible in K, then

$$((a))+((-a)) = ((4a))+((-a)) = ((3a))+((-12a^3)) = ((3a))+((-3a))\ ;$$

when $K/\mathfrak{m}$ is isomorphic to $\mathbb{Z}/3\mathbb{Z}$, you can assume $a-1\in\mathfrak{m}$ and write

$$((a))+((-a)) = ((\,(1-\mu^2a^2)a))+((-(1-\mu^2a^2)a)) \quad \text{with } \mu=(a-1)a^{-1}(a+1)^{-1}.$$

(8.ex.2) Let $\mathrm{WQ}_0(K)$ be the subgroup of $\mathrm{WQ}(K)$ generated by the Witt classes of quadratic spaces of even rank at every prime ideal of K; it is even an ideal of $\mathrm{WQ}(K)$.

(a) Explain the exact sequence

$$0\longrightarrow \mathrm{WQ}_0(K)\longrightarrow \mathrm{WQ}(K)\longrightarrow \mathrm{Ip}'(K)\longrightarrow 0.$$

(b) When K is a local ring in which 2 is invertible, we denote by $[a,b]$ the Witt class of the quadratic space $\langle a,b\rangle$. Prove that $\mathrm{W}_0(K)$ is generated as an additive group by all classes $[a,b]$ with a and b in $K^\times$; it is even generated by all classes $[a,b]$ such that a is a fixed element of $K^\times$, whereas b runs through $K^\times$.

(c) Let K still be a local ring with maximal ideal $\mathfrak{m}$, in which 2 is invertible. Prove that $\mathrm{W}_0(K)$ is the additive group generated by all symbols $[a,b]$ constrained to these relations:

(i) $[a\lambda^2,\ b]=[a,b]$ for all $a,\ b,\ \lambda\in K^\times$.
(ii) $[a,b]=[a+b,\ ab(a+b)]$ whenever $a+b\in K^\times$.
(iii) $[a,b]=[(1+ab\mu^2)a,\ (1+ab\mu^2)b]$ whenever $a+b$ and μ are in $\mathfrak{m}$.
(iv) $[a,-a]=0$ for all $a\in K^\times$.
(v) $[a,b]=[b,a]$ for all $a,\ b\in K^\times$.
(vi) $[a,b]+[c,d]=[a,c]+[b,d]$ for all $a,\ b,\ c,\ d\in K^\times$.

Besides, the relations of type (iii) are consequences of the other ones when $K/\mathfrak{m}$ is not isomorphic to $\mathbb{Z}/3\mathbb{Z}$, and also when the mapping $\mu\longmapsto\mu-\mu^2$ is surjective from $\mathfrak{m}$ onto $\mathfrak{m}$. And the relations (iv) can be replaced with the only relation $[1,-1]=0$ (see (8.ex.1)).

(8.ex.3) Let K be a local ring in which 2 is invertible. Since $((a))^2=((1))$ for all $a\in K^\times$, the relation (ii) in (8.1.8) can also be written $((1))+((ab)) = ((a+b))\,(\,((a))+((b))\,)$; the two previous relations are the beginning of an infinite sequence in which the third relation (valid when a, b, c and $a+b+c$ are invertible) may be written in this way:

$$((1))+((ab))+((ac))+((bc)) = ((a+b+c))\ \big(\,((a))+((b))+((c))+((abc))\,\big)\ .$$

To obtain the nth relation in this sequence, we take n elements $a_1,\dots,a_n$ of $K^\times$, we assume that their sum b_n is invertible, and we consider this polynomial P_n with coefficients in the ring $\mathrm{WG}(K)$:

$$P_n(X) = \big(\,((1))-((a_1))X\big)\ \big(\,((1))-((a_2))X\big)\cdots\big(\,((1))-((a_n))X\big)\ ;$$

now the announced relation is $P_n(\langle\!\langle b_n\rangle\!\rangle) = 0$. Prove that this relation actually holds.

Hint. Let $P_n(x)$ be the image of $P_n(X)$ in the quotient of the polynomial ring $\mathrm{WG}(K)[X]$ by the ideal (X^2-1); prove the existence of $w_n \in \mathrm{WG}(K)$ such that $P_n(x) = w_n(\langle\!\langle 1\rangle\!\rangle - \langle\!\langle a_1\rangle\!\rangle x)$, and establish the induction formula $w_n = w_{n-1}(\langle\!\langle 1\rangle\!\rangle + \langle\!\langle a_1 a_n\rangle\!\rangle)$ (for all $n>1$); thus $P_n(\langle\!\langle b_n\rangle\!\rangle) = w_n(\langle\!\langle 1\rangle\!\rangle - \langle\!\langle a_1 b_n\rangle\!\rangle)$; when $n=3$, and almost always when $n>3$, it occurs that $b_n - a_j$ is invertible for some $j\in\{1,2,\dots,n\}$, and you can assume that $j=n$, so that b_{n-1} is invertible (if $b_{n-1} = b_n - a_n$); deduce the equality $P_n(\langle\!\langle b_n\rangle\!\rangle) = 0$ from $P_{n-1}(\langle\!\langle b_{n-1}\rangle\!\rangle) = 0$; but when $b_n - a_j$ is never invertible, then $a_n + a_{n-1}$ and $b_{n-2} = b_n - a_n - a_{n-1}$ are both invertible, and the announced result can be deduced from $P_{n-2}(\langle\!\langle b_{n-2}\rangle\!\rangle) = 0$.

(8.ex.4) Let K be a local ring in which 2 is not invertible. As in **8.2**, the notation $\langle a,b\rangle$ means the quadratic space K^2 with quadratic form $(x,y)\longmapsto ax^2+xy+by^2$, and $\langle\!\langle a,b\rangle\!\rangle$ is its class in $\mathrm{WGQ}(K)$.

(a) Prove that $\langle\!\langle a,b\rangle\!\rangle = \langle\!\langle -a(1-4ab)^{-1},\ b\rangle\!\rangle = \langle\!\langle -a(1-4ab)^{-1},\ -b(1-4ab)\rangle\!\rangle$.

(b) Let a, b, c, d be four elements of K, and let us set $\lambda = 1-4ab$ and $\mu = 1-4cd$. Prove that the class of the tensor product $\langle a,b\rangle\otimes\langle c,d\rangle$ is equal to

$$\begin{aligned}\langle\!\langle a,b\rangle\!\rangle\,\langle\!\langle c,d\rangle\!\rangle &= \langle\!\langle 2ac,\ 2bd\rangle\!\rangle + \langle\!\langle 2ad\lambda\mu,\ 2bc\lambda^{-1}\mu^{-1}\rangle\!\rangle\\ &= \langle\!\langle 2ac,\ -2bd\mu^{-1}\rangle\!\rangle + \langle\!\langle -2ad\lambda,\ 2bc\lambda^{-1}\mu^{-1}\rangle\!\rangle\\ &= \langle\!\langle 2ac,\ 2bd\lambda^{-1}\mu^{-1}\rangle\!\rangle + \langle\!\langle -2ad\lambda^{-1},\ -2bc\mu^{-1}\rangle\!\rangle\\ &= \langle\!\langle -2ac\mu,\ -2bd\mu^{-1}\rangle\!\rangle + \langle\!\langle -2ad\lambda,\ -2bc\lambda^{-1}\rangle\!\rangle\ .\end{aligned}$$

The Witt ring $\mathrm{W}(\mathbb{Q})$ and its subrings $\mathrm{WB}(\mathbb{Z}_{(2)})$ and $\mathrm{WQ}(\mathbb{Z}_{(2)})$

(8.ex.5) Let Θ be the set of all group morphisms $\theta : \mathrm{W}(\mathbb{Q}) \to \mathbb{Z}/8\mathbb{Z}$ satisfying these properties: when the integer a is not divisible by 4, then $\theta([a])$ only depends on the image of a in $\mathbb{Z}/8\mathbb{Z}$, and $\theta([a]) = 1$ modulo 8 whenever $a\equiv 1$ modulo 8.

(a) Suppose that θ is an element of Θ and prove that $\theta([3]) = 3$ modulo 8, and that $\theta([2])$ is equal to 1 or 5 modulo 8. Conclude that the cardinal of Θ is 0 or 2.

Hint. $[1]+[1] = [2]+[2]$ and $[3]+[-1] = [2]+[-6]$; don't forget the morphism $\delta_2' : \mathrm{W}(\mathbb{Q}) \to \mathbb{Z}/8\mathbb{Z}$ derived from $\delta_2 : \mathrm{W}(\mathbb{Q}) \to \mathbb{Z}/2\mathbb{Z}$.

(b) Prove the existence of a morphism $\theta\in\Theta$ that maps $[a]$ to the following value when $a = 2^\alpha a'$ with a' an odd integer:

$$\theta([a]) = a' \text{ modulo } 8 \quad \text{if } \alpha \text{ is even,}$$
$$\theta([a]) = (-1)^{(a'-1)/2} \text{ modulo } 8 \quad \text{if } \alpha \text{ is odd.}$$

Hint. Verify that this definition of θ is compatible with the relations between the generators $[a]$ of $\mathrm{W}(\mathbb{Q})$; since the relation $[a]+[b] = [a+b]+[ab(a+b)]$ is not modified when (a,b) is replaced with (b,a) or $(a,-a-b)$, you can suppose that $a = 2^\alpha a'$ and $b = 2^\beta b'$ with a' and b' odd integers, and $\alpha < \beta$. It may be useful to notice that $(-1)^{(a'-1)/2} \equiv a'$ modulo 4.

(8.ex.6) Let θ be the group morphism $\mathrm{W}(\mathbb{Q}) \to \mathbb{Z}/8\mathbb{Z}$ defined in (8.ex.5)(b), and δ_2 the group morphism $\mathrm{W}(\mathbb{Q}) \to \mathbb{Z}/2\mathbb{Z}$ defined in (8.3.4). Here our purpose is to prove that $\mathrm{WB}(\mathbb{Z}_{(2)})$ and $\mathrm{WQ}(\mathbb{Z}_{(2)})$, considered as subgroups of $\mathrm{W}(\mathbb{Q})$, are equal to $\mathrm{Ker}(\delta_2)$ and $\mathrm{Ker}(\theta) \cap \mathrm{Ker}(\delta_2)$. We shall not use (8.3.5).

(a) Verify that $\delta_2(w) = 0$ for all $w \in \mathrm{WB}(\mathbb{Z}_{(2)})$, and that $\theta(w) = 0$ for all $w \in \mathrm{WQ}(\mathbb{Z}_{(2)})$.

(b) Let w be an element of $\mathrm{W}(\mathbb{Q})$; prove the existence of $w' \in \mathrm{WQ}(\mathbb{Z}_{(2)})$ and $k \in \mathbb{Z}$ such that $w - w'$ is equal either to $k\,[1]$ or to $k\,[1] + [2]$.

Hint. When $n \equiv 2$ modulo 8, then $[n] - [2]$ is one of the generators $[2a] + [2a(4ac - b^2)]$ of $\mathrm{WQ}(\mathbb{Z}_{(2)})$; when $n \equiv 1$ modulo 8, then $[n] + [1] = [n+1] + [n(n+1)]$ and $[n] + [1] - [2] - [2]$ is a sum of two such generators; when $n \equiv 3$ modulo 8, then $[n] - [1] - [2] - [2]$ belongs to $\mathrm{WQ}(\mathbb{Z}_{(2)})$; finally remember $[1] + [1] = [2] + [2]$.

(c) Let w be an element of $\mathrm{W}(\mathbb{Q})$ such that $\delta_2(w) = 0$; prove that $w \in \mathrm{WB}(\mathbb{Z}_{(2)})$. Now suppose that $\delta_2(w) = 0$ and $\theta(w) = 0$; prove that $w \in \mathrm{WQ}(\mathbb{Z}_{(2)})$.

(8.ex.7)

(a) It is known that $\mathrm{WB}(\mathbb{Z}_{(2)})$ is generated as a group by all $[b']$ with b' an odd integer. Consider the element $w = [2a] + [2a(4ac - b^2)]$ of $\mathrm{WQ}(\mathbb{Z}_{(2)})$ (with b an odd integer, and a and c nonzero integers), and decompose it into a sum of two or four elements like $[b']$ (with an odd b'), according to the exponent α and the odd factor a' in the equality $a = 2^\alpha a'$. When α is odd, you get immediately a decomposition $w = [b_1] + [b_2]$. Explain how to get a decomposition $w = [b_1] + [b_2] + [b_3] + [b_4]$ when α is even.

(b) Let $b_1, b_2, \ldots, b_n$ be odd integers. From (8.ex.6) deduce that $\sum_k [b_k]$ belongs to $\mathrm{WQ}(\mathbb{Z}_{(2)})$ if and only if $\sum_k b_k \equiv 0$ modulo 8.

(c) *Example.* Let $b_1$and b_2 be two odd integers such that $b_1 + b_2 \equiv 0$ modulo 8. Find a triplet of integers (a, b, c) (with b an odd integer) such that $[b_1]+[b_2] = [2a] + [2a(4ac - b^2)]$.

Fields with property $C_1(2)$

A field K is said to have the property $C_1(2)$ if for every integer $r > 2$ and for every quadratic form q on K^r there exists a nontrivial $x \in K^r$ such that $q(x) = 0$. More generally it is said to have the property $C_i(d)$ (with $i \geq 0$ and $d \geq 2$) if every homogeneous polynomial function of degree d over K^r vanishes at some nonzero element of K^r whenever $r > d^i$; and it has the property C_i if it has the property $C_i(d)$ for all degrees d. For instance, an algebraically closed field has the property C_0.

(8.ex.8) Prove that these two assertions are equivalent when K is a field of characteristic $\neq 2$:

(i) K has the property $C_1(2)$;
(ii) every nondegenerate quadratic form on K^2 is a surjective mapping $K^2 \to K$.

When K is a field of characteristic 2, prove that (i) implies (ii).

(8.ex.9) Let K be a field of characteristic $\neq 2$ and with the property $C_1(2)$. With every quadratic space (M,q) is associated an integer $\dim(M)$ in $\mathbb{N}$ and a determinant $\det(q)$ well-defined modulo the subgroup of squares $K^{\times 2}$: it is the image in $K^\times/K^{\times 2}$ of the determinant of b_q in any basis of M.

(a) Prove that every quadratic space (M,q) of dimension r contains an orthogonal basis $(e_1,\dots,e_r)$ such that $q(e_i)=1/2$ whenever $i<r$. What can you say about $2q(e_r)$ and $\det(q)$?
(b) Prove that there is a group isomorphism from $\mathrm{WG}(K)$ onto $\mathbb{Z}\times(K^\times/K^{\times 2})$ that maps the class of every quadratic space (M,q) to $(\dim(M),\ \det(q))$. This isomorphism maps the class of the hyperbolic space $\langle 1,-1\rangle$ to $(2,\omega)$, where ω is the image of -1 in $K^\times/K^{\times 2}$.
The group $\mathbb{Z}\times(K^\times/K^{\times 2})$ is here treated as an additive group, and the tensor product of quadratic spaces provides it with the following multiplication: $(r,\xi)\ (s,\zeta)=(rs,\ \xi^s\zeta^r)$.
(c) Derive from (b) an exact sequence

$$0 \longrightarrow K^\times/K^{\times 2} \longrightarrow \mathrm{W}(K) \longrightarrow \mathbb{Z}/2\mathbb{Z} \longrightarrow 0 .$$

Prove that the Witt ring $\mathrm{W}(K)$ is isomorphic to the set $(\mathbb{Z}/2\mathbb{Z})\times(K^\times/K^{\times 2})$ provided with this modified addition and this multiplication:

$$(\rho,\xi)\ \hat{+}\ (\sigma,\zeta) = (\ \rho+\sigma\ ,\ \omega^{\rho\sigma}\xi\zeta\ )\ ,$$
$$(\rho,\xi)\ (\sigma,\zeta) = (\ \rho\sigma\ ,\ \xi^\sigma\zeta^\rho\ ).$$

Conclude that the additive group $\mathrm{W}(K)$ is isomorphic to the ordinary direct product $(\mathbb{Z}/2\mathbb{Z})\times(K^\times/K^{\times 2})$ if and only if -1 is a square in K, and that in the other cases every element of $\mathrm{W}(K)$ with a nontrivial image in $\mathbb{Z}/2\mathbb{Z}$ has order 4 in the additive group $\mathrm{W}(K)$.

(8.ex.10) Let K be a field of characteristic 2 with the property $C_1(2)$.

(a) Prove that every nonhyperbolic quadratic space is the orthogonal sum of a hyperbolic quadratic space and an anisotropic quadratic space of dimension 2.
(b) From (8.ex.8) it follows that every quadratic space of dimension 2 contains a basis (e_1,e_2) such that $q(e_1)=\mathrm{b}_q(e_1,e_2)=1$. Prove that its isomorphy class is determined by the image of $q(e_2)$ in $K/\wp(K)$, where $\wp(K)$ is the subgroup of all elements $\kappa-\kappa^2$ with $\kappa\in K$. Deduce from this fact that an element $\mathrm{Arf}(M,q)\in K/\wp(K)$ is associated with every quadratic space (M,q).

Remark. This element $\mathrm{Arf}(M,q)$ coincides with the Arf invariant already met in (3.ex.25).

(c) Prove that there is a group isomorphism form $\mathrm{WGQ}(K)$ onto $2\mathbb{Z} \times (K/\wp(K))$ that maps the isomorphy class of every (M,q) to $(\dim(M),\ \mathrm{Arf}(M,q))$. Prove that it induces an isomorphism of additive groups $\mathrm{WQ}(K) \to K/\wp(K)$.

(8.ex.11) Let K be a finite field of cardinal n (a power of the characteristic of K).

(a) Prove that K has the property $C_1(2)$.
Hints. When n is even, this follows immediately from $K^{\times 2} = K^\times$; when n is odd, prove that $q(M) = K$ for every quadratic space (M,q) of dimension 2 (see (8.ex.8)); it suffices to prove that the equation $a\lambda^2 + b\mu^2 = c$, with a, b, c given in K, $ab \neq 0$, always admits a solution $(\lambda, \mu) \in K^2$; calculate the number n' of elements in the image of $\lambda \longmapsto a\lambda^2$, and the number n'' of elements in the image of $\mu \longmapsto c - b\mu^2$, and verify that $n' + n'' > n$.

(b) From (8.ex.9) and (8.ex.10) deduce a description of the rings $\mathrm{WGQ}(K)$ and $\mathrm{WQ}(K)$. The additive group $\mathrm{WGQ}(K)$ is isomorphic to $2\mathbb{Z}$ when n is even, to $\mathbb{Z} \times (\mathbb{Z}/2\mathbb{Z})$ when n is odd. The additive group $\mathrm{WQ}(K)$ is isomorphic to $\mathbb{Z}/2\mathbb{Z}$ if n is even, to $(\mathbb{Z}/2\mathbb{Z}) \times (\mathbb{Z}/2\mathbb{Z})$ if $n \equiv 1$ modulo 4, and to $\mathbb{Z}/4\mathbb{Z}$ if $n \equiv 3$ modulo 4. In all cases, the Witt ring $\mathrm{WQ}(K)$ contains a nonzero element the square of which vanishes.

(8.ex.12) Here we suppose that K is a finite ring; consequently K is semilocal, in other words, it contains finitely many maximal ideals $\mathfrak{m}_1, \mathfrak{m}_2, \ldots, \mathfrak{m}_k$. By definition the Jacobson radical $\mathfrak{r}$ is the intersection of all maximal ideals.

(a) Prove the bijectiveness of the natural ring morphism $K/\mathfrak{r} \to \prod_{i=1}^{k} K/\mathfrak{m}_i$.
Comment. This is a particular case of the "Chinese remainder theorem".

(b) Prove the existence of an integer n such that $\mathfrak{r}^n = 0$.
Hint. If K contains only one maximal ideal $\mathfrak{m}$, there exists an exponent n such that $\mathfrak{m}^n = \mathfrak{m}^{n+1}$, whence $\mathfrak{m}^n = 0$ because of Nakayama's lemma; in the general case, every localization of K is still a finite ring.

(c)* Let K momentarily be any ring, and $\mathfrak{r}$ an ideal of K; the identity mapping of K induces surjective ring morphisms $\cdots \to K/\mathfrak{r}^3 \to K/\mathfrak{r}^2 \to K/\mathfrak{r}$, whence a ring morphism from K into the projective limit of the rings $K/\mathfrak{r}^j$ (see (1.ex.29)); when this morphism $K \to \varprojlim(K/\mathfrak{r}^j)$ is bijective, K is said to be "complete for the $\mathfrak{r}$-adic topology"; this refers to the topology on K for which the ideals $\mathfrak{r}^j$ are a basis of neighbourhoods of 0. In [Baeza 1978] the following theorem is proved: if K is a semilocal ring that is complete for the $\mathfrak{r}$-adic topology determined by its Jacobson radical $\mathfrak{r}$, then the natural ring morphisms $\mathrm{WGQ}(K) \to \mathrm{WGQ}(K/\mathfrak{r})$ and $\mathrm{WQ}(K) \to \mathrm{WQ}(K/\mathfrak{r})$ are isomorphisms.
Prove that the above finite ring K is complete for the $\mathfrak{r}$-adic topology.

(d) From (a) and (8.ex.11) deduce a description of the rings $\mathrm{WGQ}(K/\mathfrak{r})$ and $\mathrm{WQ}(K/\mathfrak{r})$ which, according to (c), are isomorphic to $\mathrm{WGQ}(K)$ and $\mathrm{WQ}(K)$;

these rings depend on the number k' (resp. k'', resp. k''') of maximal ideals $\mathfrak{m}_i$ such that the cardinal of $K/\mathfrak{m}_i$ is even (resp. equal to a multiple of 4 plus 1, resp. equal to a multiple of 4 minus 1).

(8.ex.13) Let F be an algebraically closed field, and $K = F(t)$ the field of rational functions in an indeterminate t. The following statement (a classical result of elimination theory) is needed in (a) below: if $P_1, P_2, \ldots, P_m$ are nonconstant homogeneous polynomial functions on F^r, and if $0 < m < r$, there is a nonzero element of F^r at which all these m polynomials vanish.

(a) Prove that K has the property $C_1(2)$.
Hint. Let q be a quadratic form on K^3; you may assume that the coefficients of q are polynomials in t, all of degree $\leq k$ for some positive integer k; let $x_1(t)$, $x_2(t)$, $x_3(t)$ be polynomials of degree $\leq k-1$; thus the coefficients of these three polynomials determine an element $\xi \in F^{3k}$; there are polynomials $P_j(\xi)$ (the coefficients of which depend on q) such that

$$q(x_1(t),\ x_2(t),\ x_3(t)) = \sum_{j=0}^{3k-2} P_j(\xi)\, t^j\ ;$$

these $(3k-1)$ polynomials P_j all vanish at some nonzero element ξ of F^{3k}.
Comment. More generally K has the property C_1 : every homogeneous polynomial of degree d vanishes at some nonzero element of K^r whenever $r > d$. In [Greenberg 1969] it is proved that more generally the field $F(t)$ has the property C_{i+1} when the field F has the property C_i.

(b) Now suppose that 2 is invertible in F. Let G be the field $\mathbb{Z}/2\mathbb{Z}$ and $G^{(F)}$ the vector space over G freely generated by the elements of F. Deduce from (8.ex.9)(c) that $\mathrm{W}(K)$ is isomorphic to the additive group $G \times G^{(F)}$ provided with the following multiplication:

$$(\rho,\ (\xi_a)_{a\in F})\ (\sigma,\ (\zeta_a)_{a\in F}) = (\rho\sigma,\ (\rho\zeta_a + \sigma\xi_a)_{a\in F}).$$

(8.ex.14) For every ring K there are surjective group morphisms $\mathrm{WQ}(K) \to \mathcal{H}(K)$ and $\mathcal{H}(K) \to \mathrm{Q}^g(K)$. Prove that they are bijective when K is a field with the property $C_1(2)$. By restriction to the subgroups of elements with trivial image in $\mathrm{Ip}'(K)$, we get isomorphisms $\mathrm{WQ}_0(K) \to \mathcal{H}_0(K) \to \mathrm{Q}(K)$.

The group $\mathcal{H}(K)$ of classes of Clifford algebras

(8.ex.15) Let $\mathcal{H}_0(K)$ be the intersection of $\mathcal{H}(K)$ with the kernel of the canonical group morphism $\mathrm{Br}^g(K) \to \mathrm{Ip}(K)$, and $\mathcal{H}_{00}(K)$ its intersection with the kernel of the canonical morphism $\mathrm{Br}^g(K) \to \mathrm{Q}^g(K)$. Consider this commutative diagram in which all lines and columns are exact:

$$\begin{array}{ccccccccc} & & & & 0 & & 0 & & \\ & & & & \downarrow & & \downarrow & & \\ 0 & \longrightarrow & \mathcal{H}_{00}(K) & \longrightarrow & \mathcal{H}_0(K) & \longrightarrow & \mathrm{Q}(K) & \longrightarrow & 0 \\ & & \updownarrow & & \downarrow & & \downarrow & & \\ 0 & \longrightarrow & \mathcal{H}_{00}(K) & \longrightarrow & \mathcal{H}(K) & \longrightarrow & \mathrm{Q}^g(K) & \longrightarrow & 0 \\ & & & & \downarrow & & \downarrow & & \\ & & & & \mathrm{Ip}(K) & \longleftrightarrow & \mathrm{Ip}(K) & & \end{array}$$

Let $\beta : \mathcal{H}_0(K) \to \mathrm{Br}(K)$ be the mapping defined in this way: it maps the class of every $\mathrm{C}\ell(M,q)$ (with (M,q) a quadratic space of even rank) to the class of $\mathrm{C}\ell(M,q)^{ng}$ (with forgotten parity grading). Prove that the restriction of β to $\mathcal{H}_{00}(K)$ is an injective group morphism (*Hint:* (3.8.8)), and that every element of $\mathcal{H}_{00}(K)$ has order 1 or 2. Now let ξ be an element of $\mathcal{H}_0(K)$ (resp. $\mathcal{H}(K)$); observe that ξ^2 (resp. ξ^4) belongs to $\mathcal{H}_{00}(K)$, and conclude that the order of ξ is a divisor of 4 (resp. 8).

(8.ex.16) Here (M,q) is a quadratic space of constant rank 4, and there exists a graded algebra isomorphism from $\mathrm{C}\ell(M,q)$ onto some $\mathrm{End}(P)$, with P a graded finitely generated projective module of constant rank 4. Prove that (M,q) is hyperbolic when these two additional hypotheses are also fulfilled: P_0 contains a direct summand of constant rank 1, and M contains a direct summand D of constant rank 1 such that $q(D)$ generates K as an ideal. These additional hypotheses are always fulfilled when K is a local ring. You can proceed in this way:

(a) Let ε be the idempotent in $\mathrm{C}\ell(M,q)$ that is mapped to the projection $P_0 \oplus P_1 \to P_0$ by the isomorphism $\mathrm{C}\ell(M,q) \to \mathrm{End}(P)$. Prove that $\varepsilon\mathrm{C}\ell_0(M,q)$ is an algebra isomorphic to $\mathrm{End}(P_0)$. Each of the algebras $\varepsilon\mathrm{C}\ell_0(M,q)$ or $\mathrm{End}(P_0)$ is a quaternion algebra over K, and is provided with a unique standard involution; prove that the standard involution of $\varepsilon\mathrm{C}\ell_0(M,q)$ is the restriction of the reversion τ of $\mathrm{C}\ell(M,q)$, whence a norm $\mathcal{N}$ defined by $x\tau(x) = \varepsilon\mathcal{N}(x)$ for all $x \in \varepsilon\mathrm{C}\ell_0(M,q)$; and the norm associated with the standard involution of $\mathrm{End}(P_0)$ is the quadratic form $f \longmapsto \det(f)$ (see (3.6.3)).

(b) Prove the existence of an isomorphism $D \otimes D \to K$ that makes D become a discriminant module such that $q(d)$ is the image of $d \otimes d$ in K for every $d \in D$ (*Hint:* (2.5.3)). Prove that the mapping $d \otimes a \longmapsto \varepsilon da$ is a bijection from $D \otimes M$ onto $\varepsilon\mathrm{C}\ell_0(M,q)$.

(c) The norms $\mathcal{N}$ and det associated with the standard involutions of $\varepsilon\mathrm{C}\ell_0(M,q)$ and $\mathrm{End}(P_0)$ make them become quadratic spaces. Prove that the above bijections $D \otimes (M,q) \to \varepsilon\mathrm{C}\ell_0(M,q) \to \mathrm{End}(P_0)$ are isomorphisms of quadratic spaces, that $\mathrm{End}(P_0)$ is hyperbolic, and conclude.

(8.ex.17) This exercise is an application of (8.ex.16) and (3.8.2). Let K be a local ring.

(a) Suppose that every element of WQ(K) is the Witt class of a quadratic space of rank ≤ 4. Prove that the surjective morphism WQ(K) $\to \mathcal{H}(K)$ is bijective.
(b) Suppose that K is a field with property $C_2(2)$; this means that for every quadratic form q on K^r there is a nontrivial $x \in K^r$ such that $q(x) = 0$ whenever $r > 4$. Prove that every element of WQ(K) is actually the Witt class of a quadratic space of dimension ≤ 4 .

(8.ex.18)* Let K be a Dedekind ring, and L its field of fractions. The K-submodules of L (also called fractionary ideals) that are invertible inside L (see (1.ex.25)), constitute a group, and the K-submodules generated by one nonzero element (also called principal fractionary ideals) constitute a subgroup; the quotient of the former by the latter is called the "group of classes of ideals" of K and denoted by C(K).

For more information about fractionary ideals of Dedekind rings, see [Jacobson 1995], Chap. 10.

(a) If P is a finitely generated projective K-module of constant rank 1, there are isomorphisms $P \otimes L \to L$. Prove that any such isomorphism maps $P \otimes 1$ onto a K-submodule of L that is invertible inside L, and that in this way we obtain an isomorphism Pic(K) $\to$ C(K).
When L is a "field of numbers" (that is a finite dimensional field extension of $\mathbb{Q}$), and K its ring of integer elements, then K is a Dedekind ring, and in [Samuel 2003] (Chap. III, §3.4., theorem 3) it is stated that C(K) is a finite group; consequently Pic(K) is also finite.
(b) Let Q′(K) be the group of isomorphy classes of quadratic extensions of K that are free K-modules; such a quadratic extension A admits a basis $(1, z)$ with $z^2 = \beta z - \gamma$ and $\beta^2 - 4\gamma \in K^\times$. Prove that we obtain a group morphism Q′(K) $\to K^\times/K^{\times 2}$ if we map the isomorphy class of any free quadratic extension A to the discriminant $\beta^2 - 4\gamma$ modulo the subgroup of squares $K^{\times 2}$. Prove the injectiveness of this group morphism when K is a Dedekind ring.
Besides, there is an exact sequence $0 \to$ Q′(K) $\to$ Q(K) $\to$ Pic$_2$(K). When K is the ring of integers of a field of numbers, the group $K^\times/K^{\times 2}$ is finite (see [Samuel 2003], Chap. IV, §4.4); consequently the group Q(K) is finite.
(c) When K is the ring of integers of a field of numbers, it is known that the subgroup Br$_2$(K) is finite (see [Grothendieck 1968]); conclude that the group $\mathcal{H}(K)$ is also finite.

(8.ex.19)* Let p be a prime integer ≥ 2, K the ring of p-adic integers (see (1.ex.29)(d)), and L its field of fractions. It is known that $K^\times/K^{\times 2}$ is a group of order 2, whereas the order of $L^\times/L^{\times 2}$ is 4 when $p > 2$, and 8 when $p = 2$ (see [Serre 1970], chap. 2, §3.3). Besides, Br(L) is isomorphic to $\mathbb{Q}/\mathbb{Z}$ (see [Serre 1962] chap. 12, or [Blanchard 1972] chap. 5), and consequently Br$_2$(L) contains only two elements. Prove that $\mathcal{H}_0(K)$ (resp. $\mathcal{H}(K)$) is a group of order 8 (resp. 16) when $p > 2$, of order 16 (resp. 32) when $p = 2$.

Comment. K is a local ring with maximal ideal pK, and K/pK is isomorphic to $\mathbb{Z}/p\mathbb{Z}$; since K is complete for the p-adic topology (see (8.ex.12)(c)), $\mathrm{W}(K)$ is isomorphic to $\mathrm{WQ}(K/pK) = \mathrm{WQ}(\mathbb{Z}/p\mathbb{Z})$ which has been calculated in (8.ex.11). According to [Auslander, Goldmann 1960], the completeness of the local ring K also implies $\mathrm{Br}(K) = \mathrm{Br}(K/pK) = \mathrm{Br}(\mathbb{Z}/p\mathbb{Z})$, whence $\mathrm{Br}(K) = 0$.

(8.ex.20) Let (M, q) be a quadratic space; prove (by means of (6.7.10)) that the class of $C\ell(M, q)$ is trivial if and only if there exists a graded finitely generated projective module P and a linear mapping $F : M \to \mathrm{End}_1(P)$ satisfying these conditions: $\mathrm{rk}(\mathfrak{m}, C\ell(M, q)) = (\mathrm{rk}(\mathfrak{m}, P))^2$ for all maximal ideals $\mathfrak{m}$, and $F(a)^2 = q(a)\ \mathrm{id}_P$ for all $a \in M$.

(8.ex.21) Let K be a ring in which 2 is invertible, κ an element of $K^\times$, and q the nondegenerate quadratic form on K^8 that maps every $(\lambda_1, \dots, \lambda_8)$ to $\frac{1}{2}\sum_{n=1}^{8} \kappa\lambda_n^2$. Since the order of every element of $\mathcal{H}(K)$ is a divisor of 8, the graded Brauer class of $C\ell(K^8, q)$ is trivial; this exercise leads to the construction of an isomorphism $C\ell(K^8, q) \to \mathcal{M}(8, 8; K)$.

(a) Let us set $A = C\ell(\langle -2, -2\rangle)^{ng}$; it is a quaternion algebra with basis $(1, i, j, ij)$ such that $i^2 = j^2 = -1$ and $ji = -ij$; its standard involution is denoted by $x \longmapsto \bar{x}$. Verify that A with the quadratic form $x \longmapsto x\bar{x}$ is a quadratic space isomorphic to $\langle 2, 2, 2, 2\rangle$.

(b) Let A^4 be the graded free module in which the even (resp. odd) elements are those of $A \times A \times 0 \times 0$ (resp. $0 \times 0 \times A \times A$). With every $(x, y) \in A^2$ we associate this endomorphism $F(x, y)$ of A^4 :

$$F(x,y)(a,b;c,d) \;=\; \left(\, \bar{x}c + d\bar{y}\ ,\ xd - cy\ ;\ \frac{\kappa}{2}\,(xa - b\bar{y})\,,\ \frac{\kappa}{2}\,(\bar{x}b + ay) \right) .$$

Calculate $F(x, y)^2$.

(c) Deduce from (8.ex.20) a graded algebra isomorphism $C\ell(K^8, q) \to \mathrm{End}(A^4)$.

(d) Let φ be the symmetric bilinear form on A^4 such that, for all $(a, b; c, d) \in A^4$,

$$\varphi((a,b;c,d),\ (a,b;c,d)) \;=\; \frac{\kappa}{2}(a\bar{a} + b\bar{b}) \;+ (c\bar{c} + d\bar{d})\ ;$$

prove that, for all (a, b, c, d) and (a', b', c', d') in A^4, and for all $(x, y) \in A^2$,

$$\varphi\big(F(x,y)(a,b;c,d)\ ,\ (a',b';c',d')\big) = \varphi\big((a,b;c,d)\ ,\ F(x,y)(a',b';c',d')\big).$$

(8.ex.22)* Let P be the quadratic space P_8 over $\mathbb{Z}$ defined just before (2.8.9), and Q the quadratic space defined in (2.ex.23); the $\mathbb{Z}$-modules P and Q are both treated as additive subgroups of $\mathbb{R}^8$. Here we consider the real quaternion algebra $\mathbb{H}$ with basis $(1, i, j, ij)$ such that $i^2 = j^2 = -1$ and $ji = -ij$, and we identify $\mathbb{R}^8$ with $\mathbb{H}^2$. Consequently the element of $\mathbb{R}^8$ formerly represented by $(x_1, x_2, \dots, x_8)$ is now represented by (x, y) with $x = x_1 + x_2 i + x_3 j + x_4 ij$ and $y = x_5 + x_6 i + x_7 j + x_8 ij$; similarly we write (a, b) instead of $(a_1, a_2, \dots, a_8)$, and (c, d) instead of $(c_1, c_2, \dots, c_8)$.

(a) With every $(x, y) \in \mathbb{H}^2$ we associate the endomorphism $F(x, y)$ of $\mathbb{H}^4$ suggested by (8.ex.21):

$$F(x,y)(a,b;c,d) \;=\; \left(\ \bar{x}c + d\bar{y}\ ,\ xd - cy\ ;\ \frac{1}{2}\,(xa - b\bar{y})\,,\ \frac{1}{2}\,(\bar{x}b + ay)\right)\,.$$

Suppose that (x, y) belongs to P, and $(a, b; c, d)$ to $Q \oplus P$; verify that $F(x, y)(a, b; c, d)$ also belongs to $Q \oplus P$.
Comment. This verification probably needs a lot of calculation.

(b) Deduce from (8.ex.20) the existence of a graded algebra isomorphism $C\ell_{\mathbb{Z}}(P) \to \mathrm{End}_{\mathbb{Z}}(Q \oplus P)$. Here the elements of Q are even, and those of P are odd.

(c) Prove that $\mathcal{H}(\mathbb{Z})$ is a trivial group, without using the difficult theorem stating that $\mathrm{Br}(\mathbb{Z})$ is a trivial group (*Hint:* (2.8.14)).

(8.ex.23) Let X be an infinite set, and L the ring of all mappings $X \to \mathbb{Q}$. For every subset Y of X, we call e_Y the function that maps all elements of Y to 1, and all elements of $X \setminus Y$ to 0.

(a) Prove that the mapping $Y \longmapsto e_Y$ is a bijection from the set of subsets of X onto $\mathrm{Ip}(L)$.

(b) By definition the "fractional support" of an element f of L is the subset of all $x \in X$ such that $f(x)$ does not belong to $\mathbb{Z}$, and K is the subset of all elements of L with finite "fractional support" in X. Prove that K is a subring of L, and that $\mathrm{Ip}(K) = \mathrm{Ip}(L)$.

(c) Now let $\mathrm{Ip}'(K)$ be the subset of all $e \in \mathrm{Ip}(K)$ such that $2e$ is invertible in Ke; it is an ideal of the boolean ring $\mathrm{Ip}(K)$. Prove that e_Y belongs to $\mathrm{Ip}'(K)$ if and only if Y is a finite subset of X, and that $\mathrm{Ip}'(K)$ is not a principal ideal of $\mathrm{Ip}(K)$.

Bibliography

Books and booklets

Albert, A.A., *Structure of algebras*, Amer. Math. Soc. Coll. Publ. **24**, New York 1939.

Amaldi, E., (edited by), *La vita e l'opera di Ettore Majorana (*1906–1938*)*, Academia Nazionale dei Lincei, Roma 1966.

Argand, J.R., *Essai sur une manière de représenter les quantités imaginaires dans les constructions géométriques*, Hoüel, Paris 1806, 1874. Republié par A. Blanchard , Paris 1971.

Atiyah, M.F., Macdonald, I.G., *Introduction to Commutative algebra*, Oxford 1969.

Baeza, R., *Quadratic forms over semilocal rings*, Lecture Notes in Math. **655**, Springer Verlag, 1978.

Bass, H., *Topics in algebraic K-theory*, Tata Institute, Bombay 1967.

Blanchard, A., *Les corps non commutatifs*, Presses Universitaires de France, 1972.

Bombelli, R., *L'Algebra*, Bologna (G. Rossi), 1579.

Bourbaki, N., *Algèbre* Chap. 6, *Groupes et corps ordonnés*, Hermann, Paris 1964 ....

Bourbaki, N., *Algèbre* Chap. 9, *Formes sesquilinéaires et formes quadratiques*, Hermann, Paris 1959....

Bourbaki, N., *Algèbre commutative* Chap. 2, *Localisation*, Hermann, Paris 1961....

Bourbaki, N., *Algèbre commutative* Chap. 7, *Diviseurs*, Hermann, Paris 1965....

Budinich, P., Trautman, A., *The spinorial chessboard*, Springer, New York 1989.

Cartan, E., *Leçons sur la théorie des spineurs*, Hermann, Paris, 1938. English translation: *The theory of spinors*, Hermann, Paris, 1966; Dover Public. Inc., New York 1981. – *Warning:* in Cartan's original text, the title of the section **98** (resp. **106**) is: "Nombres de Clifford-Lipschitz" (resp. "Spineurs simples et ν-plans isotropes"); yet in the English translation this title has become: "The Clifford algebra" (resp. "Pure spinors and isotropic ν-planes").

Cartan, H., Eilenberg, S., *Homological Algebra*, Princeton Univ. Press, 1956.

Cayley, A., *Collected mathematical papers*, 13 volumes, Cambridge Univ. Press, 1889–1898.

Chevalley, C., *The algebraic theory of spinors*, Columbia Univ. Press, 1954. Reprinted in his Collected Works, vol. 2, Springer, 1997.

Chevalley, C., *Formes quadratiques sur un corps quelconque*, Notes (rédigées par M. Lusson) d'après un cours à l'Université de Paris, 1956.

Chisholm, M., *Such silver currents*, The Lutterworth Press, Cambridge 2002. – *Warning:* this book is devoted to the lives and works of both William K. Clifford and his wife Lucy who was a writer.

Clifford, W.K., *Mathematical papers*, London 1882. Reprinted by Chelsea Publ. Co., New York 1968. – Especially *On the classification of geometric algebras*, pp. 397–401, an unfinished paper.

Crowe, M.J., *A History of Vector Analysis*, Dover, 1985.

Crumeyrolle, A., *Orthogonal and symplectic Clifford algebras*, Math. and its applications **57**, Kluwer Acad. Press, 1990. – *Warning:* the chapters 17, 18, 19, 22 devoted to "symplectic Clifford algebras" (that are Weyl algebras, eventually enlarged) are completely unreliable.

Deheuvels, R., *Formes quadratiques et groupes classiques*, P.U.F. Paris 1981.

DeMeyer, F., Harrison, D., Miranda, R., *Quadratic forms over* $\mathbb{Q}$ *and Galois extensions of commutative rings*, Memoirs of the Amer. Math. Soc. **394**, 1989.

Dieudonné, J., *La géométrie des groupes classiques*, Springer Verlag, 1963.

Dieudonné, J., *Sur les groupes classiques*, Hermann, Paris 1967.

Ebbinghaus, H.D., Hermes, H., Hirzebruch, F., Koecher, M., Mainzer, K., Neukirch, J., Prestel, A., Remmert, R., *Zahlen*, Springer Verlag 1988. English version: *Numbers*, Springer Verlag 1990. French version: *Les nombres, leur histoire, leur place et leur rôle de l'Antiquité aux recherches actuelles*, Vuibert, Paris 1999.

Fontana, M., Huckaba, J.A., Papick, I.J., *Prüfer Domains*, Marcel Dekker, New York 1997.

Gauss, C.F., *Werke*, 12 volumes, Göttingen, 1870–1927.

Grassmann, H., *Gesammelte Werke*, 3 vol., Teubner, Leipzig, 1894–1911.

Greenberg, M.J., *Lectures on forms in many variables*, Benjamin, New York 1969.

Gross, H., *Quadratic forms in infinite dimensional vector spaces*, Progress in Math. **1**, Birkhäuser, 1979.

Hahn, A.J., *Quadratic algebras, Clifford algebras and arithmetic Witt groups*, Universitext, Springer Verlag, New York 1994.

Harrison, D.K., *Witt rings*, Lecture Notes, University of Kentucky, Lexington 1970.

Harvey, F.R., *Spinors and calibrations*, Acad. Press, Persp. in Math. **9**, 1990.

Helmstetter, J., *Algèbres de Clifford et algèbres de Weyl*, Cahiers Math. Montpellier **25**, 1982.

Helmstetter, J., *Algèbres de Weyl et* $\star$*-produits*, Cahiers Math. Montpellier **34**, 1985.

Hestenes, D., *Space-time algebra*, Gordon + Breach 1966, 1987, 1992.

Hestenes, D., Sobczyk, G., *Clifford algebra to geometric calculus*, Reidel, 1984, 1987.

Hirzebruch, F., Neumann, W.D., Koh, S.S., *Differentiable manifolds and quadratic forms*, Marcel Dekker Inc., New York, 1971.

Jacobson, N., *Basic Algebra*, vol. I and II, W.H. Freeman and Co., New York, 1974,..., 1995,....

Knebusch, M., Kolster, M., *Witt rings*, Aspekte der Math., F. Vieweg und Sohn, Braunschweig/Wiesbaden 1982.

Knebusch, M., Scharlau, W., *Generic methods and Pfister forms*, DMV Seminar **1**, Birkhäuser, 1980.

Knus, M.A., *Quadratic forms, Clifford algebras and spinors*, Seminários de Matemática **1**, IMECC-UNICAMP, Campinas 1988.

Knus, M.A., *Quadratic and hermitians forms over rings*, Grundlehren der Math. Wissenschaft **294**, Springer Verlag, 1991.

Knus, M.A., Ojanguren, M., *Théorie de la descente et algèbres d'Azumaya*, Lecture Notes in Math. **389**, Springer Verlag, 1974.

Knus, M.A., Merkurjev, A., Rost, M., Tignol, J.P., *The book of involutions*, Amer. Math. Soc. Colloq. Publ. **44**, 1998.

Lam, T.Y., *The algebraic theory of quadratic forms*, Benjamin, New York 1973.

Lam, T.Y., *Orderings, valuations and quadratic forms*, Conference Board of the Math. Sc. **52**, Amer. Math. Soc., 1983.

Lawson, H.B., Michelsohn, M.L., *Spin Geometry*, Princeton Univ. Press, 1989.

Lipschitz, R., *Untersuchungen über die Summen von Quadraten*, Max Cohen und Sohn, Bonn 1886 (147 pages). French summary in Bull. Sci. Math. (2), **10**, pp. 163–183, 1886.

Lounesto, P., *Clifford algebras and spinors*, London Math. Society Lecture Note **239**, Cambridge Univ. Press, 1997.

Micali, A., Revoy, Ph., *Modules quadratiques*, Cahiers Math. Montpellier **10**, 1970. Reprinted in Bull. Soc. Math. Fr. **63**, 1979.

Micali, A., Villamayor, O.E., *Algèbres de Clifford sur un anneau local*, duplicated notes, Instituto de Pesquisas Matemáticas, Universidade de São Paulo, 1966.

Milnor, L., Husemoller, D., *Symmetric bilinear forms*, Ergebnisse der Math. **73**, Springer Verlag, 1973.

O'Meara, O.T., *Introduction to quadratic forms*, Springer Verlag, 1973.

Orzech, M., Small, Ch., *The Brauer group of commutative rings*, Lecture Notes in pure and applied Math., **11**, Marcel Dekker Inc., New York 1975.

Peano, G., *Calcolo geometrico secondo l'Ausdehnungslehre di Grassmann, preceduto dalle operazioni della logica deduttiva*, Torino 1888.

Porteous, I.R., *Clifford algebras and the classical groups*, Cambridge studies in Adv. Math., **50**, Cambridge Univ. Press, 1995, 2000.

Revoy, Ph., *Autour des formes quadratiques*, Cahiers Math. Montpellier **13**, 1978.

Riesz, M., *Clifford numbers and spinors*, Institute for Fluid Dynamics and Applied Math., Lecture Series **38**, Univ. of Maryland, 1958. Reprinted by E.F. Bolinder and P. Lounesto, Kluwer Acad. Publishers, 1993. – Interesting historical comment by P. Lounesto.

Samuel, P., *Théorie algébrique des nombres*, Hermann, Paris, 1967,..., 2003.

Schafer, R.D., *An introduction to nonassociative algebras*, Academic Press, New York and London, 1966.

Scharlau, W., *Quadratic and hermitian forms*, Grundlehren der Math. Wissenschaft **270**, Springer Verlag, 1985.

Serre, J.P., *Corps locaux*, Hermann, Paris 1962.

Serre, J.P., *Cours d'Arithmétique*, Presses Universitaires de France, 1970,..., 1995. English translation: *A course in Arithmetic*, Springer Verlag, 1973.

Wessel, C., *Essai sur la représentation analytique de la direction*, présenté en 1797 à l'Académie Royale de Danemark, republié en traduction française en 1897 par l'Académie Royale de Danemark.

Other publications

Abłamowitz, R., Lounesto, P., *On Clifford algebras of a bilinear form with an antisymmetric part*, in "Clifford algebras with numeric and symbolic computations" pp. 167–188, edited by R. Abłamowitz, P. Lounesto, L.M. Parra, Birkhäuser, Boston, 1996.

Amitsur, S.A., *Simple algebras and cohomology groups of arbitrary fields*, Trans. Amer. Math. Soc. **90**, pp. 73–112, 1959.

Arf, C., *Untersuchungen über quadratische Formen in Körpern der Characteristik* 2, Journal für die reine und angew. Math. **183**, pp. 148–167, 1941.

Auslander, M., Goldmann, O., *The Brauer group of a commutative ring*, Trans. Amer. Math. Soc. **97**, pp. 367–409, 1960.

Atiyah, M.F., Bott, R., Shapiro, A., *Clifford modules*, Topology **3**, Suppl. **1**, pp. 3–38, 1964.

Baez, J.C., *The octonions*, Bull. Amer. Math. Soc. **39**, pp. 145–205, 2001.

Bass, H., *Clifford algebras and spinor norms over a commutative ring*, Amer. J. Math. **46**, pp. 156–206, 1974.

Bergé, A.M., Martinet, J., *Formes quadratiques et extensions en caractéristique* 2, Ann. Institut Fourier (Grenoble) **35**, pp. 57–77, 1985.

Brauer, R., Weyl, H., *Spinors in n dimensions*, Amer. J. Math. **57**, pp. 425–449, 1935.

Cartan, E., *Les groupes projectifs qui ne laissent invariante aucune multiplicité plane*, Bull. Soc. Math. France **41**, pp. 53–96, 1913.

Clifford, W.K., *Applications of Grassmann's extensive algebra*, Amer. J. Math. **1**, pp. 350–358, 1878. Reprinted in his *Mathematical papers* (see above), pp. 266–276.

Cortiñas, G.H., *Clases características para módulos cuadráticos*, PhD at the Univ. of Buenos Aires, 1988.

Craven, T.C., Rosenberg, A., Ware, R., *The map of the Witt ring of a domain into the Witt ring of its field of fractions*, Proc. A.M.S. **51**, pp. 25–30, 1975.

Dias, I., *Formas quadráticas sobre LG-aneis*, PhD at the Univ. of Campinas, 1988.

Dias, I., Micali, A., *Anneaux de Witt de LG-anneaux*, Indag. Mathem., N.S. **9**, pp. 21–220, 1998.

Dias, I., Micali, A., Paques, A., *On transversality for quadratic spaces over rings with many units*, Communications in Algebra **32** (2), pp. 551–559, 2004.

Dirac, P.A.M., *The quantum theory of the electron*, Proc. Royal Soc. A **117**, pp. 610–624, 1928.

Dixmier, J., *Sur les algèbres de Weyl*, Bul. de la Soc. Math. de France **96**, pp. 209–242, 1968.

Fernandes, N.C., *As álgebras de Grassmann-Schenberg e grande unificação*, in "Perspectivas em Física Teórica", Instituto de Física da Universidade de São Paulo, pp. 368–390, 1987.

Fernandes, N.C., *Mário Schenberg e a descoberta da supersimetria*, Boletim informativo da Sociedade Brasileira de Física **1**, ano 22, pp. 3–6, 1991.

Ferrero, M., *Galois theory and Clifford algebras*, Mathematicae Notae **23**, pp. 99–101, 1973. –– *Correction*, – **24**, p. 97, 1975.

Flanders, H., *Tensor and exterior products*, Journal of algebra **7**, pp. 1–24, 1967.

Frobenius, G., *Über lineare Substitutionen und bilineare Formen*, Journal de Crelle **84**, pp. 1–63, 1878.

Fujisaki, G., *A note on Witt rings over local rings*, J. Fac. Sc. Math. **19**, pp. 403–414, 1972.

Gabriel, P., *Degenerate bilinear forms*, J. of Algebra **31**, pp. 67–72, 1974.

Grothendieck, A., *Sur le groupe de Brauer III: exemples et compléments*, pp. 88–188 in this book: Giraud, J., et. al., *Dix exposés sur la cohomologie des schémas*, Advanced studies in pure mathematics, vol. 3, North-Holland 1968.

Helmstetter, J., *Groupe de Clifford pour des formes quadratiques dégénérées*, C.R. Ac. Sc. Paris, A **285**, pp. 175–177, 1977.

Helmstetter, J., *Systèmes de puissances partiellement divisées*, Ann. Scuola Normale Sup. Pisa, serie IV, **9** $n^o 1$, pp. 27–56, 1982.

Helmstetter, J., *Monoïdes de Clifford et déformations d'algèbres de Clifford*, J. of Algebra **111**, pp. 14–48, 1987. – *Warning:* at that time the author still ignored Lipschitz's contribution to Clifford algebras, and gave the name "Clifford monoid" to what is now called the Lipschitz monoid.

Helmstetter, J., *Clifford groups for arbitrary quadratic forms*, in "Clifford algebras and their applications..." pp. 33–38, edited by A. Micali, R. Boudet, J. Helmstetter, Kluwer Ac. Publishers, Dordrecht 1992.

Helmstetter, J., *Lipschitz monoids and Vahlen matrices*, Adv. in applied Clif. alg. **15**, pp. 83–122, 2005.

Helmstetter, J., Micali, A., *Relations between Witt rings and Brauer groups*, in "Clifford algebras and their applications..." pp. 9–12, edited by Brackx, F., Delanghe, R., Serras, H., Kluwer Ac. Publishers, Dordrecht 1993.

Hoestmaelingen, C. (that is Lefranc, C. after her marriage), *Sur l'anneau de Witt d'un anneau de Prüfer*, C.R. Acad. Sc. Paris **280**, pp. 69–71, 1975.

Kaplansky, I., *Quadratic forms*, J. Math. Soc. Japan **5**, pp. 200–207, 1953.

Karoubi, M., *Algèbres de Clifford et K-théorie*, Ann. scient. Ec. Normale Sup. (4e série) **1**, pp. 161–270, 1968.

Karoubi, M., *Localisation des formes quadratiques I*, Ann. scient. Ec. Normale Sup. (4e série) **7**, pp. 359–404, 1974.

—— *Localisation des formes quadratiques II*, — **8**, pp. 99–155, 1975.

Kientega, G., *Sur une question de Micali-Villamayor*, in "Clifford algebras and their applications.", edited by J.S.R. Chisholm and A.K. Common, D. Reidel Publishing Co., Dordrecht, pp. 109–114, 1986.

Knebusch, M., *Specialization of quadratic and symmetric bilinear forms and a norm theorem*, Acta Arithmetica **24**, pp. 279–299, 1973.

Knebusch, M., Rosenberg, A., Ware, R., *Structure of Witt rings and quotient of abelian group rings*, Amer. J. Math. **94**, pp. 119–155, 1972.

Knebusch, M., Scharlau, W., *Quadratische Formen und quadratische Reziprozitätsgesetze über algebraischen Zahlkörpern*, Math. Z. **121**, pp. 346–368, 1971.

Knus, M.A., Paques, A., *Quadratic spaces with trivial Arf invariant*, J. of Algebra **93**, pp. 267–291, 1985.

Laborde, O., *Formes quadratiques et algèbres de Clifford*, Bull. Sc. Math. (2), **96**, pp. 199–208, 1972.

Laborde, O., *Formes quadratiques, algèbres de Clifford et signatures*, C.R. Ac. SC. Paris **278**, pp. 1599–1602, 1974.

Larotonda, A.R., *Trivialidad del fibrado tangente a la n-esfera algebraica*, PhD at the Univ. of Buenos Aires, 1968.

Larotonda, A.R., Micali, A., Villamayor, O.E., *Sur le groupe de Witt*, Symposia Math. **11**, pp. 211–219, 1973.

Lazard, D., *Sur les modules plats*, C.R. Acad. Sc. Paris **258**, pp. 6313–6316, 1964.

Lefranc, C., *Formes bilinéaires non dégénérées sur un anneau de Prüfer*, D.E.A. de Math. pures, Univ. de Montpellier, 1970. – Interesting review by H. Gross in MR 45 ♯3400, 1973.

Legrand, D., *Formes quadratiques et algèbres quadratiques*, PhD at the Univ. of Paris-Sud-Orsay, 1971. Reprinted in Annales Sc. Univ. de Clermont **54**, Math. 10ème fascicule, 1975.

Lipschitz, R., *Principes d'un calcul algébrique...*, C.R. Acad. Sc. Paris **91** pp. 619–621 and 660–664, 1880. Reprinted in Bull. Sci. Math. (2), **11**, pp. 115–120, 1887.

Lipschitz, R., *Recherches sur la transformation, par des substitutions réelles, d'une somme de deux ou de trois carrés en elle-même*, J. Math. pures appl. **2**, pp. 373–439, 1886.

Lorenz, F., Leicht, J., *Die Primideale des Wittschen Ringes*, Inv. Math. **10**, pp. 82–88, 1970.

Lounesto, P., *Scalar products of spinors and an extension of Brauer-Wall groups*, Foundations of Physics **11**, pp. 721–740, 1981.

Lounesto, P., *Clifford algebras and Hestenes spinors*, Foundations of Physics **23**, pp. 1203–1237, 1993.

Micali, A., *Résultats récents sur la théorie des formes quadratiques*, Séminaire P. Dubreil (26-ème année, exp. $n^o 12$), Paris 1973.

Micali, A., Paques, A., *Le groupe des algèbres quadratiques*, Comm. in Algebra **10**(16), pp. 1765–1799, 1982.

Micali, A., da Rocha Barros, A.L., *Notice biographique sur Mário Schenberg*, in "Clifford algebras and their applications..." edited by A. Micali, R. Boudet and J. Helmstetter, Kluwer Acad. Publishers, pp. 501–503, 1992. This notice is followed by a reprint of [Schenberg, 1964] (see below) and a list of works of M. Schenberg (pp. 519–523).

Micali, A., Villamayor, O.E., *Sur les algèbres de Clifford*, Ann. Sc. Ec. Normale Sup., 4ème série **1**, pp. 271–304, 1968.

Micali, A., Villamayor, O.E., *Sur les algèbres de Clifford II*, Journal für die reine und angew. Math. **242**, pp. 61–90, 1970.

Micali, A., Villamayor, O.E., *Algèbres de Clifford et groupe de Brauer*, Ann. Sc. Ec. Normale Sup., 4ème série **4**, pp. 285–310, 1971.

Micali, A., Villamayor, O.E., *Formes quadratiques sur les corps de caractéristique* 2, Comm. in Algebra **17**, pp. 299–312, 1989.

Micali, A., Villamayor, O.E., *Algèbres de Clifford sur un corps de caractéristique* 2, in "Clifford algebras and their applications..." pp. 33–38, edited by A. Micali, R. Boudet, J. Helmstetter, Kluwer Ac. Publishers, Dordrecht 1992.

Milnor, J., *Algebraic K-theory and quadratic forms*, Inv. Math. **9**, pp. 318–344, 1970.

Nimeier, H.V., *Definite quadratische Formen der Dimension* 24 *und Determinant* 1, J. of Number Theory **5**, pp. 142–178, 1973.

Nouazé, Y., Revoy, Ph., *Sur les algèbres de Weyl généralisées*, Bull. Sci. Math. (2), **96**, pp. 27–47, 1972.

Ojanguren, M., *Quadratic forms over regular rings*, J. Indian Math. Soc. **44**, pp. 109–116, 1979.

Papa Serra, J.M., *L'àlgebra vectorial. Una història que ens cal reescriure*, Bulletí de la Societat Catalana de Matemàtiques **10**, pp. 75–120, 1995.

Paques, A., *Sobre cohomologia de formas quadráticas*, PhD at the Univ. of Campinas, 1977.

Pfister, A., *Quadratische Formen in beliebigen Körpern*, Inv. Math. **1**, pp. 116–132, 1966.

Picco, D.J., *El grupo de Brauer relativo*, PhD at the Univ. of Bahia Blanca, 1969.

Quebbemann, H.G., Scharlau, R., Scharlau, W., Schulte, M., *Formen in additiven Categorien*, Soc. Math. France, Mémoire **48**, pp. 93–101, 1976.

Revoy, Ph., *Sur les deux premiers invariants d'une forme quadratique*, Ann. Sc. Ec. Normale Sup. **4**, pp. 311–319, 1971.

da Rocha Barros, A.L., *Schenberg e a visão algébrica da realidade física*, in "Perspectivas em Física Teórica", Instituto de Física da Universidade de São Paulo, pp. 130–143, 1987.

da Rocha Barros, A.L., *Schenberg e as álgebras geométricas da teoria quântica*, Boletim informativo da Sociedade Brasileira de Física **1**, ano 22, pp. 7–9, 1991.

da Rocha Barros, A.L., Schenberg, M., *On the Clifford and Jordan-Wigner algebras*, Rev. Unión Mat. Argentina **20**, pp. 239–258, 1960.

Roy, A., *On a characterization of Clifford algebras*, Math. Zeitschrift, **85**, pp. 241–244, 1964.

Saltman, D.J., *The Brauer group is torsion*, Proc. Amer. Math. Soc. **81**, pp. 385–387, 1981.

Sato, M., Miwa, T., Jimbo, M., *Holonomic quantum fields I* , Publ. R.I.M.S., Kyoto Univ. **14**, pp. 223–267, 1978. – *Warning:* this is just a paper about Clifford algebras.

—— *Holonomic quantum fields II* , — **15**, pp. 201–278, 1979.

—— *Holonomic quantum fields IV* , — **15**, pp. 871–972 (especially the Appendix), 1979.

Scharlau, W., *Quadratic reciprocity laws*, J. of Number Theory **4**, pp. 78–97, 1972.

Schenberg, M., *Sobre uma extensão do cálculo espinorial I* , Anais da Ac. Brasil. de Ciências **13**, pp. 129–135, 1941.

—— *Sobre uma extensão do cálculo espinorial II* , — **15**, pp. 97–108, 1943.

Schenberg, M., *On Grassmann and Clifford algebras I*, Anais da Ac. Brasil. de Ciências **28**, pp. 11–19, 1956.

—— *On Grassmann and Clifford algebras II*, — **32**, pp. 299–322, 1960.

Schenberg, M., *Quantum mechanics and geometry I*, Anais da Ac. Brasil. de Ciências **29**, pp. 473–499, 1958.

—— *Quantum mechanics and geometry II*, — **30**, pp. 1–20, 1958.

—— *Quantum mechanics and geometry III*, — **30**, pp. 117–131, 1958.

—— *Quantum mechanics and geometry IV*, — **30**, pp. 259–280, 1958.

Schenberg, M., *Algebraic structures of finite point sets I*, 1964, reprinted in 'Clifford algebras and their applications..." edited by A. Micali,..., Kluwer Acad. Publishers, pp. 505–518, 1992.

Serre, J.P., *L'invariant de Witt de la forme* $\mathrm{Tr}(x^2)$, Comment. Math. Helvetici **59**, pp. 651–676, 1984.

Shale, D., Forrest Stinespring, W., *States of the Clifford algebra*, Ann. of Math. **80**, pp. 365–381, 1964.

Small, Ch., *The Brauer-Wall group of a commutative ring*, Trans. of the Amer. Math. Soc. **156** pp. 455–491, 1971.

Small, Ch., *The group of quadratic extensions*, J. pure and appl. Algebra **2**, pp. 83–105, 1972.

Small, Ch., *Normal bases for quadratic extensions*, Pacific J. Math. **50**, pp. 601–611, 1974.

Tao, D., *The generalized even Clifford algebra*, J. of Algebra **172**, pp. 184–204, 1995.

Tamagawa, T., *On quadratic forms and pfaffians*, J. Fac. Sc. Tokyo, Sect. I, **24**, pp. 213–219, 1977.

Tits, J., *Formes quadratiques, groupes orthogonaux et algèbres de Clifford*, Inventiones Math. **5**, pp. 19–41, 1968.

Tsuzuku, T., *On a conjecture of Kaplansky on quadratic forms*, J. Math. Soc. Japan **6**, pp. 325–331, 1954.

Wall, C.T.C., *Graded Brauer groups*, Journal für die reine und angew. Math. **213**, pp. 187–199, 1964.

Wall, C.T.C., *Graded algebras, anti-involutions, simple groups and symmetric spaces*, Bull. Amer. Math. Soc. **74**, pp. 198–202, 1968.

Weil, A., *Correspondence* published anonymously in Ann. of Math. **69**, pp. 247–251, 1959. Reprinted in his *Collected papers*, vol. **2**, pp. 557–561, Springer Verlag 1979.

Witt, E., *Theorie der quadratischen Formen in beliebigen Körpern*, Journal für die reine und angew. Math. **176**, pp. 31–44, 1937.

Zassenhaus, H., *On the spinor norm*, Arch. Math. **13**, pp. 434–451, 1962.

Zorn, M., *Alternativkörper und quadratische Systeme*, Abh. Math. Sem. Hamburg **9**, pp. 395–402, 1933.

Index of Definitions

Index of Notation

Here every algebra is denoted by A (or B), every module by M (or N),.... Only notations that are important, or that are used several times, are recalled here. According to the context, single letters like β, γ, ε, φ,... may also be used with another meaning than the one recalled here.